The Handbook of Global Science, Technology, and Innovation

Handbook of Global Policy Series

Series Editor
David Held
Master of University College and Professor of Politics and International Relations at Durham University

The Handbook of Global Policy series presents a comprehensive collection of the most recent scholarship and knowledge about global policy and governance. Each Handbook draws together newly commissioned essays by leading scholars and is presented in a style which is sophisticated but accessible to undergraduate and advanced students, as well as scholars, practitioners, and others interested in global policy. Available in print and online, these volumes expertly assess the issues, concepts, theories, methodologies, and emerging policy proposals in the field.

Published

The Handbook of Global Climate and Environment Policy
Robert Falkner

The Handbook of Global Energy Policy
Andreas Goldthau

The Handbook of Global Companies
John Mikler

The Handbook of Global Security Policy
Mary Kaldor and Iavor Rangelov

The Handbook of Global Health Policy
Garrett Brown, Gavin Yamey, and Sarah Wamala

The Handbook of Global Science, Technology, and Innovation
Daniele Archibugi and Andrea Filippetti

Forthcoming

The Handbook of Global Policy and Social Policy-Making in Education
Karen Mundy, Andy Green, Robert Lingard, and Antoni Verger

The Handbook of Global Science, Technology, and Innovation

Edited by

Daniele Archibugi and Andrea Filippetti

WILEY Blackwell

This edition first published 2015

Registered Office
John Wiley & Sons, Ltd, The Atrium, Southern Gate, Chichester, West Sussex, PO19 8SQ, UK

Editorial Offices
350 Main Street, Malden, MA 02148-5020, USA
9600 Garsington Road, Oxford, OX4 2DQ, UK
The Atrium, Southern Gate, Chichester, West Sussex, PO19 8SQ, UK

For details of our global editorial offices, for customer services, and for information about how to apply for permission to reuse the copyright material in this book please see our website at www.wiley.com/wiley-blackwell.

Library of Congress Cataloging-in-Publication Data

The handbook of global science, technology, and innovation / edited by Daniele Archibugi and Andrea Filippetti.
pages cm. – (Hgp – handbooks of global policy)
Includes bibliographical references and index.
ISBN 978-1-118-73906-8 (hardback)
1. Research–Industrial–Handbooks, manuals, etc. 2. Technology transfer–Handbooks, manuals, etc. 3. Technology and state–Handbooks, manuals, etc. 4. Science and state–Handbooks, manuals, etc.
I. Archibugi, Daniele. II. Filippetti, Andrea, 1977–
T175.H35 2015
338.9′26–dc23

2015004138

A catalogue record for this book is available from the British Library.

Cover image: Computer circuit board © blackred / iStockphoto

Set in 10/12.5pt Sabon by SPi Global, Pondicherry, India

1 2015

Contents

Notes on Contributors

Daniele Archibugi is a Research Director at the Italian National Research Council (CNR-IRPPS) in Rome, and Professor of Innovation, Governance and Public Policy at the University of London, Birkbeck College. He works on the economics and policy of science, technology, and innovation and on the political theory of international relations. He has worked at the Universities of Sussex, Cambridge, London School of Economics, Harvard, and Rome LUISS. In 2006 he was appointed honorary professor at Sussex University. He has chaired the European Commission's Expert Group on "A Wide Opening of the European Research Area to the World." Besides several edited books and articles in academic journals, he has authored *The Technological Specialization of Advanced Countries* (with Mario Pianta, Kluwer, 1992) and *Innovation and Economic Crisis: Lessons and Prospects from the Economic Downturn* (with Andrea Filippetti, Routledge, 2011).

Suma Athreye is Professor of International Strategy at Brunel Business School, Brunel University. Her research interests include technology development of emerging market firms and their use of international markets to gain competiveness. Her recent work has focused on the internationalization of Indian and Chinese firms, especially those in the software and pharmaceutical sectors.

Helena Barnard, since obtaining her PhD in Management from Rutgers University, has been working at GIBS, University of Pretoria, in South Africa where she is Director of Research. Her research interests are in how knowledge (and with it technology, organizational practices, and innovation) moves from more to less developed countries. She focuses both on organizational mechanisms (notably emerging multinationals) and individual mechanisms, especially the diaspora and scientific collaborations.

Sabine Brunswicker is researcher with a particular interest in open innovation and innovation ecosystems. She is an Associate Professor of Innovation and Director of the Research Center for Open Digital Innovation at Purdue University, Indiana. She is also a Visiting Professor at ESADE Business School at Ramon Llull University, Barcelona, and is a strategic advisor for open innovation at the Fraunhofer Institute for Industrial Engineering, Stuttgart. Her general research interests lie in understanding collaborative models of innovation and value creation in today's global and digital economy.

Fulvio Castellacci is Director of the TIK Centre, University of Oslo, and Research Professor at the Norwegian Institute of International Affairs (NUPI). He holds a PhD in Innovation Studies (University of Oslo, 2004) and a PhD in Economics (University of Rome La Sapienza, 2005). His main research interests are in the fields of the economics of innovation, and particularly on the relationship between innovation and economic performance.

Jongmin Choi is a doctoral student in the Department of Public Policy at the University of North Carolina, Chapel Hill. His main research interests include innovation, entrepreneurship, and economic development policy suited for regional characteristics. In particular, he focuses on industry cluster as an economic development strategy. He plans to shed light on regional characteristics that spark the initial takeoff of an industry in a region.

Alessandro Cordova is a PhD candidate in Business Administration at Bocconi University, Milan. He graduated cum laude in Economics at Bocconi University, and he obtained an MSc in Economics from the London School of Economics. In 2013, he has been awarded a research grant and worked with the Dean of the Bocconi PhD School on projects related to the economy of innovation, entrepreneurship, and public administration. Currently, his main research interests lie in new business ventures and Big Data. Alongside, he is founding a start-up in the health sector.

Simone Corsi is Programme Manager of the Lancaster China Catalyst Programme (Lancaster University) and Honorary Researcher at the Institute for Entrepreneurship and Enterprise Development (IEED) at the Lancaster University Management School. His research and work deal with R&D internationalization and collaboration in China. He holds a PhD in Management from Scuola Superiore Sant'Anna in Pisa (Italy) and he is Research Associate at the Research Center for Global R&D Management (GLORAD) at Tongji University in Shanghai (China) where he has been visiting doctoral student (2011).

Robin Cowan is Professor of Management at the University of Strasbourg, Professor of the Economics of Technical Change at Maastricht University, and Professorial Fellow at UNU-MERIT. His research focuses on the economics of networks, knowledge, and innovation.

Alberto Di Minin is an Associate Professor at the Institute of Management of Scuola Superiore Sant'Anna in Pisa (Italy), and Research Fellow at the Berkeley Roundtable on the International Economy (BRIE). He has a PhD from University of California, Berkeley. His research deals with the appropriability of innovation, open innovation, business models, intellectual property, and R&D management. His research has appeared in *Research Policy*, *California Management Review*, *R&D Management*, and *Journal of International Business Studies*.

Johanna Dolci is a financial consultant at CMC Capital in London, where she executes M&A transactions, principal investing, and valuation assignments. She completed her Master of Science in Finance at Bocconi University, Milan. Johanna started research on crowdfunding during her studies at Bocconi University.

Paolo Davide Farah is University Professor at West Virginia University, Department of Public Administration within the Eberly College of Arts and Sciences and Law School; Research Scientist and Principal Investigator at gLAWcal – Global Law Initiatives for Sustainable Development (United Kingdom); Principal Investigator and Research Team Coordinator for European Commission-funded projects at the University Institute of European Studies (IUSE) in Turin (Italy). He has been Visiting Scholar (2011–2012) at Harvard Law School, East Asian Legal Studies (EALS) and Marie Curie Fellow (2013–2014) at Peking University, Center for European Studies in Beijing (China).

Maryann P. Feldman is the Heninger Distinguished Professor in the Department of Public Policy at the University of North Carolina, Chapel Hill. Her research and teaching interests focus on the areas of innovation, the commercialization of academic research and the factors that promote technological change and economic growth.

Marta Fernandez de Arroyabe Arranz is a PhD student at the Center for Research in Economics and Management (CREA) at the University of Luxembourg and at UNU-MERIT. Her research interests include the areas of innovation roadmap, innovation planning, networks, and alliances.

Andrea Filippetti is Marie Curie Fellow at the London School of Economics and Political Science, Department of Geography and Environment, and Researcher at the National Research Council of Italy (CNR) Institute for Regional Studies. He is interested in innovation and technological change, the globalization of science and technology, and intellectual property rights. He has been Visiting Fellow at Birkbeck College, University of London, Fulbright-Schuman Post Doc at Harvard University, Center for European Studies, Visiting Fellow at the University of Queensland, Faculty of Economics, Center for Productivity Analysis, Australia, Marie Curie Visiting Fellow at the School of Government, Beijing Normal University, Beijing (China) in 2013, and Visiting Fellow at the Department of Political Science of Columbia University, New York. He has co-authored (with Daniele Archibugi) *Innovation and Economic Crisis: Lessons and Prospects from the Economic Downturn* (Routledge, 2011).

Kieron Flanagan is Senior Lecturer in Science and Technology Policy at the Manchester Institute of Innovation Research of the University of Manchester. He works on rationales for policy intervention and the construction of science policy problems; on policy dynamics in science, technology and innovation; on the international dimensions of science policy; on the place of science and technology in government; and on science and innovation policies for local and regional economic development. He has published in a range of innovation, science policy, and economic geography journals, is an active commentator on science policy issues, and is a founding contributor to the science policy blog of *The Guardian*.

Richard Florida is Director of the Martin Prosperity Institute at the University of Toronto's Rotman School of Management and Global Research Professor at New York University. He is the author of several global best-sellers, including the award-winning *The Rise of the Creative Class* (Basic Books, 2002). He previously taught at Carnegie Mellon and George Mason Universities, and has been a visiting professor at Harvard and MIT. He earned his Bachelor's degree from Rutgers University and his PhD from Columbia University

Gianfranco Gianfrate is an "Empedocle Maffia Fellow" of the Sustainability Science Program at the Harvard Kennedy School, an Assistant Professor of Finance at Bocconi University, and a Research Affiliate of Tufts University. He did both his undergraduate and postgraduate work (PhD in Business Administration) at Bocconi University. In 2014 he was awarded the Alfred D. Chandler Jr. Fellowship by the Harvard Business School. His current research focuses on the valuation, financing, and geography of entrepreneurial ventures.

Donald Gillies studied mathematics and philosophy at Cambridge University (1962–1966). In 1966 he became a graduate student in Sir Karl Popper's department at the London School of Economics, and completed his PhD with Imre Lakatos as supervisor in 1970. From 1971 until retirement in 2009, he taught at London University, and researched in the history and philosophy of science and mathematics. He is Emeritus Professor at University College London.

Frederick Guy is Senior Lecturer at Birkbeck University of London, Department of Management. He has been Visiting Scholar at the London School of Economics, University of California at Berkeley, Waikato University, Hamilton, New Zealand, Chuo University, Tokyo, and Consejo Superior de Investigaciones Científicas at Universidad Politécnica de Valencia, Spain. He is interested in international political economy and international business, labor economics and industrial relations, and inter-firm relations. He has published *The Global Environment of Business* (Oxford, 2009).

Stefan Hennemann is research associate in Economic Geography at the University of Giessen, Germany. He is interested in spatial network science (the analysis and visualization of geographical networks) and in the application of these methods to science, technology, and innovation systems and to world city networks. His research was published in the fields of economic geography (e.g., *Environment and Planning A/B*), computer and information science (e.g., *Journal of Informetrics*), and business and economics (e.g., *Journal of International Technology Management*).

Jean-Yves Heurtebise holds a doctoral degree of philosophy from Aix-Marseille University (AMU, France). He is currently Assistant Professor, Fu Jen Catholic

University, Taiwan (ROC). He is also Associate Member of the Research Center for Comparative Epistemology and Ergology at AMU and Affiliated Research Scholar of the Kozmetsky Global Collaboratory in Stanford University.

Simona Iammarino is Professor of Economic Geography at the Department of Geography and Environment of the London School of Economics and Political Science. Her main research interests lie in the following areas: multinational corporations, location and innovation strategies, and global-local interactions; geography of innovation and technological change; regional systems of innovation; regional and local economic development.

Grazia Ietto-Gillies is Emeritus Professor of Applied Economics at London South Bank University and Visiting Research Professor at Birkbeck University of London. She held Visiting Professorships at various international institutions and is one of the founders of the World Economics Association. Her main field of research has been the transnational corporation, its activities and effects. She has authored six books, co-edited two, and published many papers in international journals and as chapters in edited volumes.

Jeremiah Johnson is an Assistant Professor in Technology Leadership and Innovation at Purdue University, Indiana. He is also a Research Fellow of the Research Center for Open Digital Innovation at Purdue. He received a PhD in Information Systems from the Marshall School of Business at the University of Southern California. Prior to pursuing a doctorate, Jeremiah worked for SAIC at Sandia National Labs. His research interests include online communities, open innovation and data, collective preference, technology policy, and anonymity.

Sandeep Kapur is Professor of Economics at Birkbeck University of London. His research interests include foreign direct investment and technology diffusion especially in emerging markets. His recent publications have analyzed the phenomenon of outward FDI from China and India.

Thanh Quang Le is currently a Lecturer in Economics working in the School of Economics at the University of Queensland. He received his PhD from The Australian National University in 2007. His main research areas are growth and development economics. He is the author of several journal articles and has taught macroeconomics for many years at intermediate and graduate levels.

Jiang Li is an Associate Professor of Information Science at Zhejiang University, and a Researcher Associate at Tsinghua University, China. He received his PhD from Nanjing University in 2010, prior to which he studied at the University of Sheffield as a visiting student. The focus of his research is bibliometrics. Recently, he is attempting to apply bibliometric methodology to science and technology policies which have similar text structure to journal articles.

Ingo Liefner is a Professor of Economic Geography at the University of Giessen, Germany. He is an economic geographer with a strong interest in knowledge production and dissemination, innovation and upgrading, and regional economic change in newly industrializing countries. He has published in international journals in economic geography (e.g., *EPA*, *Applied Geography*, *Geoforum*), business (e.g., *Technovation*), science research (e.g., *Higher Education*, *Scientometrics*), and interdisciplinary (e.g., *Research Policy*, *Growth & Change*). His research has been funded with grants from the German Research Foundation, the Volkswagen Foundation, the German Academic Exchange Service, the Alexander von Humboldt Foundation, governments, and companies.

Mark Lorenzen is Professor at the Department of Innovation and Organizational Economics at the Copenhagen Business School, Denmark, director of DRUID, and editor of *Industry and Innovation*. His research is in the field of industrial dynamics, with a special focus on the relations between innovation and the economic organization of the market in networks, projects, and clusters, currently within the creative industries. He has published in journals including *Journal of Economic Geography*, *Organization Studies*, and *Economic Geography* and convened sessions at DRUID, Academy of Management, AIB, EGOS, and AAG.

Andrea Martínez-Noya is Associate Professor in the Management Department at the University of Oviedo in Spain. Her research is mainly focused on internationalization and cooperation strategies on R&D, especially on international R&D outsourcing practices by technological firms and how these firms can benefit from outsourcing. She was a visiting PhD at the Wharton School of the University of Pennsylvania and she is a Visiting Fellow at the John H. Dunning Centre for International Business, University of Reading (UK).

Philip McCann holds the University of Groningen Endowed Chair of Economic Geography. He was previously Special Adviser to EU Commissioner for Regional Policy Johannes Hahn (2010–2013), Chief Independent Economic Advisor on the EU Sixth Cohesion Report on Economic Social and Territorial Cohesion (2013–2014), and also works on range of research activities with both the OECD and the European Investment Bank. He is co-editor of *Spatial Economic Analysis* and also editor of the Edward Elgar "New Horizons in Regional Science" series.

Charlotta Mellander is Professor of Economics, Jönköping International Business School. Her research is focused on the location patterns of creative individuals and firms, as well as the interrelationship between these actors, and how this in the end shapes regional development.

Ian Miles is Professor of Technological Innovation and Social Change at Manchester Business School, University of Manchester, and Head of the Laboratory for the Economics of Innovation in Moscow. His work for this volume is an output of a research project implemented as part of the Basic Research Program at the National Research University Higher School of Economics (HSE), where this Laboratory is based. He is researching service innovation, KIBS, and other topics.

Marcela Miozzo is Professor of Economics and Management of Innovation at Manchester Business School, University of Manchester. She has published around 40 refereed journal articles and books, many focusing on the relation between internationalization and innovation of services firms. Her recent research is on collaboration for innovation of knowledge-intensive service firms and the means through which knowledge-intensive service firms capture value from innovation. She is also working on organizational forms and governance in science-based sectors.

Ram Mudambi is Professor and Perelman Senior Research Fellow at the Fox School of Business, Temple University, Philadelphia and Visiting Professor at the Henley Business School, University of Reading, UK. He previously served on the faculties of the University of North Carolina Chapel Hill and Case Western Reserve University. His current research projects focus on innovation and governance of knowledge-intensive processes. He has published over 80 peer-reviewed articles, including work in the *Journal of Political Economy*, the *Journal of Economic Geography*, the *Strategic Management Journal*, and the *Journal of International Business Studies*. He holds a Master's degree from the London School of Economics and a PhD from Cornell University.

Moritz Müller is Assistant Professor at the Faculty of Economics and Management of the University of Strasbourg and researcher at the Bureau for Economic Theory and Applications (BETA), a research laboratory of the University of Strasbourg, University of Lorraine, and the French National Centre for Scientific Research (CNRS). His research centers on the implication of human interaction in industrial as well as science systems.

Rajneesh Narula is Professor of International Business Regulation at the Henley Business School, University of Reading, UK. He is also Adjunct Professor at the Norwegian School of Business in Oslo. His research and consulting have focused on the internationalization of innovation, R&D alliances, and the role of multinational firms in industrial development.

Jose Miguel Natera is Catedras CONACYT researcher at the Universidad Autónoma Metropolitana – Unidad de Xochimilco in México and associate researcher at the Instituto Complutense de Estudios Internacionales (ICEI). He holds a PhD in Economics and Management of Innovation (Universidad Complutense de Madrid, 2014). His main research interests are linked to use of knowledge as a development tool, emphasizing the evolutionary aspects of this relationship.

Andrea Piccaluga is Professor of Innovation Management at the Institute of Management of Scuola Superiore Sant'Anna in Pisa (Italy) where he is Deputy Director and coordinator of the PhD program in Management. He holds a PhD from Scuola Superiore Sant'Anna and a Master's in Technology and Innovation Management from SPRU (University of Sussex). He has published books and papers in the field of R&D management and technology transfer.

F.M. Scherer is Aetna Professor Emeritus at the John F. Kennedy School of Government, Harvard University. His research specialties are industrial economics and the economics of technological change, leading inter alia to his books *Industrial Market Structure and Economic Performance* (Rand McNally, 1970), *Innovation and Growth: Schumpeterian Perspectives* (MIT Press, 1984), *Mergers, Sell-offs, and Economic Efficiency*, with David J. Ravenscraft (Brookings Institution, 1987), *Industry Structure, Strategy, and Public Policy* (Harper Collins, 1996), and *New Perspectives on Economic Growth and Technological Innovation* (Brookings Institution, 1999).

Paul Stoneman is Emeritus Professor, formerly Research Professor, Warwick Business School, University of Warwick, UK. He has been Visiting Professor at Stanford University and Visiting Fellow at Nuffield College, Oxford. He has been a former member of the UK Competition Appeal Tribunal. He is also an advisor to governments and private sector companies on innovation, productivity, and performance. Research interests center upon the economics of innovation and technical change, especially diffusion.

Kam Ki Tang is an Associate Professor in Economics at the University of Queensland and a research associate at the Centre for Applied Macroeconomic Analysis (CAMA) at The Australian National University. His past and current research covered development, health, education, and international economics. He is particularly interested in interdisciplinary research.

Alessio Terzi is Affiliate Fellow at Bruegel, Brussels and Research Associate at the Hertie School of Governance, Berlin. His work focuses on competitiveness, structural reforms, and determinants of long-term growth.

Riccardo Tremolada is PhD Candidate at University Federico II Naples (Italy), SJD Candidate at Shanghai JiaoTong University (China), Research Fellow, Università degli Studi del Piemonte Orientale, Dipartimento di Studi per l'Impresa e il Territorio (Italy), and a Research Associate at gLAWcal – Global Law Initiatives for Sustainable Development (UK).

Andrew Tylecote is Professor of the Economics and Management of Technological Change, University of Sheffield. His book *The Long Wave in the World Economy* (Routledge, 1992), described by Chris Freeman as a tour de force, looked at the interaction between techno-economic paradigms and socio-institutional frameworks over two centuries. *Corporate Governance, Finance and the Technological Advantage of Nations*, with Francesca Visintin (Routledge, 2007) shared the 2010 Myrdal Prize. He has been Visiting Professor at Tsinghua and Zhejiang Universities.

Ping Zhou is a Professor of Information Science at the Department of Information Resources Management, School of Public Affairs, Zhejiang University. Former Senior Researcher at the Institute of Scientific and Technical Information of China. Researcher at the Centre for R&D Monitoring, Catholic University Leuven. She holds a PhD from the University of Amsterdam. Her research interests include bibliometrics, research evaluation, and science and technology management.

Editors' Introduction: Science, Technology, and Innovation Go Global

Daniele Archibugi and Andrea Filippetti

"ideas, knowledge, art, hospitality, travel – these are the things which should of their nature be international" (John Maynard Keynes, 1933)

It is a great pleasure to present in this series dedicated to global policy a collection of fresh papers devoted to science, technology, and innovation. The realm of knowledge is, in fact, an area which is most contributing to the coming of a genuine global society. The community of learned, what in the Enlightenment was labeled the Republic of Letters, has been one of the first components of society to challenge existing geographical and political boundaries, exploring and establishing interactions across different spaces. Medieval cloisters, academic circles, and, more recently, universities and research centers, have welcomed contributions from outsiders and have often been free land for anybody willing to exchange ideas. Scientists as well as craftsmen have been early frequent travelers to learn what other communities were offering and to disseminate the outcome of their ingenuity. Driven by the passion for knowledge and in search of personal rewards, the community of learned has been one of the great vehicles for disseminating ideas, know-how, artefacts, products, and techniques. Long before the word entered into the jargon of our age, knowledgeable people have been mighty pioneers of globalization.

The relationship between science, technology, and innovation and globalization is however not unidirectional. Revolutions in transports, ICT advent, migrations, growth in international trade and in foreign direct investment, are aspects that have substantially shaped human life, including the way in which scientists and engineers work and how they organize their activities. This Handbook is a comprehensive collection of essays that attempts to answer a number of questions related to the way in which the forces of globalization have affected the generation and diffusion of science, technology, and innovation (STI). It is common sense that STI has become more important in everyday life and we know by instinct that the pills we swallow and the mobile phones in our pockets are associated with investigations and discoveries carried out in several parts of the world. But are we sure that the opportunities

The Handbook of Global Science, Technology, and Innovation, First Edition.
Edited by Daniele Archibugi and Andrea Filippetti.

offered by the new global age have been fully exploited? How can society at large better benefit on a global scale from the knowledge developed?

The rise of the so-called distance-shrinking technologies and their global diffusion, like the World Wide Web, is making communication and the exchange of information virtually instantaneous and with very negligible costs. In a thought-provoking book Thomas Friedman (2006) claims that our planet has become a level playing field as a result of globalization, and that most aspects are today dominated by a global market, which has made historical and geographical divisions increasingly irrelevant. Is this also true for STI? Has the generation of STI become "placeless" or does location continue to play a crucial role in the successful generation of knowledge?

These days, big advancements in science like the mapping of the genome, the discovery of the Higgs boson at the CERN, or the landing of the Rosetta spacecraft on a comet are all the result of big international projects. The CERN was founded by 12 countries and today has 21 member states, it employs over 2400 people, representing 608 universities and research facilities and 113 nationalities. The Human Genome Project was also established as an international, collaborative research program whose goal was the complete mapping and understanding of all the genes of human beings and it remains the world's largest collaborative biological project. Almost all of the actual sequencing of the genome was conducted at numerous universities and research centers throughout the United States, the United Kingdom, France, Germany, Japan, and China. While we are drafting this introduction, for the first time in history a spacecraft has just landed on a comet as a result of an effort of the European Union in establishing a European Research Area with a budget dedicated to collaborative cross-country research.

Collaborative research is becoming the norm also in ordinary research activity. Collaboration across national boundaries is generally increasing, as reflected in international co-authorship of scientific articles. According to the US National Science Foundation[1] in 1988, only 8% of the world's science and engineering (S&E) articles had international co-authors; by 2009, this share had grown to 23%. For the world's major advanced regions, the 2009 rate ranged from about 27% to 42%.

The production of technology and innovation has also become an activity which is more and more carried out on a global basis. It is typical for corporations in the automobile, electronics, and chemical industries to establish Research and Development (R&D) abroad, in both advanced countries and, more recently, emerging countries. Western companies develop innovations also in emerging countries that eventually are exploited in their domestic markets. Innovations are often the results of cross-country collaboration activity.

This is also extremely visible by looking at figures on the internationalization of patenting activity, one of the typical methods used by profit-seeking agents to secure returns from their inventions and innovations. For many years, inventors were exploiting their innovations in their domestic markets only and were not particularly eager to file their patents abroad. In the last decades, the situation has dramatically changed and foreign-oriented patents – patents that are granted in at least one country other than the applicant's home country – have quintupled since 1970.[2] Today, the inventions of emerging economies are becoming a fundamental component, and are growing faster than those in established countries. This witnesses a striking shift in the strategies of firms regarding the geographical scope of their innovations.

The influence of globalization is not only visible in the *production* of STI, but is also macroscopic in the *diffusion* of STI. The diffusion of knowledge has increased rapidly as a result of a number of factors. Global integration has accelerated exchanges across countries and the development of global value chains (Meliciani and Savona 2014). Further, distance-shrinking technologies have dramatically reduced the time and cost of transmitting information across the world. As a result, any new technology could reach any corner of the world, provided there is enough absorptive capacity to make sense of it. The lags with which new technologies arrive to countries have dropped dramatically over the last 200 years: "technologies invented in the nineteenth century such as telegrams or railways often took many decades to first arrive to countries. In contrast, new technologies such as computers, cell phones or the internet have arrived on average within a few decades (in some cases less than one) after their invention" (Comin and Ferrer 2013: 1).

Mobility of people has also increased dramatically, encouraging face-to-face interaction, adding another important dimension to the globalization of STI. Professor Venkatraman Ramakrishnan, who received the 2009 Nobel Prize in Chemistry for studies of the structure and function of the ribosome, was born in India and studied at Ohio University but when he received his Nobel Prize, he worked in Cambridge in the United Kingdom. This is a rather recurrent story of top scientists and engineers. While migration has historically characterized individuals with low skills and education, over the years a new phenomenon has emerged in the form of migration of highly skilled workers and talents, the so-called *global creative class* (Florida 2005). Today, when countries debate their immigration policy, one question at stake is how to attract skilled workers who foster innovation and entrepreneurship. Inventors tend to be quite mobile, with 10% of inventors worldwide showing a migratory background in 2005. The United States is by far the most popular destination for migrant inventors, hosting 57% of the world's inventors who reside outside their home country. By contrast, China and India are the largest providers of inventors (Fink, Miguelez, and Raffo 2013).

The globalization of STI has not made the world flatter but rather more fragmented. Spatial agglomeration continues to be crucial for the generation and diffusion of knowledge and it continues to exert an effect on a larger scale. The economic agents producing knowledge and technologies in specific domains continue to agglomerate in specific areas and districts, which are extending also in emerging countries. Bangalore in India is developing along lines that resemble the Silicon Valley in California, while the large concentration of hi-tech industries in a few coastal regions in China is mirroring the concentration of technology-based activities in some regions of the Western countries. The geography of technology and innovation has not disappeared with globalization; rather it has expanded its scale.

An Attempt to Categorize the Globalization of STI

The facts and trends mentioned above are explored in a growing literature and the chapters collected in this volume provide an overview of how globalization is shaping STI. The increasing evidence collected by scholars, business strategists, and policy analysts is often fragmentary and can sometimes point out different or even opposite trends. It might therefore be useful to approach such a huge literature through a categorization of the various forms of globalization of STI, looking at how knowledge is developed, disseminated, and exploited. An attempt to categorize global STI is provided in Table 0.1. It is possible to identify four principal ways according to which STI is generated and disseminated in the global sphere. This categorization has as unit of analysis the individual scientific and technological activity (i.e., a successful research project, the introduction of a product, a process, or component). It is therefore possible that the same players (e.g., a university, a company, a research lab) make use of different categories for each individual STI activity they carry out.

The first category, **nationally confined knowledge developments**, occurs when there is no knowledge transmission across borders. This category should not be included in the globalization discourse since both the knowledge development and its dissemination occurs at the local or at the national levels. We may wonder how important this category is in contemporary society, and many readers will find it difficult to visualize some knowledge components or technical innovations that have been confined inside the local space only. Of course, these inventions and innovations do exist; empirically, we may identify them through some outcomes: scientific articles written in national languages and that are not cited outside their own

Table 0.1 Categories of globalization of scientific, technological, and innovative activities.

Categories	*Public*	*Business*
Nationally confined knowledge development	Activities promoted by public institutions in the national context	Business players developing knowledge and innovations for local markets only
International dissemination and exploitation of nationally generated knowledge and innovation	Although developed predominantly in the national context, the outcome of knowledge is disseminated internationally through a variety of channels of the academic community, including scientific papers and conferences	National and multinational corporations (MNEs) develop knowledge within their home country and attempt to profit from them internationally through trade, licenses, or international production
Global techno-scientific collaborations	Preserving their national identity, this includes collaboration across scientists and institutions of different countries, generally for specific research projects	Independent companies undertake arm-length strategic technological agreements to reduce the costs and risks of specific innovative projects
Global generation of knowledge	Knowledge generated in permanent institutions financed by several national governments or by international organizations	Knowledge generated by MNEs in their headquarters and their subsidiaries, through intra-firm but international innovative networks

Source: Archibugi and Iammarino (2002), Archibugi and Michie (1995).

country; patent applications within national offices that are not extended abroad; and, perhaps more importantly, forms of tacit knowledge that serve specific local purposes. As first approximation, we can identify this knowledge as experiments but not particularly fruitful ones since, if successful, it is likely that their diffusion would also contaminate other parts of the world. Recently, there has been an increasing interest for knowledge specific to local communities, a sort of anthropology of knowledge able to bring to the fore the knowledge of indigenous communities (see Agrawal 1995). This would include, for example, the therapeutic properties of plants and flowers known to indigenous communities but not yet codified. An attempt to explore and exploit this knowledge has also been carried out by profit-seeking agents, often creating tensions concerning the intellectual property.

The second category, **the international dissemination and exploitation of nationally generated knowledge and innovation**, comprises knowledge that is developed within national contexts but with explicit attempts to disseminate it across countries. Both the public and the business sector contribute to this category. On the one hand, this includes activities promoted by the government and by the non-profit sector. The international dissemination is generally free and it occurs through the standard channels used by the academic community such as publications in scientific journals, books, conferences, personal contacts, and others.

On the other hand, the business sector also makes explicit attempts to profit from its innovations overseas. When innovations introduced into the domestic market prove to be successful and to get substantial market shares, companies have strong incentives to exploit them also in foreign markets. We can imagine these knowledge advances when the original focus and its more important elaborations are clearly identified with a national location, which is typically the nation hosting the headquarter of the innovating company. The company, however, makes explicit attempts to profit from the innovations also in foreign markets. The traditional way is through trade, exporting abroad innovative products or processes that are competitive because of the advantages associated to the innovation. But when trade is obstructed by high transport costs or by administrative restrictions, the innovating company may also find it convenient to license the know-how to foreign corporation. A third option may be to exploit the innovation in foreign markets by establishing production facilities in the host country, especially when labor costs in the host country are lower than in the home country.

The third category is represented by **global scientific and technological collaborations** among separate institutions. These collaborations have operated for centuries within the academic community: sabbatical years, student exchange, joint research programs, are established forms used by scholars to learn and to disseminate their expertise. These collaborations have for long been supported by public policies, for example through scientific academies and other support schemes, under the assumption that exchanges with other communities will make the national scientific base stronger. The key belief behind these long established forms of collaboration is that the knowledge developed is not rival and that the advantages obtained through collaboration are much larger than those provided by isolation.

It is more than 30 years since attention was devoted to explore how similar forms of collaboration also operate in the business sector. Companies are much keener than expected to share their knowledge with actual and potential competitors and to undertake strategic technological agreements to share the costs and risks of innovative projects. The conventional economic model that provides paramount importance to inter-firm competition is not particularly suited to explain the new generation of inter-firm technology agreements that came to the fore in the early 1980s and that is still progressing. It can be said that the business sector has learnt from universities that collaboration is vital to augment the knowledge base and that there is more convenience in sharing know-how with rivals rather than in keeping it secret. The legal controversy between Apple and Samsung concerning smart phones has shown how innovation can be the competitive edge of a company and that companies use all their muscles to protect their successful know-how. But this is only part of the story: behind such a striking case of commercial rivalry, the courts soon discovered that the two contending firms undertook technological strategic agreements to elaborate selected products together, and that some devices and methods were developed by third companies with whom both of them collaborated.

The fourth and last category is the **global generation of knowledge**. There are cases of important STI activities that cannot be associated to any specific national flag since scientists and engineers from different countries participated in its achievement. In the public sector this happens when genuine transnational laboratories and research centers are established. CERN is one of the best known cases of a genuine global laboratory. The European Union has promoted several other permanent research centers and devotes a substantial part of its budget to scientific and technological activities carried out in collaboration among member countries.

But the public sector has not been the first to use the global generation of knowledge. MNEs have used this form after the Second World War, creating research laboratories and technical centers also in the countries of their subsidiaries. Through intra-firm but international research networks, large MNEs often manage to develop complex products as well as incremental innovations, exploiting the expertise available in each national innovation system. MNEs' introduction of successful products and processes is frequently associated with the effectiveness of their internal networks which explore, examine, and assimilate the scientific and technological opportunities of different continents.

In the presentation of Table 0.1, we have preferred to keep separate the public and the business sectors. This might be useful analytically, but it ignores another important form of collaboration that has gained greater importance in the last two decades, namely the collaborations between the public and the business sectors in STI. These collaborations are often national, where companies collaborate effectively with the governments and the universities of their home country. But also these forms of public/private partnership are less and less confined to national boundaries, and integrate universities with companies of host countries.

Can the four categories singled out above also be considered diachronic? Not entirely, since all the four forms can be traced even in ancient times. But their relative relevance has been modified over time, and the later categories have grown in importance at the expense of the previous ones. The first category, nationally confined knowledge development, seems today residual and certainly unable to describe the most significant knowledge progresses of our age. The second category, international dissemination and exploitation of nationally generated knowledge and innovation, is probably quantitatively still the most important, but it is losing ground since both public and business players engage in transnational collaborations not only for the exploitation and dissemination of their knowledge, but also for its generation. Global techno-scientific collaborations are becoming increasingly important in both the public and the business sectors. But what will probably become the distinctive modality of the twenty-first century is the global generation of knowledge, at least for what concerns the most relevant scientific and technological advances.

An Overview of the Book

The impact of globalization on STI is first explored in terms of the ongoing trends in Part I of the book ("Global Trends"). The chapters in this part show that STI is less a domain of the most advanced countries. Until a few decades ago, North America, Europe, and Japan – haughtily self-labeled the Triad – accounted for the dominant share in the production of science and technological advancements and countries outside these regions were, at most, extras, often relying on the knowledge developed elsewhere. Measures such as R&D expenditure, scientific articles, and patents, reflected the dominance of the Triad in the generation of STI. These chapters show instead that a process of convergence in science and technological capabilities is taking place at a global scale. Technological capabilities are increasingly spreading in emerging economies; the production of science shows a tendency to move to the East. This is also mirrored in the rise of first-class universities outside Western countries and in the distribution of key technical infrastructures for basic science like supercomputers. We have also to deal with how the concept itself of innovation is evolving in an increasingly post-industrial society. While for many years it was understood mainly as "technological," it is now largely accepted that other forms of innovations, including aesthetic changes, may be of crucial importance for economic competitiveness.

Part II ("The Globalization of Technology and Innovation") reports a wealth of analyses on how these activities are carried out. Transnational corporations continue to be leading players and often create important *inter*-national but *intra*-firm networks, integrating, through a variety of channels, also emerging and developing countries. Through international R&D alliances and capital and technology flows, companies manage to exploit and to assimilate knowledge and technological expertise developed elsewhere. The service sector is no longer a passive spectator and absorber of the innovations developed in manufacturing, as it was for most of the twentieth century, but is also generating its expertise through more complex cross-border interactions. This Part also has a chapter devoted to crowdfunding, a rising form to finance innovative projects. To what extent will this form be able to overcome the traditional difficulties of financing innovation?

Part II also addresses the changing ways of organizing STI in developing countries. In the past decades we have witnessed a reverse of the international cycle of technology as traditionally set out in Vernon's (1966) early works, shifting from a scenario where innovation was only developed in advanced economies to one where global innovation also stems from inputs from emerging economies. Large firms in developing countries acquire technology through outward foreign investment, typically through acquisitions of firms with a portfolio of technology products. The rise of large markets, such as China and India, has led foreign companies to internationalize their R&D activities there. Western MNEs are more and more focusing on how the Chinese market, as well as other emerging markets, could provide these companies with new sources of innovation, a phenomenon called *innovation in reverse*. Geographical agglomerations prove to be very important also in developing countries. Clusters emerge as a viable strategy for catching up in emerging areas or even for becoming global technological leaders.

Part III ("Spaces and Flows of Knowledge") discusses the role of geography in the globalization of STI, and the importance of transnational collaboration and mobility as drivers of knowledge flows. Globalization has extended the importance of spatial agglomeration across the globe. This natural tendency of innovative activity to cluster spatially becomes a major concern for policymakers since it may lead to the exclusion or the marginalization of vast areas. When policies try to pursue economic development through innovation it becomes fundamental to understand what makes locations more or less attractive.

In the current phase of economic globalization, geography and space are increasingly important for both public and business organizations. The rise of city-regions as knowledge sources have led to increasingly differentiated geographies across all parts of the world. But locations have to interact with people: they may be more or less knowledge-friendly and innovative also according to how they support actual and prospective residents. To this, one has to add the new dimension of *creativity*. More than anything, creativity has become the key attribute for successful, sustainable, regional economic growth. A new creative class, more mobile than other population groups, is often leading economic, social, and cultural progress. Its geography, however, is profoundly uneven, not just across but within nations. The way in which companies and public institutions, research labs, and universities manage human resources becomes crucial not only for their performance but also for the economic environment in which they operate.

A great driver of knowledge flows has always been human interaction. In fact, it has been argued that the rise of the information communications technologies (ICT) have increased, rather than reduced, the need for face-to-face interaction: "Face-to-face contacts are particularly important in environments where information is imperfect, rapidly changing, and not easily codified, key features of many creative activities" (Storper and Venables 2004). This Part addresses various mechanisms by which this can occur, such as global collaboration in scientific projects, international mobility often leading to brain drain, and interpersonal ties of scholars.

Part IV ("Global Institutions and Intellectual Property Rights") addresses the importance of institutions and international regimes for the globalization of STI. The international economic landscape is periodically rattled by controversies concerning intellectual property rights (IPRs). Recent cases of the so-called smart phone war or anti-HIV drugs are just a few examples. These controversies reflect the fact that in the current global economy, knowledge and intangibles have become increasingly important both as production factors and as consumption goods. While previously IPRs were mostly a matter of nation states, they became more relevant at a global scale after the introduction of a global regime of protection of intellectual property with the establishment of the "Trade Related Aspects of Intellectual Property Rights" (TRIPS) within the World Trade Organization in 1994. This Part reframes the debate regarding the globalization of IPRs by discussing how the generation, transmission, and diffusion of knowledge is a complex phenomenon and that what IPRs can do and what they cannot do to reward inventors and innovators and to prevent imitators is limited. It then addresses the key issue of how, within this context, least affluent citizens can obtain the benefits of modern pharmaceutical therapy at affordable prices. Finally, it discusses the relationship between intangible cultural heritage protection, intellectual property rights regime, and the Information Society.

The final Part V ("The Global Governance of Science and Technology") addresses some more general questions about governance. The first relates to the importance of where knowledge is generated and how it gets diffused worldwide, and whether there is the need of some kind of governance or not. This recalls the classic case for public intervention in the case of public goods, which has also been made about knowledge since the early studies by Nelson (1959) and Arrow (1962). In addition, today, governments release governmental data that were previously hidden from the public. This democratization of governmental open data aims to increase transparency but also fuels innovation. Indeed, the release of governmental open data is a global trend, which has evolved into governmental open innovation, a new perspective on technology and innovation policy that has so far received very scant attention.

Part V then touches upon other relevant issues that are at stake. Two chapters discuss the risks associated with scientific and technological developments, for instance environmental disasters, from philosophical and economic standpoints respectively. Another chapter takes an epistemological perspective and discusses the role of serendipity and chance in scientific and technological discovery, examining the implications for the individual researcher and the policy implications for the funding and organization of research in a global society. Finally, the last chapter challenges the view of international economic integration as being necessarily *global*, by claiming that regional economic integration – in the sense of subcontinental mega-regions such as Europe and China – is likely to eclipse global integration in coming decades.

Efforts and Ambitions

The realm of knowledge is already one of the most important spaces of our society and it is an easy forecast that it will become even more important in the future. We need to get the appropriate tools to understand how it is working, how it is transforming the global landscape and, in return, how globalization is also shaping the STI community. The chapters collected in this volume provide a substantial amount of tools and make a brave attempt to identify geographical, economic, and social similarities and differences in the phenomena observed. We hope very much that this handbook will be helpful for the community of scholars working in the economics of innovation, in social studies of science and technology, economic geography, global studies, and in science and technology policy studies. We made a special effort in providing chapters that could be useful for teachers as well as for their students. Our ambition, however, is to be able to involve in this fascinating area also scholars outside the traditional boundaries of the discipline. Scientists and engineers, university researchers and company technicians, are the leading protagonists of inventions and innovations that nurture our society and that allow progress. We hope that some of them will find it useful to read the analyses here presented that ultimately try to make sense of what they do.

Acknowledgements

Editing a handbook like this one is first and foremost a collective venture. Therefore our gratitude goes in the first place to the enthusiastic and professional participation of all the authors. We are also indebted to Ben Thatcher from Wiley-Blackwell for his relentless and professional support. Daniele would like to acknowledge a grant from the School of Business, Economics and Informatics, Birkbeck College, University of London, which initially supported research on the globalization of intellectual property rights. He would also like to thank his colleagues at the Italian National Research Council, IRPPS, and at the Department of Management, Birkbeck, for their continuous support. Andrea would like to acknowledge that this project has been supported by a Marie Curie Intra-European Fellowship project under the EC Grant Agreement No. PIEF-GA-2011-298167, carried out at the Department of Geography and Environment of the London School of Economics and Political Science. He would like to thank the Department for its support, and to express a particular gratitude to his supervisor, Simona Iammarino. Part of this work has been carried out during his pleasant visiting period at the Department of Political Science at the University of Columbia, New York. Finally, he would also like to thank his colleagues at the Italian National Research Council, ISSiRFA, and particularly Stelio Mangiameli, for their continuous support.

Rome and New York, December 2014

Notes

1 Statistics available at http://www.nsf.gov/statistics/seind12/c0/c0s7.htm (accessed December 17, 2014).

2 See data and various annual reports from the WIPO at http://www.wipo.int/ipstats/en/statistics/patents/ (accessed December 17, 2014).

References

Agrawal, Arun. 1995. "Dismantling the Divide Between Indigenous and Scientific Knowledge." *Development and Change* 26(3): 413–439.

Archibugi, Daniele, and Simona Iammarino. 2002. "The Globalization of Technological Innovation: Definition and Evidence." *Review of International Political Economy* 9(1): 98–122.

Archibugi, Daniele, and Jonathan Michie. 1995. "The Globalization of Technology: A New Taxonomy." *Cambridge Journal of Economics* 19: 121–140.

Arrow, Kenneth. 1962. "Economic Welfare and the Allocation of Resources for Invention." In *The Rate and Direction of the Inventive Activity: Economic and Social Factors*, ed. R.R. Nelson. Princeton, NJ: Princeton University Press.

Comin, D.A., and M.M. Ferrer. 2013. "If Technology Has Arrived Everywhere, Why Has Income Diverged?" National Bureau of Economic Research (Cambridge, MA), Working Paper 19010.

Fink, Carsten, Ernest Miguelez, and Julio Raffo. 2013. "The Global Race for Inventors." *VoxEU.org*. http://www.voxeu.org/article/global-race-inventors (accessed December 17, 2014).

Florida, Richard. 2005. *The Flight of the Creative Class: The New Global Competition for Talent*. New York: HarperCollins.

Friedman, Thomas L. 2006. *The World Is Flat: A Brief History of the Twenty-First Century*. New York: Farrar, Straus and Giroux.

Meliciani, Valentina, and Maria Savona. 2014. "The Determinants of Regional Specialisation in Business Services: Agglomeration Economies, Vertical Linkages and Innovation." *Journal of Economic Geography* (published online).

Nelson, Richard R. 1959. "The Simple Economics of Basic Scientific Research." *Journal of Political Economy* 67(3): 297–306.

Storper, Michael, and Anthony J. Venables. 2004. "Buzz: Face-to-Face Contact and the Urban Economy." *Journal of Economic Geography* 4(4): 351–370.

Vernon, Raymond. 1966. "International Investment and International Trade in the Product Cycle." *Quarterly Journal of Economics* 80(2): 190–207.

Chapter 1

The Convergence Paradox: The Global Evolution of National Innovation Systems

Fulvio Castellacci and Jose Miguel Natera

Introduction

The investigation of the reasons behind cross-country differences in the creation and international diffusion of advanced knowledge constitutes, in a nutshell, the main interest of the national innovation systems (NIS) approach (Lundvall 1992; Nelson 1993; Edquist 1997). Studies in this tradition have become increasingly popular in the last two decades, and have greatly enriched our understanding of the functioning and evolution of nations' capabilities and competencies. In addition to the great variety of theoretical insights that the systemic perspective has made it possible to achieve, several empirically oriented studies have also tried to operationalize the approach by measuring different aspects of the process of knowledge creation and dissemination in a large sample of countries, and by investigating the extent of cross-country differences (Archibugi and Coco 2004).

Several empirical studies in this tradition are rooted in the traditional literature on technology and convergence (Abramovitz 1986; Verspagen 1991; Fagerberg 1994). Following a technology-gap Schumpeterian approach, applied econometric studies have focused on a few key variables that explain cross-country differences in the innovation ability of countries as well as their different capabilities to imitate foreign advanced knowledge, and analyzed the empirical relationship between innovation and imitation factors and cross-country differences in GDP per capita growth (Fagerberg and Verspagen 2002; Fagerberg, Srholec, and Knell 2007; Castellacci 2008; Castellacci and Archibugi 2008).

Most of this empirical research, however, has so far focused on the cross-country comparative aspect ("why growth rates differ") and mostly neglected the time

The Handbook of Global Science, Technology, and Innovation, First Edition.
Edited by Daniele Archibugi and Andrea Filippetti.

series dimension and the analysis of the dynamics of the technological catch-up and economic growth process over time. Further, an exceptional amount of research has been devoted to the study of the determinants of GDP and income per capita growth. By contrast, only a limited number of studies have empirically investigated the dynamics of innovative capability and absorptive capacity over time and the main factors that may explain their long-run evolution (Castellacci 2011; Filippetti and Peyrache 2011; Castellacci and Natera 2013). Thirdly, applied research in this field has so far provided a rather stylized view and highly simplified operationalization of the multidimensional concept of innovation systems, typically focusing on a few key variables (e.g., R&D, patents, human capital) and neglecting several other potentially relevant factors. In short, the existing literature provides only limited insights on the global evolution of national systems of innovation and the mechanisms that may explain their evolution and growth over time. This is a crucial task for future research in this field, and the present chapter intends to take a step in this direction.

Motivated by this background, this chapter presents an empirical analysis of the global evolution of national innovation systems. The objective is to carry out a general and comprehensive study aimed at describing the cross-country distribution of a large number of technological, economic, and social factors that broadly contribute to define national systems of innovation, and how these factors have evolved during the period 1980–2008. The work focuses on six main dimensions characterizing NIS, three of which are related to the techno-economic domain (innovation and technological capabilities, openness, and infrastructures), while the other three define countries' socio-institutional system (education, political institutions, and social cohesion). We adopt a large number of statistical indicators to measure these six conceptual dimensions. These indicators are from the CANA database, a newly released dataset that provides a rich set of information on a large number of countries worldwide for the last three-decade period, and enables a dynamic cross-country analysis of national systems, growth, and development (Castellacci and Natera 2011).

The results of our empirical analysis, in short, point out a contrasting pattern that we call the *convergence paradox*. On the one hand, the three dimensions defining the socio-institutional system – basic education, political institutions, and social cohesion – have experienced a process of worldwide convergence. On the other hand, however, the three dimensions related to the techno-economic system have experienced a marked divergent dynamics with increasing polarization between rich and poor country groups. The paradox we outline is therefore that national systems have progressively become more similar in terms of their socio-institutional frameworks while at the same time diverging and becoming more different with respect to their techno-economic pillars.

Admittedly, the objective and nature of this work is descriptive rather than analytic. The chapter does not intend to analyze the interactions and complex coevolutionary process that links together the techno-economic and socio-institutional variables that define national systems. Rather, as other contributions in this volume, we seek to provide a comprehensive and empirically founded overview of a key issue in the study of global science, technology, and innovation systems.

Literature and Conceptual Framework

Empirical Studies on National Innovation Systems (NIS)

NIS are key drivers of economic growth and competitiveness. The NIS perspective originally developed during the 1990s to understand the broad set of factors shaping the innovation and imitation ability of countries, and how these factors could contribute to explain cross-country differences in economic growth and competitiveness (Lundvall 1992; Edquist 1997). The study of NIS focuses on the main components of the system, such as private firms and public organizations, and investigates their mutual interactions as well as their relationships with the social and institutional framework in which the system is embedded (Freeman 1995; Lundvall 2007). Empirical studies in this tradition initially focused mostly on advanced economies in the OECD area (Nelson 1993), although the NIS literature has recently shifted the focus toward the study of innovation systems within the context of developing and less developed economies (Lundvall et al. 2009).[1]

NIS empirical literature has traditionally devoted substantial attention to the process of international knowledge diffusion and investigated the set of factors that affect the extent to which a national system is able to grow and catch up with the technological frontier by means of international learning and imitation activities. This approach was originally inspired by the work of authors such as List, Landes, Gerschenkron, and Abramovitz, which, by focusing on historical case studies of the technological catch-up process, pointed out that international knowledge diffusion is a complex and demanding process, and investigated the set of factors that are necessary for imitation-based technological development, so-called *absorptive capacity* of a country (Abramovitz 1986).

More recently, a strand of empirical research made an effort to measure some of the relevant dimensions composing a national system through aggregate data and statistical indicators, and carried out applied econometric investigations of the relationships between these dimensions, on the one hand, and income per capita dynamics, on the other. However, a well-known challenge for applied research in this field is how to operationalize the innovation system view in empirical studies and, relatedly, how to measure the complex and multifaceted concept of NIS and its relationship to countries' economic performance. Quantitative applied studies of NIS and development have so far made use of two different (albeit complementary) approaches.

The first approach is rooted in the traditional literature on technology and convergence (Abramovitz 1986; Verspagen 1991; Fagerberg 1994). Following a technology-gap Schumpeterian approach, recent econometric studies have focused on a few key variables that explain (or summarize) cross-country differences in the innovation ability of countries as well as their different capabilities to imitate foreign advanced knowledge, and analyzed the empirical relationship between these innovation and imitation factors and cross-country differences in GDP per capita growth (Fagerberg and Verspagen 2002; Castellacci 2004; Castellacci 2008; Castellacci 2011; Fagerberg et al. 2007). Since one main motivation of this type of studies is to analyze the dynamics and evolution of national systems in a long-run perspective, they typically consider a relatively long time span (e.g., from the 1970s or 1980s onward), but must for this reason focus on a more restricted sample of countries (e.g., between 70 and 90 countries). Due to the lack of statistical data for a sufficiently long period of time, therefore, a great number of developing economies and the vast majority of less developed countries are often neglected by these cross-country studies.

The second approach is based on the construction and descriptive analysis of composite indicators. In a nutshell, this approach recognizes the complex and multidimensional nature of national systems of innovation and tries to measure some of their most important characteristics by considering a large set of variables representing distinct dimensions of national systems and capabilities, and then combining them together into a single composite indicator – which may be interpreted as a rough summary measure of a country's relative position vis-à-vis other national systems. Desai et al. (2002) and Archibugi and Coco (2004) have firstly proposed composite indicators based on a simple aggregation (simple or weighted averages) of a number of technology variables. Godinho, Mendonça, and Pereira (2005), Castellacci and Archibugi (2008), and Fagerberg and Srholec (2008) have then considered a larger number of innovation system dimensions and analyzed them by means of factor and cluster analysis techniques. As compared to the first approach, the composite indicator approach has a more explicit focus on the comparison across a larger number of countries. Consequently, due to the lack of data availability on less developed countries for a sufficiently long period of time, these studies typically focus on a relatively short time span, giving a cross-section description of the sample in one point in time (e.g., the 1990s and/or the 2000s).

Most of this empirical research, however, has so far focused on the cross-country comparative aspect ("why growth rates differ") and mostly neglected the time series dimension and the analysis of the dynamics of the technological catch-up and economic growth process over time. Further, an exceptional amount of research has been devoted to the study of the determinants of GDP and income per capita growth. By contrast, only a limited number of studies have empirically investigated the dynamics of innovative capability and absorptive capacity over time and the main factors that may explain their long-run evolution (Filippetti and Peyrache 2011; Castellacci and Natera 2013). This is a crucial task for future research in this field, and the present chapter intends to take a step in this direction.

Conceptual Framework and Dimensions

Although the concept of national system was originally intended to comprise a broad and multidimensional set of capabilities, modeling and empirical exercises have often provided a stylized and simplified operationalization of it. As recently argued by Archibugi and Coco (2004), Godinho et al. (2005), and Fagerberg and Srholec (2008), it is indeed important to adopt a multifaceted description and measurement of the various factors that contribute to shape the innovative capability and absorptive capacity of nations.

Given that the concept of national systems is complex, multifaceted, and comprises a great number of relevant factors interacting with each other, our empirical analysis adopts a broad and multidimensional operationalization of it. Our stylized view, broadly in line with the previous literature, is presented in Figure 1.1.[2] The diagram provides a stylized representation of the main aggregate dimensions that define the

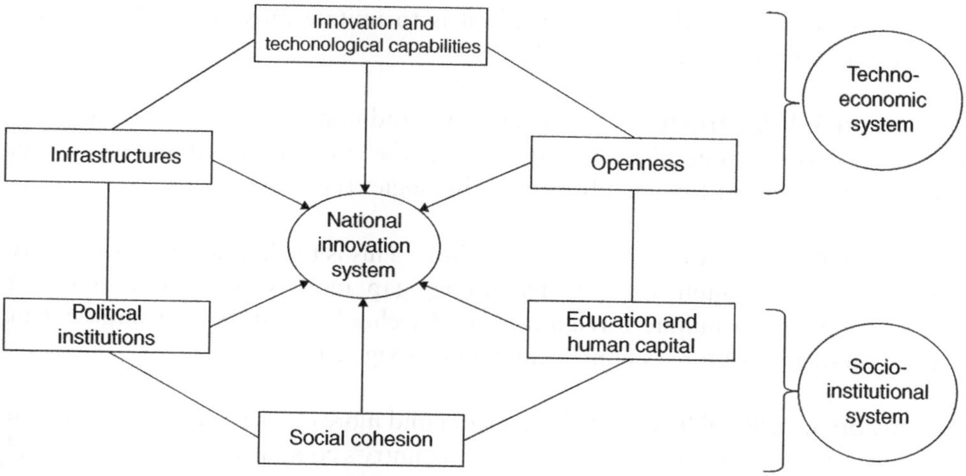

Figure 1.1 National innovation system: a stylized framework.

dynamics of a national system. Specifically, we represent NIS as composed of six main dimensions: (1) Innovation and technological capabilities; (2) Openness; (3) Infrastructures; (4) Education and human capital; (5) Political institutions; (6) Social cohesion.

The first three dimensions are broadly related to a host of technological and economic factors, which classical neo-Schumpeterian work by Chris Freeman (1987) summarized under the heading *techno-economic system*. The next three dimensions refer instead to a heterogeneous set of social and institutional factors that may be combined together under the label *socio-institutional system* (see also Freeman and Louça 2001; Castellacci 2004). The underlying idea motivating this framework is that the dynamics and complex interactions between these six dimensions represent the driving force of national systems' social and economic development, and it is therefore crucial for empirical analyses in this field to have statistical information available for a large number of indicators and country-year observations, and to study how each of these dimensions evolve in the long run.[3]

Dimension 1: Innovation and Technological Capabilities This is one of the crucial dimension on which empirical literature in this field has so far focused. Three related aspects contribute to shape each country's innovative capability. Innovative inputs represent the total efforts and investments carried out by a country for R&D and innovative activities, that is, its innovation intensity. Its scientific output denotes the result of research and innovation activities carried out by the public scientific and technical system, for example, scientific and technical publications. The technological output refers to the total output of technological and innovative activities carried out by private firms, such as in particular patents, the commercialization and export of new products.

Dimension 2: Openness This represents the openness of the national system, a proxy of the potential for exchanging knowledge and technical information with other nations and hence imitating and absorbing foreign advanced technologies. The more open the system, the more capable it is to imitate internationally available advanced knowledge (Fagerberg 1994).

Dimension 3: Infrastructures A greater level and quality of infrastructures (e.g., network, transportation, distribution) increases the country's capability to absorb, adopt, and implement foreign advanced technologies (Freeman 2004).

Dimension 4: Education and Human Capital This is the key absorptive capacity variable typically emphasized by technology-gap models, according to which countries' ability to imitate foreign advanced technologies does to a large extent depend on their workforce's human capital (Verspagen 1991).

Dimension 5: Political Institutions A better and more efficient governance system and institutional quality tends to increase the country's commitment to technological upgrading as well as its imitation capability (Fagerberg and Srholec 2008).

Dimension 6: Social Cohesion A national system with a greater level of social cohesion and within-country income equality is in general characterized by a higher degree of trust and knowledge sharing, hence supporting the pace of diffusion and adoption of advanced knowledge within the country (Arocena and Sutz 2003).

National Specificities and Cross-Country Heterogeneity

This stylized framework intends to provide a general background for our empirical analysis of the global evolution of national innovation systems. Nevertheless, it is important to emphasize that, although these six conceptual dimensions are arguably important for all countries, the way in which they combine and interact with each other greatly differs among national economies. In particular, the specific trajectory followed by distinct national systems of innovation and the specific set of key factors driving the dynamics of the system may differ across country groups characterized by different levels of development. To illustrate this, consider for instance some of the main differences between advanced (high-income), middle-income, and less developed economies, as often depicted in the literature.

Advanced countries are usually rich in terms of both absorptive capacity (infrastructures, openness, and human capital) and innovative capability. In particular, private R&D capabilities become a more central engine of growth as compared to less advanced national systems, and they are also more closely related to the development of public scientific activities. This is one of the central points made by historically oriented studies on the emergence of new world leaders such as Imperial Germany (Veblen), the United States (Abramovitz), and Japan (Freeman), as well as the seminal studies on national systems of innovation in advanced economies (Nelson 1993). Therefore, this group of countries is typically characterized by an intense and complex web of relationships among the key variables driving the system, and in particular a two-way process of coevolution between innovative capability and absorptive capacity.

Compared to the advanced group, middle-income countries often lack the innovative capability that is required to become a world technological leader. However, they have in general a well-developed absorptive capacity and may then exploit the international diffusion of advanced technologies. For instance, previous studies of East Asian innovation systems have pointed out their rapidly increasing levels of human capital, ICT infrastructures, and export specialization and competitiveness (Edquist and Hommen 2008). Compared to the East Asian group, Latin American economies are instead characterized by lower levels of human capital and technical training, as well as a more traditional (resource-based) export specialization profile (Nelson 1993; Alcorta and Peres 1998). Finally, former Soviet countries in Eurasia, in spite of their high levels of human capital, have in general lower private R&D capabilities and weak scientific and technical interactions, which can be explained as the historical heritage these countries have received from the Soviet innovation system and its heavy reliance on publicly funded defense R&D (Freeman 1987; Nelson 1993). All in all, within the broad (and heterogeneous) group of middle-income economies, we normally find weaker interactions among the innovative capability variables, because of the less central role of R&D in the national system and, for the same reason, weaker interactions between innovative capability and absorptive capacity.

Finally, in less developed countries, national systems do not only lack strong innovative and R&D capabilities, but are also characterized by a more limited absorptive capacity, for example, in terms of lower levels of human capital and technical training, infrastructural investments, and ability to reap benefits from international trade activities (Fagerberg et al. 2007; Castellacci 2008; Lundvall et al. 2009). As a consequence, for this group of countries most of the dynamic interactions and coevolution patterns among the six conceptual dimensions noted above are likely to be weaker or non-existent. Below a threshold level of development, innovative capabilities and absorptive capacity are low, and their interaction is therefore not expected to be a key driver of the economic system.

Data, Indicators and Descriptive Evidence

Our empirical analysis makes use of the CANA database, a newly released cross-country panel dataset containing a large number of indicators for the period 1980–2008 (Castellacci and Natera 2011). The novelty of the database is that it provides full information for the whole set of country-year observations, and so it contains no missing value. The dataset has been constructed by combining together indicators available from a number of existing cross-country data sources, and then applying the method of multiple imputation recently proposed by Honaker and King (2010).[4]

The CANA dataset is an attempt to provide a broad, comprehensive, and widely accessible database of a large number of indicators measuring countries' technological, social, and economic characteristics. In particular, the CANA dataset enables us to measure some of the main characteristics of national systems of innovation and carry out a cross-country comparative analysis of their differences and long-run dynamics. Based on the theoretical framework outlined in the previous section, we use this database to obtain four indicators for each of the six dimensions of NIS (see Figure 1.1).

Dimension 1: Innovation and Technological Capabilities

- R&D: R&D expenditures as a percentage of GDP (source: UNESCO, OECD).
- Scientific articles: Number of scientific articles per million people (source: NSF).
- High-tech products: Export of high-tech products as a share of manufacturing exports (source: World Bank).
- Patents: Number of patents registered at the US Patent and Trademark Office per million people (source: USPTO).

Dimension 2: Openness

- Inward FDI: Inward FDI flow, share of GDP (source: World Bank).
- Import: Imports of goods and services as a share of GDP (source: UNCTAD).
- Export: Export of goods and services as a share of GDP (source: UNCTAD).
- Openness: (Export + Import) / GDP (source: UNCTAD).

Dimension 3: Infrastructures

- Electricity: Kilowatts of electricity consumed per hour per capita (source: World Bank).
- Telephony: Sum of telephone mainlines and mobile phones per 1000 people (source: ITU, UNDP, World Bank).
- Carrier departures: Domestic takeoffs and takeoffs abroad of air carriers registered in the country, per 1000 inhabitants (source: World Bank).
- Paved roads: Paved roads as a percentage of the whole road length of the country (source: World Bank).

Dimension 4: Education and Human Capital

- Primary enrollment ratio: Share of primary students (source: UNDP, World Bank).
- Secondary enrollment ratio: Share of secondary students (source: UNDP, World Bank).
- Mean years of schooling: Average number of years of school completed in the population over 14 (Barro and Lee 2001; UNDP, World Bank).
- Tertiary enrollment ratio: Share of tertiary students in science and engineering in the population of that age group (source: UNDP, World Bank).

Dimension 5: Political Institutions[5]

- Political rights: People's free participation in the political process. It ranges from –7 (low freedom) to –1 (total freedom) (source: Freedom House).
- Civil liberties: People's basic freedoms without interference from the state. It ranges from –7 (low freedom) to –1 (total freedom) (source: Freedom House).
- Freedom of press: It reflects the degree of freedom that journalists and news organizations enjoy in each country, and the efforts made by the authorities to respect and ensure respect for this freedom. Index from –115 (no freedom) to 0 (high freedom) (source: Reporters Without Borders).

- Freedom of speech: Extent to which freedoms of speech and press are affected by government censorship, including ownership of media outlets. Index from 0 (Government censorship) to 2 (No Government Censorship) (source: Cingranelli and Richards 2008).

Dimension 6: Social Cohesion

- Corruption: Corruption Perception Index, ranging from 0 (High Corruption) to 10 (Low Corruption) (source: Transparency International).
- Inequalities: Gini index (within-country income inequalities).
- Trust: Percentage of respondents who agree with the statement "Most people can be trusted" (source: World Value Survey).
- Happiness: Index ranging from 3 (very happy) to 0 (not happy) (source: World Value Survey).[6]

As noted in the previous section, these dimensions and indicators of national systems have arguably different relevance and patterns for countries characterized by different levels of economic, institutional, and social development. It is therefore important that our empirical analysis of these indicators will not only consider the overall global patterns across all countries in the world, but also take into account the existence of cross-country heterogeneity and the specificities of different groups, or clubs, of economies. We do this by clustering countries into different groups, and then analyzing convergence and divergence patterns within each of these clubs too.

We have chosen to cluster countries in a simple hierarchical two-step manner. First, we identify three major country clubs, which are broadly in line with the existing literature: advanced economies, catching-up countries, and least developed economies (Howitt and Mayer-Foulkes 2005; Castellacci and Archibugi 2008). Second, in order to achieve a finer characterization of the widely different nature of economies within these three heterogeneous clubs, we further divide them into a few subgroups. We make this based on an exogenous and intuitive criterion: we follow broad geographical areas, which on the whole group together countries that are similar with respect to both the initial GDP per capita level (the usual clustering variable in this literature) and the overall institutional context and capitalist mode of development. All in all, we end up with a total of seven subgroups, as defined in Table 1.1.

Although it is clear that no clustering exercise is faultless, the advantage of the intuitive clustering method described here is twofold: it is broadly in line with the three-club specification adopted by most theoretical models and empirical exercises in this field, and at the same time, by working with seven fairly homogeneous subgroups, it deals in a satisfactory manner with the cross-country heterogeneity issue.

In order to illustrate that heterogeneity at a greater detail, we propose a comparative exercise for selected countries from different groups. We evaluate countries' deviation from the world average for each dimension, in order to see how changes are taking place around the globe in two points of time, 1980 and 2008. If a developing country exhibits a higher level in 2008 than in 1980, it means that its relative position has improved or, equivalently, that it has experienced a process of catching up. By

Table 1.1 Countries with similar GDP per capita by group.

Advanced Countries
OECD: Australia, Austria, Belgium, Canada, Denmark, Finland, France, Germany, Greece, Iceland, Ireland, Israel, Italy, Japan, Netherlands, New Zealand, Norway, Portugal, Spain, Sweden, Switzerland, United Kingdom, United States

Catching-up Countries
Latin America: Argentina, Bolivia, Brazil, Chile, Colombia, Costa Rica, Dominican Republic, Ecuador, El Salvador, Guatemala, Guyana, Honduras, Jamaica, Mexico, Nicaragua, Panama, Paraguay, Peru, Trinidad and Tobago, Uruguay, Venezuela
East Asia: Cambodia, China, Fiji, Indonesia, Malaysia, Mongolia, Philippines, Singapore, South Korea, Thailand, Vietnam
Eurasia: Albania, Armenia, Azerbaijan, Bulgaria, Croatia, Czech Republic, Estonia, Georgia, Hungary, Kazakhstan, Latvia, Lithuania, Moldova, Poland, Romania, Slovakia, Slovenia, Tajikistan, Ukraine, Uzbekistan

Least Developed Countries
South Asia: Bangladesh, India, Nepal, Pakistan, Sri Lanka
North African and Middle East: Algeria, Iran, Jordan, Morocco, Tunisia, Turkey
Sub-Saharan Africa: Benin, Botswana, Burkina Faso, Burundi, Cameroon, Ethiopia, Gabon, Ghana, Guinea, Kenya, Lesotho, Madagascar, Malawi, Mali, Mauritania, Mauritius, Mozambique, Namibia, Niger, Nigeria, Rwanda, Senegal, Sierra Leone, South Africa, Swaziland, Tanzania, Uganda, Zambia, Zimbabwe

contrast, if an advanced country shows a lower level in 2008 than in 1980, it means that its leadership position has weakened in relative terms.

Figure 1.2 presents some descriptive evidence on some of the indicators noted above, for a selection of some major large economies for these seven groups. Each panel in Figure 1.2 reports a radar graph, which contains the composite indicators of each national economy on the six NIS dimensions highlighted in this paper and shows the evolution of the NIS between 1980 and 2008 relatively to the world average.

The radar graphs are informative. More advanced countries have on average a much greater surface than the catching-up BRICS and other less developed economies, indicating an overall greater level of the set of relevant technological, social, and economic capabilities both at the beginning and at the end of the period. Regarding the time evolution of these national systems, South Korea is a good illustration of a country that has improved rapidly its relative position during the last three decades, particularly in terms of innovation capabilities and infrastructures. Within the group of BRICS countries, the catching-up process between the beginning and the end of the period has also been striking for China, Brazil, and South Africa, and less so for Russia and India.

Empirical Analysis

Methods

Our empirical analysis of the global evolution of national systems intends to provide a comprehensive description of cross-country differences in terms of the indicators outlined in the previous section, and how these differences have evolved over the

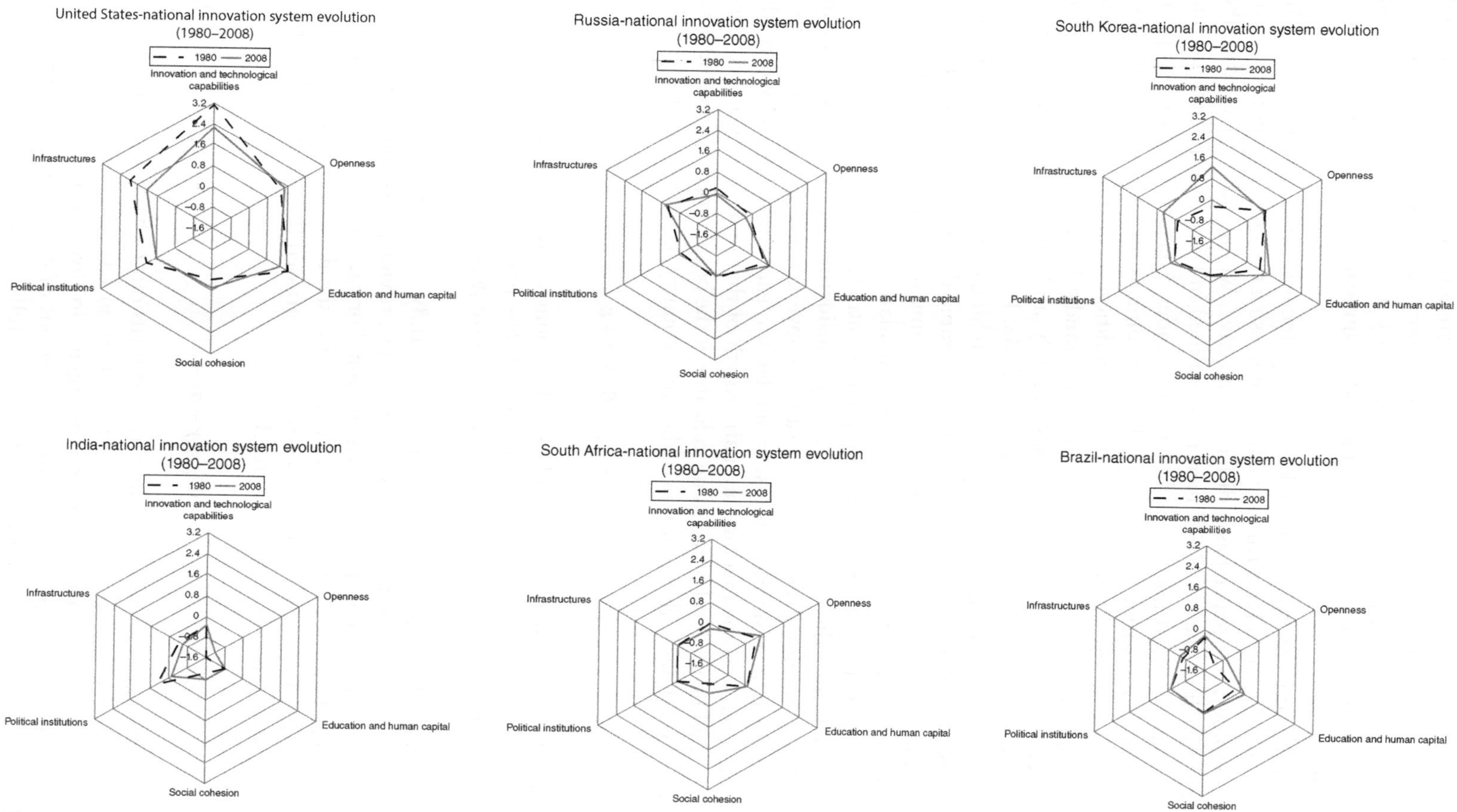

Figure 1.2 Selected countries' relative evolution to the world average 1980–2008.

period 1980–2008. We make use of a simple and commonly used method of applied economic research: convergence analysis. Cross-country convergence is one of most popular areas of research in applied growth theory and, despite some important limitations, it still provides an appealing method to summarize cross-country patterns and dynamics.

Most of the convergence literature has however focused on the dynamics of GDP per capita and the related determinants. But, while using the same convergence analysis methodology, our paper has a different objective. We intend to investigate the dynamics of each of the six NIS dimensions outlined in the previous sections, and how these have evolved during the last three decades. Since there exists no prior theory or model indicating how exactly these six dimensions evolve over time, we are not able to set up a structural model of conditional convergence, and follow instead a simpler empirical strategy. We carry out a standard analysis of (unconditional) convergence for each of these six factors separately, and study how their statistical distributions have evolved in the period 1980–2008. For each dimension, a pattern of *convergence* would indicate that a process of catching up is in place, meaning that less developed economies have experienced a more rapid dynamics than industrialized countries and hence been able to partly close their initial gap. By contrast, a finding of *divergence* would indicate the presence of a cumulative mechanism that is leading to increasing disparities between rich and poor countries.

More specifically, the analysis proceeds by considering the two most standard notions of (unconditional) convergence: β- and σ-convergence. For the β-convergence, we have estimated a simple regression model for each indicator A, where the dependent variable is the (average annual) growth of that indicator over the period 1980–2008, while the level of the same indicator at the beginning of the period is the only regressor (along with a constant). The cross-country regression model is:

$$\Delta A_i / A_i = \tau_A + \beta_A A_{i,0} + \varepsilon_i \qquad (1)$$

where $\Delta A_i/A_i$ is the growth of each indicator for country i over the period, and $A_{i,0}$ is the log of its level at the beginning of the period. The parameter of interest in these regressions is β_A, which measures the speed of convergence, or divergence, for that indicator.

By contrast, the idea of σ-convergence is to study whether the dispersion of a variable has increased or decreased over time, thus providing a synthetic measure of the dynamics of the variability of its distribution. For each indicator, σ-convergence is computed as follows. First, we have calculated the standard deviation of the cross-country distribution at each time t; then we have regressed the standard deviation's time series on a constant and a time trend:

$$SD_t = \gamma + \sigma t + \varepsilon \qquad (2)$$

If the estimated coefficient of the time trend σ turns out to have negative (positive) coefficient, this means that the indicator in question has decreased (increased) its variability over time, hence it shows σ-convergence (σ-divergence).

It is important to emphasize that the two notions of convergence, albeit related, are quite distinct from each other. β-analysis tells us whether, on average, a cross-country distribution has experienced a process of catching up – where initially poorer economies have developed more rapidly than richer countries (β-convergence) – or instead a cumulative process according to which initially richer economies have been able to reinforce their leadership position over time (β-divergence). On the other hand, σ-analysis provides information on the dynamics of the cross-country variability of an indicator: σ-convergence indicates decreasing cross-country dispersion, whereas σ-divergence points to increasing polarization between rich and less developed economies.

Further, in order to take into account the existence of substantial differences across country groups, we have undertaken our convergence analysis not only for the whole world distribution, but we have also repeated it for each of the seven geographically defined country groups in Table 1.1.

Results

We will now present the results of the convergence analysis for each dimension separately, and will then conclude by summarizing the main general results. Table 1.2 reports a summary of convergence analysis results,[7] and Figures 1.3–1.8 show geographical maps and kernel densities to illustrate visually the cross-country distribution of selected indicators and their evolution over time.[8]

Dimension 1: Innovation and Technological Capabilities Figure 1.3 shows that the cross-country distribution of innovative capabilities is very skewed, since the bulk of innovative investments and results is highly concentrated in countries within the OECD area. Such a skewed pattern has not changed substantially during the period 1980–2008, indicating that no significant worldwide improvement has taken place in this dimension (Castellacci 2011). More specifically, the convergence analysis results indicate that three out of four indicators (R&D, scientific articles, high-tech products) are characterized by a combination of slow β-convergence and visible σ-divergence. This means that, on the one hand, countries that were less technologically advanced at the beginning of the period have on average improved their capabilities somewhat faster than already developed economies; on the other hand, however, several rich countries have continued to increase their technological efforts, leading to an increasing worldwide dispersion (i.e., the right tails of these

Table 1.2 Summary of convergence analysis.

	Decreasing dispersion (σ-convergence)	*Increasing dispersion (σ-divergence)*
Catch-up process (β-convergence)	• Basic education • Social cohesion • Political institutions	• Infrastructures • Openness • Advanced education • Innovation (R&D, articles, high-tech exports)
Cumulative process (β-divergence)		• Innovation (patents)

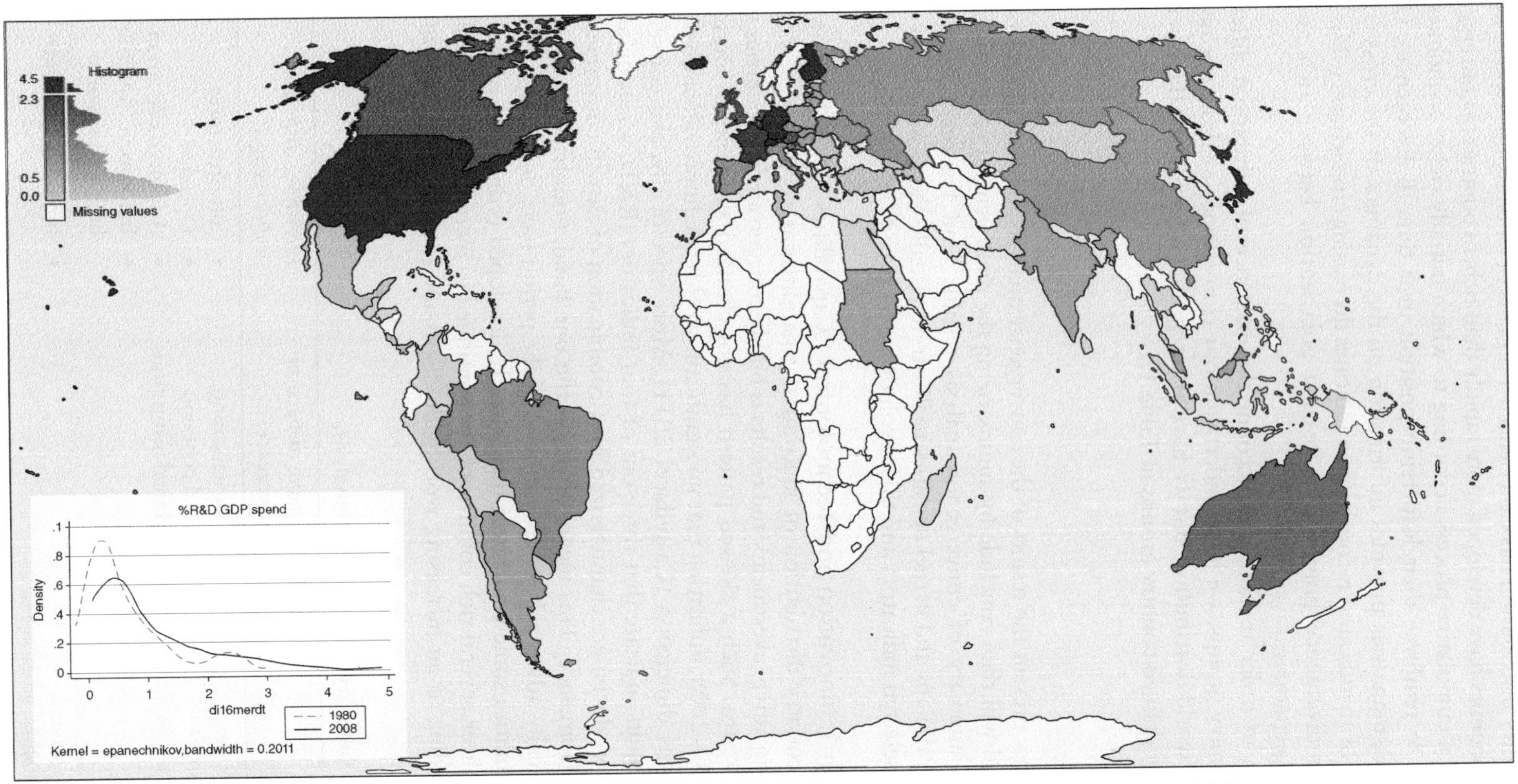

Figure 1.3 The cross-country dynamics of innovation and technological capabilities: ratio of R&D expenditure to GDP.

cross-country distributions have become longer). This pattern holds for all geographical regions, except the OECD area, which is the only club in which one observes cross-country convergence. The fourth indicator, patents, shows an even more dramatic pattern of divergence, since the variable combines at the same time β- and σ-divergence. This indicates that innovative results and performance are becoming more polarized, with an increasing gap between a small bunch of innovative OECD countries and a large group of non-innovating middle-income and less developed economies. As Lall (2003) does, we also wonder how big could be the influence of the TRIPS agreement on this pattern, in which external conditions might be fostering the divergence process across the world.

Dimension 2: Openness The four openness indicators (inward FDI, export, import, openness) show an overall pattern and dynamics that is largely similar to that of innovation indicators. The cross-country distributions of these variables are quite skewed at the beginning of the period (Figure 1.4). These distributions move rightward over time – indicating an overall worldwide progress in this dimension – but the variability is still high at the end of the period. The convergence results point out a combination of β-convergence and σ-divergence (as it is the case for the innovation indicators). This general pattern holds for most of the geographical regions, although we also observe a more pronounced σ-convergence process within less developed regions such as South Asia, Middle East, and Africa.

Dimension 3: Infrastructures The infrastructures dimension considered in Figure 1.5 does also show a cross-country pattern similar to that observed for the innovation and openness dimensions. The overall convergence dynamics is one that combines at the same time a catch-up process (poor economies' infrastructures have grown on average faster than already developed economies) with an increasing polarization, that is mostly due to the fact that some of the richest countries have continued to rapidly upgrade their infrastructures (e.g., telephony lines, road and air transportation), leading to an enlargement of the gap between the richest and the poorest economies.

Dimension 4: Education and Human Capital A worldwide and relatively rapid process of convergence is instead apparent when we shift the focus to Figure 1.6, which studies the evolution of the human capital dimension. The kernel densities show that the cross-country distributions of these indicators have visibly shifted toward the right, thus indicating an overall improvement of countries' education systems. Differently from the previous dimensions, however, this worldwide growth has not led to increasing polarization but to a rapid convergence process (both β- and σ-convergence), and hence to a more equal cross-country distribution of human capital. This is clearly the case for the three indicators of basic education (primary and secondary education). By contrast, advanced education (measured by the tertiary enrollment ratio) has a markedly different pattern, combining β-convergence and σ-divergence like the indicators of innovation, openness and infrastructures discussed above. In short, it is interesting to point out that while countries are progressively becoming more similar in terms of basic education, this is not the case with respect to advanced education and the university sector's capabilities, where the gap between OECD and less developed economies is indeed enlarging.

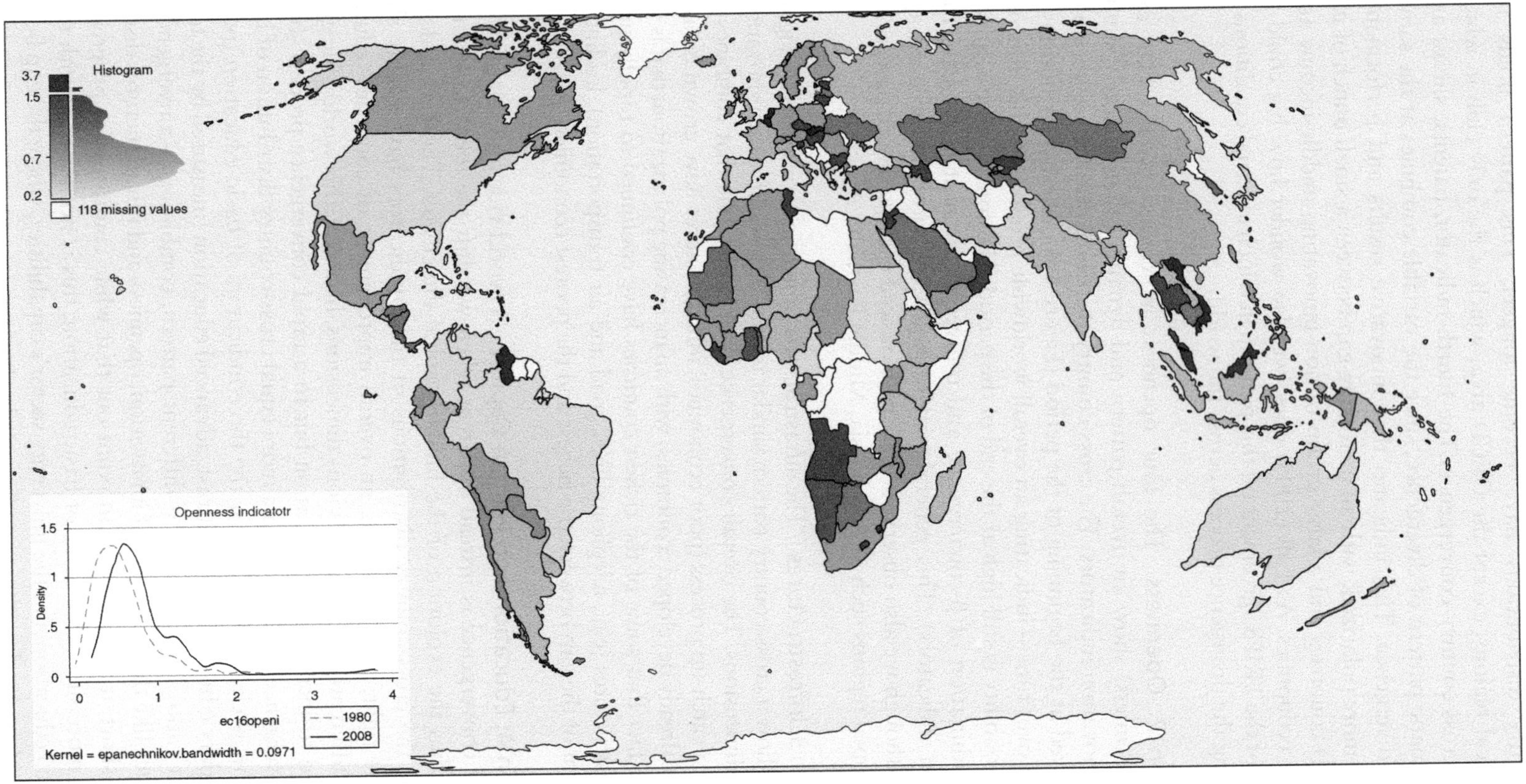

Figure 1.4 The cross-country dynamics of openess: openess 2008 (Exports + Imports / GDP).

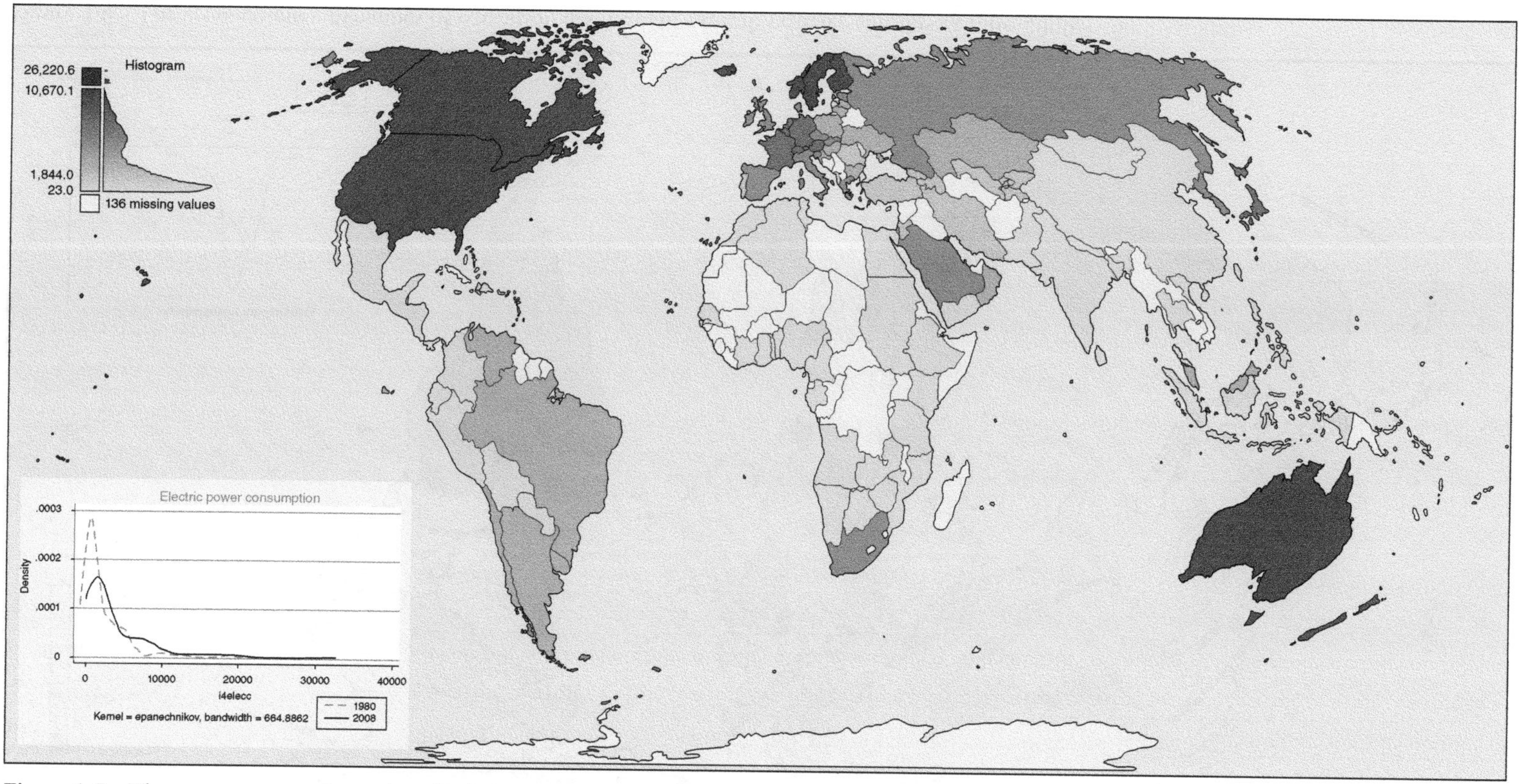

Figure 1.5 The cross-country dynamics of infrastructures: electric power consumption 2008.

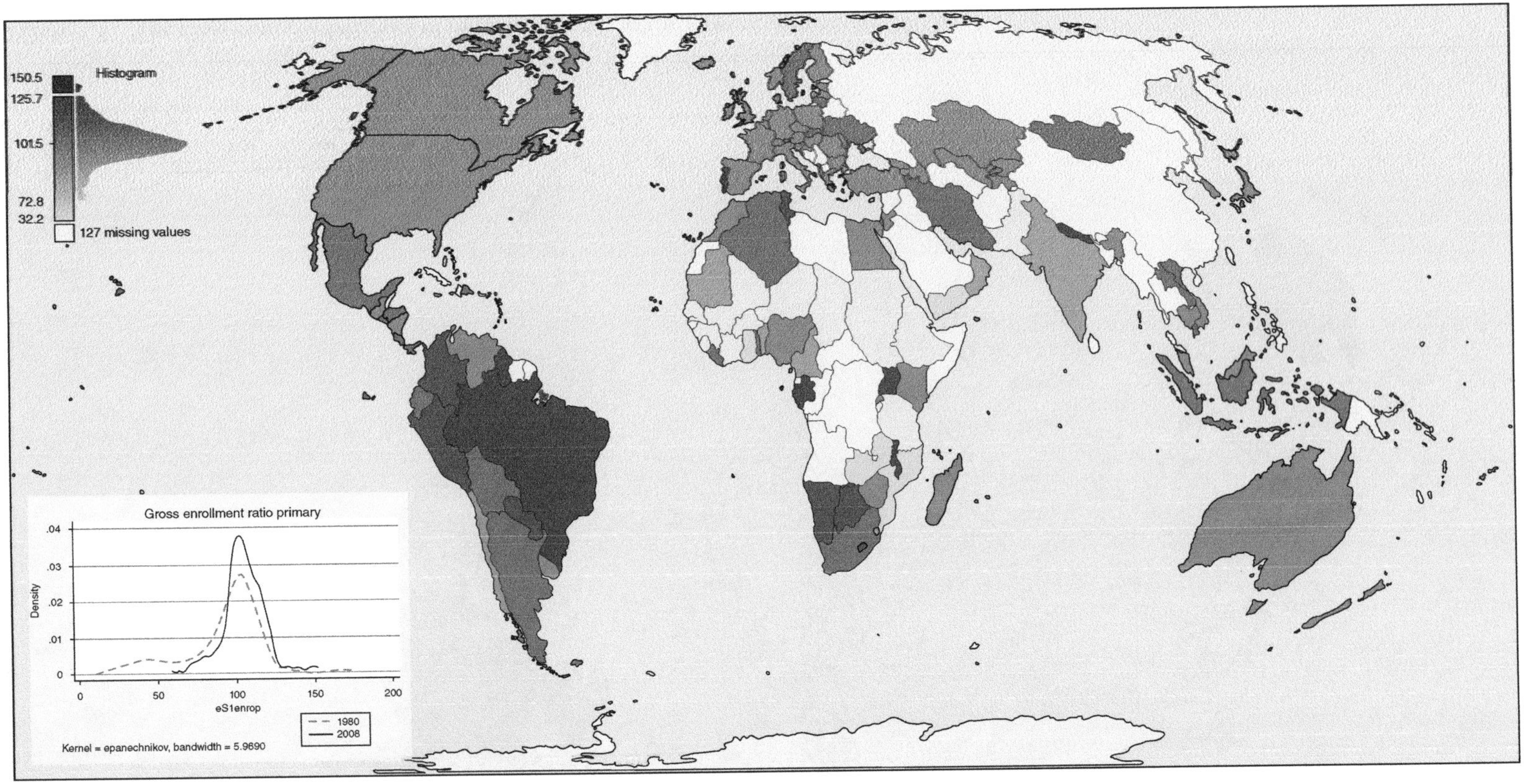

Figure 1.6 The cross-country dynamics of education and human capital: primary enrollment ratio 2000.

Dimension 5: Political Institutions The four indicators of political institutions follow a cross-country dynamics that largely resembles that of basic education variables discussed above. All of these variables – which are proxies for countries' respect for political rights, civil liberties, freedom of press, and freedom of speech – have experienced both β- and σ-convergence during the last three decades. This convergence pattern also holds within most of the seven geographical regions that we have considered, with the exception of East Asia and Eurasia, where the dispersion of the distribution has actually increased somewhat over time, arguably due to the process of institutional transition that has led to more cross-country heterogeneity within these regions. Figure 1.7, for instance, shows that the shape of the distribution of the freedom of press indicator has substantially modified during this period, shifting from a bimodal and highly polarized pattern at the beginning of the period toward a more even and slightly more concentrated (less dispersed) distribution at the end of the time span.

Dimension 6: Social Cohesion Finally, Figure 1.8 considers the social cohesion dimension. The patterns we obtain from the convergence analysis are somewhat mixed. On the one hand, the two variables that we consider more reliable because of their larger country coverage and data quality – corruption and the Gini index – show an overall process of catch-up and convergence (both β- and σ-convergence), indicating that national systems worldwide have on average become more similar in terms of internal social cohesion. On the other hand, the other two indicators, trust and happiness – which we consider a little less reliable since they are based on survey data for a somewhat smaller sample of countries – seem to have combined a process of catch-up and β-convergence with an overall increasing polarization between rich and poor economies (σ-divergence).

The *Convergence Paradox* Table 1.2 summarizes the main results of our convergence analysis; it is based on convergence results of dimension indicators across the world and for each of the seven country groups defined.[9] The table shows a quite clear overall pattern. On the one hand, the three dimensions defining the socio-institutional system – basic education, political institutions, and social cohesion – have experienced a process of convergence: less developed economies have on average improved these factors at a faster rate than more advanced countries (β-convergence), and at the same time the dispersion and polarization of these distributions have decreased over time (σ-convergence). On the other hand, however, the three dimensions related to the techno-economic system have experienced a marked divergent dynamics: most of the indicators of infrastructures, openness, advanced human capital, and innovation have in fact combined a slow β-convergence pattern (slow and hence not sufficient for less developed economies to close the gap vis-à-vis more advanced countries) with a σ-divergence trend pointing to increasing polarization between rich and poor country groups. In a nutshell, we summarize this contrasting pattern by pointing to the existence of a *convergence paradox*: during the last three decades national systems worldwide have become more similar in terms of their socio-institutional frameworks but increasingly more different with respect to techno-economic dimensions.

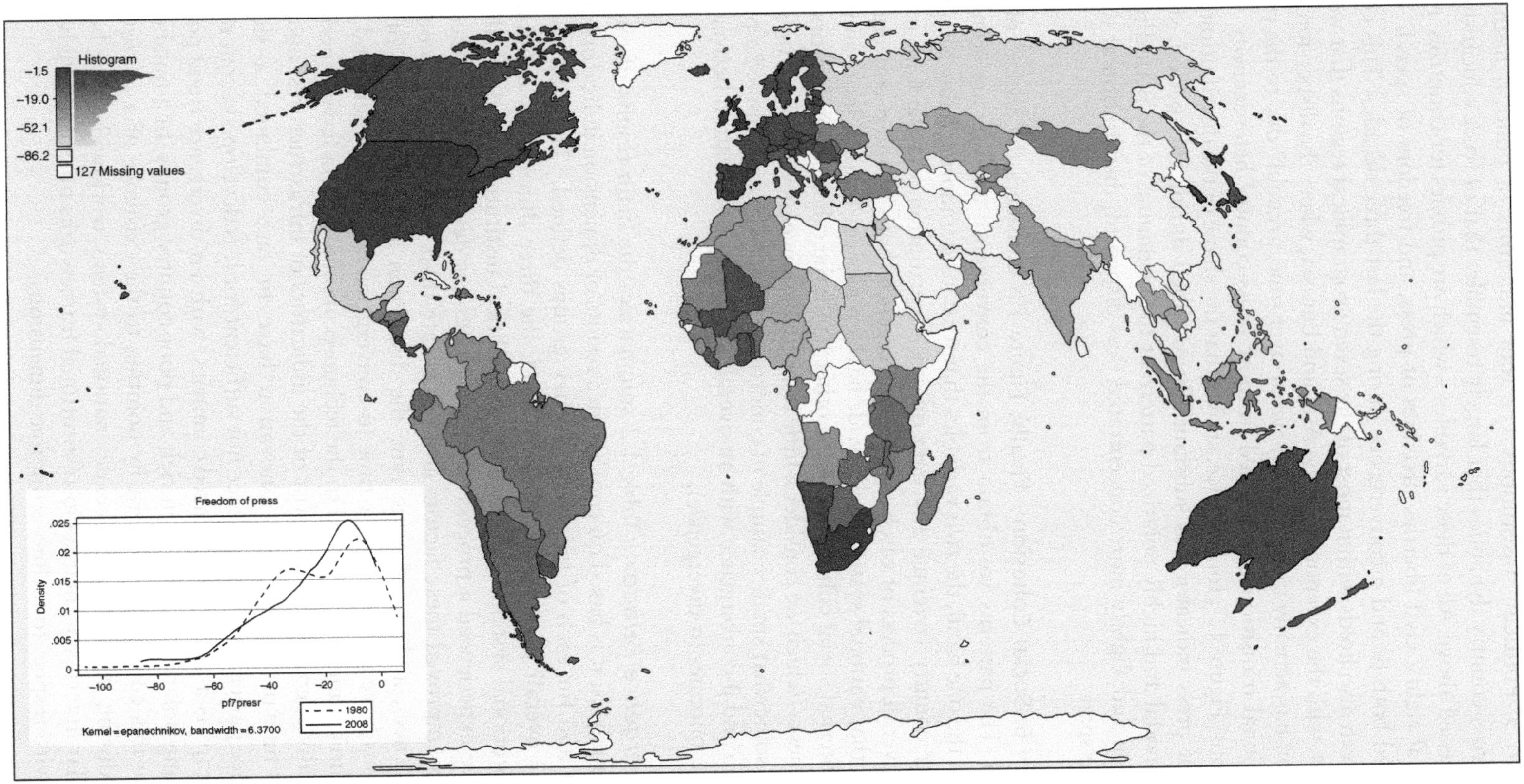

Figure 1.7 The cross-country dynamics of political institutions: freedom of press 2000.

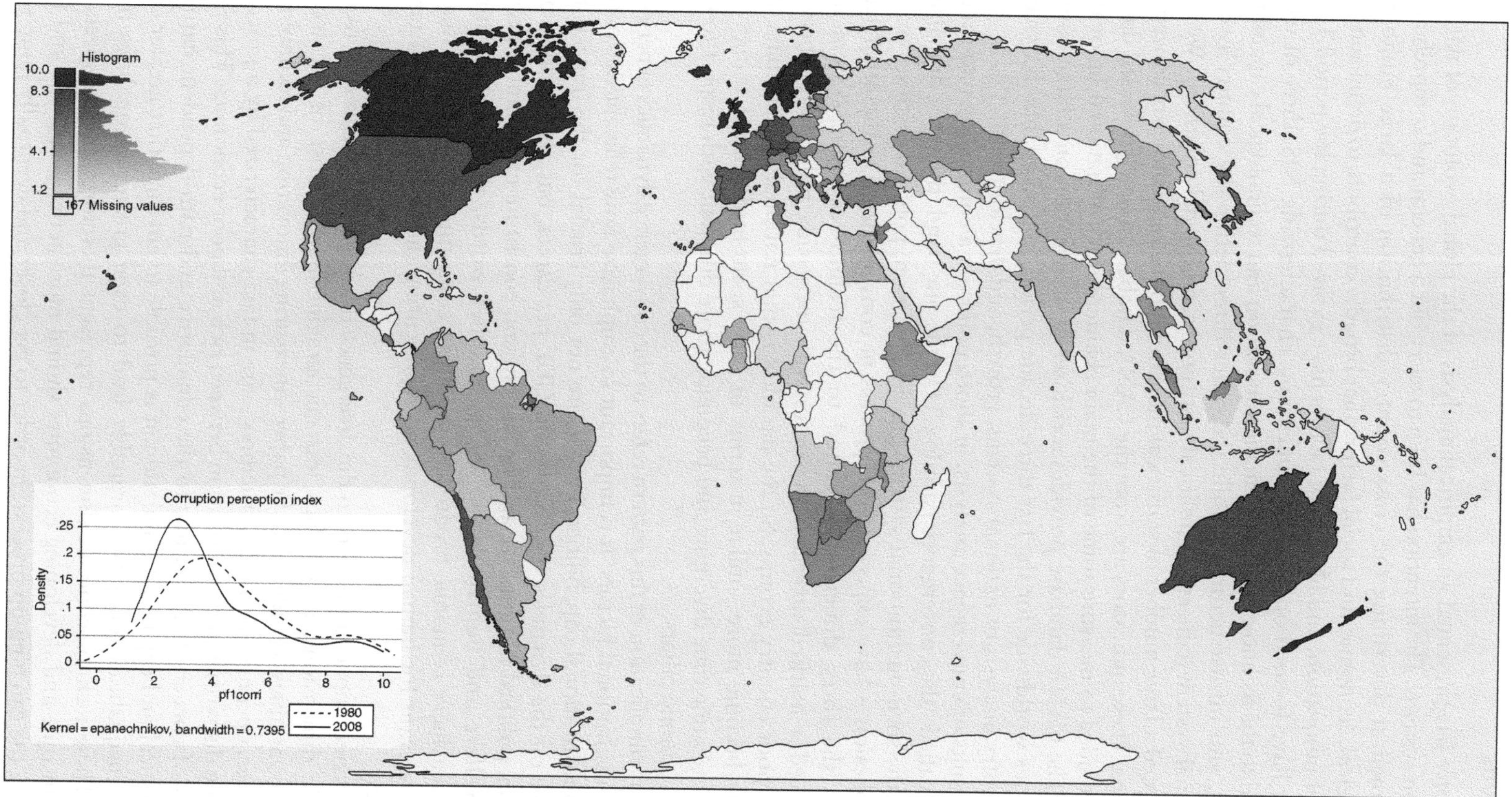

Figure 1.8 The cross-country dynamics of social cohesion: Corruption Perception Index 2008.

Conclusions

This chapter has presented an empirical analysis of the global evolution of national innovation systems. The intention has been to carry out a general and comprehensive study, aimed at describing the cross-country distribution of a large number of technological, economic, and social factors that broadly contribute to define national systems of innovation, and how these factors have evolved during the period 1980–2008. We have focused on six main dimensions characterizing NIS, three of which are related to the techno-economic domain (innovation and technological capabilities, openness, and infrastructures), while the other three define the country's socio-institutional system (education, political institutions, and social cohesion). We have then adopted a large number of statistical indicators to measure these six conceptual dimensions. These indicators are from the CANA database, a newly released dataset that is intended to measure some of the main characteristics of national innovation systems. The CANA database provides a rich set of information on a large number of countries worldwide for the last three-decade period, and so enables a dynamic cross-country analysis of national systems, growth, and development.

The results of convergence analysis, in short, point out a contrasting pattern that we have called "the convergence paradox." On the one hand, the three dimensions defining the socio-institutional system – basic education, political institutions, and social cohesion – have experienced a process of convergence. On the other hand, however, the three dimensions related to the techno-economic system have experienced a marked divergent dynamics with increasing polarization between rich and poor country groups. So the paradox is that national systems have progressively become more similar in terms of their socio-institutional frameworks while at the same time diverging and becoming more different with respect to their techno-economic pillars.

One first implication of this pattern refers to the convergence dynamics experienced by socio-institutional factors. It is important to emphasize how relevant this pattern is. Socio-institutional capabilities, in fact, are not only crucial for the impacts they may have on the dynamics of income and GDP per capita, but also because basic education, political institutions, and social cohesion constitute important aspects of social welfare and human development. In other words, they are not simply means to achieve economic progress, but do also constitute achievements that are directly relevant for human development, representing therefore important targets for policy.

The second implication refers to the divergence dynamics experienced by techno-economic dimensions, and brings less encouraging news. Most of the techno-economic variables considered in the paper – measuring innovation and technological capabilities, openness, and infrastructures – are in fact characterized by a process of increasing disparities and polarization between rich and poor economies. This fact is a reason of concern because, due to the cumulative and path-dependent nature of technological and capital accumulation, it is possible to envisage that greater cross-country inequalities today will arguably lead to even greater inequalities in the future. These techno-economic dimensions represent therefore crucial aspects that middle-income and developing economies should more actively upgrade and focus on during the catching-up process. In order to get closer and eventually jump to the innovation development stage, developing economies should therefore implement an appropriate combination of policies that takes into account the need to simultaneously develop R&D activities, openness and export performance, traditional and ICT infrastructures, and advanced human capital. At the same time, advanced countries should also take their responsibility in this divergence process: living in a globalized world, creating global economic inequalities and limiting the possibilities of latecomers to catch up might lead to unstable situations, affecting not only developing countries but also their own economies. It is in the best interest of both advanced and catching-up countries to increase the level of technological progress of the whole world in the long run. In more general terms, the finding of increasing cross-country divergence in terms of techno-economic dimensions can also be interpreted as evidence of the continuing relevance of the NIS concept: even in the context of an increasingly globalized economy, a large number of techno-economic factors persistently differ across countries, following country-specific and path-dependent trajectories of development. Global economic integration is indeed strengthening country-specific characteristics and hence the relevance of national systems and policies (Archibugi and Michie 1995).

Finally, we would like to conclude by acknowledging one important limitation of the paper and pointing out a related challenge for future research in this field. Our empirical analysis has been provocatively simple, based on a simple description of the univariate evolution and unconditional convergence dynamics of a large battery of indicators. The chapter has said nothing regarding the underlying key issue of interest: how do these factors interact with each other, and how can their coevolutionary process explain the different growth trajectories experienced by national economies in the long run? While there are several papers investigating this question in a cross-country comparative perspective, very few works have approached this issue focusing on the time series process that may explain the complex coevolutionary patterns followed by national systems (Castellacci and Natera 2013). We consider this an important area for future applied research on the evolution of national innovation systems.

Notes

1 For further references and information regarding the flourishing field of innovation systems and development, see the website of the Globelics network: www.globelics.com.

2 Other empirical exercises in the NIS literature have previously made use of (at least some of) these dimensions and indicators. See in particular Godinho et al. (2005), Castellacci and Archibugi (2008), and Fagerberg and Srholec (2008).

3 In another paper (Castellacci and Natera 2013), we study the interactions among these dimensions and carry out a time series multivariate analysis of their coevolutionary process.

4 The CANA database, along with the sources and definitions of the indicators and a description of the construction methodology, can be downloaded at: http://english.nupi.no/Activities/Projects/CANA.

5 Three indicators of this dimension have been modified from the original sources. We have multiplied the original observed data by minus one (–1) such that a higher value indicates a more favorable situation.

6 This indicator was constructed as a weighted average of the responses of the original data source.

7 Full detailed results are provided in Tables 1.3a–d at the end of this chapter. Interested readers might look at specific dimensions or geographical regions.

Table 1.3a Results of convergence analysis: techno-economic system (I).

	World		*OECD*		*Latin America*		*East Asia*		*Eurasia*	
Innovation & technological capabilities.	β	σ	β	σ	β	σ	β	σ	β	σ
%R&D GDP Spend	−0.327274 (0.234249)	0.012522*** (0.000822)	−0.029722*** (0.005152)	0.010344*** (0.001659)	−4.32027 (3.446528)	0.000085 (0.000299)	−1.234516* (0.549059)	0.02289*** (0.001011)	−0.03977*** (0.010984)	−0.003345*** (0.000865)
Scientific and technical journal articles	-808.246** (312.3217)	0.00000381*** (0.000000144)	-111.9871*** (20.66416)	-0.00000117*** (0.000000295)	-10483.06* (5345.123)	0.000000517*** (0.0000000185)	-7579.298 (6318.195)	0.00000901*** (0.000000481)	-9701.348 (5922.183)	0.0000036*** (0.000000184)
High-technology exports (% of manufactured)	−0.087992* (0.051513)	0.135567*** (0.027889)	−0.00187 (0.001358)	−0.073104*** (0.019678)	−0.010717 (0.007609)	0.191656*** (0.060321)	−0.046664 (0.047924)	0.383142*** (0.057562)	−0.001492 (0.002079)	0.054928 (0.034804)
Patent applications, nonresidents per capita	−0.00000621 (0.0000368)	−31.42311** (11.65714)	−0.0000192 (0.0000526)	10.06299*** (1.841461)	−0.001083** (0.000448)	1.520069*** (0.20066)	−0.003317 (0.004959)	20.2475*** (2.713616)	−0.0000274 (0.0000864)	−102.1013*** (28.84035)
					Openness					
Inward foreign direct investment stock	−0.00265** (0.001186)	1.773146** (0.671757)	−0.000768*** (0.000267)	−0.121833 (0.157576)	−0.00477* (0.002449)	0.568872*** (0.062479)	−0.011415 (0.013491)	1.008179*** (0.107496)	−0.018817** (0.008185)	0.958098*** (0.090196)
Imports of goods and services (% of GD)	−0.000438*** (0.0000898)	0.091432*** (0.028459)	−0.000249** (0.000106)	0.09532*** (0.021217)	−0.000118 (0.00021)	0.266961*** (0.086841)	−0.000413* (0.00022)	0.245843*** (0.029628)	−0.001298*** (0.000336)	0.167303*** (0.04581)
Exports of goods and services (% of GD)	−0.000569*** (0.000104)	0.174551*** (0.019069)	0.000000633 (0.000141)	0.287369*** (0.021758)	−0.000507*** (0.000176)	0.134757 (0.087911)	−0.000675 (0.000393)	0.308287*** (0.039623)	−0.00099** (0.000359)	0.189373*** (0.052187)
Openness Indicator	−0.026196*** (0.00792)	0.000288 (0.002768)	−0.001325 (0.006488)	0.00437*** (0.000537)	−0.023193* (0.011649)	0.002993*** (0.001055)	−0.015681** (0.006239)	−0.003522* (0.002009)	−0.15357*** (0.031581)	0.00058 (0.000854)

					Infrastructure					
% Telecommunication Revenue	-0.04868**	0.07825***	-0.024378***	0.018963***	-0.009236	0.162822***	-0.037896	0.038657***	-0.097533	0.032265***
Electric power consumption	-0.00000333*** (0.000000763)	72.6738*** (2.568785)	-0.000001** (0.00000047)	86.02092*** (6.282711)	-0.00000613 (0.0000072)	24.57857*** (1.545643)	-0.0000304* (0.0000137)	88.73484*** (2.773622)	-0.00000285 (0.0000022)	16.69627*** (1.632516)
Mobile and fixed-line subscribers	-0.000427*** (0.0000354)	17.90163*** (1.189618)	-0.000161*** (0.0000147)	2.171602*** (0.221338)	-0.001135*** (0.00016)	8.536118*** (0.860196)	-0.000886** (0.000328)	16.7257*** (1.115447)	-0.000544*** (0.000122)	17.17018*** (1.574)
Registered carrier departures worldwide	-0.007334 (0.004744)	0.200876*** (0.024771)	-0.000794 (0.000517)	0.159005*** (0.032081)	-0.022368 (0.013966)	-0.039179** (0.014345)	-0.002147 (0.003147)	0.461142*** (0.121918)	-0.024986*** (0.005216)	0.041102*** (0.011321)
% Paved Roads	-0.000929** (0.000391)	0.012855* (0.006756)	-0.000293*** (0.0000541)	-0.090784*** (0.01335)	0.000447 (0.0004)	-0.074832*** (0.018723)	-0.0000404 (0.000243)	0.27429*** (0.022949)	-0.008779*** (0.00252)	0.004101 (0.035683)

Note: Significance levels: 1%***, 5%**, 10%*. Standard errors in parenthesis.

Table 1.3b Results of convergence analysis: techno-economic system (II).

	South Asia		*North Africa and Middle East*		*South and Central Africa*	
Innovation & technological capabilities.	β	σ	β	σ	β	σ
%R&D GDP Spend	-1.249661 (0.964799)	0.004802*** (0.00103)	-1.368182* (0.645719)	0.006431*** (0.000915)	-0.15931 (0.145447)	0.007034*** (0.001377)
Scientific and technical journal articles	-21624.91* (6992.043)	-0.0000000316* (0.0000000155)	-1771.593** (751.8455)	-0.000000664*** (0.0000000917)	-12717.91 (7837.22)	0.000000487*** (0.0000000299)
High-technology exports (% of manufactured)	0.399452 (0.357574)	0.029892*** (0.00653)	-0.902152 (0.879375)	0.075679*** (0.019456)	-0.220056 (0.219991)	0.016807 (0.029601)
Patent applications, nonresidents per capita	-0.227431 (0.200529)	0.152691*** (0.022684)	-0.0000693 (0.000606)	-0.246017** (0.116292)	-0.004553 (0.003307)	-0.234901** (0.105938)
			Openness			
Inward foreign direct investment stock	-0.030409** (0.009456)	0.035936** (0.01497)	-0.003785 (0.003337)	0.699126*** (0.043861)	-0.002174* (0.001162)	3.505717** (1.339075)
Imports of goods and services (% of GD)	-0.001032 (0.000493)	-0.137868*** (0.043821)	-0.000395*** (0.0000972)	-0.480234*** (0.084992)	-0.000536*** (0.000176)	-0.079536 (0.056121)
Exports of goods and services (% of GD)	-0.002174 (0.001006)	-0.056849* (0.029188)	-0.00059*** (0.000187)	-0.361843*** (0.067983)	-0.000829*** (0.000187)	0.015189 (0.022196)
Openness Indicator	-0.070948 (0.031934)	-0.00157*** (0.000505)	-0.032903*** (0.006485)	-0.008251*** (0.000953)	-0.00514 (0.022342)	-0.000241 (0.00732)
			Infrastructure			
% Telecommunication Revenue	-0.018586	0.013823***	-0.033731***	0.044785***	-0.081684	0.078122***
Electric power consumption	-0.000274*** (0.0000448)	5.295248*** (0.252979)	-0.00000361** (0.00000129)	114.7191*** (6.646084)	-0.0000142** (0.00000529)	9.01667*** (0.745526)
Mobile and fixed-line subscribers	-0.015057 (0.00736)	4.44475*** (0.823235)	-0.000761*** (0.000127)	12.78409*** (1.240232)	-0.002976*** (0.000716)	7.800345*** (0.88737)
Registered carrier departures worldwide	-0.071967 (0.065898)	0.00012 (0.002883)	-0.001921 (0.001914)	0.2246*** (0.050897)	-0.050535 (0.045998)	-0.044017*** (0.00439)
% Paved Roads	-0.001575** (0.00049)	0.136303** (0.055325)	-0.000484*** (0.000127)	-0.206584*** (0.016199)	-0.001062*** (0.000358)	0.091045*** (0.00747)

Note: Significance levels: 1%***, 5%**, 10%*. Standard errors in parenthesis.

Table 1.3c Results of convergence analysis: socio-institutional system (I).

	World		OECD		Latin America		East Asia		Eurasia	
Education and Human Capital	β	σ	β	σ	β	σ	β	σ	β	σ
Gross Enrollment Ratio Primary	−0.000512*** (0.0000236)	−0.441884*** (0.018528)	−0.000296*** (0.0000543)	−0.089952*** (0.011067)	−0.000284*** (0.0000572)	−0.14632*** (0.017726)	−0.000281 (0.000212)	−0.136063** (0.05171)	−0.000379*** (0.0000596)	−0.212354*** (0.027181)
Gross Enrollment Ratio Secondary	−0.000797*** (0.0000587)	−0.046228* (0.024197)	−0.000627*** (0.000124)	0.183092*** (0.057361)	−0.000665*** (0.0000939)	−0.032679 (0.020521)	−0.00089*** (0.000279)	−0.167756*** (0.021944)	−0.000428*** (0.0000671)	−0.117525*** (0.028813)
Mean years of schooling	−0.004363*** (0.000316)	−0.002672*** (0.000594)	−0.003089*** (0.000409)	−0.027338*** (0.001576)	−0.003967*** (0.000899)	0.001013 (0.000752)	−0.001323 (0.001289)	0.026851*** (0.002034)	−0.001186 (0.000688)	0.013124*** (0.00112)
Gross Enrollment Ratio Tertiary	−0.016179* (0.008503)	0.503428*** (0.016155)	−0.001407*** (0.000209)	−0.146808*** (0.039278)	−0.007874*** (0.001866)	0.343581*** (0.013067)	−0.038421 (0.026216)	0.689115*** (0.018019)	−0.002927*** (0.000447)	0.33849*** (0.045999)
Political Institutions										
Political Rights	−0.004673*** (0.001083)	−0.001782* (0.000886)	−0.012937*** (0.002148)	−0.016753*** (0.001347)	−0.010838*** (0.003566)	−0.038184*** (0.002931)	−0.015772** (0.006483)	0.018313*** (0.001762)	−0.008069* (0.004191)	0.046577*** (0.004167)
Civil Liberties	−0.005172*** (0.000804)	−0.010926*** (0.001313)	−0.010575* (0.005517)	−0.005452** (0.002392)	−0.011601*** (0.002186)	−0.02237*** (0.003148)	−0.009464*** (0.002783)	−0.008044*** (0.002604)	−0.009378* (0.005035)	0.032111*** (0.002735)
Freedom of Press	−0.001029*** (0.000114)	−0.12458*** (0.017158)	−0.004122*** (0.000557)	−0.081799*** (0.010191)	−0.002038*** (0.000451)	−0.330072*** (0.038153)	−0.000702*** (0.000194)	0.27377*** (0.036365)	−0.000903*** (0.000279)	0.359378*** (0.037505)
Freedom of Speech	0.205712*** (0.03923)	−0.002253 (0.001792)	−0.083181*** (0.014662)	−0.001926 (0.001488)	−0.009549 (0.02845)	−0.003387* (0.001953)	0.385553 (0.277164)	0.003868 (0.002605)	0.047721 (0.191013)	0.012976*** (0.002998)

(Continued)

Table 1.3c *(Continued)*

	World		*OECD*		*Latin America*		*East Asia*		*Eurasia*	
Education and Human Capital	β	σ	β	σ	β	σ	β	σ	β	σ
					Social Capital					
Corruption Perception Index	−0.007296***	0.000663	−0.003508***	−0.020309***	−0.007956***	−0.018982***	−0.002494	0.002714	−0.009351***	−0.006534**
	(0.001954)	(0.000804)	(0.000594)	(0.001711)	(0.001764)	(0.002434)	(0.001947)	(0.001746)	(0.00151)	(0.002481)
Gini Index	−0.000711***	−0.145552***	−0.000987***	−0.011321	−0.000581***	−0.029548**	−0.001637***	−0.2125***	−0.000961	0.232464***
	(0.000063)	(0.010306)	(0.00026)	(0.009381)	(0.000146)	(0.012886)	(0.000384)	(0.024404)	(0.000596)	(0.026288)
Most people can be trusted	−0.056101***	0.000233***	−0.011884**	0.00028**	−0.239054***	−0.0000566	−0.078595***	−0.000459**	−0.145275***	−0.000301**
	(0.010627)	(0.0000826)	(0.005376)	(0.000135)	(0.051344)	(0.000134)	(0.01835)	(0.000183)	(0.032112)	(0.000118)
Feeling of Happiness	−0.000954***	0.000216	−0.001063	−0.000297*	−0.001251	0.00000757	−0.005749**	−0.000773*	−0.001076	0.000398
	(0.000348)	(0.000181)	(0.001013)	(0.000165)	(0.002033)	(0.000314)	(0.001697)	(0.000379)	(0.001029)	(0.00031)

Note: Significance levels: 1%***, 5%**, 10%*. Standard errors in parenthesis.

Table 1.3d Results of convergence analysis: socio-institutional system (II).

	South Asia		*North Africa and Middle East*		*South and Central Africa*	
Education and Human Capital	β	σ	β	σ	β	σ
Gross Enrollment Ratio Primary	−0.000331**	−0.331319***	−0.00038***	−0.277927***	−0.000499***	−0.47073***
	(0.0000771)	(0.038534)	(0.0000427)	(0.047844)	(0.0000478)	(0.033362)
Gross Enrollment Ratio Secondary	−0.000449	0.219976***	−0.000901***	−0.299383***	−0.002047***	0.314106***
	(0.000306)	(0.014315)	(0.000231)	(0.030695)	(0.000378)	(0.018309)
Mean years of schooling	−0.006505**	−0.012677***	−0.005934***	0.004681***	−0.009612***	0.016128***
	(0.001369)	(0.0012)	(0.000942)	(0.001113)	(0.001425)	(0.001274)
Gross Enrollment Ratio Tertiary	−0.022508	0.363008***	−0.005322***	0.061497**	−0.111863	0.091934***
	(0.039221)	(0.021236)	(0.000982)	(0.026447)	(0.085062)	(0.007456)
			Political Institutions			
Political Rights	−0.00509	−0.008136	−0.011107***	−0.000345	−0.01208***	0.015962***
	(0.004675)	(0.011311)	(0.00198)	(0.00379)	(0.001917)	(0.003084)
Civil Liberties	−0.005678**	−0.006472*	−0.004826***	−0.004037*	−0.007711***	0.006019***
	(0.00152)	(0.00354)	(0.000911)	(0.002326)	(0.001663)	(0.001769)
Freedom of Press	−0.000943*	−0.356358***	−0.000643***	−0.523354***	−0.001352***	−0.162969***
	(0.000333)	(0.102334)	(0.0000775)	(0.036265)	(0.000265)	(0.019634)
Freedom of Speech	0.215426	−0.003668	0.783036**	−0.004699*	0.128801	−0.000663
	(0.252573)	(0.005634)	(0.283735)	(0.002454)	(0.077408)	(0.002164)
			Social Capital			
Corruption Perception Index	−0.022044	−0.01011***	−0.004962**	−0.00675**	−0.029513***	−0.018602***
	(0.013992)	(0.003497)	(0.001641)	(0.002462)	(0.009261)	(0.00174)
Gini Index	−0.000371	0.244798***	−0.000622	−0.110757***	−0.000565***	−0.175183***
	(0.000597)	(0.030138)	(0.000299)	(0.031826)	(0.0000857)	(0.020744)
Most people can be trusted	0.000928	−0.000546	−0.040653	0.000629*	−0.182915**	0.0000614
	(0.049108)	(0.000383)	(0.100556)	(0.000319)	(0.078626)	(0.000179)
Feeling of Happiness	−0.022786	0.0000187	0.002167	0.001053***	−0.001165	0.000862**
	(0.022009)	(0.000377)	(0.002674)	(0.000342)	(0.000981)	(0.000361)

Note: Significance levels: 1%***, 5%**, 10%*. Standard errors in parenthesis.

8 In order to provide a better perspective of the world maps, we decided to use year 2000 in Figure 1.3, Figure 1.6, and Figure 1.7, instead of using 2008 as we did in all of the other figures. The reason is that we wanted to report only observed data in the cartograms and avoid using imputed (estimated) observations.

9 Detailed results are included in Tables 1.3a–d.

References

Abramovitz, Moses. 1986. "Catching Up, Forging Ahead, and Falling Behind." *Journal of Economic History* 46(2): 385–406.

Alcorta, L., and W. Peres. 1998. "Innovation Systems and Technological Specialization in Latin America and the Caribbean." *Research Policy* 26(7): 857–881.

Archibugi, Daniele, and Alberto Coco. 2004. "A New Indicator of Technological Capabilities for Developed and Developing Countries (ArCo)." *World Development* 32(4): 629–654.

Archibugi, Daniele, and Jonathan Michie. 1995. "The Globalization of Technology: A New Taxonomy." *Cambridge Journal of Economics* 19: 121–140.

Arocena, R., and J. Sutz. 2003. "Inequality and Innovation as Seen from the South." *Technology in Society* 25(2): 171–182.

Barro, Robert J., and Jong-Wha Lee. 2001. "International Data on Educational Attainment: Updates and Implications." *Oxford Economic Papers* 53(3): 541–563.

Castellacci, Fulvio. 2004. "A Neo-Schumpeterian Approach to Why Growth Rates Differ." *Revue Économique* 55(6): 1145–1169.

Castellacci, Fulvio. 2008. "Technology Clubs, Technology Gaps and Growth Trajectories." *Structural Change and Economic Dynamics* 19(4): 301–314.

Castellacci, Fulvio. 2011. "Closing the Technology Gap?" *Review of Development Economics* 15(1): 180–197.

Castellacci, Fulvio, and Daniele Archibugi. 2008. "The Technology Clubs: The Distribution of Knowledge Across Nations." *Research Policy* 37(10): 1659–1673.

Castellacci, Fulvio, and Jose Miguel Natera. 2011. "A New Panel Dataset for Cross-Country Analyses of National Systems, Growth and Development (CANA)." *Innovation and Development* 1(2): 205–226.

Castellacci, Fulvio, and Jose Miguel Natera. 2013. "The Dynamics of National Innovation Systems: A Panel Cointegration Analysis of the Coevolution between Innovative Capability and Absorptive Capacity." *Research Policy* 42(3): 579–594.

Cingranelli, David L., and David Richards. 2008. "The Cingranelli and Richards (CIRI) Human Rights Data Project." *Human Rights Quarterly* 32: 395–418.

Desai, Meghnad, Sakiko Fukuda-Parr, Claes Johansson, and Fransisco Sagasti. 2002. "Measuring the Technology Achievement of Nations and the Capacity to Participate in the Network Age." *Journal of Human Development* 3(1): 95–122.

Edquist, Charles. 1997. *Systems of Innovation: Technologies, Institutions and Organizations*. London: Pinter.

Edquist, Charles, and Leif Hommen. 2008. *Small Country Innovation Systems: Globalization, Change and Policy in Asia and Europe*. Cheltenham: Edward Elgar Publishing.

Fagerberg, 1994. "Technology and International Differences in Growth Rates." *Journal of Economic Literature* 32(3): 1147–1175.

Fagerberg, Jan, and Martin Srholec. 2008. "National Innovation Systems, Capabilities and Economic Development." *Research Policy* 37(9): 1417–1435.

Fagerberg, Jan, and Bart Verspagen. 2002. "Technology-Gaps, Innovation-Diffusion and Transformation: An Evolutionary Interpretation." *Research Policy* 31(8–9): 1291–1304.

Fagerberg, Jan, Martin Srholec, and Mark Knell. 2007. "The Competitiveness of Nations: Why Some Countries Prosper While Others Fall Behind." *World Development* 35(10): 1595–1620.

Filippetti, Andrea, and Antonio Peyrache. 2011. "The Patterns of Technological Capabilities of Countries: A Dual Approach Using Composite Indicators and Data Envelopment Analysis." *World Development* 39(7): 1108–1121.

Freeman, Chris. 1987. *Technology, Policy, and Economic Performance: Lessons from Japan*. London: Pinter.

Freeman, Chris. 1995. "The 'National System of Innovation' in Historical Perspective." *Cambridge Journal of Economics* 19(1): 5–24.

Freeman, Chris. 2004. "Technological Infrastructure and International Competitiveness." *Industrial and Corporate Change* 13(3): 541–569.

Freeman, Chris, and Francisco Louça. 2001. *As Time Goes By: The Information Revolution and the Industrial Revolutions in Historical Perspective*. Oxford: Oxford University Press.

Godinho, Manuel Mira, Sandro F. Mendonça, and Tiago Santos Pereira. 2005. "Towards a Taxonomy of Innovation Systems." Instituto Superior de Economia e Gestão (Lisbon), Working Papers 13.

Honaker, J., and G. King. 2010. "What to Do About Missing Values in Time-Series Cross-Section Data." *American Journal of Political Science* 54(2): 561–581.

Howitt, Peter, and David Mayer-Foulkes. 2005. "R&D, Implementation, and Stagnation: A Schumpeterian Theory of Convergence Clubs." *Journal of Money, Credit and Banking* 37(1): 147–177.

Lall, Sanjaya. 2003. "Indicators of the Relative Importance of IPRs in Developing Countries." *Research Policy* 32(9): 1657–1680.

Lundvall, Bengt-Åke. 1992. *National Systems of Innovation: Towards a Theory of Innovation and Interactive Learning*. London: Pinter.

Lundvall, Bengt-Åke. 2007. "National Innovation Systems – Analytical Concept and Development Tool." *Industry and Innovation* 14(1): 95–119.

Lundvall, Bengt-Åke, K.J. Joseph, Cristina Chaminade, and Jan Vang. 2009. *Handbook of Innovation Systems and Developing Countries: Building Domestic Capabilities in a Global Setting*. Cheltenham: Edward Elgar Publishing.

Nelson, Richard R. 1993. "National Innovation Systems: A Comparative Analysis." Social Science Research Network (Rochester, NY) Scholarly Paper ID 1496195.

Verspagen, Bart. 1991. "A New Empirical Approach to Catching Up or Falling Behind." *Structural Change and Economic Dynamics* 2(2): 359–380.

Part I Global Trends

Chapter 2

World Top University Rankings: From Distribution to Implications on National Knowledge Creation and Competitiveness

Thanh Quang Le and Kam Ki Tang

Introduction

Universities have been at the heart of human capital and knowledge creation since Plato founded the first one in 387 BCE at Athens. Nowadays, universities play a central role in driving economic development around the world. Alan Greenspan, the former Chairman of the US Federal Reserve System, speaking on the "structural change in the new economy" to the National Governors Association in 2000, emphasized that:

> In a global environment in which prospects for economic growth now depend importantly on a country's capacity to develop and apply new technologies, our universities are envied around the world. ... If we are to remain preeminent in transforming knowledge into economic value, the US system of higher education must remain the world's leader in generating scientific and technological breakthroughs and in preparing workers to meet the evolving demand for skilled labor. (Greenspan 2000)

This statement succinctly summarizes the mission of leading US universities in contributing to the economic development of the country. Universities are important initiating forces for national innovation capacity as they seed new generations of applied research, scientific breakthroughs, and streams of new products. They also play a central and strategic role in educating and training scientists, engineers, teachers, researchers, entrepreneurs, and other skilled workers. Knowledge created by universities is diffused to the society through an array of activities, ranging from educating undergraduate students, training postgraduate students, publishing research papers, to involvement in public debates, providing consultancy services, and producing patents. In this modern era, economic growth is increasingly

The Handbook of Global Science, Technology, and Innovation, First Edition.
Edited by Daniele Archibugi and Andrea Filippetti.

knowledge-intensive; as such, universities are at the heart of the economic vitality and competitiveness of their host nations. Michael Porter of Harvard Business School once commented: "Skilled human resources and knowledge resources are two of the most important factors for upgrading national competitive advantage" (Boulton 2009). Given the significant economic and social contributions of universities to their communities and the whole economy, there is a strong need to evaluate their knowledge creation capability.

Universities' knowledge creation capacity is intrinsically dependent on their stocks of academic talent. Generally speaking, academic talent is a specific type of human capital alongside other types of human capital such as athletics, art, music, or entrepreneurship (Li, Shankar, and Tang 2010, 2011). Academic talent and tacit knowledge are hard to measure directly, but they can be reflected through people's academic performance. In that sense, a university's research and publication performance is the best tool to capture the size and quality of its academic talent stock. Accordingly, universities' knowledge pool could be calculated and compared with each other.

Over the last 10 years, the rise of the middle class in many populous emergent economies has created a high demand for higher education. At the same time, the marketization of the education sector, especially the tertiary one, in many advanced economies (most noticeably the United Kingdom, Canada, Australia, and the United States) plus the globalization of education services have created strong demand for comparative information on the quality of world universities. A large number of world university league tables have been created as a result.

As a consumer good, university league tables are very "successful" as they have attracted a lot of attention from the general public, policymakers, and stakeholders, and have evolved into valuable new benchmarking tools. At the individual level, league tables act as a quality assurance although the credibility of certain scoring methods has been a subject of criticisms. At the institutional level, many universities have heavily advertised their ranking positions for marketing purposes. At the national level, being able to occupy the top spots of university league tables signifies a country's *current* innovative capacity and human capital stock. Furthermore, since academic talent is highly mobile nationally and internationally, there is a conjecture that the higher the university's position in the world rankings, the easier for it to attract good academics and therefore, the higher the country's *future* innovation capacity (Li et al. 2011).

Against this background, the main objective of this chapter is to examine the distribution of top universities across nations. However, it is necessary to say straight away that although teaching is an integral part of universities' mission, the focus of this study is on their research. This is because although teaching is more concerned about education and training, it is research that reflects more fully the knowledge creation process of an institution. To that end, we make use of the Academic Ranking of World Universities by the Shanghai Jiaotong University over the 2003–2013 period. Instead of using the rankings, we use the scores underlying the rankings to measure university performance. An advantage of the score measure over the ranking measure lies in its cardinal characteristic, which allows us to aggregate the scores of all top universities of a country into a national score. In doing so, we are able to investigate the research performance differences across countries. In addition, unlike the ranking measure, this measure can help us visualize how far apart countries are in their academic performance and whether they are catching up with each other over time.

In the data analysis, we will firstly focus on national aggregate data. In particular, we look at the number of top universities and the aggregate academic scores that each country obtains. The main purpose is to use the distribution of top universities as indicators of the disparity in science, technology, and innovation capacities across countries (see Castellacci and Natera in this volume, Chapter 1). It is a country's total stock of technology and scientific knowledge that determines its international competitiveness, regardless of population. Knowing, however, that aggregate measures tend to "benefit" larger countries, this chapter will also examine the countries' performance in relation to their size. To remove the scale factor, the aggregate score measure constructed is deflated either by population or GDP level. The comparison of countries' research output intensity is expected to provide us with a better idea on how efficient countries are in performing advanced academic research within their limited resources.

The next section studies research and innovations in universities and their characteristics, while the following section shifts the focus to the distribution of top universities. It explains how league tables of top universities are actually measured and why the distribution of top universities across countries matters. It also compares the distribution of world universities at the top end across countries and geographical regions. The last section concludes with a discussion on policy implications.

Research and Innovations in Universities

Universities as an Important Source of Technical Change

Throughout history, universities have been the birthplace of many significant innovations. For example, Penicillin was developed by scientists at Oxford University in 1939; blood preservation technology by the University of Columbia in 1940; ultrasound by the University of Vienna in 1942; Liquid Crystal Display (LCD) by Kent State University in 1967; and the Hepatitis B vaccine by the University of Pennsylvania in 1969. The number of US patents awarded to universities increased from 500 in 1982 to more than 3100 in 1998 (Lach and Schankerman 2008). Within a decade from 1990 to 2000, gross licensing revenues generated by university inventors increased by more than seven-fold from US$186 million to roughly US$1.3 billion. This rapid growth in university innovation is partly due to the passage of the Bayh-Dole Act in 1980, which gave US universities intellectual property rights on patentable inventions funded by federal government money (Lach and Schankerman 2008; Goldstein and Renault 2004). By year 2000, a large number of US universities had set up their own technological licensing offices governed by policies concerning intellectual property protection and royalty sharing arrangements for academic scientists (Lach and Schankerman 2008). By receiving funding from the government, universities to some extent have accepted the responsibility of directly contributing to national development instead of doing so merely indirectly through their traditional missions of teaching and research.

The knowledge created by universities spills over to industries through a number of channels. First, many students graduating from universities work in industries.[1] Second, companies sometimes send their staff to universities to learn new techniques or knowledge. Third, companies and universities can engage in R&D collaboration (Jaffe, Trajtenberg, and Henderson 1993). Fourth, there is movement of scientists between the academic and industrial sectors (Zucker, Darby, and Toreto 2002). Fifth, university academics publish their findings in technical and scientific papers (Jaffe et al. 1993). Sixth, universities can license their own inventions to private firms (Jensen and Thursby 2001; Thursby and Kemp 2002). The knowledge flow from universities to the private sector presumably will stimulate industrial R&D, which then increases productivity and output (Fagerber, 1994).

Characteristics of Academic Research

The conventional view places basic research at the center of academic research. According to that view, academia conducts research that comes up with fundamental insights (Sauermann and Stephan 2010). As academic research often aims at recognition, through publications in scientific journals or presentations at conferences, rather than raising commercial value, the knowledge produced in this type of research is more like a public good. The timely and widespread disclosure of research results is the key factor for accelerating the accumulation of knowledge in society over time (Sorenson and Fleming, 2004). The accumulated stock of knowledge then becomes an input for the production of many goods and services.

This view is deemed to be too narrow to capture the characteristics of contemporary academic science.[2] As discussed above, many academic institutions have had their research outputs commercialized. However, it is still correct to say that a substantial part of university research is a public good. Publications, an evidence of peer recognition from the scientific community, are still the main incentive for the continuous flow of academic research (Dasgupta and David 1994; Stephan 1996).

As compared to their industrial counterparts, academic researchers have more freedom in choosing research topics and methodologies, in deciding when to "pause" and disseminate the results. There have been claims that scientists working at universities are underpaid.[3] But this may actually create an unequivocal advantage for academic sciences as they can conduct the research at relatively lower costs (Sauermann and Stephan 2010), and that is why universities are ideal places for undertaking explorative research. On the other hand, academic researchers may pursue projects that they find interesting or prestige enhancing but generate little value to society.

Traditionally, state government is the main funding source for university research. Government funding comes in three forms: direct funding to finance infrastructure and staff salaries, research and development grants, and loans and tuition subsidies to students. As these funds are discretionary items in the state budget, they are highly susceptible to budgetary cuts (Zusman 2005). A recent survey by Moffitt (2014) reveals that due to the recent global recession (2008–2012), deep budget cuts were common worldwide. This creates concerns about the flow-on effect to the innovation capacity of the private sector due to the close connection between universities and industries.

History has shown that many important innovations ranging from medicines to information technology products have their roots from publicly funded projects conducted at universities. According to Nelson (1959) and Arrow (1962), industries have less incentive to invest in basic research, because they cannot fully appropriate the economic value associated with research ideas due to knowledge spillovers and imperfect intellectual property rights. This leaves the explorative research to mainly academic institutions. Nowadays, besides the government, universities also look to the private sector for funding. As these funding sources often have some particular research agenda, academic researchers' freedom could be constrained. In the meantime, by controlling the research funding, firms are able to direct academic researchers to projects that help maximize their economic payoffs.

A number of strategies have been employed by the universities in response to the recent cut in state funding. Zusman (2005) states that one strategy is that universities make high-return and high-demand professional programs (mainly in business, law, and medicine) self-funded or charge a higher fee on them. Some institutions introduce self-supporting part-time programs and contract education programs with specific industries. A number of public universities have become privatized, while others have found support in the form of private donations and external grants. An indication of privatization of public higher education is the number of online courses produced by private and for-profit companies (Robertson and Leumer 2014). Facing shrinking research grants, several universities opt to form for-profit collaborations with industries and make commercial technological transfer (Zusman 2005).

International Comparison of Academic Research Output

Academia is often perceived as a homogeneous sector. This view, however, is far from the reality. For any given country, there are top tier research institutions as well as lower tier ones, and there is also considerable heterogeneity among researchers within an institute in terms of research output and impacts.

As mentioned above, academic scientists have to rely on funding from the government or industry. In deciding the allocation of research grants, funding agencies typically consider the researchers' track record as well as the research environment of the academic institutions they work for. This, on the one hand, makes the competition for funding more competitive because the institutions that manage to stand out from the crowd tend to take most of the fruits. On the other hand, there is an urgent need for a way to measure the research quality of academia. To this end, there has been strong interest in ranking worldwide universities based on their achievements in both teaching and research, known as university rankings or university league tables. This benchmarking exercise is done at the institutional level as well as at the national level. At the institutional level, league tables, given their emphasis on research outputs, obviously provide a measure of a university's research output and impacts. At the national level, league tables also provide an informal measure of a country's research capacity and its competitiveness in a knowledge-based world economy (Aghion et al. 2007).

Among all the league tables, the most influential and widely cited ones are the Academic Ranking of World Universities (ARWU) by Shanghai Jiaotong University, the THE World University Rankings published by Times Higher Education, and the

QS World University Rankings published by Quacquarelli Symonds. These rating agencies nowadays provide university rankings as well as department rankings for specific subjects. Historically, the THE Rankings and QS Rankings had a period of joint publication from 2004 to 2009. In 2009, THE split from QS to form a new partnership with Thomson Reuters. While QS continues to use their preexisting methodology, THE has created and adopted a new method that takes advantage of the data on scientific paper citations supplied by its partner Thomson Reuters. Table 2.1 lists the major features of these three rankings.

Amongst these three league tables, the ARWU is commonly considered the most preferable indicator due to its high objectivity and central focus on research (Marginson 2007; Hazelkorn 2007). First published in 2003, the ARWU was originally developed to benchmark the research performance of Chinese universities but was then extended to include universities of other countries. Since the purpose of this study is to compare research capacity across different countries, we choose the ARWU ranking table over the other two for our analysis.

The ARWU measures universities' research strengths using six different indicators, namely: (1) the number of alumni that have received Nobel prizes in science and economics and Fields medals in mathematics,[4] (2) the number of staff that have received Nobel prizes and Fields medals, (3) the number of highly cited researchers in 21 broad subject categories selected by Thomson Scientific, (4) the number of papers published in *Nature* and *Science*,[5] (5) the number of papers indexed in Science Citation Index (Expanded) and Social Science Citation Index, and (6) the per capita academic performance of these indicators. Each year, more than 1000 universities[6] are surveyed, but only the rankings and scores of the top 500 (denoted as ARWU500 hereafter) are reported. The scores for the top 500 universities are normalized in a way that has the topmost university scoring 100, and other institutions are calculated as a percentage of this top score. Harvard University has maintained its leading position throughout the whole sample period (2003–2013), suggesting that it has more research talent than any other university.

Comparison Across All Fields

Over the period 2003–2013, the ARWU500 universities spanned across 45 countries. Universities in the OECD countries account for the lion's share as only 71 ARWU500 universities were from non-OECD countries (of which 29 are from China and 5 from Hong Kong). The biggest share went to Europe. America came second, followed by the Asia Pacific region. The Middle East and Africa came last with a negligible share. Throughout the period, a large number of countries did not have a single university in the ARWU500 ranks.

Figure 2.1 shows regional distribution of ARWU500 for the two endpoints of the data period. In 2003, 219 universities (43.89%) were located in 23 European countries, 193 (38.68%) in six American countries (160 of them in the United States alone), 83 (16.63%) in nine Asia Pacific countries, and only four universities (0.80%) in a Middle Eastern and African country (South Africa). In 2013, the shares of Asia Pacific and Middle East and Africa slightly improved to 19.76% (with 99 universities in 10 countries) and 1.80% (with nine universities in four countries) respectively. The expansion of these two regions implies the contraction of the others.

Table 2.1 Some major world university rankings.

	ARWU Rankings		*QS Rankings*		*THE Rankings*		*Remarks*
Indicators and their weightings	Alumni winning Nobel prizes and Fields medals	10%	Academic peer review on research output	40%	Industry income – innovation	2.5%	(i) ARWU focuses entirely on research. (ii) Both QS and THE take account of teaching, infrastructure, and international diversity. (iii) QS puts a significant weight on two subjective measures: academic peer review and employer review.
	Staff winning Nobel prizes and Fields medals	20%	Recruiter review on quality of graduates	10%	Teaching – the learning environment	30%	
	Highly cited researchers in 21 broad subject categories	20%	Faculty student ratio	20%	Research – volume, income, and reputation	30%	
	Articles published in *Nature* and *Science*	20%	International orientation	10%	Internationaloutlook– staff, students, and research	7.5%	
	Science Citation Index and Social Science Citation Index	20%	Scientific paper citations per faculty	20%	Scientific paper citations – research influence	30%	
	Per capita academic performance	10%					
Period covered	2003–2013		2004–2014		2004–2009; 2010–2014		QS and THE had a joint publication in 2004–2009 before their split in 2009.

Major rankings and number of top universities released	Aggregate rankings Subject rankings Field rankings	Top 500 Top 200 Top 200	Aggregate rankings Subject rankings Faculty rankings Asia's rankings Latin America's rankings BRICS's rankings Under 50's rankings	Top 800 Top 200 Top 400 Top 300 Top 250 Top 100 Top 50	Aggregate rankings Subject rankings	Top 400 Top 100	(i) Besides aggregate rankings and subject rankings, ARWU and QS additionally have field/ faculty rankings. (ii) QS also has rankings for different geographical regions.

Source: Academic Ranking of World Universities (2014), QS World University Rankings (2014), and THE World University Rankings (2014).

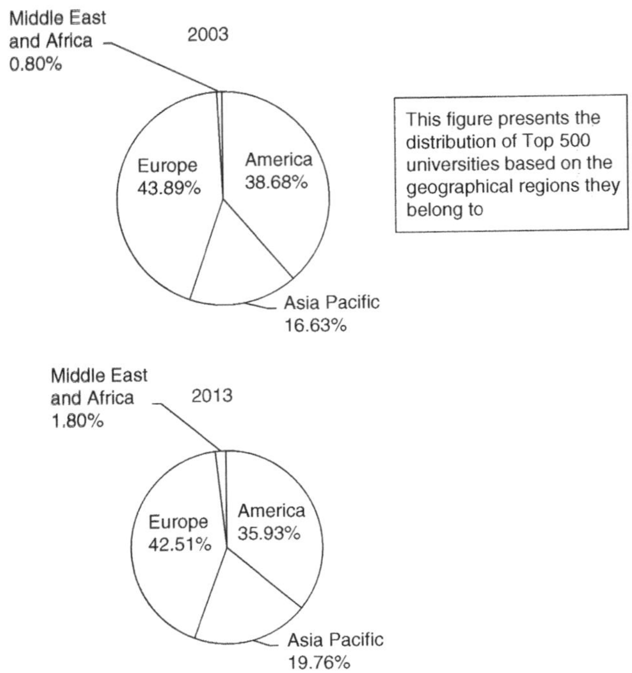

Figure 2.1 Regional distribution of ARWU500 universities.
Source: Data from Academic Ranking of World Universities (2014).

Indeed, the number of European ARWU500 universities dropped to 213 (42.51%) while the American ones shrank to 180 (35.93%).

Figure 2.2 depicts the performance of the top countries in Europe, America, and Asia Pacific over the sample period. The United States, the top country in America and also in the world, dominated the whole decade with around 31% of the ARWU500 universities on average. The number of US universities decreased slightly in recent years from the highest number of 167 in 2004 to 146 in 2013. By contrast, China, one of the top performers in the Asia Pacific region, had made impressive progress. The number of Chinese ARWU500 universities had increased nearly three-fold, from 10 in 2003 to 28 in 2012 and 27 in 2013. This improvement had allowed China to rise into the top 10 countries since 2010. Australia, another top performer in the Asia Pacific region, had made slight improvement over the years, jumping from 12 universities in ARWU500 to 18 universities in 2013. Despite having a small decrease in the number of British universities joining in ARWU500 over the years, the European champion was steady in its second place overall. These are four of the top 10 countries. Most of the other members of the top 10 are from Europe including Germany, France, Italy, and the Netherlands, leaving the other two regions with an additional top 10 country each: Canada (America) and Japan (Asia Pacific).

Although the total number of US universities in the ARWU rankings has fallen over the 2003–2013 period, they consistently dominated the top spots. As shown in Table 2.2, most of the top 10 positions are occupied by US universities. The only non-US universities that made it to the top 10 are the Oxbridge duo from the United Kingdom.

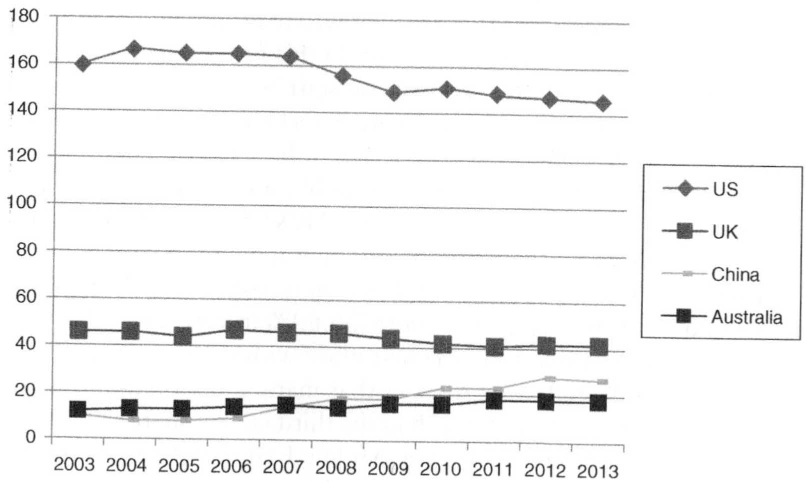

Figure 2.2 Number of ARWU500 universities by the four top countries.
Source: Data from Academic Ranking of World Universities (2014).

Table 2.2 Top 10 university rankings.

Ranking	*2003*	*2008*	*2013*
1	Harvard University	Harvard University	Harvard University
2	Stanford University	Stanford University	Stanford University
3	California Institute of Technology	University of California, Berkeley	University of California, Berkeley
4	University of California, Berkeley	University of Cambridge	Massachusetts Institute of Technology
5	University of Cambridge	Massachusetts Institute of Technology	University of Cambridge
6	Massachusetts Institute of Technology	California Institute of Technology	California Institute of Technology
7	Princeton University	Columbia University	Princeton University
8	Yale University	Princeton University	Columbia University
9	University of Oxford	University of Chicago	University of Chicago
10	Columbia University	University of Oxford	University of Oxford

Source: Elaborations on Academic Ranking of World Universities (2014).

Among the six indicators used in the ARWU rankings, the purpose of the last one is to capture per capita academic performance of all the other indicators. The main purpose of this indicator is to control for the size of universities. While this indicator is useful for comparison across institutions, it is less relevant for the comparison across countries. Hence, we exclude this indicator from computing the total research score for an individual university:

$$TotalScore = Staff + Alumni + HC + NS + PUB$$

To keep it simple, the research score of the university is the unweighted sum of five different research dimensions, which is in contrast to the original weighted sum method by the ARWU. As no arbitrary weight is given to a particular research measure, all indicators are considered equally important. In this formulation, *TotalScore* is the university's aggregate score, *Staff* is the score on staff being Nobel laureates or Fields medalists, *Alumni* is the score on alumni being Nobel laureates and Fields medalists, *HC* is the score on highly cited researchers, *NS* is the score on *Nature* and *Science* publications, and *PUB* is the score on indexed publications. In the same manner, the national measure is the total score of a country's ARWU500 universities in that year.

Figure 2.3 presents the distribution of the new national scores across regions. Despite having the largest number of ARWU500 universities over the period, Europe only came second in terms of aggregate score (39.69% in 2003 and 34.80% in 2013 respectively). It is America that took the first place with 47.06% (52.64%) share in 2003 (2013) respectively, reflecting the fact that many top scored universities were located in the United States. Asia Pacific had the third largest share (12.83% in 2003 and 12.83% in 2013 respectively), leaving Middle East and Africa in the last place with a minimal share in either year.

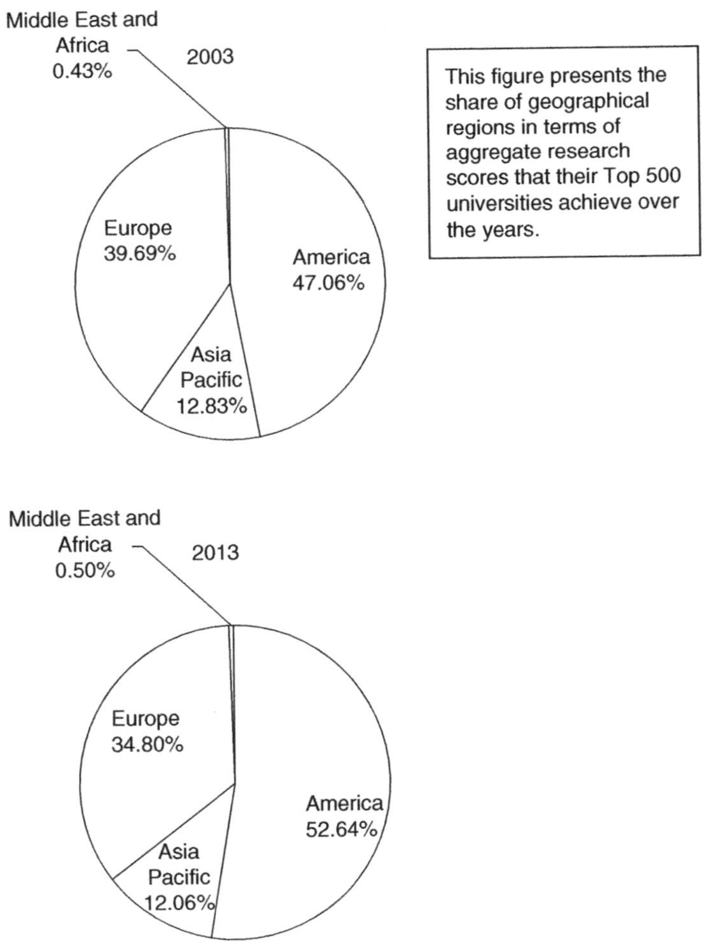

Figure 2.3 Regional distribution of research performance scores 2003 and 2013.
Source: Data from Academic Ranking of World Universities (2014).

Table 2.3 lists the top 10 countries in our sample based on the aggregated scores over the period 2003–2013. With a large number of ARWU500 universities being at the top end of the league tables, not surprisingly the United States secured its number one position every year. The United Kingdom occupied the second place while the third place belonged to Germany. While the top three seems quite stable over time, fourth and fifth places alternated between Japan and Canada. France seemed to be well secured in its sixth position. The last four places in the top 10 remained in competition between Italy, Australia, the Netherlands, Sweden, and China. The emergence of China over the last decade is stunning, with its national score increasing three-fold from 464 to 1432 over the sample period. In the last four years (2010–2013), China had overtaken many OECD countries and moved into the top 10 group.

It is worth pointing out that there is a huge gap between the United States and the rest in the top 10 group. Over the years, the United States scored more than four times the second-placed United Kingdom, and as high as 15 times the tenth-placed country. Although China has joined the top 10, its score is still more than 10 times smaller than that of the United States, as well as lagging behind all G7 countries except Italy.

As much as 30% of the original ARWU scores are allocated to association with Nobel laureates and Fields medalists.[7] The inclusion of these two indicators in measuring quality of research is controversial in the sense that it benefits long-established universities as opposed to their younger counterparts. This is evident in the fact that these prizes are often awarded to scholars in US universities and, to some extent, Western European universities (Marginson 2013). Furthermore, to the extent that Nobel laureates and Fields medalists are likely to have some of the strongest publication records, which are already included in another three indicators of ARWU (*HC*, *NS*, and *PUB*), the scores associated with Nobel laureates and Fields medalists could lead to double counting. On the other hand, one might argue that the other three indicators do not sufficiently account for the impact differentials of publications by different researchers, and that could be captured by the Nobel prize and Fields medal achievement. However, even for the impact argument, given there are numerous international awards, some of them are equally prestigious (e.g., the Wolf Prize of the Wolf Foundation of Israel, the Leroy P. Steele Prize of the American Mathematical Society in Mathematics) and some of them cover areas outside, but no less important than pure sciences (e.g., the Wolf Prize of the Wolf Foundation in Agriculture, the American Academy of Arts and Letters Gold Medal in Architecture), it is debatable why only Nobel prizes and Fields medals are counted. Put it the other way, the awarding of merely Nobel prizes and Fields medals, as compared to the publication of journal articles, by nature are extremely rare events; as such, its associated score could have much greater "noise" as a measure of human capital.

To see whether the results in Table 2.3 are sensitive to such potential noise, we exclude these two indicators from the calculation of national scores. As a result, the unweighted aggregate is computed as the following:

$$TotalScore = HC + NS + PUB$$

From Table 2.4, it can be seen that there is no dramatic change in the top 10 countries. The only difference is that China would have entered the list of top 10 countries three years earlier in 2007 instead of 2010, and it would have been ranked

Table 2.3 Top 10 country rankings in terms of aggregate scores.

	2003		*2005*		*2008*		*2011*		*2013*	
Ranking	*Country*	*Score*	*Country*	*Score*	*Country*	*Score*	*Country*	*Score*	*Country*	*Score*
1	US	15,095	US	17,417	US	16,916	US	16,627	US	19,378
2	UK	3,629	UK	4,229	UK	4,290	UK	4,124	UK	4,882
3	DEU	2,680	DEU	2,988	DEU	2,973	DEU	2,914	DEU	3,548
4	JPN	2,190	JPN	2,278	JPN	2,121	CAN	1,849	CAN	2,293
5	CAN	1,715	CAN	1,878	CAN	1,827	JPN	1,807	JPN	2,075
6	FRA	1,277	FRA	1,637	FRA	1,705	FRA	1,732	FRA	2,017
7	ITA	1,153	ITA	1,257	ITA	1,226	AUS	1,310	CHN	1,896
8	NLD	932	NLD	992	AUS	1,089	ITA	1,243	AUS	1,757
9	AUS	850	AUS	971	NLD	1,018	CHN	1,184	ITA	1,499
10	SWE	815	SWE	957	SWE	912	NLD	1,094	NLD	1,172

Source: Elaborations on Academic Ranking of World Universities (2014).
Note: US – United States, UK – United Kingdom, DEU – Germany, JPN – Japan, CAN – Canada, FRA – France, ITA – Italy, NLD – Netherlands, AUS – Australia, SWE – Sweden, CHN – China.

Table 2.4 Top 10 country rankings in terms of aggregate scores (without Nobel prizes and Fields medals).

	2003		*2005*		*2008*		*2011*		*2013*	
Ranking	*Country*	*Score*	*Country*	*Score*	*Country*	*Score*	*Country*	*Score*	*Country*	*Score*
1	US	13,497	US	13,916	US	13,462	US	13,147	US	16,094
2	UK	3,256	UK	3,263	UK	3,370	UK	3,180	UK	4,129
3	DEU	2,388	DEU	2,302	DEU	2,310	DEU	2,284	DEU	3,004
4	JPN	2,107	JPN	2,071	JPN	1,929	CAN	1,627	CAN	2,071
5	CAN	1,612	CAN	1,643	CAN	1,647	JPN	1,546	CHN	1,854
6	FRA	1,080	FRA	1,042	FRA	1,168	FRA	1,181	JPN	1,795
7	ITA	1,052	ITA	1,019	ITA	986	CHN	1,173	AUS	1,604
8	NLD	851	AUS	863	AUS	935	AUS	1,154	FRA	1,483
9	AUS	820	NLD	839	NLD	874	ITA	1,014	ITA	1,239
10	SWE	693	SWE	702	CHN	865	NLD	938	NLD	1,227

Source: Elaborations on Academic Ranking of World Universities (2014).
Note: US – United States, UK – United Kingdom, DEU – Germany, JPN – Japan, CAN – Canada, FRA – France, ITA – Italy, NLD – Netherlands, AUS – Australia, SWE – Sweden, CHN – China.

the fifth in 2013 instead of the seventh, bypassing Japan and France. In terms of score, the gap between the United States and the other countries remains as large as before. This means that the country results are very robust to the exclusion of the Nobel prize and Fields medal factor.

Comparison of Top-Quality Science and Engineering Research Output

As industrial production is more closely related to frontier research in science and engineering than in many other fields, we also exclusively consider an indicator of publications in *Nature* and *Science*, the leading journals in these fields. Figure 2.4 depicts the regional results for 2003 and 2013. The overall pattern is similar to

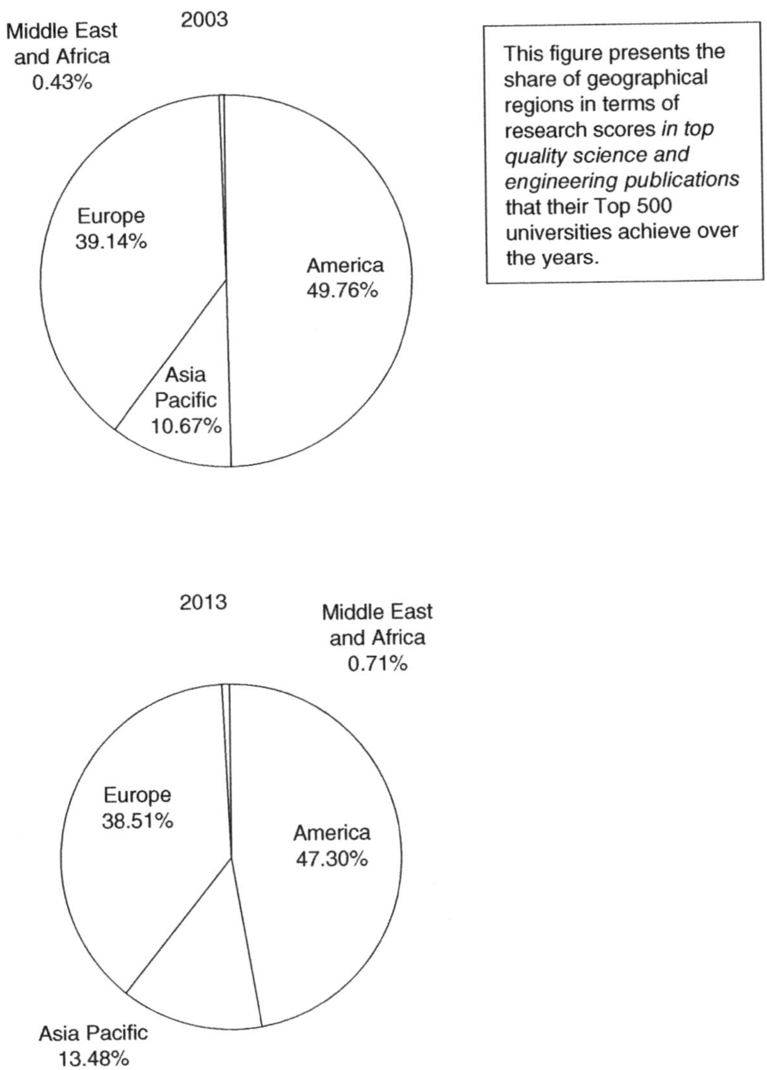

Figure 2.4 Regional distribution of top-quality science and engineering research scores. Source: Data from Academic Ranking of World Universities (2014).

Figure 2.3. However, now the shares of America and Europe fell slightly between 2003 and 2013, from 39.14% to 38.52% for Europe and from 49.76% to 47.30% for America. By contrast, the shares of Asian Pacific and Middle Eastern and African countries increased slightly over the decade, from 10.67% to 13.48% for Asia Pacific and from 0.43% to 0.71% for Middle East and Africa (see Zhou and Li in this volume, Chapter 5).

Table 2.5 shows the top 10 country rankings over the period based entirely on the publications in *Nature* and *Science*. Interestingly, the results are very similar to Table 3. The US was again far ahead of all other countries. China was not able to break into the top 10 group until 2012 and its ranking lagged behind the G7 countries (except for Italy) and Australia. This means that the ARWU500 league table weighs heavily on performance in science and engineering and hence it is a useful indicator of a countries' research capability in these fields.

Accounting for the Country Size

So far, the comparison of national research capability has been conducted without controlling for the size of the countries. An important argument is that the size of a country (either in terms of its market scale or population) matters to how universities from this country perform on the international front. Without a doubt, richer countries will be more likely to spare a larger amount of resources for tertiary education. Similarly, more highly populated countries will tend to benefit more from their larger stock of human capital. To explore these aspects, this section considers the intensity of research output. To that end, the total national aggregate scores calculated previously are deflated by either total population or GDP to give new intensity measures.

Table 2.6 lists top 10 countries based on the aggregate research scores weighted by population. Surprisingly, a completely different picture is presented. The United States and other G7 countries no longer dominated the top most places as seen in the previous section. By contrast, the table is now mainly occupied by a group of smaller countries from Northern Europe. The steadiest countries are Switzerland and Sweden that took the top two spots for the whole period of study. The next two positions were mainly competed for by Denmark and Israel. Despite being overtaken by Finland in 2010, the United Kingdom well secured its fifth place. In the bottom five, Norway and the Netherlands ranked ahead of Finland, Australia, Canada, and the United States. During 2003–2009, the United States mostly took a modest rank of number nine. However, with the progress of Australia during 2010–2012, the United States was driven out of the list. China fared even worse as it never made to the list over the whole sample period.

Table 2.7 presents top 10 country rankings in terms of research scores deflated by GDP instead of population. Again, the topmost spots are mainly occupied by countries located in Northern Europe. The United States did not even make it to this table over the sample period, let alone China. Israel topped the list for the whole period and was followed by Sweden and then Switzerland. The United Kingdom sits comfortably in the fourth place. The presence of either Denmark or New Zealand rounded up the top five. Other top 10 countries include Finland, the Netherlands, Hong Kong, Australia, and Canada.

Table 2.5 Top 10 country rankings in top-quality science and engineering research performance.

	2003		*2005*		*2008*		*2011*		*2013*	
Ranking	*Country*	*Score*	*Country*	*Score*	*Country*	*Score*	*Country*	*Score*	*Country*	*Score*
1	US	3,497	US	3,508	US	3,232	US	3,308	US	3,220
2	UK	851	UK	842	UK	848	UK	793	UK	769
3	DEU	546	DEU	521	DEU	510	DEU	533	DEU	531
4	JPN	446	JPN	436	CAN	401	JPN	362	JPN	335
5	CAN	358	CAN	337	JPN	389	CAN	338	CAN	334
6	FRA	257	FRA	264	FRA	304	FRA	331	FRA	281
7	NLD	215	ITA	201	AUS	201	AUS	235	AUS	242
8	ITA	197	NLD	191	SWZ	175	ITA	188	CHN	221
9	SWZ	196	SWZ	188	NLD	171	NLD	184	SWZ	192
10	AUS	181	AUS	183	ITA	166	SWZ	180	NLD	185

Source: Elaborations on Academic Ranking of World Universities (2014).
Note: US – United States, UK – United Kingdom, DEU – Germany, JPN – Japan, CAN – Canada, FRA – France, ITA – Italy, NLD – Netherlands, AUS – Australia, CHN – China, SWZ – Switzerland.

Table 2.6 Top 10 country rankings in terms of aggregate scores deflated by population.

Ranking	*2003*	*2004*	*2005*	*2006*	*2007*	*2008*	*2009*	*2010*	*2011*	*2012*
1	SWZ	SWZ	SWZ	SWZ	SWZ	SWZ	SWZ	SWZ	SWZ	SWZ
2	SWE	SWE	SWE	SWE	SWE	SWE	SWE	SWE	SWE	SWE
3	DNK	DNK	ISR	ISR	ISR	DNK	DNK	ISR	ISR	DNK
4	UK	ISR	DNK	DNK	DNK	ISR	ISR	DNK	DNK	ISR
5	NLD	UK	UK	UK	UK	UK	UK	FLD	UK	UK
6	ISR	NLD	NOR	NLD	NOR	NOR	NOR	UK	NLD	NLD
7	FLD	NOR	NLD	NOR	NLD	FLD	NLD	NOR	NOR	NOR
8	CAN	FLD	US	US	FLD	NLD	FND	NLD	FLD	AUS
9	US	US	CAN	FLD	US	US	US	CAN	AUS	FLD
10	NOR	CAN	FLD	CAN	CAN	CAN	CAN	AUS	CAN	CAN

Source: Elaborations on Academic Ranking of World Universities (2014).
Note: SWZ – Switzerland, SWE – Sweden, DNK – Denmark, UK – United Kingdom, NLD – Netherlands, ISR – Israel, FLD – Finland, CAN – Canada, US – United States, NOR – Norway, AUS – Australia.

Table 2.7 Top 10 country rankings in terms of aggregate scores deflated by GDP.

Ranking	*2003*	*2004*	*2005*	*2006*	*2007*	*2008*	*2009*	*2010*	*2011*	*2012*
1	ISR	ISR	ISR	ISR	ISR	ISR	ISR	ISR	ISR	ISR
2	SWE	SWE	SWE	SWE	SWE	SWE	SWE	SWE	SWE	SWE
3	SWZ	SWZ	SWZ	SWZ	SWZ	SWZ	SWZ	SWZ	SWZ	SWZ
4	UK	UK	UK	UK	UK	UK	UK	FLD	UK	UK
5	DNK	DNK	DNK	NZ	NZ	NZ	DNK	UK	NZ	NZ
6	FLD	CAN	CAN	DNK	CAN	FLD	FLD	NZ	DNK	DNK
7	CAN	HK	HK	CAN	NLD	DNK	NLD	DNK	AUS	AUS
8	HK	FLD	NLD	NLD	DNK	CAN	CAN	CAN	NLD	NLD
9	NLD	NLD	FLD	HK	AUS	NLD	NZ	AUS	FLD	FLD
10	AUS	AUS	AUS	FLD	FLD	AUS	AUS	NLD	CAN	CAN

Source: Elaborations on Academic Ranking of World Universities (2014).
Note: ISR – Israel, SWE – Sweden, SWZ – Switzerland, UK – United Kingdom, DNK – Denmark, FLD – Finland, CAN – Canada, HK – Hong Kong, NLD – Netherlands, AUS – Australia, NZ – New Zealand.

Discussions and Conclusions

This chapter looks at the tally of top universities across countries and what that means to international competitiveness. Generally speaking, universities are institutions designed to create and transfer general or specific knowledge. They are also clusters of human capital with technological as well as non-technological knowledge. Therefore, to certain extent, top university rankings reveal countries' capability of knowledge creation.

There has been a proliferation of academic rankings of world universities over the last 10 years. These rankings have cemented the notion that irrespective of the type, universities can be scored, sorted, and thus compared in a single league table (Marginson and van der Wende 2007). Although university rankings have not been in place for long, their importance has gone beyond the interests of students and university administrators and caught the attention of policymakers. More importantly, top university rankings are considered to contain economic implications in both the short and long term. In the long term, having top universities will secure global "soft power" in the future (in the sense of "knowledge is power"). The attraction and retention of international talent by top universities subsequently feeds back into their host countries' knowledge system by boosting its innovation capacity. In the short term, top universities can command a better position in the global education market, which accounts for a significant part of service exports for countries like Australia, Canada, and the United Kingdom, as well as being increasingly important for the United States and European countries. Through generating billions of dollars in revenue, academia gradually becomes a major contributor to the GDP growth of those higher education exporting countries.

Making use of the published ARWU500 data set, this chapter has taken a bold step in quantifying university output through which inference on national innovative capacity could be made. However, instead of using the rankings provided, it utilizes research scores behind those rankings to construct different aggregate research measures for each country, which are then used to compare across countries and/or geographical regions.

Aggregate data on top universities reveal that US universities dominated all tables throughout the 2003–2013 period. More importantly, they occupied almost all of the top ranking spots. For example, in 2013, the United States accounted for 35 of the top 50 universities and eight of the top 10, followed by the United Kingdom with five top 50 and two top 10. However, the United States had lost some ground in the middle and lower end of the league table over the years. This suggests that universities in other countries, especially in other OECD countries, were catching up with their US counterparts (Li et al. 2011). In proportional terms, China was the best performer as it had nearly tripled its number of ARWU500 universities within a decade and those universities that were already on the list in earlier years on average had also improved their ranking over time (although none of them had managed to break into the top 100). This was probably due to great support from the Chinese government. From 1996 to 2000, it was estimated that the Chinese government had distributed US$2.2 billion under Project 211 to about 6% of the universities in order to raise their standards (Li 2004). This is to say that establishing a top university requires a lot of resources and time because human capital may take several decades to accumulate. However, with the labor market of academics being highly globalized, it is possible for countries to build (or actually buy) up its talent stock quickly. The results on countries' academic research performance are robust with the exclusion of scores for Nobel prizes and Fields medals in aggregate research score calculation. The United States also took the lead in possessing top research in science and engineering.

Although the United States monopolized the world's top universities and China consistently outperforms other countries, their performance is less impressive when it comes to research output intensity. The dominance of the United States is strongly associated with its large population and economic size. After controlling for country size, either in the form of population or GDP level, the United States was found to lag behind many smaller countries such as Switzerland, Israel, or Australia. And China did not even make to the top 10 once over the years studied.

Clearly, results obtained are very different depending on what benchmark is used. If one wishes to compare countries' knowledge creation capability, then one should focus more on their aggregate research scores because embedded in that measure is the non-rivalrous characteristic of academic research. In this case, the United States clearly dominates the whole world in terms of innovative capacity and national competitiveness. In the meantime, China is quickly rising as a future challenger to the US hegemony. However, if examining how efficient a country is in conducting research given its limited resources is the main objective, then the research output intensity is a preferred measure. In this regard, small countries like Switzerland and Israel are truly the frontrunners.

Results of any league tables are obviously specific to the criteria used to rank the subjects, and different ranking systems have different assessment criteria and definitions of what constitutes university quality (Van Dyke 2005; Usher and Savino 2006). Another problem with rankings is that most systems evaluate universities as one unit, that is, different quality and performance characteristics are used to construct a composite index (Marginson 2007, 2012). Similarly, although the ARWU rankings are considered as the prime benchmark in their own fields, they are not without limitations and competing alternatives. For example, this league table has commonly been criticized for too much focusing on English language research output. Therefore, one should be cautious not to read too much into any snapshot results from them. Focusing on their trend results is recommended because systematic biases due to specific benchmarking criteria are likely to remain unchanged over time. Despite limitations, university league tables provide an informal measure of a country's ability to compete in a knowledge-based world economy given the emphasis on research outputs (Aghion et al. 2007).

As mentioned above, universities contribute to economic development not only through their enhancement of national knowledge creation capacity but also through their revenue from providing education services to non-nationals. Regarding this latter aspect, in our modern time of globalization, universities are mimicking multinational corporations in transforming themselves into transnational organizations. Many universities start to offer programs to learners located outside the domicile country, making education services globally tradable (Gribble and Ziguras 2003). These tradable programs are referred to as transnational education (UNESCO/Council of Europe 2002). Gribble and Ziguras observe that in the last decade, the magnitude of transnational education has grown dramatically, especially in Asia where British, American, and Australian institutions are offering education services. One common way of offering transnational education services that universities often do is setting up a satellite campus abroad. For examples, the US Carnegie Mellon University has recently established a campus in Australia and the Australian Royal Melbourne Institute of Technology has been operating a campus in Vietnam for more than a decade. The United States has the highest number of overseas campuses, with 78 in 2011 (Obst, Kuder, and Banks 2011). Another way of exporting education services is via franchising, twinning, or validating degree programs of local partner organizations such as universities, colleges, and professional associations (Clark 2012). For example, according to the UK Council for International Student Affairs (2014), the number of students studying wholly overseas for a UK qualification increased from 570,665 in 2011/12 to 598,930 in 2012/13. Out of the figure for

2013, 2% study in an overseas campus, 17% in distance, flexible, and distributed learning, and 50% in an overseas partner organization. In the United States, a number of universities, including the top brass like Harvard University or Stanford University, have started to offer online introductory courses for free and expect students to enroll in further courses on-campus at some later stage in order to complete a degree (Lewin 2013). This will shape a new direction of development for universities in the era of globalization.

Besides offering transnational education services globally, universities for long have been recruiting staff and students from different parts of the world. More than ever, the traditional state boundaries are fading in the university context. Looking at how universities are internationalizing and the impact of this process on countries' advancement in science and technology will be an exciting topic on our future research agenda.

Notes

1 Sometimes firms are set up by academics or graduate students (Sauermann and Stephan 2010).

2 According to Sauermann and Stephan (2010), a considerable number of US universities are also engaged in regional economic development through their applied work.

3 Aghion, Dewatripont, and Stein (2008) explain that because academic scientists derive utility from research freedom, they are willing to accept a lower wage.

4 The effect of a Nobel prize or a Fields medal is assumed to last for a century and its weighting diminishes 10% in each decade.

5 For institutions with high specialization in humanities and social sciences, such as the London or Paris School of Economics, this indicator is not considered. Instead, its weight is allocated to other indicators.

6 Universities are considered by the ARWU only if they have Nobel laureates, Fields medalists, highly cited researchers, papers published in *Science* or *Nature*, or significant number of papers indexed in Science Citation Index (Expanded) and Social Science Citation Index.

7 One of the main criticisms of the ARWU is that it rewards heavily those institutions whose faculty or alumni have won Nobel Prizes or Fields Medals.

References

Academic Ranking of World Universities. 2014. http://www.shanghairanking.com/ (accessed December 18, 2014).

Aghion, Philippe, Mathias Dewatripont, Caroline Hoxby, Andreu Mas-Colell, and Andre Sapir. 2007. *Why Reform Europe's Universities?* Brussels: Bruegel.

Aghion, Philippe, Mathias Dewatripont, and Jeremy Stein. 2008. "Academic Freedom, Private-Sector Focus, and the Process of Innovation." *The Rand Journal of Economics* 39(3): 617–635.

Arrow, Kenneth. 1962. "Economics of Welfare and the Allocation of Resources for Invention." In *The Rate and Direction of Inventive Activity*, ed. Richard Nelson, 609–626. Princeton, NJ: Princeton University Press.

Boulton, Geoffrey. 2009. "Global: What Are Universities for?" *University World News* 69.

Clark, Nick. 2012. "Understanding Transnational Education, Its Growth and Implications." *World Education News and Reviews*. http://wenr.wes.org/2012/08/wenr-august-2012-understanding-transnational-education-its-growth-and-implications/ (accessed December 18, 2014).

Dasgupta, Partha, and Paul David. 1994. "Toward a New Economics of Science." *Research Policy* 23(5): 487–521.

Fagerberg, 1994. "Technology and International Differences in Growth Rates." *Journal of Economic Literature* 32(3): 1167–1175.

Goldstein, Harvey, and Catherine Renault. 2004. "Contributions of Universities to Regional Economic Development: A Quasi-Experimental Approach." *Regional Studies* 38(7): 733–746.

Greenspan, Alan. 2000. "Structural Change in the New Economy." Speech to National Governors Association 92nd Annual Meeting. http://www.federalreserve.gov/boarddocs/speeches/2000/20000711.htm (accessed December 18, 2014).

Gribble, Kate, and Christopher Ziguras. 2003. "Learning to Teach Offshore: Pre-Departure Training for Lecturers in Transnational Programs." *Higher Education Research and Development* 22(2): 205–216.

Hazelkorn, Ellen. 2007. 'The Impact of League Tables and Ranking Systems on Higher Education Decision-Making." *Higher Education Management and Policy* 19: 87–110.

Jaffe, Adam, Manuel Trajtenberg, and Rebecca Henderson. 1993. "Geographic Localization of Knowledge Spillovers as Evidenced by Patent Citations." *Quarterly Journal of Economics* 108: 577–598.

Jensen, Richard, and Marie Thursby. 2001. "Proofs and Prototypes for Sale: The Licensing of University Inventions." *American Economic Review* 91(1): 240–259.

Lach, Saul, and Mark Schankerman. 2008. "Incentives and Invention in Universities." *The Rand Journal of Economics* 39(2): 403–433.

Lewin, Tamar. 2013. "Public Universities to Offer Free Online Classes for Credit." *The New York Times*, January 23.

Li, Lixu. 2004. "China's Higher Education Reform 1998–2003: A Summary." *Asia Pacific Education Review* 5(1): 14–22.

Li, Mei, Sriram Shankar, and Kam Ki Tang. 2010. "Why Does the USA Dominate University League Tables." *Studies in Higher Education* 36(8): 923–937.

Li, Mei, Sriram Shankar, and Kam Ki Tang. 2011. "Catching Up with Harvard: Results from Regression Analysis of World Universities League Tables." *Cambridge Journal of Education* 41(2): 121–137.

Marginson, Simon. 2007. "Global University Rankings: Implications in General and for Australia." *Journal of Higher Education Policy and Management* 29: 131–142.

Marginson, Simon. 2012. "Global University Rankings: The Strategic Issues." http://www.cshe.unimelb.edu.au/people/marginson_docs/Latin_American_conference_rankings_17-18May2012.pdf (accessed December 18, 2014).

Marginson, Simon. 2013. "Nobels Aside, Local Unis Punching Above Weight." *The Australian*, July 17.

Marginson, Simon, and Marijk van der Wende. 2007. "To Rank or to be Ranked: The Impact of Global Rankings in Higher Education." *Journal of Studies in International Education* 11(3–4): 306–329.

Moffitt, Ursula. 2014. "Budget Cuts in Academic Institutions (Universities, Think Tanks, NGOs)." http://blog.inomics.com/en/budget-cuts-in-academic-institutions-universities-think-tanks-ngos/ (accessed December 18, 2014).

Nelson, Richard. 1959. "The Simple Economics of Basic Scientific Research." *Journal of Political Economy* 67: 297–306.

Obst, Daniel, Mathias Kuder, and Clare Banks. 2011. *Joint and Double Degree Programs in Global Context: Report on an International Survey*. Institute of International Education.

QS World University Rankings. 2014. http://www.topuniversities.com/university-rankings (accessed December 18, 2014).

Robertson, Anne, and Bill Leumer. 2014. "Towards the Privatization of Public Education in America: Imposing a Corporate Culture." *Global Research*. http://www.globalresearch.ca/towards-the-privatization-of-public-education-in-america/5364567 (accessed December 18, 2014).

Sauermann, Henry, and Paula Stephan. 2010. "Twins or Strangers? Differences and Similarities Between Industrial and Academic Science." *NBER Working Paper Series* 16113.

Sorenson, Olav, and Lee Fleming. 2004. "Science and the Diffusion of Knowledge." *Research Policy* 33: 1615–1634.

Stephan, Paula. 1996. "The Economics of Science." *Journal of Economic Literature* 34(3): 1199–1235.

THE World University Rankings. 2014. http://www.timeshighereducation.co.uk/world-university-rankings/2013-14/world-ranking (accessed December 18, 2014).

Thursby, Jerry, and Sukanya Kemp. 2002. "Growth and Productive Efficiency of University Intellectual Property Licensing." *Research Policy* 31(1): 109–124.

UK Council for International Student Affairs. 2014. "International Student Statistics: UK Higher Education." http://www.ukcisa.org.uk/Info-for-universities-colleges--schools/Policy-research--statistics/Research--statistics/International-students-in-UK-HE/ (accessed December18, 2014).

UNESCO/Council of Europe. 2002. "Code of Good Practice in Provision of Transnational Education." http://www.coe.int/t/dg4/highereducation/recognition/code%20of%20good%20practice_EN.asp (accessed December 18, 2014).

Usher, Alex, and Massimo Savino. 2006. "A World of Difference: A Global Survey of University League Tables." *Canadian Education Report Series*.

Van Dyke, Nina. 2005. "Twenty Years of University Report Cards." *Higher Education in Europe* 30(2): 103–125.

Zucker, Lynne, Michael Darby, and Maximo Toreto. 2002. "Labour Mobility from Academe to Commerce." *Journal of Labour Economics* 20(3): 629–660.

Zusman, Ami. 2005. "Challenges Facing Higher Education in the Twenty-First Century." In *American Higher Education in the Twenty-First Century*, ed. P.G. Altbach, R.O. Berdahl, and P.J. Gumport, 115–160. Baltimore, MD: Johns Hopkins University Press.

Chapter 3

The International Race of Top Supercomputers and Its Implications

Kam Ki Tang and Thanh Quang Le

Introduction

In September 2013, the US National Aeronautics and Space Administration (NASA) achieved a landmark in space exploration. Its Voyager-1 spacecraft launched 36 years ago (September 1977), became the first human object to go beyond the Solar System.[1] Voyager-1's interstellar journey became a possibility when, back in 1961, Michael Minovitch solved the extremely difficult "three-body problem" in celestial mechanics. The problem had been described as one of the hardest in that field, and it had tested and "defeated" some of the finest scientists including Isaac Newton (Riley and Campbell 2012). Minovitch solved the problem with the help of a machine called the IBM 7090 – the fastest computer at that time.

The fact that a calculation machine assisted in solving a mathematical problem and, as a consequence, pushed the boundaries of human exploration beyond planet Earth, is a powerful demonstration of De Solla Price's (1983) thesis of "instrumentality." According to this thesis: "advances in instrumentation and experimental techniques have been of major importance in stimulating and enabling both radical theoretical advance in fundamental science, and radical innovations in practice application." De Solla Price illustrates the idea of instrumentality using another cosmic example – the development of deep-dished concave lenses. Such lenses allowed Galileo to construct ever more powerful telescopes in order to observe the moon in greater detail and clarity than ever before, during the early seventeenth century. This not only opened up a new chapter for astronomy, but also created strong interest in telescopes in the community. The boom of the telescope industry then led to the development of other optical instruments, including the microscope. The development of the microscope itself created a new way to examine biological species such as cells, and this

The Handbook of Global Science, Technology, and Innovation, First Edition.
Edited by Daniele Archibugi and Andrea Filippetti.

changed the discipline of biology forever. In summary, a scientific instrument can trigger a chain of unintentional and completely different discoveries and innovations.

The most omnipresent instrument across all fields of science and industry nowadays is no doubt the computer (both the hardware and software). What makes a supercomputer "super" amongst its peers is its speed: a supercomputer is simply a very fast computer. As such, supercomputers enable tasks that are beyond the capacity of regular computers, and this assists in accelerating the pace of scientific discovery and product innovation. One of the benefits of high-speed supercomputers is their capability in modeling and simulating large-scale, complex systems. For instance, scientists of Illinois's Argonne National Laboratory used a supercomputer to simulate how the universe evolved from the Big Bang until today (a time period spanning over 13.7 billion years). They did this by tracing the movement of trillions of "particles" in their computer model, in order to study the mysterious dark energy (Moskowitz 2012). Similarly, the need for accurate and timely weather forecasting also makes supercomputer simulation and modeling an indispensable tool for meteorological scientists. Traditionally theory and experimentation are considered the two legs of science and engineering. Nowadays computational modeling and simulation is considered to be the "third leg" (Middleton 2013), which signifies the crucial instrumentality of supercomputers.[2]

Supercomputers have also been used widely in various industries for designing and creating products. Cummins uses simulation to help design diesel engines better, faster, and cheaper; Goodyear uses it to design safer tyres much more quickly; Boeing uses it to build more fuel-efficient aircraft; and Johnson & Johnson uses it to improve the absorbency of nappies (Middleton 2013; Nuttall 2013). Even the humble light bulb has had its lifespan increased thanks to the simulation done with a supercomputer (Cray 2011). The importance of supercomputers in molecular modeling results in big pharmaceutical companies owning some of the world's largest supercomputers in the private sector. Supercomputing capability can also have implications for national security. Simulation allows scientists to conduct virtual nuclear tests and to monitor existing nuclear stockpiles.

Furthermore, the brute power of supercomputers lends them to handing "Big Data." This can range from investigating large amounts of genetic data (McBride 2013), to processing medical reports and diagrams pertaining to patients with cancer (Lenzner 2013), to decoding and analyzing vast streams of information gathered for intelligence purposes (Tsay 2013).[3] IBM's CEO Ginni Rometty calls data the world's "next natural resource." How well a company or an organization "mines" those resources will determine whether it is a winner or a loser in its field (Lenzner 2013).

With faster processing speeds, supercomputers can obviously reduce the time between the inception of an experiment and the final outcome. Aside from the time effect, supercomputers can also make qualitative differences to the outcome. By shortening the time between asking a question and seeing the answer, researchers can quickly ask follow-up questions, find the answer, and keep repeating this process. As Eccles (1989) puts it:

> When results come more slowly, we are more conservative in the questions we can ask, and we leave many more stones unturned. When the answers come back while the questions are still fresh, the train of thought can continue, and less time is spent trying to remember where one left off and what line of reasoning was being explored.

In other words, creativity increases through the stream of continuous and bolder thoughts.

To the extent that supercomputer capacity is an indicator of a country's *current* inputs in cutting-edge research and development (R&D), it therefore also projects the country's *future* scientific and technological prowess. As demonstrated by the success of Voyager-1, the United States' leading supercomputing capability back in 1961 is still giving the country an unparalleled advantage more than half a century later in the exploration of the "final frontier." The social and economic benefits of having the highest supercomputing power could be broad and deep. The former US Energy Secretary Steven Chu sums it up:

> The nation that leads the world in high-performance computing will have an enormous competitive advantage across a broad range of sectors, including national defense, science and medicine, energy production, transmission and distribution, storm weather and climate prediction, finance, commercial product development, and manufacturing. (NNSA 2012)

Against this background, the objective of this chapter is to examine the distribution of top supercomputers across nations and explore its implications for national technological and economic competitiveness. Given that technological and economic competitiveness is a very complex issue, focusing only on supercomputers may seem to be somewhat narrow. Our limited scope of study can be justified for a number of reasons. First, top supercomputers can be viewed as indicators of the frontier R&D that drives the development of the high-tech sector. Second, rankings of top supercomputers have attracted increasing attention from the general public, policymakers, and stakeholders, and have the potential to evolve into valuable new benchmarking tools. Thirdly, although there are studies providing measures of countries' technological and innovation capacities,[4] they are not explicitly focusing on high-tech and have not made extensive use of top supercomputer data yet.[5]

Measure of Supercomputer Capacity

According to the *Encyclopedia of Computer Science*, "the most powerful computers of any time have been called high-speed computers, supercomputers, high-performance computers and, most recently, high-end computers" (Padua and Hoeflinger 2003). The first supercomputer, the Control Data Corporation (CDC) 6600, released in 1964, was up to 10 times faster than the fastest computer at the time.[6] As known by any computer user, it does not take too many years for what is considered a very fast computer to become an ordinary computer if not a vintage one. The time-honored example is that a smart phone produced nowadays has more computing power than the Apollo program, which sent three men to the Moon back in 1969. What underlies this exponential growth in computer power is the fact that the number of integrated circuits on a computer chip doubles approximately every two years – known as the Moore's Law. As such, the term supercomputer is reserved for the fastest computers in the world at the present time.

Since 1993, a group of high-performance computer (HPC) experts have published a list of the 500 most powerful computers in the world, referred to as TOP500. Its purpose is to "provide a reliable basis for tracking and detecting trends in high-performance computing."[7] The TOP500 list is compiled twice a year with input from HPC experts, computational scientists, manufacturers, and the Internet community in general. To date, this is the only public document that details the configuration, purpose, and processing power of HPCs around the world, and thus it is the most authentic source of supercomputer capacities.

The project ranks computers based on their speed in solving a set of numerical problems widely used in the industry for benchmarking. It is known as the LINPACK benchmark. The speed is measured in "flops" (floating-point operations per second), which is basically the number of calculations per second. The more powerful a supercomputer, the larger its flops. The measurement scale of flops, in ascending order, is: kilo (1000), mega (1 million), giga (1 billion), tera (1 trillion), peta (1 quadrillion = 1000 trillion), and exa (1 quintillion = 1 billion trillion). To put this into perspective, the fourth generation iPad's graphic processing unit has a computing power of 76.8 gigaflops, a human brain's equivalent power is estimated to be 2.2 petaflops (Fischetti 2011),[8] while the fastest supercomputer on Earth in 2013 has a power of 33.85 petaflops.

It is, however, commonly accepted that focusing solely on the number of flops is far from the best way to measure the performance of a computer. For example, software is crucial in determining the speed of a supercomputer for running applications, but this is not accounted for in the LINPACK test. The fact that a human (still) has much stronger cognitive abilities like emotion recognition or concept formation than artificial intelligence, despite achieving a smaller LINPACK score than many supercomputers, also demonstrates clearly the limitation of one-dimensional benchmarks. Despite that, TOP500 is still a very useful benchmark for a number of reasons. First of all, using a single measure based on flops is highly transparent, and provides a focal point of discussion, especially for non-specialists. Furthermore, the performance data for each individual supercomputer in a country can be aggregated to provide a single measure of a country's supercomputer capacity. Lastly, the constancy in the benchmarking method allows comparison of a supercomputer's capability, not only across countries at a given time, but also over time for each country.

World Distribution of Supercomputer Capacity

In the following data analysis, we focus on national aggregate data without adjusting for country size. For instance, we will look in absolute terms at the number of top supercomputers that countries have, instead of their per capita figures. Obviously aggregate measures will benefit larger countries as opposed to smaller ones. We do so because our purpose is to use the distributions of top supercomputers as an indicator of the disparity in technology and innovation capacities across countries. It is a country's total stock of technology and knowledge that determines its international competitiveness, not its total population.

Figure 3.1 shows the system distribution of TOP500 supercomputers by segments in 1993 and 2013 respectively. A segment's system share is measured based on the total number of TOP500 computers hosted by that segment. The TOP500 list is updated twice a year; all data used in this chapter are from the mid-year June list

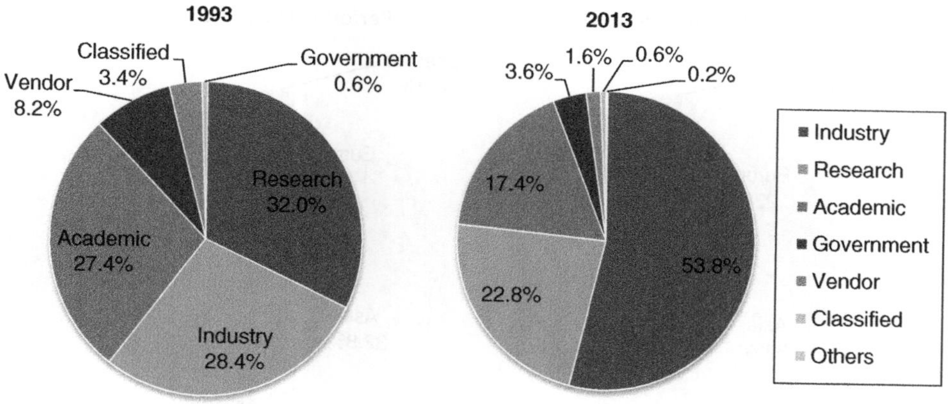

Figure 3.1 The system distribution of TOP500 by segments.
Source: Data from TOP500.org.

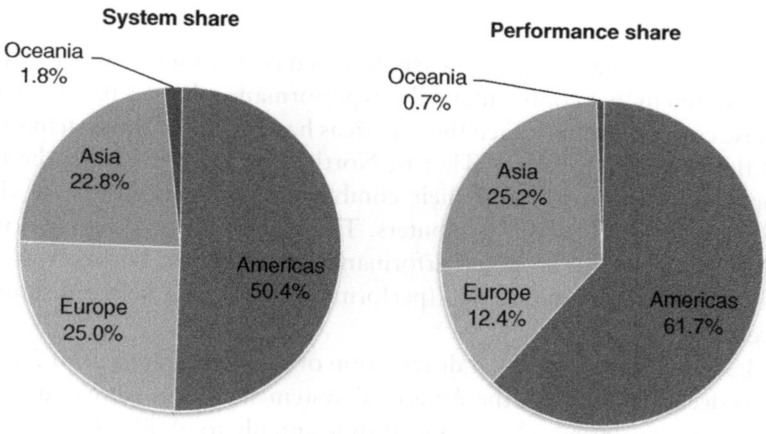

Figure 3.2 The distribution of TOP500 in 1993 by continent.
Source: Data from TOP500.org.

except for 1993, where the November list is used.[9] In 1993, 28.4% or 142 of TOP500 supercomputers were deployed by particular industries. By 2013, that number has been nearly doubled to 53.8% or 269, at the expense of all other segments. This reflects the increasing importance of HPC for industrial innovation.

However, there is limited information on what the exact applications of these supercomputers are. In 2013, 2.4% of the TOP500 machines are known to be used for weather and climate research, 1.8% for energy, 1.4% for defense, 1% for benchmarking, and less than 1% (individually) for uses in relation to environment, aerospace, information service, finance, software, life science, semiconductor, web services, Internet providers, and geophysics. All these amount to less than 10% of the total. The remainder is used for unspecified "research" (13.8%), or for applications "not specified" (76.4%). In other words, there is a great deal of secrecy about the applications of, and access to, top supercomputers.

The distribution of HPC power across regions has changed a lot since the launch of TOP500. Figure 3.2 shows the continental distribution of supercomputers as of

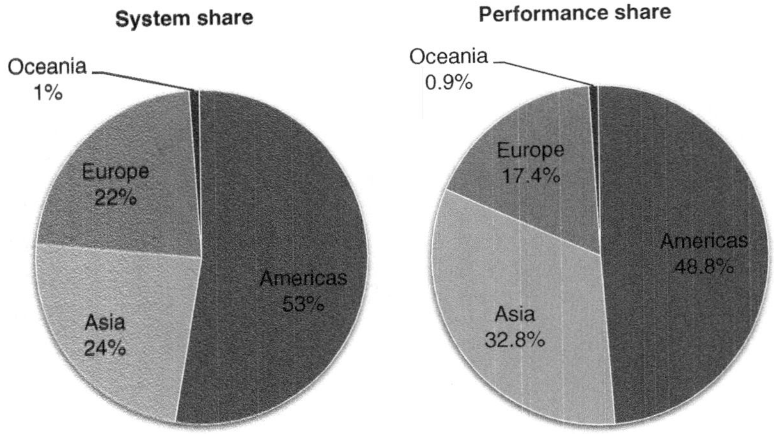

Figure 3.3 Continental distribution of TOP500 in 2013.
Source: Data from TOP500.org.

1993. A continent's system share is measured based on the total number of TOP500 computers located in that continent, while its performance share is based on the total flops of those computers. Back then the Americas had 50.4% of the system share and 61.7% of the performance share. That is, North America had 252 of the top 500 supercomputers in the world and their combined speed was 61.7% of the total combined speed of all TOP500 computers. The second biggest share belonged to Asia: 22.8% (system) and 25.2% (performance) respectively. Europe was in third place with 25% (system) and 12.4% (performance). Oceania held a distant fourth (last) place.

Figure 3.3 shows the continental distribution of supercomputers as of 2013. Over the last two decades, although the Americas' system share has increased slightly to 52.8%, its performance share has dropped substantially to 48.8%. On the contrary, Asia's system share has increased slightly to 23.8%, while its performance share has surged to 32.8%. Europe's system share also slid to 22.4%, but its performance share has increased to 17.4%. This means that while Asia nowadays owns roughly the same number of TOP500 machines as it did 21 years ago, those machines are relatively more powerful than before.

As expected, within each of the four continents there are very diverse performances among the countries themselves. Therefore, in Tables 3.1 and 3.2 we break down the continental-level data into country-level data for not just 2013, but for all the years after 1993. Table 3.1 shows how the world share of TOP500 systems by countries has changed over time, while Table 3.2 shows the corresponding results in terms of performance. To keep the figures intelligible, only the top 10 countries in 2013 (and "others") are listed.

The United States dominates throughout the whole period, with a world share of TOP500 systems ranging roughly between 45% (as occurred in 2002) to 60% (2006). As of 2013, it had 252 TOP500 supercomputers, which is just over 50%. This is almost the same as its share two decades ago. Its share of TOP500 performance has a similar story. Its performance share ranged roughly between 43% (2011) and 67% (2008). As of 2013, its share is around 48%.

Table 3.1 Share of TOP500 systems by countries 1993–2013.

%	*US*	*China*	*Japan*	*UK*	*France*	*Germany*	*India*	*Canada*	*Russia*	*Sweden*	*Others*
1993	49.0	0.0	21.6	4.2	4.0	10.6	0.0	0.6	0.0	0.2	9.8
1994	52.4	0.0	21.0	4.4	3.8	8.8	0.0	0.6	0.0	0.4	8.6
1995	54.8	0.2	13.2	3.2	5.4	10.4	0.0	1.8	0.0	0.8	10.2
1996	51.0	0.2	18.6	2.4	3.6	9.6	0.0	1.2	0.0	1.2	12.2
1997	53.0	0.0	17.4	4.8	3.8	9.0	0.0	1.4	0.2	1.6	8.8
1998	57.0	0.0	16.6	5.0	2.0	9.2	0.0	0.4	0.2	1.2	8.4
1999	58.4	0.0	11.4	5.8	3.6	9.4	0.0	1.6	0.0	1.4	8.4
2000	51.6	0.4	12.4	5.6	4.0	13.0	0.0	1.8	0.0	1.0	10.2
2001	50.6	0.0	10.8	6.2	4.2	12.8	0.0	1.8	0.0	0.8	12.8
2002	45.2	0.6	10.6	7.4	4.6	12.8	0.0	2.6	0.2	1.0	15.0
2003	49.0	1.2	8.2	7.0	3.8	11.2	0.4	1.8	0.4	0.8	16.2
2004	52.4	2.8	7.0	6.8	3.8	7.4	1.2	1.4	0.4	0.6	16.2
2005	55.4	3.8	4.6	6.4	2.2	8.0	1.6	1.4	0.6	0.6	15.4
2006	59.6	5.6	5.8	7.0	1.6	3.4	0.0	1.8	0.2	0.2	14.8
2007	56.0	2.6	4.6	8.8	2.6	4.6	1.6	2.0	1.0	2.0	14.2
2008	51.6	2.4	4.4	10.4	6.8	9.4	1.2	0.4	1.6	1.8	10.0
2009	58.2	4.2	3.0	8.6	4.6	6.0	1.2	1.6	0.8	2.0	9.8
2010	56.0	5.0	3.6	7.6	5.8	4.8	1.0	1.4	2.2	1.6	11.0
2011	51.0	12.2	5.2	5.4	5.0	6.0	0.4	1.6	2.4	1.0	9.8
2012	50.4	13.6	7.0	5.0	4.4	4.0	1.0	2.0	1.0	0.8	10.8
2013	50.4	13.2	6.0	5.8	4.6	3.8	2.2	1.8	1.6	1.4	9.2

Source: TOP500.org.

Table 3.2 Share of TOP500 performance by countries 1993–2013.

%	*US*	*China*	*Japan*	*UK*	*France*	*Germany*	*India*	*Canada*	*Russia*	*Sweden*	*Other*
1993	59.61	0.00	23.83	3.00	2.34	4.67	0.00	1.54	0.00	0.07	4.94
1994	56.56	0.00	27.01	3.93	2.50	4.62	0.00	0.99	0.00	0.09	4.31
1995	55.49	0.15	24.97	3.49	2.92	5.72	0.00	1.59	1.97	0.51	3.18
1996	47.28	0.10	31.15	2.72	2.58	6.82	0.00	1.22	0.00	0.66	7.47
1997	51.78	0.00	23.30	3.64	3.44	9.69	0.00	0.85	0.13	1.07	6.11
1998	56.82	0.00	16.46	5.99	2.31	10.45	0.00	1.14	0.09	1.07	5.68
1999	60.69	0.09	14.23	6.79	2.42	8.96	0.00	1.23	0.00	0.90	4.68
2000	56.20	0.16	15.33	6.09	3.37	11.08	0.00	1.14	0.00	0.61	6.01
2001	58.91	0.36	14.62	5.15	2.75	8.84	0.00	1.09	0.00	0.63	7.64
2002	44.66	0.29	25.79	5.35	4.19	9.05	0.00	1.04	0.25	0.70	8.68
2003	53.73	0.86	16.92	6.19	3.21	7.30	0.31	1.14	0.29	0.64	9.41
2004	56.76	3.40	11.19	7.13	2.68	5.29	0.70	1.49	0.18	0.47	10.72
2005	61.71	3.16	7.59	5.05	1.17	4.75	0.83	1.39	0.46	0.43	13.48
2006	62.21	3.30	9.10	4.83	2.36	3.29	1.29	1.31	0.23	0.17	11.91
2007	62.13	1.95	5.77	6.83	3.15	4.97	0.92	1.11	0.63	1.03	11.52
2008	61.56	1.15	4.53	6.63	5.86	8.01	1.56	0.19	1.27	2.21	7.02
2009	60.60	3.48	3.86	5.44	4.44	9.75	1.09	1.58	0.74	1.52	7.50
2010	55.25	9.30	3.86	5.50	5.41	6.93	0.87	1.25	2.51	1.28	7.82
2011	42.87	12.11	18.98	3.18	5.40	5.50	0.32	1.09	2.28	0.83	7.45
2012	48.60	9.23	14.59	5.37	5.15	6.59	0.64	1.15	1.05	0.43	7.19
2013	47.77	21.23	9.08	3.61	4.00	5.08	1.20	0.79	0.90	0.52	5.82

Source: TOP500.org.

The surge of China is nothing but spectacular. The country did not even have one machine qualify for the TOP500 list in 2000, now it has 66 (or 13.2%) in 2013, which is more than any other country bar the United States. Its world share of TOP500 performance has risen even more dramatically, from 0%, to over 21% in less than one and a half decades. The rise of China came at the expense, not of the United States, but of the previous runner-up, Japan. In 1993 Japan commanded nearly 22% of the world's TOP500 systems and 24% for performance. These figures have declined to just 6% and 10% in 2013. Germany also experienced a substantial drop in its world share of TOP500 systems from 10.6% in 1993 to 3.8% in 2013, but it has maintained its performance share of roughly 5% over the two periods.

Due to space limitations, Tables 3.1 and 3.2 do not show the individual share of countries beyond the top 10 countries. However, an examination of the full list reveals another interesting phenomenon. In 1993, there were 23 countries in the TOP500 list, all of which were contemporary or future OECD countries bar Brazil and Taiwan. One decade later in 2003, the list had increased to 34 countries. This list includes newly industrialized economies like Singapore and Hong Kong, emerging economies like Malaysia and Thailand, as well as less prominent ones like Egypt and Oman. However, one more decade later on in 2013, the number of countries in the TOP500 list has been reduced to 27, and the list once against is dominated by OECD countries; the exceptions being China, India, Russia, Saudi Arabia, Brazil, Hong Kong, and Taiwan. This finding suggests that despite the seemingly globalized nature of computer technology, the "supercomputer club" has been, and will continue to be, a very exclusive group.

High-Tech Output and Supercomputing Capacity

An argument in support of supercomputer capacity being used as a competitiveness indicator is that it is a measure of a country's science and engineering prowess. In this section, we look at countries' high-tech sector performance and examine if there is any evidence in support of this assertion. All the data regarding high-tech output and exports are drawn from SEI (2012).[10]

Figure 3.4 shows the world share of high-tech manufacturing exports (HTME) of the top five countries, based on the current value measures. Surprisingly China managed to beat the United States to take top spot. This is probably because of a shift of final reassembling lines for high-tech products after 2001 from other Asian economies and developed economies to China (SEI 2012). This means that the gross HTME value greatly overstates China's value-added. Therefore, it is necessary to consider high-tech imports as well as exports. Even so, China has been a net high-tech exporter since 1998. In 2010 China's net HTME was valued at USD 156.7 billion, making it the largest net high-tech exporter. On the contrary, the United States has been a net high-tech importer since 1995. In 2010, the United States had net high-tech manufacturing import value of USD 94.3 billion. However, these figures also need to be interpreted with caution because it is possible that China's large net HTME value is still attributed to its sheer volume of low to mid-tech processing work.

To shed more light on the issue, we show in Figure 3.5 the world share of high-tech manufacturing value-added (HTMVA) of the top five countries. At first glance, the figure shares three features with Figures 3.3 and 3.4, namely the dominance of

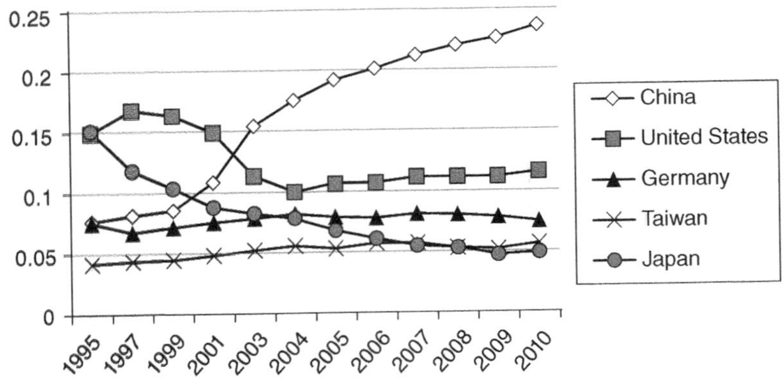

Figure 3.4 World shares of high-tech exports (%).
Source: Data from SEI (2012).

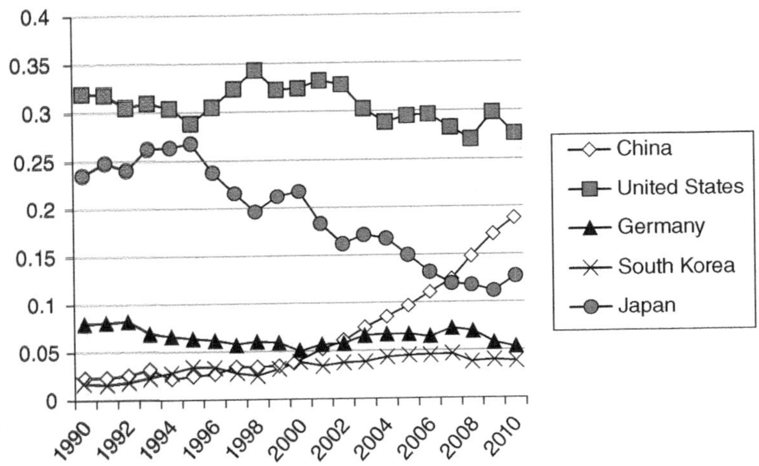

Figure 3.5 World shares of high-tech manufacturing valued-added (%).
Source: Data from SEI (2012).

the United States over the whole sample period, the rise of China since year 2000, and the decline of Japan.

To verify if there is any association between TOP500 share and HTMVA share, we provide in Figure 3.6 a scatter plot showing the world share of HTMVA and that of TOP500 system, from 1993 to 2010.[11] The figure shows that there is a strong positive correlation (0.82) between the two world shares. However, the United States is a clear outlier. If the United States is excluded, the correlation between the two variables drops to 0.68. If Japan is also considered an outlier and removed (together with the United States), the correlation drops further to 0.46. However, China seems to follow the path of Japan, suggesting that it may be inappropriate to treat Japan as an outlier after all. In fact, for consistency, if China is removed together with Japan and the United States, the correlation actually increases to 0.63 and this is depicted in Figure 3.7. In other words, the high correlation between the two world shares is robust to the exclusion of the top three countries. Yet, even a strong correlation does not imply causation, and therefore the results should be interpreted with caution.

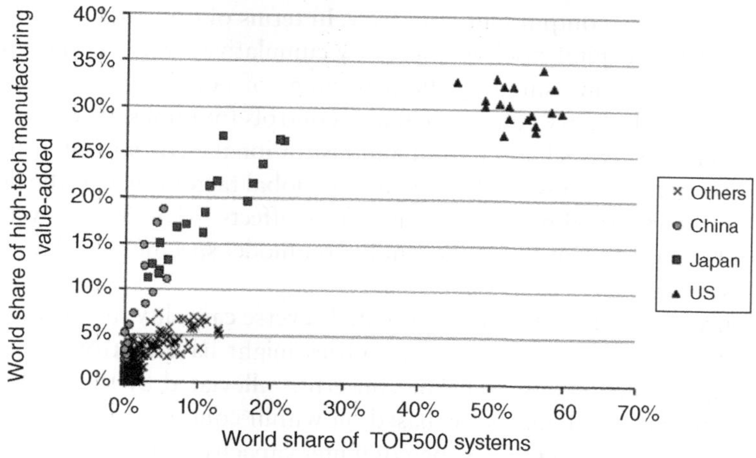

Figure 3.6 World shares of high-tech manufacturing valued-added and TOP500 system (%) 1993–2010.
Source: Authors' own calculation.

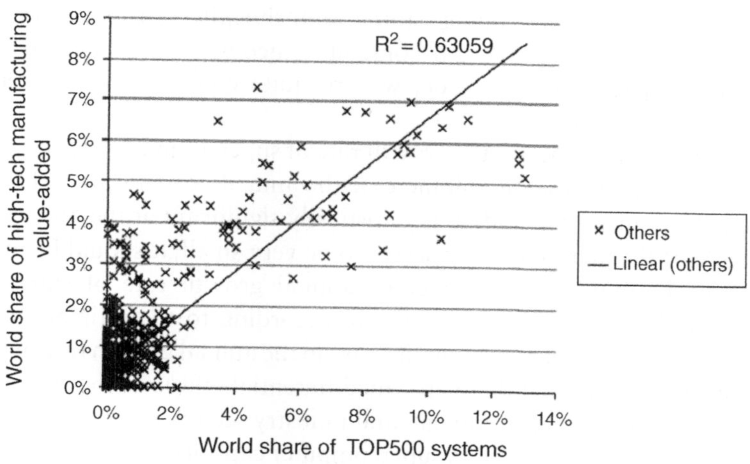

Figure 3.7 World shares of high-tech manufacturing valued-added and TOP500 system (%) 1993–2010, excluding the US, China, and Japan.
Source: Authors' own calculation.

Figures 3.6 and 3.7 only depict the correlation between HTMVA shares and TOP500 systems shares without controlling for any other convoluting factors. In a related study (Le and Tang 2014), we examine how the growth of a country's supercomputer capacity affects its growth in HTMVA using a regressional analysis. The data used in the study is very similar to the current one but with some differences. In particular, it takes into account the capacity of the supercomputers that have dropped out of the TOP500 list over time. The reason for this is that a supercomputer's operational lifespan is much longer than the average time it stays on the TOP500 list and a non-TOP500 supercomputer can still contribute to its host country's high-tech output.

An advantage of using the regressional approach is that it can control for other observed or unobserved factors, allowing us to isolate the impact of supercomputer capacity on high-tech output more precisely. In terms of observed factors, the study controls for R&D capital stock measured by cumulative R&D expenditure, as well as high-skilled labor measured by the percentage of population aged 25 and over that has completed tertiary education. It also controls for unobserved, time-invariant country heterogeneity such as cultural and institutional environment, using country fixed effects, and for unobserved, time-variant global factors such as the demand for high-tech products or oil prices using time fixed effects.[12] The study covers 27 to 28 OECD and emerging economies (depending on model specifications) and the study encompasses years 1991 to 2010.

One concern may arise about the potential reverse causality that a country with a stronger high-tech sector than other countries might have a stronger incentive to invest in supercomputers as well. The concern is alleviated by the methodology of the study. The empirical analysis is based on within-country variations of the data; that is, we look at the growth in supercomputer capacity of a country from one year to another year and see how it connects to the growth of the country's HTMVA over the same period. Even so, one might still argue that if a country's high-tech sector grows faster in a given year, it may also invest more in its supercomputer capacity in that year. But one should be aware that investment takes a considerable amount of time to materialize into functioning capital and, thus, it is inconceivable that the growth of a country's HTMVA can contemporaneously affect the growth of its supercomputer capacity. To conclude, we are quite comfortable to interpret the results as evidence of causation.

The study finds that if the annual growth rate of supercomputer capacity increases by 10 percentage points, then it will increase the annual growth rate of HTMVA by 0.27 percentage points. This finding is statistically significant at the 5% level. While the magnitude of the marginal effect seems to be very small, it should be interpreted against the background that the average annual growth rate of supercomputer capacity in the dataset is over 65%, which, according to the estimation result, is translated into a 1.76 percentage point increase in the annual growth rate of HTMVA.

The study also examines if there are any differential effects of the supercomputer capacities in the academic, government, and industry sectors respectively. It is found that growth in the academic sector's supercomputer capacity has a statistically larger impact on the growth of HTMVA than growth in the government sector's capacity. However there are no statistically discernible differences between the academic and industry sectors, or between the industry and government sectors.

Taking all the evidence into account, it seems reasonable to conclude that a country's supercomputer capacity does have a positive impact on the growth of its high-tech manufacturing sector.

A New Arms Race?

On October 4, 1957, the Soviet Union successfully launched the first Earth-orbiting satellite, Sputnik 1, beating the United States in the first round of their space race. Losing out in that round spurred the United States to set up the Advanced Research Projects Agency (ARPA) within five months in order to regain military technological supremacy.[13] Since then US politicians have been using the term "sputnik moment" to stress the need to increase efforts on science and technology development so that the United States will not be outperformed by other countries. Fast-forwarding to 2011, in his State of the Union speech President Barack Obama warned that the United States had faced another Sputnik moment: this time, losing its title of having the fastest supercomputer to China.

In late 2010, Tianhe-1A (or Milky Way-1A), a supercomputer developed by China's National University of Defense Technology, became the country's first supercomputer to reach the pinnacle of TOP500, knocking down the US Jaguar to second place. Provoked by China's technological ascension, President Obama called for higher levels of R&D spending on biomedical research, clean energy, and information technology. The writing on the wall is clear to see: the technology race between China and the United States has intensified.

In June 2012, the US Department of Energy's 16 petaflops Sequoia regained for the United States the crown of the fastest supercomputer, overtaken in November by the DOE's Titan. China responded by committing four billion Yuan (about USD 650 million) to building a 50–100 petaflops supercomputer by 2015 as part of its 12th Five-Year Plan for Science and Technology Development (2011–2015) (Zhong 2011). China even managed an impressive step toward its target much earlier. The TOP500 list released in June 2013 confirmed that China's Tianhe-2 had become the world's new number one system.[14] At a proven speed of 33.86 petaflops, it is almost twice as fast as the now number two and previous champion, Titan.

The success of China in 2010 and 2013, in taking the top prize of building the fastest supercomputer, cannot be taken as exceptional events. China's investment in HPC technology is for the long haul, and the development of Tianhe-2 signals the amount of resources that the Chinese authority is willing to put in this race. Some experts estimate that Tianhe-2 cost around USD 400 million to build, four times the cost of Titan (Tsay 2013). And the electricity bill for running could be up to USD 24 million per year (Thibodeau 2013). This level of investment is even more impressive given that China's per capita income, measured in purchasing power parity (PPP) terms, is still only one-sixth of the United States.

Although it is clear that China has become one of the world's top players in HPC and has the potential to challenge the hegemony of the United States (Service 2013a),[15] its rise should be put into perspective. On the one hand, the United States still has the largest share of TOP500 machines, 3.8 times the share of China (and 2.3 times in terms of performance). Also, the United States can arguably claim some credit for China's success. This is because the "brain" of a supercomputer is its processor chips, and Tianhe-2 was built using chips produced by Intel Corporation. On the other hand, despite Tianhe-2's brain being "Intel inside," the rest of its body, namely the operating system, software, and interconnect were all designed and built in China. Industrial experts are of the opinion that China is well on its way to developing its own processor chips so that it will soon have a full package of home-grown supercomputing technology (Thibodeau 2013). The economic implication of this is that China could soon be ready to compete with key chipmakers Intel and AMD for the commercial chip markets (Service 2013b). This scenario is not far-fetched given that China has a good track record in commercializing some of its other high-tech outputs such as satellite and rocket technology.[16]

The current generation supercomputers are of petascale; the next generation will be of exascale, one thousand times faster. When discussing which countries are committed to putting resources into building the first exascale HPC, Service (2013a) echoes the comment of Steven Chu given earlier:

> The answer will determine more than who gets bragging rights for leading high-performance computing technology into the future. Because the effort is expected to revolutionize the design of everything from computer logic and memory to the interconnections and software that make them all run, the race could determine which country's high-tech firms are likely to dominate computer technology in the decades ahead. And because scientists who take advantage of the world's top computers tend to be leaders in their own fields, the race to exascale could also affect which nation's researchers will drive developments in disciplines including materials science, alternative energy production, and climate research.

In that regard, the Chinese government has already approved the next five-year plan (2016–2020) to develop an exascale HPC (Zhong 2011). In the United States, on the contrary, despite the call by the Obama administration to increase R&D spending in 2011, the need to fix the Federal Government's debt problems has actually reduced the federal R&D expenditure in dollar terms since then. It is now down to its lowest level in 40 years as a share of Gross Domestic Product (Zambon 2013). This raised concerns that US supremacy in the technology realm might be eroded as a result. The EU, India, Japan, and Russia have also geared up their investment for the supercomputer race (Ramachandran 2013; Izvestia 2012), although that probably will only help slow down the rate at which they are falling behind China and the United States. So all the indications point to a two-horse race between China and the United States.

With so much emphasis on computer speed in public commentaries, it is important to stress that, at the end, it is not the raw HPC power that matters but what a country does with that power. In that regard, China's supercomputers have been highly underutilized compared to its US counterparts (Stone and Xin 2010; Tsay 2013). The main reason is that China's investment in software, including human capital, has lagged behind that in hardware. That is indeed a common problem in many other Chinese industries.

In their thought-provoking book, *Race Against the Machine*, Brynjolfsson and McAfee (2011) recount how the development in chess competitions involving humans and computers demonstrates that less powerful computers *working together* with humans are far more superior than a supercomputer alone. In 1996, chess world champion Gary Kasparov played against IBM's supercomputer Deep Blue, which was programmed with the help of a team of chess experts. Kasparov won that match. In 1997, the two replayed and this time an improved Deep Blue defeated the champion, raising questions about the superiority of artificial intelligence. Given that computer power grows exponentially (as always), over time the competition between human and computer chess players has become even more one-sided. The situation, however, becomes more fluid in "freestyle" competitions, where any combination of human and computer is allowed. In one of the freestyle competitions discussed in the book, a pair of amateur chess players managed to defeat their opponents of either grandmaster skills or greater computer power, by working with their computers more efficiently than others. Kasparov commented on the result: "Weak human + machine + better process was superior to a strong computer alone and, more remarkably, superior to a strong human + machine + inferior process" (Brynjolfsson and McAfee 2011: 55).

Supercomputers, like ordinary computers, are essentially tools. Their power eventually lies with the humans that utilize them. Those countries that win the game of speed could still lose every other game if the ultimate users – their people – are not able to use the tools creatively and effectively. Recognizing the limitation of speed-based benchmarking, the US administration warns that:

> While it would be imprudent to allow ourselves to fall significantly behind our peers with respect to scientific performance benchmarks that have demonstrable practical significance, a single-minded focus on maintaining clear superiority in terms of FLOPS count is probably not in our national interest. Engaging in such an "arms race" could be very costly, and could divert resources away from basic research aimed at developing the fundamentally new approaches to HPC that could ultimately allow us to "leapfrog" other nations, maintaining the position of unrivaled leadership that America has historically enjoyed in high performance computing. (PCAST 2010: 67)

To sum up, if there is an HPC race between countries, then the race is much more than the TOP500 ranking. The true race is not even in HPC per se, but in the fundamentals that support the development of top HPCs or other frontier technologies, namely research, education, and business. In other words, it is a race of being the most advanced knowledge-based nation.

Concluding Remarks

Computational modeling and simulation is described as the third pillar of scientific inquiry, especially in areas where experiment is dangerous (e.g., testing nuclear weapons), expensive (e.g., investigating aerodynamics), or simply impossible (e.g., recreating the Big Bang). Supercomputers are at the forefront of this kind of scientific inquiry. Some even describe top HPCs as "time machines" because they give people access to a capability that will be broadly available within five to 10 years.[17] The countries that have such machines are able to conduct new experiments, design new products, and develop new processes well ahead of the others (NPR 2013). In short, they will have the first mover's advantage.

This chapter looks at the tally of top supercomputers across countries and what that means in terms of their international competitiveness. Our analysis is based on TOP500, which in turn is developed using publicly available information as well as the acknowledgements of HPC experts and vendors. Certain government agencies, like the military, obviously have huge computing power but will not reveal it for national security reasons. Likewise large private enterprises like Google or Amazon will not reveal it due to issues of trade secrecy. Therefore, a country's true prowess in HPC cannot be measured just based on TOP500. Misrepresentation may also happen at the lower end of an HPC league table. This is because for nations like China, or companies like IBM, getting to the top rank of the TOP500 list can generate positive publicity for themselves. However, those countries that can only be at the bottom end may not be willing to put in the effort to participate (Jackson 2013). If this trend is persistent and widespread, it could affect the representativeness of the league table.

Also, the results of any of the league tables are specific to the criteria used to rank the subjects. While TOP500 is considered as the prime benchmark for HPC, it is not without limitations and competing alternatives. For instance, another HPC benchmark, Graph500,[18] emphasizes data-intensive applications instead of calculations. According to this benchmark, as of June 2013, the US DOE's Sequoia is ranked the first while China's Tianhe-2 is ranked only sixth (positions unchanged in November 2014). Because ranking results could change when different criteria are used, one should be cautious not to read too much into any snapshot results from any specific league table. Focusing on their trend results is recommended because systematic biases due to specific benchmarking criteria are likely to remain unchanged over time. On that note, a clear trend in the top supercomputer ranking is the emergence of China as a global player and its ability to challenge the technological supremacy of the United States. This observation is broadly consistent with the conclusions drawn from other indicators of innovation input or performance such as R&D expenditure and patents: the United States is still leading but China is closing in (see, e.g., Naude, Szirmai, and Lavopa 2013).

One reason why supercomputer rankings have drawn so much attention is that they are viewed as a proxy of countries' capability in broader advanced technology, beyond just computing. Projecting countries' technological sophistication based on their HPC capacity ranking is not an overstretch because of the wide-ranging applications of supercomputing in science discovery and product innovation. In this chapter we try to place the top supercomputer distributions within a broader picture of international competitiveness, whilst being mindful of their proportion in this picture.

The broad picture of international competitiveness obviously contains far more elements than just technology or knowledge creation. Political and social institutions are equally important, if not more so. Countries like China, India, and Russia, though commanding some of the most advanced technology, are lagging behind in some very basic areas of development. In his Hong Kong art exhibition, dissident Chinese artist Ai Weiwei made a huge map of China out of milk power tins as a reflection of a man-made tragedy in China in 2008, where 300,000 infants were poisoned by a milk formula power containing industrial chemicals, and six of them eventually died. He commented that: "A country like this can put a satellite into space but it can't put a safe bottle teat into a child's mouth. I think it's extremely absurd" (Reuters 2013).

If China's rise in the TOP500 league tables is a sign of it being a technology superpower, then its uncountable cases of social scandals and corruption are similarly a symptom of a sick nation. These two apparently polar developments are likely to coexist in the medium term. In the long run a country's advancement in science and technology requires not only the state's mobilization of resources, but also a stable and functioning institutional environment where laws and regulations are implemented impartially. It also requires investment to be governed through economic principles instead of political interests.

Last but not least, the fast pace of computer evolution is well known, and supercomputers are not an exception. Moore's Law implies that supercomputers evolve at the pace of one generation every few years. As such, the national and international landscape of HPC in a decade or two could be very different from what it is now.

Indeed, in the future, defining the high-tech landscape based on countries' borders may be insufficient. For instance, the emergence of numerous private spaceflight companies (e.g., SpaceX) competing with each other for a slice of the aerospace travel/transportation market, created partly by the termination of NASA's space shuttle program, has tested the decade-old idea that the space race must be like the Olympic Games – one nation against another. A similar situation is observed in supercomputing. As shown in Figure 3.1, in the past two decades, the industry sector has increasingly become the dominating force in HPC at the expense of government and academic sectors. To the extent that companies that can afford to invest in HPC tend to be large and multinational corporations, how to take their cross-border activities into account when measuring national technological competitiveness could be a challenging issue.

Notes

1 Voyager-1 actually had reached interstellar space in August 2012, but NASA officially confirmed the success in September 2013.

2 Even political campaigns, which are traditionally run based on individual experience and instinct, have moved to relying upon data mining and simulation. In the 2012 US Presidential election, the campaign team of Obama is reported to have run 66,000 simulations every night (without using a supercomputer, though) to determine how to allocate resources the following day in order to maximize votes and donations (Scherer 2012).

3 One should not assume that a higher HPC capacity automatically means stronger cyber warfare. For example, the computer worm "stuxnet" – allegedly created by Israel – that attacked Iran's nuclear facilities, is a completely different type of technology to supercomputing.

4 Examples include the Science and Engineering Indicators Report of the US National Science Board, the Science, Technology and Industry Scoreboard report of the Organisation for Economic Co-operation and Development (OECD), and the Global Innovation Index scores of the World Intellectual Property Organization (a specialized agency of the United Nations).

5 The 2012 Science and Engineering Indicators Report however, contains a "box" on China's progress in supercomputers. This echoes our proposition that supercomputer data could evolve into a useful benchmarking tool for high-tech development.

6 For the history of supercomputers, see Anthony (2012).

7 Details can be found in the organization's website at http://top500.org/.

8 Hawkins (2004), however, points out the fallacy of comparing a brain to a computer.

9 The June 1993 list was the first list of TOP500 ever published and it showed that the United States owned all TOP500 supercomputers in the world, but the figure dramatically dropped to 245 in the November list. This is probably because information for other countries was not collected in June 1993.

10 The details are in the appendix tables of the report, which are available at: http://www.nsf.gov/statistics/seind12/appendix.htm (accessed December 20, 2014).

11 Similar results are obtained using TOP500 performance but results are omitted here to save space.

12 Fixed effects are dummy variables and therefore do not require observed values for the factors that they represent.

13 ARPA has been renamed the Defense Advanced Research Projects Agency (DARPA). Its mission is "to prevent technological surprise to the US, but also to create technological surprise for our enemies."

14 Both Tianhe-2 and Titan retained their number one and number two positions in the releases in November 2013, June 2014, and November 2014.

15 Development in some other high-tech sectors like space technology seems to point to the same conclusion. For instance In 2003 China became the third country that independently sent its astronauts into space, and it looks set to become the second country that can put a person on the Moon within the next two decades (Shadbolt 2013).

16 China is one of the only nine countries that can provide commercial satellite launching services, and in 2012 China launched 19 satellites, ahead of the United States for the first time (China Daily 2013).

17 Based on the LINPACK benchmark, iPad 2 is as fast as Cray-2, the world's fastest supercomputer in 1985, and it could have stayed on the TOP500 list through 1994 (Markoff 2011). However, with a price tag of about USD 700 for iPad 2 and USD 32 million for Cray-2 (both in 2010 dollar), the former has a cost per megaflops of USD 0.42, compared to the latter's USD 32,000 (Olds 2012).

18 www.graph500.org.

References

Anthony, Sebastian. 2012. "The History of Supercomputers." *ExtremeTech*, April 10. http://www.extremetech.com/extreme/125271-the-history-of-supercomputers (accessed December 19, 2014).

Brynjolfsson, E., and A. McAfee. 2011. *Race Against the Machine: How the Digital Revolution is Accelerating Innovation, Driving Productivity, and Irreversibly Transforming Employment and the Economy*. Lexington, MA: Digital Frontier Press.

China Daily. 2013. "Chinese Seek Greater Share of Satellite Market." *China Daily*, June 20.

Cray. 2011. "LED Lighting Comes Out of the Dark." *Cray.com*. http://www.cray.com/Assets/PDF/products/xe/XE-NERSC-LED-0811.pdf (accessed December 19, 2014).

De Solla Price, D.J. 1983. "The Science/Technology Relationship, the Craft of Experimental Science, and Policy for the Improvement of High Technology Innovation." *Research Policy* 13(1): 3–20.

Eccles, B. 1989. "Deciding to Acquire a Powerful New Research Tool-Supercomputing." In *Supercomputers: Directions in Technology and Application*, 73–80. Washington, DC: National Academic Press.

Fischetti, M. 2011. "Computers Versus Brains." *Scientific American*, October 25.

Hawkins, J. 2004. *On Intelligence*. New York: Times Books.

Izvestia. 2012. "Russia Joins the Supercomputer Race." *Izvestia*, August 24.

Jackson, J. 2013. "Growing Inequality in Supercomputing Power." *PCWorld*, November 20.

Le, T., and K. Tang. 2014. "Impacts of Academic R&D on High-Tech Manufacturing Products: Tentative Evidence from Supercomputer Data." *Studies in Higher Education*, published online April 14.

Lenzner, R. 2013. "IBM CEO Ginni Rometty Crowns Data as the Globe's Next Natural Resources." *Forbes*, March 7.

Markoff, J. 2011. "The iPad in Your Hand: As Fast as a Supercomputer of Yore." *The New York Times*, May 9.

McBride, R. 2013. "Big Data Hits Eastern Europe as IBM Fires Up Supercomputer." *FierceBiotechIT.com*, April 5. http://www.fiercebiotechit.com/story/big-data-hits-eastern-europe-ibm-fires-supercomputer/2013-04-05 (accessed December 19, 2014).

Middleton, J. 2013. "The World's Fastest Computers." *Australian Personal Computer* 33(5): 64–69.

Moskowitz, C. 2012. "Supercomputer Recreates Universe from Big Bang to Today." *Space.com*, September 11. http://www.space.com/17530-universe-dark-energy-supercomputer-simulation.html (accessed December 19, 2014).

Naude, W., A. Szirmai, and A. Lavopa. 2013. "Industrialization Lessons from BRICS: A Comparative Analysis." IZA Working Paper 7543.

NNSA. 2012. "Titan Supercomputer Named Fastest in the World, NNSA's Sequoia Supercomputer Ranked Second." National Nuclear Security Administration, November 13. http://nnsa.energy.gov/blog/titan-supercomputer-named-fastest-world-nnsa%E2%80%99s-sequoia-supercomputer-ranked-second (accessed December 19, 2014).

NPR. 2013. "A Calculating Win for China's New Supercomputer." *NPR.org*, June 21. http://www.npr.org/2013/06/21/194230816/a-calculating-win-for-chinas-new-supercomputer (accessed December 19, 2014).

Nuttall, C. 2013. "Supercomputers: Battle of the Speed Machines." *Financial Times*, July 9.

Olds, D. 2012. "Supercomputing vs. Our Computing: iPad Owns Cray-2? Wife's Desktop Beat All?" Gabriel Consulting Group, March 8.

Padua, D., and J. Hoeflinger. 2003. "Supercomputers." In *Encyclopaedia of Computer Science*, 4th edn, ed. A. Ralston, E. Reily, and D. Hemmendinger. Chichester, UK: John Wiley & Sons Ltd.

PCAST. 2010. *Designing a Digital Future: Federally Funded Research and Development in Networking and Information Technology: Report to the President and Congress*. President's Council of Advisors on Science and Technology.

Ramachandran, T. 2013. "India Tries to Regain Lost Ground in Supercomputer Race." *The HINDU*, July 4.

Reuters. 2013. "China's Ai Weiwei Takes Inspiration from Milk Scandal." *Reuters*, March 17.

Riley, C., and D. Campbell. 2012. "The Maths That Made Voyager Possible." *BBC News*, October 23. http://www.bbc.co.uk/news/science-environment-20033940 (accessed December 19, 2014).

Scherer, M. 2012. "Inside the Secret World of the Data Crunchers Who Helped Obama Win." *Time Swampland*, November 7.

SEI. 2012. *Science and Engineering Indicators 2012*. Arlington, VA: NSB.

Service, Robert F. 2013a. "Who Will Step Up to Exascale?" *Science* 229(6117): 264–266.

Service, Robert F. 2013b. "China's Supercomputer Regains No. 1 Ranking." *Science Insider*, 18 June.

Shadbolt, P. 2013. "China Sets Course for Lunar Landing this Year." *CNN.com*, August 30.

Stone, R., and H. Xin. 2010. "Supercomputer Leaves Competition – and Users – in the Dust." *Science* 330(6005): 746–747.

Thibodeau, P. 2013. "What China's Supercomputing Push Means for the U.S." *Computerworld*, June 10.

Tsay, B. 2013. *"The Tianhe-2 Supercomputer: Less Than Meets the Eye?" SITC Bulletin Analysis,* July, University of California Institute on Global Conflict and Cooperation.

Zambon, K. 2013. "AAAS Analysis Shows Uncertain Future for Federal R&D Spending." American Association for the Advancement of Science, May 2. http://www.aaas.org/news/aaas-analysis-shows-uncertain-future-federal-rd-spending (accessed December 19, 2014).

Zhong, J. 2011. "Recent Advances of Chinese Efforts in HPC." International Exascale Software Project (IESP) Exploratory Meeting 6, San Francisco, CA, USA, April 6–7. http://www.exascale.org/mediawiki/images/b/b8/Talk25-zjin.pdf (accessed December 19, 2014).

Chapter 4

Soft Innovation and Changes in Product Aesthetics: An Omitted Dimension in Economic Analyses of Innovation Activities

Paul Stoneman

Introduction

The purpose of this contribution is to draw attention to a class of innovative activities primarily concerned with changes of an aesthetic[1] nature or in intellectual appeal, that has largely been ignored in the study of innovation prevalent in economics. This is not to downplay the extent and importance of technological innovations, in fact aesthetic and related innovations may rely upon technological innovations, but rather to indicate that the study of technological changes provides an incomplete picture of the totality of innovative activity. The inclusion of aesthetic and similar innovations means that not only is innovation more widespread than previously considered but may also take different forms than previously considered. This is not a completely new argument: Bianchi and Bortolotti (1996) draw attention to what they label "formal innovation." which is innovation that changes product form without any necessary changes in product function. They consider that the new form "exalts the aesthetic or symbolic content of the product." Postrel (2004) argues that aesthetics is an increasingly important element of our society and people are concerned not only with function but also with how things look and feel. Marzal and Esparza (2007) argue that there are a number of industries that experience aesthetic innovations when novelty is conferred on a product in terms of visual (broadly, sensory) attributes, while Swan, Kotabe, and Allred (2005) confirm that practitioners realize the importance of visual or aesthetic design in consumer choice.

The chapter will discuss the nature and extent of such activities and within the space available draw attention to related policy issues. A much fuller discussion of the topic can be found in Stoneman (2011),[2] where the term "soft innovation"

The Handbook of Global Science, Technology, and Innovation, First Edition.
Edited by Daniele Archibugi and Andrea Filippetti.

has been coined to label the type of innovation considered here and more precisely defined as: "innovation in goods and services that primarily impacts upon sensory perception and aesthetic appeal rather than functional performance" (2011: 22).

The third edition of the Oslo manual (OECD 2005) defines innovation to encompass product, process, organizational, and marketing innovations. For the purposes of this paper we set aside organizational innovations. Product innovation is then further defined as: "the introduction of a good or service that is new or significantly improved with respect to its characteristics or intended uses. This includes significant improvements in technical specifications, components and materials, incorporated software, user friendliness or other functional characteristics" (2005: 48).

This definition has two particular limitations that arise because it defines innovations as involving newness or significant improvements in terms of functional characteristics. First, we argue that this inappropriately excludes those new products that reflect changes in aesthetic characteristics, that is, changes in form rather than function encompassing appearance, taste, sound, smell, or intellectual appeal. The only way that such changes are addressed by the current definition is by the inclusion of marketing innovation, defined to be "the implementation of a new marketing method involving significant changes in product design or packaging, product placement, product promotion or pricing" (2005: 49), but this only scratches the surface of the issue. Second, the definition of product innovation rules out as innovation all new product launches (involving changes in form or function) that involve horizontal product differentiation (i.e., where the new product is different to existing products rather than better than existing products) and all those new product launches that involve vertical product quality change where the new product is of lesser performance than existing products but is proportionately cheaper. These arguments in a less clear-cut way can also be applied to the OECD definition of process innovation which involves "new or significantly improved production or delivery method" (2005: 49).

It is possible to classify innovations that are more aesthetic than functional, or encompass form rather than function, into two types. The first type are all product innovations in the creative industries[3] and the second is those innovations in other industries that are of an aesthetic or intellectual nature rather than of a functional kind (the importance of which is noted for example by WIPO 2012: 19, where it is argued that the look and feel of modern electronic devices drive consumer choice, as they determine the ease of use and influence consumer experience of a product). Examples of the former are:

1. The development and recording of a new CD
2. The development of a new computer game
3. The writing, rehearsing and staging of a new theatre production
4. The writing, production and launching of a new movie film
5. The writing and publishing of a new book
6. The development and launch of a new advertising promotion
7. Architectural activity in the generation of new built form designs

Examples of the latter are

1. The development and introduction of new food products
2. The development and launch of a new clothing line
3. The design and production of a new range of furniture

One may note that many such changes will involve the introduction of products that are different but not all potential buyers will necessarily consider to be "better." In accepting therefore that such changes are innovations one is thus accepting as well that horizontal product differentiation may be considered to be innovation.[4]

Such innovative products may be one-offs, produced in small numbers, or mass produced. Thus, for example, a new architect-designed building might be a one-off, or an architect-designed hotel room may be reproduced many times; a new film may be an art house product or a mass blockbuster; a new book or CD or DVD may be purchased by many or by very few; and a new food product may sell in large quantities or small quantities.

There are many soft innovations produced and it is of interest to separate out those that are "significant" from those that are not. With technological innovations (see above) the OECD definitions regard improvements in functionality as a measure of significance. As the innovations we are discussing do not (necessarily) improve functionality this is of little relevance here. There is a line of argument that would however offer a similar approach to significance by proposing that just as it may be judged that one functional advance is more functionally significant than another, so it may be possible to judge one aesthetic advance as more artistically significant than another. Although several internal metrics, such as, for example, influence upon others, the number of imitators, or the extent of copying have been suggested (see also Marzal and Esparza 2007 and the references therein), at the current time, there seems to be no agreed metric that an external observer would be able to employ. In fact some commentators (e.g., Carey 2005) have argued that there are no absolute criteria of value in the arts. Nor does an index of artistic contribution adequately address advances in the aesthetics of non-artistic products (e.g., is one new amusement park ride of greater aesthetic significance than another?).

Instead of attempting to judge significance by an index of aesthetic contribution therefore, we pursue a much more economic approach and use a market-based measure instead. We suggest that an indicator such as the market share attained by a new product is a useful measure of the significance of that innovation. Such a measure caters well for the horizontally differentiated nature of many aesthetic innovations as well as for those showing vertical quality changes. A market-based measure may either refer to a point in time or involve accumulation over time. Market-based measures of the importance of aesthetic and artistic innovations are not new, for example, Galenson (2005) argues that for major artists there is a correlation between prices and artistic importance and that the most valuable art is made by the greatest artists, and Cowen (2000; see also 2006) examines how economic incentives affect the artist's choice of audience rather than trying to use aesthetic criteria to order art works on a high/low spectrum. Taking a market-based approach and in particular a market share-based approach to measuring significance also enables us to judge the overall innovativeness of a particular market using changes in the shares over time of newly introduced products. However, market-based measures of significance are not without their critics (see Eltham 2013).

Production of Innovations

We have classified two main types of aesthetic innovations: product innovations in the creative industries and innovations with new aesthetic characteristics in other industries. The means by which these are produced are heterogeneous. In the creative industries we may have the model of the lone artist figuratively working in a garret and then displaying works for sale in a gallery. Similarly there may be authors working at their own expense on producing new novels or other books hoping for publication eventually. Alternatively there may be small software houses producing new computer games, or large record companies generating new music recordings, employing professional singers and orchestras. Theatre companies will invest in commissioning and then developing, rehearsing, and presenting new offerings. There may be large companies producing new films with investments in the hundreds of millions of dollars matching the investments made by drug companies in new chemical entities. Architectural practices may undertake small or large commissions meeting the expressed requirements of clients.

In other industries, although much will be spent on improving functionality, firms will also be addressing aesthetic issues. For example car companies will not only be concerned with comfort, performance, and efficiency but will spend time on appearance, smell, color choice, exhaust note, the sound made by the closing door, the shape of the grille, the angle of the headlights, and the extent of the chrome. For example, Swan et al. (2005) cite evidence that in the automobile industry one of the most important aspects for body design is how light will reflect off the surface of a car and that the aesthetics of luxury automobiles are critical to their consumer appeal. Similarly confectioners will launch new chocolate bars where taste will be a prime issue. Cosmetic firms may launch new perfumes where smell is of utmost importance. In fact appearance, taste, sound, and smell as much as functionality may well be major factors affecting the demand for many consumer products such as mobile phones, laptops, and televisions (see WIPO 2012: 19).

Geographically innovations may be generated, produced, and sold in the same country or region. More generally however they may be invented in one country, developed in another, and launched, produced, and sold in another or several different countries. Some soft innovations may be restricted to local markets because, for example, of cultural or language factors whereas others may be sold in world markets either as is, or via adaptation such as local remaking, or as with film, via dubbing or adding subtitles.

It should be noted that investments in the soft innovation production process are not generally defined as Research and Development (R&D). International standards for the measurement of R&D were first put forward 40 years ago in the Frascati manual. The latest edition of that manual states that: "The basic criterion for distinguishing R&D from related activities is the presence in R&D of an appreciable element of novelty and the resolution of scientific and/or technological uncertainty" (OECD 2002: 34). Clearly expenditure on those activities that have an aesthetic rather than a technological or scientific component, and which we have called aesthetic innovation, will to a large extent not be defined as R&D and thus not be reflected in the measures of R&D expenditure as usually collected and published. Thus R&D indicators cannot be used to measure the extent of soft innovation and as a result R&D figures will give an incomplete picture of overall innovative activity.

Intellectual Property Rights (IPR)

One means by which inventors or innovators may facilitate obtaining a return from innovative investments is to protect their intellectual property rights. One such protection mechanism is the use of patents. In many countries only advances of an industrial nature can be patented, causing them to be labeled utility patents. These cannot be granted for aesthetic improvements. This means that one cannot use (utility) patent data to measure soft innovation.

In other countries, such as the United States, there is also a form of patents called design patents. These patents correspond to various design rights mechanisms available for the protection of IPR in other countries. Such design rights apply to intellectual property in the physical appearance of a product and are not concerned with the function or operation of that product. In different countries there are several different types of design protection available. For example, in the United Kingdom, registered designs (RD) offer protection throughout the United Kingdom. Application must be made for this IP right, a fee has to be paid and it is not an automatic right. Registered Community design (RCD) offers like protection in all EU member states. UK design right is an automatic right which does not need to be applied for, and prevents others from copying a design, but it covers only the 3D aspects of the item and does not protect the surface decoration of the product or any 2D pattern such as a wallpaper or carpet design. Unregistered Community design right is also an automatic right for which one does not need to apply and offers protection from copying of the design of any item. To qualify for any of these rights, the design must be new and individual in character – which means that the overall impression the design gives the informed user must be different from any previous designs. Given their purpose, counts of design rights granted would appear to be a useful indicator of the extent of soft innovation.

Copyright is a third IP mechanism. It relates to the expression of an idea, not the idea itself, nor any process by which that idea is embodied in a physical artifact. Copyright protects sound recordings, films, broadcasts, and original artistic, musical, dramatic, and literary works, including, for example, photographs, sculptures, websites, computer programs, plays, books, videos, databases, maps, and logos. But it does not protect the names, designs, or functions of the items themselves. It is not necessary to formally apply or pay for copyright in the United Kingdom (although this is not the case in all countries). It is an automatic right. The copyright arises as soon as the work is "fixed" (e.g., written down, recorded, or stored in a computer memory) and in the United Kingdom is established once the © symbol is attached to the work with the creator's name and the date created. The owner of the copyright has the right to license or sell or otherwise transfer the copyright to someone else. One may note that the periods of copyright protection are considerably longer than the terms of even extended patent rights (see below). In Europe (and beyond) there have been attempts to harmonize copyright law dating back to the Berne Convention for the Protection of Literary and Artistic Works in 1886. However, the direction of relevant directives has been inconsistent and copyright laws still vary considerably between member states, particularly between common law jurisdictions (e.g., Cyprus, Ireland, Malta, and the United Kingdom) and civil law countries. There is still however no general requirement for registration of copyright. In the United States copyright may be registered but this is not a requirement.

Whereas patents do not cover soft innovations, copyright is particularly applicable to new products in the creative industry and also soft innovation in the non-creative sector. Thus a count of copyrights claimed might be a useful further indicator of soft innovative activity. However, as copyrights are generally not legally registered, it is particularly difficult to obtain numbers of copyrights claimed.

The fourth basic IPR mechanism is trademarks. A trademark is a sign which can distinguish a firm's goods and services from those of other traders. A sign includes, for example, words, logos, pictures, or a combination of these. Whereas patents require novelty and copyright requires originality, the counterpart for trademark is distinctiveness. Whereas patents are not available for aesthetic innovations, such innovations may be trademarked. Non-aesthetic innovations may also be trademarked. For example a rock group can trademark its name, a product with a particular aesthetic can be trademarked (e.g., the iPod), and particular products may also be trademarked (e.g., Mars bars). Registered and unregistered marks are available.

An unregistered trademark provides certain rights under common law and the owner can use the TM symbol. However, it is easier to enforce rights if the mark is registered and the owner may use the ® symbol to indicate that it is registered. A registered mark confers the right of use of that mark on the goods and services in the classes for which it is registered, and the legal right to take action against anyone who uses the mark or a similar mark on the same, or similar goods and services to those that are set out in the registration. To be registrable, the trademark must be distinctive for the goods and services (for which application is made); and not the same as (or similar to) any earlier marks on the register for the same (or similar) goods or services. In the United Kingdom application for registration is made to the Trade Marks Registry of the Patent Office. There is a need to pay a renewal fee every 10 years. European Union protection via a Community Trade Mark application is made via the Office for Harmonization in the Internal Market (OHIM). As with other IPR instruments it is necessary for the owner to police his/her rights via the courts.

Overall therefore whereas an innovative aesthetic product may not be patented, it may be copyrighted, trademarked, and even employ a design right protection as well.

Macroeconomic Indicators of Soft Innovation

In the next two sections of this chapter we consider the nature and extent of soft innovation activity using various indicators of such activity. We first consider macroeconomic indicators before moving on to consider microeconomic (i.e., industry level) indicators.

In most traditional studies of innovation, measurement relies extensively upon measures relating to R&D and/or patenting. As we have made clear above, however, neither of these indicators is particularly valid when it is soft innovation that is being discussed. There are some alternatives that might be explored, for example, a useful indicator might be numbers employed in the creative and other industries on creative tasks (such as design). Such data however, although available for some countries, is not sufficiently widely available to provide a world view. In the circumstances useful indicators to employ are measures of the extent to which non-patent IP protection measures have been used. In Table 4.1 we present some data on the worldwide use of design rights registrations and trademarks as well as, for comparative purposes, some data upon patenting and R&D activity. Unfortunately comparable copyright data is not available.

Of the three IP indicators measured, patents, trademarks, and industrial design applications, we note that patents will not cover soft innovations while trademarks will encompass soft innovations, but they may also be used to protect the more traditional technological product and process innovations. On the other hand (see WIPO 2012) design rights should primarily (and perhaps exclusively) encompass industrial designs, being legally defined thus: "an industrial design is the ornamental or aesthetic aspect of an article. The design may consist of three-dimensional features, such as the shape or surface of an article, or of two dimensional features, such as patterns, lines or color." Counts of design rights will therefore be a useful indicator of protected aesthetic innovations. The data in Table 4.1 indicates that the use of design rights (as well as patents and trademarks) around the world is extensive.

Direct comparisons of numbers across type of IP would mean little. However, although counts of design right applications may reflect changing behavior with respect to registration rather than increased levels of design activity (a problem we are unable to overcome), the growth rates presented in Table 4.1 indicate that design rights applications have grown faster than the use of the other IP mechanisms, which is consistent with the view that soft innovation is growing faster than technological product and process innovation. This view might be reinforced by the indication that, in addition, trademark applications, which include soft innovations, grew faster than patent applications which do not. One may also observe that design right applications have been growing faster than total R&D spend, suggesting again perhaps that soft innovation is growing faster than other innovation.

Table 4.1 R&D spending and IP applications worldwide 2002–2011.

Year	*R&D spending (GERD $PPP bn)*	*Patent applications*	*Trademark applications*	*Industrial design applications*
2002	787.7	1,442,500	2,389,889	322,039
2007	1155.4	1,866,700	3,328,561	520,837
2009	1276.9	1,846,700	3,253,887	586,785
2010	n/a	1,985,300	3,686,502	668,470
2011	n/a	2,140,600	4,175,987	775,631
2002–2009 growth	62%	28%	36%	82%
2010–2011 growth	n/a	7.8%	13.3%	16.0%

Sources: R&D from UNESCO Institute for Statistics, remainder from WIPO Statistics Database, October 2012.

Exploring design rights further, Table 4.2 indicates the split of design right applications across groups of countries with different income levels. The most obvious indicator in this data is the growing importance of China in the design statistics, accounting in 2011 for 68.1% of all applications, having grown from 33.4% in 2004. This data also makes clear that design rights are largely the prerogative of high and upper middle income countries. Only around 3% of design rights applications relate to lower middle and lower income countries. Although it is possible that this reflects propensities to register, it is more likely that the importance of industrial design may be much less when income levels are very low. That residents account for between two thirds and three quarters of applications in these low income countries also implies that non-residents in wealthier countries do not attempt to protect their designs in the lower income countries.

There is limited data on the sectoral breakdown of applications. However, in Table 4.3 we present data on the top 10 companies applying for design registrations in the EU in 2011 (which is similar to patterns found in other countries).

WIPO (2012) conclude that such data show that the electronics and ICT, automotive, clothing and fashion, interior design and decoration industries, and – to a lesser extent – firms in the consumer product industries use the industrial design system most intensively. This reflects our initial view that soft innovation is not restricted to the creative industries but is much more widespread.

Table 4.2 Design rights, shares of global applications (%).

Income group	*Design applications 2004–2011**		*Design count 2011*	*Resident share*
High	52.5	24.9	37.1	75.4
Upper middle	42.4	72.0	59.6	95.3
of which China	33.4	68.1	53.2	–
Lower middle	4.6	2.9	3.2	66.7
Low	0.6	0.3	0.2	75.0

Note: *An application may include more than one design.
Source: WIPO (2012).

Table 4.3 Leading applicants for design registration, OHIM, 2011.

Rank	*Company*	*Applications*	*Sector*
1	Rieker Schuh AG	947	Footwear
2	Microsoft Corporation	644	IT
3	Electrolux Home Products Corporation NV	500	Electricals
4	Sony Corporation	485	Electricals
5	Eglo Leuchten GmbH	476	Lighting
6	Pierre Balmain SA	437	Fashion
7	Creation Nelson	403	Fashion
8	Samsung Electronics Co Ltd	350	Electricals
9	Nike International Ltd	319	Sportswear
10	Koninklijke Philips Electronics NV	318	Electricals

Source: WIPO (2012).

Microeconomic Indicators of Soft Innovation

Exploring soft innovation at a lower level of aggregation provides a number of insights beyond those found at the macro level. In this section we initially explore innovation in the film industry before discussing whether the patterns displayed are also to be found in other industries.

The film industry operates in an international marketplace, characterized by a cross-border flow of products and skills. Currently, major technological changes are altering patterns of both production and consumption such that, today, most films are seen away from traditional cinemas and increasingly involve formats other than celluloid for shooting, editing, and production. The production of films can require large investments with estimates of the cost of the two most expensive films (en.wikipedia.org) in 2013 prices being $332m for *Pirates of the Caribbean* and $286m for *Titanic*. These sums match those for a number of new pharmaceutical entities and as such illustrate that some soft innovations require levels of innovative investment matching those of the technological innovations more often studied.

Cinema admissions are an indicator of the demand for film. India had the largest number of admissions in 2009, with 2917 million, more than double the US admissions (1415 million), which were more than five times those of the next country, China (264 million). In 2009, the frequency of admission in the United States was 5.2 films per year, in India 2.7. Nigeria does not show up in the statistics although as we shall see it is a major film producer. There, consumption typically does not take place in theatres but instead the norm is what are conventionally called "alternative screening venues," like video theatres or semi-private communal television sets (UNESCO Institute for Statistics 2012). However in most countries participation in cinema-going culture is in fact quite limited. A select few make up the bulk of the activity.

In the absence of full data upon expenditures for all films we use data upon the number of films produced as a measure of the extent of innovation in the movie production industry. The pattern of world film production 2005–2009 is displayed in Table 4.4.

Although there is no obvious criterion by which one may judge whether the number of films produced is large or small, the growth rate, which averages 6.5% per annum, represents an increasing rate of innovation over time. We may also observe that the leading producers of new films, India and Nigeria, are not the usual

Table 4.4 World production of feature films 2005–2009.

	2005	*2006*	*2007*	*2008*	*2009*
World production	5,658	6,255	7,071	7,083	7,233
Countries covered	95	90	95	98	100
Growth rate (%)		10.6	13.0	0.2	2.1
Top five producers					
India	1,041	1,091	1,146	1,325	1,288
Nigeria	872	1,000	1,559	956	987
United States	699	673	656	759	734
Japan	356	417	407	418	448
China	260	330	411	422	475

Source: UNESCO Institute for Statistics (2012).

candidates for heading tables of innovativeness, although the movement of China up the league table from fifth to fourth is in line with its growing activity as an innovator in many other fields.

Further data on language (although some was missing for key countries) shows that in 2009, English was by far the dominant language for film production, used in one-quarter of all films, Yoruba (11%), used in Nigerian productions, was the second most common language, followed by Spanish (8%), French (6%), Russian (5%), and Hindi (5%). These six languages together represent the majority (60%) of all films made in 2009.

In the definitions above we have placed some emphasis upon when innovation is to be considered significant. We have argued that market impact is a useful indicator of significance. That films may be of some financial importance is shown by estimates that the top two grossing films of all time, adjusted for inflation (en.wikipedia.org) were *Gone with the Wind*, grossing $3.3bn at 2010 prices and *Avatar*, grossing $2.7bn, although these are extreme. In a useful exercise the UNESCO Institute for Statistics (2012) requested countries to identify the 10 most popular feature films over three years, 2007–2009. Most countries measured this popularity by cinema admissions, though a few used gross box office revenue. Scoring a first-ranked film as 10 and a tenth-ranked film as 1, lists of the world's most popular films can be made. The data indicates that popularity declines rapidly. In 2009, whereas the most popular film merited a score of 450 the tenth ranked film scored only 57 and the twentieth only 18. It is also apparent that the top 10 most popular films in each of 2007 and 2008 were no longer in the most popular rankings for the following year. The new products that dominate the market in one year are thus superseded by later new products in the following year.

Of the top ten films in 2007, 2008, and 2009 all were in English and either sole US production or joint productions employing US producers. It is however argued by UNESCO Institute for Statistics (2013) that these highly visible, big budget, English language, franchise feature films represent only one strata or tier of popular global film culture. This top tier, produced and promoted by the largest multinational corporations, represents an international standard for the exclusive and selective environment of the movie theatre. However although these may be the most visible and most widely shared there is also an active sphere that is more localized, nationally specific, in diverse languages, and likely enjoyed in venues other than the traditional cinema house. Locally made films may well be significant locally, but are not globally significant.

Overall however we may observe certain characteristics in the film industry that we shall see to be common to soft innovations in a number of sectors. The first is the large number of new products that are launched each year. This indicates a very innovative sector. However of these launches only a very few achieve large sales and may thus be considered significant, although what is significant in one culture or country may not be significant in another and some products may be locally significant but not globally significant. The market leading products turn over very quickly and thus significant soft innovations generally have short lives, being quickly replaced by new market leading products. Soft innovations in this industry may also yield large returns although, on the other hand, they may also, but not necessarily, be expensive to produce.

In book publishing we also observe many new products being launched each year. Data upon new books published (sourced from Wikipedia, "Books published per country per year") based upon data originally collected by UNESCO, makes clear that there is evidence of innovation in even smaller countries. This data suggests that approximately 2,200,000 new books were published in total across the main 116 publishing countries in the sample in 2010/11. In Europe (Federation of European Publishers 2012) slightly more than 500,000 new books are published each year, with an active catalogue in 2011 of 8.5m titles). China is the single most highly ranked country (a recent phenomenon) publishing 369,523 new titles, closely followed by the United States with 328,259. Even Thailand, ranked thirtieth, published 13,607 titles. As with film we might expect of course that the books being published in any country will reflect the culture of that country, although the large number of new publications in the United Kingdom and the United States implies that the most common language in use for new publications will be English.

As with film, many of the new titles are irrelevant as innovations of economic significance. Using data from the *New York Times* on fiction book sales, Sorensen (2007) observes that the 205 books that entered the *New York Times* bestseller lists accounted for 84% of annual sales in his sample (the top 12 (1%) accounted for 25% of total six-monthly sales and the top 43 (3.5%) accounted for 50% of sales). Thus, like film, only a small number of the soft innovations launched actually succeed. Using weekly data from the *New York Times* bestseller lists for fiction titles from 1970 until 2004, grouped into seven five-year periods (1970–1974, 1975–1979, 1980–1984, 1985–1989, 1990–1994, 1995–1999, 2000–2004), and calculating for each period, the number of titles that appeared in the top 10 and for how many weeks these titles remained there, Stoneman (2011) shows that the number of titles that enter the top 10 per period has increased significantly since 1970–1974; that the number of weeks that a book spends in the top 10 has fallen considerably; and that on average 19% of sales derive from titles new to the top 10 each week. Such data suggests that in publishing the rate of innovation as indicated by the number of significant new products launched is not only high but appears to have increased three times over the 35-year sample period.

Although Stoneman (2011) also provides some evidence relating to the video games industry we look at recorded music as a final example of a creative industry. There are now a number of formats in which recorded music is sold (or pirated). While vinyl and cassettes are now almost defunct, CDs are still widespread although electronic downloads are now supreme. Global recorded music industry revenues were US$16.5 billion in 2012 with the United States as the largest market (with sales totalling $4.48 billion) and Japan being the world's number two market remains (figures sourced from www.billboard.com). The data indicates that in 2013, 11,625 new five-inch CD albums were released on to the Japanese market while the back catalogue of such CDs stood at 118,215 (Recording Industry Association of Japan 2013). This indicates an industry where large numbers of new products are launched on the market each year, similar to publishing. In addition Stoneman (2011) shows using UK data that only 0.7–0.8% of albums released enter the top 40 sales charts at any time, and that sales of any album, as a proportion of sales of the highest-selling album, decline sharply as one moves down the rank of sales. It is obvious that sales soon die away as one moves down the bestseller lists. Stoneman also shows that there will be complete churn of the top 40 on average every two months. The overall picture of soft innovation in the music industry is thus similar to that seen with books and films. There is very extensive soft innovation as indicated by the data on launch patterns, but many titles fail and only a few are significant in terms of sales. There is also a high rate of churn in the top titles.

As argued in WIPO (2012), innovation in terms of non-functional characteristics may well be widespread in industries other than the creative industries. In contrast to the creative industry examples it is thus informative to look at soft innovation in an industry outside the creative sector. Stoneman (2011) explores innovation in three such industries: food, pharmaceuticals, and credit cards. Here we look at just the food industry.

A food industry is found in all economies and is probably one of the oldest if not the oldest industry in the world. Food has both functional and aesthetic characteristics. In developed economies such as the United Kingdom, food is no longer considered to be just a means to avoid dying or to stave off hunger pangs; it is a product that is to be enjoyed via its taste, appearance, smell, and so on. New food products thus to a large degree will reflect changes in form rather than function. Following this argument we may thus again use indicators of the extent of product innovation in order to indicate the extent to which soft innovation is occurring.

Literature on new products in the food industry is informative. According to McNamara, Weiss, and Wittkopp (2003), thousands of new products (variants) are launched every year. In Germany, for example, 32,478 new products were introduced into the food market in the year 2000 (Madakom 2001, cited in McNamara et al. 2003). Winger and Wall (2006) cite that in the United States about 18,000 "new" products are offered to the supermarkets each year (typically in Australia/New Zealand there are between 5,000 and 10,000) and about 10% are chosen to be displayed on the shelves. New introductions to the shelves are almost always linked to the discontinuation of another product. The literature also suggests that most product innovations are failures. For example, McNamara et al. (2003) state that of the many new products introduced into the German food market in 2000 a large share did not survive beyond the first year. Winger and Wall (2006) quote findings from the literature that shows failure rates ranging from 48% to 99%. Feigl and Menrad (2008) using survey data are a bit more optimistic although they find some significant differences across countries. The picture that emerges is similar to that found with books and films. Large numbers of new products are being offered to the market each year. Although there is some dispute, many of these innovations do not succeed and/or they quickly disappear from the market.

Some further information may be gained by exploring trademark data. As argued above, trademarks may be counted to measure jointly both soft innovations and TPP innovations. WIPO (2013) estimates that trademark applications in the food and drink industries (NICE Classes 29–33 inclusive) encompass 9.2% of all applications made under the Madrid system. Also, of the total number of trademarks across all industrial sectors registered in 2005, those in the food industry represented 10%, 3%, 17%, and 8% respectively for the United Kingdom, the United States, Germany, and Korea. This suggests that the food industry is relatively very innovative. However as R&D in the Food, Drink, and Tobacco industry in Europe, for example, represented less than 2% of all Business Enterprise R&D in 2009 (OECD STAN database) it would seem that most of this considerable innovation in the food industry is beyond the (technological) innovation normally measured using R&D data. WIPO counts (http://ipstats.wipo.org) of new trademarks applications in NICE sector 29 further indicate that in terms of overall registrations Europe and China lead, with China showing massive growth between 2008 and 2011 whereas Europe shows a decline. One may also observe that there is trademark registration by non-residents, again suggesting that soft innovation is an international activity.

Across both the creative and other industry examples explored, our first observation is that many new products reflecting changes in form or intellectual content rather than function are being launched each year, indicating high rates of soft innovation. Such innovations are being introduced particularly in the United States, United Kingdom, and Western Europe although China is an increasingly important source and some locally specific important innovations are being introduced in most countries. However of the new products launched very few succeed (in terms of sales). The most successful products sell in very large quantities but sales quickly decline as the success rank falls. It may also be that product life cycles are becoming shorter and such markets are exhibiting greater and greater churn with more and more "bestsellers" each year. These findings may be interpreted as indicative of extensive and increasing rates of significant soft innovation.

Policy Issues

Limiting the Scope

There is now a large literature considering whether, and if so how, government can intervene in the innovation process (see, e.g., Steinmuller 2010). It is not the intention in this section to address that literature in anything like its entirety. Instead we attempt here to cover issues that are particular to the consideration of soft innovation and may not be encompassed by discussions more usually associated with technological innovation. In line with much of the current economic literature we first take an approach based upon market failure, it being argued that governments should only intervene if, left unaided, markets may fail. Even here however we restrict ourselves to arguments that relate to the horizontal product differentiating nature of soft innovation rather than encompassing the whole relevant literature that may also apply to soft innovation. Second, we move to consider copyrights and the needs for and probable amendments required to this IP regime in the light of the problems raised by the ease with which copying has become enabled with advances in IT. Again we note that other extensive literatures on appropriability are not addressed here. Third, we consider whether, rather than market failure, one can justify intervention on cultural grounds.

Market Failure

It has long been considered by economists that market failures are prevalent in the innovation process and as result the private incentives to innovate may not well match the social incentives to innovate. Government intervention can thus often be justified. There are a number of different lines in this argument, but the emphasis we wish to place here is upon the characterization of much of soft innovation as involving horizontal product differentiation. The question of market failure to be addressed is then one of whether unaided the market will generate too much or too little variety or will generate the welfare optimal amount.

Key issues in the analysis of markets for horizontally differentiated products concern how many product variants will be on the market, where they may be located, how the outcomes may change over time and what are the driving forces in the process. These are matters addressed in Stoneman (2011) building upon the highly influential Dixit and Stiglitz (1977) model of product variety. Several different scenarios are explored. The approach leads one to conclude that, with inviolate IPRs, the equilibrium number of firms and product variants we might expect on the market depends upon launch cost, fixed costs, marginal production costs, market structure, the price elasticity of demand, and consumer preferences. Some literature suggests that variants will be grouped in product space. Other literature does not support this view. Of the factors discussed it is clear that: (i) if the number of suppliers is predetermined, then more suppliers means more variety; (ii) lower production costs encourage higher product sales and more variety; (iii) lower costs of developing new variants encourages more variety; (iv) through its impact on production costs, R&D spending on process innovation or the buying in of new technology (which may be determined endogenously) will encourage more variety; (v) as R&D and new technology reduce the costs of generating new variants the number of new variants may increase (the link between TPP innovation and soft innovation so indicated is an important point to note); (vi) if the costs of production are falling over time this may spread sales and also variant proliferation over time; and (vii) for some products, as the market for older variants becomes exhausted so there is a greater incentive to develop newer variants.

There is however little agreement in the literature as to whether the free market provides excess incentives to horizontal product variety. Although Dixit and Stiglitz (1977) suggest not, the potential for scale economies, common pool effects, and creative destruction effects may generate excess incentives and thus the issue is not resolved. The outcome may well be situation specific. It is not possible to state a priori that intervention is or is not desirable on market failure grounds. In some scenarios expenditure on developing variety may be too low. In others it may be too great. The conclusion of the survey by Lancaster (1990) is still a useful summary of the position:

> There is much disagreement on an important policy issue – whether particular market structures produce more or less variety than is optimal. The conclusion in this regard varies from model to model, and in the more complex models, from situation to situation. A fair statement, however, is that most of the models predict that the monopolistic competition equilibrium will give more than optimal variety under most circumstances, and that protected monopoly will gives less variety than is optimal. There seems to be no clear cut answer to such a question as whether an oligopolistic structure of multiproduct firms, or a monopolist attempting to deter entry, will result in more or less than the optimal degree of variety.

As with many fields of economics, one is left in a position that says that there may be market failures but it all depends. It suggests that in some cases there may be a reason for government intervention on the grounds that there is insufficient incentive to variety but the analysis has to be situation specific and general recommendations are unlikely. Even if there are market failures it still leaves open the issue of appropriate instruments. For example should there be tax based subsidies, project support, information provision, and so on? We will not explore this issue in detail, the reader is instead referred to, for example, Steinmuller (2010).

Copyrights

The results for variety proliferation were generated in the context of an assumption of strong IPR. However we know that over time there has been a need to introduce regulatory structures that can back up IPR, for example, patents, design rights, trademarks, and copyrights. Clearly if the lack of IPR protection means that the variety and advances are less than is socially desirable then there is an argument for having stronger protection. Deciding upon the appropriate level of protection however is no simple matter. For example there is a large literature upon the ideal length of patent life. Once again, however, given the limited scope of this chapter we are not going to review the whole of this literature. Stoneman (2011) addresses it more fully. Hargreaves (2011) also provides a wide-ranging review. What we are going to do here is to look only at issues surrounding copyright protection. The reasons for this are that with the advent of the Internet there has been increasing amounts of and increasing concern with copying and unapproved downloads of especially music, films, and books, and there have been some recent changes in copyright law that are not covered in Stoneman (2011).

Hurt and Schuchman (1966) propose that one of the arguments in favor of copyrights is based on the rights of the creator of the protected object or on the obligation of society toward him/her. We, however, follow a more economic line of argument and consider that it would seem clear that if authors, musicians, and filmmakers cannot protect their intellectual property and there are unauthorized downloads of their work then the incentives to produce such work will be less than the social value of that work and a sub-optimal outcome will exist. However, as with all IPR mechanisms, copyright involves a trade-off between providing an incentive to the producer/innovator and the cost of a monopoly welfare loss. It may well be that at the welfare optimum, given low reproduction costs, music, video, and game downloads should be (all but) free, however if they are free there is no incentive to produce new music, films, or games. The excludability provided by copyrights requires that a price has to be paid for the products in order to reward the innovator for his/her creative efforts. One way to try to resolve this conflict is by choice of the length of copyright life.

Copyright in literary, musical, artistic, and dramatic work in the United Kingdom lasts for the creator's lifetime plus 70 years (basically the same as in the EU and the United States). For films it is 70 years after the death of the last of the directors, score composer, dialogue or screenplay authors and for TV and radio it is 50 years from the first broadcast. Sound recording copyright lasts for 50 years. Publisher's right, which covers the typographical layout of published editions like books or newspapers (how it is presented on the page), lasts for 25 years from creation. These lives are considerably longer than the terms of even extended patent rights. Lindsay (2002) argues on the basis of the literature that:

> Although there are good economic explanations for many of the existing legal limits on copyright protection, there is no basis for assuming that the current limits are optimal. There is even less basis for assuming that the limits should be immutable. (111)

Long lives are of course irrelevant if they cannot be enforced. The Gowers (2006) review of copyright in the United Kingdom suggests that copyright suffers from a lack of public legitimacy with little guilt or sanction associated with infringement. While criminal and civil legal sanctions against copyright infringement are tough, infringement is extremely common. The fact that the letter of the law is rarely enforced only adds to the public sense of illegitimacy surrounding copyright law. A basic problem is that the enforcement is in the hands of the owner and many owners will be individual artists, authors, or perhaps academics for whom the costs of enforcement are too great. Enforcement through the civil courts is costly, and cases are difficult to prove.

The long length of life of copyright protection has also created another problem, that of orphan works. Gowers (2006) notes that:

> the existence of such a large volume of old work protected but unavailable (estimates of up to 98 per cent of published work under copyright) means that a great amount of intellectual capital is wasted. Firms and individuals are unable to restore, rework or revive these "orphan" works to create new commercial and creative capital. (39)

In a UK government commissioned review, Hargreaves (2011) found that copyright law may be too restrictive, with the excludability that it provides having significant social costs, arguing that:

> copyright is falling behind what is needed. Copyright, once the exclusive concern of authors and their publishers, is today preventing medical researchers studying data and text in pursuit of new treatments. Copying has become basic to numerous industrial processes, as well as to a burgeoning service economy based upon the internet. The UK cannot afford to let a legal framework designed around artists impede vigorous participation in these emerging business sectors. ... creative industries too need change, in the form of more open, contestable and effective global markets in digital content and a setting in which enforcement of copyright becomes effective once more. (1)

In line with these arguments Hargreaves (2011) proposes that the UK government: (i) should establish a cross-sectoral Digital Copyright Exchange; (ii) support moves by the European Commission to establish a framework for cross-border copyright licensing; (iii) require by law that collecting societies adopt codes of practice, approved by the IPO and the UK competition authorities, to ensure that they operate in a way that is consistent with the further development of efficient, open markets; (iv) legislate to enable the licensing of orphan works; (iv) deliver copyright exceptions at national level to realize all the opportunities within an EU framework, including format shifting, parody, non-commercial research, and library archiving; (v) promote, at EU level, an exception to support text and data analytics; and (vi) lead at an EU level to develop a copyright exception enabling adaptability to new technologies, designed to allow uses enabled by technology of works in ways which do not directly trade on the underlying creative and expressive purpose of the work.

HM Government (2012) proposed responses which are largely in line with these recommendations and at the time of writing the Intellectual Property Office is in the process of consultation upon these changes. Generally the changes may be seen as weakening copyright. It is difficult to see any changes here that will increase the extent to which individual copyright holders may more effectively pursue their rights.

In other countries, change is also occurring. For example in the EU the orphan works directive (2012/28/EU) has been introduced which allows any institution (e.g., public libraries, education establishments, museums, and archives) wishing to use an orphan work to do so after first carrying out a "diligent search" in good faith from "appropriate sources." They may not exploit the works commercially (www.openrightsgroup.org). In Canada, however, such a change has already been instituted with Section 77 of the Canadian Copyright Act 1985 enabling the Copyright Board to issue a non-exclusive license, on any terms it chooses to specify, to anyone seeking a license for orphan works, as long as it satisfies the Board that they have made "reasonable efforts to locate the owner." However, despite attempts to do so, no changes have been made to US copyright law to address this issue.

The Cultured Society

The arguments above for policy intervention have largely been built upon a view of soft innovation that values such activities because of the economic welfare (the sum of consumer and producer surplus) that they generate. This economic approach is not uncontentious (Eltham 2013). Of particular importance is a set of views that value art more in terms of "art for art's sake" and embody different values, beliefs, and foundations to those implicit or explicit above. The art for art's sake view is that art and aesthetic advances cannot be valued like margarine or washing powder, and one can never measure the value of beauty or artistic achievement by price alone (and especially one cannot expect the market price to be a true valuation). Art is argued to have its own inherent aesthetic value.

Such arguments support the view that cultural activity is the sign of a civilized society and as the United Kingdom is a civilized society the government should support art and, by implication, soft innovation. Such a view may be used as a separate basis for policy intervention. Thus, for example, the Arts Council states its ambition to "put the arts at the heart of national life and people at the heart of the arts (www.artscouncil.org.uk).

In the United Kingdom, funding for such cultural reasons (although often overlaid with arguments for economic benefits) is provided to the arts via the Arts Council, while the Arts and Humanities Research Council (AHRC) supports university research within a subject domain from traditional humanities subjects, such as history, modern languages, and English literature, to the creative and performing arts and, on behalf of the Higher Education Funding Council for England, provides funding for museums, galleries, and collections that are based in, or attached to, higher education institutions in England.

In addition, perhaps with a greater eye upon the economic benefits, there are also tax reliefs to certain creative industries. Creative industry tax reliefs are a group of four Corporation Tax reliefs that allow qualifying companies to claim a larger deduction, or in some circumstances claim a payable tax credit, when calculating their taxable profits. Film Tax Relief was introduced in April 2007. Two additional reliefs, Animation Tax Relief and High-end Television Tax Relief were introduced in April 2013. A fourth relief for Video Games Development will be introduced after state aid approval (i.e., after EU approval). Although these incentives may also have economic spin-off as a target, to qualify for the creative industry tax reliefs all films, television programs, animations, or video games must pass a "cultural test" or qualify through an internationally agreed co-production treaty, certifying that the production is a "British" film, "British" program, or "British" video game. In all cases, formal certification is required to qualify.

The United Kingdom is not the only country to offer such incentives. For film production, for example, the Italian government since 2009 has given local producers a 20% tax break, outside investors a 40% tax shelter, and foreign productions a 25% tax credit. In the United States there are various state-level incentives to encourage filmmakers encompassing, for example, tax credits, cash rebates, and grants. In many other countries similar schemes are also available.

Conclusions and Implications

In this chapter it has been argued that there is a type of innovation, here labeled soft innovation, that is more concerned with changes in form and intellectual appeal rather than changes in function. This type of innovation has largely been ignored in the main definitions of innovation used in the literature. Such innovation tends not to be the result of R&D activity although it can involve extensive levels of investment, nor can it be patented, although there are a number of appropriability issues relating to such innovation that are addressed by other IPR mechanisms. These characteristics mean that the standard indicators of the extent of innovation, such as R&D spending or the numbers of patent applications, do not reflect such activity.

Here, by using alternative indicators such as counts of new product innovations and applications for design rights and trademarks, we have been able to show that such innovation is not only widespread in the creative industries but can also be observed in other industries as well. It may even be the case that it is growing faster than traditional technological innovation. It can also generate large returns to successful innovators. We have shown that such innovation is not restricted to the advanced economies. In film, for example, we found that Nigeria is second in the league table of the greatest number of films made annually. The innovations produced may well differ according to the cultural environment in which they are generated. The main innovative activity is however occurring in the upper and upper middle income countries. In addition, but not surprisingly, the data also show the growing influence of China as a source and user of soft innovation – a pattern that reflects its growing importance in technological innovation.

If utilized indicators of innovation exclude soft innovation activity, not only do we get a biased picture of overall activity, but our analysis of issues such as the returns to innovation, the incentives to innovation, and the economic impact of innovation will be unsatisfactory. In addition if we ignore changes in form and concentrate only upon changes in function then the policy recommendations that are produced will be misguided and incomplete.

Our analysis of the nature of soft innovation has indicated that in a number of cases (e.g., books, film, music, food) large numbers of new products are launched onto the market each year. Nearly all of these fail. However a small number go on to have very extensive sales and generate large revenues. It is such innovations we consider to be significant. In most cases however even these most successful innovations have short lives at the top and as such the apparent rate of innovation in many of the relevant industries is very fast. We have not here been able to explain why such "letting a thousand flowers bloom" strategies are pursued so widely.

There are a number of implications of these findings:

1. There is now a need to reconsider the OECD definitions of innovation to take fuller account of soft innovation activity beyond the current inclusion of marketing innovations. These definitions also need to be reconsidered to fully account for innovation that encompasses new horizontally differentiated products in addition to new products showing vertical quality improvement (change) as currently included.
2. The reliance upon R&D as an indicator of investment in innovation-producing activity needs to be revised with investment in such activities as, for example, design, filmmaking, games development, or book writing being considered as well.
3. The widespread reliance upon the analysis of patents as an IPR mechanism needs revisiting. More attention should be paid to design rights, copyrights, and trademarks.
4. Theoretical analysis of market failures needs to place greater emphasis upon incentives to variety than is currently the case. This may well be built upon a more satisfactory approach to explaining the widespread pursuit of innovation strategies such as letting a thousand flowers bloom.
5. Policy initiatives built on the basis of market failures and the correction thereof, need to consider soft innovation as well as the more traditional technological innovations. In addition soft innovation may merit more consideration purely in terms of cultural issues.

Notes

1 Aesthetic is considered here to encompass issues wider than the concept of beauty and artistic experience, and in particular visual beauty, to which it may be restricted. The *Online Etymology Dictionary* refers to aesthetic as popularized in English by a translation of Immanuel Kant, and used originally in the classically correct sense "the science which treats of the conditions of sensuous perception." Although today sensory is a more appropriate word than sensuous, it is this meaning that is adopted here. Aesthetic is thus taken as encompassing not only sight but also touch, smell, sound, and for current purposes extended to include the intellect.

2 This contribution provides a reprise of that book, while also updating and extending it in relevant directions.

3 Defined variously in different countries but essentially encompassing, Advertising, Architecture, Art and Antiques, Crafts, Design, Designer Fashion, Video Film and Photography, Music and the Visual and Performing Arts, Publishing, Software, Computer Games and Electronic Publishing, Radio and TV. Although there is reasonable literature considering these industries (e.g., Caves 2000), there is little that explicitly concentrates upon innovation in these industries.

4 Interestingly, recent macroeconomic literature on economic growth has allowed for horizontal product differentiation in capital inputs into the production process (Romer 1990) but has preferred models of vertical quality improvement when considering innovation in outputs (Aghion and Howitt 1992).

References

Aghion, P., and P. Howitt. 1992. "A Model of Growth through Creative Destruction." *Econometrica* 60: 323–351.

Bianchi, B., and F. Bortolotti. 1996. "On the Concept of Formal Innovation." European Regional Science Association, 36th European Congress, ETH Zurich, August 26–30.

Carey, J. 2005. *What Good Are the Arts?* London: Faber and Faber.

Caves, R. 2000. *Creative Industries, Contracts Between Art and Commerce*. Cambridge, MA: Harvard University Press.

Cowen, T. 2000. *In Praise of Commercial Culture*. Cambridge, MA: Harvard University Press.

Cowen, T. 2006. *Good and Plenty: The Creative Successes of American Arts Funding*. Princeton, NJ: Princeton University Press.

Dixit, A., and J. Stiglitz. 1977. "Monopolistic Competition and Optimum Product Diversity." *American Economic Review* 67: 297–308.

Eltham, B. 2013. "Three Arguments Against Soft Innovation: Towards a Richer Understanding of Cultural Innovation." *International Journal of Cultural Policy* 19(5): 537–556.

Federation of European Publishers. 2012. *European Book Publishing Statistics*. Brussels.

Feigl, S., and K. Menrad. 2008. "Innovation Activities in the Food Industry in Selected European Countries." Report for Truefood-Project, University of Applied Sciences Weihenstephan, Straubing, Germany.

Galenson, D.W. 2005. *Who Are the Greatest Living Artists? The View from the Auction Market*. NBER Working Paper 11644.

Gowers, A. 2006. *Gowers Review of Intellectual Property*. Norwich, UK: HMSO.

Hargreaves, I. 2011. *Digital Opportunity: A Review of Intellectual Property and Growth*. London: Department for Business, Innovation and Skills.

HM Government. 2012. *Modernising Copyright: A Modern, Robust and Flexible Framework*. Newport, UK: Intellectual Property Office.

Hurt, R.M., and R.M. Schuchman. 1966. "The Economic Rationale of Copyright." *American Economic Review* 56(1–2): 421–432.

Lancaster, K. 1990. "The Economics of Product Variety: A Survey." *Marketing Science* 9(3): 189–206.

Lindsay, D. 2002. *The Law and Economics of Copyright, Contract and Mass Market Licences*. Strawberry Hills, NSW: Centre for Copyright Studies Ltd.

Madakom. 2001. *Innovations Report 2001*. Köln: Madakom GmbH.

Marzal, J.A., and E.T. Esparza. 2007. "Innovation Assessment in Traditional Industries: A Proposal of Aesthetic Innovation Indicators." *Scientometrics* 72(1): 33–57.

McNamara, K.T., C. Weiss, and A. Wittkopp. 2003. *Market Success of Premium Product Innovation: Empirical Evidence from the German Food Sector*. Working Paper FE 0306, Department of Food Economics and Consumption Studies, University of Kiel.

OECD. 2002. *Frascati Manual: Proposed Standard Practice for Surveys on Research and Experimental Development*, 6th edn. Paris: OECD.

OECD. 2005. *Oslo Manual: Guidelines for Collecting and Interpreting Innovation Data*, 3rd edn. Paris: OECD.
Postrel, V. 2004. *The Substance of Style: How the Rise of Aesthetic Value Is Remaking Commerce, Culture, and Consciousness*. New York: HarperCollins.
Recording Industry Association of Japan. 2013. *Yearbook of the Recording Industry in Japan, 2013*. Tokyo.
Romer, P. 1990. "Endogenous Technological Change." *Journal of Political Economy* 98: 71–102.
Sorensen, A. 2007. "Bestseller Lists and Product Variety." *Journal of Industrial Economics* 55(4): 715–738.
Steinmuller, E. 2010. "The Economics of Technology Policy." In *The Economics of Innovation*, ed. B. Hall and N. Rosenberg. Amsterdam: North Holland.
Stoneman, P. 2011. *Soft Innovation: Economics, Product Aesthetics, and the Creative Industries*. Oxford: Oxford University Press.
Swan, K.S., M. Kotabe, and B.B. Allred. 2005. "Exploring Robust Design Capabilities: Their Role in Creating Global Products, and Their Relationship to Firm Performance." *Journal of Product Innovation Management* 22(2): 144–164.
UNESCO Institute for Statistics. 2012. *From International Blockbusters to National Hits*. Information Bulletin No. 8.
UNESCO Institute for Statistics. 2013. *Feature Film Diversity*. Fact Sheet No. 24.
Winger, R., and G. Wall. 2006. *Food Product Innovation: A Background Paper*. Rome: Food and Agriculture Organization of the United Nations.
WIPO. 2012. "The Rise of Design in Innovation and Intellectual Property: Definitional and Measurement Issues." In *World Intellectual Property Indicators 2012*. Geneva: WIPO Economics and Statistics Division.
WIPO. 2013. *2013 Madrid Yearly Review*. Geneva: WIPO Economics and Statistics Division.

Chapter 5

Is the World of Science Moving to the East? What Bibliometrics Says

Ping Zhou and Jiang Li

With the emergence of the BRICS countries (Brazil, Russia, India, China, and South Africa) in the world economy, the traditional leading roles of the West including the United States and EU nations have been challenged. China, for instance, is now the second largest economy in the world with $8.22 trillion GDP (World Bank 2013). A similar situation is happening in science and technology. On December 2, 2013, China successfully launched the Long March 3B rocket carrying the Chang'e-3 probe and Yutu lunar rover. In terms of world publications in science China has been the second largest producer since 2006 (Zhou and Leydesdorff 2008). Statistics of the Institute of Scientific and Technical Information of China (ISTIC 2013) show that China contributed 12.08% of publications included in the Science Citation Index (SCI) of Thomson Reuters in 2012. It is clear that China has become a major player in world science (e.g., Zhou and Leydesdorff 2006; Glänzel, Debackere, and Meyer 2008; ISTIC 2013; see also Le and Tang in this volume, Chapter 2). The performance of other Asian countries, such as Japan, South Korea, and India, is also remarkable. In fact, Leydesdorff and Zhou (2005) noted the change and the possible shift of gravity center of the world science system in 2005.

Is it true that the world of science is indeed moving to the East? The current chapter will try to answer the question from the perspective of bibliometrics. Comparative analysis will be done between the West and the East. The United States, England, Germany, and France are selected to represent the West because they produce most publications in the region. China, Japan, South Korea, and India are chosen to represent the East for the same reason. Analysis will focus mainly on investment in research and development (R&D) and scientific productivity and impact in terms of publications. Data from the OECD are used for analysis of R&D investment. Publication data from the Web of Science (WoS) of Thomson Reuters are used to map the overall performance of the selected countries. Further investigations are based on data from the Essential Science Indicators (ESI) of Thomson Reuters.

The Handbook of Global Science, Technology, and Innovation, First Edition.
Edited by Daniele Archibugi and Andrea Filippetti.

In R&D analysis, India will not be included because it is not covered by the OECD data. Data for the United Kingdom are also somewhat complex. For R&D analysis we used data of the United Kingdom. For investigation based on publication data, only England is included because both the WoS and the ESI only label publications from the four parts of the United Kingdom (i.e., England, Scotland, Wales, and Northern Ireland) separately without integrating them together under the United Kingdom. Considering England is by far the largest publication producer of the United Kingdom, we use England to represent the United Kingdom.

Investment in Research and Development

R&D includes systematic activities combining both basic and applied research. "Research" is to identify new knowledge and ideas, and "development" is to turn the ideas into tangible products or processes. As defined by the US National Science Foundation (2013), applied research is systematic study to gain knowledge or understanding necessary to determine the means by which a recognized and specific need may be met. In contrast, basic research advances the knowledge of science without an explicit, anticipated commercial outcome. R&D activities are performed by both the public and the business sectors. The R&D performed in the business sector is now larger than the R&D performed in the public sector. As an important form of R&D outcomes, scientific publications are more likely to reflect the public component of R&D. R&D can be funded by the government and performed by either/both public or industrial sectors. To a large extent, however, it continues to be true that public R&D more likely generates publications while business R&D generates patents, although public R&D is generating more patents and business R&D is generating more scientific articles than in the past. As this chapter uses publications as research output, the OECD data of public R&D expenditure will be used in the current section.

Public R&D Expenditure

R&D investment plays a significant role in innovation and economic growth. For example, Robbins and Moylan (2007) estimated that to real GDP growth R&D contributes approximately 0.20% of the average 2.9% growth, or about 7.1% of the average growth rate from 2002 to 2007 in the United States. In fact, R&D expenditure has long been an important indicator of investment in science and technology (e.g., the OECD Science, Technology, and R&D Statistics). Investment in R&D can be reflected by two aspects, namely expenditure and human talents or manpower. A nation's R&D expenditure relative to its GDP is usually considered as an important indicator for measuring the capacity and aspiration of a nation to science.

Based on public R&D expenditure of the seven countries representing the West and the East, two groups can be classified, with US$15 billion as a divide. The leading group is represented by the United States, China, Japan, and Germany. The United States takes the absolute and unshakable lead with a continuous increase (except for a slight drop in 2010). The growing investment of China is impressive. It has replaced Japan and Germany as the second largest investor since 2007. In the 10 years 2001–2010, Germany and Japan had spent the same in terms of public R&D (Figure 5.1a).

South Korea, France, and the United Kingdom are in the second group. France takes the lead with relatively stable public R&D expenditure in ten years. The growing investment of South Korean is remarkable. It surpassed the United Kingdom in 2009 and is about to catch up with France. Slight increase in the United Kingdom is visible but not as significant as that of South Korea (Figure 5.1b).

Percentage share of GDP is often used to eliminate size effect of different economies. In terms of the share of public R&D expenditure in a nation's GDP, four groups can be identified. South Korea and the United States are in the first group and take the lead. Germany and France spend similarly and are in the second group. The United Kingdom and Japan are in the third group, whereas China alone is in the last group (Figure 5.2).

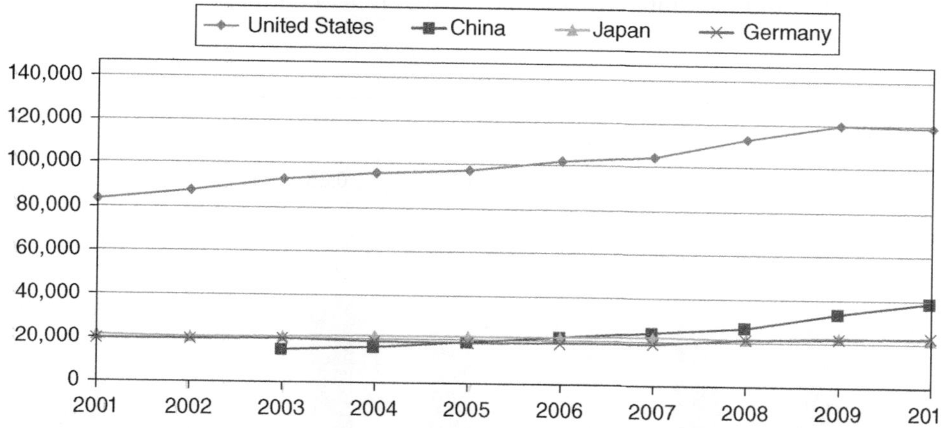

Figure 5.1a Public expenditure on R&D of countries in the first group 2001–2010 (US $ million, 2005 constant prices and PPPs).
Source: Data from OECD (2013a).

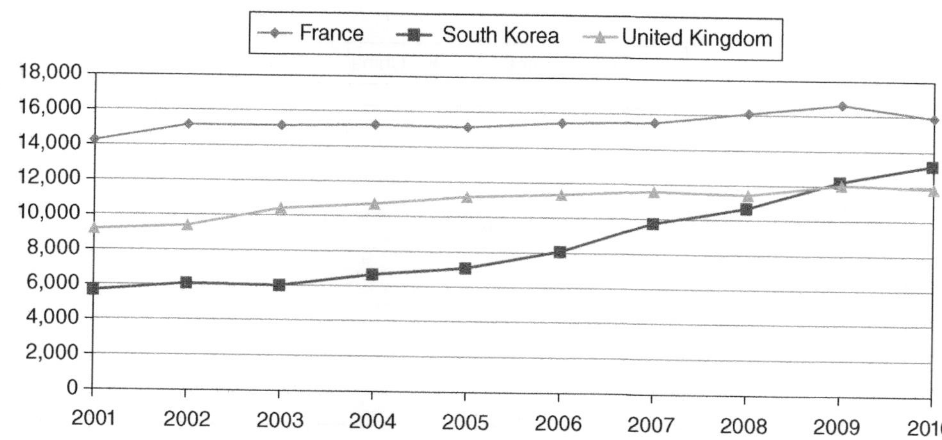

Figure 5.1b Public expenditure on R&D of countries in the second group 2001–2010 (US $ million, 2005 constant prices and PPPs).
Source: Data from OECD (2013a).

R&D Manpower

R&D activities are mainly conducted by researchers who engage in the conception or creation of new knowledge, products, processes, methods, and systems, and in the management of the projects concerned. Numbers of researchers in the seven countries can also be classified into two groups (Figures 5.3a and 5.3b) with Germany now falling into the second group. Countries in the first group (United States, China, and Japan) have over 600,000 researchers. The United States and Japan seem to have reached their manpower capacity, whereas China continued to enlarge its research team until 2009 when a significant drop happened (Figure 5.3a).

The number of researchers in all countries in the second group has grown continuously, although with significantly smaller scale than those in the first group. South Korea and Germany are especially remarkable. South Korea had already exceeded France in 2007 and the United Kingdom in 2010. France continues to grow although

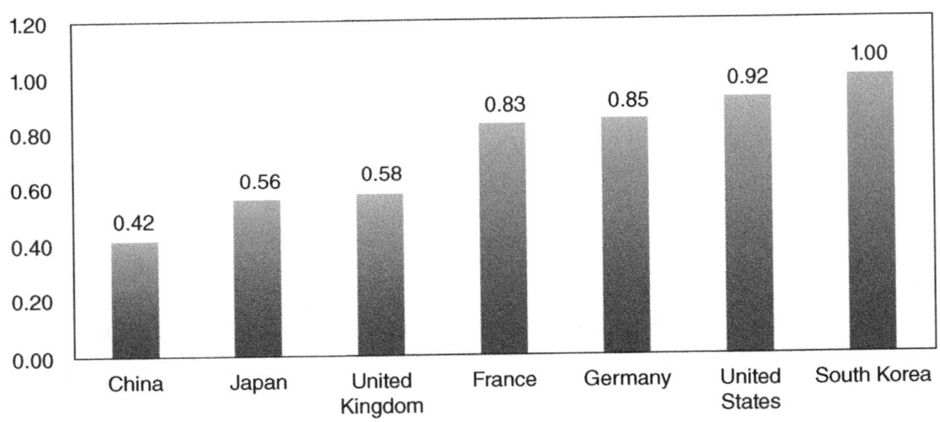

Figure 5.2 Public expenditure on R&D (as a percentage of GDP).
Source: Data from OECD (2013a).

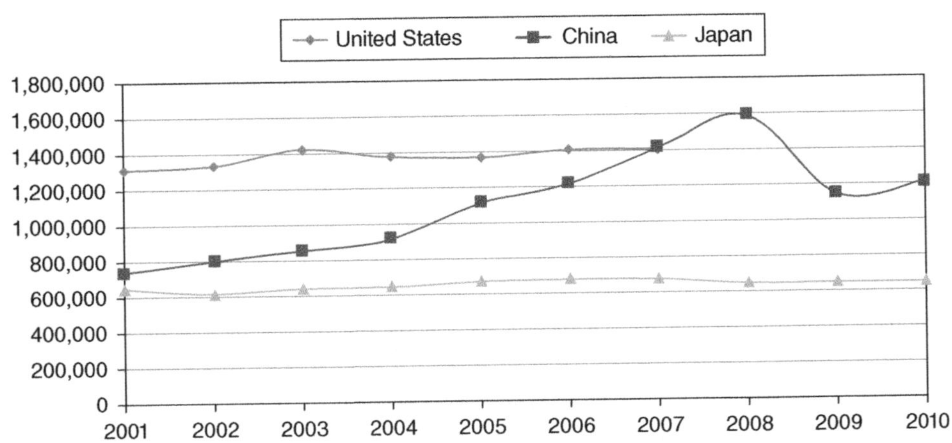

Figure 5.3a Number of researchers of countries in the first group 2001–2010.
Source: Authors' elaboration from OECD (2013b).

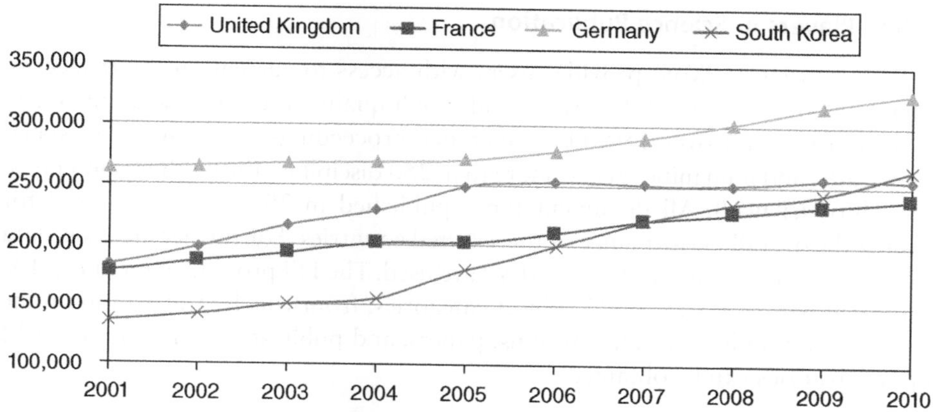

Figure 5.3b Number of researchers of countries in the second group 2001–2010.
Source: Authors' elaboration from OECD (2013b).

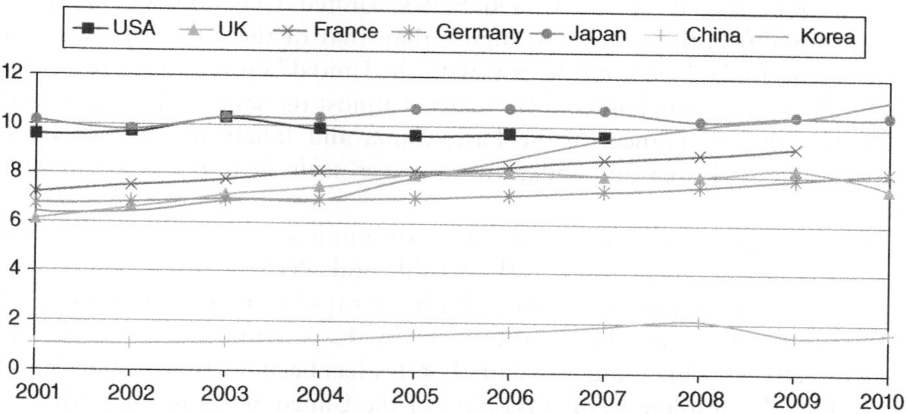

Figure 5.4 Full-time equivalent researchers per thousand employed 2001–2010.
Source: Authors' elaboration from OECD (2013b).

with smaller steps. The United Kingdom seems to have reached its saturation level since 2005 (Figure 5.3b).

Similar to R&D expenditure being dependent on a nation's GDP, the number of researchers of a nation also relies on the size of employed population. In terms of number of researchers per thousand employed, South Korea has experienced fast growth and took the first position by replacing Japan in 2010. Among the Western countries, France and Germany show growth and the United Kingdom went down slightly after five stable years. The ratio of the United States has also gone down slowly (Figure 5.4). Compared to its huge number of employed population, the number of researchers in China is the lowest (Figures 5.3a and 5.4).

In terms of absolute value (e.g., R&D expenditure and size of research team), China takes the second position. When a relative measurement is applied (e.g., ratio of R&D to GDP, number of researchers per thousand employed), China lags far behind the other six countries. The contrary performance of China implies a potential or option for China to further increase its research investment.

Leading Players in Science Publication

The Web of Science (WoS) provides users with access to current and retrospective content. It covers over 12,000 worldwide high-quality journals including Open Access journals and over 150,000 conference proceedings in the sciences, social sciences, arts, and humanities across more than 250 disciplines (http://thomsonreuters.com/web-of-science/). All document types published in 2003–2012 are used for mapping the overall performance of the selected countries. For further analysis, data from the Essential Science Indicators (ESI) are used. The ESI provides data of top 1% highly cited papers in each of 22 subject categories, from which information on the most influential individuals, institutions, papers, and publications in each of the 22 subject categories can be obtained.

An Overview

In 2003–2012, publication productivity in terms of number of publications of the West (represented by the United States, United Kingdom, Germany, and France) is significantly (1.3 times) higher than that of the East (represented by China, Japan, India, and South Korea) with the United States taking the absolute lead. Productivity of England and Germany is almost the same and is significantly higher than that of France. In the East, China and Japan are the two largest producers. South Korea and India are at lower scale in terms of productivity (Table 5.1).

From the perspective of world share of publications, the selected eight countries contribute over 66% of the world total. Performances of the United States and China are most distinctive. With the rapid increase of Chinese publications the United States lost world share by 4.7% from 2003 to 2012. The huge gap between China and the United States has been narrowed significantly. Nonetheless, the absolute leading position of the United States is still unshakable (Figure 5.5a).

In 10 years, Germany and England have kept similar and stable world shares. The world share of France has also been kept stable but significantly lower than those of Germany and the United Kingdom. In the East, Japan is a bigger loser. In 10 years, Japan has lost nearly 2% of its world share. South Korea and India grow in a similar way with similar world share of publications. (For greater clarity, see Figure 5.5b which excludes the United States and China.)

Table 5.1 Publications indexed in the Web of Science 203–2012.

The West: 8,349,174		*The East: 3,587,433*	
United States	5,195,098	China	1,646,564
England	1,174,979	Japan	1,063,209
Germany	1,174,977	South Korea	446,166
France	804,120	India	431,494

Source: Authors' elaboration on data from WoS.

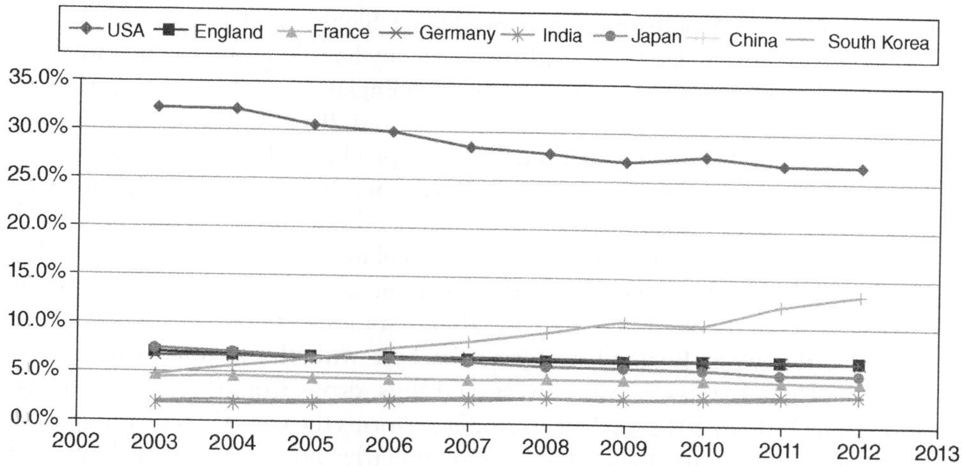

Figure 5.5a Contributions to world publications 2003–2012.
Source: Authors' elaboration from WoS.

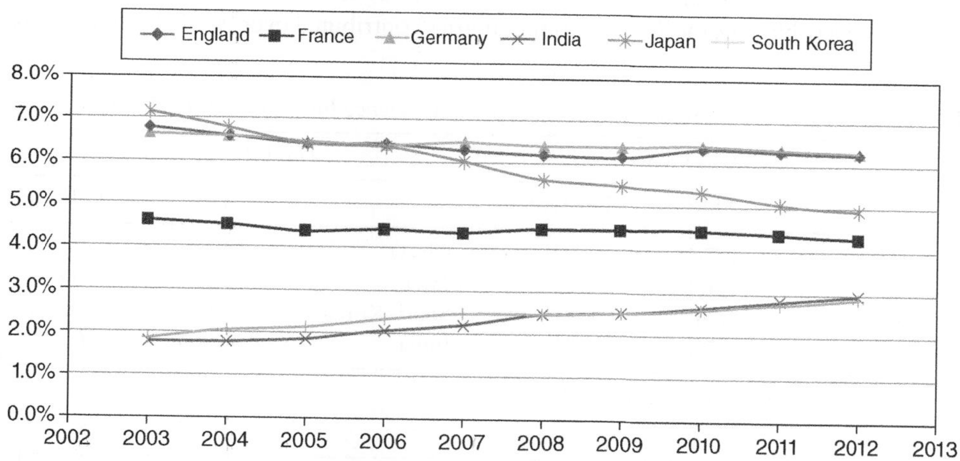

Figure 5.5b Contributions to world publications 2003–2012 (excluding United States and China).
Source: Authors' elaboration from WoS.

Highly Cited Publications

Based on publications in 2003–2012 in the ESI, the productivity of the West is 3.7 times that of the East. In terms of highly cited papers, the gap between the West and the East is much wider compared to that in overall publications included in the WoS. In the Western four countries, the United States takes the absolute lead. England contributes slightly higher than Germany. In the East, China also takes the lead but only slightly higher than France. Highly cited papers from Japan total about half of Germany's number. Each of the four Western countries produces 1.1–1.2% highly cited papers of its total in the WoS, which is twice that of the Eastern countries (Table 5.2).

In terms of annual performance in producing highly cited papers, the United States takes the absolute lead. England and Germany have kept the second and third position until 2012 when China slighly surpassed England (in 2012 China produced 2023 highly cited papers whereas England produced 2012. In the seven countries excluding the United States, China grew fastest in producing highly cited papers and has replaced Japan's leading position in the East since 2006. South Korea and India have kept their relative position in the 10 year period (Figures 5.6a and 5.6b).

To exclude size effect of publications, we calculated percentage share of highly cited versus WoS publications of the corresponding countries (Figure 5.7). A clear gap exists between the West and the East, although ratios of both groups have been growing. In the West, England performs the best with the fastest-growing rate, whereas the United States grows slowest, and thus, drops from the first to the last position from 2003 to 2012. In the East, India performs the worst in the same period. China, Japan, and South Korea progressed similarly: no improvement happened until 2010 when the ESI/WoS ratio started to rise from around 0.50.

Looking into highly cited publications among the 22 subject categories in the ESI, the West has played a major role in each field. In fields such as psychiatry/psychology, social sciences, neuroscience and behavior, economics and business, clinical medicine, as well as microbiology, the Western four countries contributed over 90% of publications

Table 5.2 Highly cited publications in the Essential Science Indicators 2003–2012.

The West: 98,863			*The East: 20,840*		
	ESI	ESI/WoS		ESI	ESI/WoS
United States	61,719	1.2%	China	10,010	0.6%
England	14,609	1.2%	Japan	6,351	0.6%
Germany	13,747	1.2%	South Korea	2,675	0.6%
France	8,788	1.1%	India	1,804	0.4%

Source: Authors' elaboration on data from ESI.

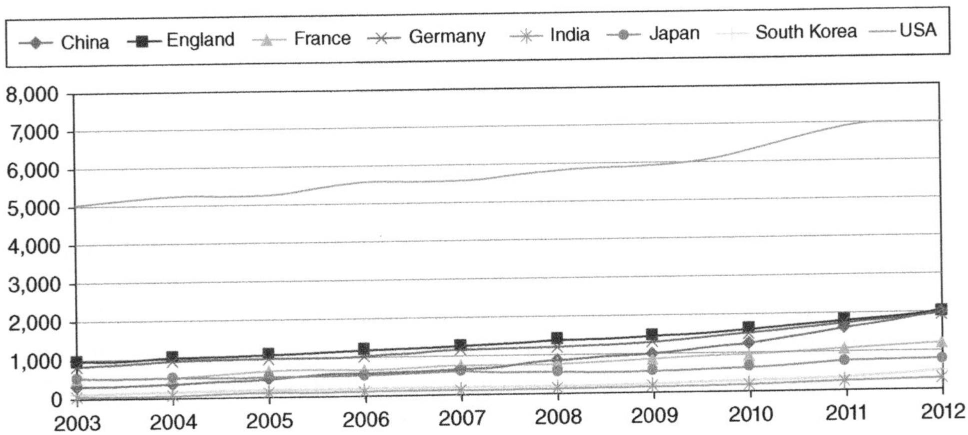

Figure 5.6a Highly cited publications of the eight countries in the Essential Science Indicators 2003–2012.
Source: Authors' elaboration from ESI.

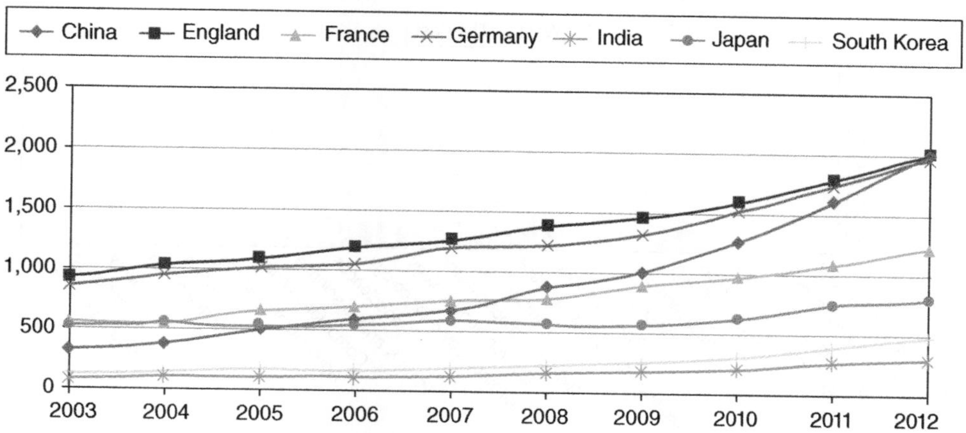

Figure 5.6b Highly cited publications of seven countries in the Essential Science Indicators 2003–2012.
Source: Authors' elaboration from ESI.

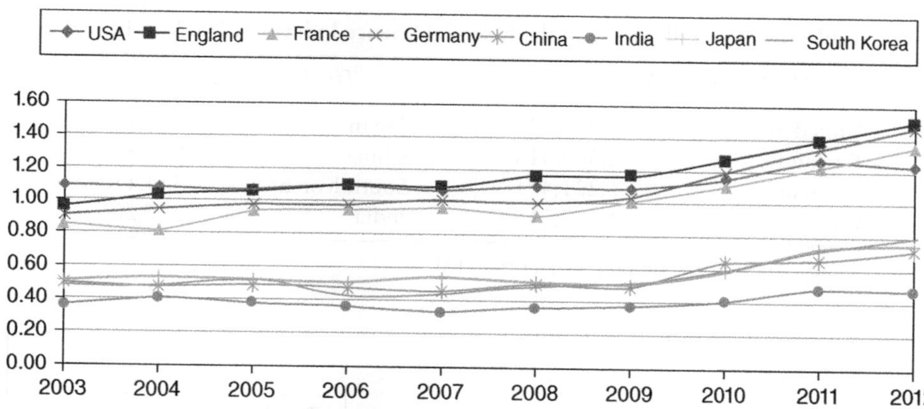

Figure 5.7 Percentage share of highly cited versus Web of Science publications 2003–2012 (ratio of Essential Science Indicators to Web of Science publications).
Source: Authors' elaboration from ESI and WoS.

from the eight countries. The East is doing relatively well both in kinds of "applied" sciences (e.g., engineering) and more basic science (i.e., mathematics and physics), but is lagging behind in the social sciences and humanities (Figure 5.8).

Publication Impact in the ESI

Number of citations received by publications is used to measure impact of publications. Overall impact is the total number of citations to publications from a country. In 2003–2012, overall impact of the West is more than six times that of the East. The United States takes the absolute lead. Furthermore, each of the selected Western countries has higher overall impact than each of the Eastern countries. Japan takes the lead among the Eastern countries. China only stands in the second position although producing the most ESI papers among the Eastern countries (Table 5.3).

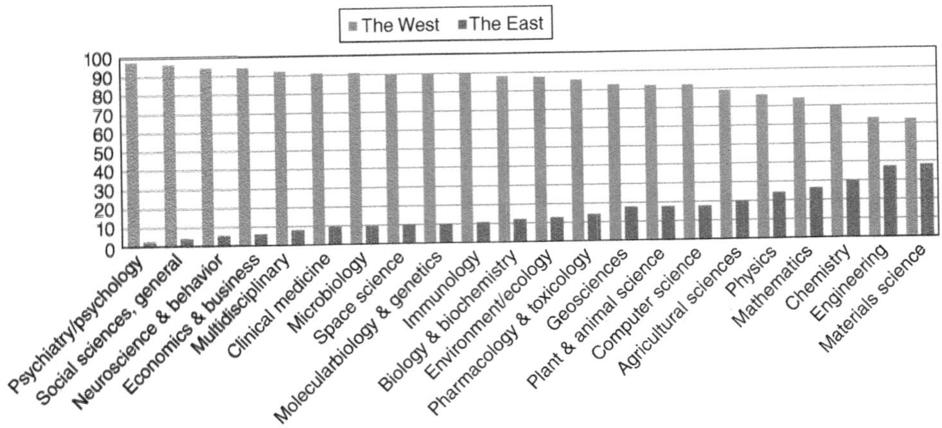

Figure 5.8 Percentage share of Essential Science Indicators publications produced by the eight countries in corresponding fields 2003–2012.
Source: Authors' elaboration from ESI.

Table 5.3 Overall impact in the Essential Science Indicators 2003–2012.

The West: 17,151,529		*The East: 2,729,359*	
United States	10,976,827	Japan	1,148,466
England	2,500,516	China	999,576
Germany	2,221,471	South Korea	354,116
France	1,452,715	India	227,201

Source: Authors' elaboration on data from ESI.

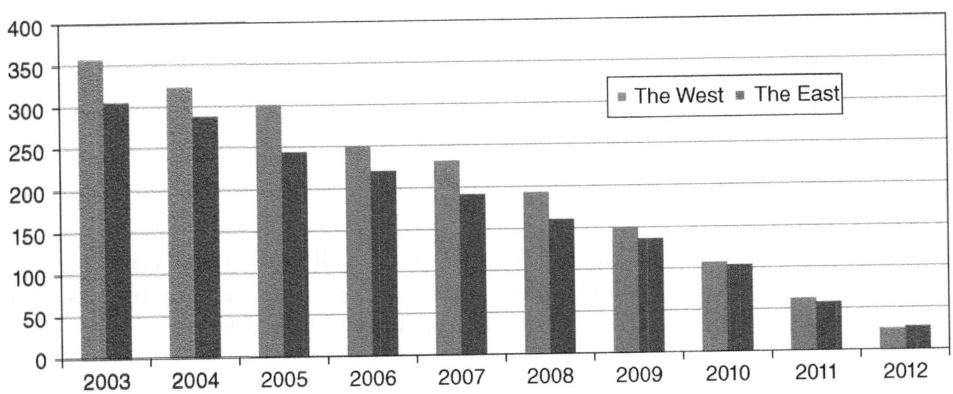

Figure 5.9 Citations per paper of the eight countries in the Essential Science Indicators 2003–2012.
Source: Authors' elaboration from ESI.

Variation exists among highly cited publications, especially among publications from the East and the West. Figure 5.9 shows citations per paper (CPP) of the West and the East. Because the ESI only provides accumulative citation counts, publications in earlier years have more chances of being cited and thus may have higher citation counts. For this reason, comparison has to be done in the same year so as to exclude the different time effect. Figure 5.9 also shows that the gap between the West and the East in terms of CPP is narrowing over time. For example, the CPP in the West and the East were respectively 26.4 and 26.6 in 2012 and in 2003 were respectively 356.6 and 305.5. No gap exists in 2012 but in 2003 the difference was 51. Nevertheless, it would be too imprudent to conclude that the gap between the West and the East has disappeared in 2012 because it takes time for citation information to be published and indexed in a database like the WoS. In general, it needs time to prove whether difference exists.

Leading Players in International Collaboration

Over the past decades, the number of internationally collaborated publications have increased significantly (e.g., He 2009; Leydesdorff and Sun 2009; Zhou and Glänzel 2010; Mattsson et al. 2010). Co-authorship in publications is widely considered as a reliable proxy for scientific collaboration (Franceschet and Costantini 2010). Various indicators have been proposed for measuring international collaboration. For instance, Garg and Padhi (2001) proposed Domestic Collaborative Index (DCI) and International Collaboration Index (ICI), Zhou and Tian (2014) classify international collaboration into domestically-led and internationally-led so as to further identify possible variations between the two types of collaboration. In exploring determinants of research collaboration Plotnikova and Rake (2014) made two conclusions: (1) geographical distance is negatively related to the intensity of international research collaborations, and (2) cognitive proximity, institutional proximity, social proximity, and cultural proximity have a positive relation with the intensity of international collaboration. In the current chapter, the Activity Index (AI) suggested by Frame (1977) and elaborated by Schubert and Braun (1986) will be used to explore bilateral collaboration based on institution names in the byline of papers.

Based on Salton's measure (Glänzel and Winterhager 1992), bilateral collaboration strength r_{ik} can be measured by the following formula:

$$r_{ik} = \frac{n_{ik}}{\sqrt{n_i \times n_k}} r_{ik} \in (0,1)$$

where n_{ik} is the number of co-authored papers with addresses of two countries with n_i and n_k total publications respectively. It is clear that r_{ik} is a relative indicator for measuring international collaboration.

In 2003–2012, the United States was the leading country in bilateral collaboration among the other seven countries, with major partners from the West (i.e., England, Germany, and France). In collaboration with the four Eastern countries, the United States collaborated mostly with Japan, with China as the second partner. For England, however, distance still matters in terms of determining partners: European countries (Germany and France) are significantly more important partners of England than Eastern countries. Western countries are the first collaboration options of Eastern countries. Unlike collaboration between European countries, bilateral collaboration between the Eastern countries is significantly weak (Table 5.4).

Table 5.4 Salton's measure for bilateral collaboration amongst the eight countries 2003–2012.

	China	*Japan*	*South Korea*	*India*	*United States*	*England*	*Germany*	*France*
China	—	0.075	0.071	0.064	0.113	0.069	0.069	0.059
Japan	0.075	—	0.083	0.059	0.117	0.093	0.102	0.092
South Korea	0.071	0.083	—	0.095	0.095	0.054	0.064	0.064
India	0.064	0.059	0.095	—	0.058	0.059	0.067	0.065
United States	0.113	0.117	0.095	0.058	—	0.196	0.190	0.160
England	0.069	0.093	0.054	0.059	0.196	—	0.210	0.210
Germany	0.069	0.102	0.064	0.067	0.190	0.210	—	0.226
France	0.059	0.092	0.064	0.065	0.160	0.210	0.226	—

Source: Authors' elaboration on data from ESI.

Conclusions and Discussion

In R&D investment, China and South Korea advance the fastest in terms of absolute R&D expenditure, whereas South Korea and Japan take the lead measured by a relative indicator (i.e., percentage share of R&D expenditure to GDP). Regarding human resources in R&D, China and South Korea also show the fastest growth. South Korea even takes the lead in terms of full-time equivalent researchers per thousand employed among the eight countries under study. The ambition of the East in science is clearly shown with South Korea's performance the most outstanding among the four Eastern countries. The advancing momentum of China is also impressive. Although China has the least value of full-time equivalent researchers per thousand employed among the eight countries, its huge population implies a potential of enlarging its R&D reservoir of manpower. As a country developed earlier than the other Eastern countries, Japan seems to have reached its extreme in R&D investment. Compared with the Eastern countries, growth momentum of R&D investment in the West is weaker. An important reason for the different momentum in R&D investment between the West and the East is the different starting point. The West has entered a relatively stable stage whereas the East, except Japan, is still in its catching-up stage.

In the total volume of publications in 2003–2012, the West contributes significantly higher than the East with the United States taking the absolute lead. Looking into the annual publications, however, the rapid growth of publications from China, together with South Korea and India, has significantly disturbed the balance of world publication, with the United States and Japan fast losing their world share. In terms of production of highly cited publications, however, the disturbance caused by the rise of China, South Korea, and India is less fierce. In other words, the gap between the West and the East is still huge. In 2003–2012, over 1.1% of WoS publications produced by the West are highly cited, whereas the ratio of the East was only 0.4–0.6%. The United States again takes the absolute lead in producing highly cited publications. After the United States are the United Kingdom and Germany. In fact, all the eight countries have produced a growing number of highly cited publications.

Field distribution of highly cited publications shows strong and balanced development among fields in the West. In the East, however, this is not the case. In psychiatry/psychology, social sciences, neuroscience and behavior, economics and business, clinical medicine, as well as microbiology, the East is extremely weak. Only in materials science, engineering, and chemistry, does the East perform relatively better, but with significantly lower share than the West. The situation in terms of publication impact is also the same.

In international collaboration, the United States has strong bilateral relations with the other countries except India. Japan is the United States' most favorite partner in the East. Collaboration between European countries is the strongest, which may be partly attributed to the EU Framework Programme. Collaboration between the Eastern countries is rather weak compared with that between the Western countries. Western countries are the first option of the East in selecting collaboration partners, partly because of the leading role of the West in science. Collaborating with the West brings more benefit than collaborating with countries from the East. With the popularization of international collaboration as well as the rising capacity of the East, both the publication productivity and impact of the East is expected to increase. The current weakness of East-East co-authorship, however, is expected to continue. In other words, it will take time for significant change to occur in East-East collaboration.

To conclude, world science is still led by the West with the United States taking the absolute lead. The East has been emerging with China and South Korea the most dynamic countries. Nonetheless, the scientific impact of the East is not yet as remarkable as it shows in R&D investment and publication production. The hysteretic effect between publishing and citation may partly explain such a phenomenon. Japan seems to have reached its potential and India has not yet woken up. To quicken the catching-up steps of the East, collaborating with the West can after all be an important option.

References

Frame, J.D. 1977. "Mainstream Research in Latin America and Caribbean." *Interciencia* 2: 143–148.

Franceschet, Massimo, and Antonio Costantini. 2010. "The Effect of Scholar Collaboration on Impact and Quality of Academic Papers." *Journal of Informetrics* 4(4): 540–553.

Garg, K.C., and P. Padhi. 2001. "A Study of Collaboration in Laser Science and Technology." *Scientometrics* 51(2): 415–427.

Glänzel, Wolfgang, and Matthias Winterhager. 1992. "International Collaboration of Three East European Countries with Germany in the Sciences, 1980–1989." *Scientometrics* 25(2): 219–227.

Glänzel, Wolfgang, Koenraad Debackere, and Martin Meyer. 2008. "'Triad' or 'Tetrad'? On Global Changes in a Dynamic World." *Scientometrics* 74(1): 71–88.

He, Tianwei. 2009. "International Scientific Collaboration of China with the G7 Countries." *Scientometrics* 80(3): 571–582.

ISTIC. 2013. "Statistical Data of Chinese S&T Papers." Institute of Scientific and Technological Information of China, http://www.istic.ac.cn/portals/0/documents/kxpj/2%E5%9B%BD%E9%99%85.pdf (accessed December 21, 2014).

Leydesdorff, Loet, and Yuan Sun. 2009. "National and International Dimensions of the Triple Helix in Japan: University-industry-government versus International Coauthorship Relations." *Journal of the American Society for Information Science and Technology* 60(4): 778–788.

Leydesdorff, Loet, and Ping Zhou. 2005. "Are the Contributions of China and Korea Upsetting the World System of Science?" *Scientometrics* 63(3): 617–630.

Mattsson, Pauline, et al. 2010. "What Do European Research Collaboration Networks in Life Sciences Look Like?" *Research Evaluation* 19(5): 373–384.

National Science Foundation. 2013. "Definitions of Research and Development." http://www.nsf.gov/statistics/randdef/fedgov.cfm (accessed December 21, 2014).

OECD. 2013a. "Expenditure on R&D." In *OECD Factbook 2013: Economic, Environmental and Social Statistics*. Paris: OECD.

OECD. 2013b. "Researchers." In *OECD Factbook 2013: Economic, Environmental and Social Statistics*. Paris: OECD.

Plotnikova, Tatiana, and Bastian Rake. 2014. "Collaboration in Pharmaceutical Research: Exploration of Country-Level Determinants." *Scientometrics* 98: 1173–1202.

Robbins, Carol A., and Carol E. Moylan. 2007. "Research and Development Satellite Account Update." *Survey of Current Business* 87(10): 49–64.

Schubert, András, and Tibor Braun. 1986. "Relative Indicators and Relational Charts for Comparative Assessment of Publication Output and Citation Impact." *Scientometrics* 9: 281–291.

World Bank. 2013. "China." http://www.worldbank.org/en/country/china (accessed December 21, 2014).

Zhou, Ping, and Huibao Tian. 2014. "Funded Collaboration Research in Mathematics in China." *Scientometrics* 99(3): 695–715.

Zhou, Ping, and Wolfgang Glänzel. 2010. "In-depth Analysis on China's International Cooperation in Science." *Scientometrics* 82(3): 597–612.

Zhou, Ping, and Loet Leydesdorff. 2006. "The Emergence of China as a Leading Nation in Science". *Research Policy* 35(1): 83–104.

Zhou, Ping, and Loet Leydesdorff. 2008. "China Ranks Second in Scientific Publications since 2006." *ISSI Newsletter* (13): 7–9.

Part II The Globalization of Technology and Innovation

Chapter 6

Innovation, Internationalization, and the Transnational Corporation

Grazia Ietto-Gillies

Introduction[1]

Innovation and internationalization are closely linked. Causality between them can go in either and, indeed, in both directions in a dynamic sequence. There are strong theoretical arguments why causation could go either way: from innovation to internationalization and/or from internationalization to innovation. It is, in fact, very likely that the two phenomena are linked by a cumulative causation mechanism. At the micro level, more innovative firms can better compete and thus become more internationalized. At the macro level, countries with a high innovation base will have a better performance on their international business activities compared to less innovative countries.

Conversely, internationalization can impact positively on innovation by setting in train a variety of behavior and learning mechanisms. At the ex-ante level, firms that plan to become involved in foreign markets, be it via exports or foreign direct investment (FDI), are likely to try and improve their innovation performance (Wagner 2007). Moreover, internationalized firms and countries are exposed to diverse cultures, knowledge, and innovation environments from which they can learn. The ex-ante strategic behavior as well as the possible ex-post learning from international activities, lead to improved innovation performance. Through time these processes are likely to cumulate, leading to the enhancement both of firms' innovation and of internationalization performances for firms as well as countries. A virtuous circle may set in, such that innovative firms and countries can compete successfully in international markets. Similarly the process could become vicious, with poor innovation performance affecting negatively the internationalization performance and the latter, in turn, impacting negatively on innovation.

The Handbook of Global Science, Technology, and Innovation, First Edition.
Edited by Daniele Archibugi and Andrea Filippetti.

From the beginning of the twentieth century, a particular type of institution – the multinational company (MNC) – has increasingly been involved in the international business activities of countries. As will be argued below, the very existence of the MNC is due to technological and organizational innovations. The involvement by MNCs in a variety of international activities has become very relevant since the Second World War. Indeed, in the last three to four decades the transnational company (TNC) has come to dominate international business activities. Moreover, its involvement has led to new modalities of activities and it has increased the geographical scope of such involvement. It has also led to the internationalization of the production process. These are the reasons why, in analyzing the connection between internationalization and innovation, this chapter focuses mainly – but not exclusively – on the TNC.

The chapter proceeds as follows. The next section introduces the MNC and discusses the role of technological and organizational innovation in its emergence and development. This is followed by a discussion of the close involvement of TNCs in innovation activities with specific reference to the internationalization of the production process and to the role of the TNCs' networks in the diffusion of knowledge and innovation. The chapter then considers two specific modalities of internationalization: exports and imports and their relationship to innovation. They are the oldest modes of cross-border activities and their existence well pre-dates the birth of TNCs. In contemporary economies trade can be undertaken by all sorts of enterprises and not just TNCs. Nonetheless the latter are the institutions responsible for the largest share of world trade. A section is devoted to issues related to the possible extension from the micro to macro economy in considering the impact of internationalization on innovation. The last section summarizes and concludes.

Innovation and the Transnational Corporations

The transnational is a corporation that owns assets and operates *direct* business activities in at least two countries. There are many adjectives and nouns designating a company with direct business activities in several countries. The most used nouns are corporation or company though firm and enterprise are also used. The most used adjective in common parlance as well as in academic works is "multinational"; "international" is also used often in the literature. The term transnational used in this chapter is often preferred by researchers – including those working at the UNCTAD[2] – because it best represents the ability that these companies have developed since 1945 to plan, manage, and control across borders. This is an ability that gives them power vis-à-vis actors that cannot do so, or not to the same extent. Among the latter are: labor; small and medium size suppliers; and governments (see Ietto-Gillies 2012: ch. 14).

Direct investment and related production is a relatively new mode of conducting business activities across countries. The oldest modality of cross-border business activity, going back millennia rather than centuries, is trade[3] (i.e., imports and exports). Companies with direct business activities across borders have existed for the last few centuries: from the Medici Bank in fifteenth-century Florence to the Dutch and British trading companies dating back some four centuries. Steven Hymer, the researcher who first developed a theory of "The International Operations of

National Firms" (1976; originally written 1960), disagrees with the view that any of these institutions can be taken as the real precursor of the modern TNC. He writes:

> But neither these firms, nor the large mining and plantation enterprises in the production sector, were the forerunners of the multinational corporations. They were like dinosaurs, large in bulk, but small in brain, feeding on the lush vegetation of the new worlds (the planters and miners in America were literally *Tyrannosaurus rex*). (Hymer 1971: 115–116)

Hymer, following Chandler (1962) sees the forerunner of the international firm in the nineteenth-century joint stock company. The first half of the twentieth century saw the growth of direct business across borders of firms from developed countries investing in developing countries – often colonies – to secure raw materials. However, the last seven decades have seen a huge change not only in the number of TNCs but also in the quantum and quality as well as in the industrial and geographical pattern of these operations. The modern transnational corporation (TNC) has developed after the Second World War.

How do innovation and transnationality interact? What is the position of TNCs – these most internationalized of all business actors – in relation to innovation development and diffusion? There are several ways in which transnationality and innovation interact and, in particular, the following.

For a start, the very existence and development of the modern TNC is the product of innovation. Then there is the interaction of innovation with a variety of TNCs' activities and, in particular, with the various modalities of internationalization such as FDI and trade. There are also a variety of collateral issues considered in the literature from the contribution of TNCs to R&D to the location of R&D laboratories to the growth of partnerships in R&D. These are not discussed in this chapter.[4]

Innovation is at the very basis of the existence and growth of the modern transnational corporation. Two types of related innovations have given the TNC the ability to plan, manage, and control across borders and across space: (a) technological innovations in communications and transportations; and (b) organizational innovations. The reason why these two types of innovation are so important has to do with control.

The distinguishing way of doing business abroad, the one that characterizes the transnationals compared with other companies, is *direct* production and generally direct business activities abroad. In order to engage in these direct activities, the TNCs establish affiliates abroad and acquire the ownership and control of their assets via FDI. This gives them a long-term interest in the strategies and management of the foreign enterprises which they control.

In the context of direct activities abroad by transnational corporations, control has two connotations. First is its relationship with the equity stake in the foreign enterprise. What percentage of the foreign assets must be owned by the main company for it to have control? This issue is far from simple because there is no single percentage of ownership – other than the 50+ percentage – that can definitely ensure control to a single owner or a group of associate owners over the company whose shares are being acquired. It all depends on how widespread the ownership is. Whenever ownership is very widespread even a relatively small percentage of ownership may be enough to give control. This is not the case for companies in which the ownership is very concentrated. The concentration of ownership varies across companies, industries, and countries. The International Monetary Fund (IMF) guidelines set a minimum of 10 percent of share ownership for the main company to be considered to have control (IMF 1977).

But is a controlling share of the equities enough for the main company to exercise control? Equity control is a necessary condition but not a sufficient condition to ensure control. Equity control by itself does not lead to strategic managerial control if the means of exercising such control are not available. Thus the second connotation of control relates to managerial and strategic control. The latter type of control is not possible if the system of communications and the organization of the business across countries are not suitable for the exercise of managerial control. This was indeed the case of much foreign business prior to the First World War. Technological innovation in communications from the wireless onward and in transportation technologies is one of the keys to the development of the modern TNC. The other one relates to organizational innovation.

Organizational innovation plays a key role in the ability of headquarters to control at a distance. Several authors and in particular Chandler (1962) and Williamson (1975, 1981, 1984) analyze how the firm's strategic objectives – be they growth or efficiency led – drive changes in internal organizations. Historically, such organization has moved from the "unitary form" dominated by the single owner and overseer of all activities to a departmentalized structure – in firms operating large manufacturing projects, particularly the building of railways, during the nineteenth century – to a multidivisional structure. The latter gave the flexibility to move from product division to geography-based divisions and, later, to mixed products and geography divisions. The flexibility for moving into different countries was therefore created as organizational constraints to multi-country expansion were removed.

In conclusion, the modern transnational corporation is characterized by both the equity ownership/control and the ability to manage strategically the foreign affiliates at a distance. The latter characteristic is the product of two relevant and interconnected innovations both of which form the *sufficient conditions* for the exercise of control: first, the technological innovation in personal communications which started with the telegraph and telephone and led, more recently, to electronic communications; and second, organizational innovations which were made possible (or strongly facilitated) by the communication technologies.

Transnationality and Innovation

TNCs, Innovation, and the Internationalization of the Production Process

Technological and organizational innovations are therefore at the basis of the very existence and growth of the modern TNC. However innovation affects and is affected by TNCs' activities in many other ways. To understand the complex relationship between innovation and TNCs let us look first at the range of activities TNCs are involved in. International business activities can take a variety of routes and modalities. The main modalities are: trade – both imports and exports – and foreign direct investment (FDI) – both inward and outward.[5] Non-equity contractual relationships such as joint ventures and sub-contracting activities are also important internationalization modalities. The TNCs are the main actor in all these modalities. However, the primary and distinguishing activities of TNCs is foreign direct investment. UNCTAD (2002: Annex B, 275)[6] writes on this:

> *Foreign direct investment* (FDI) is defined as an investment involving a long-term relationship and reflecting a lasting interest and control by a resident entity in one economy (foreign direct investor or parent enterprise) in an enterprise resident in an economy other than that of the foreign direct investor (FDI enterprise or affiliate enterprise or foreign affiliate). FDI implies that the investor exerts a significant degree of influence on the management of the enterprise resident in the other economy.

Though FDI is the main and defining business activity by TNCs it is not the only one. Trade – both imports and exports – figures very prominently. In fact, TNCs are responsible for some 80 percent of world trade. Moreover, a third of world trade is intra-firm trade, trade that is internal to the firm and external to countries (UNCTAD 2013: 135–136, Fig. IV.14 and Box IV.3). TNCs are responsible for all intra-firm trade, that is, for all the exchange of goods and services between a unit of a TNC – say headquarters or an affiliate – and another unit located in a different country. The growth in intra-firm trade is part of the expansion of TNCs and their activities as well as part of innovation in their location strategies made possible by technological innovations. The digital technologies have made possible the development of vertically-integrated international production or globalized value chains.[7] This means that the production process is split into segments according to a variety of elements such as the degree of skills required, the relative cost of labor and the cost of transportation. The strategy is to locate segments requiring cheap low-skills labor in developing countries and those requiring high-skill and high-cost labor in developed countries. One outcome of this process is the movement of components from country to country for further processing. This movement often takes place within units of the same TNC and is, therefore, intra-firm. Increasingly international vertical integration involves outsourcing to firms with which the links are contractual rather than equity. In such cases the resultant international trade may be inter-firm.

The overall strategy of international vertical integration of the production process has been applied largely to manufacturing from the 1960s onward and has led to the so-called New International Division of Labor (NIDL).[8] This describes the fact that large segments of manufacturing are located in developing countries, contrary to before the Second World War when the direct foreign investment was resource-seeking and manufacturing was located almost exclusively in developed countries.

More recently the digital information and communication technologies (ICTs) have led to further strategies which affect mainly disembodied, service products. The ICTs have, in effect, made possible the development of a new internationalization mode in which service components of a final product in the form of data or documents can be made available in real time to a distant location – whether in the same country or in a foreign one – at a very low cost. We are talking about the processing of data and documents relating to accounting, booking, or the editing of manuscripts for publishing companies. This new form is neither international production nor trade of the traditional type (Ietto-Gillies 2002). It is a new modality of international activity made possible by the ICTs. Nonetheless, some FDI is likely to take place in order to set up the business structure – as either a single or joint venture with a local partner – that will carry on the processing activity.

Both internationalization strategies just described relate to the production process in manufacturing (the first case) and in services (the second case). However, the internationalization of activities by TNCs is increasingly affecting other functional activities from R&D to Sales and Marketing to Business Services (Crescenzi, Pietrobelli, and Rabellotti 2013).

Strategies of international vertical integration – whatever the function or the sector involved – lead to increase in both FDI and trade; they are strategies that generate complementarities between FDI and trade. Resource-seeking FDI also stimulates trade and thus generates complementarity between these two modalities of international business. If a petrol company from a developed country invests abroad in an oil-rich country, its FDI generates trade as the oil is exported to another country for refining and eventually from the latter to several other countries for final consumption.

Nonetheless, the relationship between international production set in action by FDI and trade is not always one of complementarity.[9] Whenever FDI aims at producing abroad for the local market (or/and for exports from the host country to others), the relationship between exports and FDI is likely to be one of actual or potential substitution: foreign markets can be sourced via exports from the home country or via direct production abroad. By choosing to source them via direct production, the company substitutes FDI for actual or potential exports.[10]

Multinationality and Knowledge Diffusion

The impact of innovation on the internationalization activities of TNCs – and particularly on their FDI and exports – is well documented and figures directly or indirectly in several theories of FDI and/or the TNC (Vernon 1966; Dunning 1977, 1980; Cantwell 1989). Innovation within TNCs is likely to affect positively both exports and outward FDI (UNCTAD 2002): those TNCs that innovate can successfully compete in both exports and FDI activities. However, the TNCs' link to innovation is also one of reverse causality: one in which the TNC and its internal organization and its wide geographical reach play a big role in the development and diffusion of innovation.

There is a specific question to be asked regarding the impact of multi- and transnationality vis-à-vis innovation. Are TNCs in a special position regarding knowledge and innovation? In other words does *multinationality per se* and the fact that TNCs operate in many countries have an effect on the development and diffusion of knowledge and innovation? The theoretical underpinnings to the links between the activities of TNCs and innovation can be found in the evolutionary theory of the firm (Nelson and Winter 1982; Nelson and Rosenberg 1993). This theory led to developments and applications to the TNCs in which the behavior and performance of the latter is linked to their capability for the development, absorption, and diffusion of innovation activities (Cantwell 1989; Kogut and Zander 1993, 2003).[11] These works, as well as those related to networks theory (Forsgren, Holm, and Johanson 2005; Hedlund 1986; Bartlett and Ghoshal 1989; Ghoshal and Nohria 1997; Hedlund and Rolander 1990) rely considerably on the role of TNCs' networks for knowledge transmission.

Two types of networks involving the TNC are discussed in the literature and in relation to knowledge diffusion:[12] internal and external networks. A TNC's *internal network* is constituted by headquarters and all the affiliates, many of them scattered in various countries of the world characterized by diverse business and organizational cultures as well as by diverse innovation environments. Each unit of the TNC can transmit and receive knowledge to/from other parts of the company via the internal network. The mobility of managerial and technical staff across units of the TNC supports knowledge transmission.

Moreover, each unit is part of various *external networks* within the environment in which it operates. These networks range from contacts with customers or suppliers or distributors and other business partners or local universities and research centers. The range and extent of external networks the TNC is involved in vary according to the types of modality it uses to operate in foreign countries: from FDI to trade to licensing or franchising to sub-contracting to joint ventures. TNCs are likely to use different modalities for different activities and/or host countries. Whichever the modality, the external networks it gives rise to can become channels for the acquisition of knowledge whose diffusion across the various units of TNCs and the countries in which they operate will then be facilitated by the company's internal networks.

In this scheme, each unit of the TNC acquires knowledge from its environment and then transmits all or some of it to other parts of the company often located in other countries. Moreover, knowledge from the unit – whether it is self-generated or acquired via the internal or external networks – spills over to the local environment via the same transmission mechanisms which led to the acquisition of knowledge by the unit. The transfer of knowledge and innovation can operate in both directions: the firm transfers to the local environment and receives from it. The transfers can be deliberate and planned or they can be accidental.

Various questions emerge in relation to this framework. How is knowledge transferred within internal or external networks and between them? What are the mechanisms that allow/facilitate the transfer? What are the possible constraints to internal or external knowledge transfer? As regards the first question we must distinguish between *codified* and *uncodified – tacit – knowledge* (Polanyi 1966, 1967). The former is the type of knowledge that can be written down in clear instructions and codes, thus transmittable to people in different localities and countries. Tacit knowledge cannot be written down in clear instructions because much of it is embodied in what workers do in their everyday tasks and in the way they work together as a group. The latter point forms the basis of Kogut and Zander's (1993) analysis of the TNC as a social community.

People and their expertise are therefore essential for the development, use, and transfer of both types of knowledge and the innovations linked to it. In the case of uncodified knowledge, transmission via people is the only effective way for it to spread from one business unit to another. The mobility of skilled labor – be it managerial or technical – is an excellent vehicle for knowledge and innovation transfers. Such mobility can take place on an intra- or intercompany basis via internal or external labor markets. In the former the skilled labor moves from subsidiary to subsidiary within the same country or between different countries. However, labor mobility – sometimes on a temporary basis – is not the only mechanism for the transmission of knowledge and innovation. The mobility of products, be they manufactures or services, equipment or final consumption, can also be facilitators of learning.

As regards constraints to – and facilitators of – the effectiveness of internal and external networks in knowledge diffusion, the following elements are relevant. First, the degree to which the subsidiary of the company is embedded in the locality. The degree to which knowledge spills over from the unit of the TNC to the local environment depends on the strength of its external networks, and, thus, on the degree of embeddedness[13] of the unit in the locality. For example, to what extent are its suppliers chosen from the local community? If they are, this facilitates the spillover process as the suppliers learn from the company while the company learns about the knowledge environment and about the requirements of local markets and production facilities. However, Uzzi (1997) finds that, beyond a certain threshold, embeddedness can have negative effects on performance and knowledge transmission by insulating the business unit from information external to the local environment, and, therefore, making it more vulnerable to external shocks.

Second, the *degree of autonomy* that the subsidiary has in dealing with the local environment, for example in dealing with suppliers and distributors. A relatively decentralized structure with little control from the center may favor local embeddedness and therefore external knowledge spillovers. However, a more centralized structure may favor the internal transfer of knowledge.

What is the evidence for positive causality link from FDI to innovation? Wagner (2007) and ISGEP (2007) – further discussed in the next section – finds evidence of self-selection mechanisms by exporters which seem to be valid also for the outward FDI of TNCs and, indeed, in a stronger way. This means that firms that plan a strategy of internationalization via exports and specially via FDI engage in innovation activities. In terms of innovation performance – proxied by productivity – there appears to be a hierarchy in relation to international activities: exports are associated with productivity increase; FDI have also this association at a higher level of performance; firms that engage in both exports and FDI appear to be the best performers of all (Kimura and Kiyota 2006). These results are compatible with the fact that exports and FDI tend to be complementary more than substitute as discussed above and this may lead to cumulation in any mechanism of knowledge diffusion.[14]

International Trade and Innovation

This section deals more specifically with the relationship between international trade and innovation. As already mentioned, most trade is now the responsibility of TNCs. Nonetheless many national companies contribute to trade. Indeed export is often the first modality of internationalization by companies that had, so far, operated only on the domestic arena. They are usually small to medium size companies and often they may be introduced to international markets and production locations via initial collaboration with a larger company that is already transnational. Once companies learn about internationalization via trade, they may branch out into other modalities.[15]

The impact of innovation on internationalization has been explored in various studies. Posner (1961) and Hufbauer (1966) found that trade performance and, specifically, exports are related to the technological gap between countries. Posner's work formed the background to Vernon (1966) in which the innovation performance of firms and countries determines their exports performance, then – in a time sequence – their propensity to FDI and, eventually, both their exports and imports propensities.[16]

The impact of trade on innovation performance is less intuitive and less researched. The evidence for these effects is also more problematic. For these reasons this section concentrates on the latter effects: the causality from trade to innovation and, in general, the impact of trade on the international diffusion of knowledge and innovation.[17] Let us start with an analysis of exports and then go on to consider imports.

Exports

The possible impact of exports on innovation has been the subject of many studies. In 1995 Bernard and Jensen published a paper which was to have a great follow-up for many years to come and still has. Using a large database of US official statistics, the authors give a systematic analysis of the performance of exporters versus non-exporters. One of the performance variables used was productivity and this was taken from then on – and in similar researches – to be a proxy for innovation. Since then, many studies have been devoted to unpicking the role of internationalization via imports, exports, or FDI on innovation. Most of these studies are at the micro level and refer either to firms or to establishments/ plants.[18] The wealth of studies, their analogies and differences, have led to meta studies. Wagner (2007) analyses 54 pieces of research related to 34 countries and finds definite evidence of a positive relationship between exporting and innovation proxied by productivity.

Most studies use productivity growth and productivity premia between exporters and non-exporters to reach conclusions on the impact of internationalization modes on innovation. There are considerable problems with using productivity as a proxy for innovation.[19] Even allowing for correspondence between productivity increase and innovation, we must take account of the fact that the relationship between exporting and productivity/innovation can come in a variety of ways. The first one is self-selection: the more productive/innovative firms become exporters because they can better compete. Moreover, firms that plan to become exporters will invest and innovate in preparation for entering foreign markets. Thus exporting affects innovation in an ex-ante way and via the strategic behavior of firms. This pre-entry innovation performance leads to a positive relationship and to a direct link between exporting and innovation.

A different link can come about via the process of learning-by-exporting (post-entry performance). Firms learn from contacts with customers and competitors and they also benefit – in terms of productivity increases – from the larger scale of activity made possible by exporting. In the 54 studies analyzed by Wagner (2007) there is evidence for all three types of positive relationship between exports and productivity with the third one (learning-by-exporting) less strong than the other two. Wagner is cautious about his conclusions, which were followed up by a comparable study[20] of 14 countries (ISGEP 2007). In this study the export premia between exporters and non-exporters are found to be consistently positive and high. They appear to be unrelated to the level of development of the exporting country. The self-selection hypothesis is confirmed; the learning-by-exporting hypothesis is less so. However, on the latter issue the conclusions cannot be fully accepted because there are problems with the paucity of data.

Imports

Let us now consider the imports component of internationalization. Firms can learn through importing in a variety of ways. Access to foreign suppliers provides access to specialized intermediate and capital goods; importing firms may have to adapt their own equipment to be able to fully use – and benefit from – the imported ones. This requires investment. The human resources component is also crucial. Whether learning is via internal or external processes (Iammarino and McCann 2013: ch. 4; Malerba 1992), investment in staff training may be necessary. There may also be elements of reverse engineering as importing firms may try to find out how the new product was arrived at.

The relevance of imports for learning is explored in several studies. MacGarvie (2006) analyzes patents citations of French firms and finds that the importing firms are more likely to be influenced by the technology of the country they import from, than firms that do not engage in imports from that country. The effect is less significant for firms that export: exporting firms do not cite significantly more patents from their destination country. Liu and Buck (2007) finds evidence that firms learn by importing. Grossman and Helpman (1991) highlight the fact that the import of technology acts as a channel through which domestic firms may "reverse engineer" the products of their foreign rivals.[21]

Damijan and Kostevc (2010) study the sequence "imports – innovation – exports – innovation" in a sample of Spanish firms. Theirs is one of the few studies that do not rely on productivity as a measure of innovation performance; they use data on product and process innovation. There is strong evidence of the sequence running from imports to innovation and then to exports. The sequence from exports to innovation to imports is less strong. Evidence of a positive impact from exports to innovation in both products and processes is in Damijan, Kostevc, and Polanec (2010). Bertschek (1995) in a sample of micro German data finds that imports and inward FDI have significant positive effects on product and process innovation. The stronger competition from foreign firms may encourage innovation in domestic firms.[22] Thus the positive impact of imports on innovation may be linked to learning mechanisms or to behavioral ones.

From Micro to Aggregate Effects

Most of the studies that analyze the impact of internationalization on innovation rely on firms' level data; some relate to establishments or plants; a few to industry level. There are also studies relating to countries. At the meso and macro levels the extent to which internationalization – whichever the modality – affects innovation depends on the following. (a) The size of imports or exports or FDI and their immediate effect on the firms directly involved in them. (b) The extent to which the effects spillover from one firm to others and from public research institutes – such as universities – to firms. (c) The type of spillovers, that is, whether they relate to production or to market access knowledge. Both affect productivity – via new production methods in the first case and via market/scale extension in the second – though only the first one leads to innovation. In other words, innovation and its spillovers lead to productivity increase. However, productivity increases should not be taken as evidence of innovation and possible spillovers from it.

Learning is not automatic or inevitable. Whether it occurs or not depends not only on the knowledge endowment of the institution – be it firm or university or research institute – or country and their ability/willingness to let their knowledge leak away from their boundaries. It also depends on the capacity of the potential receivers, that is, on the extent to which the latter have the relevant absorptive capacity and the appropriate learning culture (Laursen and Salter 2006; Gunawan and Rose 2014).

Blomstrom, Globerman, and Kokko (1999) review the literature on determinants of host country spillover from FDI.[23] They list the main determinants as technological complementarities between home and host country and a series of elements related to the host country such as: strength of intellectual property (IP) rights; competition; size and wealth of host country; technical competence; government policies. Spillovers tend to be higher when the technological gap between the trading countries is low; this means that countries at a similar stage of technological development can better absorb foreign knowledge. It could also be argued that countries may learn more from those with higher knowledge frontiers than themselves. Whether this is the case or not may depend on whether the receiving country has the necessary absorptive capacity (AC) to capture knowledge.

The concept of AC goes back to Cohen and Levinthal (1990).[24] The AC of the locality is usually seen as the main determinant of the ability of firms, sectors, or countries to capture spillovers. The AC has many dimensions ranging from the human skills required to use and develop innovative products and processes to the physical infrastructure and to the knowledge infrastructure related to the build-up of specific knowledge and research capacity. The AC is likely to be the result of cumulative processes; the persistence of research and innovative activities are likely to have a positive impact on the AC at any given point in time.[25]

In relation to TNCs and their activities, high AC allows the locality to take advantage of spillovers from local subsidiaries; indeed, it may be a condition for spillovers from a foreign TNC to local firms to take place. Many components of AC – be they in relation to knowledge or to human resources or to physical infrastructure – may also become attractive locational advantages which encourage foreign firms to invest in that particular area.[26] Agglomeration effects set in motion cumulative processes. Therefore the initial location advantages in terms of elements of AC may lead to further location advantages[27] and to the enhancement of AC.

Summary and Conclusions

This chapter deals with the relationship between innovation and internationalization focusing, in particular, on the TNCs and their activities. The TNCs are, currently, the main actors in international business activities. The chapter considers issues of causality between internationalization and innovation and issues of both development and diffusion of knowledge and innovation. The historical trajectory and current organization and activities of TNCs are discussed. The chapter then considers the relationship between innovation and TNCs. It traces the role of innovation in the very existence and birth of the TNC; in the contemporary international fragmentation of the production process and the related impact on trade and on the international division of labor. The organization of TNCs and the related internal and external networks are of relevance for the diffusion of knowledge across countries.

The chapter then analyzes the role of innovation on trade with particular emphasis on the impact of trade – both exports and imports – on innovation. Lastly, there is a discussion of problems and issues related to the application of analysis of impact of internationalization on innovation from the micro to the macro sphere. The passage from one to the other involves analysis of possible spillovers whose capture requires relevant absorptive capacity at firm, industry, and country levels.

Innovation and internationalization are both old concepts. However they are also very new and present specific characteristics in the contemporary era. This is now dominated by the activities of TNCs. The domination is in terms of number of modalities with which they operate; depth, quantity, and relevance of such activities; and in terms of geographical extension. Moreover, it is in terms of the TNCs' ability to plan, organize, and control across borders to an extent not yet achieved by other economic actors. All these elements point to a position of dominance in today's world by TNCs in world economics, politics, and power. Studying the links between – and the impact of – TNCs and their activities on innovation is therefore important for a proper understanding of contemporary economic systems. Some of the theoretical and empirical works cited above are not fully conclusive, interesting though they may be. There is a great need for further research in the field and I hope this chapter encourages young researchers to contribute to the advancement of knowledge in it.

The analysis and conclusions in this chapter point also to possible policy implications and specifically toward those policies that facilitate learning from different business and innovation environments and increase the absorptive capacity of countries and regions.

Notes

1 Some parts of this chapter draw on Ietto-Gillies (2012: chs. 1 and 17) and Filippetti, Frenz, and Ietto-Gillies (2013b).

2 The Geneva-based United Nations Conference on Trade and Development has had for many years a section researching into transnational companies and their activities, particularly in relation to policy issues toward the developing countries. The name of the relevant section has changed through the years. Currently it is the Division of Investment and Enterprise that conducts research on the TNCs. They publish several journals and reports including the annual *World Investment Report* and the quarterly *Transnational Corporations*.

3 The expression "international" trade should, strictly speaking, be used only since the advent of nation-states.

4 On the first of these issues see UNCTAD (2005) and Narula and Zanfei (2004); on R&D location, Patel and Pavitt (1994). The growth of research related inter-firm partnerships is in Hagedoorn (1996) and Cohen (1995).

5 A main component of cross-border transactions is portfolio investment. This chapter deals with issues related to the real sector and, therefore, portfolio investment is not considered. Though digital innovation has been one of the key elements in the global growth of portfolio investment, the overall impact of portfolio investment on innovation in the real sector is not substantial.

6 The UNCTAD definition of FDI is based on IMF (1993) and OECD (1996).

7 See UNCTAD (2013) for a detailed analysis and empirical information on Global Value Chains (GVCs). The theory of vertical and horizontal integration is developed in Caves (1971).

8 See Frobel, Heinricks, and Kreye (1980).

9 On the relationship between trade and FDI see Cantwell (1994).

10 It may, however, be the case that the international market could only be developed via a direct production presence into the host country and not via exports.
11 Cantwell's and Kogut and Zander's works are summarized and commented on in Ietto-Gillies (2012: ch. 11).
12 Several works on innovation have used the concepts of internal and external networks. See Castellani and Zanfei (2004, 2006); Frenz, Girardone, and Ietto-Gillies (2005); Frenz and Ietto-Gillies (2007, 2009); Filippetti, Frenz, and Ietto-Gillies (2011, 2013a); Zahra, Ireland, and Hitt (2000).
13 On the concept of embeddedness see Granovetter (1985).
14 See also Javorcik (2004) and Poole (2010). The impact of foreign TNCs on the productivity and innovation performance of Italy has been detailed in Imbriani and Reganati (1997) and in Castellani and Zanfei (2006).
15 The modalities of internationalization sequence have been well argued theoretically and evidenced in the Scandinavian School model of international business (Johanson and Wiedersheim-Paul 1975; Johanson and Vahlne 1977).
16 More recent works exploring the technological gap theory include Soete (1981), Dosi, Pavitt, and Soete (1990), Wakelin (1998). Trade performance, in general, has been linked positively to innovation in many works including the following: Hughes (1986); Amendola, Dosi, and Papagni (1993; Cantwell (1989, 1994); Cantwell and Sanna Randaccio (1993); Krugman (1995); Fagerberg (1996); and Cassiman and Golovko (2011).
17 Keller (2004) finds that foreign sources of technology are of dominant importance for productivity growth.
18 Lileeva and Trefler (2010) find that access to foreign markets by Canadian firms improves their productivity at the plant level.
19 The problems are discussed in Filippetti et al. (2013b). Alternative measures of innovation are patents filing or counts of products and process innovation deriving from surveys.
20 The consistency refers mainly to the methodology used. There are many differences in the 14 countries' sample: some studies refer to firms, others to establishments; the years covered vary and so does the size of firms or establishments.
21 Balcet, Wang, and Richet (2012) discuss reverse engineering in the case of the Chinese automotive RNC Geely.
22 Castellani and Zanfei (2004) in a study of France, Italy, and Spain conclude that a positive impact of inward FDI on the productivity of the host country's firms partly depends on the technological gap between the home and host country.
23 See also Kokko (1992) and Görg and Strobl (2005). A critique of the concept and empirics of knowledge spillovers is in Breschi and Lissoni (2001).
24 There have, since, been further elaborations and refinements particularly in Zahra and George (2002). Mahroum et al. (2008) consider and estimate a wide range of indicators of AC.
25 The three dimensions are developed in Filippetti et al. (2013a). The indicators used to represent them are considered as stock data rather than flows to capture the relevance of history and cumulation in determining the AC of a country at any given point in time.
26 See Cantwell and Iammarino (2003) and Driffield and Love (2003).
27 See Iammarino and McCann (2013: ch. 5, as well as this volume, Chapter 14).

References

Amendola, G., G. Dosi, and E. Papagni. 1993. "The Dynamics of International Competitiveness." *Review of World Economics* 129(3): 451–471.

Balcet, G., H. Wang, and X. Richet. 2012. "Geely: A Trajectory of Catching Up and Asset-Seeking Multinational Growth." *International Journal of Automotive Technology and Management* 12(4): 360–375.

Bartlett, C.A., and S. Ghoshal. 1989. *Managing Across Borders: The Transnational Solution.* Boston, MA: Harvard Business School Press.

Bernard, A.B., and J.B. Jensen. 1995. "Exporters, Jobs and Wages in US Manufacturing: 1976–1987." *Brookings Papers on Economic Activity: Microeconomics* 1995: 67–119.

Bertschek, I. 1995. "Product and Process Innovation as a Response to Increasing Imports and Foreign Direct Investment." *Journal of Industrial Economics* 43(4): 341–357.

Blomstrom, M., S. Globerman, and A. Kokko. 1999. "The Determinants of Host Country Spillovers from Foreign Direct Investment: Review and Synthesis of the Literature." *The European Institute of Japanese Studies Working Paper* 76.

Breschi, S., and F. Lissoni. 2001. "Knowledge Spillovers and Loxcal Innovation Systems: A Critical Survey." *Industrial and Corporate Change* 10(4): 975–1005.

Cantwell, J. 1989. *Technological Innovation and Multinational Corporations*. Oxford: Blackwell.

Cantwell, J. 1994. "The Relationship Between International Trade and International Production." In *Surveys in International Trade*, ed. D. Greenway and L.A. Winters, 303–328. Oxford: Blackwell.

Cantwell, J., and S. Iammarino. 2003. *Multinational Corporations and European Regional Systems of Innovation*. London: Routledge.

Cantwell, J., and F. Sanna Randaccio. 1993. "Multinationality and Firm Growth." *Weltwirtschaftliches Archiv* 129(2): 275–299.

Cassiman, B., and E. Golovko. 2011. "Innovation and Internationalization Through Exports." *Journal of International Business Studies* 42(1): 56–75.

Castellani, D., and A. Zanfei. 2004. "Choosing International Linkage Strategies in the Electronic Industry: The Role of Multinational Experience." *Journal of Behaviour and Organization* 53: 447–475.

Castellani, D., and A. Zanfei. 2006. *Multinational Firms, Innovation and Productivity.* Cheltenham: Edward Elgar.

Caves, R.E. 1971. "International Corporations: The Industrial Economics of Foreign Investment." *Economica* 38: 1–27.

Chandler, A.D. 1962. *Strategy and Structure: Chapters in the History of the Industrial Enterprise*. Cambridge, MA: MIT Press.

Cohen, W. 1995. "Empirical Studies of Innovative Activity." In *Handbook of the Economics of Innovation and Technological Change*, ed. P. Stoneman. Oxford: Basil Blackwell.

Cohen, W.M., and D.A. Levinthal. 1990. "Absorptive Capacity: A New Perspective on Learning and Innovation." *Administrative Science Quarterly* 35: 128–195.

Crescenzi, R., C. Pietrobelli, and R. Rabellotti. 2014. "Innovation Drivers, Value Chains and the Geography of Multinational Corporations in Europe." *Journal of Economic Geography* 14(6): 1053–1086.

Damijan, J.P., and C. Kostevc. 2010. "Learning from Trade Through Innovation: Causal Link Between Imports, Exports and Innovation in Spanish Microdata." Leuven, Licos Centre for Institutions and Economic Performance Discussion Paper 264/2010.

Damijan, J.P., C. Kostevc, and S. Polanec. 2010. "From Innovation to Exporting or Vice Versa?" *The World Economy* 32(2): 374–398.

Dosi, G., K. Pavitt, and L. Soete. 1990. *The Economics of Technological Change and International trade*. Brighton: Harvester Wheatsheaf.

Driffield, N., and J.H. Love. 2003. "Foreign Direct Investment, Technology Sourcing and Reverse Spillovers." *The Manchester School*, 71(6): 659–672.

Dunning, J.H. 1977. "Trade, Location of Economic Activity and the MNE: A Search for an Eclectic Approach." In *The International Allocation of Economic Activity*, ed. B. Ohlin, P.O. Hesselborn, and P.M. Wijkman, 395-431. London: Macmillan.

Dunning, J.H. 1980. "Explaining Changing Patterns of International Production: In Defense of the Eclectic Theory." *Oxford Bulletin of Economics and Statistics* 41(4): 269–295.

Fagerberg, J. 1996. "Technology and Competitiveness." *Oxford Review of Economic Policy* 12(3): 39–51.

Filippetti, A., M. Frenz, and G. Ietto-Gillies. 2011. "Are Innovation and Internationalization Related? An Analysis of European Countries." *Industry and Innovation* 18(5): 437–459.

Filippetti, A., M. Frenz, and G. Ietto-Gillies. 2013a. "The Impact of Internationalisation on Innovation: An Analysis of Patent Activities in 42 Countries." Workshop on Internationalization and Innovation, Centre for Innovation Management Research (CIMR), Department of Management, Birkbeck University of London, September 28.

Filippetti, A., M. Frenz, and G. Ietto-Gillies. 2013b. "The Role of Internationalization as a Determinant of Innovation Performance: An Analysis of 42 Countries." Birkbeck College University of London, CIMR Research Working Paper 10.

Forsgren, M., U. Holm, and J. Johanson. 2005. *Managing the Embedded Multinational: A Business Network View*. Cheltenham: Edward Elgar.

Frenz, M., and G. Ietto-Gillies. 2007. "Does Multinationality Affect the Propensity to Innovate? An Analysis of the Third UK Community Innovation Survey." *International Review of Applied Economics* 21(1): 99–117.

Frenz, M., and G. Ietto-Gillies. 2009. "The Impact on Innovation Performance of Different Sources of Knowledge: Evidence from the UK Community Innovation Survey." *Research Policy* 38(7): 1125–1135.

Frenz, M., C. Girardone, and G. Ietto-Gillies. 2005. "Multinationality Matters in Innovation: The Case of the UK Financial Services." *Industry and Innovation* 12(1): 1–28.

Frobel, F., J. Heinricks, and O. Kreye. 1980. *The New International Division of Labour*. Cambridge and Paris: Cambridge University Press and Éditions de la Maison des Sciences de l'Homme.

Ghoshal, S., and N. Nohria. 1997. *The Differentiated MNC: Organizing Multinational Corporations for Value Creation*. San Francisco, CA: Jossey-Bass.

Görg, H., and E. Strobl. 2005. "Spillovers from Foreign Firms Through Worker Mobility: An Empirical Investigation." *Scandinavian Journal of Economics* 107: 693–709.

Granovetter, M. 1985. "Economic Action and social Structure: The Problem of Embeddedness." *American Journal of Sociology* 78(3): 3–30.

Grossman, G., and E. Helpman. 1991. *Innovation and Growth in the Global Economy*. Cambridge, MA: MIT Press.

Gunawan, J., and E.L. Rose. 2014. "Absorptive Capacity Development in Indonesian Exporting Firms: How Do Institutions Matter?" *International Business Review* 23(1): 45–54.

Hagedoorn, J. 1996. "Trends and Patterns in Strategic Technology Partnering since the Early Seventies." *Review of Industrial Organization* 11: 601–616.

Hedlund, G. 1986. The Hypermodern MNC – A Heterarchy? *Human Resource Management* 25(1): 9–35.

Hedlund, G., and D. Rolander. 1990. "Action in Heterarchies: New Approaches to Managing the MNC." In *Managing the Global Firm*, ed. C.A. Bartlett, Y. Doz, and G. Hedlund, 1–15. London: Routledge.

Hufbauer, G.C. 1966. *Synthetic Materials and the Theory of International Trade*. London: Duckworth.

Hughes, K. 1986. *Exports and Technology*. Cambridge: Cambridge University Press.

Hymer, S.H. 1971. "The Multinational Corporation and the Law of Uneven Development." In *Economics and World Order*, ed. J.W. Bhagwati, 113–140. London: Macmillan.

Hymer, S.H. 1976. *The International Operations of National Firms: A Study of Direct Foreign Investment*. Cambridge, MA: MIT Press.

Iammarino, S., and P. McCann. 2013. *Multinationals and Economic Geography: Location, Technology and Innovation*. Cheltenham: Edward Elgar.

Ietto-Gillies, G. 2002. "Internationalization and the Demarcation Between Services and Manufactures: A Theoretical and Empirical Analysis." In *Internationalization, Technology and Services*, ed. M. Miozzo and I. Miles. Cheltenham: Edward Elgar.

Ietto-Gillies, G. 2012. *Transnational Corporations and International Production: Concepts, Theories and Effects*. Cheltenham: Edward Elgar.

Imbriani, C., and Reganati, F. 1997. "Spillovers internazionali di efficienza nel settore manifatturiero italiano – International Efficiency Spillovers into the Italian Manufacturing Sector." *Economia Internazionale* 50(4): 583–595.

IMF. 1977. *Balance of Payments Manual*, 4th edn. Washington, DC: International Monetary Fund.

IMF. 1993. *Balance of Payments Manual*, 5th edn. Washington, DC: International Monetary Fund.

ISGEP. 2007. "Exports and Productivity – Comparable Evidence for 14 Countries." The International Study Group on Exports and Productivity. Leuven, Licos Centre for Institutions and Economic Performance Discussion Paper 192/2007.

Javorcik, B. 2004. "Does Foreign Direct Investment Increase the Productivity of Domestic Firms? In Search of Spillovers Through Backward Linkages." *American Economic Review* 94: 605–627.

Johanson, J., and J.-E. Vahlne. 1977. "The Internationalization Process of the Firm: A Model of Knowledge Development and Increasing Foreign Market Commitment." *Journal of International Business Studies* 8(1): 23–32.

Johanson, J., and F. Wiedersheim-Paul. 1975. "The Internationalization of the Firm: Four Swedish Cases." *Journal of Management Studies* 12(3): 305–322.

Keller, W. 2004. "International Technology Diffusion." *Journal of Economic Literature* 42: 752–782.

Kimura, F., and K. Kiyota. 2006. "Exports, FDI, and Productivity: Dynamic Evidence from Japanese Firms." *Review of World Economics* 142(4): 695–719.

Kogut, B., and U. Zander. 1993. "Knowledge of the Firm and the Evolutionary Theory of the Multinational Corporation." *Journal of International Business Studies* 24: 625–645.

Kogut, B., and Zander, U. 2003. "A Memoir and Reflection: Knowledge and an Evolutionary Theory of the Multinational Firm 10 Years Later." *Journal of International Business Studies* 34: 505–515.

Kokko, A. 1992. *Foreign Direct Investment, Host Country Characteristics and Spillovers*. Stockholm: Stockholm School of Economics, Economic Research Institute.

Krugman, P. 1995. "Technological Change in International Trade." In *Handbook of Economics of Innovation and Technological Change*, ed. P. Stoneman. Oxford: Blackwell.

Laursen, K., and A. Salter. 2006. "Open for Innovation: The Role of Openness in Explaining Innovation Performance among UK Manufacturing Firms." *Strategic Management Journal* 27: 131–150.

Lileeva, A., and D. Trefler. 2010. "Improved Access to Foreign Markets Raises Plant-Level Productivity … for Some Plants." *The Quarterly Journal of Economics* 125(3): 1051–1099.

Liu, X., and T.W. Buck. 2007. "Innovative Performance and Channels for International Technology Spillovers: Evidence from Chinese High-Tech Industries." *Research Policy* 36(3): 355–366.

MacGarvie, M. 2006. "Do Firms Learn from International Trade?" *The Review of Economics and Statistics* 88(1): 46–60.

Mahroum, S., R. Huggins, N. Clayton, K. Pain, and P. Taylor. 2008. *Innovation by Adoption: Measuring and Mapping Absorptive Capacity in UK Nations and Regions*. London: National Endowment for Science, Technology and Arts.

Malerba, F. 1992. "Learning by Firms and Incremental Technical Change." *The Economic Journal* 102: 845–859.

Narula, R., and A. Zanfei. 2004. "Globalization of Innovation: The Role of Multinational Enterprises." In *The Oxford Handbook of Innovation*, ed. J. Fageberger, D.C. Mowery, and R.R. Nelson, 318–346. Oxford: Oxford University Press.

Nelson, R.R., and N. Rosenberg. 1993. "Technical Innovation and National Systems." In *National Innovation Systems: A Comparative Analysis*, ed. R.R. Nelson. Oxford: Oxford University Press.

Nelson, R.R., and S.G. Winter. 1982. *An Evolutionary Theory of Economic Change.* Cambridge, MA: Harvard University Press.

OECD. 1996. *Detailed Benchmark Definition of Foreign Direct Investment*, 3rd edn. Paris: OECD.

Patel, P., and K. Pavitt. 1994. "Uneven (and Divergent) Technological Accumulation among Advanced Countries: Evidence and a Framework of Explanation." *Industrial and Corporate Change* 3: 759–787.

Polanyi, M. 1966. "The Logic of Tacit Inference." *Philosophy* 41: 1–18.

Polanyi, M. 1967. *The Tacit Dimension*. London: Routledge.

Poole, J. 2010. "Knowledge Transfer from Multinational to Domestic Firms: Evidence from Worker Mobility." Working Paper, Department of Economics, University of California, Santa Cruz.

Posner, M.V. 1961. "International Trade and Technical Change." *Oxford Economic Papers* 13: 323–341.

Soete, L. 1981. "A General Test of Technological Gap Trade Theory." *Weltwirtschaftliches Archiv* 117: 638–659.

UNCTAD. 2002. *World Investment Report 2002: Transnational Corporations and Export Competitiveness*. Geneva: United Nations.

UNCTAD. 2005. *World Investment Report 2005: Transnational Corporations and the Internationalization of R&D*. Geneva: United Nations.

UNCTAD. 2013. *World Investment Report. Glabal Value Chains: Investment and Trade for Development.* Geneva: United Nations.

Uzzi, B. 1997. "Social Structure and Competition in Interfirm Networks: The Paradox of Embeddedness." *Administrative Science Quarterly* 42: 35–67.

Vernon, R. 1966. "International Investment and International Trade in the Product Cycle." *The Quarterly Journal of Economics* 80: 190–207.

Wagner, J. 2007. "Exports and Productivity: A Survey of the Evidence from Firm-Level Data." *The World Economy* 30(1): 60–82.

Wakelin, K. 1998. "The Role of Innovation in Bilateral OECD Trade Performance." *Applied Economics* 30: 1335–1346.

Williamson, O. 1975. *Markets and Hierarchies: Analysis and Anti-trust Implications.* New York: Free Press.

Williamson, O.E. 1981. "The Modern Corporation: Origins, Evolution, Attributes." *Journal of Economic Literature* 19: 1537–1568.

Williamson, O.E. 1984. "Efficient Labour Organization." In *Firms, Organization and Labour: Approaches to the Economics of Work Organization*, ed. F.H. Stephens, 87–118. London: Macmillan.

Zahra, S.A., and G. George. 2002. "Absorptive Capacity: A Review, Reconceptualization, and Extension." *Academy of Management Review* 27(2): 185–203.

Zahra, S.A., R.D. Ireland, and M.A. Hitt. 2000. "International Expansion by New Venture Firms: International Diversity, Mode of Market Entry, Technological Learning and Performance." *Academy of Management Journal* 43: 925–950.

Chapter 7

International R&D Alliances by Firms: Origins and Development

Rajneesh Narula and Andrea Martínez-Noya

Introduction

One of the hallmarks of globalization has been the growing interdependence of firms across borders, and the greater variety of organizational modes deployed by companies (see this volume, Ietto-Gillies, Chapter 6 and Guy, Chapter 28). Globalization has not always been a force for positive outcomes: Firms have greater opportunities, but these are matched by commensurate challenges. The liberalization of markets, the decline of barriers to trade and investment, and the greater cross-border enforceability of contracts has expanded the de facto markets available to firms, while also increasing cross-border competition. New systemic technologies and the growing need for firms to have a greater breadth and depth in a variety of scientific and technological areas have also changed the competitive landscape. The dynamism and complexity of the marketplace and the intertwining of such technological and economic forces has meant that it is difficult for firms to effectively compete when relying entirely on their own resources.

Globalization and technology are deeply interconnected and concatenated at multiple levels, and the growing international, cross-border aspect of collaboration is regarded as one of the primary categories of the well-known tripartite taxonomy proposed by Archibugi and Michie (1995). The greater interaction between users and producers of scientific knowledge has naturally led to greater collaboration and interaction between a variety of actors, and it should be stressed that this collaboration is not limited to commercial, for-profit economic actors. There is a considerable degree of exploitation and generation of knowledge that is facilitated through informal and formal collaboration, inter alia, through migration, expatriation, student exchanges, scientific cooperation, conferences, and so on (Archibugi and

The Handbook of Global Science, Technology, and Innovation, First Edition.
Edited by Daniele Archibugi and Andrea Filippetti.

Iammarino 2002; see also in this volume: Flanagan, Chapter 17; Florida and Mellander, Chapter 15; Barnard et al., Chapter 18). In addition, the large role of non-firms (which we take to mean actors that do not have a primary economic motivation, such as universities, colleges, public institutes, regulatory authorities, and so on) should not be underestimated (Narula 2003). However, this chapter intentionally focuses on collaboration between entities that have at their heart a commercial motivation behind their activities.

It is not an accident, then, that the past few decades have seen a rapid growth of collaborative agreements between firms (Narula 2003). This trend is particularly evident in those industries where consumption patterns are more homogeneous across countries, and which show a high level of capital intensity as well as knowledge intensity in terms of investment in innovation and technology (Narula and Duysters 2004).

Collaboration in innovation does not necessarily mean that innovation is the primary intention, but may be a consequence of exploitation. Indeed, some have described the current era as being the "age of alliance capitalism" where flexible economic arrangements are finding increasing favor (Dunning 1995, 1997). It should be noted that growing collaboration has not led to the death of the "traditional" firm. Instead, cooperative arrangements have, in many cases proven to be supplementary to hierarchical, fully internalized activities within the boundaries of a vertically or horizontally integrated firm (Dicken 2011). It is not entirely new either – economic actors have relied upon cooperation to gain competitive advantage since time immemorial.[1]

Nonetheless, it is now an especially large and pervasive phenomenon, and the increased use of collaborative agreements has been observed by all firms independent of their size and country of origin. Firms have always needed external partners (suppliers, competitors, customers, universities, technological research centers, or institutes) to collaborate with on different, specific activities along their value chain (Narula and Duysters 2004), but the extent to which firms in the twenty-first century systematically exploit cooperative networks and agreements is quite breathtaking.

In this chapter we focus on R&D alliances, or strategic technology partnering (STP) (these terms are used as synonyms here). We intend to highlight two specific characteristics of the current growth of cooperative activity that are novel to the current era. First, that the last three decades have seen cooperative agreements increasingly being used for high-value and knowledge-based activities, such as those related to research and development (R&D). It turns out that the growing complexity and multidisciplinary nature of the innovation process has implied a greater need for firms to be flexible and responsive to external partners in order to access complementary resources, to take advantage of more business opportunities, achieve lower costs, or reduce time-to-market (Duysters and De Man 2003; Hagedoorn 1993; Martínez-Noya, García-Canal, and Guillén 2012). In effect, technological change and global competition have forced firms to search for external sources of knowledge through a wide diversity of alliances (Hagedoorn and Osborn 2002).

Second, we pay specific attention to *international* alliances, which are agreements that are international in their span and partner selection. Theoretical work over much of the twentieth century has argued that, due to the strategic and usually tacit nature of R&D, firms have tended to centralize this activity in the home country and to internalize it (i.e., in-house) to maintain control over the assets. Firms preferred, wherever possible, to establish wholly-owned subsidiaries, and where this was not possible for whatever reason, to maintain a controlling stake in an affiliate. This was considered to be particularly important when engaging in foreign markets, so as to protect their technology and avoid appropriability hazards (Kogut 1988; Oxley 1997). However, over the past few decades this preference for full internalization has changed, even for R&D intensive firms (Martínez-Noya and García-Canal 2011).

The great paradox of globalization is evident in the analysis of R&D alliances: countries, governments, and nation-states matter more, even as economic activity is increasingly cross-border, and national borders matter less (Narula 2003). There remains heterogeneity of country-specific resources, and these are geographically bound. The external resources needed by a firm may not be available within its home country, and these cross-country differences in resource endowment may drive the firm to seek such resources internationally (offshore), searching for location-specific advantages (Dunning 1998). Nowadays, knowledge-intensive firms from both advanced and developing countries are globally dispersing and disintegrating their value chains to control costs and leverage their capabilities (Mudambi 2008). Through STP firms have found a way not only to be more efficient or flexible, but also to benefit from the distinctive capabilities of specialized partners located worldwide (Chen 2004; Graf and Mudambi 2005) even from emerging countries (UNCTAD 2005).

Defining Strategic Technology Partnering (STP)

This chapter is focused on (international) STP, which represents a particular subset of cooperative agreements. STP refers to inter-firm cooperative agreements where R&D is a significant part of the collaborative effort, and which are intended to affect the long-term product-market positioning of at least one partner (Hagedoorn 1993). Compared to quasi-external, vertical solutions such as outsourcing and customer-supplier networks (which tend to involve lower levels of joint activity), STPs are usually horizontal agreements which tend to reflect a more complex strategic intent, and require closer collaboration (Narula and Hagedoorn 1999; Narula 2001).[2] The word "strategic" in STP suggests that such agreements are aimed at the longer-term objective of enhancing the value of the firm's assets.

This definition is important to note because there are a variety of cooperative agreements that do not have a primary strategic intent, but are primarily cost-economizing: for instance those associated with global value chains and production networks, where the primary objective is to reduce costs along a vertical chain. A strong strategic motive implies that the firm is often willing to forego (some degree of) cost-savings in order to strengthen its portfolio of assets or its competitive position, or at least to defend its current position. Needless to say, most agreements have elements of both, but horizontal agreements tend to have a greater strategic aspect.

STP differs from other kinds of alliances in several ways. First, STP is often designed to undertake specific tasks and is generally terminated at the completion of these tasks, and so is by definition short- and often fixed-term in nature, unlike non-R&D

alliances and networks which may have a long-term and formal aspect which link aspects of their businesses (Porter and Fuller 1986).

Second, STP also differs from other strategic alliances in that while firms engaged in asset-exploiting activities such as production or sales have a broader choice of options, that include wholly-owned subsidiaries and arms-length technology acquisition, some of these options are simply not available to firms that are seeking to undertake R&D. On the one hand, R&D is partly tacit, surrounded by high levels of uncertainty and has a long-term horizon. As such, the value of the research cannot be estimated ex-ante. On the other hand, arms-length transactions are simply not as effective, particularly in technology-intensive sectors or new, "emerging" sectors, even if markets for these technologies were to exist. The further away these technologies are from commercialization, the less likely that technology can be obtained through market mechanisms. Besides, the partly public good nature of these activities prevents prospective selling firms from making technologies available for evaluation, and without doing so the prospective buyer is unable to determine its worth. Thus full internalization may simply not be a choice available to the MNE, and the use of mergers or acquisitions (M&As) or greenfield investments may not be advisable either. First, because it is known that the use of M&As is not a viable option where the technology being sought is a small part of the total value of the firm (Kogut 1988). Second, due to the high dynamism of the technological market, a greenfield investment may not be viable either, because the time and costs of building new competencies from scratch may be prohibitive. In relation to this, it should be noted that in some instances alliances are used as a precursor to M&As (Hagedoorn and Sadowski 1999). Therefore, it is no surprise that STP has grown fastest in high-technology sectors where market options are less well developed. In general though, partnering activity has steadily declined as a percentage of all agreements in all sectors (Hagedoorn 2002); see Figure 7.1.

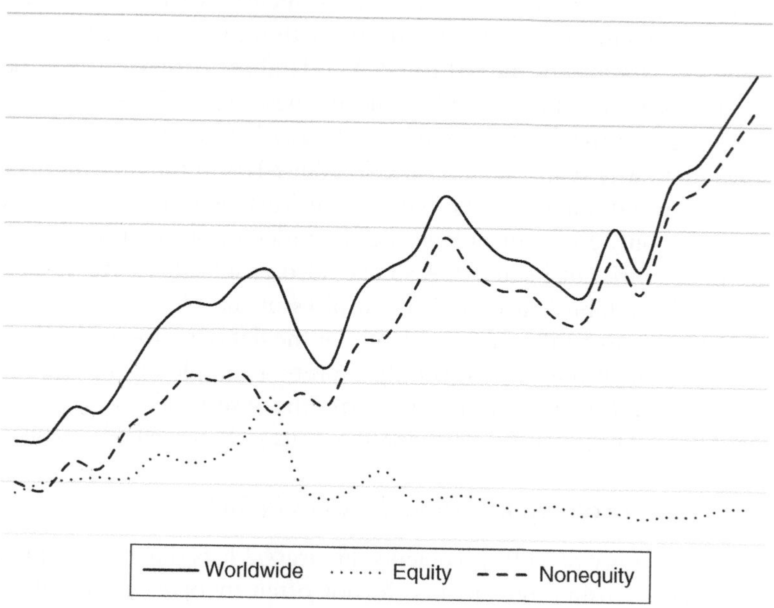

Figure 7.1 Worldwide equity and non-equity technology alliances 1980–2006.
Source: Data from MERIT-CATI database, as published in NSF (2010).

Databases on STP

Although often complemented with additional data sources stemming from survey data or news retrieval searches, studies on research and technology alliances tend to utilize large alliance databases such as Securities Data Company (SDC), Bioscan, MERIT-CATI, Recombinant Capital (RECAP), or CORE. Three are multisector databases (SDC, MERIT-CATI, and CORE) and two are specific to the biotechnology sector (RECAP and Bioscan). Each of these databases has its unique advantages and disadvantages: see Schilling (2009) for an extensive review on their differences. Therefore it is important to know them so as to use the most appropriate one depending on the type of research to be undertaken. Given that SDC and MERIT-CATI databases have been found to be the most appropriate for large-scale analyses on STP, their main features are described here.

SDC is a division of Thomson Financial and is the most commonly used database covering the widest range of sectors and details over 138,000 worldwide joint ventures and strategic alliances. Specifically, SDC tracks cooperative agreements by two or more separate entities that may result in the formation of a third entity, either as a start-up or through the merger of assets, or acquisition of stakes; as well as agreements involving licensing, manufacturing, research and development, mining and exploration, supply services. Thus, although R&D alliances are just a subset of the data available through the mergers and acquisitions section of the database, the main advantage of SDC is its extensive searchability. It offers over 200 data search elements including the name, SIC code, and nationality of participants, the terms of the deal, and deal synopsis for each alliance agreement.

On the other hand, the MERIT-CATI database, administered at Maastricht University in the Netherlands, is focused exclusively on STP, which includes any alliance that entails the transfer of technology or the undertaking of joint research, such as joint research pacts, joint development agreements, R&D contracts, (mutual) second sourcing agreements, and joint ventures with technology sharing or an R&D program. The key strength of the MERIT-CATI database is that it tracks data since 1960 which makes it very valuable for looking at long-range STP historical trends. The disadvantages are that the availability of this dataset is restricted (2006 was the last year for which data was available published in NSF 2010, at the time of this chapter), and that it only includes agreements that have at least two industrial partners, thus alliances involving only government labs or universities, or one industrial partner with a university or lab, are not included. Also excluded is subsidized R&D cooperation such as agreements established under the auspices of the EU framework programs. Overall, it should be noted that independently of the dataset chosen, Schilling (2009) found that even though the databases only capture a sample of alliance activity, all these databases exhibit strong symmetries in patterns of sectoral composition, alliance activity over time, and geographic participation, thus yielding reliable results.

The Growing Variety of Organizational Modes in R&D Alliances

There are a wide variety of STP agreements. Figure 7.2 lists the most common types. Strictly speaking, licensing agreements do not often involve active collaboration, and knowledge flows are largely one-way (with the exception of sporadic flows due for instance to legal clauses that require the licensee to feedback to the licensor

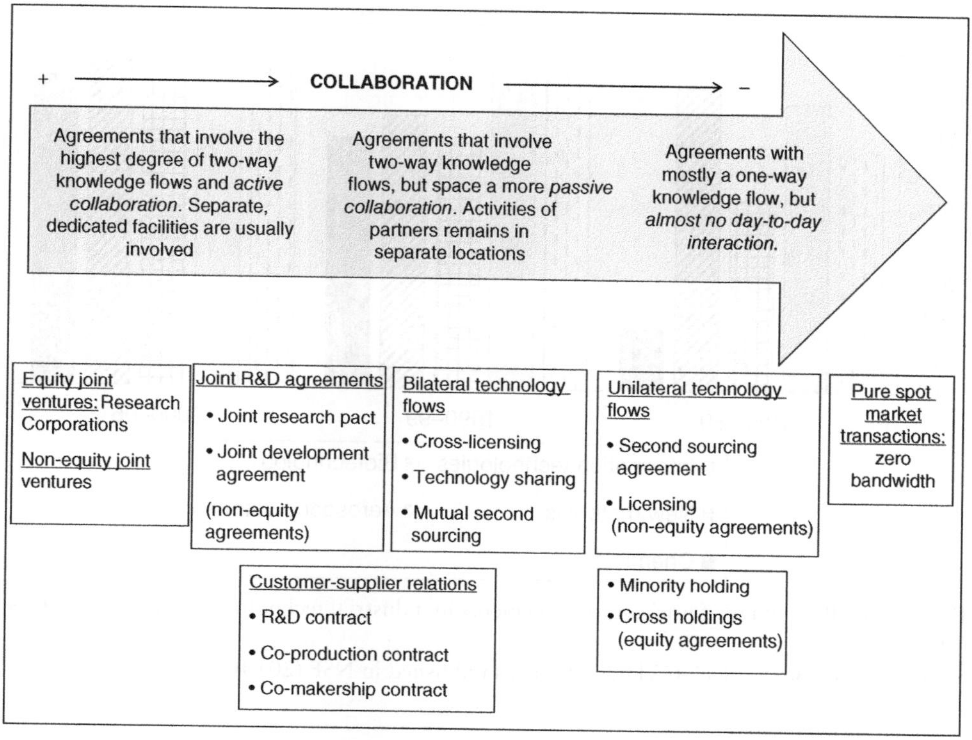

Figure 7.2 Organizational modes of STP agreements and extent of inter-firm collaboration.

improvements and innovations they may have made to the licensed technology). Licensing agreements are therefore passive conduits for knowledge flows, and do not normally qualify as alliances *by themselves*. Similarly, two-way licensing agreements, mutual second sourcing, and cross-licensing involve knowledge flows, but have a minimal degree of collaboration. They are also – by definition – utilized where knowledge has been codified. Nonetheless, they are likely to have a strategic element to them, because they may be a precursor for more complex and intense collaboration, and may also be utilized in parallel with more interactive collaboration. Indeed, there are few instances where more complex joint R&D alliances do not also involve an exchange of technology licenses, and other lesser forms of collaboration.

Thus, the extent and form of interaction between learning activities depend upon the organizational form of cooperation, and this varies considerably by industry and by the relative importance of specific technologies to the firm's portfolio of competencies. On the one hand, non-equity forms of agreements tend to be more efficient for undertaking activities in more research-intensive industries, and where technological change is rapid since they promote negotiation and can lead to more intensive cooperation than equity forms. However, where firms seek to learn and transfer tacit knowledge back to the parent firm, such as market-specific knowledge when entering a new market, or are engaged in production as well as research, equity forms of agreement may be more appropriate. Equity agreements are preferred in relatively mature sectors, while non-equity agreements are utilized in high-tech sectors (Hagedoorn and Narula 1996). Figure 7.3 shows the differences in the preference for non-equity agreements by major sectors. By and large, there has been a growing

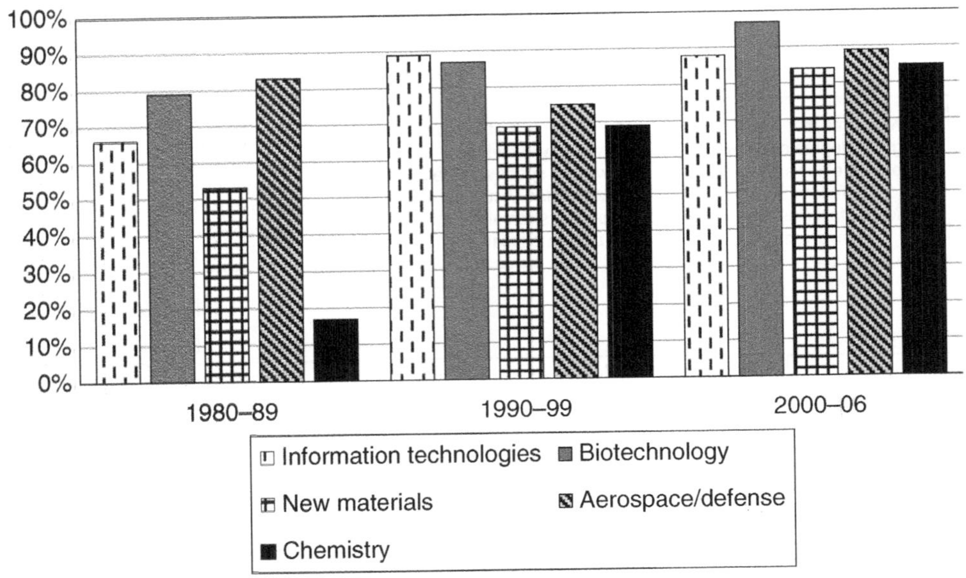

Figure 7.3 Percentage of non-equity agreements in industrial technology alliances by sectors 1980–2006.
Source: Data from MERIT-CATI database, as published in NSF (2010).

preference for non-equity agreements across all sectors, and this reflects the declining coordination costs of STP, as well as their ability to monitor and enforce their intellectual property rights within the growing framework of international and national regulations governing intellectual property rights.

On the one other hand, the preference for alliances relative to other modes is also determined by the relative importance of specific technologies to their portfolio of competencies. The strategic importance of these technologies determines to what extent firms will use alliances or outsourcing in their development. This, in turn, is determined by the extent to which the technology is tacit, the extent to which collaboration is required to utilize it, and to what extent the partner's activities need to be monitored (Narula 2001). In addition, the maturity of the technology, and its characteristics, determines the extent to which the innovation process can be internalized (Teece 1986, 1996). Furthermore, not all new technological developments are equally important to a firm's technological portfolio, and this influences the choice of organizational mode.

The facilitating role of globalization has expanded firms' use of external resources to reduce, inter alia, time spans, costs, and risks associated with innovation, as well as provide firms with greater flexibility in their operations (Hagedoorn 1993). The increased knowledge content of products in general, the cross-fertilization between previously distinct technological areas, coupled with the multiple technological competencies that modern products, services, and processes require of firms put growing pressure on their in-house resources, which are limited. So firms must collaborate if they are to overcome this resource bottleneck. Falling coordination, monitoring, and enforcement costs have made it easier for firms of all sizes to monitor, identify, and establish collaborative ventures than previously had been the case (Narula 1999;

Rangan 2000). In other words, hierarchical control and full internalization is no longer always a first-best option to MNCs, especially where innovatory activities are concerned.

Internalizing activities under conditions of rapid technological change impose inflexibility precisely when flexibility is most needed (Poppo and Zenger 1998). In fact, the literature analyzing strategic technology partnering has found that whenever firms need quick responses to changes in technological leadership, non-equity agreements are preferred to joint ventures because they provide firms with greater strategic flexibility (Osborn and Baughn 1990). Due to the fact that investments in technology are often quite specialized, rapid technological change may increase the likelihood of technological investments in knowledge and routines being rendered obsolete (Balakrishnan and Wernerfelt 1986). For this reason, previous research has shown that greater use of non-equity agreements may deliver more flexibility, which may help firms to respond quickly to unanticipated threats and market opportunities (Hitt, Keats, and De Marie 1998).

As Figure 7.3 shows, the fraction of non-equity STPs has been growing over the last 25 years. One can interpret this as evidence of the fact that low-commitment intensive agreements are more effective as a mechanism to gain timely and extensive access to rapidly evolving technology across borders. From this perspective, STP may represent a "first-best" option to MNEs (Narula 2003), especially where innovatory activities are concerned. In other words, firms do not necessarily resort to these strategies because they cannot have access to more effective and more profitable channels of technology transfer (as uncertainty is too high or institutional barriers constrain "internal" strategies); on the contrary, STPs, especially non-equity agreements, can be preferred as a tool that is both more flexible and more apt for knowledge development and learning.

Does STP Substitute for In-House R&D?

Considerable research has sought to understand how much a firm can substitute STP for in-house R&D. The attempt to understand the reasons behind a firm's choice between non-internal and internal technological development is not new. The work of Teece (1986) presents a pioneering analysis of this issue, which builds on Abernathy and Utterback (1978), and Dosi (1982) among others, and further developed by Pisano (1990), Arora and Gambardella (1990), Henderson and Clark (1990), Granstrand, Patel, and Pavitt (1997), Buckley and Chapman (1998), Nagarajan and Mitchell (1998), Veugelers and Cassiman (1998), Croisier (1998), Lowe and Taylor (1998), Tidd and Trewhella (1997), Gambardella and Torrisi (1998), Nooteboom (1999), Narula (2001) and Brusoni, Prencipe, and Pavitt (2001).

Cooperative agreements act as complements rather than as substitutes for traditional hierarchical modes. This is because the excessive use of non-internal R&D entails considerable risks and costs. As a general rule, firms find it costly and difficult to access competencies from other firms or organizations in technological fields which are unrelated to their own capabilities, and with which they have little initial familiarity. The point is that the internalization advantages of undertaking a variety of activities in-house derives (inter alia) from the technological coherence of these activities (Teece et al. 1994). Where such technological coherence does not exist, non-internal activities become feasible options. However, this feasibility is a function of the firm's "absorptive capacity." When the firm acquires knowledge from its external environment or where one knowledge-creating part of the firm interacts with another, the recipient needs to have some innovative potential of its own to be able to learn and effectively adapt the technologies to which it may wish to have access (Cohen and Levinthal 1989). Thus, STP tends to develop in areas in which partner companies share some complementary capabilities, and these alliances create a greater degree of interaction between the partners' respective paths of learning and innovation (Mowery, Oxley, and Silverman 1998; Cantwell and Colombo 2000; Santangelo 2000).

STP is not an alternative to in-house R&D, but complementary to it. STP and other collaboration modes do not replace the need for firms to undertake internal R&D activities, but enhance it. Indeed, Veugelers (1997) demonstrates that there is a positive relationship between external technology sourcing and internal R&D. Firms must know considerably more about what they need (and do not have) both to monitor their partners and also to be able to efficiently perform the role of systems integrator (Brusoni et al. 2001). Such a threshold level of expertise provides them with the absorptive capacity to benefit from external sources of knowledge (Cohen and Levinthal 1990). STP tends to be highly correlated with large firms with ample resources in technology-intensive sectors, and is most frequently the means firms use to keep up with the technological frontier. By associating complementary resources and competencies STP makes it possible for firms to explore and exploit new technological opportunities and expand the boundaries of their knowledge base.

Theoretical Perspectives to Explain Growth of STP

The growth of STP has been explained taking either an economizing perspective through the lens of transaction cost theory, or a more strategic one through the lens of a number of different theoretical perspectives, such as the resource-based theory of the firm (Barney 1991; Das and Teng 2000; Wernerfelt 1984), the dynamic capabilities approach (Teece, Pisano, and Shuen 1997; Zollo and Winter 2002), the knowledge-based view and organizational learning theories (Kogut and Zander 1993), as well as social network theory (Gulati 1995; Powell and Grodal 2005). Overall, underlying the difference between the economizing and the more strategic perspectives is a fundamentally different view of the way firms make decisions. Transaction cost theory assumes that firms' internalization decisions are driven by their desire to seek protection from opportunism, while the strategic theories highlight the fact that firms are boundedly rational and undertake decisions based on the need to enhance their organizational and technological capabilities; that is, because they need to enhance the value of the firm.

The transaction cost or internalization perspective, which derives from Williamson (e.g., 1975) and Coase (1937), explains the behavior and organizational mode and the mode of entry of firms based on their need to *minimize net transaction costs* faced by the firm. This body of literature has been expanded by others including Buckley and Casson (1976), Hennart (1993), and Rugman (1980). From this economic perspective, one of the fundamental reasons for the growth in STP (and in alliances in general) is the reduction of transaction costs faced by firms as a result of globalization of markets. On the one hand, recent improvements of information and communication technologies (ICTs) have reduced the costs of communicating and coordinating with external partners, which have also facilitated management of international arrangements. On the other hand, economic liberalization of markets has also led to a harmonization of regulations and barriers across countries, which has reduced the costs of enforcing and monitoring international alliances and contracts. In this sense, the establishment of supra-national regional and inter-regional agreements such as the EU in Europe and NAFTA in North America have all helped to further lower transaction costs within firms located in those regions.

However, as we have emphasized throughout this chapter, the reduction of transaction costs is only a partial explanation for the growing use of STP as a viable option to in-house R&D. As argued by Narula and Dunning (1998) four primary characteristics differentiate collaborative activity in the era of alliance capitalism from those in earlier periods. They observed that in this new era: (1) alliances are not primarily made to overcome market failure; (2) alliances are increasingly undertaken to achieve not only vertical integration, but also horizontal integration; (3) the use of alliances has been expanded worldwide and became typical of most advanced industrialized economies; and finally (4) while alliances were previously made primarily to enhance or achieve market entry or presence (i.e., asset-exploitation motives), in recent decades they are increasingly being made to protect or enhance their technological assets (i.e., asset-creation or acquisition motives). It is for these distinctive characteristics that a growing number of recent studies assert that firms' decisions to internalize are subject not only to a cost minimization strategy, but also determined by value-enhancing considerations.

Indeed, there are seven factors of a more strategic nature that explain much of the growth of STP in the past few decades.

1. The emergence of new technological sectors (such as biotechnology) and the growing technological convergence between sectors (such as computers and automobiles, or new materials with transportation) have played an important role. The cross-fertilization of technological areas has meant that firms need to have an increasing range of competencies (Granstrand et al. 1997), which encourages the use of STP to seek complementary assets rapidly.
2. Firms do not always have recourse to patenting as a means to protect their intellectual property, and must rely on secrecy or co-invention instead (Levin et al. 1987, Arundel and Kabla 1998).
3. By co-invention, alliances allow firms to monitor competitors. In certain cases, firms may also engage in STP in order to co-opt the competition (Narula and Dunning 1998). For instance, in those industries in which there is a high likelihood of selecting a standard, competitors may decide to cooperate in the development stage so as to prevent rivalry in the commercialization stage. Thus, by sharing costs and benefits through STP firms can maximize their chances of becoming the winning standard.
4. Where the expected benefit of the new area of research is still unknown, the firm will not be interested in investing large internal resources, until the potential benefits are more tangible. Under such circumstances, a risk-reduction strategy through collaboration is most often viable (Mitchell and Singh 1992).

5. Firms are path-dependent, and find it costly to break away from existing routines toward radically new or different concepts. There are additional costs involved in switching trajectories which may impede organizational change and exacerbate the level of uncertainty and therefore economic risk.
6. Non-internal activities, apart from the obvious benefits of exploring new areas and instigating radical change, have the advantage of being a "reversible" form of investment (Gambardella and Torrisi 1998) which creates a strong motivation to undertake alliances, no matter how much firms may prefer to go it alone. The capital needed is smaller, and the risks are substantially reduced, and in case of failure or organizational crisis, limited damage is inflicted on the primary operations of the firm.
7. Lastly, there are the game-theoretic considerations. As Kay (1997: 215) explains, "it is necessary to engage in networks with certain firms not because they trust their partners, but in order to trust their partners." In addition, there is the follow-my-leader strategy, as originally highlighted by Knickerbocker (1973). Firms seek partnerships in response to similar moves made by other firms in the same industry, not always because there is sound economic rationale in doing so, but in imitation of their competitors.

MNCs need to seek a variety of technological inputs, and this means partnering with not just "technology leaders." If this were the case, asset-augmenting activities would remain the exclusive domain of only a handful of firms, and all industries would become oligopolies quite rapidly. It is clear, therefore, that technology leaders actively engage with "followers" as well. There are several reasons why an MNC should wish to collaborate with a partner which has limited or as-yet-undemonstrated resources to offer. First, because of the nature of innovation, the only way to determine the nature of a potential partner's research efforts is to examine them. One way it can do so is by engaging in some form of mutual hostage exchange, which an alliance provides. Second, even where the partner's resources prove to be of a limited or inappropriate nature, and the alliance is terminated prematurely, information about its former partner's competencies are then available to either firm in future periods, should it require competencies similar to those on offer by its ex-partner. Third, as Hagedoorn and Duysters (1997) have argued, while selecting partners that are well-established players in existing technologies may represent a profit maximizing situation, it is optimal only in a static environment. In a dynamic environment, where there is a possibility of technological change (or even a change in technological trajectories), having ties to a wide group of companies, including companies that have yet to demonstrate their value, represents a higher learning potential. At the technology frontier where dominant technological designs have not yet been determined and several potential options exist, it pays to have a number of overlapping, redundant agreements. It may be optimal to partner with all sorts of companies, even those without a demonstrated track record.

In conclusion, both transaction cost-minimizing and value-enhancing reasons underlie most of the behavior of firms, so the two schools may be regarded as complementary to each other (see Madhok 1997 for an in-depth analysis of this debate). Firms would prefer to increase short-term profits through cost-economizing as well as long-term profit maximizing through value enhancement, although this is not always possible. For this reason very few agreements are distinctly driven by one motivation or the other. However, it should be noted that agreements that are established with primarily short-term cost efficiencies in mind (with cost-minimizing objectives) are generally customer-supplier networks, while agreements with a more strategic intent where a long-term value enhancement is the primary objective underlies our understanding of STP.

Strategic alliances can be thought of as "an attractive organizational form for an environment characterized by rapid innovation and geographical dispersion in the sources of know how" (Teece 1992: 20). In other words, the need for a timely and effective knowledge access may well overcome short-term, static (transaction and organizational) cost minimization. Consistent with the view that there needs to be a complementarity between internal and external competence accumulation, there is a mutually reinforcing relationship between intra-firm and inter-firm networks. The relevant implication here is that greater multinational experience can be expected to expand the exploration potential and hence lead to a greater recourse to international STP.

Relating STP to Internationalization

The international aspect of R&D alliances, broadly speaking, reflects the gradual internationalization of firms. In general, while production activities have gradually been increasingly internationalized, R&D tends to stay significantly "at home" due to its "sticky" nature (see, e.g., Kumar 2001; Narula 2002a). Nevertheless, it is worth noting that there has been some growth in the technological development activities of MNEs relative to its level 30 years ago, and its increased internationalization is now due not only to demand factors, generally associated with adaptive R&D in response to specific market conditions, but mainly to supply-side ones (Narula and Zanfei 2005).

Given that R&D activities are knowledge-based, and knowledge tends to be location-bound, locations may offer specialized know-how or capabilities within a specific technological domain (Calderini and Scellato 2005; Cantwell and Santangelo, 1999). Indeed, it has been shown that a key driver for firms' geographic distribution of R&D activity is the access to knowledge spillovers (Feinberg and Gupta 2004; Lahiri 2010). STP, like other asset-augmenting activities, is largely driven by supply-side issues. That is, firms seek to utilize immobile assets, which may be either firm-specific and/ or location specific. It is well acknowledged that both location advantages and ownership advantages are idiosyncratic and path-dependent, and the nature of innovatory activities of firms in a given location is associated with its systems of innovation. When a firm wishes to benefit from location-bound assets, it can establish an affiliate in that location, because benefits generally accrue from physical proximity to the firm or cluster. However technology spillovers through collocation can be a highly costly, uncertain, and random procedure that requires a long-term horizon, because linkages need to be developed over time. In fields where innovation is very dynamic, a wholly-owned subsidiary may not provide a fast enough response. The use of M&A is even less attractive where the area where the complementary resources sought only covers a small area of the firm's interests, and is generally not possible to do, except in rare circumstances. As a result, in order to tap these foreign external resources and access this specialized technological expertise, firms may find it convenient to ally with a partner located within such economies.

The majority of high-end product development and engineering activities tend to be carried out in advanced Western economies and Japanese firms, because world leaders in knowledge and technology are typically located within developed economies (Arora, Fosfuri, and Gambardella 2001). This is also true for STP. Technological paradigms are defined on a global basis, because competition is global in nature, and this affects the way in which firms sustain their competitiveness. Firms seek to emulate the technological advantages of leading competitors in the same industry, regardless of their national location (Cantwell and Sanna-Randaccio 1990). Likewise, firms seek to engage in STP with technological leaders in the same industry, irrespective of their national origins (Narula and Hagedoorn 1999).

On the other hand, recent studies argue that innovation activities are increasingly being offshored to developing economies searching for talent, due to the emergence of new geographical technological clusters in developing economies, such as in India or China (Lewin, Massini, and Peeters 2009; Manning, Massini, and Lewin 2008; Ricart et al. 2011). However, much of this activity is being undertaken through wholly-owned affiliates of major MNEs, rather than through alliances. Besides, a majority of the non-Triad firms are still some distance behind the technology leaders, with a few exceptions. According to NSF (2010), the 50 US alliances with Asia-Pacific companies, excluding Japan, were driven by collaborative agreements with companies headquartered in India (15), China (12), and South Korea (11). As domestic firms from these emerging economies build up technological capability, it is likely we will see a growth in activities. However, the participation of developing countries is rather modest. In 2006 (the latest year for which MERIT-CATI data is available) Japan accounted for as many R&D alliances with US firms as does the rest of Asia combined. The pattern reflects the increasing, if still modest, role of developing countries as hosts for US-owned R&D (NSF 2010). Firms from the East Asian newly industrialized countries (mainly Korea and Taiwan) and Eastern Europe account for the bulk of the non-Triad partners. Indeed, when one examines the industry and country distribution of STP involving developing country firms, these largely reflect the industrial structure of their economies (Narula and Sadowski 2002).

Even smaller technology-based MNEs are involved in a web of such agreements, and their growing significance raises numerous conundrums (Narula 2002b). Firms – regardless of size – must maintain the appropriate breadth of technological competencies, and to do this they must maintain complex international internal and external networks. Such increasingly complex linkages, both of networks internal to the firm, and those between external networks and internal networks, require complex coordination if they are to provide optimal benefits (Zanfei 2000). Such networks are not only difficult to manage, but also require considerable resources (both managerial and financial). Small and medium enterprises (SMEs) have historically tended to rely on non-internal sources more than their larger counterparts and must be more skillful at managing their portfolio of technological assets, but have limited resources (Narula 2002b). It is no surprise, therefore, that external technology development is primarily the domain of larger MNCs with greater resources, and more experience in transnational activity (Hagedoorn and Schakenraad 1994). Managing a web of different types of agreements across borders is not without its price, and highlights the role of transaction-type ownership advantages in the success of the MNE. A dispersion of activities across the globe also requires extensive coordination between them – and particularly with headquarters – if they are to function in an efficient manner with regard to the collection and dissemination of information. Indeed, the management of intangible assets is potentially a core, inimitable advantage of the firm (Madhok and Phene 2001).

The Effect of Geographic Distance on STP Formation

Although R&D alliances may involve lower sunk costs when compared with other governance modes such as M&As, they are also subject to other transaction costs due to the usually uncertain and tacit nature of innovation activities. First, firms face high information costs due to information asymmetries (i.e., high costs of searching and evaluating alliance partners). Second, firms searching for an R&D alliance partner are subject to the risk of adverse selection (i.e., this is the risk of not selecting the optimal partner) (Reuer and Lahiri 2014). Previous research has demonstrated that firms tend to respond to high information costs and the risk of adverse selection by searching for nearby partners and avoiding partners in distant locations (Rangan 2000). Selecting spatially proximate partners also offers the advantage of facilitating control, which becomes critical in STP, where misappropriation hazards are high (Li et al. 2008). As a consequence, it has been demonstrated that R&D alliance formation tends to decline with geographic distance (Reuer and Lahiri 2014).

Nevertheless, there are firm-level characteristics as well as those associated with specific dyads that shape the degree to which geographic distance matters to R&D alliance formation (Reuer and Lahiri 2014). The value of both direct and indirect prior ties between the exchange partners reduces the negative effect of geographic distance on R&D alliance formation. This is so because prior direct ties between the exchange partners, due to collaborations in the past, as well as indirect ties that they may have through common partners, both help reduce information asymmetries as well as the risk of adverse selection because they will have access to better information on the real resources and capabilities of the potential partner (Zaheer, Hernandez, and Banerjee 2010). Indeed, it should be noted that a firm's belonging to a particular network (i.e., both the direct and indirect linkages a firm has with other organizations) may act as a signal of its quality or reputation, thus reducing the risk of adverse selection. Furthermore, both product market relatedness of firms as well as their similarity in technological resources can also reduce information costs as well as the risks of adverse selection, and thus reduce the negative effect of geographic distance on STP.

There is also an extensive literature on the important role that geographical proximity plays in the propensity of firms to form technological alliances from social network theory, economic geography, and innovation systems literature (Narula and Santangelo 2009). These approaches share the emphasis that firms innovative activities show a "spatial stickiness," and for this reason location is a primary determinant of the competencies a firm possesses (Iammarino and McCann 2006). These literatures suggest that location (or collocation) has an indirect effect on the choice of partner because of the role of informal institutions in collaborations, which results in firms becoming embedded in relationships that firms have a natural tendency to perpetuate with other collocated firms and organizations. This is so because firms belonging to the same spatially localized social network are "likely to have a greater awareness of the rules, routines, and procedures each follows" (Gulati 1998: 304), which improves knowledge transfer and reduces the risk of opportunistic behaviors within the network. Indeed, there is evidence that shows that MNEs locate asset-augmenting R&D facilities in offshore locations mainly with the intention of exploiting the benefits that derive from collocation (Blanc and Sierra 1999; Criscuolo, Narula, and Verspagen 2005). Despite this, it is important to note that not all firms like proximity.

Research has found that there are some firms that, depending upon their competitive position and industry, tend to avoid collocation with the purpose of minimizing undesired knowledge spillovers and leakage of valuable technological assets (Alcácer 2006, Narula and Santangelo 2009; Shaver and Flyer 2000). Alcácer (2006) found that despite the higher concentration of R&D facilities compared to manufacturing or sales, the more capable firms collocate less than less capable ones, regardless of the activity because more capable firms have more to lose than to gain from clustering. For instance, more technologically advanced firms prefer to locate close to universities, and are less interested in locating proximate to rivals, whereas less competitive firms are more willing to collocate with other firms in the same industry (Alcácer and Chung 2007). Similarly, it has been found that in industries where the knowledge being exchanged is highly tacit, ceteris paribus, the firms' propensity to cluster is higher than in industries where the knowledge is codifiable (Cantwell and Santangelo 1999, 2000; Iammarino and McCann 2006). This is so because although the marginal cost of transmitting codified knowledge across geographical space does not depend on distance, the marginal cost of transmitting tacit knowledge does increase with distance (Criscuolo and Verspagen 2008). Finally, empirical evidence has demonstrated that the firms' preference for clustering or not is highly subject to the nature of the industry structure (Cantwell and Kosmopoulou 2002) and for firms operating in oligopolistic industries collocating with rivals may not be the preferred option.

STP can act either as a complement or a substitute for collocation. STP has the potential to act as a substitute for collocation where firms are not located in the same cluster, while at the same time, these alliances enable firms to directly monitor knowledge exchange with their collocated partners and to access complementary capabilities, thus STP acts as a complement to collocation. Using data for the European ICT industry, Narula and Santangelo (2009) found that STP is complementary to prior collocation of firms' R&D labs. They found that in that type of oligopolistic industry, instead of using STP as a mechanism to promote knowledge flows, firms strategically use STP as a mechanism to limit knowledge flows and protect competencies.

The Challenges of STP

Although this chapter has described the potential benefits of engaging in STP particularly with regard to innovation performance (Duysters and De Man 2003), they are notoriously costly in terms of resources, and suffer from a high failure rate. Common reasons for their failure include: unfulfilled expectations, lack of trust, asymmetry of learning, incompatibility of organizational cultures, unfair or unclear division of control and responsibilities, as well as certain government policies and regulations. Indeed, most alliances have a 70% failure rate, and even where successful, the majority of them last less than four years. In a study analyzing terminated research alliances in the biotech industry, Reuer and Zollo (2005) found that only 15% of the terminated alliances examined were successful, in the sense that they achieved their intended outcome (although an alliance may be a success from one partner's perspective and a failure from the other, because firms may learn asymmetrically). Given the strategic, uncertain, and knowledge-based nature of the activities involved, these high rates of failure are not surprising. Firms involved in STP face a critical dilemma because they have to maintain the necessary knowledge exchange to achieve the alliance objectives, while at the same time avoid the unintended leakage of valuable technology (Mudambi and Tallman 2010; Oxley and Sampson 2004). This dilemma is common to most non-internalized modes of governance, and is more challenging in horizontal alliances. Although horizontal agreements provide opportunities for economies of scale and scope there are also increased possibilities for the leakage of intellectual property from one partner to the other. Obviously, these managerial challenges become greater if the agreements are of an international nature. The higher the differences in their national and organizational cultures, language, or institutional environments, the higher the resources the partners need to invest in order to assure communication and control within the alliance.

One of the main managerial challenges firms face when doing STP is associated with how to minimize the risk of opportunistic behavior by the partner. It should be noted that given the characteristics of innovative activities, partners cannot afford complete contracts and strong formal governance mechanisms, which increases the room for appropriability hazards. By appropriability hazards we mean the risk of inadequate uses or modifications of the technology and knowledge transferred, not intended in the contract, and injurious to the transferor. Specifically, the transferor can be worse off under two circumstances. First, when the partner becomes a future competitor; and second, when the knowledge gained by the partner benefits other competitors that are actually clients of the same supplier (Kogut 1988; Oxley 1997). Indeed, STP can evolve into a situation that Hamel (1991) calls learning races, in which each partner tries to speed up its learning rate to reach their individual goals first before defecting on the other. Although partners can develop complex contracts to protect their knowledge when self-enforcing safeguards are not effective (Reuer and Ariño 2007), the partner's notion of the other's ability to exploit the information exchanged beyond the scope of the agreement might deviate its behavior from the theoretically optimal one aimed at maximizing the value of the focal transaction (Khanna, Gulati, and Nohria 1998). For this reason, despite the positive effects that the development of relationship-specific investments (both tangible and intangible) may have on the performance of the alliance, managers of the firms involved in the STP may opt to under-invest in these investments to avoid misappropriation issues, even at the cost of a lower alliance performance. In other words, as argued by Martínez-Noya, García-Canal, and Guillén (2013), managers have to be careful with how much they invest in the relationship, because although these relationship-specific investments may improve communication and coordination among partners, they may also be a platform for undesired technology transfers and asymmetric learning among partners. However, in relation to this, another factor to be considered is the underlying motivational or strategic orientation of the potential partner. For instance, compared to business firms, non-profits have different motivational or strategic orientation (Das and Kumar 2011; Li, Wei, and Liu 2010) which means that they have lower incentives to apply the knowledge gained via the alliance to markets or products outside of the agreed-upon scope. This lack of complementary capabilities to commercially exploit the acquired knowledge lowers appropriability hazards, and thus may make them a more suitable partner for those STP in which a high level of tacit and firm-specific knowledge is exchanged (Martínez-Noya et al. 2013).

On the other hand, another important challenge of STP is related to the higher preference for nearby and known partners, which in some circumstances may limit a firm's innovativeness. As it was previously explained, because firms tend to be location- or network-bounded, and thus either physically or strategically interconnected in a particular organizational network, they may be reluctant to take full advantage of the gains from switching to newer and potentially more valuable partners. This preference for nearby and known may lead to over-embedded relationships (Uzzi 1996, 1997). As a consequence, instead of embracing new, valuable exchanges with new partners (whose propensity to cooperate is uncertain), firms may prefer to preserve recurring ties with familiar actors (Portes and Sensenberger 1993). However, it should be noted that this preference for the familiar partner can be especially detrimental in a novel context because it has been found that overcoming the limitations of contextually localized search (Rosenkopf and Almeida 2003), and thus accessing novel resources, is particularly positive for innovation performance when the activities undertaken within the alliance involve exploration such as in the case of radical innovation projects (Nooteboom et al. 2007).

In conclusion, what is clear from previous studies is that external technology development is primarily the domain of larger firms with greater resources, and more experience in transnational activity (Castellani and Zanfei 2003; Martínez-Noya and García-Canal 2011). In this sense, Reuer and Zollo (2005) found partner-specific experience to have a greater favorable impact for non-equity alliances than for equity structures affording stronger formal governance mechanisms. Their findings also indicated that alliance complexity adversely influences firms' termination outcomes in these alliances. For this reason, in order to extract the maximum value from STP, especially those being international, firms need to have strong technological capabilities so as to be able to act as an efficient system integrator (Brusoni et al. 2001).

Policy Issues Associated with STP

Within policy forums, increasingly there is an acceptance that alliances are complementary to in-house R&D, whether undertaken by firms, or public research organizations (PROs). As discussed in this chapter, R&D cooperation does not replace the need for firms and PROs to undertake internal R&D activities, but it enhances and supports internal R&D. After a period in the 1970s and 1980s where international R&D cooperation was viewed with some suspicion, it is now largely accepted that by and large R&D cooperation across borders can have net positive effects on the economy (see, e.g., Vonortas 1997).

Most policymakers recognize that there are cognitive limits to the resources available to any given firm and the costs of maintaining national innovation systems to

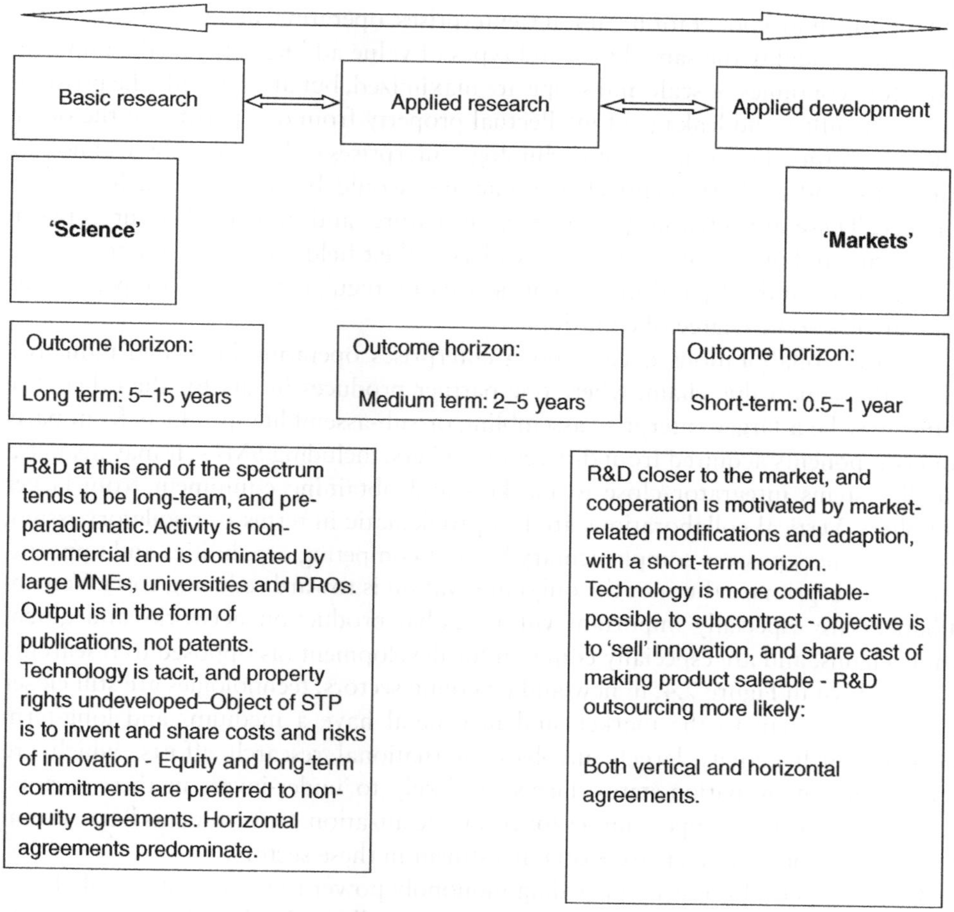

Figure 7.4 Distance-to-market and how it affects STP.
Source: Adapted and modified from European Commission (2012).

acquire world-class expertise in all the different knowledge bases needed in multi-technology products and services is prohibitive. Locations, on the other hand, such as knowledge clusters maintain their overall strengths and specialization over long periods of time, because countries' innovation systems do not evolve rapidly which is why firms are less willing to relocate their R&D abroad (Narula 2002a).

In general, firms and PROs in location A often cooperate with firms and PROs in location B on an ongoing basis as a means to upgrade or even maintain their technological assets. Globalization means that the geographical location or nationality of the potential partner in location B is now largely irrelevant, what matters is the quality of their assets. Furthermore, cooperation can act as an alternative to collocation, and sometimes as a substitute to locating R&D abroad. From a policy perspective this is good, because it counters the dangers of "hollowing out" (Narula 2003).

Policies to intervene or promote R&D cooperation vary by the distance-to-market of the R&D activity, as well as whether they are horizontal or vertical collaborations (Figure 7.4).

In horizontal cooperation (among enterprises operating in the same industry, engaged in roughly the same kinds and types of value adding activity) the opportunities for economies of scale and scope are maximized, but also provide the possibilities for conflict and leakage of intellectual property from one partner to the other. The cooperation between two biotechnology enterprises or between a biotechnology enterprise and a pharmaceutical manufacturer would be considered a horizontal alliance. These are often largely strategic in nature, and in general occur between large firms and organizations that are leaders in their field. They are commonly used to establish standards, and may often be seen by regulators as anticompetitive, as they involve some degree of collusion.

Vertical collaborations occur among enterprises operating in related industries along the same value chain, where one partner produces inputs for the other. The latter may be a larger enterprise assembling or sub-assembling products from parts and components acquired from different suppliers, including SMEs. It may also be a small systems integrator close to markets and obtaining equipment from larger suppliers. Vertical collaborations are less problematic in terms of regulatory issues, as the partners possess complementary but not competing capabilities and opportunities. Their primary (but not the only) motivation is toward reducing costs. Vertical alliances are especially important within global production networks and global value chains, and are especially common for development (as opposed to research).

As shown in Figure 7.4, in new and emerging sectors, technologies are still closer to "science" than to the market, and in general have a medium- and long-term horizon. Such activity largely involves international research efforts, which are mainly driven by nation-states. Firms are likely to under-invest in these sectors, unless there are clear opportunities for commercialization. Governments have several policy instruments to increase R&D investment in these sectors.

One means to do so is by providing monopoly power to the inventors/collaborators so that they may continue to innovate at a socially optimal level through patents or other means. This can be achieved by multilateral collaborative mechanisms, such as consortiums and joint ventures which have the advantage of sharing goals and strategies among the partners. However, where society values maximum diffusion and availability of products at the lowest possible costs, firms will under-invest in R&D when they are uncertain of appropriating sufficient returns. Challenge-driven research is less attractive for firms, especially where the intention is to generate public goods. Key actors from the private sector will only invest resources and take risks of project failure if they can expect a reasonable return in the case of success. On the other hand, swift diffusion of new products and processes is crucial to have a significant impact on global challenges. Diffusion may be hampered if prices for innovations are too high and there are no funding mechanisms to make them broadly available. Another instrument is for governments to directly affect the returns to the innovator by creating a market for the product. Thus the second instrument is for governments to subsidize basic initial research, or to require patent pooling or compulsory cross-licensing of outputs. The difficulty in so doing is that governments may not necessarily select ex ante what the most superior technology is, and it requires a suspension of anti-trust regulations in most cases.

The other instrument is to invest in multi-user research facilities. In this way, the state becomes a participant, and has property rights. Governments can engage as a participant in R&D alliances through PROs or universities. This is especially common in basic research projects, as public research institutes and universities have the human and capital resources to undertake fundamental R&D, or what is referred to as pre-competitive research by the EU. An additional advantage of such participation is that it is better able to monitor the utilization of the resources and act as an honest broker, and prevent the misallocation of funds by commercial (and profit-oriented) partners (European Commission 2012).

In new and emerging sectors, there are often a number of alternative scientific solutions, but firms may not see a viable market (and therefore are reluctant to invest). Governments play the role of market-maker, by creating incentives for private actors to exchange complementary technologies, and by creating a market for the final product that will provide a sufficient return on investment. This may involve PROs who can act as a knowledgeable intermediary between the various parties. Universities play an important role in such alliances.

Although governments are unable to prevent alliances from being unstable, or indeed, from reducing the inherent risk of R&D activity, whether collaborative or otherwise, there is a role for governments in providing information to help identify synergies, complementarities, and opportunities, since there are market imperfections in the market for partners. Even where governments do not own the technologies, they can play an important role in match-making firms. Governments can help diffuse the results of basic research output produced either by government research institutes or private establishments to interested parties by creating a sort of "market place" where potential partners can meet and exchange information. Good government intervention acts to provide "brokering" organizations specific to industries. Identifying the relevant knowledge producers and knowledge users and bringing them together in a comprehensive and communication-rich network is an important role. Epistemic communities – researchers who share a similar approach or a similar position on an issue and maintain contact with each other across their various locations and fields – create new channels for information and discussing new perspectives (European Commission 2012).

In the case of short- and medium-term initiatives which are closer to the market (see Figure 7.4), these focus on activities that are close to commercialization or already commercialized. In such areas, collaboration has a significant timing element, because commercialization of new products and processes must occur before or close to that of the main competitors. As such the window to establish collaboration and complete the joint innovation activity is quite small – as low as six months. Establishing collaboration is a management- and resource-intensive affair, and for short-term collaborations firms prefer to rely on non-equity modes of cooperation which have an important trust element (equity arrangements are slow and cumbersome to establish).

An important role of governments is to assist standard setting, and STP forms a primary means to do this. There is the need to reduce cross-border duplication of activity, especially in terms of multiple (and not necessarily compatible) standards which can be potentially sub-optimal in terms of expenditure on a global basis. However, successful governmental initiatives in standard setting are, in general, the exception rather than the rule, and voluntary standards developed by the main commercial and technological players are more likely to be successful.

Vertical collaborations (customer-supplier) are built on the principle of redundancy, given the high failure rate of collaborations. "Central" firms within a global production network will make specifications available, and collaborate with a number of similarly qualified suppliers on the same project. The principle is that competition will increase the options of selecting the best technology, given that several partnerships will fail to produce optimal results.

Especially where the technological fields reflect existing research competencies and capacities and maintaining or catching-up competiveness in crucial, a much stronger bottom-up approach is required, because short life cycles means that time-to-market is important. Delays due to bureaucratic issues and high transaction costs can make a huge difference. Here, the priority for firms tends to be on the de facto technological competencies of the partners, as judged by the actors themselves, rather than geographical or national level issues, which are shaped by political imperatives, and can result in sub-optimal outcomes from a technological perspective.

Lastly, most policymakers pay special attention to SMEs. One of the policy concerns for states is to increase the participation of SMEs in innovation activities, and specifically to STP, to reduce transaction costs for SMEs to participate. There are high barriers to participation due inter alia to high transaction costs of participating in international STP. These are exacerbated by the bureaucracy of engaging in state-subsidized R&D consortiums. Given that R&D time frames operate on a relatively small window – especially development activities closer to the market, and in fast-moving sectors – delays can make the purpose of the collaboration redundant.

Notes

1 Intricate linkages between economic entities that create informal and formal networks to undertake value added activity dates back to before the seventeenth century, where production of goods was undertaken by "putting out." Rapid Japanese industrial growth over the last century has been partly attributed to the cooperation between interlinked firms with limited equity cross-holdings within industrial groups.

2 *Horizontal collaborative agreements* are those among firms operating in the same industry, engaged in roughly the same kinds and types of value adding activity. The opportunities for economies of scale and scope are here maximized, but also provide the possibilities for conflict and leakage of intellectual property from one partner to the other. *Vertical collaborations* occur among firms operating in related industries along the same value chain, where one partner produces inputs for the other. These agreements tend to be less problematic, as a priori the partners possess complementary but not competing capabilities and opportunities. Their primary (but not the only) motivation is reducing costs.

References

Abernathy, W., and J. Utterback. 1978. "Patterns of Industrial Innovation." *Technology Review* 80: 97–107.

Alcácer, J. 2006. "Location Choices Across the Value Chain: How Activity and Capability Influence Collocation." *Management Science* 52(10): 1457–1471.

Alcácer, J., and W. Chung. 2007. "Location Strategies and Knowledge Spillovers." *Management Science* 53(5): 760–776.

Archibugi, D., and S. Iammarino. 2002. "The Globalization of Technological Innovation: Definition and Evidence." *Review of International Political Economy* 9(1): 98–122.

Archibugi, D., and J. Michie. 1995. "The Globalisation of Technology: A New Taxonomy." *Cambridge Journal of Economics* 19: 121–140.

Arora, A., and A. Gambardella. 1990. "Complementarity and External Linkages: The Strategies of the Large Firms in Biotechnology." *Journal of Industrial Economics* 37: 361–379.

Arora, A., A. Fosfuri, and A. Gambardella. 2001. *Markets for Technology: The Economics of Innovation and Corporate Strategy*. Cambridge, MA: MIT Press.

Arundel, A., and I. Kabla. 1998. "What Percentage of Innovations are Patented? Empirical Estimates for European Firms." *Research Policy* 27: 127–141.

Balakrishnan, S., and B. Wernerfelt. 1986. "Technical Change, Competition and Vertical Integration." *Strategic Management Journal* 7: 347–359.

Barney, J.B. 1991. "Firm Resources and Sustained Competitive Advantage." *Journal of Management* 17(1): 99–120.

Blanc, H., and C. Sierra. 1999. "The Internationalization of R&D by Multinationals: A Tradeoff Between External and Internal Proximity." *Cambridge Journal of Economics* 23: 187–206.

Brusoni, S., A. Prencipe, and K. Pavitt. 2001. "Knowledge Specialization and the Boundaries of the Firm: Why Do Firms Know More Than They Do?" *Administrative Science Quarterly* 46: 597–621.

Buckley, P., and M. Casson. 1976. *The Future of the Multinational Enterprise*. New York: Holmes & Meier.

Buckley, P., and M. Chapman. 1998. "The Management of Cooperative Strategies in R&D and Innovation Programmes." *International Journal of the Economics of Business* 5: 369–381.

Calderini, M., and G. Scellato. 2005. "Academic Research, Technological Specialization and the Innovation Performance in European Regions: An Empirical Analysis in the Wireless Sector." *Industrial and Corporate Change* 142: 279–305.

Cantwell, J.A., and M.G. Colombo. 2000. "Technological and Output Complementarities, and Inter-firm Cooperation in Information Technology Ventures." *Journal of Management and Governance* 4: 117–147.

Cantwell, J., and E. Kosmopoulou. 2002. "What Determines the Internationalization of Corporate Technology?" In *Critical Perspectives on Internationalization*, ed. M. Forsgren, H. Håkanson, and V. Havila, 305–334. London: Pergamon Press.

Cantwell, J., and F. Sanna-Randaccio. 1990. "The Growth of Multinationals and the Catching up Effect." *Economic Notes* 19: 1–23.

Cantwell, J.A., and G.D. Santangelo. 1999. "The Frontier of International Technology Networks: Sourcing Abroad the Most Highly Tacit Capability." *Information Economics and Policy* 11: 101–123.

Cantwell, J.A., and G.D. Santangelo. 2000. "Capitalism, Profits and Innovation in the New Techno-Economic Paradigm." *Journal of Evolutionary Economics* 10(1–2): 131–157.

Castellani, D., and A. Zanfei. 2003. "Technology Gaps, Absorptive Capacity and the Impact of Inward Investments on Productivity of European Firms." *Economics of Innovation and New Technology* 12: 555–576.

Chen, S. 2004. "Taiwanese IT Firms' Offshore R&D in China and the Connection with the Global Innovation Network." *Research Policy* 33(2): 337–349.

Coase, R. 1937. "The Nature of the Firm." *Economica* 4(16): 386–405.

Cohen, W.M., and D.A. Levinthal. 1989. "Innovation and Learning: The Two Faces of R&D." *Economic Journal* 99: 569–596.

Cohen, W.M., and D.A. Levinthal. 1990. "Absorptive Capacity: A New Perspective on Learning and Innovation." *Administrative Science Quarterly* 35: 128–152.

Criscuolo, P., and B. Verspagen. 2008. "Does It Matter Where Patent Citations Come From? Inventor Versus Examiner Citations in European Patents." *Research Policy* 37(10): 1892–1908.

Criscuolo, P., R. Narula, and B. Verspagen. 2005. "Measuring Knowledge Flows Among European and American Multinationals: A Patent Citation Analysis." *Economics of Innovation and New Technologies* 14: 417–433.

Croisier, B. 1998. "The Governance of External Research: Empirical Test of Some Transaction-Cost Related Factors." *R&D Management* 28: 289–298.

Das, T.K., and R. Kumar. 2011. "Regulatory Focus and Opportunism in the Alliance Development Process." *Journal of Management* 37(3): 682–708.

Das, T.K., and B.-S. Teng. 2000. "A Resource-Based Theory of Strategic Alliances." *Journal of Management* 26(1): 31–61.

Dicken, P. 2011. *Global Shift: Mapping the Changing Contours of the World Economy*. New York: Guilford Press.

Dosi, G. 1982. "Technological Paradigms and Technological Trajectories: A Suggested Interpretation of the Determinants and Directions of Technical Change." *Research Policy* 11: 147–162.

Dunning, J. 1995. "Reappraising the Eclectic Paradigm in the Age of Alliance Capitalism." *Journal of International Business Studies* 26: 461–491.

Dunning, J. 1997. *Alliance Capitalism and Global Business*. London: Routledge.

Dunning, J.H. 1998. "Location and the Multinational Enterprise: A Neglected Factor." *Journal of International Business Studies* 29(1): 45–66.

Duysters, G., and A. De Man. 2003. "Transitory Alliances: An Instrument for Surviving Turbulent Industries?" *R&D Management* 33: 49–58.

European Commission 2012. *International Cooperation in Science, Technology and Innovation: Strategies for a Changing World*. Luxembourg: Publications Office of the European Union.

Feinberg, S., and A. Gupta. 2004. "Knowledge Spillovers and the Assignment of R&D Responsibilities to Foreign Subsidiaries." *Strategic Management Journal* 25(8–9): 823–884.

Gambardella, A., and S. Torrisi. 1998. "Does Technological Convergence Imply Convergence in Markets? Evidence from the Electronics Industry." *Research Policy* 27: 445–463.

Graf, M., and S.M. Mudambi. 2005. "The Outsourcing of IT-Enabled Business Processes: A Conceptual Model of the Location Decision." *Journal of International Management* 11: 253–268.

Granstrand, O., P. Patel, and K. Pavitt. 1997. "Multi-Technology Corporations: Why They Have 'Distributed' Rather Than 'Distinctive Core' Competencies." *California Management Review* 39(4): 8–25.

Gulati, R. 1995. "Does Familiarity Breed Trust? The Implications of Repeated Ties from Contractual Choice in Alliances." *Academy of Management Journal* 38: 85–112.

Gulati, R., 1998. "Alliances and Networks." *Strategic Management Journal* 19: 293–317.

Hagedoorn, J. 1993. "Understanding the Rationale of Strategic Technology Partnering: Inter-Organizational Modes of Cooperation and Sectoral Differences." *Strategic Management Journal* 14: 371–385.

Hagedoorn, J. 2002. "Inter-Firm R&D Partnerships: An Overview of Major Trends and Patterns since 1960." *Research Policy* 31: 477–492.

Hagedoorn, J., and G. Duysters. 1997. "Satisficing Strategies in Dynamic Inter-Firm Networks: The Efficacy of Quasi-Redundant Contacts." MERIT Working Paper Series 97–016.

Hagedoorn, J., and R. Narula. 1996. "Choosing Modes of Governance for Strategic Technology Partnering: International and Sectoral Differences." *Journal of International Business Studies* 27: 265–284.

Hagedoorn, J., and R.N. Osborn. 2002. "Interfirm R&D Partnerships: Major Theories and Trends Since 1960." In *Cooperative Strategies and Alliances*, ed. F.J. Contractor and P. Lorange, 517–542. Oxford: Pergamon Press.

Hagedoorn, J., and B. Sadowski. 1999. "Exploring the Potential Transition from Strategic Technology Partnering to Mergers and Acquisitions." *Journal of Management Studies* 36: 87–107.

Hagedoorn, J., and J. Schakenraad. 1994. "The Effect of Strategic Technology Alliances on Company Performance." *Strategic Management Journal* 15: 291–311.

Hamel, G. 1991. "Competition for Competence and Interpartner Learning Within International Strategic Alliances." *Strategic Management Journal* 12: 83–103.

Henderson, R., and K. Clark. 1990. "Architectural Innovation: The Reconfiguration of Existing Product Technologies and the Failure of Established Firms." *Administrative Sciences Quarterly* 35: 9–30.

Hennart, J. 1993. "Explaining the Swollen Middle; Why Most Transactions are a Mix of Market and Hierarchy." *Organization Science* 4: 529–547.

Hitt, M.A., B.W. Keats, and S.M. De Marie. 1998. "Navigating in the New Competitive Landscape: Building Strategic Flexibility and Competitive Advantage in the 21st Century." *Academy of Management Executive* 12(4): 22–42.

Iammarino, S., and P. McCann. 2006. "The Structure and Evolution of Industrial Clusters: Transactions, Technology and Knowledge Spillovers." *Research Policy* 35(7): 1018–1036.

Kay, N. 1997. *Pattern in Corporate Evolution*. Oxford: Oxford University Press.

Khanna T., R. Gulati, and N. Nohria. 1998. "The Dynamics of Learning Alliances: Competition, Cooperation, and Relative Scope." *Strategic Management Journal* 19: 193–210.

Knickerbocker, F. 1973. *Oligopolistic Reaction and the Multinational Enterprise*. Cambridge, MA: Harvard University Press.

Kogut, B. 1988. "Joint Ventures: Theoretical and Empirical Perspectives." *Strategic Management Journal* 9(4): 319–332.

Kogut, B., and U. Zander. 1993. "Knowledge of the Firm and the Evolutionary Theory of the Multinational Enterprise." *Journal of International Business Studies* 24: 625–646.

Lahiri, N. 2010. "Geographic Distribution of R&D Activity: How Does It Affect Innovation Quality?" *Academy of Management Journal* 53(5): 1194–1209.

Levin, R., A. Klevorick, R. Nelson, and S. Winter. 1987. "Appropriating the Returns from Industrial Research and Development." *Brookings Papers on Economics Activity* 3: 783–820.

Lewin, A., S. Massini, and C. Peeters. 2009. "Why Are Companies Offshoring Innovation? The Emerging Global Race for Talent." *Journal of International Business Studies* 40(6): 901–925.

Li, D.T., L. Eden, M.A. Hitt, and R.D. Ireland. 2008. "Friends, Acquaintances, or Strangers? Partner Selection in R&D Alliances." *Academy of Management Journal* 51: 315–334.

Li, Y., Z. Wei, and Y. Liu. 2010. "Strategic Orientations, Knowledge Acquisition, and Firm Performance: The Perspective of the Vendor in Cross-Border Outsourcing." *Journal of Management Studies* 47: 1457–1482.

Lowe, J., and P. Taylor. 1998. "R&D and Technology Purchase Through License Agreements: Complementary Strategies and Complementary Assets." *R&D Management* 28: 263–278.

Madhok, A. 1997. "Cost, Value and Foreign Market Entry Mode: The Transaction and the Firm." *Strategic Management Journal* 18: 39–61.

Madhok, A., and A. Phene. 2001. "The Co-evolutional Advantage: Strategic Management Theory and the Eclectic Paradigm." *International Journal of the Economics of Business* 8: 243–256.

Manning, S., S. Massini, and C. Lewin. 2008. "A Dynamic Perspective on Next-Generation Offshoring: The Global Sourcing of Science and Engineering Talent." *Academy of Management Perspectives* 22(3): 35–54.

Martínez-Noya, A., and E. García-Canal. 2011. "Technological Capabilities and the Decision to Outsource/Outsource Offshore R&D Services." *International Business Review* 20: 264–277.

Martínez-Noya, A., E. García-Canal, and M. Guillén. 2012. "International R&D Service Outsourcing by Technology-Intensive Firms: Whether and Where?" *Journal of International Management* 18: 18–37.

Martínez-Noya, A., E. Garcia-Canal, and M.F. Guillen. 2013. "R&D Outsourcing and the Effectiveness of Intangible Relationship-Specific Investments: May Proprietary Knowledge Be Walking out the Door?" *Journal of Management Studies* 50(1): 67–91.

Mitchell, W., and K. Singh. 1992. "Incumbents' Use of Pre-Entry Alliances Before Expansion into New Technical Subfields of an Industry." *Journal of Economic Behaviour and Organisation* 18: 347–372.

Mowery, D.C., J.E. Oxley, and B.S. Silverman. 1998. "Technological Overlap and Interfirm Cooperation: Implications for the Resource-Based View of the Firm." *Research Policy* 27(5): 507–524.

Mudambi, R. 2008. "Location, Control and Innovation in Knowledge-Intensive Industries." *Journal of Economic Geography* 8(5): 699–725.

Mudambi, S., and S. Tallman. 2010. "Make, Buy or Ally? Theoretical Perspectives on Knowledge Process Outsourcing Through Alliances." *Journal of Management Studies* 47: 1434–1456.

Nagarajan, A., and W. Mitchell. 1998. "Evolutionary Diffusion: Internal and External Methods Used to Acquire Encompassing, Complementary, and Incremental Technological Changes in the Lithotripsy Industry." *Strategic Management Journal* 19: 1063–1077.

Narula, R. 1999. "Explaining Strategic R&D Alliances by European Firms." *Journal of Common Market Studies* 37(4): 711–723.

Narula, R. 2001. "Choosing Between Internal and Non-Internal R&D Activities: Some Technological and Economic Factors." *Technology Analysis & Strategic Management* 13: 365–388.

Narula, R. 2002a. "Innovation Systems and 'Inertia' in R&D Location: Norwegian Firms and the Role of Systemic Lock-in." *Research Policy* 31: 795–816.

Narula, R. 2002b. "R&D Collaboration by SMEs: Some Analytical Issues and Evidence." In *Cooperative Strategies and Alliances*, ed. F.J. Contractor and P. Lorange. Oxford: Pergamon Press.

Narula, R. 2003. *Globalisation and Technology: Interdependence, Innovation Systems and Industrial Policy*. Cambridge: Polity Press.

Narula, R., and J. Dunning. 1998. "Explaining International R&D Alliances and the Role of Governments." *International Business Review* 7: 377–397.

Narula, R., and G. Duysters. 2004. "Globalisation and Trends in International R&D Alliances." *Journal of International Management* 10(2): 199–218.

Narula, R., and J. Hagedoorn. 1999. "Innovating Through Strategic Alliances: Moving Towards International Partnerships and Contractual Agreements." *Technovation* 19: 283–294.

Narula, R., and B. Sadowski. 2002. "Technological Catch-up and Strategic Technology Partnering in Developing Countries." *International Journal of Technology Management* 23: 599–617.

Narula, R., and G.D. Santangelo. 2009. "Location, Collocation and R&D Alliances in the European ICT Industry." *Research Policy* 38: 393–403.

Narula, R., and A. Zanfei. 2005. "Globalisation of Innovation: The Role of Multinational Enterprises." In (Eds.), *Oxford Handbook of Innovation*, ed. J. Fagerberg, D. Mowery, and R. Nelson. Oxford: Oxford University Press.

NSF. 2010. *Science and Engineering Indicators 2010*. Arlington, VA: National Science Foundation.

Nooteboom, B. 1999. *Inter-Firm Alliances: Analysis and Design*. London: Routledge.

Nooteboom, B., W.V. Haverbeke, G. Duysters, V. Gilsing, and A.V. Oord. 2007. "Optimal Cognitive Distance And Absorptive Capacity." *Research Policy* 36(7): 1016–1034.

Osborn, R.N., and C. Baughn. 1990. "Forms of Interorganizational Governance for Multinational Alliances." *Academy of Management Journal* 33(3): 503–519.

Oxley, J.E. 1997. "Appropriability Hazards and Governance in Strategic Alliances: A Transaction Cost Approach." *Journal of Law, Economics and Organization* 13(2): 387–409.

Oxley, J., and R. Sampson. 2004. "The Scope and Governance of International R&D Alliances." *Strategic Management Journal* 25: 723–749.

Pisano, G. 1990. "The R&D Boundaries of the Firm: An Empirical Analysis." *Administrative Science Quarterly* 35: 153–176.

Poppo, L., and T. Zenger. 1998. "Testing Alternative Theories of the Firm: Transaction Cost, Knowledge-Based, and Measurement Explanations for Make-or-Buy Decisions in Information Services." *Strategic Management Journal* 19(9): 853–877.

Porter, M., and M. Fuller. 1986. "Coalitions and Global Strategy." In *Competition in Global Industries*, ed. M. Porter. Boston, MA: Harvard Business School Press.

Portes, A., and J. Sensenberger. 1993. "Embeddedness and Immigration: Notes on the Social Determinants of Economic Action." *American Journal of Sociology* 98: 1320–1350.

Powell, W.W., and S. Grodal. 2005. "Networks of Innovators." In *The Oxford Handbook of Innovation*, ed. J. Fagerberg, D. Mowery, and R. Nelson, 56–85. Oxford: Oxford University Press.

Rangan, S. 2000. "Search and Deliberation in International Exchange: Microfoundations to Some Macro Patterns." *Journal of International Business Studies* 31: 205–222.

Reuer, J., and A. Ariño. 2007. "Strategic Alliance Contracts: Dimensions and Determinants of Contractual Complexity." *Strategic Management Journal* 28: 313–330.

Reuer, J.J., and N. Lahiri. 2014. "Searching for Alliance Partners: Effects of Geographic Distance on the Formation of R&D Collaborations." *Organization Science* 25(1): 283–298.

Reuer, J., and M. Zollo. 2005. "Termination Outcomes of Research Alliances." *Research Policy* 34: 101–115.

Ricart, J.E., P. Agnese, N. Pisani, and T. Adegbesan. 2011. *Offshoring in the global Economy: Management Practices and Welfare Implications*. Bilbao: Fundación BBVA.

Rosenkopf, L., and P. Almeida. 2003. "Overcoming Local Search Through Alliances and Mobility." *Management Science* 49: 751–766.

Rugman, A. 1980. "Internalisation as a General Theory of Foreign Direct Investment: A Reappraisal of the Literature." *Weltwirtschaftliches Archiv* 116: 365–379.

Santangelo, G. 2000. "Corporate Strategic Technological Partnerships in the European Information and Communications Technology Industry." *Research Policy* 29: 1015–1031.

Schilling, M. 2009. "Understanding the Alliance Data." *Strategic Management Journal* 30: 233–260.

Shaver, J., and F. Flyer. 2000. "Agglomeration Economies, Firm Heterogeneity and Foreign Direct Investment in the United States." *Strategic Management Journal* 21(12): 1175–1193.

Teece, D. 1986. "Profiting from Technological Innovation: Implications for Integration, Collaboration, Licensing and Public Policy." *Research Policy* 15: 285–305.

Teece, D.J. 1992. "Competition, Cooperation and Innovation: Organizational Arrangements for Regimes of Rapid Technological Progress." *Journal of Economic Behavior and Organization* 18: 1–26.

Teece, D. 1996. "Firm Organisation, Industrial Structure and Technological Innovation." *Journal of Economic Behavior and Organization* 31: 193–224.

Teece, D., G. Pisano, and A. Shuen. 1997. "Dynamic Capabilities and Strategic Management." *Strategic Management Journal* 18(7): 509–533.

Teece, D., R. Rumelt, G. Dosi, and S.G. Winter. 1994. "Understanding Corporate Coherence: Theory and Evidence." *Journal of Economic Behavior and Organization* 23: 1–30.

Tidd, J., and M. Trewhella. 1997. "Organizational and Technological Antecedents for Knowledge Creation and Learning." *R&D Management* 27: 359–375.

UNCTAD. 2005. *Transnational Corporations and the Internationalization of R&D*. Geneva: United Nations.

Uzzi, B. 1996. "The Sources and Consequences of Embeddedness for the Economic Performance of Organizations: The Network Effect." *American Sociology Review* 61: 674–698.

Uzzi, B. 1997. "Social Structure and Competition in Interfirm Networks: The Paradox of Embeddedness." *Administrative Science Quarterly* 42(1): 35–67.

Veugelers, R. 1997. "Internal R&D Expenditures and External Technology Sourcing." *Research Policy* 26: 303–315.

Veugelers, R., and B. Cassiman. 1999. "Make and Buy in Innovation Strategies: Evidence from Belgian Manufacturing Firms." *Research Policy* 28: 63–80.

Vonortas, N.S. 1997. *Cooperation in Research and Development*. Dordrecht: Kluwer Academic Publishers.

Wernerfelt, B. 1984. "A Resource-Based View of the Firm." *Strategic Management Journal* 5(2): 171–180.

Williamson, O. 1975. *Markets and Hierarchies: Analysis and Anti-Trust Implications*. New York: Free Press.

Zaheer, A., E. Hernandez, and S. Banerjee. 2010. "Prior Alliances with Targets and Acquisition Performance in Knowledge-Intensive Industries." *Organization Science* 21(5): 1072–1091.

Zanfei, A. 2000. "Transnational Firms and the Changing Organisation of Innovative Activities." *Cambridge Journal of Economics* 24: 515–542.

Zollo, M., and S.G. Winter. 2002. "Deliberate Learning and the Evolution of Dynamic Capabilities." *Organization Science* 13(3): 339–351.

Chapter 8

The Globalization of Knowledge-Intensive Services

Ian Miles and Marcela Miozzo

Introduction

The globalization of services (especially knowledge-intensive services) not only involves the geographical extension of services economic activity across borders, but also is an intrinsic part of wider processes of change in the international organization of production and division of labor. Liberalization and the application of new information and communication technologies have transformed services, not simply improving their growth processes or increasing their trade, but altering how industrial activities are conducted, integrating services and goods, and contributing to the unbundling of services activities.

This chapter discusses the many factors involved in the globalization of services. The first two sections explore the particular modes (and barriers) of international "trade" in services. The following section examines the characteristics of services multinationals. The next section explores the role of services in globalization and international division of labor. The chapter then turns to knowledge-intensive services, defining knowledge intensity and knowledge-intensive business services, and the particular issues associated with the internationalization of knowledge-intensive business services (KIBS), including their outsourcing and offshoring.

Liberalization and International "Trade" in Services

Internationalization of services is encountered by many of us on a daily basis. We may drink coffee in a branch of a British coffee shop chain, connecting to the Internet through a French telecommunications company, executing web searches through an American search engine, and ordering goods from an online retail store that operates out of just about any country. The high street could be populated by

The Handbook of Global Science, Technology, and Innovation, First Edition.
Edited by Daniele Archibugi and Andrea Filippetti.

global banks, offices of international accountancy and consultancy firms, and hotels, the brands of which are encountered in most capital cities. Is this so surprising? Most of the cars we see and phones we use are produced for a world market – so why not services?

Service sectors have grown to take major shares of the economy in industrially advanced countries such as members of the EU and OECD. Economists often view services as simply "intangible goods"; but this term is now increasingly used to cover digital content, software, and the like, as distinct from more traditional services. Many researchers stress that service activities are not just "intangible," they are also very much a matter of service relationships: customer input often makes a major contribution to the activity, so that we often see reference to the "co-production" of services. The features of service products vary considerably across different classes of services: TV broadcasts, psychiatric consultation, veterinary treatment, freight transport, and land remediation are just a few of the vast range of activities included as services. But the features of intangibility and interactivity mean that many services are (largely) produced and consumed at the same time and same place, that they are hard to store, that supplier and consumer need to be collocated, and so on. Not surprisingly, these features mean that service trade is not strictly similar to trade in manufactures, which are physical artifacts that can usually be transported (by transport services!). Often a service company will need to set up branches in an overseas country if it wants to internationalize, rather than follow traditional exporting strategies.

Thus services trade is strikingly low, compared to the share of services in economic activity. Cross-border trade in services has stood at only around 20% of total world trade, and has been since the 1970s, despite the growing prominence of services (World Bank 2014). In contrast, foreign direct investment (FDI) was dominated by services over the first part of this century (it averaged 50% of FDI over 2005–2007), but nosedived after the economic crisis set in, down to 39% in 2010 and 40% in 2011, compared to 50% and 46% for manufacturing, respectively (UNCTAD 2012). UNCTAD figures show a shift away from international trade in services to FDI and especially mergers and acquisitions (rather than greenfield FDI). However, it is not necessarily the case that trade and investment are competing strategies – one may complement the other.

The liberalization of international trade and FDI in services has been underpinned by the introduction in 1995 of the General Agreement on Trade in Services (GATS). The GATS was advanced despite resistance from less developed countries such as Brazil and India, who argued that services were primarily a matter of domestic regulation (and in many cases, public provision). Indeed, many service sectors recently opened to internationalization are sensitive because of their distributional, cultural, or strategic significance. Following long negotiations, the GATS was the first multilateral agreement to cover trade in services and involved two elements: the framework of agreement containing rules and regulations; and each country's schedule of "bound commitments" in individual sectors with respect to national treatment and market access for services in each mode (World Trade Organization 2014). The level and nature of commitments generally differ across modes.

Recognizing the greater diversity of international transactions in services compared to goods, the GATS followed earlier researchers (e.g., Bhagwati 1984;

Sampson and Snape 1985) differentiating between four modes of international services "trade" (see Table 8.1) (cf. Baker, Miozzo, and Miles 2002):

- Mode 1, cross-border trade. This is most similar to trade in physical goods, with services being produced in one country and consumed in another. Cross-border communications are employed so that neither the producer nor the consumer moves physically: instead they interact through a postal or a telecommunications network (e.g., exporting software on a computer disc/CD-ROM, sending business reports or architectural blueprints by fax and computer-to-computer transfer; or providing educational or call center services by telecommunications).
- Mode 2, consumption abroad. The consumer moves from one country to another to acquire foreign services, as in tourism or travel for educational or medical purposes. This may also include activities such as aircraft maintenance abroad, where only the property of the consumer moves (there are even cross-border laundry services in Scandinavia!).
- Mode 3, commercial presence. Here the service producer sets up an establishment in a foreign country where consumers are located; this can involve corporations, joint ventures, partnerships, representative offices, and branches, and is evident in international restaurant and hotel chains and in some professional services. Note that commercial presence usually involves investment in the trading partner country. It is also liable to involve intra-firm trade in services.
- Mode 4, trade by temporary presence of natural persons. This involves the temporary movement of service personnel from one country to another, as in the case of lawyers traveling abroad to provide legal services, engineers moving to work on overseas operations, or intra-corporate transfers.

As noted earlier, with respect to modes 1 and 3, the different modes need not be straightforward substitutes one for another. There may be synergies between them, and specific firms may move from one combination of modes to another over time. The entwining of trade and investment issues in the case of services means that the

Table 8.1 Modes of international services trade.

Mode	*Service provider (based in country A)*	*Service customer (based in country B)*
1 Cross-border trade	Provider remains in country A; Service is produced in country A; Service crosses borders	Customer remains in country B; Service crosses borders; Service consumed in country B
2 Consumption abroad	Provider remains in country A; Service remains in country A	Customer crosses border to country A; Service consumed in country A
3 Commercial presence	Provider extends to country B; Service produced in country B	Customer remains in country B; Service consumed in country B
4 Temporary presence	Provider remains in country A; Employee moves to country B; Service produced in country B	Customer remains in country B; Service consumed in country B

GATS is as much an investment agreement as one concerned with cross-border trade. Indeed, commercial presence is the mode of supply where there have been the greatest number of bound commitments (countries guaranteeing levels of market access, national treatment, and that they would not impose new measures restricting entry into the market or the operation of the service) (Zampetti and Sauvé 2007).

"Barriers" to Internationalization of Services

Despite the introduction of the GATS, there remains much perception that services trade is problematic. Indeed, there is some evidence that services trade is substantially more expensive than trade in manufactures. Miroudot, Sauvage, and Shepherd (2012) examine just modes 1 and 2 services trade (due to issues of data availability for the other modes), across 12 services and 17 other sectors. Trade costs are difficult to assess, but here they are estimated by considering the extent to which a country trades with itself, rather than with others; this should take account of all factors that influence trade costs, not just protectionism or regulation. (Though they are unable to identify which specific factors are involved). Trade costs in services are much higher than for goods – often two to three times higher.

These authors estimate, further, that trade costs in services remained relatively steady (indeed, slightly increasing) over the first decade of the present century, whereas trade costs for goods declined overall by more than 15%. Despite the internal market, EU countries displayed quite varied experiences, with different overall levels of trade costs, and different trends over time. Substantial reductions in service trade costs between 1995 and 2007 were noted in Germany, Ireland, and Poland, but over the same period France displayed substantial increase in these costs. All four economies registered substantial declines in trade costs for goods over this period.

Thus services trade seems to be costly, which would be expected to affect the scope for economies of scale and other developments and innovation trajectories. The European Commission (2002) analyzed these issues in the course of preparing for the Services Directive (it instituted this in 2006, with the aim of further liberalizing service trade within Europe). Following the conventions of trade policy, it portrayed the issue here as one of "barriers" to trade, though the protectionist connotations of this term are not strictly accurate. It grouped barriers into legal and non-legal types, with numerous specific difficulties within each type. The main non-legal barriers were simply lack of information about regulations, competent authorities, etc., and broader cultural/language "barriers" affecting formal issues (like the need to have documents translated) as well as less formal ones (such as understanding local values and habits). As regards legal barriers, these were helpfully grouped according to the stage in the service process that was concerned, with a startlingly long list of impediments discussed under each heading. (The document also goes into some detail about the typical treatment of different services, under several of these headings.) A flavor of these legal "barriers" is provided in the following list:

1. Difficulties relating to the establishment of service providers. There may, for example, be "red tape" requirements to comply with local regulations concerning licenses, authorizations, etc., which can consume significant time and financial costs.

2. Difficulties relating to the use of inputs necessary for the provision of services. For example, rules governing the hiring and qualifications of staff, the mobility of personnel, pensions, and working hours.
3. Difficulties relating to the promotion of services. For example, what sorts of advertising are allowed (in some services, advertising is even forbidden in some countries).
4. Difficulties relating to the distribution of services. For example, cross-border supply of some services may be rendered problematic by national rules governing what sorts of professional are entitled to supply services in the country.
5. Difficulties relating to the sale of services. For example, because contracts tend to play a more important role in services' sales than in those of goods, variations in the specifications for contracts across countries constitute obstacles, as do rules and practices related to invoicing, payment, and value-added tax.
6. Difficulties relating to after-sales aspects of services. For example, variations in practice concerning legal (and criminal) liabilities, the validity of post-sales obligations, warranties, etc.

Such issues lead to the EU's "single market" being much less integrated in the sphere of services than in that of goods. This varies, to be sure, across different types of services. Cuadrado-Roura (2010) notes that even within KIBS there are substantial variations across activities – the advertising and marketing industry is relatively highly integrated in Europe, as compared to many traditional professional services (like legal services), but also some more technology-related services, like engineering and R&D.

Globally, the situation is even more diverse. Table 8.2 reproduces a recent assessment of the key regulatory issues confronting international service operations across a range of activities. All of the services encounter difficulties of one kind or another.

Services Multinationals

Services sectors' multinational firms have become very evident in the last few decades. International fast food and retail chains are prominent in many world cities, along with hotels, banks, insurance, and travel firms; the logos of many other service corporations are familiar ones. While the best known are probably American and European, we can see the emergence of prominent Asian transport, hotel, and food chains (and the restaurant chain Nando's is of South African origin). As well as consumer services, we can also see prominent global professional service companies (accountancy, consulting, computer services) and others engaged in international freight transport and business finance. Services multinationals come from all continents, and have come to play a substantial role in most metropolitan centers.

International business scholars have debated whether the international operations of services firms are different from that of manufacturing firms. In the run-up to GATS, many argued that theories developed for manufacturing firms are equally applicable to the international operations of services (e.g., Boddewyn, Halbrich, and Perry 1986; Dunning 1989; Enderwik 1989). Even though research has not kept up with the growth of services multinationals, their international behavior has also been

Table 8.2 Key regulatory issues and modes of supply in selected services sectors.

Type of service	*Key regulatory issues*	*Modes of supply*
Business	Human capital and skills; telecommunications infrastructure Institutions that affect contract enforcement Labor mobility and foreign direct investment regulations; outward policies Trading partners' policies	Cross-border, commercial presence, and presence of natural persons
Financial	Prudential regulations and regulations affecting board members and management Electronic infrastructure and regulations on personal data protection and transfer Innovation of new products, access to payments system Effect on social security through links to health insurance and pension assets management Regulations that limit scope of services (e.g., firewalls between banking, securities, and insurance)	Cross-border, consumption abroad, commercial presence, and temporary movement of labor
Professional	Nationality and residency requirements that limit access to market Labor laws and regulations affecting professionals Differential treatment of applications from foreign and domestic suppliers, including criteria relating to education, experience, examinations, and ethics; competence of applicants; and the need for in-country experience examinations	Presence of natural persons was traditional mode of delivery, but cross-border professional services are increasingly becoming a substitute or complement to this mode
Tele-communications	Electronic (virtual) delivery of services, especially on a cross-border basis Terms, conditions, quality, and reliability of physical infrastructure, absence of which limits electronic delivery of services Barriers to new entrants by incumbents and other limitations to competition	Cross-border and commercial presence
Transportation	Intensive use of physical infrastructure Customs and border management Regulations dealing with international cargo and passengers, sabotage, and non-discriminatory access to and use of basic infrastructure Regulations on drivers, pilots, and crew	Cross-border and commercial presence
Travel	Physical infrastructure and natural endowments Customs and border management (e.g., entry requirements) Other services, such as transportation and health and security standards	Consumption abroad and commercial presence

Source: Molinuevo and Sáez (2014: 12, Table PF1).

viewed as mirroring that of manufacturing firms (see, e.g., Li and Guisinger 1992 on determinants of internationalization and Capar and Kotabe 2003 on multinationality and firm performance). But other commentators argue that studies on the international operations of service firms illuminate some subtleties peculiar to service multinationals – or phenomena that have been neglected in studies of manufacturing firms – that may well aid in the better understanding of internationalization processes in general (Lovelock 1983; Carman and Langeard 1980; Gebauer and Kowalkowski 2012).[1]

Compared to studies of manufacturing multinationals, empirical examination of the international operation of services firms continues to be grossly unrepresented. We know relatively little about the international strategy, organization, and evolution of service multinationals. Despite fragmentation in the literature, there is some consensus that the international expansion of services has followed three "waves" (Kundu and Merchant 2008). A first wave involved finance and business services firms following internationalizing manufacturing firms before the 1980s. A second wave involved the internationalization of services firms in telecommunications and other sectors seeking new markets abroad. A third wave has been due to liberalization and privatization of service sectors, especially in less developed countries.

One controversial topic concerns the extent to which service multinationals can achieve scale economies. Campbell and Verbeke (1994) argued that the characteristics of services (as reviewed above) mean that scale economies are less relevant for service multinationals, though they do apply to some operations such as marketing (branding, investment in corporate image, etc.). Lovelock and Yip (1996) argued that information-based service firms can achieve economies of scale by providing global customers with standardized services and by centralizing upstream value chain activities. In one empirical study, however, Katrishen and Scordis (1998) reported that service firms actually suffer diseconomies of scale with increasing multinationality.

Another controversial topic is whether the acknowledged restructuring of (manufacturing) multinationals, from being loosely coupled organizations to ones that are more integrated and centrally coordinated (Nolan, Sutherland, and Zhang 2002; Rugman 2005), also applies to services multinationals. Miozzo and Yamin's (2012) study suggests that such restructuring may not have been necessary for service multinationals (the internationalization of which has lagged behind that of manufacturing multinationals), but that service multinationals have a fairly centralized structure from the beginning. Instead, information and communication technology (ICT) has speeded up the internationalization process of services firms and facilitated the integration of subsidiaries through global clients, corporate processes and global suppliers rather than restructuring. Other determinants, however, account for variations among service multinationals in the autonomy of subsidiaries. A first set of factors relates to the characteristics of the different sectors, including whether the multinational serves global or local customers and the scale and diversity of subsidiary operations. A second includes the influence of institutions and regulations in the host country (such as restrictions on land or media ownership), sometimes curtailing the scope of local subsidiaries, but other times leading to joint ventures with local firms, enhancing the role of local subsidiaries of services firms.

A further issue of debate is whether the factors influencing the choice of entry modes by manufacturing firms are generalizable to services firms. Some suggest that this is the case (see Agarwal and Ramaswami 1992; Terpstra and Yu 1988; Weinstein 1977); but others stress how such characteristics of service firms as low capital intensity and the inseparability of production and consumption affect entry modes. For example, Erramilli (1990) argued (as GATS was to conclude) that the inseparability of production and consumption in services differentiates entry modes for services and manufacturing firms. While some services can be exported (mode 1), others require close physical proximity for simultaneous production and consumption, and are limited to contractual entry, licensing or franchising, and FDI. In turn, Erramilli and Rao (1990) argue that services firms prefer to start with full-control entry modes. They show variation in choice of entry mode when firms that are characterized by low asset specificity give up control in response to the rising costs of integration or the diminishing ability to integrate. Given that services firms need fewer resources to establish a wholly-owned subsidiary (and have greater flexibility to move those resources), service firms may have greater capability and latitude than manufacturing firms to establish integrated modes for the purpose of reducing transaction costs or implementing what are called "ethnocentric" strategies (Erramilli 1991).

The development of ICT is widely seen to have challenged the traditional inseparability of production and consumption in many services (Miozzo and Soete 2001), which increases the entry and internationalization options available for services. ICT's role in separating the location and timing of (some) service production from its delivery and consumption, and thus in helping reshape the "mode of presence" of service organizations in foreign markets, was early on spotted by Vandermerwe and Chadwick (1989). New ICT allows for some services to be traded, and some management control procedures to be conducted, remotely. These practices compete with and may substitute for traditional physical modes of presence.

Often the initial impetus of such service firms to internationalize is in order to follow multinational corporations that are their major long-term clients, which is then followed by extending operations in the new host economy (Roberts 1998). Such client-following is one of the three paths that Toivonen et al. (2009) see internationalizing services firms as pursuing. They see some other firms as simply expanding markers through independent gradual internationalization (here we might expect services to follow something close to the "Uppsala model"; see Johanson and Vahlne 2009). A third path is that of the "born globals," rapidly internationalizing new service firms, often relying on ICT-enabled exports. Despite all these contributions, we still know a lot less about the entry modes and international strategy and organization of services as compared to manufacturing multinationals.

The Role of Services in Globalization and the International Division of Labor

Internationalization of services (and, as we will see below, of KIBS) is one of the most important parts in the general process of the globalization of production, distribution, and innovation, having implications for the wider international division of labor and for the competitiveness of firms, regions, and countries. One of the problems with understanding these wider processes, however, is that different sets of literature address different issues. International trade issues are usually treated separately from the operation of multinationals, with each of these lines of research drawing upon different traditions of economic thought (see Chesnais 1992), although the issues have become increasingly closely intertwined since the mid-1970s.

Developments in ICT and liberalization have facilitated the reconfiguration of organizations in all sectors of the economy. ICT systems with unprecedented flexibility and reach have facilitated radical transformations of the organization of production, including not only an increase in services trade but also the development of new (international) services activities, new service suppliers, and new business models for service firms. Later, we argue that these developments apply especially to knowledge-intensive service firms.

The activities of service and other multinationals may simply follow the logic of expanding their sales in wider markets; they may involve establishing roughly similar systems of production and/or delivery in more countries (in the case of services operations, it has often been argued that their growth is as much a matter of replication of branches as a greater and more complex divisions of labor); or they may involve new configurations of production and consumption across borders. An example is the blurring of services and goods, with manufacturing firms responding to the commoditization of products by adopting strategies that recast these as service offerings (e.g., Rolls Royce selling power by the hour or Apple products' links to iTunes). Another is the use of the Internet as a platform for delivery of services (note the evolution of shared services and of Cloud Computing, through which organizations in all sectors can store and access data without building their own data centers).

ICT has transformed services by enabling the formalization of activities requiring rules-based logic, which allows for the "unbundling" of business functions, which can be recombined and purchased from third parties (Breznitz and Zysman 2013). Thus the back-office service operations of many companies have been relocated to lower wage areas, often initially in the country of origin and only later in other and less costly countries. The offshoring of office services was already widely discussed in the 1980s, while, more recently, offshoring of many more knowledge-intensive services has been contentious. The more immediate point concerns multinational firms, whose operations span continents. Their operations can be restructured more readily through advances in ICT (and in transport and other services), and this is facilitated by the liberalization of cross-border transactions through GATS and related initiatives. Those parts of the business processes that do not require such close physical location to other parts can be located where new strategies suggest is most profitable (taking into account factors like long-term security and business and political partnerships). Low wages are not the only driver, there may be scarce skills concentrated in some locations, and and in some types of business the ability to span several time zones may be advantageous. Organizations in all sectors are reconsidering how each stage of their production processes (and business functions) are performed, establishing and reconfiguring global value chains, with business activities located in, and intermediate inputs derived from, several countries (Nicita, Ognivtsev, and Shirotori 2013).

Knowledge Intensity

The concept of "knowledge-intensive services" can be seen as rooted in early discussions of the service economy and post-industrial society, where the eminent theorist Daniel Bell placed much emphasis on knowledge as an "axial principle" in the emerging economy, and one that was transforming the nature of work (Bell 1974). But it became prominent toward the turn of the century, having been pioneered by researchers such as Edvardsson (1990) and Edvardsson, Edvinsson, and Nystrom (1993) who discussed knowledge-intensive service firms, such as management consultancies, often in the context of their international expansion. As statisticians turned their attention increasingly to services, there was demand for some parallel to the classification of manufacturing industries into high- and low-tech sectors. This was partly to do with the desire to get more insight into those sectors that are more innovative, and those that are more likely to display higher levels of growth and to carry competitive advantage and benefit national economies. High-tech manufacturing could be defined (if problematically) in terms of R&D intensity, but services firms typically conduct much more limited R&D, organizing their innovative efforts in other ways (Miles 2007).

A knowledge-intensive activity is likely to be produced by workers who are capable of deploying considerable knowledge and understanding in the service work. This can be assessed (again, problematically) in terms of levels of educational attainment. Educational attainment data for international comparisons are available in terms of three categories, the highest being university graduate or equivalent. As an operational definition, Eurostat considers sectors to be knowledge-intensive if more than 33% of the total employment in the sector is tertiary educated (ISCED97, levels 5+6). On this basis, Eurostat has concluded that around half of Europe's service economy can be regarded as knowledge-intensive. NACE rev. 1.1 sectors G (Hotels, Restaurants and Catering), H (Transport, Storage), K (Trade and Repair), and L (Public Administration and Defence, Compulsory Social Security) are generally seen as not, while I (Financial Intermediation), J (Real Estate, Renting, Business Activities), M (Education), and N (Health and Social Work) are considered knowledge-intensive.[2] Eurostat's definition of knowledge-intensive services features four broad categories, as outlined in Table 8.3.

This approach defines a large number of services sectors as knowledge-intensive. Perhaps if around half of Europe's service workforce belongs to such sectors, this just confirms that we live in a knowledge economy. The large number and wide range of these services means that a thorough analysis of internationalization of all knowledge-intensive services would need to be very extensive.

Some service sectors, however, do have much higher levels of graduates in the workforce than most others, in particular education, health, and KIBS. Setting a higher bar to determining which sectors (i.e., which broad sets of core service transformation) are knowledge-intensive would focus just on these. Sectors like telecommunications and financial services are important in the internationalization processes, and they undoubtedly require high levels of knowledge to function, and employ many highly educated staff. But they also typically have many staff performing more routine tasks (e.g., telephone line installation and repair, and bank teller services). Health and education services are also subject to much internationalization pressure,

Table 8.3 Knowledge-intensive services categories.

NACE (rev. 1.1) codes	*Categories*
	Knowledge-Intensive High-Technology Services
64	Post and telecommunications
72	Computer and related activities
73	Research and development
	Knowledge-Intensive Market Services (excl. Financial Intermediation and High-Tech Services)
61	Water transport
62	Air transport
70	Real estate activities
71	Renting of machinery and equipment without operator and of personal and household goods
74	Other business activities
	Knowledge-Intensive Financial Services
65	Financial intermediation except insurance and pension funding
66	Insurance and pension funding, except compulsory social security
67	Activities auxiliary to financial intermediation
	Other Knowledge-Intensive Services
80	Education
85	Health and social work
92	Recreational, cultural, and sporting activities

Source: Authors' own elaboration, based on Eurostat.

but they are often organized as collective public services, and this status led to these two types of service being the focus of much debate about limitations that should be placed on the GATS. Thus we focus particularly on KIBS in what follows.

What Are Knowledge-Intensive Business Services?

Business services are services that are sold to (or undertaken within) firms and other organizations in the formal economy to support their business processes. (The broader category of business-related services includes services that are also supplied on a large scale to consumers, but that are vital to business, e.g., telecommunications, finance, and transport).

In sectoral terms, NACE (rev. 1.1) section J covers Divisions 70 Real estate; 71 Renting equipment; and 72–74 Business services. Divisions 70 and 71 are not usually considered to be KIBS, not featuring such high shares of graduate employment. Thus many studies identify the industries that comprise Divisions 72–74 as KIBS.

Division 72 consists of computer-related services: 72.1 Hardware consultancy; 72.21 Publishing of software; 72.22 Other software consultancy and supply; 72.3 Data processing; 72.4 Database activities; 72.5 Maintenance and repair of office, accounting and computing machinery; and 72.6 Other computer related activities. These are new technology-related activities, usually in the business of supplying and applying specialized knowledge that supports information systems and related business processes, often to customers from many other economic sectors. The workforce has a high share of people with science and engineering training.

Division 73 covers Research and experimental development services (R&D), consisting of two groups: 73.1 R&D on natural sciences and engineering; and 73.2 R&D on social sciences and humanities. This is a relatively small division in terms of employment and value-added. It is distinguished because of its significance for innovation studies, though how far R&D should be segregated here from activities like design, testing, and even market research is arguable.

The largest set of services in Section J is Division 74, where KIBS include: 74.11 Legal activities; 74.12 Accounting, book-keeping and auditing activities, tax consultancy; 74.13 Market research and public opinion polling; 74.14 Business and management consultancy activities; 74.15 Management activities of holding companies; 74.2 Architectural and engineering activities and related technical consultancy; 74.3 Technical testing and analysis; and 74.4 Advertising.[3]

Numerous studies demonstrate that KIBS are: typically small businesses, though a few large firms operate on a worldwide basis in many sectors; they report high levels of technological and organizational innovation; they tend, of course, to have high shares of highly qualified workers, and some are thus seen as "professional services." While many services are known to customize their products for individual customers and to engage in "co-production" with these customers, KIBS are particularly prone to do so (Martinez-Fernandez, Miles, and Weyman 2011). Most service branches offer mixtures of relatively standardized and more customized/specialized service products, and some KIBS are relatively standardized (i.e., scanning the business environment for many clients). But many KIBS require high levels of contact between service producers and clients, which naturally has implications for internationalization.

Issues Affecting the Internationalization of KIBS

The need for interaction often implies physical proximity, so it is not surprising that many KIBS follow mode 3 patterns, and "trade" through FDI or through setting up professional relationships with firms operating in the host countries. As well as physical proximity, this may also be driven by factors such as language, knowledge of local cultures, and access to local social networks.

Additionally, the internationalization of KIBS may be influenced by regulatory frameworks, less deriving from conventional protectionist trade barrier motivations, and more from public interest considerations connected with these services. A recent IBRD report (Molinuevo and Sáez 2014) has examined the trade difficulties encountered by a number of KIBS in considerable detail. This review indicates that different KIBS have to operate under very different regulatory as well as market frameworks.

Long-established professional services often operate under conditions that can pose challenges to internationalization; national systems of accreditation and self-regulation of professions have grown up more or less independently. Despite many recent efforts to achieve comparability, these may make it hard to practice across different national contexts. More recently, emerging KIBS, such as market research and ICT services, are often less heavily regulated; ICT services, with many transactions mediated through telecommunications, can be particularly hard to monitor and control.

The regulations covering many traditional professional services may determine who is entitled to provide the service (entry regulation: e.g., what qualifications should be held), and what norms they should follow (conduct regulation: e.g., rules about advertising and pricing). There is a public interest rationale for such regulations: there are liable to be market failures, for example, because it is hard for consumers to assess service quality and value for money. There may also be doubts about the effectiveness of self-regulation, and suspicions that service suppliers may seek to provide premium services even when customer requirements would be met by more basic services. Thus, consumers may need to be provided with reliable information, and/or market entry should be limited by application of professional standards (education and qualification requirements; clear specification of what tasks specific types of professional should undertake).

One example of professional services that Molinuevo and Sáez (2014) examine in some depth is accountancy; reliable accountancy services are essential to the performance and stability of financial markets. Such KIBS may be subject to qualitative requirements, such as restrictions on access to the profession (where professional associations may exert a monopoly over training institutions), and multiple certification requirements. A country may have laws governing banking and insurance industries that require that companies use approved auditors (generally affiliates of one of the major companies, making it hard for smaller suppliers to access the market). Some national differences may be attenuating, however, as International Standards on Auditing are being adopted. Another long-established professional KIBS sector is legal services, where Molinuevo and Sáez (2014) describe three models of regulation (varying mainly in terms of the extent to which exclusive rights are granted to legal professionals). They point to price regulations, restrictions on advertising and on business structures (some sorts of intra-professional partnership and multidisciplinary activity may be prohibited to ensure the independence of professional advice and to reduce conflicts of interest).

In contrast, ICT-related services (among which they include back-office services, ICT maintenance, and software development), generally feature relatively open regulatory regimes. One reason for this is that cross-border trade relies on telecommunications platforms, which are difficult for regulators to monitor and therefore regulate. The novelty of these services may also mean that there has been less time for long-term institutionalization processes to have diverged across countries. There may be issues connected with data protection and similar regulations. The most important barriers in this sector thus relate less to regulations, more to constraints that affect services and infrastructures that are essential inputs, such as telecommunications, financial services, and professional services. (These constraints may impact the demand side as well as the supply side: for example, limitations on cross-border provision of professional services affect ICT services related to that professional activity.)

Outsourcing and Offshoring of KIBS

The general processes of trade and investment liberalization and advances in ICT have significantly increased the scope for segmentation of organizations' activities in all sectors, through modular production networks in manufacturing (Sturgeon 2002) and unbundling of services (Breznitz and Zysman 2013). These developments have enabled multinationals to widen the scope of coordination, so that it includes not only their own subsidiaries, but also their network of external suppliers

(Rugman and D'Cruz 2003; Nolan et al. 2002). Greater reliance by headquarters on externalization includes the unbundling of service functions, allowing multiple points of entry for new types of organizations and for innovation. The growing significance of services in adding value to products (e.g., marketing, distribution, and after sales maintenance), and the emergence of new markets for business services, have contributed to this trend.

As a result of such dynamics, many internal functions of firms and other organizations have been unbundled; they can be produced within the firm at new locations, or secured from a domestic or foreign company. *Outsourcing* refers to the decision to buy services previously produced internally from another (domestic or offshore) company. Researchers have examined how organizations are shedding important business functions such as human resources, ICT, accounting and finance, and R&D (Howells 1990; Miozzo and Grimshaw 2005; Sako and Tierney 2005). These processes of subcontracting and outsourcing are sometimes accompanied by the consolidation and geographic expansion of global suppliers. In the case of services, some very large suppliers are emerging in ICT services, personnel or human resources, and engineering services. These are growing in a different way from contract manufacturers or autoparts producers, whose strategy is to capture higher value-added activities in the same industry as their clients. Instead, services outsourcing firms provide novel combinations of services and business functions and develop functional knowledge and market position in a different industry from their clients (Miozzo and Grimshaw 2011).

Offshoring refers to a domestic company obtaining services from a foreign-based company, be that an overseas subsidiary (captive or international in-sourcing), or an independent service provider, in which case we have outsourcing and offshoring combined (offshore outsourcing). Nicita et al. (2013) note that from a very low base, services offshoring has mushroomed in the present century, though precise data are lacking. Estimates for 2010 (Gereffi and Fernandez-Stark 2010) suggest that some US$250–300 billion is spent on offshore services, and that there is increasing offshoring of more sophisticated services (Figure 8.1).

Data processing (and data entry), basic software, and other ICT services were the first services to be outsourced and offshored to a significant degree. In the 1980s, some US firms sent credit card processing to, and established call centers in, the Caribbean, while PC manufacturers established software centers in Malaysia. A substantial software offshoring industry emerged in the 1990s. India, Singapore, Ireland, Israel, and Hungary were all early entrants in the offshoring business and benefited from first-mover advantages. One driver of offshoring was labor shortages in the United States during the 1990s, especially connected to solving the "Y2K problem" and the speculative development of new Internet products and services during the dot-com boom. Other countries, too, faced similar pressures and were sourcing software development overseas. The burst of the dot-com bubble at the beginning of the century came after momentum had been established; offshoring was now seen as the transfer of jobs abroad, rather than the supplementing of an insufficient US labor market. (Likewise, in Europe longstanding concerns about ICT skills shortages were voiced less often.) Firms in India and the other countries moved up the value chain through specialization, deepening their expertise and building relations with clients which would eventually demand progressively higher value-added tasks for them

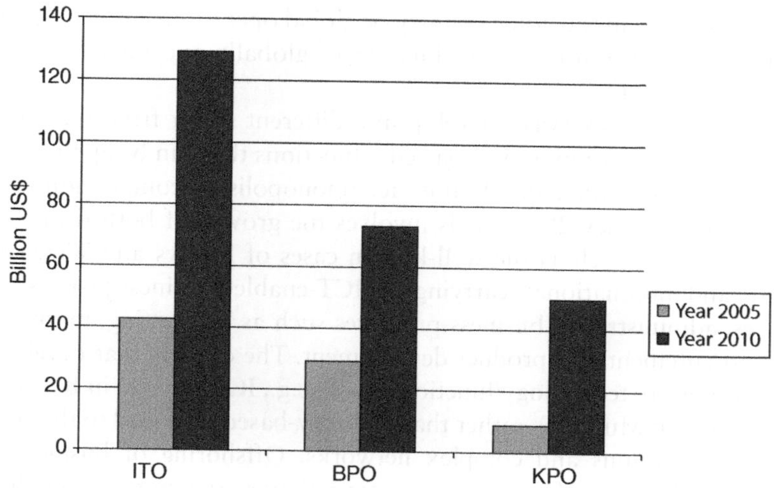

Figure 8.1 Offshored services by segment 2005 and 2010.
Source: Gereffi and Fernandez-Stark (2010), based on OECD data.
Note: Information technology outsourcing (ITO) for Gereffi and Fernandez-Stark covers software development, applications and infrastructure management, and ICT consulting. Business process outsourcing (BPO) includes enterprise, human and customer resource management. What Gereffi and Fernandez-Stark call knowledge process outsourcing (KPO) is regarded as the more sophisticated by them and covers business consulting, market intelligence, and legal services.

to do. There were often substantial wage differentials for programmers of equivalent capability in emergent economies like India as compared to industrially advanced countries such as the United States. In India the process began with body-shopping, mode 4 service provision, with trained programmers sent to work for a few months in the client firm's premises in another country. Later, following visa problems (Heeks 1996), a blended strategy was pursued: some of the work was done on the client's site and some at the vendor's site in India. US firms pioneered the relocation of operations to countries such as India, activities such as call center operations as well as software. Some of the service suppliers from less developed countries moved up the value chain, adding ICT-enabled knowledge-processing to more basic functions. Firms like TCS, Infosys, and Wipro have become prominent service suppliers, competing in the provision of outsourced business functions with large multinational suppliers like Accenture and IBM (Dossani and Kenney 2007).

Although outsourcing and offshoring are not altogether new phenomena, the contemporary outsourcing and offshoring of KIBS involves a number of new features. Massini and Miozzo (2012) document the trend in the offshoring of administrative services, call centers, ICT services, procurement, and product development from the United States and Europe to less developed countries. They outline the differences between the present phase of outsourcing and offshoring of KIBS and previous phases of outsourcing and offshoring. Among the peculiarities is that firms outsourcing and offshoring KIBS today involve not only large multinationals, but also less internationalized companies and small and medium-sized firms. This is a big transformation. Also, the activities outsourced or offshored are not designed to serve the local market of the host country, but instead to serve those activities based in the home country or global operations of the organization. This demands a great effort of coordination of globally and inter-firm dispersed knowledge and activities.

Suppliers of business services develop in a different sector from the outsourcer/ offshorer, and develop know-how in specific functions that can be applied to clients in a broad range of sectors, and do not face (monopolistic) competition from their clients (Sako and Tierney 2005). This involves the growth of both firms from less developed countries (such as the well-known cases of Infosys and Wipro in India cited above) and multinationals carrying out ICT-enabled business processes such as ICT services, administrative business processes such as accounting, marketing, and sales, and procurement and product development. The clusters that develop in offshoring locations are technology/function-based (e.g., ICT services in Bangalore and financial services in Mumbai), rather than industry-based, and lead to the emergence of hybrid organizations and complex networks. Offshoring of business services involves rapid relocation of existing domestic activities to countries including India and China starting in the 1990s and expanding to other Asian countries, Eastern Europe, and Latin and Central America in the 2000s, through hybrid forms of organization and complex networks.

The developments in the outsourcing and offshoring of KIBS raise important questions about the potential erosion of knowledge-based competitive advantage of developed countries. While some less developed countries show upgrading in a number of KIBS, these activities are still very "footloose" (Miozzo and Grimshaw 2008). Also, most of the more creative and knowledge-intensive activities, and most of the functions of technological integration and coordination of lead firms, still remain rooted in dynamic regions in developed economies.

Conclusions

The globalization of knowledge intensive services thus involves a wide variety of processes. Alongside the overseas expansion of service production and delivery, we see new international service activities and the emergence of new (knowledge-intensive) service suppliers. These processes interact with other dimensions of the globalization of innovation (Archibugi and Iammarino 2002); they require efforts of coordination and management of international flows of knowledge and affect regions in the world to different extents and in various ways (e.g., incorporating countries such as China and India into new divisions of knowledge-intensive labor).

Understanding these processes is made difficult because, as has long been recognized, data on services remain relatively underdeveloped compared to statistics on manufacturing and extractive industries; the peculiarities of services "trade" compound the problem. Trade and FDI are intricately entangled, and both theory and measurement are in need of elaboration to assess trends and explore dynamics. The problems are, if anything, growing as a result of liberalization and the ways in which new ICT enables services to be unbundled and relocated. Both organizational and geographical change is underway, as firms (and even some public service bodies) reconsider how and where each stage of their production process is performed. The development of new KIBS and (more generally) knowledge-intensive services are also prompted by new technologies, as well as by social change associated with economic growth and the associated shifts in markets. Interest in service internationalization is thus bound to grow, even if our abilities to monitor and assess the processes here lag behind.

Much of the literature that we have reviewed examines services' internationalization through a "barriers" perspective, following the widespread presumption that trade is necessarily beneficial to all parties.[4] Future research could do well to adopt a more nuanced view here. Service internationalization can give rise to "nostalgic" regret, affronted aesthetic sensibilities, or a sense of threatened community, as, for example, one high street after another around the world becomes dominated by the same set of global shop fronts. But there are also concerns about the international cloning of knowledge-intensive services. Global businesses are emerging in community service fields like education, health, and even security services, and here profit orientation may not always align with such services' other social goals. Less evidently, there are concerns that some international KIBS (e.g., advertisers, consultancies, and ICT service firms) transpose solutions originally designed for one cultural context to very different contexts, with problematic consequences. It is often pointed out that service clients find it hard to assess the suppliers' offerings in advance of purchase, which may render their purchase of international knowledge-intensive services more prone to be influenced by factors such as prestige than on alignment of the service offering with local circumstances. What are taken to be barriers to service trade may really reflect the fact that the service does not mean the same thing or take the same form in different cultural and economic contexts: caution should be exercised in treating knowledge-intensive services as if they were simply intangible goods.

This caution applies particularly when we are considering the globalization of science, technology, and innovation. Knowledge-intensive services, and particularly KIBS, are not just supplying component goods that can be put together to make finished products for global markets. They are often playing important roles in shaping and integrating knowledge, in linking together various actors and communities, in contributing to innovation systems of all kinds. There remains much to be known about how they contribute to globalization, as well as how they respond to this phenomenon.

Notes

1 See Miozzo and Miles (2002), who argue that the views that goods and services are essentially similar, or fundamentally different, obscure the scope for understanding economic activities of all types in an integrated fashion.

2 NACE rev. 2 substantially reorganized the services sectors, but most available statistics are still in the earlier framework, which also happens to group most KIBS together (rev. 2 has distributed them across two broad groups, though does helpfully separate them from more operation business services).

3 Some other services in the NACE rev. 1.1 Division 74 are likely to involve few KIBS firms (based on their workforce skills): 74.5 Labour recruitment and provision of personnel; 74.6 Investigation and security activities; 74.7 Industrial cleaning; 74.81 Photographic activities; 74.82 Packaging activities; 74.85 Secretarial and translation activities (some of these may be KIBS); 74.86 Call centre activities; and 74.87 Other business activities. In NACE rev. 2 there is a major revision of the treatment of KIBS.

Many KIBS are located in Section M, Professional and scientific services, while the activities just mentioned are now located in section N (Administrative and support services).

4 See Hines (2013), who formulates the critique of neoliberal globalization in terms of "progressive protectionism."

References

Agarwal, Sanjeev, and Sridhar N. Ramaswami. 1992. "Choice of Foreign Market Entry Mode: Impact of Ownership, Location and Internalization Factors." *Journal of International Business Studies* 23: 1–27.

Archibugi, Daniele, and Simona Iammarino. 2002. "The Globalization of Technological Innovation: Definition and Evidence." *Review of International Political Economy* 9(1): 98–122.

Baker, Paul, Marcela Miozzo, and Ian Miles. 2002. "The Internationalisation of European Services: What Can Data on International Services Transactions Tell Us?" In *Internationalisation, Technology and Services*, ed. Marcela Miozzo and Ian Miles. Cheltenham, UK: Edward Elgar.

Bell, D. 1974. *The Coming of Post-Industrial Society: A Venture in Social Forecasting*. London: Heinemann.

Bhagwati, Jagdish N.1984. "Splintering and Disembodiment of Services and Developing Countries." *The World Economy* 7(2): 133–144.

Boddewyn, Jean J., Marsha B. Halbrich, and Anne C. Perry. 1986. "Service Multinationals: Conceptualisation, Measurement and Theory." *Journal of International Business Studies* 17(3): 41–57.

Breznitz, Dan, and John Zysman. 2013. *The Third Globalization: Can Wealthy Nations Stay Rich in the Twenty-First Century?* Oxford: Oxford University Press.

Campell, Alexandra, and Alain Verbeke. 1994. "The Globalization of Service Multinationals." *Long Range Planning* 27(2): 95–102.

Capar, Nejat, and Masaaki Kotabe. 2003. "The Relationship Between International Diversification and Performance in Service Firms." *Journal of International Business Studies* 34(4): 345–355.

Carman, James M., and Eric Langeard. 1980. "Growth Strategies of Service Firms." *Strategic Management Journal* 1: 7–22.

Chesnais, François. 1992. "National Systems of Innovation, Foreign Direct Investment and the Operations of Multinational Enterprises." In *National Systems of Innovation*, ed. Bengt-Åke Lundvall. London: Pinter.

Cuadrado-Roura, Juan R. 2010. "Internationalisation of Services." *Paradigmes: economia productiva i coneixement* 4: 13–26, available at http://www.raco.cat/index.php/Paradigmes/article/viewFile/219635/300261 (accessed December 24, 2014).

Dossani, Rafiq, and Martin Kenney. 2007. "The Next Wave of Globalization: Relocating Service Provision to India." *World Development* 35: 772–791.

Dunning, John. 1989. "Multinational Enterprises and the Growth of Services: Some Conceptual and Theoretical Issues." *The Service Industries Journal* 9(1): 5–39.

Edvardsson, Bo. 1990. "Management Consulting: Towards a Successful Relationship." *International Journal of Service Industry Management* 1(3): 4–19.

Edvardsson, Bo, Leif Edvinsson, and Harry Nystrom. 1993. "Internationalisation in Service Companies." *The Service Industries Journal* 13(1): 80–97.

Enderwick, Peter. 1989. *Multinational Service Firms*. London: Routledge.

Erramilli, M. Krishna. 1990. "Entry Mode Choice in Service Industries." *International Marketing Review* 7(5): 50–62.

Erramilli, M. Krishna.1991. "The Experience Factor in Foreign Market Entry Behaviour of Service Firms." *Journal of International Business Studies* 22(3): 479–501.

Erramilli, M. Krishna, and C.P. Rao. 1990. "Choice of Foreign Market Entry Modes by Service Firms: Role of Market Knowledge." *Management International Review* 30(2): 135–150.

European Commission. 2002. *Report on the State of the Internal Market for Services – COM 441*. Brussels: Commission of the European Communities.

Gebauer, Heiko, and Christian Kowalkowski. 2012. "Customer-Focused and Service-Focused Orientation in Organizational Structures." *Journal of Business and Industrial Marketing* 27(7): 527–537.

Gereffi, Gary, and Karina Fernandez-Stark. 2010. *The Offshore Services Global Value Chain*. Durham, NC: Center on Globalization, Governance and Competitiveness, Duke University.

Heeks, Richard. 1996. *India's Software Industry*. London: Sage.

Hines, Colin. 2013. *Progressive Protectionism*. London: Compass.

Howells, Jeremy. 1990. "The Location and Organisation of Research and Development: New Horizons." *Research Policy* 19(2): 133–146.

Johanson, Jan, and Jan-Erik Vahlne. 2009. "The Uppsala Internationalization Process Model Revisited: From Liability of Foreignness to Liability of Outsidership." *Journal of International Business Studies* 40(9): 1411–1431.

Katrishen, F.A., and N.A. Scordis. 1998. "Economies of Scale in Services: A Study of Multinational Insurers." *Journal of International Business Studies* 29(2): 305–323.

Kundu, Sumit K., and Hemant Merchant. 2008. "Service Multinationals: Their Past, Present, and Future." *Management International Review* 48(4): 371–377.

Li, Jiatao, and Stephen Guisinger. 1992. "The Globalization of Service Multinationals in the Triad Regions: Japan, Western Europe and North America." *Journal of International Business Studies* 23(4): 675–696.

Lovelock, Christopher H. 1983. "Classifying Services to Gain Strategic Marketing Insights." *Journal of Marketing* 47: 9–20.

Lovelock, Christopher H., and George S. Yip. 1996. "Developing Global Strategies for Service Business." *California Management Review* 38: 64–86.

Martinez-Fernandez, Cristina, and Ian Miles, with Tamara Weyman (eds.). 2011. *The Knowledge Economy at Work: Skills and Innovation in Knowledge Intensive Service Activities*. Cheltenham, UK: Edward Elgar.

Massini, Silvia, and Marcela Miozzo. 2012. "Outsourcing and Offshoring of Business Services: Challenges to Theory, Management and Geography of Innovation." *Regional Studies* 46(9): 1219–1242.

Miles, Ian. 2007. "R&D Beyond Manufacturing: The Strange Case of Services' R&D." *R&D Management* 37(3): 249–268.

Miozzo, Marcela, and Damian Grimshaw. 2005. "Modularity and Innovation in Knowledge-Intensive Business Services: IT Outsourcing in Germany and the UK." *Research Policy* 34(9): 1419–1439.

Miozzo, Marcela, and Damian Grimshaw. 2008. "Service Multinationals and Forward Linkages with Client Firms: The Case of IT Outsourcing in Argentina and Brazil." *International Business Review* 17(1): 8–27.

Miozzo, Marcela, and Damian Grimshaw. 2011. "Capabilities of Large Services Outsourcing Firms: The 'Outsourcing Plus Staff Transfer Model' in EDS and IBM." *Industrial and Corporate Change* 20(3): 909–940.

Miozzo, Marcela, and Ian Miles. 2002. "The Relation Between the Internationalization of Services and the Process of Innovation: A Research Agenda." In *Internationalisation, Technology and Services*, ed. Marcela Miozzo and Ian Miles. Cheltenham, UK: Edward Elgar.

Miozzo, Marcela, and Luc Soete. 2001. "Internationalization of Services: A Technological Perspective." *Technological Forecasting and Social Change* 67(2): 159–183.

Miozzo, Marcela, and Mo Yamin. 2012. "Institutional and Sectoral Determinants of Headquarters-Subsidiary Relations: A Study of UK Service Multinationals in China, Korea, Brazil and Argentina." *Long Range Planning* 45: 16–40.

Miroudot, Sebastien, Jehan Sauvage, and Ben Shepherd. 2012. "Trade Costs and Productivity in Services Sectors." *Economics Letters* 114(1): 36–38.

Molinuevo, Martin, and Sebastian Sáez. 2014. *Regulatory Assessment Toolkit: A Practical Methodology for Assessing Regulation on Trade and Investment in Services*. Washington, DC: World Bank.

Nicita, Alessandro, Victor Ognivtsev, and Miho Shirotori. 2013. *Global Supply Chains: Trade and Economic Policies for Developing Countries*. Geneva: UNCTAD, Policy Issues in International Trade and Commodities Study Series 55.

Nolan, Peter, Dylan Sutherland, and Jin Zhang. 2002. "The Challenge of the Global Business Revolution." *Contributions to Political Economy* 21(1): 91–110.

Roberts, Joanne. 1998. *Multinational Business Service Firms: The Development of Multinational Organisational Structures in the UK Business Services Sector*. Aldershot, UK: Ashgate.

Rugman, Alan. 2005. *The Regional Multinationals: MNEs and "Global" Strategic Management*. Cambridge: Cambridge University Press.

Rugman, A., and Joseph R. D'Cruz. 2003. *Multinationals as Flagship Firms: Regional Business Networks*. Oxford: Oxford University Press.

Sako, Mari, and Abigail Tierney. 2005. "Sustainability of Business Service Outsourcing: The Case of Human Resource Outsourcing (HRO)." Advanced Institute of Management Working Paper 019-June-2005.

Sampson, Gary P., and Richard H. Snape. 1985. "Identifying Issues in Trade in Services." *The World Economy* 8(2): 171–182.

Sturgeon, Timothy J. 2002. "Modular Production Networks: A New American Model of Industrial Organization." *Industrial and Corporate Change* 11(3): 451–496.

Terpstra, Vernon, and Chwo Ming Yu. 1988. "Determinants of Foreign Investment of US Advertising Agencies." *Journal of International Business Studies* 19(1): 33–46.

Toivonen, Marja, Tiina Tuominen, Anssi Smedlund, and Irma Patala. 2009. "Models and Paths of Internationalisation in KIBS." Paper prepared for the 9th International RESER Conference, Budapest, Hungary.

UNCTAD. 2012. *World Investment Report 2012*. Geneva: UNCTAD.

Vandermerwe, Sandra, and Michael Chadwick. 1989. "The Internationalisation of Services." *Services Industries Journal* 9(1): 79–93.

Weinstein, Arnold K. 1977. "Foreign Investments by Service Firms: The Case of Multinational Advertising Agencies." *Journal of International Business* 8: 83–91.

World Bank. 2014. World Development Indicators. World Bank, http://data.worldbank.org/data-catalog/world-development-indicators (accessed December 26, 2014).

World Trade Organization. 2014. "Guide to Reading the GATS Schedules of Specific Commitments and the List of Article II (MFN) Exemptions." WTO, http://www.wto.org/english/tratop_e/serv_e/guide1_e.htm (accessed December 26, 2014).

Zampetti, Americo Beviglia, and Pierre Sauvé. 2007. "International Investment." In *Research Handbook on International Economic Law*, ed. Andrew T. Guzman and Alan O. Sykes. Cheltenham, UK: Edward Elgar.

Chapter 9

Capital and Technology Flows: Changing Technology Acquisition Strategies in Developing Countries

Suma Athreye and Sandeep Kapur

Traditional Modes of Technology Acquisition in Developing Countries

Acquisition of technology is a vital element of the competitive business strategy of firms in developing countries. Technology acquisition can reduce the productivity gap between developed and developing countries, and even more so if acquired technologies diffuse into wider use within developing countries. To the extent that the income gap across countries is sensitive to the technology gap, understanding the channels for technology flow are of great policy relevance.

Acquisition of technology involves transfer of knowledge. This transfer might be in embodied forms, with technology incorporated in plant and machinery, or even embodied in the form of technical personnel. Alternatively, the transfer could be in disembodied forms, such as the provision of intangible technical services. While some elements of technology are codifiable in blueprints and designs, or articulated in patents, other crucial elements may be tacit, captured only in the general know-how and experience of those who operate the technology. These different types of technological knowledge lend themselves in varying extents to conventional market exchange. Technology embodied in plant and machinery or codified into a patent may be acquired quite readily through arms-length market transactions. However, where technology takes the form of tacit knowledge or general know-how, market transactions are poor channels for technology transfer.

There are well-known reasons why markets for technology do not always work well. First, to the extent that technology represents knowledge, sometimes the mere description of an idea, or sale of an object that incorporates that idea, amounts to a complete transfer of that knowledge itself. Arrow termed this the "paradox of disclosure." This characteristic can interfere with conventional market

The Handbook of Global Science, Technology, and Innovation, First Edition.
Edited by Daniele Archibugi and Andrea Filippetti.

transactions unless supported by the enunciation of intellectual property rights (IPR). Even then, not all elements of technology can be protected, and even when patent protection is used, it may be hard to enforce it. Second, and partly as a corollary of the above, those hoping to acquire technology might have only a poor idea of which out of a competing set of technologies will meet their needs and/or where it might be available. Think of this as the problem of ignorance. A third problem comes from the non-rival nature of information – that its use by one person does not diminish the amount available for others to use. The ability of others to copy or duplicate technology makes it hard to prevent the resale of transferred technology: this problem of appropriability may inhibit the initial transfer of technology. Lastly, there is the problem of absorptive capacity, in that even those who acquire technologies may struggle to deploy it productively. All these features limit the extent to which technology can be transferred across firms and countries through arms-length market transactions.

Where market-based technology transactions across firms do not work, technology could always be transferred across countries but within the boundaries of a firm. For instance, a multinational enterprise (MNE) from a developed country could invest in a manufacturing facility in a developing country. This direct investment might circumvent the problems of disclosure and ignorance, but the transfer would remain vulnerable to the risk of imitation from local rivals. If the host country's institutional environment afforded greater protection for IPR, so that leakage is restrained, an MNE parent would be more willing to transfer more technology to its overseas subsidiary. A porous regime would reduce the incentive to transfer the most valuable technologies, even though there is greater diffusion of that which is transferred.

Thus, conventional wisdom suggested two major channels for the acquisition of technologies of developing countries. In some situations, and for some forms of technology, there might be scope for outright arms-length technology acquisition, through purchase of technology licenses or capital that embodies technology. Where this is not a viable channel, a developing country eager to acquire technology could adopt a more permissive attitude to inward foreign investment, and especially in technology-intensive sectors.

The traditional view is well encapsulated in Archibugi and Pietrobelli (2003) who distinguish between three sources of global technology generation (and potentially technology acquisition) for developing countries. Developing country firms could participate in the international exploitation of nationally produced technology such as occurred in the case of Korean component manufacturers and Japanese cars or Indian software producers and US firms. Both would involve some sort of international licensing activity as well as production on contract for the foreign firm. Alternatively, they could participate in the global generation of innovation (by MNEs) by providing tax incentives for their location and an amenable infrastructure for knowledge-based work. Lastly, developing countries could increase their share of global technological collaborations. They argue that globalization provided huge opportunities for participation by developing country firms in all three forms of technology generation (by developed economy firms). In turn, such participation could provide fertile ground for technology acquisition activities by developing country firms.

Limitations of Technology Acquisition Through Licensing and FDI

Licensing proved to be a poor channel for technology acquisition in the early decades after the Second World War, and largely failed to bridge the technology gap between developed and developing countries. A large number of studies identified why this might have been so. First, developing country firms lacked information about the set of available technologies and did not always choose the technology most appropriate to their needs (Fransman 1985). Second, firms could buy a technology license but, in the absence of appropriate domestic R&D, they often lacked the complementary infrastructure of intermediate goods and services needed to operationalize the acquired technologies (Dahlman 1978; Lall 1983; Westphal, Kim, and Dahlman 1985). Third, the "know-how" to make the necessary adaptations required tacit understanding of the technology (the "know-why"), which could not be fully contracted for. Firms that made an effort to invest in technology (e.g., through internal R&D) were better able to learn and master acquired technologies but some saw acquired technology as a substitute for their own R&D (Bell and Pavitt 1997; Archibugi and Pietrobelli 2003). As a result, successful technology transfer was limited to firms and countries with not too dissimilar technological capabilities, or at least where the technology gap was not too large relative to the absorptive capacity of the firms acquiring the technology. Where the technology gap was large, licensing failed as a mode of technology acquisition in the developing world.

Of course, international labor mobility may well have aided the transfer of technology. Historical evidence for an earlier era, namely the 1920s and 1930s, documents the role played by the movement of scientific personnel as carriers of know-how and tacit knowledge in the transfer of technology from Europe to the United States (Mowery and Rosenberg 1989; Athreye and Godley 2009). These studies revealed that the informal aspects of technology transfer, such as knowledge and networks that were associated with particular people, were as important as the formal elements. Reduced mobility of scientists in the post-war period made this type of people-embedded technology transfer less common, especially to developing countries.[1] To some extent, international joint ventures and strategic technology partnering could have filled this gap. Narula and Sadowski (2002) analyze data on technological collaborations to find that developing countries accounted for only 6.9% of all technology agreements in 1997 and that over 90% of these agreements involved only developed economy firms.

In principle, the mobility of capital could help transfer technology even in the face of low labor mobility. Where developing countries acquire new technologies through inward FDI, the parent MNE would have standard ownership incentives to ensure the effective transfer of technology and essential know-how to their overseas venture. Thus technology-imbued FDI could transfer technology where arms-length transfers were weak. Although it is now commonplace to see developing countries embrace FDI, attitudes to foreign capital were somewhat ambivalent in the immediate post-war period. A long history of colonial stifling of indigenous industrialization had left newly-independent developing countries with a reflexive suspicion of foreign capital. MNEs were seen as foreign monopolies that would foreclose investment opportunities for domestic industry and, at the macroeconomic level, repatriation of profits would strain precarious foreign exchange balances.

Economic policy in India epitomized this approach: Athreye and Kapur (2001) discuss the vicissitudes of FDI policy that tried to restrain foreign ownership even as it sought to enhance the arms-length technology acquisition necessary for establishing a domestic industrial base. However, such attempts to disentangle technology transfer from foreign ownership failed. As policymakers found that foreign firms would not transfer advanced technologies without large ownership shares, a more permissive regime for MNEs came to be regarded as a vital compromise to enable technological catch-up. With the growing realization that previous restrictions of foreign investment had choked off access to technology and also stifled competition in many sectors, the earlier ambivalence had been discarded by the late 1980s. Increasingly, developing countries came to the view that the presence of MNEs would expose local firms to advanced technologies and production processes through demonstration effects, leading to spillovers that could boost the productivity of firms in their proximity, be it proximate in a geographic or industrial sense.

Blomström and Kokko (1998) identified three channels for information and knowledge transfer between foreign MNEs and domestic firms. First, the inevitable mobility of domestic workers from more productive foreign firms to host-country firms would generate technological spillovers (Fosfuri, Motta, and Rønde 2001; Glass and Saggi 2002; Görg and Strobl 2005). Second, vertical linkages whereby domestic firms competed to supply inputs to foreign firms can raise the productivity and technological capabilities of domestic suppliers (Barba Navaretti and Venables 2004; Crespo and Fontoura 2007; Smarzynska Javorcik and Spatareanu 2008; Giroud, Jindra, and Marek 2012). Automobile manufacture provided a classic example of this second channel, as the manufacture of ancillary automobile components is usually subcontracted to local firms. Zanfei (2012) identifies pecuniary externalities where increased demand for output of a particular firm can generate scale-related productivity advantages for the local industry. Third, there may be demonstration effects and imitative behavior through reverse engineering, when domestic firms observing a superior management practice or technology at work may be induced to adopt it themselves: this represents the purest externality effect of foreign presence.

Empirical evidence suggests that this view of technological externalities was simplistic. Several factors can limit beneficial spillovers from foreign to domestic firms. First, while vertical linkages could trigger technological improvements among domestic input suppliers, the precise nature of these linkages is important. For instance, many foreign firms in the dragon economies of Asia had strong linkages with domestic firms, but this was less so in other countries and in some industrial sectors. Second, the process of local technological diffusion is sensitive to the intensity of local competition vis-à-vis foreign firms, and its success depends on the absorptive capacity of local firms. If domestic firms are very far behind the technology frontier, then they are both less able to absorb the superior technology brought in by MNEs and less able to compete with them. Over time such firms may even lose their market shares and help establish foreign monopoly enclaves in the industrial sector. Third, the motives behind the investment are important: where FDI is purely of a market-seeking intent, it might involve primarily the establishment of distribution and sales outlets in the developing country, with few technological linkages or knowledge exchange.

For these reasons, the expected gains of FDI-induced technology acquisition turned out to be over-optimistic. Notably, the factors that inhibit technology transfer associated with FDI are not dissimilar to the factors that inhibited technology transfer through licensing in the post-war decades. Where the technological gap between foreign and domestic firms was large relative to the absorptive capacity of domestic firms, technology acquisition was weak, whether through licensing or through anticipated spillovers from foreign firms. Low technology capability and inadequate technological infrastructure are common features of firms located in countries behind the technology frontier: the very outcome that inward FDI and technology licensing hoped to change. In some cases, large technology gaps created fertile conditions for the emergence of foreign-owned monopolies, which were likely to be more unpopular than domestic monopolies.

That both licensing and foreign direct investment turned out to be poor channels for technology transfer is not surprising if we consider the motivation of foreign firms involved in the process. Many of these firms viewed the licensing of technology as a strategy to "enter" overseas markets, alongside FDI or simply exporting their product (Telesio 1979; Calvet 1981; Porter 1986). Licensing and direct investment were not always substitutes; sometimes they complemented each other with common factors driving both. For instance, the difficulty of operating production facilities in remote developing countries might generate a preference for licensing but some accompanying equity participation may be necessary to protect proprietary technology from the risk of leakage. Host-country intellectual property regimes also influenced this choice: in countries where IPR protection was considered to be weak, MNEs preferred to establish subsidiaries rather than license their technology (Contractor 1981).

Changes in Modes of Technology Acquisition in the 1990s

Amidst the general pessimism about difficulties of technology acquisition by developing countries, the 1990s saw two new trends. First, there was an increase in the global volume of trade in technology services, with much greater involvement of developing countries as buyers and sellers. Second, more surprising, was the emergence of outward FDI from developing countries to developed countries; the latter especially has offered a new channel for technology acquisition by developing countries. Both of these developments represent a departure from the traditional modes of technology acquisition. We explore these trends in this section, relying on some case studies.

The Increase in Cross-Border Technology Trade

As the OECD (2011) categorizes it, trade in technology comprises four types of transactions: transfer of techniques through patents and licenses and the disclosure of know-how; transfer of designs; services with a technical content, including technical assistance; and industrial R&D. Not all of this trade is amenable to measurement. The volume of royalty and license fees (RLF) associated with technology trade has been reported on a regular cross-country basis since 1996, as these transfers are reported in balance-of-payments data. The use of these data as a proxy

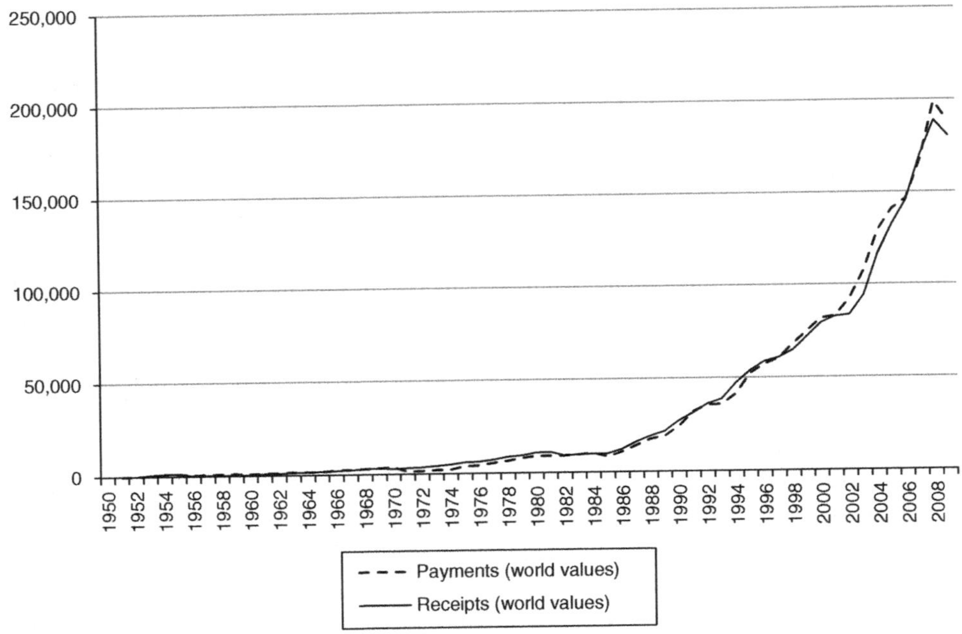

Figure 9.1 Growth in international royalty and licensing payments and receipts 1950–2009, in US$ millions.

Source: data for 1950–1970 from IMF Balance of Payments Yearbook (various years), which reports royalty and licensing fees in current USD by country. Data for 1970–2009 from the World Bank World Development Indicators database which collates royalty and licensing revenues in current USD separately for each country.

for trade in technology requires caveats. The value of RLF payments, especially intra-firm transactions, is distorted by transfer-pricing that aims to minimize the global tax liability of MNEs.[2] Figure 9.1 shows how cross-border RLF flows have accelerated since 1990, after a relatively flat trajectory in the post-War decades.

Over the period 1990–2009, computations using World Development Indicators data show that global royalty and licensing receipts and payments grew at an average rate of 9.9% per annum. To put this surge into perspective, this is higher than the rate of growth for world merchandise exports during 1992–2009, which at 7% per annum (WTO 2010) in nominal terms, is regarded as evidence of increased globalization.[3] And this growth was quite universal: WIPO (2011) found that in almost all countries for which data are available, transactions involving royalties and license fees grew faster than their gross domestic product (GDP). While some of the rise in RLF flows may be overstated due to under-reporting or measurement issues in the pre-1990 period, even for the more recent period since 1999, global flows have risen at an average of 7.7% per annum in real terms. For the Russian Federation, China, and India, international flows of royalties increased by more than 20% per annum between 1997 and 2009.

The list of countries participating in this trade also grew. In 1990, 62 countries made licensing payments but by 2007, this number had increased to 147 countries. Similarly, in 1990, only 43 countries received any international royalty or license fees, but by 2007, this number had increased to 143 countries. This suggests the

Table 9.1 Royalty and license fees receipts and payments by country groups in US$ millions. Shares in global totals and average annual rates of growth 1999–2009.

Country group	*1999*	*2009*	*Share in 1999*	*Share in 2009*	*Average annual rate of growth*
All countries					
RLF receipts	72,710	153,190			7.74
RLF payments	77,373	153,138			7.07
High-income countries					
RLF receipts	71,959	151,119	98.97	98.65	7.70
RLF payments	70,370	135,162	90.95	88.26	6.74
Middle-income countries					
RLF receipts	737	2,055	1.01	1.34	10.80
RLF payments	6,931	17,942	8.96	11.72	9.98
Low-income countries					
RLF receipts	14	16	0.02	0.01	1.04
RLF payments	72	34	0.09	0.02	–7.26

Source: Data from World Development Indicators database.
Note: Values deflated using GDP deflator provided in the World Development Indicators. Country groups used are World Bank categories.

gradual emergence of a large market in international licensing, perhaps facilitated by the growth of new industrializing economies of the BRIC countries (Brazil, Russia, India, China) and also the harmonization of IPR systems/regimes due to agreements like the Trade-Related Intellectual Property Rights (TRIPS).

Table 9.1 offers a more disaggregated picture, categorizing country by income groups. High-income countries are involved in the bulk of these flows, accounting for over 98% of all RLF receipts and over 90% of payments. The share of middle-income countries remains quite small but is rising briskly, at around 10% per annum for both receipts and payments. Although not reported here, within the middle-income group, the so-called transition economies show higher growth of RLF receipts (20% per annum in nominal terms) relative to payments (13% per annum); the BRIC countries collectively show a higher rate of growth of RLF payments (21% per annum) relative to that of receipts (16%).[4] The data also show that low-income countries are largely excluded from such trade in technology.

Looking more closely at the countries that increased their share of international RLF payments between 2005 and 2009, we find Ireland and China increased their share by 4.9% and 2.1%. Other new countries gaining shares in international licensing payments are the BRIC economies, and former East European nations like Hungary and Poland. Many of the BRIC countries also increased their shares in international licensing receipts, but large gains in exporting shares were made by European countries and South Korea.

The evidence on increased RLF flows matches up quite well with findings on national R&D expenditures which may be seen as an indicator of national investment in technological capability. WIPO (2011) finds worldwide R&D spending is skewed toward high-income countries which still account for around 70% of the world total, although their share dropped by 13 percentage points between 1993 and 2009. The share of middle- and low-income countries more than doubled between 1993 and 2008; the most spectacular increase in the share of world R&D has been that of China, now the second largest R&D spender in the world. Other middle-income countries that registered an increase in R&D spending include Brazil, Russia, and South Korea. Recalling our earlier discussion, as firms in countries become technologically more able, they are also better able to search for technologies that can be licensed to their advantage.

Nevertheless, the growth in international licensing was confined to a few industrial sectors in almost every country. Data on industrial composition for this trade are hard to come by and researchers have instead looked at data from technology agreements of firms. The Thomson Financial SDC Platinum database covers over 6235 technology licensing agreements involving 7006 firms between 1976 and 2009, and provides detail of the industrial sector, contract details (e.g., extent of exclusivity, lump-sum versus fee elements), and information on licensor-licensee pairs, details not found in aggregate data.[5]

Gambardella and Torrisi (2010) provide some detail on the sectoral composition of licensing trade. They find that the majority of the licensing contracts in their sample occurred in IT-semiconductors-electronics, chemicals-pharmaceuticals-biotech, and engineering technological classes. They also show that licensing tends to correspond to the country's technology specialization indices as computed from patent databases. The sectors of licensing activity that emerge from the firm database are quite remarkably congruent and similar to those from previous studies on licensing (Caves, Crookell, and Killing 1983; Anand and Khanna 1996). The theoretical literature suggests that such concentration in particular sectors occurs when horizontal contracts prevail. This can happen when the technology is general purpose and applicable in different uses.

Gambardella and Torrisi (2010) also use these data to shed light on the technology flows between sectors and this information is presented in Table 9.2. As can be seen from Table 9.2, the largest flows of technology through licensing are, in fact, within the same technological sectors, although related sectors (such as chemicals and drugs and computers and electronic equipment sectors) benefit from licensing arrangements. In addition, sectors like instrumentation and knowledge-intensive business services (KIBS) sell to a range of other sectors.

Outward FDI and Technology Acquisition

A surprising feature of more recent globalization since the 1990s has been the emergence of outward FDI from many developing countries, especially the BRIC countries. By 2010 these countries accounted for 9% of global outward FDI flows. Figure 9.2 plots the trends in outward investments for the BRIC countries, with data drawn from the UNCTAD database. While outflows are not large in absolute terms, they display a growing trend, which for China and Russia persists even after the onset of the financial crisis.

While a fair chunk of these investment flows went to the developing countries, a significant amount of outward investment also went from these developing countries to developed countries. UNCTAD (2013) finds that between 2000 and 2008 as

Table 9.2 Licensing flows across industrial sectors (percentage of total number of agreements).

	Licensee							
Licensor	*Drugs*	*Chemicals*	*Computers*	*Electrical/ electronic equipment*	*Transport*	*Instruments*	*KIBS*	*Number of agreements*
Drugs	64.78	3.65	0.37	0.22	0.07	4.62	11.69	1,343
Chemicals	16.85	42.82	1.93	3.31	2.49	4.42	9.39	362
Computers	0.16	1.63	27.08	22.35	3.10	5.55	27.73	613
Electrical/ electronic equipment	0.75	2.13	17.00	46.38	1.00	4.88	20.50	800
Transport	1.96	6.86	7.84	12.75	27.45	5.88	24.51	102
Instruments	18.99	2.79	6.42	10.61	1.68	29.89	13.97	358
KIBS	10.56	2.41	9.81	10.43	1.17	2.65	45.62	1,620

Source: Adapted from Gambardella and Torrisi (2010: Table 4 and Table B.1).

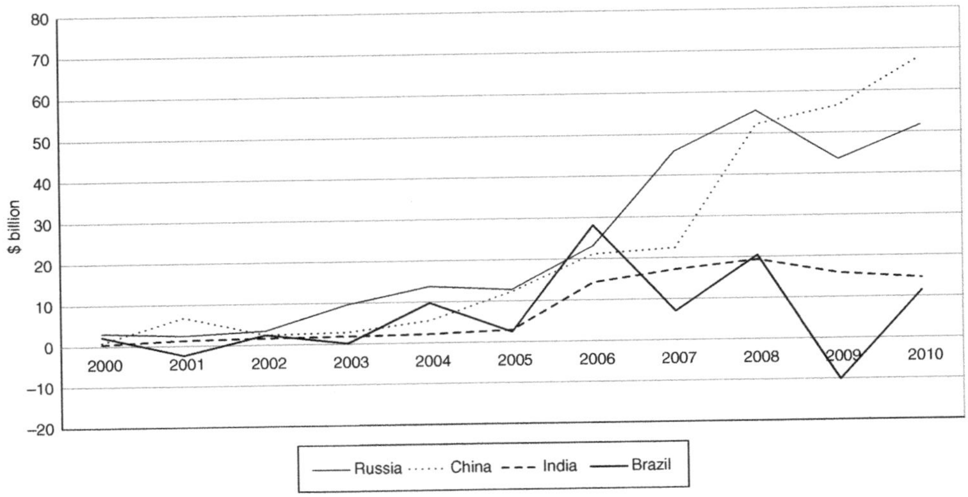

Figure 9.2 Value of outward FDI from BRIC countries, in US$ billions.
Source: Based on data from UNCTAD Statistics (http://unctadstat.unctad.org/EN/Index.html).

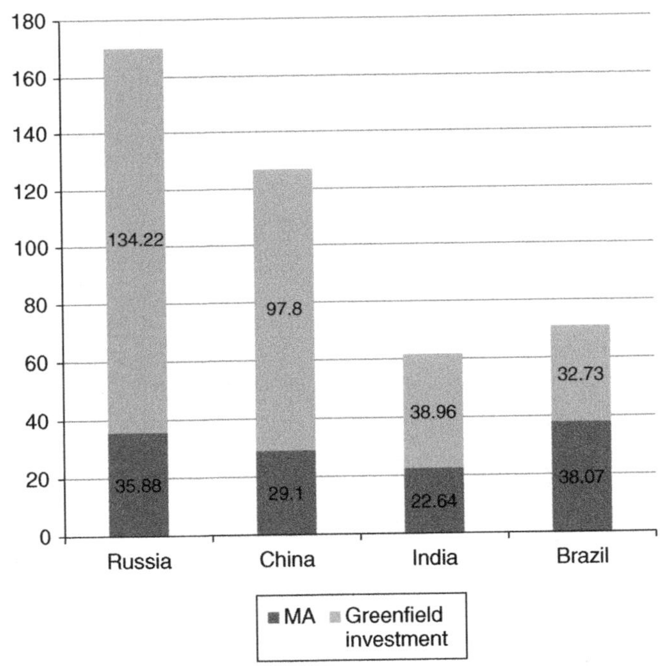

Figure 9.3 Form of outward FDI from BRIC countries 2000–2008, in US$ billions: greenfield investment vs. mergers and acquisitions.
Source: Based on data from UNCTAD Statistics (http://unctadstat.unctad.org/EN/Index.html).

much as 42% of all investment from the BRIC countries went to "northern" (i.e., developed) countries. Figure 9.3 shows the distribution of BRIC outflows by form of investment, "greenfield" vs. mergers and acquisitions (M&A). UNCTAD (2013) reports that while most of the BRIC investment outflows to Africa were of the greenfield variety, M&A accounted for 22% of FDI into developed countries.

Direct investment flows from developing to developed countries present a conundrum. Conventional economic theory suggests that capital should flow from capital-rich developed countries to capital-scarce emerging economies. Indeed, that was the prevailing pattern for much of the post-war period. And as capital-rich developed countries are technologically more advanced than developing ones, capital and technology flowed down together along the same gradient, from advanced to developing economies. In keeping with this scenario, from the viewpoint of developing countries, it was *inward* FDI that was expected to serve as a channel for technology transfer. *Outward* FDI from developing economies amounts to what has been described as the "uphill flow of capital." Outward FDI (OFDI) may have a variety of motivations, as outlined in Athreye and Kapur (2009), ranging from a quest for critical natural resources, the desire to develop distribution channels to improve access to export markets, to simply geographic diversification of corporate assets. However, we are going to argue, at least some of this uphill flow was motivated by the desire to gain access to overseas technological assets.

If uphill flows are to serve as a channel for technology acquisition, technology must flow in a direction opposite to that of capital. In well-functioning technology markets, such opposing flows of technology and financial resources could simply be an outcome of arms-length technology trade, in the form of licenses and royalty payments. In less-than-perfect technology markets where technology-seeking firms lack the ability to identify the technological assets they need, or where the assets exist in tacit form, firms from developing countries must resort to the outright acquisition of technology-rich firms overseas. In acquiring these assets, firms gain access not only to a body of knowledge codified as patents, but also the human capital in established R&D facilities or networks, or simply embodied in production facilities or processes that cannot be easily replicated. Amighini, Rebellotti, and Sanfilipo (2010) survey the empirical evidence that OFDI has become a channel for technological catch-up.

Where overseas investment is motivated by technology acquisition, it should be evident in the form and characteristics of the investment. First, if the aim is to acquire quick access to production or research facilities or associated human capital, investment will take the form of acquisition rather than the organic growth of a kind associated with greenfield investment. Furthermore, technology acquisition as a motivation for OFDI makes commercial sense only if the acquiring firm has significant interests in the relevant (or a related) industrial sector, either in the form of existing production capacity or the desire to set up new production capacity. Third, the outward investment must be in firms typically in regions which are seen to have a significant technological edge.

Child and Rodrigues (2005) assess the extent to which desire to acquire strategic assets motivates the physical and organizational expansion of Chinese firms in overseas locations. While much of Chinese OFDI aims to secure access to raw materials, and some merely to improve access to overseas markets, they argue that at least one strand of such investment aims to correct for the competitive disadvantages of late development, notably the need to catch up with technology and know-how if they are to become global players. While joint ventures with foreign partners provided Chinese original equipment manufacturers with access to technology, including some tacit knowledge, typically these left the Chinese collaborators as the junior partners in the venture.[6] Foreign acquisitions have allowed Chinese firms to get improved access to technology. For instance, in 2001 China's Holly Group, which had previously specialized in the production of energy meter equipment, acquired from Philips Semiconductors in the United States its operational assets for CDMA handset reference designs used in mobile telephony. This acquisition gave Holly access to know-how and IPR, including an exclusive license for the CDMA software protocol. In 2004, the Chinese state-owned Shanghai Automotive Industry Corporation acquired the rights to engine and transmission technology from the UK's MG Rover Group, although it later held off from outright acquisition of the group. The Chinese firm Quianjiang acquired a motor cycle manufacturer, Benelli, in Italy, mostly for its R&D facilities. Lenovo's acquisition of IBM's personal computer division in 2005 gave it control over a key brand for marketing purposes, but also provided access to its technological assets. It then built on its earlier acquisition to make a bid for the IBM x86-server business: not only did this give Lenovo access to IBM's servers and associated software, it also gave them control over a workforce of 7500 employees based in the United States and other locations.

India's OFDI provides similar examples. Pradhan and Singh (2009) study the overseas investment of Indian firms in the automotive sector. Consider, for example the acquisition by the Tata Group of UK-based Jaguar Land Rover in 2008. Tata Motors, as part of the Tata Group, had been manufacturing automobiles in India since 1945. Its early focus was on commercial vehicles but in the 1990s it entered the market for passenger cars and light commercial vehicles by developing indigenous designs. These had only limited success but established the infrastructure for engineering design and development. In 2003 Tata announced a project to produce the Nano, an ultra-low-cost car based on domestic design and engineering. But crucially, this project was accompanied by what Bruche (2010) calls complementary upstream internationalization. Tata set up the Tata Motors European Technical Centre (TMETC) in Coventry to work on what Tata identified as "critical gap" areas, and did so in close cooperation with Tata Motors' own research facilities in India.[7] At the same time, one of the Tata subsidiaries that focused on automotive design-engineering acquired UK-based design services company INCAT International, a provider of engineering services to the automotive and aerospace industry, with 3000 employees in North America, Germany, and India. Both entities were closely involved with Tata's indigenous technological efforts, which included 37 patent applications in India. These technological precursors were followed by the purchase of Jaguar Land Rover (JLR) from the Ford Group in 2008. This acquisition represented a diversification into the market for premium vehicles but had a clear technological dimension, in the form of Jaguar's two engineering design centers in the United Kingdom.

Similar initiatives by Indian firms to acquire technology were evident in other sectors. Pradhan (2008) reports evidence on overseas acquisitions by Indian pharmaceuticals in their quest for technology. Historically, the domestic innovation activity of Indian pharmaceutical firms was limited to cost-effective process development rather than product development. Accordingly, a significant fraction of early Indian OFDI in this sector went to developing countries, especially Africa, largely to extend and secure markets. Since 2000, OFDI has been used to acquire strategic assets in the developed world, most notably in the United States and United Kingdom. For instance, Ranbaxy Laboratories has made significant investments in Europe, Japan, and the United States: not all of this was in the generic drug segments, long regarded as the strength of Indian firms. Another large pharmaceutical firm from India, Dr Reddy's Laboratories, has made large investments in the United Kingdom and Germany. Clearly, acquisitions have had a variety of motivations, from pure market capture to improved ability to engage with host country drug-approval regimes through acquired proxies. Nonetheless, the technology-acquisition motivation is quite evident in contemporary corporate announcements.[8]

Nayyar (2008) cites the acquisition of Hansen Transmissions International (a Belgium-based manufacturer of turbine gearboxes) by Suzlon Energy from India, motivated by the desire to extend its capabilities in the manufacture of wind turbines,[9] and other acquisitions in the telecommunications sector. Niosi and Tschang (2009) point to the importance of overseas acquisitions in the early development of software firms from India and China.

While these case studies provide robust examples of technology acquisition through OFDI from developing countries, the salience of this motivation must not be overstated. Clearly, technology is an important consideration in many overseas acquisitions, but realistically it is not the only one, and in many cases not even the primary one. Buckley et al. (2007) found that the annual patent registration of the host countries – as a proxy for their technological intensity – was not a significant determinant of the level and direction of Chinese FDI outflows over the period 1990–2001. A lot of the investment had more direct and immediate objectives, such as securing access to key raw materials (oil, gas, or minerals) or simply acquiring market shares. One could speculate on whether the motivation for internationalization evolved in more recent years – with the technology objective acquiring greater primacy – but it is quite likely that resource-seeking or market-seeking flows swamp the few investments in search of technology.

However, the picture may be somewhat nuanced. Even when technology was not the primary motivation for acquisition, ex post it may have delivered technological dividends. For instance, in Lenovo's acquisition of IBM personal computer segments, its established market brand was a major consideration, as it allowed it to convert its existing manufacturing capacity into a larger market share. But even here, while it may not have been the initial objective, access to IBM's production facilities served as a bridge to IPR in the server segment, and gave Lenovo control over IBM's technology-rich workforce.

Aggregate data show a close relation between the pace of outward investments and the hunger for technology through international licensing. Figure 9.4 plots the rise in licensing payments from the BRIC countries and compares it with the rise in technology payments by EU transition economies which globalized (albeit less aggressively than the BRIC countries). From 1999 to 2009 the share of international licensing payments of the BRIC countries increased from 4.2% to 10.2%. In the same period OECD transition economies increased their share of international licensing payments from 1.5 to 2%.

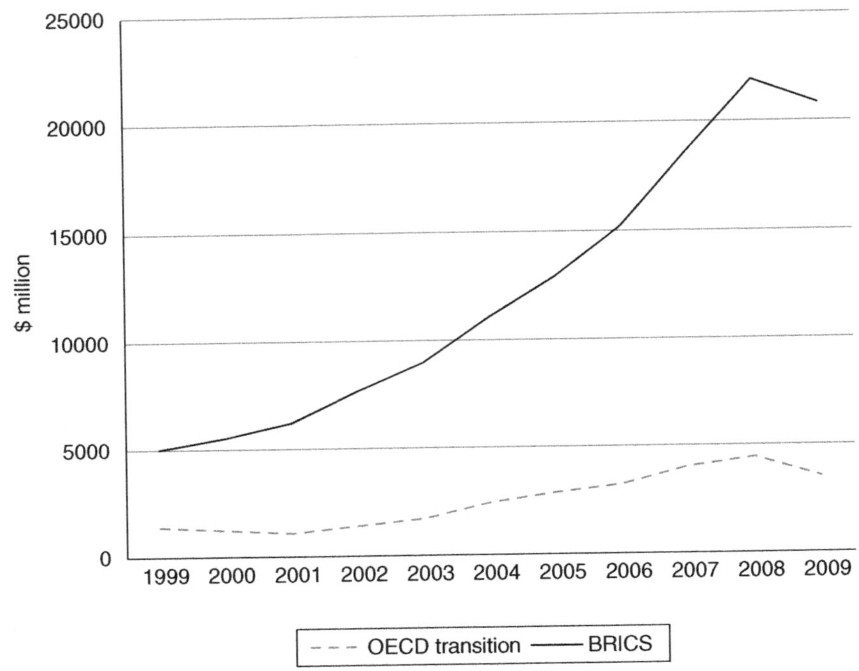

Figure 9.4 Value of arms-length technology purchases, BRIC vs. OECD transition countries 1999–2009, in US$ millions.
Source: Data for 1950–1970 from IMF Balance of Payments Yearbook (various years), which reports royalty and licensing fees in current USD by country. Data for 1970–2009 from the World Bank World Development Indicators database which collates royalty and licensing revenues in current USD separately for each country.

Impact of Globalization on Technology Acquisition Strategies

The increase in international trade in technology and the emergence of OFDI as a new tool of technology acquisition raises the question of what factors lay behind these trends. We contend that the deep globalization that started in the late 1980s contributed to both trends. Globalization saw a marked increase in the mobility of labor and capital, and the revolution in communications technologies served to reduce the knowledge gap across countries. By the 1990s, decades of post-war development had increased the technological capabilities and absorptive capacity of many firms in developing countries, also reflected in the increased R&D spending by these firms (WIPO 2011). Globalization offered an array of opportunities for the more able firms that were already actively looking for technology.

Some writers have drawn comparisons between the globalization of the 1980s and the globalization that occurred a century earlier, starting in the 1880s. This previous globalization saw the economic emergence of the Atlantic economies, particularly the United States. While Rosenberg (1976) and Standage (1998) have highlighted the importance of technological changes – the growth of the railways, emergence of machine tools and mechanical engineering – in enhancing productivity growth, O'Rourke and Williamson (1999) place far more emphasis on the impact of falling trade barriers which led to an exceptional mobility of capital and labor. As we noted earlier, the free movement of people was an important mechanism for technology transfer in the 1920s and 1930s and this technology transfer directly contributed to productivity increases.

One could seek similar gains from the globalization of the late 1980s, which went hand in hand with the spectacular growth of information flows, especially via the Internet. The Internet had a dramatic impact on shrinking distance, making information more widely available and enabling the globalization of ideas. When scarcities in human capital threatened the pace of technological progress in the emerging technological sectors, many developed countries lowered the barriers to the immigration of highly skilled professionals (reminiscent of the way the repeal of the Corn Laws enabled the availability of cheaper grain to support workers during the Industrial Revolution). These freer flows of information and skilled workers facilitated technology transfer to an extent unimaginable in the immediate post-war decades.

Many of the skilled workers who filled the emerging gaps in human capital for the technology-intensive sectors came from middle-income countries such as India, China, Ireland, Israel, and Taiwan. Arora and Gambardella (2005) show that these countries, having invested in human capital in excess of that warranted by the growth of their domestic economies, were typically at the forefront of the nascent software industry in emerging countries. These migrant workers created diasporic populations in technology clusters, while still retaining close ties to their countries of origin. Saxenian (1994, 2006) argues that those who gravitated to technology clusters such as Silicon Valley in California were able to export the model of technology entrepreneurship to satellite cities around the world, especially in India, China, Taiwan, and Israel.

As we argued above, one major impediment to trade in technology was asymmetric information between technology suppliers and buyers. Fransman (1985) had found that one of the key factors constraining technology transfer had been the ability of firms in developing countries to search for and pick the right technologies for their purposes. It was hard to acquire when buyers did not know which technologies were available, the best source from which to acquire their chosen technology, and when they lacked the tacit knowledge – the absorptive capacity – to deploy acquired technologies. The information gap across international borders was bridged through increased mobility of labor, a dramatic reduction in telecommunication costs, and global expansion of the Internet as a communication tool. Increased investment in domestic R&D by developing country firms, along with internationalization of R&D, enhanced the absorptive capacity allowing firms to adapt foreign technology. Examples include Tata, Reliance who built two petrochemical complexes in India with licensed technology, and Hyundai who manufactured their first car with licensed automotive technology from the United Kingdom.

Case studies of technology entrepreneurship offer insights into the use of knowledge networks. Consider, for instance, the case of Suntec, a Chinese company that produces solar panels. Its founder was exposed, as a student in Australia, to cutting-edge technology for photovoltaic cells. His start-up firm then licensed technology from Australia and the United States in order to produce solar panels in China. Entrepreneurship in technology-intensive sectors often involved the use of proprietary technology accessed through licensing, but even that access relied on the emergence of formal and informal networks.

There were other elements of globalization in the 1990s that may explain the rising trend in international licensing. First, during the mid-1990s, the rapid expansion of global trade had led to multilateral reforms such as the harmonization of TRIPS. This agreement forced countries to strengthen their IPR regimes, increasing the willingness of technology suppliers to license more valuable technology across countries. Branstetter, Fisman, and Foley (2006) show that moving to stronger IPR regimes was associated with greater intra-MNE technology transfers across 19 IP-reforming countries; however, they could not discern any effect of strong IP regimes on trade between independent parties. Kanwar (2012) exploits the changes in the different components of the IP index to show that developing countries received greater imports of technology following stronger IP protection, mainly as a result of an expansion in the sectors of economy that were covered due to stronger IP.

Stronger IPR regimes also boosted capital movements, both inward and outward. Hassan, Yaqub, and Diepeveen (2010) present evidence to show that TRIPS increased FDI inflow into developing countries. Strong IPR regimes favored the internationalization of R&D activity of developed country firms. Many set up knowledge centers and research labs in developing countries to take advantage of pools of trained scientists. Consider, for instance, the research facilities set up by Microsoft in India, Ireland, and Israel. Firms behind the technology frontier often have problems assessing the full value of a technology. In this context, strong IPR may induce more technology-based M&A by giving control over the residual rights in technology investments (Athreye and Godley 2009).

Summary and Policy Implications

Traditional thinking on technology transfer identified technology licensing and inward FDI as the two main channels for technology acquisition by developing country firms. Empirical evidence suggests that the extent of technology transferred through these channels was limited. In the presence of the well-known asymmetries of information and other imperfections in the market for technology, licensing remained a weak channel. FDI from developed to developing economies could have strengthened technology transfer, but the precise technologies transferred were not the most appropriate for developing country needs, especially as developing countries were relatively passive recipients of technology transfer. In many ways both channels were adversely affected by similar factors: for instance weak IPR regimes in developing countries made owners of technology reluctant to transfer, whether through licensing or through FDI.

In recent decades, firms in middle- and low-income developing countries have increased their technological capability and have become active seekers of the technology they need. A broader appreciation of technologies on offer and complementary investments in domestic R&D to better absorb acquired technologies, has made licensing a more viable channel than it used to be. With the wave of globalization that started in the late 1980s, trade in disembodied technology has boomed: more countries are now involved in cross-border licensing arrangements with increasing participation from developing regions.

More recently, some large firms from developing countries have acquired technology through OFDI, typically via acquisitions of firms with a portfolio of technology products. Technology-motivated corporate acquisitions involve active search for the right technologies – firms must carefully select their acquisition targets and invest resources to integrate acquired technologies into their domestic operation. As such, OFDI is likely to prove a more promising channel for technology acquisition than the relatively passive inflow of technologies that accompany inward FDI.

What are the policy implications of the emergence of these new channels for technology acquisition? Developing countries' governments need to recognize the new modes of technology transfer and devise policies that enable their firms to take advantage of the new channels.

One set of policies must focus on improving the functioning of technology markets. Clearly, stronger IPR regimes and enforcement of those regimes are a necessary condition if potential licensors are to be persuaded to transfer their best technologies. But stronger protection for licensors needs to be combined with policies to improve the bargaining position of potential licensees. When it comes to pricing, technology buyers are often at a disadvantage relative to monopoly suppliers. If governments could maintain and publish registers on the value of technology transactions, the price transparency could benefit potential buyers. Publicly funded intermediaries who would act as brokers between technology buyers and sellers, as well as providing legal advice on typical license contract forms, could lower the high costs of transactions in technology markets. For instance, the Chinese government has actively sponsored international joint ventures to improve technology transfer. Chinese experience in this field merits academic scrutiny: key policy insights into how the Chinese were able to persuade joint venture partners to transfer technology would be invaluable to other countries.

Policy should also be informed by a better understanding of the link between capital flows and technology acquisition. Where inward FDI has failed to transfer significant or appropriate technologies, there may be a case of dismantling fiscal regimes that encourage such FDI for its technology transfer. There may well be a case for encouraging inward FDI for other reasons, such as promoting competition in domestic markets, but that case should not be overstated to include the gains from technology transfer that may not materialize. In some cases, it may be more productive to encourage or even subsidize OFDI through sovereign guarantees. In many cases this might only involve the leveling of the fiscal regime between outward and inward FDI, correcting the inherited policy bias in favor of inward FDI.

There are policy lessons for developed countries too. Policymakers in these countries have campaigned hard in international forums for the reinforcement of IPR regimes, but this is often motivated by narrow support for national technology champions rather than to improve the functioning of technology markets. Better-functioning markets for technology are especially desirable in "green technologies" where the global environmental benefits of faster technological diffusion should trump any narrow national calculus of cost and benefit. Similar maturity may be necessary in the pharmaceuticals sector to reconcile the conflict between preserving incentives for R&D in discovering new drugs and to offer access to life-saving generics for poor developing countries.

Developed countries have also shown considerable ambiguity in their policy toward corporate acquisitions by overseas firms. At one level, many developed country regimes remain eager to encourage inflows of FDI, especially when a combination of public austerity and unemployment places a premium on projects that generate domestic jobs. But foreign corporate acquisitions are quite often regarded with public hostility and political objections. Tata's takeover of Jaguar and Corus, or the acquisition of the European steel giant Arcelor by Mittal faced a cacophony of complaints about potential job losses. Although these events are relatively recent to generate reliable evidence, the limited UK experience suggests that Indian and Chinese companies have managed to turn around failing companies, often by commercializing their technology assets to sell in their home markets. If this is a more general trend then technology-seeking OFDI from developing countries may be a boon for developed countries. It has long been said that Europe and the United Kingdom in particular are good at invention but poor at commercializing those inventions into innovation. Indian and Chinese firms may not know as much about technology as the European firms they are buying, but they may have the marketing savvy to use these inventions to satisfy the growing demand in their own economies.

Notes

1 Trends in mobility changed again after the 1980s when many countries began to pursue a policy of selective immigration, with notions of *brain gain* and *brain circulation* gaining currency. Archibugi and Pietrobelli (2003) report that, with greater funding for scholar exchanges, there has been a small increase in the share of scientific papers with developing country co-authors, from 15.8% of all co-authored papers in 1986–1988 to 19.2% in 1995–1997.

2 Madeuf (1984) and OECD (1995: Box 12.1) contain a detailed discussion of the problems and limitations of using royalty and license fee data to infer technology transfer.

3 Data from International Monetary Fund and World Development Indicators suggest that world exports stood at US$4.2 trillion in 1992 but steadily increased to US$14.2 trillion by 2010, implying an average growth rate of 7% per annum.

4 These rates of growth are for nominal values. Whilst we have used the GDP deflator in reporting real values, such deflators do not work very efficiently for estimating the value of technology.

5 The data draws upon voluntary disclosure by firms of their technology agreements, which may also be driven by strategic concerns.

6 Van Reenan and Yeuh (2012) argue that Chinese joint-venture policy was a huge success as it conferred a total factor productivity growth of 1% on all participating firms.

7 As Ravi Kant, the managing director of Tata Motors Ltd (TML) put it, TMETC provided a window on advanced technologies allowing "TML to plug into the expertise that is available in Britain and be in on the key developments in automobile manufacturing." See http://www.tata.com/article/inside/q63wqXlx5rc=/TLYVr3YPkMU= (accessed December 27, 2014).

8 For instance, as Pradhan (2008) cites, the corporate announcement at Reddy's acquisition of Dowpharma UK noted: "The proprietary chiral and bio-catalysis technology at the Cambridge site and the scale-up capability in the Mirfield site will add significant value to the company. This acquisition will also bring strengths in industrial synthesis of complex prostaglandins and carbohydrate chemistry. These newer capabilities will add to our existing R&D and commercial infrastructure."

9 The corporate statement at that acquisition offered the following motivation: "The acquisition of Hansen gives us technological leadership and will make Suzlon a leading integrated wind turbine manufacturer in the world. Although the company will be run as an independent business unit, the acquisition of Hansen will allow us to integrate gearbox technology into the total turbine solution enabling a more reliable and competitive product ... with this acquisition Suzlon has truly emerged as

a global player with significant market presence, manufacturing base and R&D centres across North America, Europe, India, China, South Korea and Australia." http://www.apax.com/our-news/apax-news/2006/march/suzlon-energy-signs-definitive-agreement-for-strategic-acquisition-of-hansen-transmissions/ (accessed December 27, 2014).

References

Amighini, Alessia, Roberta Rebellotti, and Marco Sanfilipo. 2010. "Outward FDI from Developing Country MNEs as a Channel for Technological Catch-Up." *Seoul Journal of Economics* 23(2): 239–261.

Anand, B.N., and T. Khanna. 1996. "Intellectual Property Rights and Contract Structure." Yale School of Management Working Paper H-3.

Archibugi, Daniele, and Carlo Pietrobelli. 2003. "The Globalisation of Technology and Its Implications for Developing Countries: Windows of Opportunity or Further Burden?"

Arora, Ashish, and Alfonso Gambardella. 2005. *From Underdogs to Tigers: The Rise and Growth of the Software Industry in Brazil, China, India, Ireland, and Israel.* Oxford: Oxford University Press.

Athreye, Suma, and A. Godley. 2009. "Internationalization and Technological Leapfrogging in the Pharmaceutical Industry." *Industrial and Corporate Change* 18(2): 295–323.

Athreye, Suma, and Sandeep Kapur. 2001. "Private Foreign Investment in India: Pain or Panacea?" *The World Economy* 24(3): 399–424.

Athreye, Suma, and Sandeep Kapur. 2009. "Internationalization of Firms from China and India: An Introduction." *Industrial and Corporate Change* 18(2): 209–221.

Barba Navaretti, G., and A.J. Venables. 2004. *Multinational Firms in the World Economy.* Princeton, NJ: Princeton University Press.

Bell, Martin, and Keith Pavitt. 1997. "Technological Accumulation and Industrial Growth: Contrasts Between Developed and Developing Countries." In *Technology, Globalisation and Economic Performance*, ed. Daniele Archibugi and Jonathan Michie, 83–137. Cambridge: Cambridge University Press.

Blomström, Magnus, and Ari Kokko. 1998. "Multinational Corporations and Spillovers." *Journal of Economic Surveys* 12(3): 247–277.

Branstetter, L., R. Fisman, and C.F. Foley. 2006. "Do Stronger Intellectual Property Rights Increase International Technology Transfer? Empirical Evidence from U.S. Firm-Level Panel Data." *Quarterly Journal of Economics* 121(1): 321–349.

Bruche, G. 2010. "Tata Motor's Transformational Resource Acquisition Path." Institute of Management Berlin Working Paper 55.

Buckley, P., J. Clegg, A. Cross, X. Liu, H. Voss, and P. Zheng. 2007. "The Determinants of Chinese Outward Foreign Direct Investment." *Journal of International Business Studies* 38(4): 499–518.

Calvet, A.L. 1981. "A Synthesis of Foreign Direct Investment Theories and Theories of the Multinational Firm." *Journal of International Business Studies* 12(1): 43–59.

Caves, R.E., H. Crookell, and J.P. Killing. 1983. "The Imperfect Market for Technological License." *Oxford Bulletin of Economics and Statistics* 45(3): 249–267.

Child, John, and Suzana B. Rodrigues. 2005. "The Internationalization of Chinese Firms: A Case for Theoretical Extension?" *Management and Organization Review* 1(3): 381–410.

Contractor, F. 1981. *International Technology Licensing: Compensation, Costs, and Negotiation*. Lexington, MA: Lexington Books.

Crespo, N., and M.P. Fontoura. 2007. "Determinant Factors of FDI Spillovers: What Do We Really Know." *World Development* 35(3): 410–425.

Dahlman, C.J. 1978. "From Technological Dependence to Technological Development: The Case of the Usiminas Steel Plant in Brazil." IDB/ECLA Research Programme in Science and Technology Working Paper 21, Buenos Aires.

Fosfuri, A., M. Motta, and T. Rønde. 2001. "Foreign Direct Investment and Spillovers Through Workers' Mobility." *Journal of International Economics* 53: 205–222.
Fransman, Martin. 1985. "Conceptualizing Technical Change in the Third World in the 1980s: An Interpretative Survey." *Journal of Development Studies* 21(4): 572–652.
Gambardella, Alfonso, and Salvatore Torrisi. 2010. "Heterogeneity of Technology Licensing Patterns Across Europe." Working Paper, GlobInn Project EC FP7 Cooperation Work Programme.
Giroud, A., B. Jindra, and P. Marek. 2012. "Heterogeneous FDI in Transition Economies: A Novel Approach to Assess the Development Impact of Backward Linkages." *World Development* 40(11): 2206–2220.
Glass, Amy J., and Kamal Saggi. 2002. "Multinational Firms and Technology Transfer." *Scandinavian Journal of Economics* 104(4): 495–513.
Görg, Holger, and Eric Strobl. 2005. "Spillovers from Foreign Firms Through Worker Mobility: An Empirical Investigation." *Scandinavian Journal of Economics* 107(4): 693–709.
Hassan, E., O. Yaqub, and S. Diepeveen. 2010. "Intellectual Property and Developing Countries: A Review of the Literature." Santa Monica, CA: Rand Corporation for UK Intellectual Property Office and UK Department for International Development.
Kanwar, S. 2012. "Intellectual Property Protection and Technology Licensing: The Case of Developing Countries." *Journal of Law and Economics* 55(3): 539–564.
Lall, Sanjaya. 1983. "Determinants of R&D in an LDC: The Indian Engineering Industry." *Economics Letters* 13(4): 379–383.
Madeuf, B. 1984. "International Technology Transfers and International Technology Payments: Definitions, Measurement and Firms' Behaviour." *Research Policy* 13(3): 125–140.
Mowery, David, and Nathan Rosenberg. 1989. *Technology and the Pursuit of Growth*. Cambridge: Cambridge University Press.
Narula, Rajneesh, and Bert Sadowski. 2002. "Technological Catch-Up and Strategic Technology Partnering in Developing Countries." *International Journal of Technology Management* 23(6): 599–617.
Nayyar, D. 2008. "The Internationalization of Firms from India: Investment, Mergers and Acquisitions." *Oxford Development Studies* 36(1): 111–131.
Niosi, Jorge, and Ted Tschang. 2009. "The Strategies of Chinese and Indian Software Multinationals: Implications for Internationalisation Theory." *Industrial and Corporate Change* 18(2): 269–294.
O'Rourke, Kevin, and Jeffery Williamson. 1999. *Globalisation and History: The Evolution of a Nineteenth-Century Atlantic Economy*. Cambridge, MA: MIT Press.
OECD. 1995. *Industry and Technology: Scoreboard of Indicators 1995*. Paris: OECD.
OECD. 2011. *OECD Science, Technology and Industry Scoreboard 2011*. Paris: OECD.
Porter, M.E. 1986. *Competition in World Industries*. Cambridge, MA: Harvard Business Press.
Pradhan, Jaya Prakash. 2008. "Overcoming Innovation Limits Through Outward FDI: The Overseas Acquisition Strategy of Indian Pharmaceutical Firms." Munich Personal RePEc Archive Paper 12362, http://mpra.ub.uni-muenchen.de/12362/ (accessed December 26, 2014).
Pradhan, Jaya Prakash, and Neelam Singh. 2009. "Outward FDI and Knowledge Flows: A Study of the Indian Automotive Sector." *International Journal of Institutions and Economies* 1(1): 156–187.
Rosenberg, Nathan. 1976. *Perspectives on Technology*. Cambridge: Cambridge University Press.
Saxenian, A. 1994. *Regional Advantage: Culture and Competition in Silicon Valley and Route 128*. Cambridge, MA: Harvard University Press.
Saxenian, A. 2006. *The New Argonauts: Regional Advantage in a Global Economy*. Cambridge, MA: Harvard University Press.
Smarzynska Javorcik, Beata, and M. Spatareanu. 2008. "To Share or Not to Share: Does Local Participation Matter for Spillovers from FDI?" *Journal of Development Economics* 85(1–2): 194–217.

Standage, T. 1998. *The Victorian Internet*. New York: Berkley Publishing Group.
Telesio, P. 1979. *Technology Licensing and Multinational Enterprises*. New York: Praeger.
UNCTAD. 2013. *The Rise of BRICS FDI and Africa: Global Investment Trends Monitor*. Geneva: UNCTAD.
Van Reenan, J., and L. Yeuh. 2012. "Why Has China Grown So Fast? The Role of International Technology Transfer." CEP Discussion Paper 1121, London School of Economics.
Westphal, Larry E., Linsu Kim, and Carl Dahlman. 1985. "Reflections on Korea's Acquisition of Technological Capability." In *International Technology Transfer: Concepts, Measures and Comparisons*, ed. Nathan Rosenberg and Claudio Frischtak, 167–221. New York: Praeger.
WIPO. 2011. *The Changing Face of Innovation*. Geneva: World Intellectual Property Organization.
WTO. 2010. *Market Access for Products and Services of Export Interest to Least Developed Countries*. Geneva: World Trade Organization.
Zanfei, A. 2012. "Effects, Not Externalities." *European Journal of Development Research* 24(1): 8–14.

Chapter 10

Clusters and Global Innovation: The Role of Connectedness and Connectivity

Mark Lorenzen and Ram Mudambi

Introduction[1]

Pointing to the continued role of geographical *clusters*[2] (agglomerations of specialized economic activity) in innovation processes, economic geographers have been a pervasive voice of dissent from the last decades' claims that the world is becoming "flat" (Iammarino and McCann, this volume, Chapter 14). Concomitantly, international business scholars have also paid attention to clusters, focusing on multinational enterprises (MNEs) and how these span geographical distance in order to search for new markets, resources, and innovation opportunities, and reaching the same conclusion (Ghemawat 2007). Now, the two disciplines are converging around an understanding of cluster *connectedness* (the quantitative volume of a cluster's global connections) and *connectivity* (the nature and scope of these global connections). This nascent theoretical framework can address some pertinent questions related to global science, technology, and innovation. For instance, why do some clusters benefit more than others from being integrated in the global economy? More specifically, what makes particular clusters able to leverage their outward trade and collaboration to catch up to global technological frontiers or even become global technological leaders – and what determines how the fruits of such innovation are distributed among individuals and organizations in these clusters?

While the internal processes of innovation in clusters are relatively well understood (Maskell and Malmberg 1999; Gertler 2003; Maskell and Lorenzen 2004; Storper and Venables 2004), it is a relatively new recognition that globalization has increased both inter-cluster competition and the innovation-related "gains from trade" that arise from connections between firms located in different clusters (OECD2004; Klagge and Martin 2005; Zademach 2009; Lorenzen and Mudambi

The Handbook of Global Science, Technology, and Innovation, First Edition.
Edited by Daniele Archibugi and Andrea Filippetti.

2013). Having been studied under diverse headings such as "global value chains" (Dicken et al. 2001; Gereffi, Humphrey, and Sturgeon 2005; Mudambi 2008), "commodity chains" (Pratt 2008), "production networks" (Coe, Dicken, and Hess 2008) "flows" (Hudson 2005), and "linkages" (Giuliani, Rabellotti, and Van Dijk 2005), the study of connections between clusters is still a nascent research field. In the following sections, we provide an overview of this field well as some pointers of how to promote our understanding of the impact of innovation through analyzing cluster connectedness and connectivity.

Global Connections: Connectedness and Connectivity

Global connections are channels for resource flows in and out of clusters (Amin and Thrift 1992; Bathelt, Malmberg, and Maskell 2004). While such connections facilitate harsher inter-cluster competition (Amin 2002; Humphrey and Schmitz 2002; Sheppard 2002; Morgan 2004; Pratt 2008) they also hold potential for innovation in clusters through infusing them with knowledge, technology, and capital from worldwide sources (Amin 2002; Davenport 2005; Gertler and Levitte 2005). Crossing the boundaries of a cluster can lead to global connections with cutting-edge technologies and knowledge – lead clusters, "centers of excellence" (Cantwell and Janne 1999) or "knowledge hotspots" (Bathelt et al. 2004). As noted by Florida (2005), such hotspots are the "spikes" in the global knowledge landscape: Highly specialized Marshallian clusters with focused innovative activity or global cities with diversified innovation profiles. Innovation systems in industry verticals are composed of globally dispersed, interlinked knowledge hotspots that are hierarchically ordered (Meyer, Mudambi, and Narula 2011). Lower order hotspots service higher order ones and all are orchestrated by apex or lead clusters. Silicon Valley is an archetypical example of a lead cluster in information technology. Other examples include Hollywood (filmed entertainment), London (financial services), Milan (fashion), and Stuttgart (automotives). MNEs' networks co-evolve with the global knowledge and innovation landscape, so that their subsidiaries in lead clusters typically serve as centers of excellence for their worldwide operations.

On the one hand, the potential for a cluster to innovate on the basis of global flows of knowledge depends on its sheer number of global connections – the scale of the cluster's *connectedness*. On the other hand, the ability of the cluster to benefit, in innovation terms, from these connections, is moderated by the connections' decision-making locus and network structure – the scope of the cluster's *connectivity*. Below, we shall develop a typology of connectivity and discuss how different types of connectivity impact a cluster's innovation potential.

The Decision-Making Locus within Connections: Personal Relationships vs. Organizational Pipelines

Within connections, we draw a fundamental distinction between two different decision-making loci: individual-based and organization-based. The first and most basic type, *personal relationships*, is operationalized through autonomous individuals, and includes family relations, friendships, and acquaintanceships. Individuals often leverage the social proximity provided by such connections to seek professional opportunities, and when personal relationships span globally, they facilitate both labor mobility and entrepreneurship across clusters. The best examples of how personal relationships may constitute global connections are those between members of global diasporas (Agrawal, Cockburn, and McHale 2006; Saxenian 2006; Zaheer, Lamin, and Subramani 2009). Affiliated to multiple geographical localities by birth, these individuals leverage not just their family and friends, but also cultural similarities and shared national backgrounds, in building "swift trust" (Meyerson, Weick, and Kramer 1996) that facilitates personal as well as professional relationships across several clusters.

The second type of decision-making locus is operationalized through organizations and has been referred to by Bathelt et al. (2004) as *pipelines*. These include strategic alliances, joint ventures, or ownership spanning different organizations or organizational branches. Even though pipelines, like all connections, are embedded in the individuals who build and operate them (indeed, organizations sometimes hire individuals because of the valuable connections they hold), they differ from personal relationships in important ways. The organizational hierarchy provides particular governance and incentive mechanisms. Consequently, organizational routines are designed to ensure that employees do not act autonomously when operating pipelines. The leverage of such pipelines is aligned with the strategy of the parent organization, not with the interests of individuals.

Organizations maintain pipelines in order to move resources across geographical distance. Mature clusters in advanced market economies are characterized by having a multitude of pipelines built by indigenous organizations. By contrast, in clusters in less developed economies, pipelines are typically brought in by foreign MNEs and operated by local subsidiaries. International business scholars have studied the latter type of pipelines for decades (Mudambi and Navarra 2004; McCann and Mudambi 2005; Beugelsdijk, McCann, and Mudambi 2010). Compared to strategic alliances, subsidiary ownership reduces risk and spatial costs, but also incurs higher setup costs (Gupta and Govindarajan 1991). The general potential of an MNE subsidiary to spur innovation in a cluster depends, on the one hand, on the type of its mother organization. Leading MNEs in the global competitive environment typically undertake more innovative activities in their subsidiaries, while lagging MNEs often undertake routine innovative activities in their subsidiaries by replicating activities performed elsewhere in their organization (Cantwell and Janne 1999). On the other hand, the potential of a subsidiary to spur innovation also depends on the subsidiary's own type. The literature distinguishes between subsidiaries with competence-exploiting mandates focused on routine replication and local adaptation and unlikely to spark a great deal of local innovative activity and subsidiaries with competence-creating mandates charged with developing new competencies for use beyond the local market, likely to be associated with significant inflows and outflows of knowledge (Cantwell and Mudambi 2005).

The Network Structure of Connections: Centralized vs. Decentralized

A cluster's configuration of global connections can be seen as a particular *social network* where local and non-local individuals and organizations are the nodes and their connections are the ties. The structure of this social network impacts how a cluster benefits from local individuals' and organizations' global connections (Lorenzen and

Mudambi 2013). Of particular importance is the connection network's degree of *centralization* (Barabasi and Albert 1999; Wasserman and Faust 1994; Watts and Strogatz 1998). In a social network, the more relations a node has to other nodes, the higher that node's centrality.[3] When a few nodes with high centrality dominate a network, the structure of the network is centralized. By contrast, when all nodes have a comparable number of ties, so that none is central, the network is decentralized.

Using this terminology, when flows into and out of a cluster are mediated by just a few gatekeepers, the structure of the connection network is *centralized* (Bell and Albu 1999; Bathelt et al. 2004; Giuliani 2007; Graf and Krüger 2010; Malecki 2010). Strong local individuals or organizations that hold many global connections typically also come to dominate the network of local connections. This makes peripheral individuals or organizations in the cluster depend on the central individual or organization for access to each other as well. For example, "industrial complex" type clusters with dominant local organizations (McCann and Mudambi 2005) are likely to progress into centralized connection network structures. This is because the locally dominant organizations receive most of the overtures from organizations outside the cluster seeking to establish inward connections and they concomitantly develop experience in creating and maintaining all types of connections (Wolfe and Gertler 2004).

Conversely, when the structure of a connection network is *decentralized*, flows occur directly between many nodes, all with comparable centrality. Decentralization of global connections is likely to co-evolve with direct relations among local individuals and organizations as well. The example to consider here is non-hierarchical, trust-based clusters without dominant resident organizations, such as the "pure agglomeration" (McCann and Mudambi 2005) or "industrial district" clusters (Markusen 1996).

Figure 10.1 illustrates these two different stylized extremes. As the figure shows, in a completely centralized connection network, all external nodes are connected to a central local node and all flows to and from the cluster nodes must take place through brokerage facilitated by this node (Burt 1992). This individual or organization can effectively operate as a gatekeeper or "guru" (Mudambi, Oliva, and Thomas, 2009), shaping the nature of resource flows into and out of the cluster, as well as capturing the bulk of the rents that arise from these flows. In contrast, in a completely decentralized connection network, every local node is connected directly to external nodes.

Global Connectivity: Archetypes and Examples

A cluster's *connectivity* denotes its global connections' particular configuration of decision-making loci and network structure (Lorenzen and Mudambi 2013). The stylized types of connections' decision-making loci and network structure developed above combine into four archetypes of cluster connectivity. These are sketched out in Figure 10.2.

The first connectivity archetype is *centralized personal relationships*. A real-world example is clans – person-based relationships revolving around a few central elders or other leaders (Boisot and Child 1996). Clans may connect different clusters and facilitate the transfer of knowledge and other resources to dominant clan members.

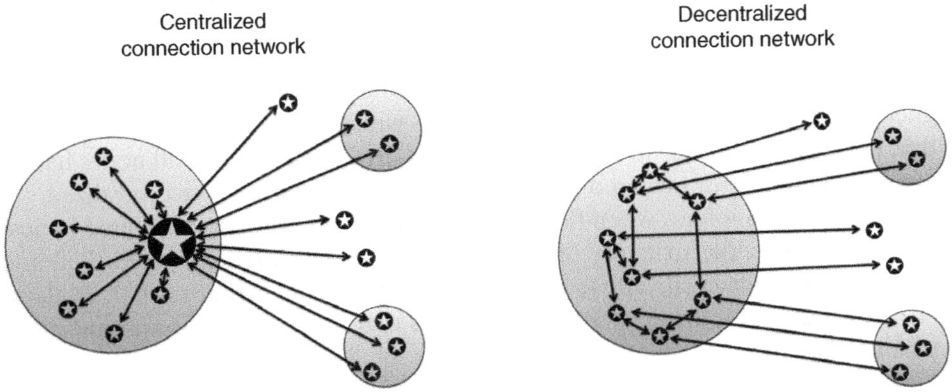

Figure 10.1 Centralized and decentralized connection networks.
Source: Lorenzen and Mudambi (2013).

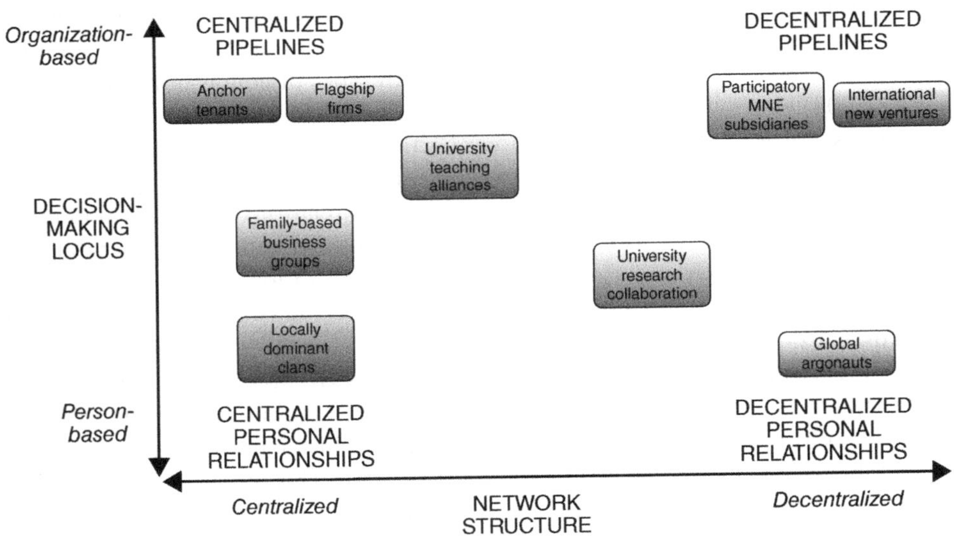

Figure 10.2 Types and examples of connectivity.
Source: Lorenzen and Mudambi (2013).

Examples of locally dominant clans are often found in East Asia, particularly within Chinese business networks where extended family connections across cities or national borders facilitate privileged access to labor, capital, information, credit, and markets (Todeva 2006; Yeung 1997). Even if this connectivity type has a very long history, it is still prevalent in clusters located in countries with strong family traditions.

Decentralized personal relationships represent the second connectivity archetype. A good example is the relations maintained between a diaspora of migrants from a cluster and their families and friends who remain in the home country. "Global Argonauts" (Saxenian 2006) are important operators of this connectivity archetype; they use their commonalities of language and culture as well as experience and contacts from one cluster in order to start up business ventures in another. Decentralized personal relationships facilitate the transfer of resources across clusters, both knowledge and remittances, often fueling business activity in a home cluster with resources generated abroad (Vaaler 2011). This connectivity type has traditionally been driven by waves of immigration. During the last half century, it has become prevalent in still more clusters due to new transportation and communication technologies allowing diasporas to stay in touch.

The third connectivity archetype is *centralized pipelines*. This may arise when "flagship firms" (Rugman and D'Cruz 1997) act as gatekeepers between other cluster organizations and global markets, or when "anchor tenants" (Pashigian and Gould 1998; Agrawal and Cockburn 2003), attract customers or investments to a cluster from global markets.[4] In mature clusters in advanced market economies, flagship firms and anchor tenants are often resource-rich indigenous organizations. However, in nascent clusters, flagships and anchor tenants are likely to be local subsidiaries of foreign MNEs that have a strong and dominant role due to the lack of development of other local organizations. It is worth noting that network position does not always follow from firm size. Rather, it is often a result of a dominant technological capability. As noted by Bell and Albu (1999), smaller organizations or public institutions can function as local technology gatekeepers. This connectivity type is limited to clusters that participate in global trade and production. The most connected such clusters are those home to the most active MNEs.

The fourth archetype, *decentralized pipelines* arises, for example, when high-technology indigenous SMEs set up "international new ventures" (Mudambi and Zahra 2007), or when MNE subsidiaries collaborate broadly with peers in a cluster, for instance, through R&D partnerships. In this case, MNE subsidiaries do not occupy central network positions. This may be a result of a strategy to be present on an equal footing with local firms in order to gain access to horizontal local knowledge-sharing (Lorenzen and Mahnke 2004), or it may be a result of the cluster being home to a large number of equally big MNE subsidiaries, so that none of them is able to take on a central role as flagship firm or anchor tenant. Since each of the MNE subsidiaries provides a pipeline to global markets, the structure of the connection network is decentralized. Driven by globalization of production and innovation activities, this connectivity type is spreading to more and more clusters that develop small-scale high-tech activities.

Of course, what we observe in the real world are mixed forms rather than pure archetypes. Real-world configurations typically include aspects of both individual-based and organization-based decision loci, and network structures with greater or lesser extents of centralization. For example, a family-based business group is a centralized organizational structure in which individual decision locus is crucial; thus, centralized personal relationships play a key role (Khanna and Palepu 2000; Yeung 1997). Universities provide another example. Their two main mission and activity areas, knowledge creation (research), and knowledge dissemination (teaching), have different types of connectivity. When teaching activities span between clusters, this is largely accomplished through bilateral alliances between universities at the organizational level, that is, through pipelines, which may well be centralized. In contrast, cross-cluster research collaboration takes place, in the main, between individual researchers, that is, through decentralized personal relationships.

As mentioned, a cluster's connectedness denotes the scale of its global connections, while its connectivity denotes the scope of a particular configuration of these connections. While connectedness represents a potential for innovation, it is connectivity that determines whether and how this potential will be realized. This is because the decision-making loci and network structure of the connections of a cluster signify how local individuals and organizations "hook on to" the global production and innovation system. Before we discuss how different forms of connectivity facilitate innovation, let us consider some general ways in which decision-making locus and network structure influence innovation.

The Decision-Making Locus within Connections and Innovation: Emergent vs. Strategic

Innovation through personal relationships is likely to be emergent and span diverse (but often related) technologies and industries, while innovation through pipelines is likely to be strategic and focused in terms of technology and industry.

Creative individuals are more focused on knowledge outcomes than firm performance outcomes (Mudambi and Swift 2009), and hence, connections with individual-based decision-making facilitate innovation activities that are *emergent* from the bottom up through interactions of many autonomous individuals who experiment, collaborate, merge, learn, spin-off, and even steal from each other. Innovation through personal relationships is often "about not recognizing boundaries or limitations – exploring for its own sake" (Gurteen 1998). Because individuals have diverse objectives, backgrounds, and knowledge bases (Doz, Santos, and Williamson 2004), innovation through personal relations is likely to cross the boundaries between organizations, industries, and technology fields. Such boundary spanning has been demonstrated to be a key aspect of exploratory innovative activities that involve distant rather than local search in technological space (Todeva 2006; Lamin and Dunlap 2011). Hence, personal relationships offer a high scope for diversity and knowledge recombination.

In contrast, connections with organization-based decision-making are likely to target organizational performance objectives, such as the development of marketable products and processes with relatively short time horizons. Thus, innovation emerging through pipelines tends to be *strategic*. Strategic innovation activities managed and owned by profit-oriented organizations with an explicit innovation strategy are often relatively narrow in scope. Successful organizations tend to link closely to suppliers, buyers and partners, that is, they tend to be very industry-focused (Christensen 1997). This implies that innovation activities through pipelines are likely to be characterized by relatively less technological diversity (Feldman and Audretsch 1999).

The Network Structure of Connections and Innovation: Appropriation vs. Participation

The main share of innovation through centralized connection networks is likely to be appropriated by central individuals or organizations, while innovation through decentralized connection networks is likely to have spillovers because individuals or organizations participate in innovation processes with a comparatively higher degree of reciprocity.

Economic geography as well as international business studies often refer to the innovation effects of social networks, but rarely apply insights from the social network analysis literature.[5] A general insight in the latter literature is that in any network, central nodes have privileged access to resources (e.g., knowledge and capital) flowing through network ties. On the one hand, this means that central nodes are equipped to perform tasks of benefit to all other nodes in the network, and to allow other nodes to enjoy some innovation-related advantages of being in a "small world" (Watts and Strogatz 1998). In the context of a cluster, central individuals or organizations can, for example, coordinate local value chains, bringing down transaction costs for all participants. They may also undertake activities of risk-taking and the attraction of capital and labor (Feldman 2003). Furthermore, if central individuals or organizations participate to knowledge sharing, knowledge may spill over to other individuals and organizations through network ties (McCann and Mudambi 2005). On the other hand, central nodes are also in a position to appropriate resources in a network, excluding other nodes from sharing. This is particularly relevant for knowledge, being the basis of intellectual property and most value creation by individuals and organizations. In a cluster, all knowledge-bearing local individuals and organizations weigh the fears that local knowledge spillovers may find their way to powerful competitors, against potential future gains from the local upgrading of current (and potential future) partners.

Hence, network structure is a key determinant of whether knowledge and other resources become appropriated or are allowed to spill over. There are theoretical arguments and empirical evidence that centralized network structures are likely to be associated with the *appropriation* of resources by a few central individuals (Granovetter 1973; Burt 1992; Uzzi and Spiro 2005) and organizations (Uzzi 1997). In the case of global connections with centralized network structure, cluster-wide spillovers are likely to be minimized by central individuals or organizations, because they have both the incentives and the ability to exclude others from most of the rents that arise from the connections that they construct. Thus, the lion's share of rents arising from innovation is likely to be captured by central individuals or organizations. Decentralized social networks lack strong and dominant coordinating individuals or organizations, but are able to facilitate *participation* in the sharing of resources among a broad category of individuals and organizations without rendering resources proprietary to any of them, and ensuring that the rents generated through network ties are distributed over a wide variety of actors. In a cluster, collaboration in decentralized networks may suffer from high coordination costs (Foss and Lorenzen 2009). However, as resources can be shifted around and have high value in alternative uses, this network structure propagates flexible specialization in value chains (Press 2008), entrepreneurship (Delgado, Porter, and Stern 2010). The associated innovation processes have many spillovers and a high degree of experimentation (Foss and Foss 2002; Lorenzen and Frederiksen 2005). In the case of global connections with decentralized network structure, both inflows and outflows of knowledge, and the fruits of innovation, occur to serve a multiplicity of interests, rather than the sole interests of the central individuals or organizations.

Connectivity and Innovation: Breadth, Depth, and Spillovers

After these general observations on how innovation relates to decision-making locus and network structure, we shall now discuss how each of the four archetypes of connectivity impacts innovation.

First, the connectivity type *centralized personal relationships* is likely to facilitate innovation across diverse technology areas and industries. Such in-breadth innovation, however, will typically be appropriated by central individuals and will have limited spillovers inside a cluster. Centralized personal relationships in the form of clans and family-based business groups may serve to coordinate local value chains as well as channel resources to local innovation activities (Todeva 2006; Yeung 1997). A clan or family may spend decades nurturing global connections with adherents or kin in valuable clusters around the world in order to extract value from the structural holes in the networks it spans. Consequently, it has obvious incentives to exclude other local individuals who are not family or clan members, maintaining tight control over the knowledge and other key resources that flow through the global connections it holds, and appropriating most of the innovation-based rents arising through them. As a result, with centralized personal relationships, access to critical resources is likely to be limited to members of the dominant group, reducing both the scope for interaction and the diversity of knowledge. Centralized personal relationships are unlikely to thrive in industries that require constant infusions of new knowledge, since family-based business groups and clans are likely to avoid becoming reliant on outsiders. Hence, they typically operate in traditional industries (Yeung 1997; Khanna and Palepu 2000), where they are likely to leverage global connections to enhance and perhaps extend their existing businesses through management innovations. This is unlikely to enhance knowledge spillovers in a cluster.

Second, the connectivity type *decentralized personal relationships* is likely to facilitate in-breadth innovation across diverse technology areas and industries with substantial spillovers to the multiple individuals participating inside a cluster. When a cluster is connected by numerous autonomous relationships of migrants (e.g., highly educated Global Argonauts), none of these can act as a gatekeeper. The only way for an individual to profit from his or her connections is to maximize their throughput of knowledge and resources, with the resulting innovation spilling over to other participants in the connection network. The results of such connectivity are, for example, knowledge transfer between diaspora members employed in different clusters (Song, Almeida, and Wu 2003; Hoisl 2007) and high levels of technological entrepreneurship (Saxenian 2002, 2006). Innovation driven by decentralized personal relationships can span broadly, across technology areas, because the personal relationships held by global migrants are typically based on education and professional experience (Saxenian 2006), or simply shared national culture. In contrast to the power-based relationships held by clan elders or heads of family-based business groups, decentralized personal relationships are able to tap into a much wider range of expertise and knowledge, and are more likely to operate in high-technology, asset-light industry environments and to implement technological innovations.

Third, the connectivity type *centralized pipelines* is likely to facilitate innovation within specialized technology areas and industries. This in-depth innovation will typically be appropriated by central organizations and will have limited spillovers inside a cluster. Centralized pipelines typically involve large sunk investments in facilities and training, which the organizations that built them seek to recoup. Hence, while a centralized pipeline can be effective in facilitating innovation, particularly in high-tech industries where R&D is scale-intensive, the organization will also design its pipelines to capture, to the extent possible, the rents from the resource flows through them. For example, if a flagship firm or anchor tenant MNE subsidiary undertakes repeated interactions with endogenous cluster organizations, fears of spillovers will lead it to retain specialized innovative activities within its wholly controlled local subsidiary and limit its local partners to routinized or "commodity" activities (Andersson, Forsgren, and Holm 2002; McCann and Mudambi 2005). This limits the innovative capabilities of the wider population of organizations in the cluster and ensures that it remains at a lower rung in the hierarchy of the global clusters (Meyer et al. 2011).

Fourth and finally, the connectivity type *decentralized pipelines* is likely to facilitate in-depth innovation within specialized technology areas and industries with substantial spillovers to the multiple organizations participating inside a cluster. We may find decentralized pipelines when several equally central competing organizations "plug into" a cluster (Lorenzen and Mahnke 2004), or where many strong indigenous SMEs each reach out and establish new international ventures. This connection network structure facilitates flexible and shifting R&D collaborations and innovation projects facilitating trial-and-error experimentation. This type of innovation is particularly prominent in consumer goods and entertainment ("cultural") industries, which is why we see such industries thriving in well-connected, non-hierarchical, trust-based clusters with a diversity of related and interconnected activities and industries (Maskell and Lorenzen 2004; Lorenzen and Frederiksen 2005).

Concluding Remarks

In this chapter, we have provided an overview of the nascent theoretical framework of cluster connectedness and connectivity. This framework is created by fusing the economic geography and IB literatures and applying insights from the literature on social networks. The framework enables us to develop testable predictions regarding the nature of innovative activity and outcomes in different cluster configurations. We have summed up this argument in Table 10.1.

We contribute to the newly arising literature of microfoundations by explicitly recognizing the role of individuals in macro outcomes. However, we go further by incorporating the fact that individuals may or may not be ensconced within organizations. Organizational pipelines may arise from personal relationships, as when corporations hire people because of their connections. Such personal relationships may then be leveraged for strategic purposes. When individuals operate within firms, individualistic decision-making is both constrained and empowered through the routines and resources of the organization. Thus, while decisions in organizations are made by individuals, they are significantly modulated by the organizational context. Thus, the reverse process of individuals leveraging existing organization-based connections for their personal advantage may be limited by routines such as rules and competition clauses.

Table 10.1 Connectivity and global innovation.

Connectivity type	*Definition*	*Impact on innovation*	*Examples*	*Evolution*
Centralized personal relationships	Connections with person-based decision locus revolving around one or few central persons	In-breadth innovation across diverse technology areas and industries. Limited spillovers appropriated by central individuals	Clans, family-based business groups (Boisot and Child 1996; Todeva 2006; Yeung 1997)	Historical and still prevalent in clusters with strong family traditions, e.g., in China, India, Iran, and Italy
Decentralized personal relationships	Connections with person-based decision locus with no dominant central persons	In-breadth innovation across diverse technology areas and industries. Substantial spillovers to the multiple individuals participating	Diasporas, global Argonauts (Saxenian 2006; Vaaler 2011)	Dependent on waves of immigration, spreading due to new transportation and communication technologies, e.g., to clusters in Taiwan, Korea, and Africa
Centralized pipelines	Connections with organization-based decision locus revolving around one or few central organizations	In-depth innovation within specialized technology areas and industries. Limited spillovers appropriated by central organizations	Flagship firms, anchor tenants (Rugman and D'Cruz 2000; Pashigian and Gould 1998; Agrawal and Cockburn 2003)	Spreading with globalization of trade and production, typically driven by well-connected clusters in United States, Europe, and East Asia
Decentralized pipelines	Connections with organization-based decision locus with no dominant central organizations	In-depth innovation within specialized technology areas and industries. Substantial spillovers to the multiple organizations participating	International new ventures, participatory MNEs (Mudambi and Zahra 2007; Lorenzen and Mahnke 2004)	Spreading with globalization of innovation, driven by high-tech MNEs from clusters in a broad range of countries

Finally, our analysis presents an analytical framework within which we can understand the innovative potential and trajectory of specific clusters. Perhaps our most important prediction is that a centralized network structure may have short-term benefits in terms of the rapid establishment of global connections, but these may be outweighed by the longer-term costs in terms of a limited extent and range of local spillovers.

In this chapter we have laid out a theoretical roadmap for studying global innovation at the intersection of economic geography and international business studies. This is a fertile intersection with numerous research opportunities. On the one hand, economic geographers have an extremely comprehensive and fine-grained approach to space, proximity, and place (Laurer, Lorenzen, and Staber 2012). In contrast, international business scholars tend to treat space in a fairly simplistic manner, often using simplistic domestic/foreign binary approaches. On the other hand, over the last several decades, international business scholars have developed extremely thorough and detailed models of global economic organization at both the macro (country) and micro (firm) levels. Economic geographers, with some exceptions, tend to treat firms as homogeneous (or eschew firm-level analysis altogether). This discussion suggests two avenues for significant advances: (1) introducing a more sophisticated view of space, proximity, and place into international business models and (2) explicitly incorporating the organization of the MNE into economic geography models.

We point to a few specific opportunities for understanding the nature of global innovation within these general research avenues. The sub-national context is only recently receiving attention in the international business literature (e.g., Beugelsdijk and Mudambi 2013; Lamin and Livanis 2013). We suggest that much more can be done by integrating the large body of work on clusters and city regions into international business models. Going in the opposite direction, the literature on clusters and city regions would be significantly advanced by recognizing the critical role of MNEs in the establishment, growth, and decline of geographic locations.

Notes

1 This chapter draws on Lorenzen and Mudambi (2013).

2 The precise nature and definition of clusters is the subject of some debate in the literature (Porter 2000; Martin and Sunley 2003). However, this debate is not central to our objectives here. For the purposes of this chapter, we adopt the economic geographers' view that defines clusters in terms of geographic space and typically identifies them with city regions (Scott 2001; Storper 2013).

3 This measure of node centrality is called degree centrality. While there are more sophisticated measures, degree centrality captures well the extent to which the central node is able to bridge "structural holes" and capture resources in the network (Freeman 1979; Burt 1992).

4 Pashigian and Gould (1998) define anchor tenants as well-known business enterprises with recognizable names that attract commercial counter-parties to a location. Rugman and D'Cruz (1997) define flagship firms as hubs of local economic activity. Less well-known enterprises can free-ride off the traffic generated by anchor tenants and flagship firms. In knowledge hotspots, firms like Google and IBM take on this role, attracting high skill human capital and a constellation of supplier and service firms.

5 For some exceptions, see Eisingerich et al. (2010), Cantwell and Mudambi (2011), and Eapen (2012).

References

Agrawal, A., and I. Cockburn. 2003. "The Anchor Tenant Hypothesis: Exploring the Role of Large, Local, R&D-Intensive Firms in Regional Innovation Systems." *International Journal of Industrial Organization* 21: 1227–1253.

Agrawal, A., I. Cockburn, and J. McHale. 2006. "Gone but not Forgotten: Knowledge Flows, Labor Mobility, and Enduring Social Relationships." *Journal of Economic Geography* 6: 571–591.

Andersson, U., M. Forsgren, and U. Holm. 2002. "The Strategic Impact of External Networks: Subsidiary Performance and Competence Development in the Multinational Corporation." *Strategic Management Journal* 23(11): 979–996.

Amin, A. 2002. "Spatialities of Globalization." *Environment and Planning A* 34: 385–399.

Amin, A., and Thrift, N. 1992. "Neo-Marshallian Nodes in Global Networks." *International Journal of Urban and Regional Research* 16(4): 571–578.

Barabasi, A.L., and Albert, R. 1999. "Emergence of Scaling in Random Networks." *Science* 286: 509–512.

Bathelt, H., A. Malmberg, and P. Maskell. 2004. "Clusters and Knowledge: Local Buzz, Global Pipelines and the Process of Knowledge Creation." *Progress in Human Geography* 28(1): 31–56.

Bell, M., and M. Albu. 1999. "Knowledge Systems and Technological Dynamism in Industrial Clusters in Developing Countries." *World Development* 27(9): 1715–1734.

Beugelsdijk, S., and R. Mudambi. 2013. "MNEs as Border-Crossing Multi-Location Enterprises: The Role of Discontinuities in Geographic Space." *Journal of International Business Studies* 44: 413–426.

Beugelsdijk, S., P. McCann, and R. Mudambi. 2010. "Place, Space and Organization: Economic Geography and the Multinational Enterprise." *Journal of Economic Geography* 10(4): 485–493.

Boisot, M., and J. Child. 1996. "From Fiefs to Clans and Network Capitalism: Explaining China's Emerging Economic Order." *Administrative Science Quarterly* 41: 600–628.

Burt, R.S. 1992. *Structural Holes*. Cambridge, MA: Harvard University Press.

Cantwell, J., and O. Janne. 1999. "Technological Globalization and Innovative Centres: The Role of Corporate Technological Leadership and Locational Hierarchy." *Research Policy* 289: 119–144.

Cantwell, J., and R. Mudambi. 2005. "MNE Competence-Creating Subsidiary Mandates." *Strategic Management Journal* 26(12): 1109–1128.

Cantwell, J., and R. Mudambi. 2011. "Physical Attraction and the Geography of Knowledge Sourcing in Multinational Enterprises." *Global Strategy Journal* 1(3–4): 206–232.

Christensen, C. 1997. *The Innovator's Dilemma: When New Technologies Cause Great Firms to Fail*. Boston, MA: Harvard Business Press.

Coe, N., P. Dicken, and M. Hess. 2008. "Global Production Networks: Realizing the Potential." *Journal of Economic Geography* 8: 271–295.

Davenport, S. 2005. "Exploring the Role of Proximity in SME Knowledge-Acquisition." *Research Policy* 34: 683–702.

Delgado, M., M.E. Porter, and S. Stern. 2010. "Clusters and Entrepreneurship." *Journal of Economic Geography* 10(4): 495–518.

Dicken, P., P.F. Kelly, K. Olds, and H.W. Yeung. 2001. "Chains and Networks, Territories and Scales: Towards a Relational Framework for Analyzing the Global Economy." *Global Networks* 1(2): 89–112.

Doz, Y., J. Santos, and P. Williamson. 2004. "Is Your Innovation Process Global?" *Sloan Management Review* 45(4): 31–37.

Eapen, A. 2012. "Social Structure and Technology Spillovers from Foreign to Domestic Firms." *Journal of International Business Studies* 43(3): 244–263.

Eisingerich, A.B., S.J. Bell, and P. Tracey. 2010. "How Can Clusters Sustain Performance? The Role of Network Strength, Network Openness, and Environmental Uncertainty." *Research Policy* 39: 239–253.

Feldman, M.P. 2003. "The Locational Dynamics of the U.S. Biotech Industry: Knowledge Externalities and the Anchor Hypothesis." *Industry and Innovation* 10(3): 311–328.

Feldman, M.P., and D. Audretsch. 1999. "Innovation in Cities: Science-Based Diversity, Specialization and Localized Competition." *European Economic Review* 43(2): 409–429.

Florida, R. 2005. "The World Is Spiky: Globalization Has Changed the Economic Playing Field, But Hasn't Leveled It." *Atlantic Monthly* 296(3): 48.

Foss, K., and N.J. Foss. 2002. "Organizing Economic Experimentation." *Review of Austrian Economics* 15(4): 297–312.

Foss, N., and M. Lorenzen. 2009. "Towards an Understanding of Cognitive Coordination: Theoretical Developments and Empirical Illustrations." *Organization Studies* 30(11): 1201–1226.

Freeman, L.C. 1979. "Centrality in Social Networks: Conceptual Clarification." *Social Networks* 1: 215–239.

Gereffi, G., J. Humphrey, and T. Sturgeon. 2005. "The Governance of Global Value Chains." *Review of International Political Economy* 12(1): 78–104.

Gertler, M.S. 2003. "Tacit Knowledge and the Economic Geography of Context, or the Undefinable Tacitness of Being (There)." *Journal of Economic Geography* 3: 75–99.

Gertler, M.S., and Y.M. Levitte. 2005. "Local Nodes in Global Networks: The Geography of Knowledge Flows in Biotechnology Innovation." *Industry and Innovation* 12(4): 487–507.

Ghemawat, P. 2007. "Why the World Isn't Flat." *Foreign Policy* 159: 54–60.

Giuliani, E. 2007. "The Selective Nature of Knowledge Networks in Clusters: Evidence from the Wine Industry." *Journal of Economic Geography* 7(2): 139–168.

Giuliani, E., R. Rabellotti, and M.P. Van Dijk (eds.). 2005. *Clusters Facing Competition: The Importance of External Linkages*. Aldershot: Ashgate.

Graf, H., and J.J. Krüger. 2010. "The Performance of Gatekeepers in Innovator Networks." *Industry and Innovation* 18(1): 69–88.

Granovetter, M. 1973. "The Strength of Weak Ties." *American Journal of Sociology* 78(6): 1360–1380.

Gupta, A., and V. Govindarajan. 1991. "Knowledge Flows and the Structure of Control within Multinational Corporations." *Academy of Management Review* 16(4): 768–792.

Gurteen, D. 1998. "Knowledge, Creativity and Innovation." *Journal of Knowledge Management* 2(1): 5–13.

Hoisl, K. 2007. "Tracing Mobile Inventors: The Causality Between Inventor Mobility and Inventor Productivity." *Research Policy* 36: 619–636.

Hudson, R. 2005. *Economic Geographies: Circuits, Flows and Spaces*. London: Sage.

Humphrey, J., and H. Schmitz. 2002. "How Does Insertion in Global Value Chains Affect Upgrading in Industrial Clusters?" *Regional Studies* 36: 1017–1027.

Khanna T., and P. Palepu. 2000. "Is Group Affiliation Profitable in Emerging Markets? An Analysis of Diversified Indian Business Groups." *Journal of Finance* 55: 867–891.

Klagge, B., and R. Martin. 2005. "Decentralized versus Centralized Financial Systems: Is There a Case for Local Capital Markets?" *Journal of Economic Geography* 5: 387–421.

Lamin, A., and D. Dunlap. 2011. "Complex Technological Capabilities in Emerging Economy Firms: The Role of Organizational Relationships." *Journal of International Management* 17(3): 211–228.

Lamin, A., and G. Livanis. 2013. "Agglomeration, Catch-Up and the Liability of Foreignness in Emerging Economies." *Journal of International Business Studies* 44: 579–606.

Laurer, I., M. Lorenzen, and U. Staber. 2012. "Space and Inter-Organizational Relations." *Industry and Innovation* 19: 181–186.

Lorenzen, M., and L. Frederiksen. 2005. "The Management of Projects and Product Experimentation: Examples from the Music Industry." *European Management Review* 2(3): 198–211.

Lorenzen, M., and V. Mahnke. 2004. "Governing MNC Entry in Regional Knowledge Clusters." In *Knowledge Flows, Governance and the Multinational Enterprise*, ed. V. Mahnke and T. Pedersen, 211–225. London: Palgrave Macmillan.

Lorenzen, M., and R. Mudambi. 2013. "Clusters, Connectivity and Catch-Up: Bangalore and Bollywood in the Global Economy." *Journal of Economic Geography* 13(3): 501–534.

Malecki, E.J. 2010. "Global Knowledge and Creativity: New Challenges for Firms and Regions." *Regional Studies* 44(8): 1033–1052.

Markusen, A. 1996. "Sticky Places in Slippery Space: A Typology of Industrial Districts." *Economic Geography* 72: 292–313.

Martin, R., and P. Sunley. 2003. "Deconstructing Clusters: Chaotic Concept or Policy Panacea?" *Journal of Economic Geography* 3(1): 5–35.

Maskell, P., and A. Malmberg. 1999. "Localised Learning and Industrial Competitiveness." *Cambridge Journal of Economics* 23(2): 167–186.

Maskell, P., and M. Lorenzen. 2004. "The Cluster as Market Organization." *Urban Studies* 41(5–6): 991–1009.

McCann, P., and R. Mudambi. 2005. "Analytical Differences in the Economics of Geography: The Case of the Multinational Firm." *Environment and Planning A* 37(10): 1857–1876.

Meyer, K., R. Mudambi, and R. Narula. 2011. "Multinational Enterprises and Local Contexts: The Opportunities and Challenges of Multiple Embeddedness." *Journal of Management Studies* 48(2): 235–252.

Meyerson, D., K.E. Weick, and R.M. Kramer. 1996. "Swift Trust and Temporary Groups." In *Trust in Organizations: Frontiers of Theory and Research*, ed. R.M. Kramer and T.R. Tyler, 166–195. Thousand Oaks, CA: Sage Publications.

Morgan, K. 2004. "The Exaggerated Death of Geography: Learning, Proximity and Territorial Innovation Systems." *Journal of Economic Geography* 4: 3–21.

Mudambi, R. 2008. "Location, Control and Innovation in Knowledge-Intensive Industries." *Journal of Economic Geography* 8(5): 699–725.

Mudambi, R., and P. Navarra. 2004. "Is Knowledge Power? Knowledge Flows, Subsidiary Power and Rent-Seeking within MNCs." *Journal of International Business Studies* 35(5): 385–406.

Mudambi, R., and T. Swift. 2009. "Professional Guilds, Tension and Knowledge Management." *Research Policy* 38(5): 736–745.

Mudambi, R., and S. Zahra. 2007. "The Survival of International New Ventures." *Journal of International Business Studies* 38: 333–352.

Mudambi, S., T. Oliva, and E. Thomas. 2009. "Industrial Marketing Firms and Knowledge Transfer: Toward a Basic Typology of Community Structures." *Industrial Marketing Management* 38(2): 181–190.

OECD. 2004. *Global Knowledge Flows and Economic Development*. Paris: OECD.

Pashigian, B.P., and E.D. Gould. 1998. "Internalizing Externalities: The Pricing of Space in Shopping Malls." *Journal of Law and Economics* 41(1): 115–142.

Porter, M.E. 2000. "Location, Competition, and Economic Development: Local Clusters in a Global Economy." *Economic Development Quarterly* 14(1): 15–34.

Pratt, A.C. 2008. "Cultural Commodity Chains, Cultural Industry Clusters, or Cultural Production Chains?" *Growth and Change* 39(1): 95–103.

Press, K. 2008. "Divide to Conquer? Limits to the Adaptability of Disintegrated, Flexible Specialization Clusters." *Journal of Economic Geography* 8(4): 565–580.

Rugman, A., and J. D'Cruz. 1997. "The Theory of the Flagship Firm." *European Management Journal* 15(4): 403–412.

Saxenian, A. 2002. "Transnational Communities and the Evolution of Global Production Networks: Taiwan, China and India." *Industry and Innovation* 9(3): 249–265.

Saxenian, A. 2006. *The New Argonauts: Regional Advantage in the Global Economy*. Cambridge, MA: Harvard University Press.

Scott, A. 2001. "Globalization and the Rise of City-Regions." *European Planning Studies* 9(7): 813–826.

Sheppard, E. 2002. "The Spaces and Times of Globalization: Place, Scale, Networks and Positionality." *Economic Geography* 78: 307–330.

Song, J., P. Almeida, and G. Wu. 2003. "Learning-by-Hiring: When Is Mobility More Likely to Facilitate Inter-Firm Knowledge Transfer?" *Management Science* 49: 351–365.

Storper, M. 2013. *The Keys to the City: How Economics, Institutions, Social Interaction and Politics Shape Development*. Princeton, NJ: Princeton University Press.

Storper, M., and A.J. Venables. 2004. "Buzz: Face-to-Face Contact and the Urban Economy." *Journal of Economic Geography* 4: 351–370.

Todeva, E. 2006. *Business Networks: Strategy and Structure*. London: Routledge.

Uzzi, B. 1997. "Social Structure and Competition in Interfirm Networks: The Paradox of Embeddedness." *Administrative Science Quarterly* 42(1): 35–67.

Uzzi, B., and J. Spiro. 2005. "Collaboration and Creativity: The Small World Problem." *American Journal of Sociology* 111: 447–504.

Vaaler, P. 2011. "Immigrant Remittances and the Venture Investment Environment of Developing Countries." *Journal of International Business Studies* 42: 1121–1149.

Wasserman, S., and K. Faust. 1994. *Social Network Analysis: Methods and Applications*. Cambridge: Cambridge University Press.

Watts, D.J., and S.H. Strogatz. 1998. "Collective Dynamics of 'Small-World' Networks." *Nature* 393(6684): 440–442.

Wolfe, D.A., and M. Gertler. 2004. "Clusters from the Inside and Out: Local Dynamics and Global Linkages." *Urban Studies* 41(5–6): 1071–1093.

Yeung, H.W. 1997. "Business Networks and Transnational Corporations: A Study of Hong Kong Firms in the ASEAN Region." *Economic Geography* 73(1): 1–25.

Zademach, H.-M. 2009. "Global Finance and the Development of Regional Clusters: Tracing Paths in Munich's Film & TV Industry." *Journal of Economic Geography* 9(5): 697–722.

Zaheer, S., A. Lamin, and M. Subramani. 2009. "Cluster Capabilities or Ethnic Ties? Location Choice by Foreign and Domestic Entrants in the Services Offshoring Industry in India." *Journal of International Business Studies* 40(6): 944–968.

Chapter 11

New Product Development in Emerging Economies: Innovation in Reverse from China

Simone Corsi, Alberto Di Minin, and Andrea Piccaluga

Introduction

The aim of this chapter is to present, discuss, and empirically show evidence of a new theoretical framework that highlights the significance of emerging countries as a source of innovation and hence influences the direction of strategic Research & Development (R&D) investment of multinational corporations (MNCs) based in advanced economies. The theoretical contribution of this chapter is to connect the theory of R&D maturation (Asakawa 2001; Almeida and Phene 2004) with the discussion on reverse innovation (Immelt, Govindarajan, and Trimble 2009; Govindarajan and Trimble 2012) and innovation from emerging economies (Hang, Chen, and Subramian 2010). We do so by shedding light on the host country factors that can affect foreign subsidiaries in China and that can trigger innovation globally.

We identify a new configuration of R&D evolution in emerging economies, which prioritizes the need to absorb and reinterpret innovation stimuli from the local market rather than the final aim simply being to tap into indigenous science and technology.

China gained the attention of foreign direct investment (FDI) in 1978 with Deng Xiaoping's Open Door Policy. Since then, foreign MNCs have triggered internationalization in China involving a wide range of activities with increasing added value (OECD 2008). Foreign R&D-oriented FDI in China has only been examined in a few studies (Gassmann and Keupp 2008) and, to date, a clear understanding of the dynamics that govern subsidiary evolution toward such kinds of activities has been neglected.

Despite the growing importance of China as a "producer" of highly qualified scientists and engineers (von Zedtwitz et al. 2007), and the emergence of Chinese

The Handbook of Global Science, Technology, and Innovation, First Edition.
Edited by Daniele Archibugi and Andrea Filippetti.

MNCs that are disrupting global markets (Hang et al. 2010), the idea of the nascent giant as a global innovator remains stuck in a low-cost innovation approach. However, China has witnessed incredibly high GDP and market growth rates together with a fast technological catch-up. Innovation is not just being stimulated, it is being generated.

Through the analysis of four case studies of Italian MNCs who have internationalized R&D activities in China, this chapter contributes to the existing literature and highlights the relevance of R&D internationalization in China where host country factors trigger innovation at a corporate level and boost the evolution of subsidiaries.

The next section overviews the literature on R&D internationalization and foreign R&D in China. Then we present our cases and discuss empirical evidence. We finish off with conclusions and suggestions for future research.

Literature Review

R&D Internationalization and the Classic Evolution of Foreign R&D Labs

International trade and MNC theories have widely investigated the reasons behind cross-border economic relations. While classical economists have used factor endowments and the cost of production to explain international trade (Smith 1776; Ricardo 1817; Hecksher 1933), scholars focusing on MNC activities have explored a wide range of factors affecting the decision to invest abroad (Hymer 1976; Buckley and Casson 1985; Vernon 1966; Dunning 1980; Johanson and Vahlne 1977).

Most of these studies concentrate on FDI by adopting a parent company perspective (Birkinshaw, Hood, and Jonsson 1998). FDI is mostly seen as a way to reproduce the specific advantages of firms abroad, without considering subsidiaries as potential contributors to the enhancement of this competitive advantage or to the development of new ones.

Since the mid-1980s extensive research on the role of subsidiaries and its evolution has spawned new perspectives on the configuration of MNCs, sources of their competitive edge, and factors affecting such an evolution. Ghoshal and Bartlett's (1990) and Gupta and Govindarajan's (1991) network models, Hedlund's (1986) heterarchy model, and Ghoshal and Nohria's (1989) differentiated organization are just four of the several studies that have questioned the parent company perspective.

The role of subsidiaries in different areas has been investigated by defining the concept of centers of excellence (Chiesa 1996; Birkinshaw 1998; Frost, Birkinshaw, and Ensign 2002), focusing on the strategic roles of subsidiaries (Jarillo and Martinez 1990; Anderson, Forsgren, and Holm 2002), and finally by identifying the dynamics of reverse knowledge transfer (Buckley, Clegg, and Tan 2003; Ambos, Ambos, and Schlegelmilch 2006; Frost and Zhou 2005).

The main factors affecting the role of subsidiaries can be divided into three categories (Birkinshaw et al. 1998; Tavares 2001): (1) *internal*: the activity and power exercised by the parent company and sister subsidiaries; (2) *subsidiary*: the endogenous forces of the subsidiary itself; and (3) *external*: the host environment.

R&D is one of the most debated business areas when it comes to internationalization. Several authors have categorized foreign R&D labs and R&D subsidiaries as *efficiency seeking* (Quan and Chesbrough 2010; Brusoni, Prencipe, and Pavitt 2001), *market seeking* (Kuemmerle 1997; Håkansson and Nobel 1993), and *knowledge-resource seeking* (Pearce 1989; Håkansson and Nobel 1993; Chiesa 1994). Kuemmerle's (1997) distinction between *home-base exploiting* and *home-base augmenting* foreign R&D centers has indeed paved the way for a rich generation of categorizations and classifications.

While it is still fair to say that R&D is the least internationalized corporate activity both at the macro (UNCTAD 2005) and industry levels (Macher, Mowery, and Di Minin 2007), empirical research suggests that the globalization of R&D is rising. Following a line drawn by more than one model such as the Product Life-Cycle Theory (Vernon 1966, 1979) or the Uppsala Model (Johanson and Vahlne 1977), it remains unclear as to what extent R&D can ultimately be internationalized. For example, issues connected with Intellectual Property (IP) management and the appropriability of research activities (Teece 1986) have been proposed as barriers to the internationalization of crucial R&D (Patel and Pavitt 1991; Di Minin and Bianchi 2011).

Most empirical investigations so far have looked at the localization of R&D FDI in other industrialized countries. Adopting Kuemmerle's dichotomy of *home-base augmenting* and *home-base exploiting* R&D sites abroad, scholars have identified a classic pattern of foreign R&D labs that usually shifts from the exploitation of parent company knowledge of foreign markets to the exploration of new knowledge, tapping into local scientific contexts once the subsidiary has become embedded (Asakawa 2001; Florida 1997).

The ability of foreign R&D centers to develop new knowledge strongly thus depends on the increasing number and intensity of technological relations that they are able to set up during their operations. Once an R&D subsidiary has reached full maturity, reverse technology transfer (Håkanson and Nobel 2000, 2001) or reverse knowledge transfer (Ambos et al. 2006) – bringing technology or knowledge that has been generated in the foreign subsidiaries back to the headquarters – are likely, thus interdependence is created between internationalization and innovation as described by Grazia Ietto-Gillies, this volume, Chapter 6.

China as a Host Country for Foreign R&D and New Product Development

Most studies concerning the mandate and evolution of foreign subsidiaries in the last stage of maturity (R&D) or on R&D internationalization have analyzed the operations of companies from advanced economies in other advanced economies (Birkinshaw and Hood 1997, 2000; Frost 2001; Frost and Zhou 2005). Yet, as mentioned above, there is still no clear understanding of the business climate and market dynamics underlying the motivations of R&D internationalization.

When it comes to China, an understanding of R&D internationalization is even foggier. Only a few studies have considered this phenomenon (von Zedtwitz 2004; Gassmann and Han 2004; Gassmann and Keupp 2008; Walsh 2007; Quan and Chesbrough 2010). If we refer to Kuemmerle's dichotomy, *home-base exploiting* tends to be considered as the most common option for foreign companies that locate their R&D in China.

The role of China as a host of foreign *explorative* R&D labs aimed at conducting "state of the art" research is thus still limited to a few anecdotal cases.

China as a location for advanced or basic R&D FDI has only been sporadically documented (Govindarajan and Ramamurti 2011). Scholars tend to identify one hindering factor in the low level of IP protection, however we believe that this is only one dimension of the explanation.

The idea of an emerging country is very often related to the image of a not fully developed customer base. The latter tends to be characterized by needs that can be fulfilled with skillful modifications of Western products, while there is little to learn from the *adaptation* to these markets and business environments that can be usefully transferred back to the headquarters for global competitive advantage.

A growing interest in a reverse cycle of innovation originating from China has been suggested by practitioner-oriented journals (Immelt et al. 2009; Zeschky, Widenmayer, and Gassmann 2011; Hang et al. 2010). These studies explain how increasingly innovation is generated in developing countries by both local and foreign firms at the expense of the decreasing importance of advanced countries as a locus of idea generation or product development.

The past 40 years have thus witnessed a shift from a scenario where innovation was developed for advanced economies based only on their local inputs, to one where global innovation also stems from inputs from emerging economies, reversing the international cycle of technology as traditionally set out in Vernon's early works (1966, 1979).

Table 11.1 shows this shift and highlights the various contributions along the shifting process. Output Quadrant 1 (local, top) represents the original, and first phase of, Vernon's product life-cycle theory, where an innovation was most likely to be conceived and developed in an advanced economy (specifically the United States) because of the contingent factors that favor it. The product would also be marketed there thanks to a mature market able to "absorb" it. In the second phase of the life-cycle, an innovation, still designed and developed in an advanced country, is marketed in other international markets in virtue of the price reduction and maturation of those markets (represented by European countries in Vernon's original model). Quadrant 2 (global, top) also represents innovation developed by subsidiaries of MNCs that are located in advanced countries (Cantwell 1995), a trend which paved the way for different contributions regarding subsidiary autonomy and capacity for innovation (Birkinshaw and Hood 1998).

Table 11.1 The shift in the international flow of innovation.

		OUTPUT	
		Local	*Global*
INPUT	*Advanced Economies*	Vernon's Product life-cycle Theory (1966: first phase)	Vernon's Product life-cycle Theory (1966: following phases; 1979) Cantwell, 1995; Birkinshaw & Hood, 1998)
INPUT	*Emerging Economies*	Frugal Innovation (Zeschky et al. 2011) Innovation at the Bottom of the Pyramid (Prahalad, 2004)	Reverse Innovation (Immelt et al. 2009; Govindarajan & Trimble, 2012)

In more recent times, the activity of both local and MNCs that have developed and marketed innovations in emerging economies from inputs from those markets has been reported (Quadrant 3, local, bottom). This phenomenon has been referred to using several terms, including frugal innovation (Zeschky et al. 2011), and innovation for the bottom of the pyramid (Prahalad 2004).

The shift from Quadrant 3 to 4 (global, bottom) has only been identified in the last few years. Companies operating in emerging economies source the inputs for innovation which are initially commercialized only in local markets but which, along with their evolution, find their way to advanced markets either by disrupting existing technologies or opening new market segments (Immelt et al. 2009; Govindarajan and Ramamurti 2011; Govindarajan and Trimble 2012). The ideas developed in this chapter fit in this latter quadrant, in an attempt to address the following question: How do the particular features of the host country affect the R&D activities of foreign MNCs in China and trigger reverse innovation?

The shift from Quadrant 1 to Quadrant 4 is evident not only in terms of the geographical location where an innovation is generated (shifting from a Western-centered to an emerging country-inclusive model), but also with regard to the role of the headquarters and the subsidiaries of an MNC. While the traditional technology life-cycle theory identifies the headquarters as the only innovation generator, more recent models (Quadrant 2) have already acknowledged that foreign subsidiaries could be the creators of innovation to be exploited later on by their headquarters. What is completely new is that this important role in terms of knowledge and innovation creation is now taken over by foreign subsidiaries located in emerging economies. These units can thus generate new value on which the headquarters can build or renew their competitive advantage in global markets.

Empirical Base

The aim of this chapter is to provide a more comprehensive perspective of the evolution of foreign R&D firms in China with particular emphasis on the contextual factors arising from the indigenous business environment that could affect R&D activities and drive them toward global product development.

We regard Italian companies in China as the "talking pig" (Siggelkow 2007) to discuss the relevance of what we have claimed could be usefully labeled as the most upgraded phase of R&D maturation: centered on the analysis and reinterpretation of innovative inputs coming from the indigenous business environment and context for global production.

The strength and sense of urgency of state regulations that guide the interventions of the public sector in China on private enterprises, associated with the relevance of the Chinese market, makes China an obvious region for this sort of maturation. Given the novelty of the topic addressed and its exploratory nature, we adopted a case study methodology (Eisenhardt 1989; Gibbert, Ruigrok, and Wicki 2008) in order analyze the dynamics behind such a phenomenon.[1]

All the cases are Italian companies that have set up R&D operations in China and are successfully developing products and adaptations for the local market. All companies are present in various national markets and claim that their experience in China is helping them to develop a global competitive advantage. All the companies also share a strong commitment toward increasing their R&D investment in China in order to reinforce their local development skills and to build the foundations for a subsidiary with a role as a global contributor.

Esaote SpA

Esaote is an Italian designer and manufacturer of medical diagnostic systems. The company operates along three main business lines, Ultrasound Diagnostic Imaging, Dedicated Magnetic Resonance, and Non-Imaging Cardiology, and also develops information technology healthcare solutions. Since 1992, Esaote's investments in China have been driven by the growing importance of China both as a market and supplying source. The company has production facilities with R&D activities and a sales office in mainland China, a sales office in Hong Kong, and a representative office in Beijing. The R&D lab was established to adapt Esaote's portfolio to the Chinese market and support local production. Esaote's corporate management is now reinforcing the R&D strategies in China in the light of three main considerations pushing Esaote to develop technological solutions that are also commercially exploitable in other, more advanced, markets.

Characteristics of the Chinese Health Sector Ultrasound medical care in China is managed directly by physicians specifically specialized in this area, while in the rest of the world there are medical doctors with broader specializations such as radiology. These highly specialized physicians provide companies with more informed and technologically focused feedback from users. Furthermore the very large Chinese population offers a number of clinical cases that is not available in any other country. Thus Esaote is exploiting the possibility of developing innovative technological solutions in environmental conditions that are not replicable elsewhere. The widespread distribution of wi-fi technologies in China and its adoption for ultrasound machines are two other important characteristics. Thus China is a unique setting for the application of new ICT systems for health and telemedicine. Esaote is developing ultrasound machines that are able to communicate with smartphones or other similar machines through wi-fi. This enables ultrasound machinery to be accessed at a low cost in rural areas and some of the features can be used in isolated care centers or in private homes. The same technological solution would also help overcome the problem, especially in Europe or the United States, of the often lower medical qualification of the technician who delivers the ultrasound, by putting him/her in contact with qualified physicians or with the best performing machines over a wi-fi connection.

Chinese Competition Esaote's most important competitors in China are foreign MNCs (mostly North American and Japanese), however there are some Chinese companies that are very strong on a local level thanks to their embeddedness and the result of a business climate characterized by a high level of *guanxi*[2] (Park and Luo 2001). These companies are increasingly gaining visibility and market share abroad. Their peculiarity is to offer low-performing low-cost technological solutions to the local market, and to develop updated products for advanced countries that are challenging incumbent firms. For example, a well-known Chinese manufacturer of ultrasound machines sells black and white imaging machines, which do not compete with color machines. This will probably have an impact on the incumbents' sales when this particular Chinese company is able to develop color ultrasound machines of an acceptable quality and, most probably, at a lower price. They might also be able to develop entry-level color ultrasound machines that fit the needs of a specific market segment which the incumbents are not currently serving well in advanced countries. Thus the need for on-site R&D capability to interpret such trends and anticipate their moves with ad hoc technological solutions is urgent.

China's Health Reform An interesting element Esaote is addressing is the fact that the Chinese government is creating a potentially very large market but with particular and unique characteristics. The strong role of the Chinese government in controlling and guiding the local economy is well known (Hsueh 2011) and is clearly visible in its major health reform. With a $125 billion investment plan between 2009 and 2012, the government aimed to provide equal access to healthcare for all the Chinese population, mixing government action and market mechanisms to support efficiency and improve quality (Yip and Hsiao 2009). The reform includes the construction of 30,000 new rural hospitals, care centers, or clinics across the country (Rein 2009). Clearly, the government's stimulus package offers considerable market opportunities to companies that operate in the healthcare sector, although the commercially successful products are those that better address the specific needs of a resource constrained rural hospital. Thus machines will need to be cheaper, to be operable in the harshest climatic conditions, and be simpler to use due to the lack of qualified personnel.

In summary, for Esaote investing in Chinese R&D is a priority in terms of the technology and market opportunities explained above.

Magneti Marelli SpA

Magneti Marelli is one of the largest suppliers of automotive systems with revenues in 2012 of about 5.8 billion euros. The company first started operations in China in 1996 and now operates there with productive and R&D facilities in three locations and in four main business lines: powertrain, exhaust systems, lighting, and electronics.

Here we will focus on electronics, which is responsible for three business lines: instrument clusters, body vehicles, and navigation. Work related to the development and industrialization of these lines is carried out by a local R&D team made up of 100 engineers, with the assistance of basic technologies available in the European R&D units. Particular features of the Chinese market provide Magneti Marelli with new sources of innovation for products that are not found elsewhere.

Specifically, three main conditions, two of which are ascribable to government intervention, have pushed Magneti Marelli to foster its innovative effort in China through the implementation of local projects for worldwide technological solutions.

Future Compulsory Introduction of Tyre Pressure Monitoring Systems (TPMS) in Passenger Vehicles As for other more advanced markets, state intervention making TPMS compulsory for vehicles determined the market success of such a technology. The Chinese case represents a special situation given the relatively high price of a

TPMS and the relatively low purchasing power of Chinese customers. One of Magneti Marelli's R&D labs in China has thus been forced to study new solutions to overcome this obstacle and is developing a new product that integrates TPMS software algorithms with Access Vehicle Systems, a smart electronic key to access and start the vehicle. The integration of these two functions, which is new in Magneti Marelli's global product portfolio, would be revolutionary due to its cost reduction estimated at around 30% for consumers. This solution is globally marketable and fits Chinese market features by responding to synergies between low-cost needs and governmental regulations.

Environment The Chinese government is actively promoting investments that lead to more environmentally friendly vehicles. Magneti Marelli has found some interesting opportunities in this respect. Electronics developed within Magneti Marelli's R&D labs offer good intermediate solutions for energy-saving control or energy management (e.g., the start and stop system, efficient battery monitoring), and our interviewees remarked that the strong pressure on environmental control in China is a unique scenario in which to develop new solutions for global markets.

Consumer Behavior Environmental concerns in China are also providing strong incentives for innovation in navigation systems. Dynamic GPS solutions are thus being developed to integrate vehicle distribution and traffic information in major Chinese cities with itinerary suggestions in order to lower fuel and energy consumption. Technological navigation solutions are strongly affected by Chinese consumer preferences, for example, for touchscreen interfaces. This trend is mainly related to the introduction of GPS and navigation systems in parallel with the diffusion of touchscreen smartphones in China. Navigation systems in China thus include a much wider number of applications than those in Europe or the United States. Once developed for the Chinese market, applications can be added to products in other markets at no additional expense for the company.

Brembo SpA

Brembo is a global leader in the design and manufacturing of braking systems for high-performance cars and motorcycles, as well as commercial vehicles. The company first set up a subsidiary in China in 2000, and its current business lines in China are: brake calipers, disc brakes, and drum brakes. Setting up an R&D lab in China was motivated by the increasing growth rates of the local market and the need to dynamically interact with customers, both foreign and domestic. When Brembo started new product development in China, its management identified two main areas of investment focus and global potential.

Drum Brakes "Advanced" carmakers usually limit drum brakes to small capacity vehicles and trucks, and only in their *simplex* (front wheels) version. On the other hand, Chinese car manufacturers tend to opt for *duplex* (both front and rear wheels) drum brakes on a large range of vehicles. The moderate growth rates of the automotive market in advanced countries does not enable Brembo to invest much in the development of new technological solutions for "low profile" segments such as those which require drum brakes. Therefore the company is using the commercial potential of the Chinese market for drum brakes to develop innovative solutions which will be implemented in advanced markets later on.

Environment Given that sustainability is an area of major interest for the Chinese government, like Magneti Marelli, Brembo is also developing solutions that respond to the strong sustainability requirements of the Chinese market. Brembo's Chinese subsidiary has ongoing projects that will reduce the weight of brake systems by 10% (a lower weight means lower consumption), a reduction in residual torque (no friction during braking also means lower consumption and emissions), and an improvement in the wear and disposal of components. The decision to respond to this market input with a strong local R&D presence highlights the importance of China as a global "think tank" for the development of environmentally-related innovations.

Carel Industries SpA

Carel is a mid-sized Italian firm operating in three main B2B industries: air-conditioning, commercial refrigeration (OEM and retail), and humidification systems. The company has been operating in China since 1997 and is now present with a productive WOFE supported by an R&D lab as well as commercial branches. Having realized the importance and the needs of the local market, Carel learned from Chinese competitors how to shape local product innovations in order to tackle the low-cost challenge that characterizes the Chinese domestic market. The company quickly learned that the development work of the Chinese R&D lab was also of great value in Western markets. Chinese subsidiary's contribution concerned two products, subsequently commercialized in advanced markets.

Cheap and User-Friendly The Chinese R&D lab analyzed local market requirements for electronic controllers for the air-conditioning industry and defined the characteristics of the product demanded by the Chinese customers and provided by local competitors, a controller for temperature and humidity. The inputs from the Chinese market provided the basis for Carel to develop a new product, relying on its existing technology platform but responding to local requirements. The new design and characteristics were offered with a 25% price reduction and a more user-friendly interface competitive with both Chinese and European markets. In the European and North American markets, the new product extended Carel's customer portfolio and enabled them to offer far more to their previous clients. The new product also gave Carel the chance to penetrate the residential air-conditioning business in Europe and North America. In the past this business field was not widely covered by the company because its products did not meet the ease of use required by residential users.

Environment China's demand for energy is well known. Carel's Chinese subsidiary detected a need from the Chinese branch of one of the most important global producers and distributors of soft drinks. This big player asked Carel China to develop a solution for an electronic controller for a bottle cooler that would meet energy-saving requirements. The aim was to save energy in a process that accounts for a considerable portion of the total cost of the product. Based on their previous experience in the refrigeration and energy-saving industry and relying on headquarters knowledge, in 2009 Carel developed a new technological solution. The new product is particularly strong in energy saving thanks to two alternative functions developed for the first time in global markets by Carel. Once this product had been adopted by the Chinese branch of the soft-drink producer, it was then adopted globally by the same company for the same purposes. This technological solution is now in Carel's product portfolio and it is offered on a global basis leading to, in accordance with customer data in different environmental and working conditions, up to 50% energy saving compared to the other solutions available.

Discussion

The four cases have some similarities in the way they entered the Chinese market via an internationalization process that essentially seems in line with the Uppsala stage model (Johanson and Vahlne 1977, 2009). An increased commitment to the foreign market has been spurred on by the collection and analysis of additional knowledge and experience of the competitive context. Despite the growing knowledge and development capabilities within the Chinese subsidiaries, favored by the establishment of ad hoc R&D labs, local engineers have to rely on existing knowledge at a corporate level in order to detect technology opportunities and develop new products.

This is even the case with locally marketable products, although our companies stated that China provides them with a large number of highly qualified engineers and technicians. If anything the problem lies in the relatively immature level of industrialization of the country which is not providing experience or accumulated knowledge in process or product innovation in several industries. Chinese R&D personnel hired by foreign companies thus need to be trained both in China and at the headquarters in Europe. In the latter case, week- or month-long training periods are organized by parent companies when a pool of Chinese engineers are required to focus on the development of new products whose concept stems from Chinese market inputs. This approach not only increases the skills of Chinese personnel and the amount of information on corporate technologies and products, but also represents a strong tool for consolidating corporate identity and reducing cultural gaps in the Chinese branch. The same objective is pursued via recruiting Chinese personnel who grew up or have studied in Europe.

The managers of Chinese subsidiaries' have a relatively low autonomy for new product development without the approval of the parent company. Despite this, all our cases confirm that R&D labs need to be set up in China and local managerial and technological skills need to be strengthened in order to develop future products autonomously.

As for host country factors behind the evolution of R&D investment in China, our cases enabled us to identify three primary reasons for this phenomenon: (1) *State Intervention*; (2) *Local Competition*; and (3) *Local Market Features*. Chinese governmental action, be it for example the promotion of new rules and laws or special fiscal or FDI policies, is especially effective in creating the premises for foreign R&D to identify new product development opportunities. Despite the socialist market economy seeming to share many characteristics with Western capitalism, the Chinese government is certainly invisibly, but strongly, managing the local economy and thus affecting the technological paths and strategies of foreign players. While foreign companies commonly have the highest technological competence, local companies struggle to reduce the technological gap. In the meantime, local companies compete, generally speaking, over what they do best: low cost. They mostly provide products with a lower quality at a much lower price than Western companies. This strategy proves successful in an economy where awareness of product quality or performance still needs refining.

The counter attack of foreign companies cannot neglect this aspect. They thus develop new products locally that can compete with local firms. Eventually the same products (may) find market success in advanced markets, maybe in a segment ignored to date, thus giving rise to low-end disruptive innovation (Christensen 1997) or disruptive innovation from emerging economies (Hart and Christensen 2002).

The cost approach is not the only difference that drives Chinese innovation. Another key factor is cultural differences. Familiarity or diffusion of a certain technology can affect the development of complementary technologies or products as in the case of wi-fi for Esaote and Magneti Marelli. On the other hand, low competence or experience related to early stages of industrialization can create the effects presented for Carel: customers need simple and user-friendly products. These factors are summarized in Table 11.2.

To date, the literature has mostly asserted that the learning curve of foreign subsidiaries in China is driven almost exclusively by cost (Zeng and Williamson 2007). The discussion of our cases suggests that this view should be extended. The sheer size of the Chinese market empowers companies to develop innovations that are targeting wider sections of the population. The focus is on lower costs and profiting from the

Table 11.2 Innovation stimuli from the Chinese market.

Company	*State Intervention/ Regulatory Issue*	*Local Competition*	*Local Market Features*
Esaote SpA	Health reform	Risk of erosion of global incumbents' share through low cost entry level technology	Physicians specialized in ultrasound; Very high number of clinical cases; Large diffusion and familiarity with wi-fi technologies
Magneti Marelli SpA	TPMS compulsory on all passenger vehicles; Environmental policies	Local firms compete in terms of low cost low quality products and customers are very price sensitive	High number of applications for navigation systems; Touchscreens
Brembo SpA	Environmental policies		Focus on drum brakes. Exploit Chinese large market as a development field
Carel Industries SpA	Environmental policies	Local firms compete over low cost and product simplicity in order to penetrate Western markets	High cost of energy implies strong energy saving incentive; Need for user friendly products

mass adoption of innovations rather than from higher margins on products and technologies that are developed for the richest, and smallest, segments of the market. This approach might eventually grant them access to market segments in advanced countries which are currently not served or exploited for low profitability reasons, as we have seen for Brembo.

We are thus far from a framework where the host country affects the innovation of MNC subsidiaries based on its technological richness and diversity (Almeida and Phene 2004; Phene and Almeida 2008; Frost 2001; Frost and Zhou 2005). Chinese subsidiaries can instead be read as interpreters of local market characteristics whose inputs configure unique innovation sources. The contextual factors are thus classified as especially important in shaping the evolution of subsidiaries. In order to fully exploit these inputs, subsidiaries need to be guided by parent companies through a technological evolution. A shift in R&D activities from home-based to a more explorative-oriented development (Asakawa 2001; Florida 1997) is thus confirmed. Strong technological knowledge must be held by parent companies, but China is the right place to develop it on the basis of new market inputs once Chinese R&D labs have filled the knowledge gap.

Implications and Future Research

In this chapter we have analyzed four cases of foreign R&D activities in China. We have seen how the evolution of foreign subsidiaries in China is affected by the host country. These are not determined by the possibility of tapping into local knowledge or technology pockets, but rather by innovative inputs stemming from local market-related factors. We divided these factors into three categories: (1) *State Intervention*; (2) *Local Competition*; (3) *Local Market Features*, and we have seen how the traditional evolution path of foreign R&D activities has moved from being exploitative to explorative. The combination of corporate knowledge and technology with innovation inputs from the Chinese market highlights the technological paths needing to be explored. The exploration of new avenues is moving from the headquarters to the subsidiaries given the characteristics of the context in which the latter operate.

We believe that future research should address two issues, which are driven respectively by the policy and managerial implications of our study. The first concerns Chinese FDI policies. Scholars have already drawn a tentative Chinese developmental state approach (Breslin 1996) and spillover effects of foreign R&D investments have been assessed from different perspectives (Cheung and Lin 2004; Wei and Liu 2006). The effects of the Chinese developmental state approach on foreign MNCs at a corporate level still needs to be clarified from a technological point of view. Can a foresight developmental state influence how the competitive advantage of foreign MNCs can spread a global level? Given the current Chinese FDI policies aimed at attracting technologies that Chinese companies still lack, by inviting foreign companies to cooperate with local ones, it might be beneficial if the Chinese government consider fostering FDI in technological areas in which the Chinese market can provide unique opportunities in order to trigger innovation and facilitate its adoption.

The link connecting the subsidiary's role and disruptive innovation from emerging economies is also worth investigating. In their work on disruptive innovation, Christensen (1997) and Christensen and Raynor (2003) discuss the importance of an independent spin-off company in taking advantage of innovation inputs from emerging markets or unserved market segments. Chinese subsidiaries of foreign MNCs can thus be seen as those independent spin-off companies responsible for the development of disruptive innovations from Chinese market inputs which could then have disruptive implications on global markets. The corporate management of foreign MNCs should then increase the autonomy of Chinese subsidiaries in order for them to overcome what we would define as "the innovator's dilemma of foreign companies in China." By granting autonomy to Chinese subsidiaries, they will be able to develop innovations based on local market inputs that could otherwise not be foreseen by managers in their country of origin.

In a world in which innovation has been traditionally driven by market inputs (Myers and Marquis 1969; Vernon 1966, 1979) rather than the push of technology, we believe that China could play the role that originally belonged to the US market. The oriental giant could represent a strong source of inspiration and generation for new products, thus reversing the traditional view of industrialized countries as innovation leaders.

Notes

1 Data collection is particularly crucial in case studies in order to avoid narration effects and respondent bias. For this reason we collected different kinds of data through different sources. We identified three different phases of the data collection process. First, we gathered secondary data through different sources such as companies' websites, Lexis Nexis Academic, Il Sole 24 Ore Database, One Source – Business Information, internal and miscellaneous documents (including balance sheets). Second, we conducted semi-structured interviews with company managers at the headquarters. Data collected in the first two phases were triangulated to avoid misinterpretation and to improve the validity of data (Eisenhardt 1989; Yin 2003). In the process, a report based on the collected information and material was produced for each company. In order for researcher triangulation, the reports were circulated amongst other researchers who are partners of the authors in a wider project on similar topics and were discussed in order to obtain a more objective reading of the data. In the third phase, we conducted on-site interviews with Chinese subsidiary managers in China. The secondary data were triangulated again with these further contributions and single case integrated reports were produced. Reports were circulated again amongst the authors and the "external" researchers for triangulation and sent back to companies for feedback and validation. Interviews were conducted face to face, in Italian for headquarter managers and in Italian or English for subsidiaries, depending on the nationality of the interviewee.

2 The term *guanxi*, usually translated as "relation," "refers to the concept of drawing from a web of connections to secure favors in personal and organizational relations" (Park and Luo 2001: 455). Companies are thus motivated to build their own *guanxi* networks, made by both public and private companies/individuals, in order to facilitate their business activities.

References

Almeida, Paul, and Anupama Phene. 2004. "Subsidiaries and Knowledge Creation: The Influence of the MNC and Host Country on Innovation." *Strategic Management Journal* 25: 847–864.

Ambos, Tina C., Bjorn Ambos, and Bodo B. Schlegelmilch. 2006. "Learning from Foreign Subsidiaries: An Empirical Investigation of Headquarters' Benefits from Reverse Knowledge Transfers." *International Business Review* 15: 294–312.

Andersson, Ulf, Mats Forsgren, and Ulf Holm. 2002. "The Strategic Impact of External Networks: Subsidiary Performance and Competence Development in the Multinational Corporation." *Strategic Management Journal* 23: 979–996.

Asakawa, Kazuhiro. 2001. "Organizational Tension in International R&D Management: The Case of Japanese Firms." *Research Policy* 30: 735–757.

Birkinshaw, Julian. 1998. "Foreign Owned Subsidiaries and Regional Development: The Case of Sweden." In *Multinational Corporate Evolution and Subsidiary Development*, ed. Julian Birkinshaw and Neil Hood, 268–298. London: Macmillan.

Birkinshaw, Julian, and Neil Hood. 1997. "An Empirical Study of Development Processes in Foreign-Owned Subsidiaries in Canada and Scotland." *MIR: Management International Review* 37(4): 339–364.

Birkinshaw, Julian, and Neil Hood. 1998. *Multinational Corporate Evolution and Subsidiary Development*. London: Macmillan.

Birkinshaw, Julian, and Neil Hood. 2000. "Characteristics of Foreign Subsidiaries in Industry Clusters." *Journal of International Business Studies* 31: 141–154.

Birkinshaw, Julian, Neil Hood, and Stefan Jonsson. 1998. "Building firm-Specific Advantages in Multinational Corporations: The Role of Subsidiary Initiative." *Strategic Management Journal* 19: 221–241.

Breslin, Shaun G. 1996. "China: Developmental State or Dysfunctional Development?" *Third World Quarterly* 17: 689–706.

Brusoni, Stefano, Andrea Prencipe, and Keith Pavitt. 2001. "Knowledge Specialization, Organizational Coupling, and the Boundaries of the Firm: Why Do Firms Know More Than They Make?" *Administrative Science Quarterly* 46: 597–621.

Buckley, Peter J., and Mark Casson. 1985. *The Economic Analysis of the Multinational Enterprise: Selected Papers*. London: Macmillan.

Buckley, Peter, Jeremy Clegg, and Hui Tan. 2003. "The Art of Knowledge Transfer: Secondary and Reverse Transfer in China's Telecommunications Manufacturing Industry." *Management International Review* 43: 67–93.

Cantwell, J. 1995. "The Globalisation of Technology: What Remains of the Product Cycle Model?" *Cambridge Journal of Economics* 19(1): 155–174.

Cheung, Kui-yin, and Ping Lin. 2004. "Spillover Effects of FDI on Innovation in China: Evidence from the Provincial Data." *China Economic Review* 15: 25–44.

Chiesa, Vittorio. 1994. "International Technology Development Organisation: A Framework." In *La globalizzazione delle attività di ricerca e sviluppo*, ed. Umberto Bertelè and Vittorio Chiesa. Milano: Consorzio MIP-Politecnico di Milano.

Chiesa, Vittorio. 1996. "Managing the Internationalization of R&D Activities." *IEEE Transactions on Engineering Management* 43: 7–23.

Christensen, Clayton M. 1997. *The Innovator's Dilemma: When New Technologies Cause Great Firms to Fail*. Cambridge, MA: Harvard Business Press.

Christensen, Clayton M., and Michael E. Raynor. 2003. *The Innovator's Solution: Creating and Sustaining Successful Growth*. Cambridge, MA: Harvard Business Press.

Di Minin, Alberto, and Mattia Bianchi. 2011. "Safe Nests in Global Nets: Internationalization and Appropriability of R&D in Wireless Telecom." *Journal of International Business Studies* 42: 910–934.

Dunning, John H. 1980. "Towards an Eclectic Theory of International Production: Some Empirical Tests." *Journal of International Business Studies* 11: 9–31.

Eisenhardt, Kathleen M. 1989. "Building Theories from Case Study Research." *Academy of Management Review* 14: 532–550.

Florida, Richard. 1997. "The Globalization of R&D: Results of a Survey of Foreign-Affiliated R&D Laboratories in the USA." *Research Policy* 26: 85–103.

Frost, Tony S. 2001. "The Geographic Sources of Foreign Subsidiaries' Innovations." *Strategic Management Journal* 22: 101–123.

Frost, Tony S., and Changhui Zhou. 2005. "R&D Co-Practice and Reverse Knowledge Integration in Multinational Firms." *Journal of International Business Studies* 36: 676–687.

Frost, Tony S., Julian M. Birkinshaw, and Prescott C. Ensign. 2002. "Centers of Excellence in Multinational Corporations." *Strategic Management Journal* 23: 997–1018.

Gassmann, Oliver, and Zheng Han. 2004. "Motivations and Barriers of Foreign R&D Activities in China." *R&D Management* 34: 423–437.

Gassmann, Oliver, and Markus Keupp. 2008. "The Internationalisation of Western Firms' R&D in China." *International Journal of Entrepreneurship and Small Business* 6: 536–561.

Ghoshal, Sumantra, and Christopher A. Bartlett. 1990. "The Multinational Corporation as an Interorganizational Network." *The Academy of Management Review* 15: 603–625.

Ghoshal, Sumantra, and Nitin Nohria. 1989. "Internal Differentiation within Multinational Corporations." *Strategic Management Journal* 10: 323–337.

Gibbert, Michael, Winfried Ruigrok, and Barbara Wicki. 2008. "What Passes as a Rigorous Case Study?" *Strategic Management Journal* 29: 1465–1474.

Govindarajan, Vijay, and Ravi Ramamurti. 2011. "Reverse Innovation, Emerging Markets, and Global Strategy." *Global Strategy Journal* 1: 191–205.

Govindarajan, Vijay, and Chris Trimble. 2012. *Reverse Innovation: Create Far from Home, Win Everywhere.* Boston, MA: Harvard Business Review Press.

Gupta, Anil K., and Vijay Govindarajan. 1991. "Knowledge Flows and the Structure of Control within Multinational Corporations." *Academy of Management Review* 16: 768–792.

Håkanson, Lars, and Robert Nobel. 1993. "Determinants of Foreign R&D in Swedish Multinationals." *Research Policy* 22: 397–411.

Håkanson, Lars, and Robert Nobel. 2000. "Technology Characteristics and Reverse Technology Transfer." *MIR: Management International Review* 40: 29–48.

Håkanson, Lars, and Robert Nobel. 2001. "Organizational Characteristics and Reverse Technology Transfer." *MIR: Management International Review* 41: 395–420.

Hang, Chang-Chieh, Jin Chen, and Annapoormina M. Subramian. 2010. "Developing Disruptive Products for Emerging Economies: Lessons from Asian Cases." *Research Technology Management* 53: 21–26.

Hart, Stuart L., and Clayton M. Christensen. 2002. "The Great Leap: Driving Innovation from the Base of the Pyramid." *MIT Sloan Management Review* 44: 51–56.

Hecksher, Eli F. 1933. *Interregional and International Trade.* Cambridge, MA: Harvard University Press.

Hedlund, Gunnar. 1986. "The Hypermodern MNC – A Heterarchy?" *Human Resource Management* 25: 9–35.

Hsueh, Roselyn. 2011. *China's Regulatory State: A New Strategy for Globalization.* Ithaca, NY: Cornell University Press.

Hymer, Stephen H. 1976. *The International Operations of National Firms: A Study of Direct Foreign Investment.* Cambridge, MA: MIT Press.

Immelt, Jeffrey R., Vijay Govindarajan, and Chris Trimble. 2009. "How GE Is Disrupting Itself." *Harvard Business Review* 87: 56–65.

Jarillo, J. Carlos, and Jon I. Martinez. 1990. "Different Roles for Subsidiaries: The Case of Multinational Corporations in Spain." *Strategic Management Journal* 11: 501–512.

Johanson, Jan, and Jan-Erik Vahlne. 1977. "The Internationalization Process of the Firm: A Model of Knowledge Development and Increasing Foreign Market Commitments." *Journal of International Business Studies* 8: 23–32.

Johanson, Jan, and Jan-Erik Vahlne. 2009. "The Uppsala Internationalization Process Model Revisited: From Liability of Foreignness to Liability of Outsidership." *Journal of International Business Studies* 40: 1411–1431.

Kuemmerle, Walter. 1997. "Building Effective R&D Capabilities Abroad." *Harvard Business Review* 75: 61–70.

Macher, Jeffrey T., David C. Mowery, and Alberto Di Minin. 2007. "The 'Non-Globalization' of Innovation in the Semiconductor Industry." *California Management Review* 50: 217–242.

Myers, Sumner, and Donald G. Marquis. 1969. *Successful Industrial Innovations*. Washington, DC: National Science Foundation Report No. 69–17.

OECD. 2008. *Review of Innovation Policy: China*. Paris: OECD.

Park, Seung H., and Yadong Luo. 2001. "Guanxi and Organizational Dynamics: Organizational Networking in Chinese Firms." *Strategic Management Journal* 22: 455–477.

Patel, Pari, and Keith Pavitt. 1991. "Large Firms in the Production of the World's Technology – An Important Case of Non-Globalization." *Journal of International Business Studies* 22: 1–21.

Pearce, Robert D. 1989. *The Internationalization of Research and Development by Multinational Enterprises*. London: Palgrave Macmillan.

Phene, Aanupama, and Paul Almeida. 2008. "Innovation in Multinational Subsidiaries: The Role of Knowledge Assimilation and Subsidiary Capabilities." *Journal of International Business Studies* 39: 901–919.

Prahalad, C.K. 2004. *The Fortune at the Bottom of the Pyramid: Eradicating Poverty through Profits*. Philadelphia, PA: Wharton School Publishing.

Quan, Xiaohong, and Henry W. Chesbrough. 2010. "Hierarchical Segmentation of R&D Process and Intellectual Property Protection: Evidence from Multinational R&D Laboratories in China." *IEEE Transactions on Engineering Management* 57: 9–21.

Rein, Shaun. 2009. "Health-Care Reform, China Style." *Bloomberg Businessweek*, August 21, http://www.businessweek.com/globalbiz/content/aug2009/gb20090821_005732.htm (accessed December 29, 2014).

Ricardo, David. 1817. "Principles of Political Economy." In *The Works and Correspondence of David Ricardo*, ed. Piero Sraffa (1951). Cambridge: Cambridge University Press.

Siggelkow, Nicolaj. 2007. "Persuasion with Case Studies." *Academy of Management Journal* 50: 20–24.

Smith, Adam. 1776. *An Inquiry into the Nature and Causes of the Wealth of Nations*, ed. E. Cannan. London: Methuen.

Tavares, Ana Teresa. 2001. "Strategic Management of Multinational Networks: A Subsidiary Evolution Perspective." University of Reading Discussion Papers in International Investment and Management.

Teece, David J. 1986. "Profiting from Technological Innovation: Implications for Integration, Collaboration, Licensing and Public Policy." *Research Policy* 15: 285–305.

UNCTAD. 2005. *World Investment Report: Transnational Corporations and the Internationalization of R&D*. New York and Geneva: United Nations.

Vernon, Raymond. 1966. "International Investment and International Trade in the Product Cycle." *Quarterly Journal of Economics* 80: 190–207.

Vernon, Raymond. 1979. "The Product Cycle Hypothesis in a New International Environment." *Oxford Bulletin of Economics and Statistics* 41: 255–267.

von Zedtwitz, Maximilian. 2004. "Managing Foreign R&D Laboratories in China." *R&D Management* 34: 439–452.

von Zedtwitz, Maximilian, Tadashi Ikeda, Li Gong, Richard Carpenter, and Seppo Hämäläinen. 2007. "Managing Foreign R&D in China." *Research – Technology Management* 50: 19–27.

Walsh, Kathleen A. 2007. "China R&D: A High-Tech Field of Dreams." *Asia Pacific Business Review* 13: 321–335.

Wei, Y., and X. Liu. 2006. "Productivity Spillovers from R&D, Exports and FDI in China's Manufacturing Industry." *Journal of International Business Studies* 37: 544–557.

Yin, Robert K. 2003. *Applications of Case Study Research*. Thusand Oaks, CA: Sage Publications.

Yip, Winnie, and William Hsiao. 2009. "China's Health Care Reform: A Tentative Assessment." *China Economic Review* 20: 613–619.

Zeng, Ming, and Peter J. Williamson. 2007. *Dragons at Your Door: How Chinese Cost Innovation Is Disrupting the Rules of Global Competition*. Boston, MA: Harvard Business School Press.

Zeschky, Marco, Bastian Widenmayer, and Oliver Gassmann. 2011. "Frugal Innovation in Emerging Markets: The Case of Mettler Toledo." *Research Technology Management* 54: 38–45.

Chapter 12

Crowdfunding: Toward the Democratization of Innovation Financing

Alessandro Cordova, Johanna Dolci, and Gianfranco Gianfrate

Introduction

Who is best suited to finance innovation is a long debated issue in both academia and policymaking. Usually, large corporates (Schumpeter 1943), venture capital (Gompers and Lerner 2001), and the State (Mazzucato 2013) are pointed to as likely candidates, but a new one has recently emerged: the crowd. The collective efforts of individuals who network and pool their money via the Internet to support innovative projects – namely, crowdfunding – is in fact becoming a potentially disruptive channel to raise capital for new ventures.

Early-stage financing is of fundamental importance to make new projects and ventures succeed (Gompers and Lerner 2004; Gorman and Sahlman 1989; Kortum and Lerner 2000) and to foster innovation in the economy (Cosh, Cumming, and Hughes 2009). Since large corporations often feel pressure to preserve existing markets rather than introduce disruptive products, innovators tend to work in small new companies and these need financing to fund their ideas (Riedl 2013). However, because of its riskier nature, the supply of this type of financing is often insufficient compared to the demand from entrepreneurs. Such capital shortage for young entrepreneurial companies is usually referred to as "equity gap."[1] In fact, because of their lack of collateral, limited cash flows, and absence of past track record, start-ups obtain bank financing with more difficulty than larger and more mature firms (Ang 1991; Berger and Udell 1995; Chittenden, Hall, and Hutchinson 1996; Carpenter and Petersen 2002; Cassar 2004; Schwienbacher and Larralde 2010).

Despite the availability of funds from angel investors and venture capitalists, the equity gap seems far from being closed. In most cases, entrepreneurial initiatives that require small amounts of funding mainly rely on friends and family (Ang 1991;

The Handbook of Global Science, Technology, and Innovation, First Edition.
Edited by Daniele Archibugi and Andrea Filippetti.

Agrawal, Catalini, and Goldfarb 2010) or own savings (Paul, Whittam, and Wyper 2007). In this scenario and thanks to the evolution of the Web 2.0 technologies (Agrawal et al. 2010), crowdfunding platforms have emerged as an innovative channel to raise capital for new ventures.

What is Crowdfunding?

Schwienbacher and Larralde (2010)[2] were among the first to study crowdfunding: they defined it as "an open call, essentially through the Internet, for the provision of financial resources either in form of donation or in exchange for some form of reward and/or voting rights in order to support initiatives for specific purposes." Crowdfunding allows initiatives of cultural, social, or for-profit nature – which are advertised on the Internet by individuals or group of individuals in search of financial support (Mollick 2013) – to gather financial resources from a large pool of small-scale investors rather than a very small group of sophisticated ones (Belleflamme, Lambert, and Schwienbacher 2014; Riedl 2013). Proposed projects range from one-time events such as parties, holidays, weddings,[3] to the starting of new ventures. More precisely, the Massolution (2013) report states that in 2012 "Social Causes" were the most active (close to 30% of all crowdfunding activity), followed by "Business & Entrepreneurship" (16.9%), "Films & Performing Arts" (11.9%), "Music & Recording Arts" (7.5%), "Energy & Environment" (5.9%). Of course, crowdfunding campaigns do not necessarily aim at financing innovative products or services: in many cases they are a mere channel for "standard" private or public fundraising purposes without any innovation-related purposes or features (i.e., financing of parties, events, or donating to charities). Nevertheless, the labeling currently used by analysts and by platforms themselves to classify crowdfunding activities fails to fully capture the innovative content of many campaigns. For example, in the above-mentioned Massolution report, projects with an innovative content are not just among the ones in the "Business & Entrepreneurship" area but highly innovative (even high-tech) elements can be found in projects presented in each of the listed categories. We will later present, for example, the case of Kite, a successfully crowdfunded project which has lead to the realization of an innovative mosquito-blocking patch. Kite was a "social cause" project which has involved advanced research from the University of California delivering a product for undeveloped countries which may potentially be commercialized in the developed ones as well. More importantly, if we assume a broader definition of innovation which encompasses "soft" innovation – namely, all the innovative activities that "primarily impact upon sensory perception and aesthetic appeal" (see the contribution by Paul Stoneman in this volume, Chapter 4) – a certain number of "artistic" projects financed via crowdfunding are "innovative" even when they have no apparent commercial intent. Therefore, current data and analyses do underestimate the potential direct and indirect contribution of crowdfunding to innovation in the economy.

To better understand how crowdfunding works, let's consider the Gerber, Hui, and Kuo (2013) description of the five steps (see Figure 12.1) that typically accompany the launch of a crowdfunding campaign:

1. Crowdfounders prepare the campaign material which requires setting up the project profile, which includes a title, video, description of planned use of funds,

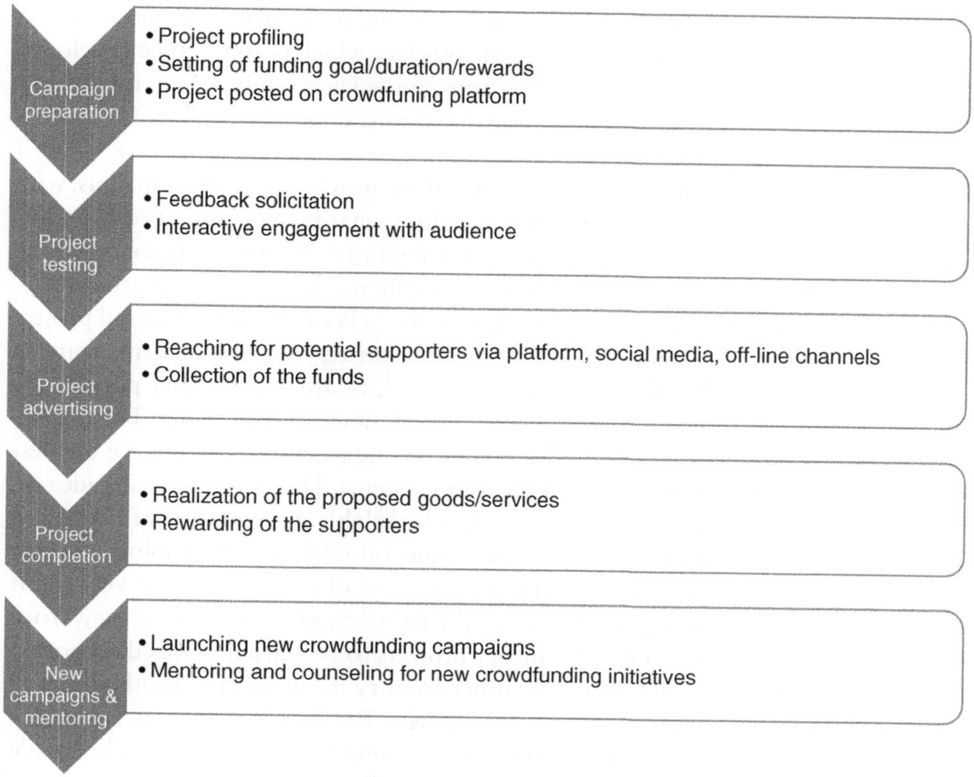

Figure 12.1 Crowdfunding campaigns' life-cycle.

funding goal, campaign duration, and reward descriptions. Creators complete these requests online and wait for the approval of the platform. Once approval is received, the project is presented on the platform website in a pre-formatted page where visitors can choose to donate;

2. Crowdfounders test their campaign material by soliciting feedback on their video and campaign description and engage their audience in deciding their preferred design direction;
3. Crowdfounders publicize the project and reach out to potential supporters via the crowdfunding platform itself, email, online social media, offline communication technologies and in-person requests;
4. Once the campaign is over, they produce and deliver their promised rewards which range from the new crowdfunded product to receiving a simple "thank you" email from the founders;
5. Finally, the fifth stage sees many creators contributing back to the community by providing advice and funding other projects.

Crowdfunding projects differ in terms of requested investment amount and kind of compensation promised to investors. One-time events usually require a lower amount of financing, at times even less than \$100,[4] and are either donation-based or offer some kind of reward (e.g., a gadget) upon contribution. The magnitude of funding implied by new business projects, instead, more likely resembles the one made by business angels and venture capitalists in the seed/early stage and investors are typically offered either rewards (e.g., the product which is going to be developed by the crowdfounder if the fundraising campaign is successful)[5] or equity shares.[6] Other platforms are instead based on pure donations, some others allow for lending money peer-to-peer.

At this point, it might be worth citing some of the most successful campaigns, with the double aim of helping the reader better picture what type of projects can be carried out on crowdfunding platforms and give a flavor of the potential economic and also social benefits that can be achieved through the realization of these projects.

A very interesting initiative started in 2013 is the "+Pool" New York-based project which managed to raise about $270,000 on Kickstarter, receiving contributions by about 5,000 people. The aim is to build a filtered, floating swimming pool in the middle of the river, allowing New Yorkers to swim in clean river water. The team is now hard at work developing the full water-cleansing assembly and finding the best structural, mechanical, electrical, and filtration systems. This is a fascinating idea on several grounds: it is innovative, it is social, it is scalable. It is innovative because it positions itself in the group of the new environmental-improving technologies which can be both beneficial for society and for the success of a new for-profit venture; it is social because the public good nature of the river brings benefits to all New York citizens and can give the impulse to further projects to be started in this direction; it is scalable because reproducible in any other country in the world. Another project running on the same lines is the "Kite" project: Kite Patch is an innovative, powerful approach to protecting humans from mosquitoes using a small, colorful sticker made of non-toxic compounds that block mosquitoes' ability to track humans for up to 48 hours. The idea was launched on Indiegogo, on the most famous US crowdfunding platform together with Kickstarter, by a team from the University of California and was support by the Bill and Melinda Gates Foundation and National Institutes of Health. Since exceeding its funding goal (it achieved a fundraising of about $550,000 with the support of an impressive 11,000 backers), the founders entered a partnership with Pilgrim Africa with the aim to distribute the patch to people living in Uganda, a country heavily affected by mosquito-borne illnesses.[7]

Among the most successful technological products we recall the Pebble Watch developed by Pebble Technology and released in 2013. It was funded via the crowdfunding platform Kickstarter. Initially designed as a watch that could display messages received on one's smartphone, several features were added later on such as the possibility to download dedicated apps ranging from music control (useful while driving, running, or across the room), sports management (checks your pace, calories burnt, etc.), game-playing. The Pebble watch, which has been in the top of the most-funded crowdfunding projects ranking for quite a long time, has been so much appreciated by the public ($10 million funding received from almost 70,000 backers), that Samsung has launched a competitor product soon after the crowdfunded smartwatch's release. But there are many other intriguing projects of this kind. For instance, Health (2013) names the following: Structure Sensor, a device that runs an iPad or Android tablet into a mobile 3D scanner which, with 25 days left before the funding period runs out, was able to raise more than $870,000; the life-logging camera Narrative, a one-inch square camera, that snaps and stores a five megapixel photo every 30 seconds, which raised more than $550,000 and later on received

$3 million from San Francisco-based True Ventures; Fuel3D, a handheld 3D scanning system capable of capturing high-resolution images which was originally developed at Oxford University for use in a 3D medical imaging system and managed to raise $325,000, well above the initial goal of $75,000.

Even from this very brief overview of crowdfunding, it is possible to glance at the potential benefits of such a new and revolutionary form of financing, especially for the funding of new business ventures. So far we have focused strictly on the crowdfunding convenience for entrepreneurs (or willing entrepreneurs), in the subsequent sections we will discuss crowdfunding convenience from the investors' point of view.

Crowdfunding Numbers at a Glance

Crowdfunding now encompasses about 450 platforms in the world and is growing at a very rapid pace: $1.5 billion of capital was raised in 2011, 2.7 billion in 2012 (more than 1 million projects were funded), and an expected 5.1 billion will be raised in 2013 (Massolution 2012, 2013). To appreciate the importance of crowdfunding as a source of financing for entrepreneurial endeavors, in 2012, according to the National Venture Capital Association (NVCA) total seed financings provided by US venture capital investors amounted to more than $752 million.

In 2012, Massolution reports that Donation and Reward-based crowdfunding platforms counted for $1.4 billion in volumes (an 85% increase from 2011), Lending-based ones for $1.4 billion (a 111% increase), while Equity-based ones still account for only a small share equal to $116 million (a 30% increase). Although crowdfunding offers a growing number of countries opportunities to use this innovative vehicle of financing, North America and Europe so far count for most of the share of total crowdfunding volumes. Precisely, in 2012 crowdfunding volumes were about $1.6 billion (a 105% increase from the year earlier) in North America and $945 million (a 65% increase) in Europe (Massolution 2013). In terms of number of platforms (see Table 12.1), the 2013 data show (expectedly) a predominance of the United States followed by Europe. However, the phenomenon does not appear to be a Western one: platforms have been launched in a number of emerging and even undeveloped countries. Given the rapid and somewhat disorderly growth of crowdfunding, data about platforms and campaigns are not fully reliable and outdate quickly, but the evidence seems to point toward an increasing number of crowdfunding projects (posted either on local or foreign platforms) from disadvantaged countries. In this sense, crowdfunding appears as a truly global infrastructure for innovation financing.

In 2011, Kickstarter appears to have provided about 10% of all angel funding for that year (Greenwald 2012). Moreover, a survey conducted by the credit agency Experian in 2012 reports that more than two-thirds of SMEs are unaware of crowdfunding, a signal of the potential development this new means of financing can yet achieve. Besides, although the reward-based platforms (such as Eppela, Ulule, Starteed, Indiegogo, and Kickstarter) are currently the most widespread, equity-based ones (such as Assob, Seedrs, and GrowVc) are rapidly growing in importance (Massolution 2012; Karabell 2013) and are attracting a great deal of attention in media and among policymakers, especially after crowdfunding was included by President Obama in the JOBS Act of 2012.

Table 12.1 Number of platforms per country in 2013.

Country	*Platforms*	*Country*	*Platforms*
United States	151	Japan	3
United Kingdom	37	Sweden	3
France	27	Austria	2
Netherlands	23	Finland	2
Germany	19	Hungary	2
Brazil	18	Israel	2
Spain	16	Mexico	2
Canada	13	Chile	1
Australia	11	Cyprus	1
Switzerland	6	Estonia	1
Italy	5	Hong Kong	1
China	4	Ireland	1
Czech Republic	4	Kenya	1
India	4	Latvia	1
New Zealand	4	Norway	1
Poland	4	Philippines	1
Portugal	4	Romania	1
Argentina	3	Russia	1
Belgium	3	South Africa	1
Denmark	3	Uganda	1
Total			388

An interesting analysis of the growing attention that crowdfunding has been receiving is performed by McGlashan and Voelker (2013) which show how Google searches for the term Kickstarter began to ramp-up in 2010 and surpassed small business loan searches in 2010, venture capital searches in 2011, and entrepreneur searches in 2012. To date, Kickstarter is in fact the most famous and successful global crowdfunding platform, having raised about $1 billion since its launch in 2009 by funding about 54,000 projects. As such, Kickstarter statistics give useful indications about crowdfunding structure and development: Figure 12.2 shows the funds raised per category of project. Overall the success rate of the posted projects is 43.8%. In particular, out of the 3416 strictly technology projects launched 35% have been successfully funded, as reported in Figure 12.3.

Most successfully funded projects on Kickstarter raise less than $10,000, but a growing number have reached six and even seven figures (see Table 12.2). Interestingly technology and gaming are the leading categories in raising more than $100,000, therefore even though less successful in general compared to other categories technology has raised relatively high amounts of money on certain successful projects. This evidence suggests that crowdfunding allows technology entrepreneurs to raise significant amounts of capital that typically is obtained by professional investors (i.e., Venture Capital and Angels).

Cordova, Dolci, and Gianfrate (2013) compiled a dataset containing 1127 cases of technology projects on four different crowdfunding platforms (Kickstarter, Ulule, Eppela, Indiegogo): the data show that, similar to the results reported in Mollick (2014), failures – namely, projects unable to gather the requested money amount – "happen by large amounts, successes by small amounts." On average, failed

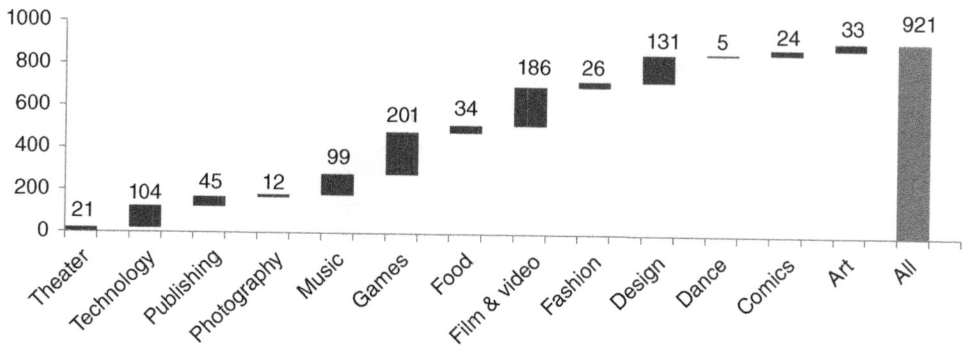

Figure 12.2 Kickstarter total amount in US$ raised by category of project.
Source: Kickstarter.com (18.12.2013).

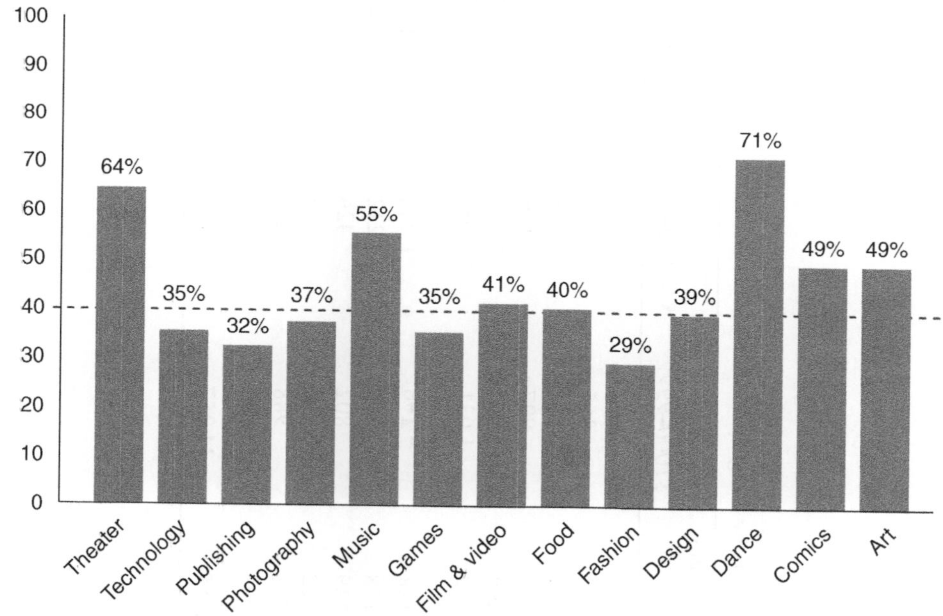

Figure 12.3 Kickstarter success rate (%) by category of project.
Source: Kickstarter.com (18.12.2013).

products miss their fundraising goal by 82%; on the other hand, successful ones obtain only 13% more than their financing target. In the remaining sections of this chapter, we will specifically focus on technological crowdfunding projects, as they are the ones most likely to turn into full-fledged innovative firms.

The Advantages of Crowdfunding

Crowdfunding platforms give the possibility to every person who owns a potentially valuable idea to test its market value. This implies more innovation is possibly brought into the open (and not necessarily just in strictly technology products) with the potential of being or becoming the new "frontier." This is especially true due to the fact that crowdfunding projects can be launched with very little upfront costs. There is no need to prepare

Table 12.2 Kickstarter's successfully funded projects by category and amount raised.

Category	*Successfully funded projects*	*Up to $1000 raised*	*$1000–9999*	*$10,000–19,999*	*$20,000–99,999*	*$100,000–999,999*	*$1 m+*
Theater	3,426	458	2,578	255	129	6	0
Technology	1,111	49	365	157	320	209	11
Publishing	4,788	739	314	581	324	24	0
Photography	1,383	214	907	176	85	0	1
Music	14,299	1,399	10,603	1,658	612	28	1
Games	267	139	1,031	509	730	232	29
Food	1,931	89	957	500	365	20	0
Film & Video	12,364	1,184	7,479	1,948	1,604	146	3
Fashion	1,352	138	769	212	200	32	1
Design	2,274	139	802	426	669	231	7
Dance	1,141	92	952	75	22	0	0
Comics	1,642	211	988	236	178	28	1
Art	5,332	1,004	3,686	441	190	11	0
Total	**53,713**	**5,855**	**34,257**	**7,152**	**5,428**	**967**	**54**

Source: Kickstarter.com (18.12.2013).

revenue forecasts for professional investors, to come out with a precise request of financing (given the crowd will eventually decide how much the initiative is worthwhile in terms of fundraising), to set up the legal structure of a new venture, to make upfront investments for product testing. Any would-be entrepreneur just needs to have an idea, to think about a way to present it, and to be willing to transform it into reality.

Another very intriguing aspect of crowdfunding (because it is on the Internet) is that these platforms allow many entrepreneurs to come together at the same place. Belonging to such a community can produce very positive spillovers for its members: the possibility for new crowdfounders to contact those who have successfully concluded a campaign in order to receive feedback and advice on their project and even obtain referrals;[8] the chance for crowdfounders to enter partnerships among each other as they match complementary skills, expertise, products/services; even the mere exchange of ideas that arise from watching others' projects can be of extreme value as sequential projects may arise.

Benefits also accrue to crowdfounders' supporters as they have access to lots of information, from the product/service to the history and character of the entrepreneur; besides, information is constantly produced and revealed on a dynamic basis as questions can be asked by investors through comments and answered back by the founders through updates. In this sense, the format of crowdfunding platforms (i.e., their informative character) helps reduce asymmetric information between investors and founders, eventually leading to the financing of the most valuable projects (more efficient resources allocation). The advantages of crowdfunding are not limited to those above, and there are clear limitations as well. In the next section, we will walk the reader through the main findings that the empirical research on crowdfunding has highlighted to date.

The Early Evidence on Crowdfunding

By referring to crowdfunding as an open call on the Internet, Schwienbacher and Larralde (2010) importantly made many authors see crowdsourcing, the outsourcing of a given task to a large group of people, in the form of an open call (Howe 2006), as the antecedent to crowdfunding (Dell 2008; Howe 2008; Kleemann et al. 2008; Belleflamme, Lambert, and Schwienbacher 2010; Rubinton 2011; Poetz and Schreier 2012); the only difference between the two being that instead of pooling labor resources, crowdfunding pools another factor of production: capital (Harms 2007).

The open call takes place on online platforms which provide the way for crowdfounders and investors to connect without standard financial intermediaries (Mollick 2013). In this direct interaction with crowdfounders, such that potential investors can see the level of support from other project backers, authors suggested that social information could have a role in the ultimate success of a crowdfunded project (Kuppuswamy and Bayus 2013).

Another early-acknowledged feature of this new financing phenomenon is that crowdfunding platforms, which provide all the means for investment transactions to take place – legal groundwork, pre-selection, the ability to process financial transactions, and so on (Ahlers et al. 2012)[9] – not only have the potential to help crowdfounders (the entrepreneurs) satisfy their financing needs, which makes crowdfunding alike *micro and social finance* (Harms 2007), but also to test new products and run new marketing campaigns (Lambert and Schwienbacher 2010; Mollick 2013). In this sense, crowdfunding draws inspiration from *social networking*, where consumers actively participate in online communities to share information and provide suggestions about new initiatives and/or brands (Ordanini et al. 2011). Moreover, when crowdfunding is used as a mean to demonstrate demand for a proposed product, successful initiatives become a signal to venture capitalists of a potential good long-term investment, possibly leading to additional future financing for crowdfounders (Mollick 2013).

From the very early attempts to define and draw the boundaries of it, research on crowdfunding has developed very quickly, either from a qualitative and quantitative point of view. Mainly focusing on reward-based crowdfunding, three main streams of literature have emerged: one, focusing on the crowdfounders' side, is centered on the reasons behind their decision to use crowdfunding platforms; the second investigates the reasons that motivate investors to support crowdfunders; finally the third looks at the determinants of projects' success and the challenges crowdfounders face in managing their initiatives. We will analyze each of these next, while leaving the more challenging analysis of the future of equity crowdfunding for later on in the chapter.

Crowdfounders' Side

Belleflamme et al. (2010) found that *raising money, getting public attention and obtaining feedback on product/service*, are all relevant factors in motivating the launchers of initiatives on crowdfunding platforms. Seemingly, by conducting a grounded-based research, Gerber et al. (2011) found that the main reasons why crowdfounders use these platforms are: to raise funds while maintaining full control over the project, to *receive validation*, to *connect with others*, to *replicate successful experiences* of others, and to *expand awareness* of work through social media. Adams (2013) also points to the role of crowdfunding as a means for *reducing risk*: by obtaining funds from the crowd, project founders who aim at starting a new business don't have to mortgage their house or dip into their kid's college fund in order to withstand the unanticipated expenses that accompany the initial phases of a start-up. Also, crowdfunding appears to founders as a more *flexible means of financing* with respect to Venture Capital. VC funds typically seek start-ups with provable and protectable intellectual property (Akin 2011) and also tend to be "geographically constrained," that is Venture Capitalists prefer being located close to their investment so as to reduce screening (ex-ante) and monitoring (ex-post) costs; finally, VC investment is highly cyclical and, therefore, the availability of capital for start-ups tightens at times of economic downturn (Christensen 2011). Research on start-ups also points to the fact that entrepreneurs with more limited knowledge of funding systems routinely make use of less beneficial funding opportunities (Seghers, Manigart, and Vanacker 2012); in this sense crowdfunding can *reduce the required knowledge of funding systems* as it provides start-ups with a single platform which to pitch several investors with. Finally, Belleflamme et al. (2014) noticed that, when used to invite consumers to pre-order a product, crowdfunding allows entrepreneurs to *price discriminate*: consumers who enjoy higher utility will pre-order the product and pay more with respect to later consumers, who will wait until the product is offered on the market at a lower price.[10]

Investors' Side

A first part of the literature has focused on what motivates investor to contribute to crowdfunding projects. Harms (2007) conducted a questionnaire-based research which led him to conclude that, in addition to *self-expression*[11] and *enjoyment*,[12] also the overall benefit investors derive with respect to their contribution (*economic value*), the presence of a guaranteed tangible output of the project (*certainty effect*), the degree to which the functional benefits of the project outcome serves a functional need of the individual consumer (*personal utility*), all significantly drive funders' intention to invest. Van Wingerden and Ryan (2011) distinguished between intrinsic motivations – *control of use of an innovation, improvement of current circumstances, enjoyment, and sense of involvement* – and extrinsic motivations – *financial reward*. Finally, Ordanini et al. (2011) added *public recognition* and *patronage* to the list.

The Geographical Dimension of Crowdfunding

A further aspect of crowdfunding that received particular attention from scholars is the one related to the role of geography, in terms of where investors are located and how much attention they pay to where new business ventures are founded. As to the former, while theory predicts that investors in early-stage firms will tend to be local, because the costs of gathering information, monitoring progress, and providing input are sensitive to distance (Tribus 1970; Florida and Kenney 1988; Florida and Smith 1993; Lerner 1995; Sorenson and Stuart 2001; Powell et al. 2002; Zook 2002; Mason 2007), Agrawal et al. (2011) show that the availability of *online platforms tends to eliminate most distance-related economic frictions*. In fact, a mean distance between entrepreneur and investor of approximately 3000 miles is found, as opposed to an average of 70 miles between lead VC and target firm (Sorenson and Stuart 2001), which suggests there is no significant difference between the number of local and distant investors. As to the latter, by comparing the distribution of the total amount invested and the number of deals across US high-tech areas made by VC funds with respect to crowdfunding investors, Cordova et al. (2013) find that looking at technological crowdfunding projects, *both the total amount invested and the number of investments made via crowdfunding are less concentrated in technologically strong*[13] *countries/regions* than those financed through VC investment. In other words, crowdfunding projects do not necessary agglomerate around highly technologically vibrant areas, such as California, but tend to be more evenly distributed.[14]

Besides, by performing regression analyses, Cordova et al. (2013) also find that the success rate of a given project is not significantly different across more or less technologically strong US States, that is, there is no possibility to predict ex-ante the higher success of a project depending on whether it was launched in California or Kentucky. Seemingly, their results show that, *within a given State, investors indifferently finance projects whose referring industry/sector is the same or not to the one that country is specialized into*. Put differently, if California is specialized into computer equipment manufacturing,[15] then we don't observe California-based crowdfunding projects of the computer equipment manufacturing type to be more successful than others.

That said, because within a single round of financing, local investors contribute relatively earlier than distant ones and the latter are more responsive to others' investment decisions, *geography is still found to play a role in so far as early contribution by family and friends can determine the pattern of subsequent contributions* by far-away investors. Such a micro-pattern of early investments opens up interesting and challenging research questions.

Focusing on a comparison between crowdfunding and venture capital early-stage financing, Cordova et al. (2013) show that both the total amount invested and the number of investments made via crowdfunding are less concentrated in more technological developed areas than those financed through venture capital investments. This suggests that crowdfunding investors are less "dependent" on geography than venture capital investors, arguably because of the online nature of the former financing vehicle. Moreover, the paper has further analyzed whether venture capital and crowdfunding are complements, substitutes, or uncorrelated types of investment. Evidence shows that crowdfunding is rather independent from the supply of credit made available to new ventures by venture capital firms, and that it does prosper in areas which are abundant or scarce with venture capital funds.

Moving to a cross-country level, the early evidence tends to confirm that projects from non-US countries are not less likely to be successfully financed on US crowdfunding platforms than the US ones. Of course, the current international data about crowdfunding is relatively limited and unhomogeneous, but crowdfunding seems poised to become a genuinely global channel for innovation financing. In fact, early-stage financing via crowdfunding by investors located in other continents are becoming more and more frequent.

The intrinsic globalized nature of crowdfunding may create an unprecedented international "level playing field" of innovation with ideas and funding freely circulating across borders, with access to the Internet, the financial infrastructure, and the national regulations and restrictions the only apparent barriers. Interestingly, no clear prediction about the direction of such flows can be drawn at the moment: entrepreneurs located in developed countries may be crowdfunded by investors located in less advantaged areas/countries but the opposite may happen as well.

While it is hard to foresee now how the map of crowdfunding will look like in a few years or even months from now, a clear suggestion seems to stem from the early international anecdotal evidence: if policymakers intend to improve economic conditions of less technologically developed areas, crowdfunding is to be encouraged.

Determinants of Success in Crowdfunding Initiatives

As to the elements of success, Cordova et al. (2013) show that higher probability of success and the extent of overfunding (funding above the threshold initially asked for) depends on the project characteristics (project funding goal and duration of the fundraising campaign) and on the behavior of later-contributing investors (contribution frequency). More specifically, the authors find that an increase in the project funding goal is correlated with a lower probability and extent of success, that project duration increases the chances of success, and that chances of success are positively related to the dollar amount contributed per day.

Mollick (2014) shows that the degree to which founders took the time and effort to ensure that project pitches conformed to standards for successful pitches and the size of the social network (measured in terms of number of Facebook friends of the crowdfounder) influences the success of crowdfounders' financing efforts. Kim and Viswanathan (2013) studied the impact of the way crowdfounders present their project on their probability of success, establishing that creators must be half-way between making credible claims and refraining from disclosing too much information. Similarly, Marom and Sade (2013) show that in Kickstarter fundraising, entrepreneurs' descriptions do matter – projects which substantially highlighted their entrepreneurs enjoyed higher rates of success[16] (this being particularly true for artistic projects as opposed to technological ones), controlling for other relevant variables.

Qiu (2013) argues that advertising efforts, such as getting advertised on Kickstarter's main web page and producing updates on one's project pitch (especially in the last week of fundraising), tend increase the likelihood of the project's success. Mollick (2014) has also found that high-quality projects attract backers who may promote the project to other potential backers, or external media, thus increasing the draw of the project.[17] On the same line of reasoning, Qiu (2013) suggested that crowdfounders could search for ways to tie rewards to investors' advertising effort – through referrals – so as to increase the visibility of their project.

Finally, as to the challenges crowdfounders face in managing their projects, one often cited argument is that as the new product/service is launched, this becomes visible to the competition (Adams 2013; Riedl 2013), thus discouraging entrepreneurs from launching their project. However, despite the fact this may induce entrepreneurs to postpone the launch of their initiative on a crowdfunding platform, it may turn positive for investors as they will be more likely to finance projects which crowdfounders are more confident about and better prepared for a successful launch.

In a SWOT analysis, Jegeleviciute and Valanciene (2013) additionally mentions administrative and accounting challenges as one of the most challenging obstacles crowdfounders might face. Because recording contributions and sending rewards to a large number of investors is time-consuming, it is to be expected that as we move toward equity-based crowdfunding and a large number of investors become shareholders, business will face even bigger administrative and accounting challenges, especially to precisely calculate the share of profits each investor is entitled to (Sigar 2012).

As a confirmation of the above, Mollick (2014) points to the fact the vast majority of founders attempt to deliver products promised to funders, but relatively few do so in a timely manner, a problem exacerbated in large or overfunded projects (thus requiring a larger scale of production/service provision). This opens issues relative to how to effectively manage manufacturing problems, manage the complexity of shipping, changes in scale, changes in scope, and unanticipated certification issues which come with unexpected project success.

Policy Implications

Despite the previously mentioned limitations, to date these appear to be outweighed by the benefits of crowdfunding, which seems positioned to play a relevant role in the financing chain of new ventures. This is also why policymakers (i.e., the Obama administration, the European Commission) have started to support crowdfunding enablers. In fact, legislators' assumption is that, similar to more traditional forms of capital investments in SMEs (angel investing and venture capitalist funds), these platforms are likely to lead to the funding of highly innovative ventures,[18] potentially making crowdfunding a relevant contributor to innovation, employment, and ultimately economic growth.

Especially when comparing crowdfunding to VC investing, one can see the potential of further developing the former. Mollick (2014) shows two main findings. Firstly, signals of quality that are used by VCs to evaluate the potential of new ventures are also used by crowdfunders (e.g., studying the background of founders, the degree of care that is put in the preparation of the project pitch, endorsements from and alliances with relevant third parties), despite possible differences in terms of expertise in evaluating projects and the connection to Internet networks specifically designed to identify and nurture start-ups (such as connections to angel investors and seed incubators). Following Mollick, these finding are a very important signal for policymakers in that they suggest that crowdfunding has the ability to distinguish quality potential projects from less promising ones. Secondly, as Cordova et al. (2013) pointed out, the fact that crowdfunding investment is less geographically concentrated implies that the original problem of whether an institutionalized selection process is biased in a way that prevents some subset of valuable technologies and promising firms from achieving success (Ferrary and Granovetter 2009), as it tends to be geographically constrained, is relaxed. In other words, even projects which are launched in more remote or less economically vibrant areas have the chance to be financed, limiting the potential drag on the incentive to start entrepreneurial initiatives imposed by regional disparities in entrepreneurial opportunities. This may happen within the same country when projects launched in less developed urban/rural areas get funded by investors located in more advantaged areas. Interestingly, crowdfunding may allow projects launched in emerging markets or undeveloped countries to raise money from the developed countries, thus resulting in a genuinely global innovation financing tool.

Interestingly, crowdfunding appears to have the potential to be a substitute and complementary means of financing to VC investing at the same time, and this should be beneficial for the level of financing of start-ups. There are in fact cases in which projects were not accepted by venture capitalists but then obtained a great success in crowdfunding platforms (an instance in which VC and crowdfunding appear to be substitutes, i.e., crowdfunding platforms capture part of the demand for financing which is not satisfied by the VC industry, the banking sector, and quotation in alternative markets).[19] Similarly, it is reasonable to expect instances in which the success obtained by some crowdfunding projects would later on allow crowdfounders to receive additional rounds of financing from VCs. The success of the initiative would in fact convey to VC investors the message that the firm is either selling a service and/or product which is liked by the market and that the firm is able to successfully carry on all of the necessary planning, marketing, and communications activities which are fundamental for start-ups to be successful. In other words, crowdfunding and VC funding are not mutually exclusive. In this regard, do you remember the description of the Pebble Watch given earlier? Well, that is a good example of what we mean by "not mutually exclusive" means of financing. With his revolutionary watch, Eric Migicovsky was embraced by Y Combinator, a quite prestigious business incubator in the United States. Once he completed the program, he was able to raise up to $375,000 from four angel investors, including Paul Buchheit, a partner at Y Combinator, and Tim Draper of venture capital firm Draper Fisher Jurvetson. Nevertheless, despite the very promising beginning, Migicovsky couldn't raise any more money from traditional investors. For this reason, the company requested crowdfunding in April 2012. Today Pebble has become the third most highly crowdfunded project to date with $10,266,844 pledged by 68,928 people.[20]

But can equity investment represent the new frontier for crowdfunding, such to make this revolutionary form of early-stage financing truly comparable in scale to venture capital funding, or even go beyond the latter by overcoming its geographical restrictions? First of all, it makes sense to explain why such an evolution would be desirable. As Miglietta et al. (2013) point out, equity-based crowdfunding has the potential to become an element of economic policy capable of strengthening the competitiveness of a country through the creation of new high-quality jobs, a more efficient allocation of capital, directly flowing to the development of the real economy and to companies which face a higher equity gap (OECD 2006). This is because more entrepreneurial ventures could be financed and funds would be more directly channeled to the real economy than having to pass through the banking sector. In addition crowdfunding may especially be important for start-up financing in countries which are highly dependent on financing from the banking sector, such as Italy, especially in a historical moment where banks are under severe stress and lending is contained (especially toward new enterprises, where risk connected to the absence of collateral and the need for more stable and prolonged financing is higher). Finally, equity crowdfunding platforms could be very important in providing consulting support to new initiatives. For instance, the new legislations on equity crowdfunding approved by the Italian Financial Market Authority (the first full-fledged equity crowdfunding regulation in Europe), has provided that crowdfunding platform operators can provide strategic and operational advice to start-ups, an activity similar to that played by incubators.[21] Start-ups making use of these consulting activities will therefore be able to further signal their ability to succeed through earlier stages of their development to the market.

As to whether such a development for crowdfunding is possible, it is harder to give a definitive answer. On the one hand, if we analyze the high innovative caliber of technology crowdfunding projects, we would tend to say that investing in such projects could have appealing expected returns. Jeffries (2013) has indeed put in evidence that most of the successful consumer electronics products launched in 2013 have been funded on crowdfunding platforms, including 3D printers and electronic watches. In more general terms, because projects presented on crowdfunding platforms often arise from consumers who, unable to find the supply of specific products in the market, initiate their own project with the aim of satisfying that need, crowdfunding ventures can be very intriguing investments as they fill in gaps in the ability of the market to satisfy particular demands.

On the other hand, there are challenges that may not be easy to overcome. First, it is not clear yet what would be the incentive for institutional investors to channel funds to start-ups on crowdfunding platforms. More precisely, while it is clear how crowdfunding platforms can represent a way for institutional investors to scan several new business ventures, yet, once the investor has individuated a viable business to invest in, they would be both better off by engaging directly in the equity transaction.

This is because they would avoid paying commissions to the platform, not to mention that the institutional investor might be better off making a specific investment contract with the firm owners. At this point either the start-up leaves the platform and it is entirely financed by the institutional investor or it receives some investment outside of the platform but remains on it to gather additional funds. In either case, the platform manager is worse off as it loses clients and/or profits. Of course, the platform will undertake some legal expedients in order to avoid such moral hazard behavior, but complete contacts (such that the platform is always protected from misbehavior) are hard to define. This may discourage over time, as the initial excitement leaves room for considerations on whether it is profitable to open a platform, and the opening and success of equity-based platforms. In addition, the main reason why VC funds are geographically proximate to start-ups is that they give great value to the face-to-face assessment of start-up teams before making their investment, and to the monitoring of the business operations overtime, once the investment has been made. In this sense, crowdfunding platforms offer no solution to such a problem. For this reason, Vass (2013) has proposed to add Internet webinars to crowdfunding projects or to consider an event registration calendar and event management function. Nevertheless, it is hard to expect that even in the presence of these functions, the role that trust plays for such investment decisions to take place would be solved. Notice additionally that the reason why most of the time venture capitalists are mentioned as an example of institutional investor is that they resemble more the institutions which are likely to engage in such a risky investment as start-up equity financing. Therefore, if venture capital has the aforementioned disincentive to go through crowdfunding platforms, even more stricter conditions would apply to pension funds and insurance companies, which are by far less accustomed to this asset class, have less expertise in evaluating such projects, and less time to spend in screening crowdfunding platforms.[22]

The Future of Crowdfunding

Initially perceived as a threat by venture capitalists, crowdfunding is likely to become more and more a tool for them to scout and test potential innovative projects to invest in. Also established corporates are likely to search for ways to integrate crowdfunding mechanisms in the launch and development of new products. In fact, the ability to pre-test innovative products/services, gathering collaborative feedback from potential clients willing to invest in the new commercial endeavor, could prove a powerful marketing tool for many companies. States and governmental entities seeking to promote innovation and entrepreneurship may as well eventually exploit crowdfunding schemes to screen for promising entrepreneurial ventures worth being supported via subsidies, tax incentives, or bureaucratic/compliance facilitations. Against the backdrop of this financing revolution, the limitations of crowdfunding are emerging clearly as well. Crowdfunding seems not appropriate for projects easy to be replicated by competitors. Paradoxically, the publicity of innovative ideas posted on open platforms is at the same time the main strength and weakness of crowdfunding. Along this line, the financial disintermediation implied by crowdfunding will not be unbounded. Professional venture capital typically based on face-to-face relationships and on strong legal/managerial/financial support will remain mostly unchallenged in cases of very complex projects or of initiatives needing a certain degree of secrecy in their initial stages. All those scenarios open up interesting research and policy questions, including whether and to what extent crowdfunding will displace angels, venture capital, and other conventional innovation financing sources. Also, since crowdfunding may not fully finance a new venture over time, its impact and signaling effect on subsequent (possibly conventional) financing should be explored.

On the other hand, the regulation and the competition among crowdfunding platforms (and among projects themselves) are going to shape the scope of this new phenomenon. New solutions and new actors (i.e., specialized investors and financial vehicles investing in crowdfunding projects) are likely to emerge as a response to the apparent flood of new projects inundating the existing and forthcoming platforms. Research questions in this area are likely to touch upon the mechanisms to minimize fraud, the differences in crowdfunding platforms affecting the type and nature of projects being proposed, and the extent to which regulation and governance influence the utility and outcomes associated with crowdfunding. So far the scholarly evidence has mostly focused on reward-based crowdfunding. If (as most expect) equity-based crowdfunding booms, the research and policy questions are likely to focus on how it will influence the rate, geography, and direction of global innovation, and to what extent it will increase the number and types of innovations that are actually funded.

Only time will provide an answer about to what extent crowdfunding will actually substitute or complement the existing innovation financing tools and channels. Nevertheless, the crowd can clearly contribute a considerable amount of funds to valuable ideas as it is motivated, not just by tangible rewards, but also by the desire to contribute to the birth of new ventures, feeling involved in a vibrant entrepreneurial community. Therefore, crowdfunding is expected to play an important role in encouraging and spreading global entrepreneurial innovation.

Notes

1 The existence of such equity gap – referred to as the lack of capital investments into innovative, high-tech start-ups – has been debated for decades, ever since the Macmillan Report (1931) and most recently by Cruickshank (2000) and the European Commission (2005).
2 Who extended the one firstly provided by Kleemann, Voss, and Rieder (2008).
3 A more comprehensive list of sectors can be found on http://crowdfunding.com/.
4 Out of more than 50,000 projects advertised on Kickstarter from 2009 to 2012, 225 of these had a goal of less than $100, while 25 had above $1 million (Mollick 2013).
5 In this case, we refer to pre-ordering: crowdfunders pay for their product before they actually receive it. Just to give some numbers, TikTok+LunaTik raised $941,718 from 13,512 individuals in the form of product pre-ordering of its multi-touch watch kit. Even more strikingly, the Pebble watch (a product of EInk) managed to raise $10,266,844 under the pledge of 68,928 people in 2012 (Belleflamme et al. 2014).
6 Crowdfunded platforms can even use a peer-to-peer lending scheme (Lin and Viswanathan 2013), where crowdfunders actually lend to crowdfounders beyond the definition of a repayment scheme.
7 These two initiatives have been mentioned in the Goodnet platform at http://www.goodnet.org.
8 Mentions appear to be particularly important in artistic projects.
9 For the services they provide, each platform charges a fee, which is usually between 5 and 12% of the amount raised (De Buysere et al. 2012). Note that the latter depends on whether the platform is of "All or Nothing" type or of "Keep it All" type. In the former case, the platform distributes the funds raised if and when the initial requested amount by the crowdfounder is achieved; in the latter case, the

crowdfounder will keep any amount raised through the crowd, regardless of whether the financing goal was achieved.

10 In their analysis, they concluded that this strategy is proved profitable as long as initial capital requirement remains relatively small, in contrast with crowdfunding through profit-sharing, where the benefits are higher when capital requirements are large.

11 When engagement in crowdfunding activities can be used to shape one's online identity and to present oneself.

12 Derived from supporting a crowdfunding initiative.

13 Technologically strong areas are defined as areas which have a high percentage of high-tech employment with respect to the US total.

14 Also Mollick (2014) finds that crowdfunding is more democratically distributed than VC funding.

15 NAICS 3341.

16 Surprisingly the number of previous successes of the entrepreneur contributes to the number of investors and to the sum pledged relative to the goal, but not significant to the chances of success.

17 The stream of literature concerning how investors react to earlier investment by peers was perhaps started by Kuppuswamy and Bayus (2013) and Burtch, Ghose, and Wattal (2013) who find evidence that additional backer contribution is negatively related to past backer support, a phenomenon labeled in the public goods literature as *substitution effect* (Shang and Croson 2009), as opposed to the reinforcement effect, which takes place when future contributions by investors depend positively on previous pledges. By using a dataset of crowdfunding initiatives for mobile apps, Kim and Viswanathan (2013) further study the peer effect by distinguishing between two types of investor: reputable funders (app developer investors and experienced investors) and the rest of the crowd. They show the former tend to contribute earlier than the latter and, consequently, their investment serves as a signal of project quality for later investors.

18 An example of steps ahead computed by countries to develop crowdfunding in its equity form is provided by the United States. In April 2012, President Obama signed into law the Jumpstart Our Business Startups (JOBS) Act to carry on the regulation of "equity crowdfunding," despite some issues still to be solved, e.g., opportunities are currently limited to "accredited investors."

19 Jeffries (2013).

20 The most funded project is Star Citizen, a space combat video game launched on Kickstarter which managed to raise $34 million, beyond the initial fundraising request of $500,000. The second in the ranking is called Ubuntu Edge, raising on Indiegogo up to $12 million but without reaching the requested amount of $32 million – the highest funding threshold set on a crowdfunding platform up to now. The Edge is a hybrid device, which functions as a high-end smartphone or, when used with a monitor, keyboard, and mouse, operates as a conventional desktop PC running Ubuntu.

21 Miglietta et al. (2013) report that in 2011 alone US incubators have provided advice and assistance to 49,000 start-ups, which in turn have created full-time jobs for some 200,000 workers, generating an annual turnover of about $15 billion. The same figures show an aspect of particular interest from the point of view of the national fiscal revenue: every dollar invested in business incubators produces about 30 times the revenue (National Business Incubation Association, business incubation FAQ, http://www.nbia.org/resource_library/faq/index.php).

22 Such considerations are partly an answer to why stock exchanges focusing on the financing of small and medium enterprises have had a hard time taking off.

References

Adams, Prince R. 2013. *Crowd Funding Made Simple: Plan, Publicize, Get Crowd Funding!* FundMeFundYou.com.

Agrawal, A., C. Catalini, and A. Goldfarb. 2011. "The Geography of Crowdfunding." NBER Working Paper 16820.

Ahlers, G.K.C., D. Cumming, C. Gunther, and D. Schweizer. 2012. "Signaling in Equity Crowdfunding." SSRN Working Paper.

Akin, M. 2011. "Does Venture Capital Spur Patenting? Evidence from State-Level Cross-Sectional Data for the United States." *Technology and Investment* 2(4).

Ang, J.S. 1991. "Small Business Uniqueness and the Theory of Financial Management." *Journal of Small Business Finance* 1(1): 11–13.

Belleflamme, P., T. Lambert, and A. Schwienbacher. 2010. "Crowdfunding: An Industrial Organization Perspective." Paper prepared for the workshop Digital Business Models: Understanding Strategies.

Belleflamme, P., T. Lambert, and A. Schwienbacher. 2014. "Crowdfunding: Tapping the Right Crowd." *Journal of Business Venturing* 29(5): 585–609.

Berger, A., and G. Udell. 1995. "Relationship Lending and Lines of Credit in Small Firm Finance." *Journal of Business* 68(3): 351–381.

Burtch, G., A. Ghose, and S. Wattal. 2013. "An Empirical Examination of the Antecedents and Consequences of Investment Patterns in Crowd-Funded Markets." *Information Systems Research* 24(3): 491–519.

Carpenter, R.E., and B.C. Petersen. 2002. "Is the Growth of Small Firms Constrained by Internal Finance?" *The Review of Economics and Statistics* 84(2): 298–309.

Cassar, G. 2004. "The Financing of Business Start-Ups." *Journal of Business Venturing* 19(2): 261–283.

Chittenden, F., G. Hall, and P. Hutchinson. 1996. "Small Firm Growth, Access to Capital Markets and Financial Structure: Review of Issues and an Empirical Investigation." *Small Business Economics* 8(1): 59–67.

Christensen, J.L. 2011. "Should Government Support Business Angel Networks? The Tale of Danish Business Angels Network." *Venture Capital* 13(4).

Cordova, A., J. Dolci, and G. Gianfrate. 2013. "The Bearable Lightness of Crowdfunding: Evidence From Technology Projects." Working Paper, Bocconi University.

Cosh, A., D. Cumming, and A. Hughes. 2009. "Outside Entrepreneurial Capital." *Economic Journal* 119(540): 1494–1533.

Cruickshank, D. 2000. *Competition in UK Banking*. London: HMSO.

De Buysere, K., O. Gajda, R. Kleverlaan, and D. Marom. 2012. "A Framework for European Crowdfunding." www.crowdfundingframework.eu (accessed December 30, 2014).

Dell, K. 2008. "Crowdfunding: Got $10? Then You Too Can Be a Venture Capitalist." *Time*, September 4.

European Commission. 2005. *Best Practices of Public Support for Early-Stage Equity Finance*. Final report of the expert group, Directorate-General forEnterprise and Industry.

Ferrary, M., and M. Granovetter. 2009. "The Role of Venture Capital Firms in Silicon Valley's Complex Innovation Network." *Economy and Society* 18: 326–359.

Florida, R.L., and M. Kenney. 1988. "Venture Capital, High Technology and Regional Development." *Regional Studies* 22(1).

Florida, R., and D. Smith. 1993. "Venture Capital Formation, Investment and Regional Industrialization." *Annals of the Association of American Geographers* 83(3): 434–451.

Gerber, E.M., J.S. Hui, and P.Y. Kuo. 2011. "Crowdfunding: Why People Are Motivated to Post and Fund Projects on Crowdfunding Platforms." Northwestern University Creative Action Lab, Evanston, IL.

Gompers, P., and J. Lerner. 2001. *The Money of Invention: How Venture Capital Creates New Wealth*. Boston, MA: Harvard Business School Press.

Gompers, P., and J. Lerner. 2004. *The Venture Capital Cycle*. Cambridge, MA: MIT Press.

Gorman, M., and W.A. Sahlman. 1989. "What Do Venture Capitalists Do?" *Journal of Business Venturing* 4(4): 231–248.

Greenwald, T. 2012. "10 Breakthrough Technologies." http://www.technologyreview.com/article/427675/crowdfunding/ (accessed December 30, 2014).

Harms, M. 2007. "What Drives Motivation to Participate Financially in a Crowdfunding Community?" Master Thesis in Marketing, University of Amsterdam.

Health, N. 2013. "10 Fascinating Tech Projects That Crowdfunding Has Made Possible." http://crowdfundbeat.com/2013/10/15/10-fascinating-tech-projects-that-crowdfunding-has-made-possible/ (accessed December 30, 2014).

Howe, J. 2006. "The Rise of Crowdsourcing." *Wired Magazine.*

Howe, J. 2008. *Crowdsourcing: Why the Power of the Crowd is Driving Future of Business.* New York: Three Rivers Press.

Jeffries, A. 2013. "How Kickstarter Stole CES: The Rise of the Indie Hardware Developer." *The Verge.com*, January 10.

Jegeleviciute, S., and L. Valanciene. 2013. "Valuation of Crowdfunding Benefits and Drawbacks." *Economics and Management* 18(1): 39–48.

Karabell, Z. 2013. "The Kickstarter Economy: How Technology Turns Us All into Bankers." *The Atlantic*, February.

Kim, K., and S. Viswanathan. 2013. "The Experts in the Crowd: The Role of Reputable Investors in a Crowdfunding Market." Available at SSRN.

Kleemann, F.G., G. Voss, and K. Rieder. 2008. "Un(der)paid Innovators: The Commercial Utilization of Consumer Work Through Crowdsourcing." *Science, Technology & Innovation Studies* 4(1): 5–26.

Kortum, S., and J. Lerner. 2000. "Assessing the Contribution of Venture Capital to Innovation." *RAND Journal of Economics* 31(4): 674–692.

Kuppuswamy, V., and B.L. Bayus. 2013. "Crowdfunding Creative Ideas: The Dynamics of Project Backers in *Kickstarter*." Available at SSRN.

Lambert, T., and A. Schwienbacher. 2010. "An Empirical Analysis of Crowdfunding." Available at SSRN.

Lerner, J. 1995. "Venture Capitalists and the Oversight of Private Firms." *Journal of Finance* 50(1): 301–318.

Lin, M., and S. Viswanathan. 2013. "Home Bias in Online Investments: An Empirical Study of an Online Crowdfunding Market." Available at SSRN.

Marom, D., and O. Sade. 2013. "Are the Life and Death of an Early Stage Venture Indeed in the Power of the Tongue? Lessons from Online Crowdfunding Pitches." Available at SSRN.

Mason, C. 2007. "Venture Capital: A Geographical Perspective." In *Handbook of Research on Venture Capital*, ed. H. Landström, 86–112. Cheltenham: Edward Elgar.

Massolution, Crowdsourcing LLC. 2012. *Crowdfunding Industry Report: Market Trends, Composition and Crowdfunding Platforms.*

Massolution, Crowdsourcing LLC. 2013. *Crowdfunding Market Outlook Report.*

Mazzucato, M. 2013. *The Entrepreneurial State: Debunking Public vs. Private Sector Myths.* London: Anthem Press.

McGlashan, R., and T.A. Voelker. 2013. "What Is Crowdfunding? Bringing the Power of Kickstarter to Your Entrepreneurship Research and Teaching Activities." *Small Business Institute Journal* 9(2): 11–22.

Miglietta, A. et al. 2013. "Gli strumenti finanziari a supporto delle start-up innovative: le opportunità offerte dal CrowdFunding." In XXV Convegno annuale di Sinergie, "L'innovazione per la competitività delle imprese," Ancona.

Mollick, E. 2013. "Swept Away by the Crowd? Crowdfunding, Venture Capital and the Selection of Entrepreneurs." Available at SSRN.

Mollick, E. 2014. "The Dynamics of Crowdfunding: An Exploratory Study." *Journal of Business Venturing* 29(1): 1–16.

OECD. 2006. *The SME Financing Gap*. Vol. I, *Theory and Evidence*. Paris: OECD.

Ordanini, A., L. Miceli, M. Pizzetti, and A. Parasuraman. 2011. "Crowd-Funding: Transforming Customers into Investors Through Innovative Service Platforms." *Journal of Service Management* 22(4): 443–470.

Paul, S., G. Whittam, and J. Wyper. 2007. "The Pecking Order Hypothesis: Does It Apply to Start-Up Firms?" *Journal of Small Business and Enterprise Development* 14(1): 8–21.
Poetz, M., and M. Schreier. 2012. "The Value of Crowdsourcing: Can Users Really Compete with Professionals in Generating New Product Ideas?" *Journal of Product Innovation Management* 29(2): 245–256.
Powell, W.W., K.W. Koput, J.I. Bowie, and L. Smith-Doerr. 2002. "The Spatial Clustering of Science and Capital: Accounting for Biotech Firm-Venture Capital Relationships." *Regional Studies* 36(3): 291–305.
Qiu, C. 2013. "Issues in Crowdfunding: Theoretical and Empirical Investigation on Kickstarter." Available at SSRN.
Riedl, J. 2013. "Crowdfunding Technology Innovation." *Computer* 46(3): 100–103.
Rubinton, B.J. 2011. "Crowdfunding: Disintermediated Investment Banking." MPRA Paper 31649.
Schumpeter, J.A. 1943. *Capitalism, Socialism and Democracy*. London: Allen & Unwin.
Schwienbacher, A., and B. Larralde. 2010. "Crowdfunding of Small Entrepreneurial Ventures." In *The Oxford Handbook of Entrepreneurial Finance*, ed. D.J. Cumming. Oxford: Oxford University Press.
Seghers, A., S. Manigart, and T. Vanacker. 2012. "The Impact of Human and Social Capital on Entrepreneurs; Knowledge of Finance Alternatives." *Journal of Small Business Management* 50(1): 63–86.
Shang, J., and R. Croson. 2009. "A Field Experiment in Charitable Contribution: The Impact of Social Information on the Voluntary Provision of Public Goods." *The Economic Journal* 119(540): 1422–1439.
Sigar, K. 2012. "Fret No More: Inapplicability of Crowdfunding Concerns in the Internet Age and the JOBS Act's Safeguards." *Administrative Law Review* 64(2); 474–505.
Sorenson, O., and T.E. Stuart. 2001. "Syndication Networks and the Spatial Distribution of Venture Capital Investments." *American Journal of Sociology* 106(6): 1546–1588.
Tribus, M. 1970. "Panel on Government and New Business Proceedings." Venture Capital and Management, Management Seminar, Boston College, Boston, MA, May 28.
Van Wingerden, R., and J. Ryan. 2011. "Fighting for Funds: An Exploratory Study into the Field of Crowdfunding." Master's Thesis, Lund University School of Economics and Management.
Vass, T.E. 2013. "Adding Webinar Presentations to Equity Internet Crowdfunding Projects: Building Trust Relationships Online." Available at SSRN.
Zook, M.A. 2002. "Grounded Capital: Venture Financing and the Geography of the Internet Industry, 1994–2000." *Journal of Economic Geography* 2(2): 151–177.

Chapter 13

Harnessing the Geography of Innovation: Toward Evidence-Based Economic Development Policy

Maryann P. Feldman and Jongmin Choi

Introduction

Policies to promote entrepreneurship and innovation within geographically defined industrial concentrations have become a dominant economic development strategy. A myriad of initiatives attempt to capture the higher than average wages and productivity associated with agglomeration economies. The Obama Administration, the European Union, and the Chinese government all have national policy initiatives with specific territorial objectives (Dühr, Colomb, and Nadin 2010; Kostka and Mol 2013; Yu and Jackson 2011). More local levels of government, recognizing that the locus of innovative activity is decidedly smaller scale, also attempt to create technology-based economic development to secure their community's economic future (Feldman, Lanahan, and Lendel 2013). Despite good intentions and the investment of large sums of public funds, empirical evidence on the impact of government cluster-based initiatives is disappointing (cf. Lerner 2009). There is an urgent need to clarify the underlying nature of agglomeration economies, the logic of economic development, and the role of government investment. The counterintuitive policy implication is that rather than specific targeted initiatives, broad investment gives rise to agglomeration economies and creates the conditions that allow industries and innovation to flourish.

This chapter will review the theoretical basis for economic development policy based on the natural tendency of innovation to cluster spatially. Certain places become the locus of creative, inventive activity at specific times when individuals are in close contact, exchanging ideas, and creating economic and social value. Consider Florence under the Medici, Paris in the 1920s, Great Britain during the Industrial Revolution, or Silicon Valley in more recent times. The question of why certain

The Handbook of Global Science, Technology, and Innovation, First Edition.
Edited by Daniele Archibugi and Andrea Filippetti.

places prosper and achieve a higher standard of living has preoccupied economists since Adam Smith's *Wealth of Nations*. Krugman (1991) observed that rather than converging as predicted by neoclassical economics, national economies were becoming more divergent over time, which counters the theoretical predictions of neoclassical growth theory.

This prompted his series of lectures published as *Geography and Trade*. Others, notably Robert Lucas and Paul Romer, challenge the classic assumption of constant or decreasing return to scale by pointing out that knowledge is subject to increasing returns due to externalities inherent in the creation and application of knowledge. Rather than diminish, the value of knowledge actually increases with use due to network effects and path dependencies. Empirical work has documented these effects but offers little guidance for policymakers.

Michael Porter's (1990) *Competitive Advantage of Nations* introduced geographic considerations to the field of management, providing a model of conditions that supported industry, emphasizing the benefits of localized competition. This work, however, highlights preconditions that influence innovation and ignores the social process inherent in the creation of institutions that support creative activity and the genesis of spatially defined innovative industries. To better guide public investment, this chapter draws on recent work for the US Department of Commerce Economic Development Administration that provides a definition of economic development (Feldman et al. forthcoming). This more expansive view of economic development articulates a new role for government as the agent of collective investment in capacity and suggests that businesses that benefit from knowledge spillovers may be instrumental partners in institution building. The best economic development policy may be predicated on a longer-term and more capacious perspective. Continuously working toward measurable increases in regional capacity will best harness the natural tendency of innovative activity to cluster spatially and led to greater prosperity.

Geography as a Platform to Organize Innovative Activity

Just as firms are one means to organize economic activity, geography provides a platform to organize resources and relationships for economic activity. Beyond the natural advantages of resource endowments, proximity to markets, or climate, certain places have internal dynamics that increase the productivity of investments and result in higher innovation and creativity (Romanelli and Feldman 2007; Rosenthal and Strange 2003; see also in this volume Iammarino and McCann, Chapter 14 and Lorenzen and Mudambi, Chapter 10). These internal dynamics are socially constructed and involve a wide variety of actors including business owners, entrepreneurs, researchers, venture capitalists, policymakers, and politicians. Most importantly, since it is difficult to predict future technological change and market evolution, the greater the number of individuals who are able to participate in creative endeavors the higher the probability that a place, be it a city, region, or nation, is able to capture the resulting benefits.

The resources required to produce innovation are typically not confined to the boundaries of a single firm. While firms contract for external resources, often important considerations, such as unexpected results or unintended findings, which might be valuable insights, will not be part of the product stipulated by the contract and an important source of new ideas may be lost. Pisano (1996) observed this for the pharmaceutical industry and predicted a loss of innovative capacity, which has proven true (Pisano 2012).

Geography and place-specific interactions can shape industries. If you enjoy coffee or fine wine, then you know that there is something about the soil, the climate, the angle of the sun, the age of the trees, and the growing and harvesting traditions that create something unique. Even the best vineyards experience different vintages, reflecting the numerous variations that determine quality. While quality winemaking is diffusing around the world with product now exported from Chile, Argentina, Australia, New Zealand, and South Africa, wines have become more complex and differentiated rather than homogeneous. Connoisseurs talk about *terroir*, a French term used to denote the special characteristics that geography bestows. The term can be translated literally as "dirt" but more poetically as a "sense of place." The term captures the total effect that the local environment has on the product, when the total effect is more than the sum of its parts.

While outsourcing allows firms to lower production costs, technologically sophisticated firms compete on the basis of differentiated performance and innovation. While firms are the entities that take ideas to the market and realize value from innovation, even the largest multinationals are embedded in ecosystems that support and sustain their activity (see Ietto-Gillies, this volume, Chapter 6). These systems are globally connected but typically are focused in certain locations – collections of firms within one specialized industry or technology, concentrated within the same geographic area.

Marshall (1890) noted this tendency, citing three reasons: an infrastructure of related and supporting industries, the presence of deep, specialized skilled labor pools, and the presence of non-pecuniary externalities due to proximity to a strong knowledge base that facilitates knowledge exchange. Marshall maintains that a specialized industry and related firms are likely to agglomerate together because they can rely on a deep pool of skilled labor and can share knowledge, either through locally mediated market-based value chain transaction or though non-market mediated knowledge spillovers. Economic actors whether firms, entrepreneurs, scientists, or workers can search for solutions to their problem through formal and informal channels more easily due to geographical proximity. Experience with a technology or industry increases the stock of available knowledge locally, yielding better ideas. That is, economic agents benefit from easy communication, sharing ideas and serendipity – the unexpected but highly relevant chance occurrence. In addition, industrial agglomeration reduces transaction costs.

Examining Britain at the height of its industrial power, Marshall tapped into a phenomenon that can be observed throughout human history as certain locations become the locus for certain activity – the right place at the right time. Economists, of course, tend to focus on the localization of industrial activity. Shortly after Marshall wrote *The Principles of Economics*, the Twelfth Census of the United States in 1900 discusses the universal characterization of the localization of industries and references a medieval English manuscript from 1250 that documents industry localization (US Census Office 1902: ccix–ccx).

Within an industrial cluster, it may be easy to find suppliers, so that firms may reduce shipping costs. A good example is automobile manufacturing in Japan. Many components suppliers are located nearby so automobile firms do not need to purchase components from distant areas. Moreover, Pisano and Shih (2009) argue that, while outsourcing at great distance has become a popular corporate strategy, industry localization creates collective operational capabilities that underpin new product and process development and hold the seeds for the next generation of innovative products.

Arrow (1962) and Romer (1986) extend Marshall's externalities argument into what has become known as Marshall-Arrow-Romer (MAR) externalities. Arrow argues that workers improve their capability from regularly doing the same type of work – *learning by doing*. When many similar firms are located together, workers may easily move between jobs, doing similar tasks and gaining greater expertise. Firms improve their productivity by employing workers with high capabilities. As workers move they transfer know-how between firms. Romer maintains that similar firms are willing to agglomerate due to benefits from increasing returns to scale. While most physical goods are subject to constant or decreasing returns, one of the unique characteristics of knowledge is that it increases rather than decreases with use. The evidence is that firms benefit (Dekle 2002; Glaeser et al. 1992) and that workers receive higher wages derived from higher productivity and gains as a result of increasing returns (Wheaton and Lewis 2002).

Of all economic activity, innovation benefits most from location. Innovation is the ability to blend and weave different types of knowledge into something new, different, and unprecedented that has economic value. Similar to art, innovation is a creative expression. However, unlike art, the measure of innovation is not in the eye of the beholder, but in acceptance within the marketplace that brings commercial rewards to the innovating entities and returns to society in terms of economic well-being, prosperity, and growth.

The importance of geographic location to innovative activity in a world increasingly dominated by instant messaging, mobile telephones, and email may seem surprising and even paradoxical. After all, new telecommunications technologies have triggered a virtual spatial revolution as geographically diverse activities may be linked electronically in almost real-time transactions. However, there are geographical boundaries of knowledge spillover among firms due to tacitness of knowledge, and innovative activities tend to be concentrated in a geographical boundary due to the geographical boundaries of knowledge spillover (see Guy, this volume, Chapter 28 for a similar argument from an institutional perspective). Audretsch and Feldman (1996) find that after controlling for the geographical concentration in production, there is a greater propensity for innovative activity to cluster in industries where knowledge spillovers are dominant. Since knowledge spillovers are likely to be geographically concentrated in a region, innovative activities are also likely to cluster in the region at the same time. This compelling result suggests that the key element for innovation is knowledge spillover. In other words, in spite of the fact that firms still benefit from agglomeration, innovative activities are likely to be vibrant more in an area where knowledge spillovers are prevalent than in an area where knowledge spillovers are weak (Feldman et al. 2014). For instance, even if physical agglomerations of firms exist, innovation may be weak if there are no knowledge spillovers among firms.

Innovation is more geographically concentrated than invention. Invention is the first stage of the innovation process. Due to the creation of large patent databases there are many studies that focus on invention, which should not be confused with innovation. The limitations of patents are well known. Even though patents are geographically concentrated, reflecting a concentration of research and development (R&D) activity, this does not necessarily translate into economic advantage for those locations. Innovation is more geographically concentrated than production. Even after controlling for the geographic distribution of production, innovation exhibits a pronounced tendency to cluster spatially (Audretsch and Feldman 1996).

On the other hand, important ingredients of new knowledge creation and innovation may be ignored if externalities from knowledge spillover among related firms are primarily emphasized (Feldman and Audretsch 1999). Jacobs (1969) argues that complementary knowledge across diverse firms is one of the main sources for innovation, helping firms create new knowledge. The most important source of knowledge spillover, according to Jacobs, derives from industrial diversity rather than industrial specialization, and cities are the greatest place for innovation since cities have the most diverse source of knowledge. That is, diverse firms in a city produce new knowledge that creates vibrant innovative activities, helping other firms gain and harness new knowledge. Of course, the difficulty lies in predicting how diverse pieces will be assembled to solve the innovation puzzle. Our definition of genius is something that once articulated become obvious – the elegance of a solution that defied prior thinking and was difficult to anticipate.

The question of whether specialization or diversity of economic activities is better for innovation has been a fierce debate that has policy implications. Specialization of economic activities implies MAR externalities, while diversity of economic activities indicates Jacobs's externalities (see Table 13.1).

Table 13.1 Various forms of externalities.

	MAR externalities	*Jacobs's externalities*
Authors	Marshall (1890), Arrow (1962), Romer (1986)	Jacobs (1969)
Main arguments	Focused on benefits from agglomeration 1. An infrastructure of related and supporting industries 2. The presence of deep, specialized skilled labor pools 3. The presence of non-pecuniary externalities due to proximity to a strong knowledge base 4. Improving workers capacity – *learning by doing* 5. Benefits from increasing returns to scale	Focused on the source of innovation 1. Diverse firms in a city produce new knowledge that creates vibrant innovative activities, helping other firms gain new profits by harnessing new knowledge. 2. Cities are the greatest place for innovation since cities have the most diverse source of knowledge
Examples	Silicon Valley, Research Triangle Park, Hong Kong Science Park	Large cities where diverse economic activities are vibrant (e.g., New York, Beijing, Seoul)

If the specialization argument by Marshall, Arrow, and Romer is correct, policymakers should promote specialized economic activities (e.g., focusing on one industry such as Bio-tech, Nano-tech, or Information and Communication Technology). On the other hand, if the diversity argument by Jacobs is true, the way in which diverse economic activities can be accelerated should be the overriding concern among policymakers. Glaeser et al. (1992) analyze the relationship between employment growth of large industries in 170 Standard Metropolitan Areas between 1956 and 1987 and both externalities. They find strong evidence that diversity in a city contributes to growth while MAR externalities do not encourage employment growth. This result is consistent with Jacobs's argument that knowledge spillovers between industries promote innovation and growth. Feldman and Audretsch (1999) also examine which types of economic activities promote innovative activities, focusing on six groups of industries: Agri-business, Chemical engineering, Office machinery, Industrial machinery, High-tech computing, and Biomedical. They reveal that diversity across complementary economic activities sharing a common science base generates more innovative activities. However, the result does not support the specialization thesis. In other words, diverse economic activities in a region help firms create more knowledge spillovers, which in turn are conductive to innovative activities. As diversity across complementary economic activities increases in a given location, the location generates knowledge spillovers, which help boost economic growth in the location. On the other hand, specialization implies that the industrial structure may have become too tailored to an industrial activity or that the industry has become concentrated because it has become mature.

These research findings offer intuitive policy ideas to policymakers who have tried just to mimic successful industrial clusters, focused primarily on industrial specialization and targeted at a core industry such as biotech or information and communication technology without considering the context of their location. As previous literatures illustrated, just following the trend may not be a good strategy. The important fact is that knowledge spillovers among firms are a conduit for innovation that is one of the pivotal sources of capacity building. In other words, what surely matters may not be industrial specialization or diversity, but how to promote knowledge spillovers among firms. In this sense, understanding the context of location certainly matters.

Policies to build industrial clusters often fail due to lack of understanding of the context and history of the location. Successful industrial clusters are socially constructed and take a long time to realize their potential. Given this reality, policymakers should first examine the context of location, such as industry structure and infrastructure in their region. By better understanding their geographical area, policymakers know what their strengths and weaknesses are and can consider them as they think about overcoming disadvantages and reinforcing advantages. They should then consider creating connections among existing firms so that firms can make positive externalities, generating knowledge spillovers across industries. In case of poor regional capacity, policymakers should focus mainly on investing in building regional capacity so that diverse economic activities can be promoted. As the research conducted by Feldman and Audretsch (1999) shows, fostering conditions that help create diverse economic activities is crucial. If a region is too specialized in a particular industry, policymakers should attract complementary industries in order to generate more knowledge spillovers across industries rather than within industries. While policymakers search for the appropriate *recipes* the best advice is to focus on local ingredients.

Harnessing the Natural Tendency of Innovative Clustering

Some economists are skeptical of place-based economic development strategies, arguing that a tradeoff exists between local gains and national welfare (Einiö and Overman 2012). The controversy is that resources are simply being redistributed from one local economy to another and to the detriment of overall national welfare. A 2009 World Bank report advocates for a "spatially blind" (or people-based) approach rather than place-based, as the "most effective way of generating efficiency, guaranteeing equal opportunities, and improving the lives of individuals where they live and work" (World Bank 2009). The report asserts that encouraging people-mobility enables people to live in places where they will likely be more economically productive which, in turn, increases individual incomes, productivity, and aggregate growth, and leads to a more even geographical distribution of wealth.

On the other hand, proponents of place-based approaches to economic development argue that it is necessary to fully understand the local and regional context in order to create development policies that will succeed in a particular area. The place-based approach asserts that one-size-fits-all policies, that do not consider the regional context of the area they are trying to assist, may have unanticipated (and potentially negative) consequences (Barca, McCann, and Rodríguez-Pose 2012).

Enterprise zones or enterprise initiatives are one policy for place-based economic development that has been extensively evaluated. To attract firms within enterprise zones, government provides many incentives, such as property or income tax exemptions for firms. Once firms are located in enterprise zone, they can qualify for additional subsidies. Policymakers expect that enterprise zones may result in mitigating the unemployment rate as well as helping a region boost economic growth in the long term. In spite of such expectations, however, little evidence supports the impact of enterprise zones. Bondonio and Engberg (2000) and Elvery (2009) focus on enterprise zones in the United States and find that the impact on employment is not significant. Gobillon, Magnac, and Selod (2012) focus on French enterprise zones implemented in 1997. They find a significant impact, but only in the short run. Einiö and Overman (2012) investigate the Local Enterprise Growth Initiative (LEGI) introduced in 2005 in the United Kingdom. The result indicates that LEGI has a positively significant impact on employment and the number of businesses as well as unemployment rate. However, they find evidence that these impacts displace activity from the areas surrounding the targeted areas. The conclusion of these studies is that enterprise zone policies are not effective in aggregate but only shift the location of firms and jobs between jurisdictions. These results show that artificially defined place-based strategies for economic development are inefficient and do not lead to industrial development. However, given that the immediately available policies for the development of industrial clusters do not yield any outcome, there is a need for profound rethinking of the theoretical basis of economic development policymaking.

Theoretical work tends to focus on the default, mathematically tractable, assumption of constant returns to scale. However, the major contribution of the new growth theories is to recognize that knowledge benefits from increasing returns to scale rather than the constant or decreasing returns associated with physical commodities. Activities that create knowledge and the sharing of knowledge create increasing returns that would lead to increased national welfare. But at this point neither theoretical nor empirical economics can adequately address this question. Policymakers cannot afford to wait. As Kline and Moretti (2013: 34) conclude, "Second best may, in practice, be very attractive relative to the status quo."

When considering the development of industrial clusters there are two diametrically opposing models. One model, practiced in China, relies on government dictating the growth of designated science cities. This is a very top-down approach to economic development that has been successful in Singapore and Taiwan: the central government dictates that a specific location will have a concentration of R&D and accomplishes this in a relatively short period of time. The verdict is still out as to whether these locations will be successful at creating a sustained competitive advantage given that innovation is more complex than simply conducting R&D.

The other model occurs in the United States and other market economies and relies on self-organization and local initiative. In market economies the central government cannot dictate the actions of private companies, but may only offer incentives to encourage company location decisions and investments in R&D. The closest that we have to a government-induced clusters is Research Triangle Park (RTP) in North Carolina, which was the result of state and local government actions. RTP was a very long undertaking beginning in the 1920s and is now the largest research park in the world (Link 1995). But what is most critical are the processes that took place subsequently as the industrial landscape developed (Feldman and Lowe 2014). Through the articulation of a vision and consistent policy efforts a successful cluster was built. There are many other examples of government trying to build clusters in market economies, but the results typically look very different from what was originally intended (Leslie and Kargon 1996).

While economic development officials and government planners want to define long-term strategies, it is difficult, if not impossible, to predict scientific discoveries, new technologies, and new opportunities. IBM, an industry leader, underestimated the potential of the computer industry, creating an opportunity for new firms to create personal computers. Few people predicted the potential of the Internet and how it would change the way we access information and communicate. Moreover, successful entrepreneurs make their own luck, adjusting and adapting to survive. Instead of wisely considered, far-sighted solutions, entrepreneurial activity is by necessity messy, adaptive, and unpredictable. Economic development strategies need to be equally adaptive.

The biggest problem is that it is impossible to predict which technologies are going to yield any pay-off. By the time a new industry, for example, biotech or nanotechnology, has a defined name and is on its way to becoming a household name, it is probably too late for other places to decide that they will participate as major centers. Creating a cluster in a market economy is a messy social process. Designing an effective economic development strategy may be the ultimate local innovation.

Evaluation Challenges for Economic Development Policies and Projects

Along with challenges of creating effective economic development strategy, policymakers have been faced with policy evaluation challenges. This section enumerates some of the measurement problems inherent in evaluating economic development programs as a prelude to a new approach, which is introduced in the following section. Calls for government accountability require program evaluation efforts that typically fall short of rigorous standards (GAO 2012). Despite the significant efforts devoted to evaluating economic development programs, the inherent complexity of innovation complicates research design.

The highest scientific standard is to discern causality. In medical studies there are no patients to randomly assign to receive medication while others get a placebo while in educational evaluations students may be randomly assigned to classrooms or schools. There are examples of experiments that examine the impact of subsidized medical insurance on health outcomes or quasi-experiments that assess the effects of individual choices such as attending college on lifetime earnings and job satisfaction. Evaluation of outcomes is easiest when comparing similar programs or inputs. Economic development cluster initiatives engage a wide range of activities from building infrastructure to incubating businesses. This variety is warranted because the projects serve heterogeneous communities and different technologies. Evaluation would be facilitated if the projects offered identical services, but that would not advance the needs of local entrepreneurs and businesses nor the objective of local economic development officials.

With economic development programs it is difficult to attribute a specific outcome, such as an increase in sales or employment, to one specific program or intervention. There are no pure treatment effects for economic development programs and initiatives and the idea of random assignment is not politically feasible. The best that we can do is to try to find a matched set of control firms that are similar, in terms of their characteristics. Feldman, Kemeny, and Lanahan (2014) provide an example in an examination of the Trade Adjustment Assistance for Firms program (TAAF). The control group was predicated on a five times match of firms in the same three-digit industry and zip codes. The results suggest that the TAAF program funded firms that were initially worse off than the control group in terms of sales. Two years after the program the treated firms had higher sales. No results were found for employment – a concern for economic development programs. Interviews revealed that firms were reluctant to lay off employees when sales were down because of the close relationships that developed in small firms. For the same reason the treated firms were reluctant to hire new employees until sales could be sustained over a longer period of time. Thus, the returns to economic development are affected by multiple individual decisions, and intervening influences that come from multiple motivations.

While it is possible to examine individual firms the analysis become less tractable when we consider regions. Case studies of the development of regional economies reveal an extremely complex process in which public investment is an important element, but only one of many important elements. Complex systems are notoriously difficult to model. There is no reason to believe that optimizing the performance of any one component of a complex system will optimize or even necessarily improve the performance of the system overall.

Current thinking is that economic development projects are not simply a series of transactions, but instead contribute to building an ecosystem (Hwang and Horowitt 2012). The building of ecosystem capacity can reduce transaction costs and increase knowledge flows, resulting in multiple unexpected outcomes. These are considered to be functional impacts rather than pure economic impacts that accrue as a function of undertaking a project or making an investment at a scale significantly larger than the original investment. It is necessary to move beyond economic impact studies to more fully capture the returns to a wide range of public economic development investments.

Moreover, the amount of funding provided for economic development initiatives, while important to recipients, is minuscule in relation to the size of a regional economy. The attribution of good outcomes to specific programs, investments, or events is probably more about good luck, publicity, and hype, rather than true causality demonstrated by sound economic analysis. Even when programs are effective, changes in the macro economy, while exogenous, may wipe out any hard-earned gains. Macroeconomic conditions certainly affect regional outcomes in ways that are not entirely predictable and certainly beyond control at a regional level.

Often successful efforts to build new companies results in an infusion of venture capital, a merger or acquisition, or some other change to ownership that induces relocation. As a result the economic development effects on, for example, local employment are negatively impacted. The result of the efforts that built a successful company can then – by commonly used metrics – appear as failure. In reality, the local area has benefited from the example of what is possible for an entrepreneurial company. This might incite others to start companies or invest in early-stage activity. The founders may stay in the region and become serial entrepreneurs. Key employees will have learned skills that will benefit their next employer. As a result of the economic development effort that was mistakenly judged to be a failure on employment, capacity in the region for the next round will actually have increased.

Finally, there are significant time lags involved in realizing the benefits of government investments that do not conform to political election or budgetary cycles. Mansfield (1991) noted that the time lag between an academic research discovery and new product innovation was seven years. Many economists have followed, trying various approaches to estimate the lag associated with the realization of benefits from investment in R&D. The consensus is that the time lags between investment in research and realized commercial advances are lengthy, uncertain, and vary significantly among fields (Merrill and Olson 2011). In fact, a National Academy of Sciences publication (2001) observed that "History ... shows us how often basic research in science and engineering leads to outcomes that were unexpected or took many years or even decades to emerge." Perhaps the best outcome for economic development is simply making progress on a variety of measures, with attention to correcting deficits and remediating problems.

A New Definition of Economic Development

Economic development is often referenced but rarely defined. This section draws on work by Feldman et al. (forthcoming), that was inspired by a request from the US Department of Commerce. In particular, when it comes to economic development, the policy debate is often hijacked toward outcomes, because of lack of a clear definition of economic development. Furthermore, the term

Table 13.2 Economic growth and economic development.

	Economic Growth	*Economic Development*
Measurement	Easily quantified as an increase in aggregate output	More qualitative and hard to measure
Process	Occurs when output increases per unit of input through productivity, enhancement and technology	Fundamental transformation of an economy including the industrial structure, the educational and occupational characteristics of the population, and the entire social and institutional framework
Relevance of distinction for the geography of innovation	Tied to macroeconomic condition such as employment rate, income, or GDP Focused on an increase in output in a given region Innovation or technology can be a source of regional economic growth	Tied to the conditions that influence the microeconomic function of the economy, affecting the quality of inputs Focused on quality improvement, innovation, and the expansion of capacities in a given region Aims to bring about fundamental transformation in a given region as well as a whole country
Necessary conditions	More inputs such as capital, labor, and technology (innovation) generate economic growth, but they may not guarantee returns to a region	Effective institutions grounded in norms of openness, tolerance for risk, appreciation for diversity, and confidence in the realization of mutual gain for the public and the private sector

"economic development" is often conflicted with economic growth or seen to reflect a professional practice (see Table 13.2). In order to better understand economic development policies this section builds a definition of economic development based on Feldman et al. (forthcoming).

Unlike economic development, the definition of economic growth is based upon strong theoretical grounding. Consider argument by David Ricardo (1891), Robert Solow (1956), and many others. Economy is a kind of apparatus that produces economic output, which is a function of inputs such as capital, labor, and technology. The mechanism is simple. The more inputs are added the greater outputs will be expected. Given this, one may expect that growth occurs when outputs increase. In this sense, economic growth is associated with an increase or decrease in outputs, so economic growth is easily measured and quantified. On the other hand, defining economic development may not be as simple as economic growth.

Rather it is very complicated since economic development is often qualitative and hard to measure. Economic development influences the microeconomic function of the economy, affecting the quality of inputs, bringing the business opportunity to firms, and in turn creating the conditions that enable long-run economic growth. Based on a review of the literature, Feldman et al. (forthcoming), inspired by Sen (1999), argue that economic development is defined as the development of capacities that expand economic actors' capabilities. These actors may include individuals, firms, or industries that are likely to exert their potentials based on the development of capacities. Rather than simple counts of jobs, economic development is concerned with the quality of jobs, the caliber of business practices, and the density of social capital.

Development can be regarded as fortifying autonomy and substantive freedom, which promotes individuals' participation in economic life (Sen 1999). Thus, economic development occurs when individuals have the opportunity to actively engage and contribute to society and are likely to realize their potential. This promotes the advancement of the whole society. In this sense, the expansion of capacities provides the basis for the realization of individual, firm, and community potential, which, in turn, contributes to the advancement of society.

According to Schumpeter (1934), economic development involves relocating capital from already established methods to new and innovative methods, which enhances productivity. In Schumpeter's view, economic development entails a fundamental transformation of an economy including the industrial structure, the educational and occupational characteristics of the population, and the entire social and institutional framework. For instance, not only did mass production in the textiles industry drive the industrial revolution, but it also influenced other complementary sectors and in turn diffused widely, increasing quality of life. While economic growth is measured by putting inputs to an existing economic framework, the aim of economic development is to change the economic framework so that people work more productively and shift their positions from simple repeat work to higher value-added activities. Thus, economic development is realized through innovation, lowered transaction costs, and the utilization of capabilities toward the responsible production and diffusion of goods and services. Here the specification of the objective of economic development as increased prosperity and quality of life seem to be guiding principles for any democratically elected government.

Rodrik, Subramanian, and Trebbi (2004) argue that the development of high-quality institutions plays a major role in economic growth. They promote productive activities, capital accumulation, skill acquisition, invention, and technology transfer (North and Thomas 1973) because they can provide well-regulated conditions for economic development. In other words, the reason why effective institutions matter is that they help individuals and businesses make investment decisions, since high-quality institutions can reduce uncertainty and risk and provide stable and predictable systems. Thus, to further build the definition, economic development requires effective institutions grounded in norms of openness, tolerance for risk, appreciation for diversity, and confidence in the realization of mutual gain for the public and the private sector. Thus, innovation is a vehicle for economic development rather than an end itself.

It is often found that many countries, especially sub-Saharan Africa, Central and South America, and Oceania, have some experience of significant increases in economic output without development due to either natural population growth or large-scale resource extraction (Acemoglu, Johnson, and Robinson 2002; De Soto 2000; Moyo 2009). In spite of their tremendous economic growth, these countries with insufficient economic development suffer from significant income inequality and limited educational attainment (Wolfson 1997) and have experienced low health outcomes such as mortality rate and life expectancy much below the average values in developed nations (Vandemoortele 2009). What this implies is that long-term outcomes such as quality of life and widespread prosperity cannot be achieved without sufficient economic development. That is, weak economic development restrains capacity-building that leads to economic growth in the future. Economic development can be a foundation for future economic growth, stimulating the agents' potentials and in turn enabling long-run economic growth. Finally, economic development is essential to creating the conditions for economic growth and ensuring our economic future.

In sum, economic development is defined as the development of capacities that expand economic actors' capabilities. Economic development is defined as the expansion of capacities that contribute to the advancement of society through the realization of individuals', firms', and communities' potential. It is a sustained increase in prosperity and quality of life realized through innovation, lowered transaction costs, and the utilization of capabilities toward the responsible production and diffusion of goods and services. Economic development requires effective institutions grounded in norms of openness, tolerance for risk, appreciation for diversity, and confidence in the realization of mutual gain for the public and the private sector. Economic development is essential to creating the conditions for economic growth and ensuring our economic future.

The question then becomes how to best invest resources in order to generate economic development. This requires delving into the process of generating clusters or what Braunerhjelm and Feldman (2006) term cluster genesis.

Cluster Genesis

Scholars and practitioners have tried to find the factors that play a critical role in building clusters of firms and revealed some main ingredients such as high-quality institutions, research universities, venture capitalists, and a strong supply of local entrepreneurs. Yet, these attributes are the result of a fully functioning industrial cluster and do not inform how policymakers might go about creating vibrant local industry. Many attempts to establish industrial clusters have ended in failure since the presence of the key factors themselves does not guarantee prosperous industrial clustering. For instance, the Johns Hopkins hospital in Baltimore, thought of as one of the top medical schools in the world, was not an early leader in the biotechnology industry (Feldman and Desrochers 2003). The question arises: in spite of well-prepared factors such as venture capitalists, research centers, and large amounts of public and private investment for building industrial clusters, why does this happen? When it comes to the cluster strategy, thus, the most important point is not about ingredients for industrial clustering but to understand how these ingredients reciprocate the dynamic process of creating industrial clusters. While fully formed industrial clusters may look alike, what truly matters is the process: the way institutions are built and social capital is created during the creation of industrial clusters. Cluster development is a sequential and temporal process rather than just a static event.

Assume that the same amount of resources is invested in certain places that are seemingly identical, why do some places prosper while other places fail to flourish? This question is a fundamental question of economists who seek to find out the driving force of technological change. In spite of the importance, only few studies offer an answer. Some triggering events can spark the emergence of industrial clusters. Koo and Choi (2013) argue that the continuous emergence of successful venture firms after some initial seeding event can accelerate cluster development. As more successful firms emerge in a given region, there will be increases in local entrepreneurship. Considering successful industrial clusters, the actions of entrepreneurs can be a possible answer.

Entrepreneurs discover opportunities, take risks, mobilize resources, create new firms, and may bring prosperity to a region. They play a pivotal role in the creation of institutions and building regional capacity that will enable regions to sustain economic growth (Feldman, Francis, and Bercovitz 2005; James 1998; Rodrik 2003). Entrepreneurs not only benefit from location but also influence the transformation of their local communities. Prosperity in a region is not deterministic, but efforts from entrepreneurs can affect the prosperity of places. Entrepreneurs can be considered *local champions* – individuals who they live and work in a region and have strong dedication to their region.

Feldman (2014) considers the case of Greenwood located in the Mississippi Delta, a small city that about 100 years ago was well known for its cotton industry. After experiencing mechanization and globalization, it became the poorest region in the poorest state in the United States. Greenwood experienced a lack of tax revenue and limited subsequent investment. It seemed that no federal or state government program could be a magical remedy. However, Fred Carl, who is the owner of Viking Range Corporation, a cutting-edge professional kitchen appliance company, brought new hope to Greenwood. While Carl was working as a building contractor, he realized that consumers wanted high-quality residential stoves that looked and cooked like commercial stoves. He identified this opportunity and created an entire new industry segment. Against the trend of offshoring production, Carl located manufacturing operations in his hometown of Greenwood, gathering financing from local investors. His efforts resulted in 1500 workers being hired with numerous benefits and educational opportunities – employing more workers than the local hospital, the typical largest employer in most small and medium-sized cities.

Fred Carl's story is not a special case but can be seen as part of a general trend. Often the story of successful places is strongly connected with the story of an individual who played a role in creating institutions and building the capacity of a local economy. For instance, Fred Terman is widely regarded as the founding father of Silicon Valley. Although he earned his Sc.D. degree at MIT, he returned to his home and served as the Dean of Engineering at Stanford University. He led the creation of Stanford Industrial Park in the 1950s. Many companies such as Hewlett-Packard and General Electric moved into Stanford Industrial Park, making the area innovative. Eventually Palo Alto became "Silicon Valley" as the hotspot for digital innovation. Like Fred Terman, George Kozmetsky, who was the founder of Teledyne, is one of the local champions in Texas. He established the Institute for Innovation, Creativity, and Capital (IC2) at the University of Texas. IC2 served as a mentor for over 260 computer companies. Ewing Marion Kauffman provides another example of a local champion. He was born and raised in Missouri and lived in Kansas City. After working as a salesman for a pharmaceutical company, Kauffman established his own pharmaceutical company in Kansas City, an unfavorable place in the 1950s, rather than in Philadelphia-New Jersey corridor, where the industry was concentrated. When he sold his company to Merrell Dow in 1989, the company had grown to become one of the global companies with $1 billion in sales and over 3400 employees. The company provided lots of benefits, which included educational and training benefit, profit-sharing plans, and employee stock options, to employees. Kauffman established the Kauffman Foundation in 1966. The Foundation is now dedicated to improvement of communities, education, the arts, and social programs in Kansas City (Feldman and Graddy-Reed 2014). Burlingham (2007) calls this phenomenon *Small Giants: Companies That Choose to Be Great Instead of Big*.

Local champions are motivated by objectives that extend beyond profits. But they take responsibility for the stewardship of the place and have a dedication to their local community. In other words, they are not seeking short-term profit maximization, but dedicated to prosperity in their home community and discovering new opportunities that may bring about new profits. These stories indicate that regions can become prosperous when entrepreneurs actively engage in extra-market activities. Moreover, local champions can advocate for the types of government interventions that will help their individual firms but can also promote an industry and a place. Of course, the reality of too great a reliance on the private sector can lead to an imbalance that favors profits over citizens' rights. Moreover, businesses change ownership and management, fail or relocate. While local champions may be a critical catalyst for developing clusters in local economies there is a need to define a role for government and to change the conversation about economic development.

While industrial clustering is a natural outcome it relies on the existence of underlying capacities. Rather than being the result of market forces, the underlying capacities require long-term investment in capacity building.

Reconsidering the Role of Government in Economic Development

Capacity-building in a jurisdiction that requires economic development is beyond the mandate of any private firm, industry associations, or other economic institution. The principal inclusive vehicle for organizing economic, social, and civic life is government. Government, most simply, is a vehicle for collective action: an agent for whom the principal is its citizens, both residents and businesses. Even the most devoted libertarian recognizes the limits of the market, while the most ardent free-marketer recognizes that government was the only entity capable of saving the financial sector from free-fall. The focus on capacity-building harkens back to an older American tradition that began with Alexander Hamilton, defining a role of government as the provider of resources and incentives that enable private enterprise to flourish. The difficult balance is for government to provide for the realization of potential while maintaining incentives; to provide scaffolding for economic transactions, while not over-regulating; and to make investments that advance the public interest and encourage the full participation of private individuals and organizations while not prying into special interests.

In contrast to a resource economy, where geographically uneven endowments predetermine place-based development strategies, contemporary economic development depends upon public and private investments in individual and group capacity. This capacity is constructed over time and requires a consistency of action that stays true to the objectives of promoting the citizen's well-being. Government investment provides skills, capabilities, programs, and incentives that private sector investment can build upon. Thus, virtuous self-reinforcing cycles of economic development yield the desired social and economic outcomes of prosperity and more sustainable economic growth.

The Reagan Thatcher Agenda and its corollary the Washington Consensus have run their course, yet with little agreement on the policies to implement and investments for government to make. While macroeconomic considerations have dominated the policy agenda there is increased urgency to address the microeconomic foundations of innovation and production. Investment in capacity, rather than austerity policies, provides the basis for economic development. Over the past 50 years, discussions of economic development have moved from a preoccupation with lagging regions and eradication of poverty to a new focus on innovation and international competitiveness. These concerns are universally relevant to a full range of jurisdictions and communities.

The increased contemporary emphasis on innovation and entrepreneurship as a source of economic growth redefines the role for government and provides rationale for government investment. Innovation comprises the complex and multifaceted process under which creativity leads to practical application, commercialization, and ultimately economic and societal gain. Considerable effort has been put toward understanding the process of innovation and identifying the factors that increase economic growth and prosperity for a region. While there is broad consensus that innovation serves as an integral catalyst in leading the trajectory of an economy and even society forward, the emphasis in economic development policy remains on traditional attraction and retention incentives, often directed at specific businesses. This is largely a zero-sum game with little or no broader effects for economic development. In addition, local governments tend to do more of the same policies over time, adding incremental changes to preexisting strategies, rather than a wholesale reconsideration of their investment strategy.

There is growing evidence to suggest that public institutions have chosen to take on an expanded role within the innovation process (Block and Keller 2009; Schrank and Whitford 2009). The nature of scientific research has changed due to increasing demands to solve society's most pressing problems. This has led to the decentralization of industrial networks, or open innovation. Rather than confined to the R&D labs of large corporations, collaborative activity is now embedded in networks between both public and private institutions, large and small firms. This degree of decentralization encourages more organizations to work in concert with one another and also fosters a greater dependence on government programs to coordinate the operations of these networks (Schrank and Whitford 2009). Evidence suggests that at a time when market fundamentalism has come to guide policy debates, government has actually become more and more immersed in the economy through technology policies in particular (Block and Keller 2009).

As innovation and entrepreneurship move to center stage as critical elements for economic and societal progress, there is a need to redefine the role of the public sector. Rather than relying on the market-based rationales for public investment it is important to consider the potential role of the public sector in building and bolstering capacity. Rather than viewing individuals, or even firms, as objects on the receiving end of public initiatives, it is important to view them as active creative agents. After all, we can never predict when genius might emerge. The best economic development strategy is to enable as many actors to productively participate in the economy to the fullest of their ability. This prioritizes improving quality of life and well-being by enhancing capabilities and ensuring that agents have freedom to achieve.

The paradox of place-specific economic development policy is that broad-based government investments in education and infrastructure are critical to future economic growth. Targeting certain sectors, specific industries, or isolated components of the innovation ecosystem is unlikely to succeed if basic capacity is lacking. Government is the only entity with the long-term perspective and command of resources to engage in the economic development activities that promote industrial agglomeration and ultimately economic growth.

Conclusion

This chapter has reviewed the theoretical basis for economic development policy based on what the literature informs about the clustering of innovation. Table 13.3 provides definitions for the key concepts discussed. Innovative activity has a natural and pronounced tendency to cluster that provides the basis for creating new firms,

Table 13.3 Key definitions.

Invention	Invention is the first stage of the innovation process. It typically manifests as a patent.
Innovation	Innovation is the realization of value for a new idea or invention. Innovation is the ability to blend and weave different types of knowledge into something new, different and unprecedented that has demonstrated economic value. Innovation may manifest as product, process, or organizational reconfiguration.
Economic Growth	Economic growth is an increase in economic output, which is a function of inputs such as capital, labor, and technology. The mechanism is simple. The more inputs are added the greater output is expected. Given this, one may expect that growth occurs when outputs increase. In this sense, economic growth is associated with an increase or decrease in outputs, so economic growth is easily measured and quantified.
Economic Development	Economic development is defined as the expansion of capacities that contribute to the advancement of society through the realization of individuals', firms', and communities' potential. Economic development is a sustained increase in prosperity and quality of life realized through innovation, lowered transaction costs, and the utilization of capabilities toward the responsible production and diffusion of goods and services. Economic development requires effective institutions grounded in norms of openness, tolerance for risk, appreciation for diversity, and confidence in the realization of mutual gain for the public and the private sector. Economic development is essential to creating the conditions for economic growth and ensuring our economic future.
Entrepreneurs	Entrepreneurs are individuals who discover opportunity, take any risks, mobilize resources, and create new firms.
Local Champions	Local champions take responsibility for the stewardship of a place. They have a dedication to place that extends beyond profit maximization. They are dedicated to promoting prosperity in their home community. Entrepreneurs can become local champions once they take responsibility and have a strong dedication to their community.
Genius	The elegance of a solution that defied prior thinking and was difficult to anticipate but obvious once articulated.
Knowledge Spillover	Knowledge spillover is an exchange of knowledge including idea, know-how, technological skills, and management skills through formal and informal channels among individuals and firms. Knowledge spillover stimulates technological improvement and innovation.
Agglomeration Economies	Agglomeration economies imply that firms gain benefits by physical location. The concept is that external economies of scale accrue to space due to proximity. As firms locate near each other, they are likely to obtain benefits such as low transaction costs, skilled labor pool, and knowledge.

new industries, and widespread prosperity and economic growth. Unfortunately, there is little theoretical guidance for policymakers responsible for making investment that will lead to economic development.

For too long the conceptualization underling place-based industrial policy relied on creating a set of preconditions such as venture capital. This emphasis has not yielded the desired results. The argument promoted here is that the social construction of the conditions conducive to the creation and genesis of spatially defined innovative industries is a process that requires a longer time horizon and investments in basic capacity that enable participation by diverse actors in a local economy.

While the concept of economic development preoccupies our collective imagination the term is often not well defined or defined in a limited manner that does not accommodate the situation of the full range of places faced with restructuring and economic uncertainly. All too often the emphasis is on innovation as an end in itself rather than as a mechanism that can create prosperity and greater well-being. This requires paying attention to the distribution of risk and returns. The chapter draws on recent work for the US Department of Commerce that provides a definition of economic development (Feldman et al. forthcoming). In the absence of a definition there is a tendency to conflate economic development with economic growth or to rely on private sector constructs such as rate of return that are inappropriate for government investments. This more expansive view of economic development articulates a new role for government as the agent of collective investment in capacity and suggests that businesses that benefit from knowledge spillovers and local capacity may be an instrumental partner in institution-building. The best economic development policy may be predicated on a longer-term and more expansive perspective that continuously works toward measurable increases in regional capacity. The best policies to harness the natural tendency of innovative activity to cluster may be policies and investments that allow economic agents the capacity to be creative and fully engaged in the economy and society.

References

Acemoglu, Daron, Simon Johnson, and James A. Robinson. 2002. "Reversal of Fortune: Geography and Institutions in the Making of the Modern World Income Distribution." *Quarterly Journal of Economics* 107(4): 1231–1294.

Arrow, Kenneth J. 1962. "The Economic Implications of Learning by Doing." *The Review of Economic Studies* 29(3): 155–173.

Audretsch, David B., and Maryann P. Feldman. 1996. "R&D Spillovers and the Geography of Innovation and Production." *The American Economic Review* 86(3): 630–640.

Barca, Fabrizio, Philip McCann, and Andrés Rodríguez-Pose. 2012. "The Case for Regional Development Intervention: Place-Based Versus Place-Neutral Approaches." *Journal of Regional Science* 52(1): 134–152.

Block, Fred, and Matthew R. Keller. 2009. "Where Do Innovations Come from? Transformations in the US Economy, 1970–2006." *Socio-Economic Review* 7(3): 459–483.

Bondonio, Daniele, and John Engberg. 2000. "Enterprise Zones and Local Employment: Evidence from the States' Programs." *Regional Science and Urban Economics* 30(5): 519–549.

Braunerhjelm, Pontus, and Maryann P. Feldman (eds.). 2006. *Cluster Genesis: Technology-Based Industrial Development*. Oxford: Oxford University Press.

Burlingham, Bo. 2007. *Small Giants: Companies That Choose to Be Great Instead of Big.* New York: Portfolio.

Dekle, Robert. 2002. "Industrial Concentration and Regional Growth: Evidence from the Prefectures." *Review of Economics and Statistics* 84(2): 310–315.

De Soto, Hernando. 2003. *Mystery of Capital: Why Capitalism Triumphs in the West and Fails Everywhere Else.* New York: Basic books.

Dühr, Stefanie, Claire Colomb, and Vincent Nadin. 2010. *European Spatial Planning and Territorial Cooperation.* London: Routledge.

Einiö, Elias, and Henry Overman. 2012. "The Effects of Spatially Targeted Enterprise Initiatives: Evidence from UK LEGI." ERSA Conference Paper.

Elvery, Joel A. 2009. "The Impact of Enterprise Zones on Resident Employment: An Evaluation of the Enterprise Zone Programs of California and Florida." *Economic Development Quarterly* 23(1): 44–59.

Feldman, Maryann P. 2014. "The Character of Innovative Places: Entrepreneurial Strategy, Economic Development, and Prosperity." *Small Business Economics* 43(1): 9–20.

Feldman, Maryann P., and David B. Audretsch. 1999. "Innovation in Cities: Science-Based Diversity, Specialization and Localized Competition." *European Economic Review* 43(2): 409–429.

Feldman, Maryann P., and Pierre Desrochers. 2003. "Research Universities and Local Economic Development: Lessons from the History of the Johns Hopkins University." *Industry and Innovation* 10(1): 5–24.

Feldman, Maryann P., and Alexandra Graddy-Reed. 2014. "Local Champions: Entrepreneurs' Transition to Philanthropy and the Vibrancy of Place." In *Handbook of Research on Entrepreneurs' Engagement in Philanthropy: Perspectives*, ed. Marilyn L. Taylor, Robert J. Strom, and David O. Renz, 43–72. Cheltenham: Edward Elgar.

Feldman, Maryann P., and Nichola Lowe. 2014. "Circling the Triangle: Understanding Dynamic Regional Economies." NSF Research Project.

Feldman, Maryann P., Johanna Francis, and Janet Bercovitz. 2005. "Creating a Cluster While Building a Firm: Entrepreneurs and the Formation of Industrial Clusters." *Regional Studies* 39(1): 129–141.

Feldman, Maryann P., Tom Kemeny, and Lauren Lanahan. 2014a. "Evaluating Trade Adjustment Assistance for Firms: Application of a Quasi-Experimental Design."

Feldman, Maryann P., Lauren Lanahan, and Iryna Lendel. 2013. "Experiments in the Laboratories of Democracy: State Scientific Capacity Building." *Economic Development Quarterly* 28(2): 107–131.

Feldman, Maryann P., Alfonso Gambardella, Dietmar Harhoff, and Myriam Mariani. 2014. "On the Benefits of Geographical Proximity for Knowledge Spillovers." Working Paper.

Feldman, Maryann P., Theodora Hadjimichael, Tom Kemeny, and Lauren Lanahan. Forthcoming. "The Logic of Economic Development: A Definition and Model for Investment." *Environment and Planning C: Government and Policy.*

GAO. 2012. *Trade Adjustment Assistance: Commerce Program Has Helped Manufacturing and Services Firms, But Measures, Data, and Funding Formula Could Improve.* Washington, DC: Government Accountability Office.

Glaeser, Edward L., Hedi D. Kallal, Jose A. Scheinkman, and Andrei Shleifer. 1992. "Growth in Cities." *Journal of Political Economy* 100(6): 1126–1152.

Gobillon, Laurent, Thierry Magnac, and Harris Selod. 2012. "Do Unemployed Workers Benefit from Enterprise Zones? The French Experience." *Journal of Public Economics* 96(9): 881–892.

Hwang, Victor W., and Greg Horowitt. 2012. *The Rainforest: The Secret to Building the Next Silicon Valley.* Los Altos, CA: Regenwald.

Jacobs, Jane. 1969. *The Economy of Cities.* New York: Random House.

James, Valentine Udoh. 1998. *Capacity Building in Developing Countries: Human and Environmental Dimensions.* Westport, CT: Praeger.

Kline, Patrick, and Enrico Moretti. 2013. "Place Based Policies with Unemployment." *The American Economic Review* 103(3): 238–243.
Koo, Jun, and Jongmin Choi. 2013. "The Rise of the Biomedical Cluster in Wonju, Korea." In *Cluster and Economic Growth in Asia*, ed. Sören Eriksson. Cheltenham: Edward Elgar.
Kostka, Genia, and Arthur P.J. Mol. 2013. "Implementation and Participation in China's Local Environmental Politics: Challenges and Innovations." *Journal of Environmental Policy & Planning* 15(1): 3–16.
Krugman, Paul R. 1991. *Geography and Trade*. Cambridge, MA: MIT Press.
Lerner, Josh. 2009. *Boulevard of Broken Dreams: Why Public Efforts to Boost Entrepreneurship and Venture Capital Have Failed – and What to Do About It*. Princeton, NJ: Princeton University Press.
Leslie, Stuart W., and Robert H. Kargon. 1996. "Selling Silicon Valley: Frederick Terman's Model for Regional Advantage." *Business History Review* 70(4): 435–472.
Link, Albert N. 1995. *A Generosity of Spirit: The Early History of the Research Triangle Park*. Research Triangle Park, NC: Research Triangle Foundation of North Carolina.
Mansfield, Edwin. 1991. "Academic Research and Industrial Innovation." *Research Policy* 20(1): 1–12.
Marshall, Alfred. 1890. *Principles of Economics*, vol. 1. London and New York: Macmillan.
Merrill, Stephen, and Steve Olson. 2011. *Measuring the Impact of Federal Investments in Research*. Washington, DC: National Academy Press.
Moyo, Dambisa. 2009. *Dead Aid: Why Aid Is Not Working and How There Is a Better Way for Africa*. London and New York: Allen Lane.
National Academy of Sciences. 2001. *Implementing the Government Performance and Results Act for Research: A Status Report*. Washington, DC: National Academies Press.
North, Douglass C., and Robert Paul Thomas. 1973. *The Rise of the Western World: A New Economic History*. Cambridge: Cambridge University Press.
Pisano, Gary P. 1996. "Learning-Before-Doing in the Development of New Process Technology." *Research Policy* 25(7): 1097–1119.
Pisano, Gary P. 2012. *Creating an R&D Strategy*. Boston, MA: Harvard Business School Press.
Pisano, Gary P., and Willy C. Shih. 2009. "Restoring American Competitiveness." *Harvard Business Review* 87(7–8): 114–125.
Porter, Michael E. 1990. *The Competitive Advantage of Nations*. New York: The Free Press.
Ricardo, David. 1891. *Principles of Political Economy and Taxation*. London: G. Bell and Sons.
Rodrik, Dani. 2003. "Introduction: What Do We Learn from Country Narratives." In *In Search of Prosperity*, ed. Dani Rodrik, 1–20. Princeton, NJ: Princeton University Press.
Rodrik, Dani, Arvind Subramanian, and Francesco Trebbi. 2004. "Institutions Rule: The Primacy of Institutions Over Geography and Integration in Economic Development." *Journal of Economic Growth* 9(2): 131–165.
Romanelli, Elaine, and Maryann P. Feldman. 2006. "Anatomy of Cluster Development: Emergence and Convergence in the US Human Biotherapeutics, 1976–2003." In *Cluster Genesis: Technology-Based Industrial Development*, ed. P. Braunerhjelm and M. Feldman, 87–112. Oxford: Oxford University Press.
Romer, Paul M. 1986. "Increasing Returns and Long-Run Growth." *Journal of Political Economy* 94(5): 1002–1037.
Rosenthal, Stuart S., and William C. Strange. 2003. "Geography, Industrial Organization, and Agglomeration." *Review of Economics and Statistics* 85(2): 377–393.
Schrank, Andrew, and Josh Whitford. 2009. "Industrial Policy in the United States: A Neo-Polanyian Interpretation." *Politics & Society* 37(4): 521–553.
Schumpeter, Joseph Alois. 1934. *The Theory of Economic Development: An Inquiry into Profits, Capital, Credit, Interest, and the Business Cycle*. Cambridge, MA: Harvard University Press.
Sen, Amartya. 1999. *Commodities and Capabilities*. Oxford: Oxford Unversity Press.

Solow, Robert M. 1956. "A Contribution to the Theory of Economic Growth." *The Quarterly Journal of Economics* 70(1): 65–94.
US Census Office. 1902. *Twelfth Census, Manufactures Part 1*. Washington, DC: US Government Printing Office.
Vandemoortele, Milo. 2009. "Growth Without Development: Looking Beyond Inequality." Overseas Development Institute (London) Briefing Paper 47.
Wheaton, William C., and Mark J. Lewis. 2002. "Urban Wages And Labor Market Agglomeration." *Journal of Urban Economics* 51(3): 542–562.
Wolfson, Michael C. 1997. "Divergent Inequalities: Theory and Empirical Results." *Review of Income and Wealth* 43(4): 401–421.
World Bank. 2009. *World Development Report 2009: Reshaping Economic Geography*. Washington, DC: World Bank.
Yu, Junbo, and Randall Jackson. 2011. "Regional Innovation Clusters: A Critical Review." *Growth and Change* 42(2): 111–124.

Part III Spaces and Flows of Knowledge

Chapter 14

Multinational Enterprises Innovation Networks and the Role of Cities

Simona Iammarino and Philip McCann

Introduction

One of the crucial aspects of the current phase of economic globalization lies in the new modes of creating and diffusing new knowledge and technology. The central role played by contemporary multinational enterprises (MNEs) in such processes has been analyzed in a variety of scholarly perspectives, and emphasis has been placed on the metamorphosis of the MNE from mere "vehicle" of technical knowledge to "creator" of new technology (e.g., Cantwell 1994; Archibugi and Iammarino 2002; Ietto-Gillies, this volume, Chapter 6). MNEs are today the largest source of technology generation, transfer, and diffusion in the world. In comparison with all other firm types, the share of new technologies produced globally by MNEs is increasing; they are associated with a higher probability of entry into new and cutting-edge technological fields; they account for the great bulk of expenditure on knowledge-creating and skill-enhancing activities, and of trade in technology and technology-intensive products (UNCTAD 2009, 2013). MNE access to a broad variety of sources of new knowledge, both intra- and inter-firm, provides immense opportunities to acquire new competitive advantages for both the firm itself and all the actors involved in its networks.

On the other hand, such MNE networks of internal and external sources of knowledge and information have obvious geographical manifestations. Recent years have seen economic debates directing increasing attention to the phenomena of spatial concentration, and one result has been an increasing body of evidence on the spatially bounded character of knowledge creation and diffusion processes. Such a boundedness is a major explanation for the emergence and reinforcement of core-periphery forces and regional divergence, especially in the face of processes of

The Handbook of Global Science, Technology, and Innovation, First Edition.
Edited by Daniele Archibugi and Andrea Filippetti.

economic integration and globalization (e.g., Fagerberg, Verspagen, and Caniëls 1997; Rodríguez-Pose 1998; Verspagen and Caniëls 2001; Rodríguez-Pose and Crescenzi 2008; Crescenzi and Rodríguez-Pose 2009; Paci and Usai 2009; Rodríguez-Pose and Fitjar 2013). Indeed, the spatial proximity of firms and people within individual regions has become progressively more important in scholarly thinking as a source of knowledge spillovers and agglomeration economies in general (see Lorenzen and Mudambi, this volume, Chapter 10). Dominant urban agglomerations appear to become increasingly specialized within the global economic system (e.g., Gordon 2002; Storper 2013), so that different localities are likely to be progressively more differentiated in terms of their competitive advantages.

This chapter argues that, in the current phase of economic globalization, geography and space are increasingly important for MNEs, and in turn MNEs are progressively more important for geography: the pivot on which this relationship turns is the creation, diffusion, and management of new knowledge and technology. The emergence of a new set of relationships between MNE organization and control of intra-firm and inter-firm innovation networks, and the role of city-regions as knowledge sources, have led to increasingly differentiated geographies across all parts of the world. The chapter is divided into five sections below. The first section revises some seminal contributions that have, *ante litteram*, addressed issues of MNEs, innovation and technology, and geographical hierarchies of places in the world economy. The second section highlights the organizational changes that have occurred in the contemporary multinational corporation and its internal and external networks, while the third section connects such changes with different spatial configurations through the lenses of knowledge and technological properties. The fourth section critically considers the role of cities, in particular the so-called global cities, in MNE innovation networks. The final section summarizes the main argument and offers some concluding remarks.

Classical MNE Theories: Hymer and Vernon on Organizational and Urban Structures

The major research questions in the study of MNEs have typically focused on determinants – *why*, *where*, and *how* firms become multinational – and effects – changes both in the sending and receiving economies involved in MNE operations, and in the MNE firm itself. Although one of the core questions regarding MNEs is precisely that of *where* their different operations take place, the explicit spatial dimension of MNEs has so far been largely neglected in the literature (Iammarino and McCann 2013). The increasing significance of specific geographies at various scales of analysis, from world regions to nations and, even more importantly, subnational regions, cities, or other typologies of spatially localized systems, has also rendered problematic the traditional scholarly separation between causes and impacts of multinational activities, as the valuation of MNE effects is inherently connected to the determinants of their existence (Ietto-Gillies 2005, 2012). Rather surprisingly, in spite of the ground-breaking insights provided long ago by scholars such as Stephen Hymer and Raymond Vernon, economics, international business, and economic geography theories have not explicitly and systematically identified the distinctive geography of MNEs activity, whether with respect to the process of becoming multinational, or with reference to the outcome of their investments (McCann and Mudambi 2004). We will therefore recall here the major intuitions of these two authors – nowadays perceived as *classical* in this field of study – that should have paved the way for a more careful observation of multinational corporations and space, which emerged only much more recently in scholarly thinking.

Power, Control and Geography

In Hymer's first argument (published in 1976, though presented in 1960) the focus of the analysis is on the investor's control afforded by a common ownership and organizational system, which ensures the exploitation of a unique set of firm-specific advantages. Although the initial analytical approach makes no serious attempt to explain the geographical spread of production, a relevant implication for location is found in Hymer's pioneering recognition that foreign direct investment tends to be concentrated in certain industries across countries, rather than in specific countries across industries. Later on, drawing both on the historical account given by Alfred Chandler (1959, 1962) and on the "creative destruction" arguments of Josef Schumpeter (1934), Hymer formulates his "law of increasing firm size" (1970), identifying the evolutionary path of the modern corporation: from the competitive Marshallian firm, with a single function, industry, and ownership, to the large national oligopolistic corporation, vertically integrated with dispersed ownership, to the modern conglomerate enterprise, multidivisional with cross-country affiliates all coordinated by a centralized "corporate brain" (Hymer 1970: 442). Rehearsing Chandler and his co-author Fritz Redlich (1962), Hymer explains that during the evolutionary process of the firm toward the worldwide enterprise – the modern MNE – a pyramidal structure of three levels of business administration and decision-making could be detected. Level III, the lowest and widest, dealing with ordinary daily operations which are connected to production activities; the intermediate Level II, generated by the separation of the head office functions from the field offices of the firm, and controlling Level III as well as coordinating the management at that lower level; Level I, the top level of management, responsible for setting the goals and growth strategies – therefore research and innovation investments – of the whole firm. Thus, while in the case of the (uni)national corporation Level II and I are split off from Level III, in the modern MNE Level I and II become completely separated, with the top Level I acting as the "corporate brain" responsible for the firm's overall strategy.

This hierarchy of corporate decision-making is further discussed in subsequent work, in which the "law of increasing firm size" is merged with the "law of uneven development" (see also Pitelis 1991, 2005). In particular, Hymer (1972) suggests a "correspondence principle," that is, the existence of a direct relationship between the centralization of power and control within the MNE and the centralization of power and control in the world economy. The "correspondence principle" thereby highlights the correlation between corporate growth and evolution, on the one hand, and uneven development across different spatial locations, on the other. It is here, in the "correspondence principle," that the critical connection between the MNE's ownership advantages, strategic behavior, and specific geographies lies.

Hymer suggests that, by applying the location theory to the Chandler-Redlich model, the "spatial dimension of the corporate hierarchy" could be effectively analyzed (1972: 122). The pyramidal structure of corporate control centralization translates directly into a *hierarchical structure of geographical locations*. Some of these locations are heavily dependent on others, underlying the uneven spatial structure of economic development. Therefore, Level III activities will tend to be relatively evenly spread and distributed worldwide, according to the attractiveness of resources such as labor, markets, and raw materials; it is this phenomenon which diffuses industrialization across less advanced economies. In contrast, Level II activities will tend to be concentrated in large city-regions, as they require primarily white-collar labor, as well as sufficient communication and information systems. At this level, MNEs from different industries will locate in the same cities, thereby giving rise to a strong spatial and functional concentration at the subnational level. Level I activities will be even more geographically clustered, as they need to be close to sufficient supplies of highly specialized human capital and services, capital markets, media, and governments: the provision and exploitation of these high-profile service relationships require strong connectivity, both locally and with the rest of the world.

Hence, the highest-level functions of the leading MNEs will almost all be located in the world's major *global cities*, which themselves will be "surrounded by regional subcapitals" (Hymer 1970: 446). Hymer argues that, eventually, the major global cities which will be the home to the core headquarter functions and high-level strategic planning of the world's MNEs will be New York, London, Paris, Bonn, Tokyo, Moscow, and Beijing (1972: 124). From these highest levels, Hymer envisages that a geographical hierarchy will characterize the spread of MNE operations, with "intermediate" and "lower-level" activities distributed across lower-tier urban centers and regions. In his words: "geographical specialization will come to reflect the hierarchy of corporate decision making, and the occupational distribution of labor in a city or region will depend upon its function in the international economic system" (1972: 124).

Much of the later international business literature has explicitly and implicitly borrowed from Hymer's analysis (see Dunning and Pitelis 2008), and his arguments have also been somewhat influential in some geography and location studies (e.g., Dicken 1977; Norton 1979, 1987). Curiously, however, in spite of a few studies (e.g., Young, Hood, and Peters 1994; Bailey and Driffield 2002; Pitelis 2002a, 2002b, 2005; Kottaridi 2005) explicitly acknowledging Hymer's key importance in explaining the relationship between MNE activity and subnational regional development, the international business literature itself has largely overlooked the *ante litteram* relevance attributed by Hymer to the interplay between spatial (particularly urban), organizational, and industrial structures, and information and communication systems, all of which underlie the locational choice and the economic geography of MNEs.

Industry Cycles and Spatial Shifts

To some extent as in Hymer's thought, in Vernon's product life-cycle (PLC) analysis the geographical location of international production follows a hierarchical pattern which is strictly linked to the ownership advantages that characterize each stage of the oligopolistic structure (Ietto-Gillies 2012). In the innovation-based oligopoly, the location of production is naturally placed in the economy where the innovation process firstly initiated. In this early stage, when technology is unstable, the demand for the new product is uncertain, technological capacity is vital, and location occurs in the metropolitan centers of advanced countries. This observation stems from the earlier work of Vernon on agglomeration economies in metropolitan regions such as New York (Vernon 1957, 1960). Critically, and in parallel with Hymer, in this work he stresses that, among the main causes underlying the growth of the core urban areas, a crucial role is played by the availability of a variety of business services, which previously were provided internally by the firm's business functions (Vernon 1957: 17). In contrast, in mature oligopolies advantages based on economies of scale, transport, and marketing lead to a strong concentration of investment either in the home country of the MNE or in similarly advanced economies, spreading the location of the firm's investments out of the initial agglomerations (Vernon 1966, 1974, 1979). The mature or standardized stage of the product cycle – characterized by stable technology, long production runs, strong price competition, and high intensity of unskilled labor – offers developing economies the opportunity to successfully enter the international competitive environment by attracting MNEs from advanced locations into their emerging local industries (see also Hirsch 1967).

The PLC model has also seen wide applications in regional and urban analysis, basically predicting that the concentration of early stages of product development will be in urban areas, where high-skilled labor, external economies, and top management are easily available, followed by subsequent shifts of standardized production to more rural and lower-cost regions (e.g., Norton and Rees 1979). However, in this literature the structural and technological aspects of Vernon's arguments as they relate to MNEs are largely ignored. On the other hand, the same simplistic skilled–unskilled dichotomy of the international product cycle model is frequently applied also at the subnational scale, thereby attracting criticism that these approaches have mainly failed to explain the processes of historical change in industry spatial and organizational structures (e.g., Storper 1985; Taylor 1986, 1987; Mack and Schaeffer 1993).

Among the criticisms of the PLC type of MNE behavior, two in particular seem to be most relevant here. First, the models tend to display a simplified and deterministic view of the innovation process, which is assumed to be a linear and uniform sequence from invention to a marketable product to standardization, and where skills and innovative capabilities are seen as effortlessly transferable across geographical and institutional contexts (Cantwell 1995). Second, the locational issues arising in each of the stages of the PLC model following the introduction of the new product have been disregarded. In many PLC models the spatial determinants of foreign and local firms' behavior are reduced mainly to labor cost differences, thereby largely ignoring the evolutionary transformation of both the firm's internal relationship between power, ownership, and control, and the external connections with the economic environment. This "ambiguous enterprise context" (Taylor 1987: 82) does not allow us to distinguish between global and local competition; nor does it shed light on the intra-firm versus inter-firm relationships which are crucial to identify the specific spatial dimension of both industry and technological cycles.

Yet, although the PLC model has been given a space-specific characterization by authors who have followed Vernon's own extensive work on location and urban and regional economics (e.g., Vernon 1957, 1959, 1960, 1966, 1991), such aspects have been surprisingly overlooked in the international trade and business literature, as well as in the economic geography of the MNEs. In fact, in spite of Vernon's path-breaking analysis of the causes of urban concentration and geographical spread of multiplant activities which takes place across boundaries of product lines, the subsequent Vernon-inspired literature still tended to treat MNE location behavior using the highly unspecific geographical dichotomy between skilled capital-abundant advanced (national) economies and unskilled labor-abundant developing countries. Vernon's emphasis on the spatial shifts involved in industry development cycles, and the changing relevance of external economies during such cycles, should have called for a more explicit incorporation of locational issues in the explanation of MNEs' operations (Iammarino and McCann 2013).

The New Context: A Fast-Changing World and the Evolving MNE

In spite of the criticisms attracted over time, the work of both Hymer and Vernon has been considered seminal in the analysis of MNEs for two reasons. First, they both introduce market imperfections and dynamic elements into the analysis, such as changes in demand structures, innovation and technological advantages, and information lags. Second, they both recognize the importance of ownership comparative advantages – particularly those based on superior knowledge – in explaining multinational behavior. As discussed above, both Hymer's and Vernon's pioneer works have strong micro-foundations, and as such provide crucial connections between the structure and behavior of MNEs and the subnational spatial organization of their economic and innovative activities.

In sharp contrast with the visions offered by Hymer's and Vernon's theories, since the 1990s and the acceleration of economic globalization the "flat world" hypothesis – the well-known claim that the world has become in some economic and social sense mostly *flat* due to the disappearance of barriers and transaction costs across space – has gained support in both scholarly and political circles (e.g., O'Brien 1992; Ohmae 1995; Cairncross 1997; Friedman 2005). Such a claim predicates that advances in technology and business organization have enabled actors anywhere in the world, be they consumers, investors, entrepreneurs, or firms, to link up and do business with others anywhere with an ease and accessibility that obviates the advantages that once accrued to those who were attached to favored localities, regions, or nations (Guy 2009; Iammarino and McCann 2013). The "flat world" thesis implies that economic activities can be done just as well in India as America, by small firms as by large firms, because we live in the age of the network organization, and these networks are global.

The hypothesis of flatness – harshly criticized in the last decade or so, but far more in academic circles than in policy ones – rests on two big assertions, one of which is geographical and the other one organizational. Many have claimed that distance is becoming unimportant, and many others have argued that big MNEs are dinosaurs while networked small and medium-sized enterprises (SMEs) are the mammals and flowering plants that will take their place (Iammarino and McCann 2013). However, there is ample evidence that the removal of barriers to international trade, investment, and communication can actually make geographical proximity more, rather than less, important. Many economic activities benefit – still, if not more – from co-location, whether for reasons of simple logistics, shared inputs, or benefits of face-to-face contact (Arita and McCann 2000). When barriers to cross-border or long-distance trade are reduced, some agglomerations – local, regional, or even national – may find that their comparative advantages are strengthened, producing fast economic growth. The integrated world becomes a more differentiated one, and proximity to places and cities such as Silicon Valley, the City of London, Hong Kong, or Guangzhou becomes more important than ever. Rather than becoming flatter, in many aspects globalization is nowadays shaping the world more unevenly, more curved (McCann 2008), or more spiky (e.g., Florida 2005; Rodríguez-Pose and Crescenzi 2008; Rodríguez-Pose and Fitjar 2013). In formal economics these changes are represented by the increasing returns models of the new trade theory and the new economic geography (see, for all, Krugman 1991; Fujita, Krugman and Venables 1999).

Turning to the organizational claim, it is true that networks have grown, as the reduction of transaction costs has made it possible for firms to outsource many functions. Yet, innovation and flexibility can thrive under administrative guidance just as mass production did: rumors of the death of the large corporation have been greatly exaggerated (Piore 1994). The picture of flat global networks is problematic for the simple fact that flatness and global reach are often in conflict. The global reach of production and innovation networks is to a substantial extent an administrative accomplishment: distance has been shrunk not only by improved technology, but also by corporate organization. Managerial hierarchies – or, according to the recent trends in corporate structure, *heterarchies* – and standardized procedures are, among their other functions, distance-spanning devices. Conversely, the most robust and innovative *flat* networks involve disproportionately local actors. One role played by large MNEs is to link such local clusters and networks to the larger world by acting as "gatekeepers" – extending them beyond national boundaries (e.g., Amin and Thrift 1992; Enright 2000; Giuliani 2011). This is an increasingly important role because the reductions of transactions, communication, and transportation costs, where they are evident, are not accessible everywhere and to any firm, while they are relatively readily available to large MNEs.

On the other hand, the stable relationship between ownership and control, which has long been understood as being problematic when looked at in terms of the divide between investor and executives, has been disturbed both along the supply chain and within the corporation (Ietto-Gillies 2005). In outsourcing and offshoring strategies, ownership changes but control of the value chain activities is largely retained through various means of pressure on the suppliers and their competitive bidding (e.g., narrow transfers of technology, strict product specifications, tight supplying schedules, etc.) (UNCTAD 2013). Conversely, in vertical integration strategies ownership is not altered, but the distribution of control within the MNE can vary greatly, with different degrees of autonomy of the affiliates and subsidiaries that can lead to intra-firm competition and even to various degrees of restraint in the powers of the central headquarters of the MNE (e.g., Birkinshaw and Hood 1998, 2000; Birkinshaw, Hood, and Young 2005).

Such organizational changes have had huge implications for the location choices and innovation advantages of MNEs, which are increasingly dependent on the balance between technological competencies and capabilities *within* and *outside* the firm, or on the integration of various sources of knowledge which are internal and external to the firm. Also, different geographies have emerged in relation to different types of multinational hierarchical versus heterarchical structural forms and internal organization, such as integration versus externalization through outsourcing and off-shoring strategies, the unbundling of headquarters and core functions such as R&D, or centralized/decentralized control of strategic assets within the firm (Desai 2009).

The identification of the MNE's spatial location, and its characteristics with respect to geography, industry structure, knowledge, and technology flows, is a result of complex interactions between firm(s), industry, organizational, and knowledge characteristics. The simple host–home country dichotomy has long become insufficient. This is particularly so not only with respect to the economically advanced parts of the world, but also increasingly in all the main cities and urban regions, where "host" and "home" may actually overlap to a great extent: the locations which are host to a major presence of MNEs are also those which are most connected by outward linkages. On the other hand, the location of corporate headquarters of large MNEs nowadays has little geographical connection with the location base for specific business units and operations: a new set of relationships between the location of knowledge assets, the role of city-regions as knowledge sources, and the spatial behavior of MNEs has emerged.

Innovation, Spillovers and Spatial Typologies

Both Hymer and Vernon depart from the prevailing neoclassical tradition in that they treat innovation as an endogenous phenomenon conditioned on structural issues – MNE innovation-based ownership advantages – rather than as an exogenous phenomenon dependent on the efficiency of market clearing. Hymer points out that multinational behavior implies the existence of international flows of groups of activities and resources, including technology, capabilities, skills, and entrepreneurship. An important qualification he made, also preempting Krugman (1991), is that while growth in the hinterland or periphery means growth in the centre, the reverse does not necessarily hold.

On the other hand, the two major criticisms of the PLC-based explanations of international production mentioned above have important spatial implications for our argument here. First, on the basis of the evolution of the contemporary world economy, ownership advantages are to be attributed to firms, rather than countries, thus making the *geographical origin* of MNEs much less predetermined by the national level. This point, which was seemingly captured in Vernon's original framework, has been highlighted by both regional economics (e.g., Taylor 1986, 1987) and international business (e.g., Cantwell 1995) scholars who attribute the limitations of the PLC model to an inadequate conceptualization of both the firm and technological progress. Second, observation suggests that agglomeration forces have attracted MNE activities – even high value-added ones – to particular locations in both advanced and emerging economies, thus making the *geographical destination* of MNEs far less dependent on purely cost-based and intrinsically static hierarchical patterns. Indeed, this latter observation seems to point to the opposite side of the globalization argument: skills formation and knowledge accumulation processes are ever more dependent on sources that are external to any single firm (however large and multinational it may be) and are highly spatially situated (Storper 1997, 2013).[1]

In the global economic race of the twenty-first century MNEs compete primarily on the basis of their internalized knowledge and technological assets as well as of how these are employed to engage with the intangible assets of the geographical localities in which they operate. In line with Hymer's and Vernon's predictions, one of the dominant features of the current wave of globalization is the emergence of a new set of relationships between MNE knowledge assets and the role of particular city-regions as knowledge sources. Although the emergence of these relationships has been taking place gradually over the last decades, in the most recent years their nature and ordering have become more apparent, as their manifested trends appear to be accelerating.

In terms of the geography of MNEs, the reality is presumably not far from that envisaged by Hymer, in his emphasis on a locational hierarchy based on ownership and control, and by Vernon (1991) in his later vision of multi-product and multi-technology industry dynamics; and possibly even more diverse and in rapid evolution than what is suggested by the "world cities" (Friedmann 1986) or "global cities" (Sassen 2001, 2002) hypotheses in urban studies,[2] applied also in the international business literature (e.g., Dunning and Norman 1983; Nachum and Wymbs 2005; Goerzen, Asmussen, and Nielsen 2013). Importantly, both these strands of urban research stress that it is not simply a matter of locating MNE facilities in the largest cities or urban regions: other factors are just as important for economic growth as is urban scale, and different spatial configurations, regions, cities, or localized production systems, which are open repositories of knowledge creation and exchanges, will offer advantages for different firms, whether MNEs or not.

By reinterpreting a classification of spatial typologies based on transaction costs (Gordon and McCann 2000; McCann, Arita, and Gordon 2002; McCann and Sheppard 2003) in terms of technology, knowledge, and innovation features (Table 14.1), we have elsewhere shown (Iammarino and McCann 2006, 2010, 2013) that it is possible to uncover more subtle and detailed aspects of firm location behavior. In particular, in the case of MNEs, any simply assumed correlations between localized knowledge spillovers, spatial concentration of activity, and the advantages of a particular location for firm investment, do not necessarily hold. For example, in the case of the first spatial configuration of our technology-based classification reported in Table 14.1 – *pure agglomeration*, which corresponds by and large to the spatial model of global city – the knowledge base is primarily explicit and codified, available to any local actor and organization, and generated outside firms' boundaries, being largely created in other private and public organizations. Variety and promiscuity are distinctive features of cities, particularly global ones, in the sense of there being an absence of long-term loyalty between agents. The combination of different streams of knowledge therefore occurs across a broad range of sectors (Jacobs-type externalities), and individual and organizational innovation linkages or relations are unpredictable due to the low degree of knowledge cumulativeness. In other words, while the overall links between innovation and cities are undoubtedly very strong (e.g., Acs 2002), the arguments relating MNEs to knowledge spillovers in such contexts are rather more complex: the knowledge objectives and the innovation performance of MNEs are not necessarily centered on cities (Simmie and Sennet 1999).

Table 14.1 Technological classification of spatial types.

			Social network	
	Pure agglomeration	*Industrial complex*	*Competence-based SN*	*Trust-led SN*
Nature of technical knowledge	Codified, explicit and mobile, transmitted by way of information	Specific, systemic, routinized, R&D-intensive; based on non-transferable experience	Tacit, generic, non-systemic, sticky and leaky; transmitted within cognitive networks	Codified, sticky, mature; transmitted within geographically localized networks
Technological trajectory	Oriented to processes, problem-solving, organizational innovation	Oriented to complex products, cost-cutting	Oriented to radically new products	Oriented to processes, customer-driven
Dynamics	Stochastic	Strategic	Disruptive, uncertain	Incremental
Sources of innovation	External to the firm, multi-sectoral	Internal to the firm and the sector, institutionalized R&D in incumbents, relevance of basic science	External to the firm and the sector, relevance of applied science and cooperative R&D	External to the firm, internal to the sector, relevance of machinery acquisition
Appropriability of innovation returns (and market structure)	Low (perfect or monopolistic competition); based on standards, copyright, design	High (oligopolistic competition); based on patents, secrecy	Low (low entry barriers); based on patents, copyright	Low (collaborative monopolistic competition); based on trademarks
Technological opportunities	Variable	Medium/low	Very high, uncertain	Low
Cumulativeness	Low	High	Low	High
Knowledge base	Broad, diversified	Narrow, targeted, specialized	Broad, research-based	Narrow, specialized along the filière

(*Continued*)

Table 14.1 *(Continued)*

			Social network	
	Pure agglomeration	*Industrial complex*	*Competence-based SN*	*Trust-led SN*
Modes of governance	Market	Corporate hierarchies	Relational and cognitive networks	Social and historical networks
Examples of industrial specialization	Finance, banking, insurance, business services, retailing	Steel, chemicals, automotive, pharmaceuticals, machine tools, medical instruments, ICT hardware	High-tech, general purpose technologies	Customized traditional goods, textiles, footwear, furniture, tourism
Example of cluster	Global cities; Silicon Valley (California) since the 1990s	Silicon Glen (Scottish Electronics Industry)	Silicon Fen (Cambridge, UK)	Italian industrial districts in *made in Italy* industries
Pavitt classification	Information Intensive, Supplier Dominated Firms	Production Intensive Firms (Scale Intensive and Specialized Suppliers)	Science-Based Firms	Supplier Dominated Firms

In contrast, the second spatial type – the *industrial complex* – is often characterized by regions or city regions with a highly specialized industrial structure. For modes of innovation which are primarily based on major R&D and capital investment and long development lead times, the industrial complex may provide the most suitable environment for technology creation and experimentation. Where evident, the composite and advanced local knowledge-production basis is also likely to display a strong concentration at the spatial level. The industrial complex structure is also often a suitable structure for the development of multilateral networks of dissimilar but complementary relations between MNEs and local actors (Patrucco 2003). Yet, in the industries in which the host region is technologically strong, the major presence of indigenous firms tends to deter MNEs in the same industry from carrying out considerable innovation activities in the primary technological fields related to the local industry strengths (Cooke 2001). As such, when MNEs choose to invest in these types of regions it generally implies that the focus of the local relations is mostly related to supply chain issues, rather than technology or knowledge sourcing. However, MNEs operating in other industries will still be attracted there for innovation and knowledge purposes in those technologies whose lines of development represent diversification from the primary technologies of their own industry (Cantwell and Kosmopoulou 2002).

There is also a fundamental difference in the particular modes of innovation governance between the two social network spatial types reported in Table 14.1 (Simmie et al. 2004; von Tunzelmann 2003, 2009). In the *competence-based social network* model, high technological opportunities come primarily from sources outside the firm and the industry, such as university academic research. In this kind of technological environment the type of knowledge tends to be both generic and non-systemic, with high rates of market entry and exit, a strong degree of volatility of market shares, and low levels of market concentration. In such an environment, the tacit and sticky nature of knowledge requires geographical proximity, particularly in the early stages of the industry or cluster life-cycle. The openness of the local system, the relatively "leaky" character of new knowledge and consequently the high potential for spillovers, and the related emergence of new rules, standards, blueprints, and verification procedures, all point to the importance of external sources of technical knowledge. Innovations therefore mainly come from knowledge that does not have a routinized nature, with new firm start-ups playing an important role, and SMEs accounting for a large share of innovative activity. There is normally either no role or at best only a limited presence of large MNEs *in situ* in the first stages of regional development. In later stages, however, where both industry and spatial patterns have reached a certain stability, such locations may become highly attractive to MNEs. In the *trust-led social network*, instead, knowledge is largely codified and mature and develops along stable trajectories which are mainly oriented toward process innovation. Knowledge is transmitted essentially by way of informal personal contacts which are strongly embedded in a common cultural matrix, comprised of elements of social and political lobbying, and intense backward and forward linkages, sometimes intensified further by the presence of MNEs. As such, the structure of vertical and horizontal linkages between MNEs and local firms may be crucial in determining the nature and extent of local spillovers. Even in the same industry, regions may be characterized either by highly vertically integrated firms, or by stage production with significant subcontracting linkages. The trust relationships and the role of reputation will differ considerably in such different spatial and industrial settings (Guy 2009).

Some empirical analyses have shown that inter-industry knowledge spillovers are likely to become more intense in centers of technological excellence where spillovers seem to operate mainly through exchanges in and around core technological systems (Crozet, Mayer, and Mucchielli 2004). This is the case where spillovers are primarily rooted in "general purpose technologies" such as, for instance, background engineering, mechanical methods, new materials, electronics, and ICT, and the additional spillovers which these create between actors in quite separate alternative fields of specialization. These types of centers of excellence exhibit inter-industry technology spillovers, and are most likely to be classified either as pure agglomerations or competence-based social networks. However, as we have seen, these two different spatial types tend to offer different possibilities with regard to the role of MNEs, and the contribution that they make to local spillovers. Yet, it is these two typologies which tend to experience a faster process of convergence between old and new technologies, and potentially therefore also a greater degree of competitiveness.[3]

In addition to the characteristics of knowledge based on industry and technological regimes – on which our simple taxonomical exercise is based – there is new substantial evidence of increasing spatial agglomeration at the functional level, particularly visible in service industries which are far more affected by functional than by sectoral features (e.g., Defever 2006; Crescenzi, Pietrobelli, and Rabellotti 2013; Goerzen et al. 2013; Ascani, Crescenzi, and Iammarino 2014). This may provide an additional angle to look in dynamic terms at the taxonomy described above.

MNEs and Cities in the World

During the twentieth century there were qualitative changes in the role of cities in the industrialized world which favored the competitive advantages associated with cities being centers of knowledge; these changes are now occurring increasingly in developing parts of the world (see Florida and Mellander, this volume, Chapter 15). The clues as to why particular cities and regions are highly productive lie in the types of innovation system that characterize them. The vast literature on innovation systems, with its strong emphasis on the links between the actors, organizations, and institutions involved in knowledge creation and diffusion – business firms, universities and research institutes, industry associations, government agencies, etc. – has discussed the main determinants of system variety at various levels of geography (see Iammarino and McCann 2013 for a review).

In order to generate the required returns to their knowledge investments, firms – and the other actors and institutions within the urban or regional system – must capture markets which extend well beyond the borders of their own country. Traditionally these returns were generated by exports, but the increasing engagement facilitated by international investment has become far more important. International flows, and particularly flows of knowledge, are increasingly understood as being bi-directional or multi-directional, with concepts such as "openness" and "connectivity" coming to replace terms such as "inward" and "outward" flows.

While clear trends toward global regionalism, rather than simply globalization, are emerging, the integration processes taking place between larger groups of both rich and also emerging economies offer greater rewards to MNEs than ever before from exploiting the possibilities for better affiliate-specific location matching, as well as coordination between dispersed subsidiaries. At the same time, the higher demands for timeliness, the greater requirements for higher frequency transactions, and the increased preferences for customization and variety, all tend to raise the distance costs associated with knowledge-related transactions (McCann 2007) and thereby the opportunity costs of sub-optimal MNE affiliate locations.

Large cities and city-regions provide growing opportunities for a wide range of sharing, sorting, and matching mechanisms and institutions, which increase the overall efficiency of the local activities. Although there is still clearly a very important role for large cities in the industrialized world in terms of driving productivity and innovativeness (Rosenthal and Strange 2004), when we consider the new context of globalization and the potential advantages of different spatial types for MNEs, the relationship between city size and productivity – the main, and often only, indicator available for measuring economic growth and attractiveness – becomes much less straightforward.

Interestingly, the majority of the world's highest productivity cities are not what the OECD (2006) classifies as "mega-cities" of over 7 million inhabitants. For OECD cities of over 1.25 million inhabitants, there is only a very weak cross-sectional link between city per capita productivity and population, which if anything is slightly negative (OECD 2006). However, the productivity advantages of very large cities appear to be relatively more important for lower-income and transition economies within the OECD than for the most advanced ones (OECD 2006). More generally, while estimates of the relationship between city size and productivity within individual countries make some sense, estimations across countries are far more complex because of the spatial, structural, and institutional heterogeneity at the national level. As well as an ∩-shaped relationship between city size and productivity, across the OECD it is generally the case that the higher-income cities are actually outgrowing lower-income cities, irrespective of population scale (OECD 2006). This suggests that other characteristics of cities are also just as important for economic performance and attractiveness as scale, and in particular the crucial role that cities or urban regions can play as centers of knowledge (e.g., Storper 2013), creativity (e.g., Florida 2005), and innovation (e.g., Acs 2002).

Interdisciplinary seminal studies on cities and economic growth (e.g., Hall 1966; Jacobs 1969; Hymer 1972; Friedmann 1986; Sassen 1991; Henderson, Kuncoro, and Turner 1995) have shown that economic growth at the international scale is being increasingly dominated by networks of particular major urban centers and regions (e.g., Borja et al. 1997; Sassen 2001; Scott 2001; Scott et al. 2001; Derudder et al. 2003; Button et al. 2006). As mentioned above, there is also abundant evidence which examines the role played by these global cities as the principal location bases for the largest MNEs (e.g., Sassen 1994, 2002; Knox and Taylor 1995; Enright 2000; Taylor 2004; Bel and Fageda 2008; Verbeke, Li, and Goerzen

2009). The analysis of global cities suggests that, in the current phase of globalization, the links among such global cores are a key determinant of the city-region's performance.

The global urban centers are locations which not only exhibit significant agglomeration advantages, but which also primarily interact with other similar globally-oriented cities in other countries, rather than with smaller urban centers and regions within their national boundaries or even within the same macro-region. In sectors such as financial services (COL 2009; MasterCard 2008) there is clear evidence that global markets are increasingly dominated by power networks of a few global urban centers such as London, Paris, Tokyo, Sydney, and New York (Sassen 2002). The increasing relative dominance of these global cities appears to be associated with strong transnational regulatory institutions and high density of information technology assets (e.g., Sassen 2002; Derudder et al. 2003; Button et al. 2006; Taylor et al. 2011); their importance as major nodes (Limtanakool, Schwanen, and Dijst 2007) within international networks is also reinforced by their role as hubs within the global air (Burghouwt 2005), rail, and marine transportation systems (Leinbach and Capineri 2007). These observations are also supported by the findings of Ni and Kresl (2010) who found that the most important element in the competitiveness rankings for global cities is connectivity, rather than urban size or structure.

There are a few points that it is worth stressing about the correspondence between global cities and MNE networks. First, as shown by the urban literature, the definition of "global cities" is not related to the city size, but rather to its degree of "connectivity," "openness," and "accessibility." As yet, and despite the emergence of global hubs in some emerging economies, most of the world's largest cities located in developing countries still do not exhibit the same information, financial, transportation, and management bi-directional flows – together with comparable local institutional settings – that the established global cities in the world exhibit.

Second, the emphasis on the role played by the presence of MNEs and their innovation networks in making cities "global" has probably been overstated, largely neglecting the crucial evolution of the relationship between corporate power and control. The centralization of political and economic power is certainly concentrated in the few cities of Hymer's Level I (1972): changes over less than a decade (2006–2014) in the location of the headquarters of Fortune Global 500 corporations show a substantial mobility in the global cities hosting more than five of such companies (McKinsey Global Institute 2013).

However, as argued above, MNE intra-firm and inter-firm networks display very different geographies according to organizational structure, degree of integration versus externalization, functional coordination, balance between control and autonomy of company's units. Such complex networks – and the variety of spatial typologies that can arise by combining one or more of those types stylized in our taxonomy above – seem to offer a rather different picture of the geography of the MNE than that offered on the basis of the location of headquarters of large corporations. Indeed, some recent evidence from the empirical literature on the economic geography of MNEs indicates that, in the cross-borders co-location of the different stages of the value chain of MNE affiliates in the context of the European Union, MNE headquarters do not display any pull effect over the location of any other

MNE function (Defever 2006; Ascani et al. 2014). Goerzen et al. (2013) have shown that *competence-exploiting* and *competence-creating* (Cantwell and Mudambi 2005) activities of MNEs follow very different spatial patterns: while the first tend to agglomerate in global cities, the latter, far more valuable for local economic development, tend instead to concentrate in metropolitan (or smaller) *core* regions.

These observations cast further doubt on the identification of the connection between global cities as the main (or sole) geographical structure of MNE innovation and production networks. It is interesting to note that, among the top 25 cities listed by the McKinsey Global Institute in 2013 as locations of the world's largest MNE headquarters (Table 14.2), five correspond to the top 25 cities of emerging economies (Table 14.3), and others in the latter list – from Mumbai to Bangkok, Johannesburg, and Buenos Aires – are very likely to enter the world top 25 in the near future. At the same time, though, the ranking of emerging markets' cities as locations of foreign subsidiaries of large MNEs shows rather different positions, with eight cities in the top 25 for MNE large subsidiaries not included in the ranking of locations for headquarters.

Table 14.2 Top 25 cities ranked by number of large companies' global HQs 2010.

Rank	*City*	*Large companies' global HQs*	*Total revenue ($bn)*	*Average revenue per company ($bn)*
1	Tokyo	613	5,231	8.5
2	New York	217	1,964	9.0
3	London	193	1,924	10.0
4	Osaka	174	1,028	5.9
5	Paris	168	2,785	16.6
6	Beijing	116	2,503	21.6
7	Moscow	115	709	6.2
8	Seoul	114	1,150	10.1
9	Rhein-Ruhr (DE)	107	1,220	11.4
10	Chicago	105	695	6.6
11	Hong Kong	96	468	4.9
12	Taipei	90	472	5.2
13	Los Angeles	82	422	5.1
14	Zurich	79	770	9.8
15	Sydney	75	466	6.2
16	Stockholm	74	360	4.9
17	Houston	74	661	8.9
18	Nagoya	70	481	6.9
19	Randstad (NL)	67	1,516	22.6
20	Singapore	64	343	5.4
21	Dallas	63	804	12.8
22	Washington DC	62	655	10.6
23	Toronto	61	436	7.2
24	Munich	61	581	9.5
25	Melbourne	58	289	5.0

Source: Adapted from McKinsey Global Institute (2013).

Table 14.3 Top 25 emerging countries' cities ranked by number of large companies' global HQs and number of large foreign subsidiaries 2010.

Rank	*City*	*Global HQs*	*Rank*	*City*	*Large foreign subsidiaries*
1	Beijing	116	1	Singapore	118
2	Moscow	115	2	São Paulo	58
3	Hong Kong	96	3	Mexico City	42
4	Taipei	90	4	Moscow	33
5	Singapore	64	5	Buenos Aires	31
6	Mumbai	57	6	Hong Kong	27
7	Shanghai	54	7	Kuala Lumpur	26
8	São Paulo	49	8	Bangkok	23
9	Bangkok	45	9	Prague	22
10	Tel Aviv-Jaffa	43	10	Warsaw	22
11	Mexico City	40	11	Istanbul	21
12	Johannesburg	37	12	Budapest	18
13	Istanbul	36	13	Santiago	17
14	Kuala Lumpur	35	14	Beijing	16
15	Buenos Aires	31	15	Taipei	14
16	Santiago	30	16	Jakarta	13
17	Shenzhen	27	17	Rio de Janeiro	13
18	Delhi	27	18	Bucharest	10
19	Jakarta	25	19	Lima	10
20	Guangzhou	22	20	Bogota	9
21	Hangzhou	22	21	Caracas	8
22	Rio de Janeiro	20	22	Shanghai	7
23	Riyadh	19	23	Manila	7
24	Warsaw	18	24	Johannesburg	6
25	Hanoi	17	25	Curitiba	6

Source: Adapted from McKinsey Global Institute (2013).

Concluding Remarks

Institutional and technological environments have changed radically since the late 1980s, and multinational corporations, the supposedly inflexible dinosaurs, have changed accordingly, to a much larger extent and rather more quickly than many other types of firm. The modes of international investment, the organization and management of intra-firm vertical and horizontal relationships for production and knowledge generation, the types of affiliation linkages, the diversification and distribution of functions, the integration of subsidiary objectives into the overall goals and strategy of the MNE, have all gone through substantial and rapid changes. These new organizational modes have occurred within MNEs, and are also seen in terms of their external relationships. Increasingly complex and systemic forms of integration of international operations have involved external firms, often SMEs, which are connected through contractual relations to the global production and innovation networks led by large corporations: many of these SMEs have actually become MNEs themselves, in advanced as in emerging economies. Also, the growing degree of complexity in MNE forms and organization has made clear that the boundaries and definitions of multi-product and multi-technology MNEs with respect to individual industrial sectors and technological fields are increasingly blurred (Iammarino and McCann 2013).

While Hymer's intuition of the top Level I acting as the "corporate brain" responsible for the firm's overall strategy (Hymer 1970: 442) has proved to be an essential feature of the contemporary multi-divisional multi-locational conglomerate enterprise, what we may observe in more recent times is actually a further splitting off of Level I functions into power and control. The "correspondence principle" suggested by Hymer (1972) has seemingly evolved in the establishment of a direct relationship between the centralization of political and financial power within the MNE and in a few truly global cities in the world economy. At the same time, in response to the organizational and technological requirements of global value chains and knowledge networks management, control has increasingly been decentralized within the corporation and across geography, involving a wider range of spatial locations. In other words, while power (political and institutional linkages, finance, lobbying, alliances, etc.) has certainly tended to agglomerate in global cities, control over MNE functions and operations – including strategic ones such as the generation of new technology and innovation – is gradually delocalized in second, third, and even lower tier cities and regions. The globalization of innovation (Archibugi and Iammarino 2002) is thus characterized by both geographical dispersion and concentration, arguably targeting a range of city-regions that extend beyond global cities.

In the light of both Hymer's and Vernon's legacy, the location behavior of MNEs cannot be understood without recourse to detailed and bundled considerations of organization, technology and innovation, institutional context, and, most importantly of all, firms' ability to access and exploit internal, local, and global knowledge sources. Indeed, Hymer's "correspondence principle" is still a crucial connection between the centralization of power and decentralization of control within the modern corporation, and the increasingly differentiated spatial hierarchy in contemporary economic systems. On the other hand, Vernon's crucial observation of the strong relationship between industry life-cycles and spatial shifts helps explain different geographical orders of multi-product and multi-technology MNEs, and the increasing specialization of places.

MNE internal (intra-firm) networks have become relatively flatter over time, with increasing decentralization of control, decision-making, and functions; at the same time, MNE external (inter-firm) networks have spurred spikier geographies and uneven regional development, depending on the variation across urban and regional innovative and institutional capabilities to cash in on the presence of global "gatekeepers" to build new localized absolute and comparative advantages. Such changes could arguably alter the connectivity of cities and regions around the world: the configuration of geographical networks calls definitely for further research efforts, both theoretical and empirical.

Notes

1 Some of these limitations are actually acknowledged by Vernon himself (1979, 1991), who recognizes that the expansion of MNEs' multi-product networks worldwide would render locational issues much more complex than what was depicted in the original formulation of his model.

2 Whilst the first approach, following more closely the MNE classical literature, uses empirically the locations of parents and subsidiaries of the world's largest MNEs to produce a matrix of "world cities," the second approach considers the locations of global, regional, and local offices of a set of producer services firms to build a matrix of "global cities." The conceptual differences to distinguish between the two variations are most clearly articulated by Derudder (2006).

3 A typical example is that of Silicon Valley, classified in Table 14.1 under pure agglomeration. Indeed, although the early post-war features of semiconductors and electronics in Silicon Valley were mainly typical of the trust-led social network model (Saxenian 1994; Hall 1998), this industry developed during the 1970s along the lines of a competence-based social network, and has now emerged into something which is akin to a pure agglomeration (Arita and McCann 2000; Saxenian 1994), exhibiting the supplier dominated characteristics of Pavitt's classification (see Iammarino and McCann 2013, for the evolutionary paths of spatial configurations).

References

Acs, Z.J. 2002. *Innovation and the Growth of Cities*. Cheltenham: Edward Elgar.

Amin, A., and N. Thrift. 1992. "Neo-Marshallian Nodes in Global Networks." *International Journal of Urban and Regional Research* 16: 571–587.

Archibugi, D., and S. Iammarino. 2002. "The Globalization of Technological Innovation: Definition and Evidence." *Review of International Political Economy* 9(1): 98–122.

Arita, T., and P. McCann. 2000. "Industrial Alliances and Firm Location Behaviour: Some Evidence from the US Semiconductor Industry." *Applied Economics* 32: 1391–1403.

Ascani A., R. Crescenzi, and S. Iammarino. 2014. "Economic Institutions and MNE Location Strategies: EU15 MNEs in Europe and the Neighbourhood." SEARCH Project, LSE mimeo.

Bailey, D., and N. Driffield. 2002. "Hymer and Uneven Development Revisited: Foreign Direct Investment and Regional Inequalities." *Contributions to Political Economy* 21(1): 55–68.

Bel, G., and X. Fageda. 2008. "Getting There Fast: Globalization, Intercontinental Flights and Location of Headquarters." *Journal of Economic Geography* 8: 471–495.

Birkinshaw, J., and N. Hood. 1998. "Multinational Subsidiary Evolution: Capability and Charter Change in Foreign-Owned Subsidiary Companies." *Academy of Management Review* 23(4): 773–795.

Birkinshaw, J., and N. Hood. 2000. "Characteristics of Foreign Subsidiaries in Industrial Clusters." *Journal of International Business Studies* 31(1): 141–154.

Birkinshaw, J., N. Hood, and S. Young. 2005. "Subsidiary Entrepreneurship, Internal and External Competitive Forces, and Subsidiary Performance." *International Business Review* 14(2): 227–248.

Borja, J., M. Castells, M. Belil, and C. Benner. 1997. *Local and Global: Management of Cities in the Information Age*. London: Earthscan.

Burghouwt, G. 2005. *Airline Network Development in Europe and Its Implications for Airport Planning*. Utrecht: Utrecht University Press.

Button, K., R. Stough, M. Bragg, and S. Taylor. 2006. *Telecommunications, Transportation and Location*. Cheltenham: Edward Elgar.

Cairncross, F. 1997. *The Death of Distance: How Communications Revolution Will Change Our Lives*. London: Orion Business Books.

Cantwell, J. (ed.). 1994. *Transnational Corporations and Innovatory Activities*. London: Routledge.

Cantwell, J. 1995. "The Globalisation of Technology: What Remains of the Product Cycle Model?" *Cambridge Journal of Economics* 19(1): 155–174.

Cantwell, J., and E. Kosmopoulou. 2002. "What Determines the Internationalisation of Corporate Technology?" In *Critical Perspectives on Internationalisation*, ed. V. Havila, M. Forsgren, and H. Håkanson, 305–334. Oxford: Pergamon Press.

Cantwell, J., and R. Mudambi. 2005. "MNE Competence-Creating Subsidiary Mandates." *Strategic Management Journal* 26(12): 1109–1128.

Chandler, A.D. 1959. "The Beginnings of 'Big Business' in American Industry." In *The Essential Alfred Chandler*, ed. T.K. McCraw (1988). Boston, MA: Harvard Business School Press.

Chandler, A.D. 1962. *Strategy and Structure: Chapters in the History of Industrial Enterprise.* Cambridge, MA: The MIT Press.

COL. 2009. *The Global Financial Centres Index 6*. London: Corporation of London.

Cooke, P. 2001. "Regional Innovation Systems, Clusters, and the Knowledge Economy." *Industrial and Corporate Change* 10(4): 945–974.

Crescenzi, R., and A. Rodríguez-Pose. 2009. "Systems of Innovation and Regional Growth in the EU: Endogenous vs. External Innovative Efforts and Socio-Economic Conditions." In *Growth and Innovation of Competitive Regions: The Role of Internal and External Connections*, ed. U. Fratesi and L. Senn, 167–192. Berlin: Springer-Verlag.

Crescenzi, R., C. Pietrobelli, and R. Rabellotti. 2013. "Innovation Drivers, Value Chains and the Geography of Multinational Corporations in Europe." *Journal of Economic Geography* (published online).

Crozet, M., T. Mayer, and J. Mucchielli. 2004. "How Do Firms Agglomerate? A Study of FDI in France." *Regional Science and Urban Economics* 34: 27–54.

Defever, F. 2006. "Functional Fragmentation and the Location of Multinational Firms in the Enlarged Europe." *Regional Science and Urban Economics* 36(5): 658–677.

Derudder, B. 2006. "On Conceptual Confusion in Empirical Analyses of a Transnational Urban Network." *Urban Studies* 43(11): 2027–2046.

Derudder, B., P.J. Taylor, F. Witlox, and G. Catalano. 2003. "Hierarchical Tendencies and Regional Patterns in the World City Network." *Regional Studies* 37(9): 875–886.

Desai, M.A. 2009. "The Decentering of the Global Firm." *The World Economy* 32: 1271–1290.

Dicken, P. 1977. "A Note on Location Theory and the Large Business Enterprise." *Area* 9(2): 138–143.

Dunning, J.H., and G. Norman. 1983. "The Theory of the Multinational Enterprise: An Application to Multinational Office Location." *Environment and Planning A* 15(5): 675–692.

Dunning, J.H., and C.N. Pitelis. 2008. "Stephen Hymer's Contribution to International Business Scholarship: An Assessment and Extension." *Journal of International Business Studies* 39(1): 167–176.

Enright, M. 2000. "The Globalization of Competition and the Localization of Competition: Policies Toward Regional Clustering." In *The Globalization of Multinational Enterprise Activity and Economic Development*, ed. N. Hood and S. Young. London: Macmillan.

Fagerberg, J., B. Verspagen, and M. Caniëls. 1997. "Technology, Growth and Unemployment Across European Regions." *Regional Studies* 31(5): 457–466.

Florida, R. 2005. *Cities and the Creative Class*. London: Routledge.

Friedman, T.L. 2005. *The World Is Flat: A Brief History of the Twenty-First Century.* New York: Farrar, Straus, and Giroux.

Friedmann, J. 1986. "The World City Hypothesis." *Development and Change* 17: 69–83.

Fujita, M., P.R. Krugman, and A.J. Venables. 1999. *The Spatial Economy: Cities, Regions, and International Trade*. Cambridge, MA: The MIT Press.

Giuliani, E. 2011. "Role of Technological Gatekeepers in the Growth of Industrial Clusters: Evidence from Chile." *Regional Studies* 45(10): 1329–1348.

Goerzen, A., C.G. Asmussen, and B.B. Nielsen. 2013. "Global Cities and Multinational Enterprise Location Strategy." *Journal of International Business Studies* 44: 427–450.

Gordon, I. 2002. "Global Cities, Internationalisation and Urban Systems." In *Industrial Location Economics*, ed. P. McCann, 187–206. Cheltenham: Edward Elgar.

Gordon, I.R., and P. McCann. 2000. "Industrial Clusters, Complexes, Agglomeration and/or Social Networks?" *Urban Studies* 37: 513–532.

Guy, F. 2009. *The Global Environment of Business*. Oxford: Oxford University Press.
Hall, P. 1966. *The World Cities*. London: Weidenfeld & Nicolson.
Henderson, J.V., A. Kuncoro, and M. Turner. 1995. "Industrial Development in Cities." *Journal of Political Economy* 103: 1067–1085.
Hirsch, S. 1967. *Location of Industry and International Competitiveness*. Oxford: Clarendon Press.
Hymer, S. 1970. "The Efficiency Contradictions of Multinational Corporations." *American Economic Review* 60: 441–448.
Hymer, S. 1972. "The Multinational Corporation and the Law of Uneven Development." In *Economics and World Order: From the 1970s to the 1990s*, ed. J.N. Bhagwati, 113–140. New York: The Free Press.
Hymer, S. 1976. *The International Operations of National Firms: A Study of Direct Foreign Investment*. Cambridge, MA: The MIT Press.
Iammarino, S., and P. McCann. 2006. "The Structure and Evolution of Industrial Clusters: Transactions, Technology and Knowledge Spillovers." *Research Policy* 35(7): 1018–1036.
Iammarino, S., and P. McCann. 2010. "The Relationship Between Multinational Firms and Innovative Clusters." In *The Handbook of Evolutionary Economic Geography*, ed. R. Boschma and R.L. Martin. Cheltenham: Edward Elgar.
Iammarino, S., and P. McCann. 2013. *Multinationals and Economic Geography: Location, Technology, and Innovation*. Cheltenham: Edward Elgar.
Ietto-Gillies, G. 2005. *Transnational Corporations and International Production: Concepts, Theories and Effects*. Cheltenham: Edward Elgar.
Ietto-Gillies, G. 2012. *Transnational Corporations and International Production: Concepts, Theories and Effects*, 2nd edn. Cheltenham: Edward Elgar.
Jacobs, J. 1969. *Economy of Cities*. New York: Vintage.
Knox, P.L., and P.J. Taylor. 1995. *World Cities in a World-System*. Cambridge: Cambridge University Press.
Kottaridi, C. 2005. "The 'Core-Periphery' Pattern of FDI-Led Growth and Production Structure in the EU." *Applied Economics* 37(1): 99–113.
Krugman, P. 1991. *Geography and Trade*. Cambridge, MA: The MIT Press.
Leinbach, T.R., and C. Capineri. 2007. *Globalized Freight Transport: Intermodality, E-Commerce, Logistics and Sustainability*. Cheltenham: Edward Elgar.
Limtanakool, N., T. Schwanen, and M. Dijst. 2007. "Ranking Functional Urban Regions: A Comparison of Interaction and Node Attribution Data." *Cities* 24(5): 26–42.
Mack, R.S., and P.V. Schaeffer. 1993. "Nonmetropolitan Manufacturing in the United States and Product Cycle Theory: A Review of the Literature." *Journal of Planning Literature* 8: 124–139.
MasterCard. 2008. *Worldwide Centers of Commerce Index*.
McCann, P. 2007. "Sketching Out a Model of Innovation, Face-To-Face Interaction and Economic Geography." *Spatial Economic Analysis* 2(2): 117–134.
McCann, P. 2008. "Globalization and Economic Geography: The World Is Curved, Not Flat." *Cambridge Journal of Regions, Economy and Society* 1(3): 351–370.
McCann, P., and R. Mudambi. 2004. "The Location Behaviour of the Multinational Enterprise: Some Analytical Issues." *Growth and Change* 35(4): 491–524.
McCann, P., and S. Sheppard. 2003. "The Rise, Fall and Rise Again of Industrial Location Theory." *Regional Studies* 37(6–7): 649–663.
McCann, P., T. Arita, and I.R. Gordon. 2002. "Industrial Clusters, Transactions Costs and the Institutional Determinants of MNE Behaviour." *International Business Review* 11: 647–663.
McKinsey Global Institute. 2013. "Urban World: The Shifting Global Business Landscape." McKinsey & Company, http://www.mckinsey.com/insights/urbanization/urban_world_the_shifting_global_business_landscape (accessed December 31, 2014).
Nachum, L., and C. Wymbs. 2005. "Product Differentiation, External Economies and MNE Location Choices: M&As in Global Cities." *Journal of International Business Studies* 36: 415–434.

Ni, P.-F., and P.K. Kresl. 2010. *The Global Urban Competitiveness Report*. Cheltenham: Edward Elgar.

Norton, R.D. 1979. *City Life Cycles and American Urban Policy*. New York: Academic Press.

Norton, R.D. 1987. "The Role of Services and Manufacturing in New England's Economic Resurgence." *New England Economic Indicators* Q II: iv–viii.

Norton, R.D., and J. Rees. 1979. "The Product Cycle and the Spatial Decentralization of American Manufacturing." *Regional Studies* 13(2): 141–151.

O'Brien, R. 1992. *Global Financial Integration: The End of Geography*. New York: Council on Foreign Relations Press.

OECD. 2006. *Competitive Cities in the Global Economy*. Paris: OECD.

Ohmae, K. 1995. *The End of the Nation-State: The Rise of Regional Economies*. New York: Simon and Schuster.

Paci, R., and S. Usai. 2009. "Knowledge Flows Across European Regions." *The Annals of Regional Science* 43(3): 669–690.

Patrucco, P.P. 2003. "Institutional Variety, Networking and Knowledge Exchange: Communication and Innovation in the Case of the Brianza Technological District." *Regional Studies* 37(2): 159–172.

Piore, M.J. 1994. "Corporate Reform in American Manufacturing and the Challenge to Economic Theory." In *Information Technology and the Corporation of the 1990s: Research Studies*, ed. T.J. Allen and M.S. Scott Morton. Oxford: Oxford University Press.

Pitelis, C.N. 1991. "The Effects of Advertising (and) Investment on Aggregate Profits." *Scottish Journal of Political Economy* 38(1): 32–40.

Pitelis, C.N. (ed.). 2002a. *The Growth of the Firm: The Legacy of Edith Penrose*. Oxford: Oxford University Press.

Pitelis, C.N. 2002b. "Stephen Hymer: Life and the Political Economy of Multinational Corporate Capital." *Contributions to Political Economy* 21(1): 9–26.

Pitelis, C.N. 2005. "Stephen Herbert Hymer and/on (the Theory of) the MNE and International Business." *International Business Review* 15(2): 103–110.

Redlich, F. 1962. "Approaches to Business History." *Business History Review* 36(1): 61–86.

Rodríguez-Pose, A. 1998. *The Dynamics of Regional Growth in Europe: Social and Political Factors*. Oxford: Clarendon Press.

Rodríguez-Pose, A., and R. Crescenzi. 2008. "Research and Development, Spillovers, Innovation Systems, and the Genesis of Regional Growth in Europe." *Regional Studies* 42: 51–67.

Rodríguez-Pose, A., and R.D. Fitjar. 2013. "Buzz, Archipelago Economies and the Future of Intermediate and Peripheral Areas in a Spiky World." *European Planning Studies* 21(3): 355–372.

Rosenthal, S.S., and W.C. Strange. 2004. "Evidence on the Nature and Sources of Agglomeration Economics." In *Handbook of Urban and Regional Economics*, vol. 4, ed. V. Henderson and J.F. Thisse. Amsterdam: North-Holland.

Sassen, S. 1991. *The Global City: New York, London, Tokyo*. Princeton, NJ: Princeton University Press.

Sassen, S. 1994. *Cities in a World Economy*. Thousand Oaks, CA: Pine Forge.

Sassen, S. 2001. *The Global City: New York, London, Tokyo*, 2nd edn. Princeton, NJ: Princeton University Press.

Sassen, S. (ed.). 2002. *Global Networks: Linked Cities*. London: Routledge.

Saxenian, A. 1994. *Regional Advantage: Industrial Adaptation in Silicon Valley and Route 128*. Cambridge, MA: Harvard University Press.

Schumpeter, J. 1934. *The Theory of Economic Development*. Cambridge, MA: Harvard University Press.

Scott, A.J. (ed.). 2001. *Global City-Regions: Trends, Theory, Policy*. Oxford: Oxford University Press.

Scott, A.J., J. Agnew, E.W. Soja, and M. Storper. 2001. "Global City-Regions." In *Global City-Regions: Trends, Theory, Policy*, ed. A.J. Scott, 11–32. Oxford: Oxford University Press.

Simmie, J., and J. Sennet. 1999. "Innovative Clusters: Theoretical Explanations and Why Size Matters." *National Institute Economic Review* 4/99(170): 87–98.

Simmie, J., C. Siino, J.M. Zuliani, G. Jalabert, and S. Strambach. 2004. "Local Innovation System Governance and Performance: A Comparative Analysis of Oxfordshire, Stuttgart and Toulouse." *International Journal of Technology Management* 28(3–6): 534–559.

Storper, M. 1985. "Oligopoly and the Product Cycle: Essentialism in Economic Geography." *Economic Geography* 61(3): 260–282.

Storper, M. 1997. *The Regional World: Territorial Development in a Global Economy.* New York: Guilford Press.

Storper, M. 2013. *Keys to the City: How Economics, Institutions, Social Interaction, and Politics Shape Development.* Princeton, NJ: Princeton University Press.

Taylor, P.J. 2004. *World City Network: A Global Urban Analysis.* London: Routledge.

Taylor, P.J., P. Ni, B. Derudder, M. Hoyler, J. Huang, and F. Witlox (eds.). 2011. *Global Urban Analysis: A Survey of Cities in Globalization.* London: Earthscan.

Taylor, R.S. 1986. *Value-Added Processes in Information Systems.* Norwood, NJ: Ablex.

Taylor, R.S. 1987. "New Thought in the Twenties: The Case of Springfield, Illinois." *Historian* 49: 329–347.

UNCTAD. 2009. *World Investment Report 2009: Transnational Corporations, Agricultural Production and Development.* New York and Geneva: United Nations.

UNCTAD. 2013. *World Investment Report 2013: Global Value Chains: Investment and Trade for Development.* New York and Geneva: United Nations.

Verbeke, A., L. Li, and A. Goerzen. 2009. "Toward More Effective Research on the Multinationality-Performance Relationship." *Management International Review* 49: 149–162.

Vernon, R. 1957. *The International Patent System and Foreign Policy: [Study] of the Committee on the Judiciary, United States Senate, Eighty-Fifth Congress, First Session, Pursuant to S. Res. 55.* Washington, DC: US Government Printing Office.

Vernon, R. 1959. *The Changing Economic Function of the Central City.* New York: Committee for Economic Development.

Vernon, R. 1960. *Metropolis.* Cambridge, MA: Harvard University Press.

Vernon, R. 1966. *The Myth and Reality of Our Urban Problems.* Cambridge, MA: Harvard University Press.

Vernon, R. 1974. *Big Business and the State: Changing Relations in Western Europe.* Cambridge, MA: Harvard University Press.

Vernon, R. 1979. *Opportunities and Challenges for Multinational Firms from Developing Countries.* Seoul: Korea International Economic Institute.

Vernon, R. 1991. "The Longer View: The Coming Global Metropolis Cities of the Next Century." *Journal of the American Planning Association* 57(1): 3–6.

Verspagen, B., and M.C.J. Caniëls. 2001. "Barriers to Knowledge Spillovers and Regional Convergence in an Evolutionary Model." *Journal of Evolutionary Economics* 11(3): 307–329.

von Tunzelmann, N. 2003. "Historical Coevolution of Governance and Technology in the Industrial Revolutions." *Structural Change and Economic Dynamics* 14: 365–384.

von Tunzelmann, N. 2009. "Regional Capabilities and Industrial Regeneration." In *Technological Change and Mature Industrial Regions: Firms, Knowledge and Policy*, ed. M. Farshchi, O. Janne, and P. McCann, 11–28. Cheltenham: Edward Elgar.

Young, S., N. Hood, and E. Peters. 1994. "Multinational Enterprises and Regional Economic Development." *Regional Studies* 28(7): 657–677.

Chapter 15

The Rise of the Global Creative Class

Richard Florida and Charlotta Mellander

Introduction

The world economy is in the midst of an epochal transformation, from industrial capitalism to a new age of knowledge-based or creative capitalism. The canonical factors of production identified by the classical economists – land, labor, and capital – are no longer the principal drivers of economic growth. National and regional economies increasingly grow and thrive to the extent that they are able to harness science and technology, innovation, and human creativity as sources of economic and social progress.

Creativity differs in fundamental ways from more traditional, tangible factors of production. Unlike stocks of things that can be depleted or worn out, it is an infinitely renewable resource that is continually enlarging via education, job experience, and the spontaneous, often fortuitous combinations and recombinations of ideas that occur as people collaborate and compete.

Just as industrial capitalism gave rise to a new socioeconomic class of blue-collar workers, the creative economy has given rise to a new class of laborers who work with their minds and creativity. This creative class, which makes up between a third to nearly half of the workforce in the advanced economies, includes scientists and technologists; innovators and entrepreneurs; designers, musicians, entertainers, media workers, academics, and artistic and cultural creatives; and knowledge-based professionals in business, education, and health-care.

Like Marx's (1887) working class, which was comprised of a number of very different occupations that all had physical skills in common, from skilled tradesmen to assembly-line workers, the diverse occupations that make up the creative class all draw upon an underlying set of creative skills.

The Handbook of Global Science, Technology, and Innovation, First Edition.
Edited by Daniele Archibugi and Andrea Filippetti.

This chapter, which draws off our previous work,[1] summarizes our rankings for 82 nations on a metric that we call the Global Creativity Index (GCI), a composite measure based on technology, talent, and tolerance, the 3Ts of economic development (Florida 2002, 2012).

Technology is the first T. As Marx (1887) and Schumpeter (1934, 1942, 1947a, 1947b) noted long ago, technology and technological innovation enable capitalism to generate new industries and spur new growth. Solow's (1956) now classic residual formalized the role of technology in economic growth. An enormous body of recent literature confirms and elaborates on it (Nelson 1959, 1977; Nelson and Winter 1982; Rosenberg 1983; Mowery and Rosenberg 1989; Rosenberg and Nelson 1994).

Talent is the second T. All the way back to Adam Smith, economists have noted the role of human capital and skill in economic growth, which Smith defined as the "acquired and useful abilities of all the inhabitants or members of the society" (Spengler 1977). Becker (1964) and Mincer (1974) identified a direct link at the individual level between knowledge and skill on one hand and productivity levels on the other (captured by wage levels). There is an extensive literature linking human capital to national and regional growth (Barro 1991; Rauch 1993; Simon and Nardinelli 1996; Simon 1998; Glaeser 1999; Glaeser and Saiz 2003). Moreover, these links are most easily seen and felt in cities, which act as the key economic engines for the contemporary economy (Glaeser 2011).

Drucker (1969, 1993) and Machlup (1962) long ago noted the growing economic importance of knowledge workers and the knowledge economy. Knowledge workers not only invent new machines and processes that turn out old products more efficiently, they come up with completely new products that create whole new markets. Romer (1986) formalized the role of knowledge and connected it with technology in his theory of endogenous growth.

Tolerance is the third T. Both technology and talent are best conceptualized not as stocks but flows. Talent and technology flow not just to places with great universities or robust industrial structures but places which are open and have low barriers to entry (Florida 2002). Tolerance thus operates as a non-market factor that increases the efficiency of technology and talent. Page (2007) has shown that innovation is closely associated with diversity. Silicon Valley, perhaps the most innovative place in the world, exemplifies this. Between one-third and one-half of its new technology-based enterprises have a foreign-born person on their founding team (Wadhwa et al. 2007). Detailed cross-national studies (Noland 2005; Noland and Pack 2004) found that the economic performance of nations was positively associated with more open attitudes toward homosexuality, which are also correlated with globalization, controlling for other factors. Openness to diversity is also in line with the broad cultural shift from materialist values about money and things to newer "post-materialist" values, which favor self-expression and a wider quest for happiness and well-being, as identified by Inglehart (1989, 1997).

Places that are open to new ideas, and that attract talented and creative people from across the globe, broaden both their technology and their talent capabilities, gaining an additional economic edge over and above those of technology and talent themselves. These places are increasingly found in and around major cities. Indeed, the economic activity produced within the world's largest metropolitan areas accounts for a much greater economic value than their population sizes would suggest (Florida, Mellander, and Gulden 2010). Smaller urban areas that are rich in amenities and tolerance also punch well above their weight when it comes to technology-fueled growth.

The remainder of this chapter proceeds as follows. The next section outlines the data, variables, and methods used in our analysis. We then discuss the overall rankings of nations on the GCI and summarize the associations between the GCI and key indicators of economic and social performance. We conclude with a short summary of our key findings.

Methodology, Data, and Variables

This section discusses the methodology used in our analysis. The data covers 82 nations for the period 2000–2009. We sometimes use different years for different variables, and also utilize running averages, depending on data availability. The following describes the main variables and data sources used to construct the GCI.

Technology

We use three variables for technology: Global R&D Investment, Global Researchers, and Global Innovation (patents), which we then combine into our Global Technology Index.

Global R&D Investment This measures R&D spending as a share of GDP. It is adapted from World Development Indicators of the World Bank, and is defined as "current and capital expenditures on creative work undertaken systematically to increase the stock of knowledge, including knowledge of humanity, culture, and society, and the use of knowledge to devise new applications. R&D covers basic research, applied research, and experimental development."

Global Researchers This variable measures professional researchers engaged in R&D per million capita. It is adapted from World Development Indicators and covers the years 2000–2005. Professional researchers are defined as "professionals engaged in conceiving of or creating new knowledge, products, processes, methods, and systems and in managing projects concerned. Postgraduate students at the doctoral level (ISCED97 level 6) engaged in R&D are considered researchers." The World Development Indicators are published annually by the World Bank and the data is reported for 127–146 different countries, depending upon the year. However, since countries do not always report on an annual basis, we use averages for several years. This generates higher numbers of observations and also helps smooth extreme values.

Global Innovation This variable measures patents granted per capita. It is adapted from the US Patent and Trademark Office (USPTO) and covers the years 2001–2008. US patents are a reasonable proxy for global innovation as inventors from around the world file for patent protection in the United States and the USPTO tracks inventors' national origins. We count the number of *granted* US patents for each nation in the world.

The Global Technology Index combines these three variables in a single measure and is based on a principal component analysis, where the correlations between the overall index and the three constituent measures are as follows: Global R&D Investment (.88), Global Researchers (.89), and Global Innovation (.94). In other words, the overall technology score is based on the value for each variable, and not its ranking. We estimate the index for countries with missing values by running regressions based on the variables for which we do have values. The R^2s for these regressions are as follows: 0.535 for Global R&D Investment, 0.588 for Global Researchers, and 0.702 for Global Innovation.

Talent

We employ two measures for talent: the occupationally-defined creative class, and the conventional measure of human capital based on educational attainment, which we then combine in our Global Talent Index.

Creative Class The creative class variable is calculated as the share of a country's labor force that is engaged in creative work that requires a high degree of problem-solving in their everyday work. This includes the following occupations: computer science and mathematics; architecture, engineering; life, physical, and the social sciences; education, training, and library science; arts and design work, entertainment, sports, and media; and professional and knowledge work occupations in management, business and finance, law, sales management, and healthcare. The data are from the International Labour Organization (ILO) and cover the years 2004–2007.

Human Capital The human capital variable is based on the rate of enrollment in tertiary or post-high school education from the World Development Indicators. The data is reported to the UNESCO Institute for Statistics by national education agencies. Tertiary education is defined as "a wide range of post-secondary education institutions, including technical and vocational education, colleges, and universities, whether or not leading to an advanced research qualification, that normally require as a minimum condition of admission the successful completion of education at the secondary level." The data cover the years 2004 and 2006 and are based on annual school surveys, normally conducted in the beginning of the year, and do not therefore reflect dropouts or actual attendance.

The Global Talent Index combines these two variables in a single index based on a principal component analysis, where the correlations are 0.87 for both the Creative Class and the Human Capital measures. Again, the overall score is based on the value for each variable and not its ranking. We estimate missing values through a regression analysis, which generates an R^2 value of 0.501.

Tolerance

We employ two measures of tolerance to construct the Global Tolerance Index: tolerance toward ethnic and racial minorities and tolerance toward gays and lesbians. Both are from the Gallup Organization's World Poll (2010), which covers approximately 150 nations for the year 2009. The first is based on the survey question: "Is your city or area a good or bad place to be in for ethnic and racial minorities?" The second is based on the survey question: "Is your city or area a good or bad place to be in for gay and lesbian people?" Our variable reflects the share of the respondents who said theirs was a good place.

The Global Tolerance Index combines these two measures. The two are equally weighted into a factor where both correlate at .92. We estimate missing values based on a regression analysis, which generates an R^2 value of 0.432.

The Global Creativity Index (GCI)

The overall GCI Index is based on a principal component analysis of the three key indexes for global technology, talent, and tolerance. Each of the three indexes is based on the actual performance of each variable. We ranked nations on each of them, with the highest ranking going to the best performer. We then added the ranks together and divided by three. In the cases where we had a value for just two of the three variables, these two were added and divided by two. The average scores of the three indices were divided by the number of observations overall to get the overall GCI score.

Correlates for Economic Performance and Related Measures

We use the following variables to examine the relationships between the GCI and the economic and social performance of nations.

Economic Output We employ the conventional measure of economic output: GDP per capita. The data are from World Development Indicators for the year 2005.

Global Competitiveness Index Developed by Porter, Sachs, and Warner (2000) for the World Economic Forum, it is based on the following categories: basic requirements (including institutions, infrastructure, macroeconomic stability, and health and primary education), efficiency enhancers (including higher education and training, goods market efficiency, labor market efficiency, financial market sophistication, technological readiness, and market size), and innovation factors (including business sophistication and innovation).

Global Entrepreneurship Index This variable is based on the Global Entrepreneurship Index developed by Acs and Szerb (2010). The index consists of several measures of entrepreneurial attitudes, activity, and aspiration, and covers the years 2004–2008.

Human Development Index This variable is based on the United Nations Human Development Index, a composite measure which aims to capture three dimensions of human development: health and measured life expectancy, education level, and standard of living. We employ the 2009 index, which is based on data from 2007.

Happiness/Life Satisfaction This variable is from the Gallup Organization's 2009 Gallup World Poll. The measure is a ranking from 1 to 10, where 10 represents the highest possible level of life satisfaction. Our variable is the national average rank of life satisfaction.

Income Inequality This variable is based on the Gini Coefficient, the standard measure of Income Inequality, which measures the distribution of incomes in a nation on a range from 0 to 100 where 0 represents absolute equality and 100 absolute inequality. This variable is from the World Bank's World Development Indicators for the year 2007.

The Global Creative Class

We now turn to our findings for the global creative class. The creative class is essentially a measure of skill based on occupations or the kinds of work people do. It thus differs from the more conventional measure of human capital that is based on educational attainment.

While there is considerable overlap between the two, they are not the same. In the United States, for example, nearly three-fourths of adults with college degrees are members of the creative class, but less than 60% of the people whose occupations qualify them as members of the creative class have college degrees (Currid-Halkett and Stolarick 2013). In Sweden, 90% of the highly educated hold creative class jobs while only one-fourth of the creative class have university degrees (Mellander 2009).

A significant body of research shows that the occupation-based creative class measure operates in addition to and through other channels than the standard education-based human capital variable. A large-scale study found that the creative class has a bigger effect on wages – a key element of regional productivity – whereas education tends to have a greater effect on income (Florida, Mellander, and Stolarick 2008). Independent research by Gabe shows that the creative class continues to have a substantial effect on regional economic growth when controlling for the effects of education and other factors (Gabe 2009). Having a creative class job brings economic benefits that extend beyond those of going to college. A college graduate working in the same occupation as a non-college graduate earns approximately 50% higher wages. But having a creative class job adds another 16%, about the same as another 1.5 years of additional education, according to Gabe's research (Gabe 2009). A 2012 study used advanced statistical models to compare the effects of the creative class and human capital across the 257 EU regions. "Our results," it concluded, "indicate that highly educated people working in creative occupations are the most relevant component in explaining production efficiency" (Marrocu and Paci 2012: 369).

McGranahan and Wojan (2007) used sophisticated statistical techniques to gauge the effects of the creative class versus human capital on regional growth. To do so, they used systems of simultaneous equations rather than the conventional simple regression models to control for population and employment growth as well as influences from a range of other local conditions and attributes. Their key findings overwhelmingly confirm the "strong independent influence on employment growth from both the initial share employed in the recast creative class occupations and its growth over the decade. By contrast, the statistical association with human capital variables is quite weak." And they add: "the econometric test of the creative class thesis provides strong support for the notion that creativity has an effect on growth independent of the endowment of human capital" (McGranahan and Wojan 2007).

Another detailed study, this one investigating regional development in the Netherlands, also found that the creative class considerably outperformed the standard education-based human capital measure in accounting for employment growth. This led its authors to conclude that the creative class measure sets a "new standard" for measuring skill and talent, especially when considering regional labor productivity (Marlet and van Woerkens 2004). "With our Dutch data set we do find evidence that Florida's creative class is a better predictor of city growth than traditional education standards," they wrote. "Therefore we conclude that Florida's major contribution is his successful attempt to create a population category that is a better indicator for levels of human capital than average education levels or amounts of highly educated people. The point is, as Florida stated, not which or how much education people can boast of, but what they really do in working life" (Marlet and van Woerkens 2004: 24).

Figure 15.1 shows how the nations of the world stack up on the creative class shares of their workforces. The range on the map is quite large, from a high of nearly half to a low of just over 2%. Fourteen countries have 40% or more of their workforce in the creative class.

With nearly half (47.3%) of its workforce in the creative class, Singapore takes the top spot. Aside from Singapore and Australia in fourth place, the top-ranked countries are overwhelmingly Northern European and Scandinavian: the Netherlands (46.3%), Sweden (43.9%), Switzerland (44.8%), Belgium (43.8%), Denmark (43.7%), Finland (43.4%), Norway (42.1%), and Germany (41.7%). Ten of the top 15 countries for the creative class are European. With 35% of its workforce in the creative class the United States ranks 27th, below Slovakia (38.4%). Of the BRIC countries, Russia ranks highest at 20th (38.6%). Brazil is 57th (18.5%), and China 75th (7.4%).

The Creative Class in Global Cities

The shift to a creative economy entails a major geographic shift as well. As a growing body of research shows, cities have become the main social and economic organizing unit of the creative economy. But whether one is measuring population density, education levels, creative class share, innovation, or economic output, the world is increasingly spiky (Florida 2005b): some cities rank much higher than others. Cities act as drivers for the disparities between national competitiveness rankings. Larger, more globally connected cities are increasingly the winners in the global competition for talent.

Though no single organized source of comparable data on the creative class in global cities exists, a significant body of studies have examined the geography of the creative class in cities and metro areas within nations outside the United States, including the United Kingdom, Germany, the Netherlands, Sweden, Denmark, Finland, Norway, and China. Overall, they confirm the clustering force of the creative class and of technology assets and innovation in large cities and metro areas.

A large-scale effort to collect and organize data for Canada, Western Europe, and the Nordic countries was carried out under the auspices of the European Science Foundation (Asheim 2009; Boschma and Fritsch 2009; Lorenzen and Andersen

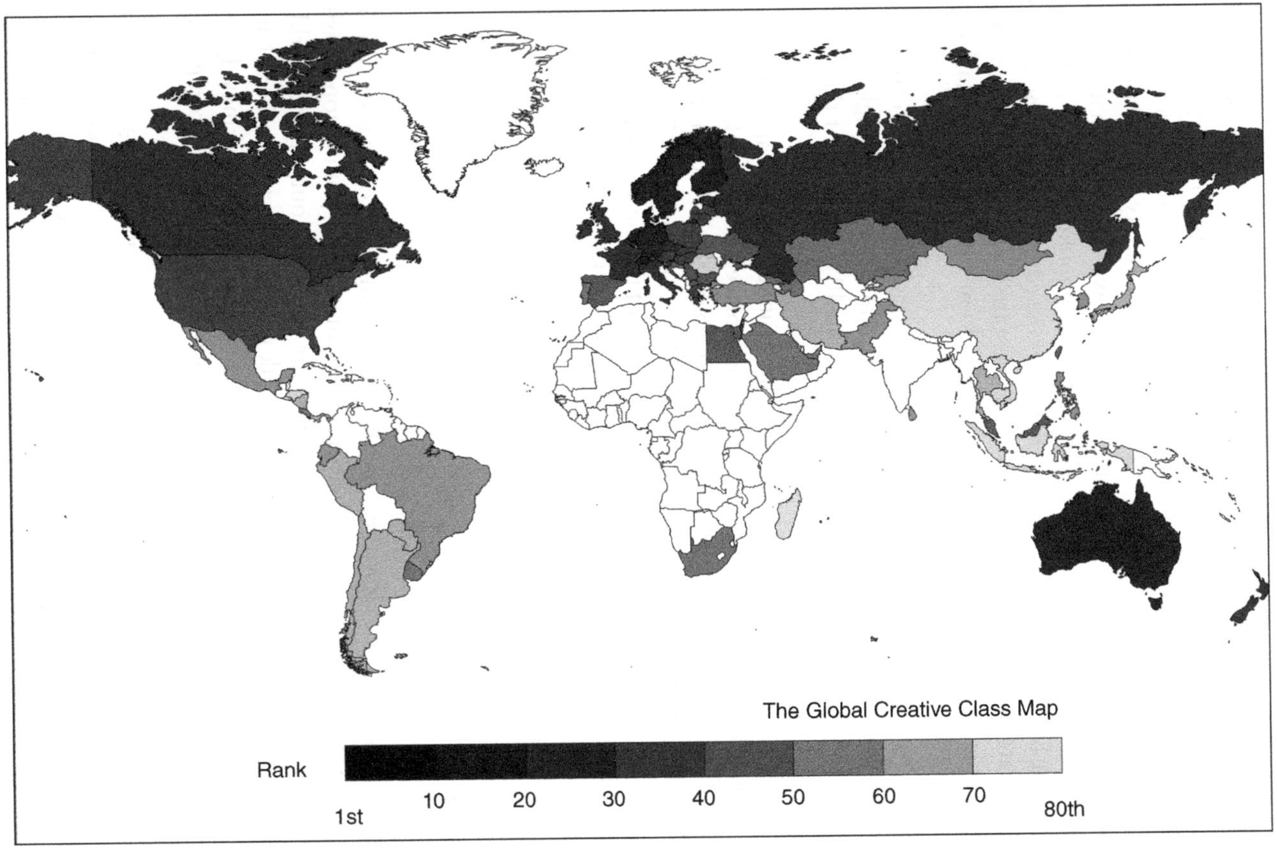

Figure 15.1 The global creative class map.
Source: Florida, Mellander, and Stolarick (2011).

Table 15.1 The creative class in global cities.

Amsterdam, Netherlands	46.0%	Hannover, Germany	37.8%
Stockholm, Sweden	46.0%	Oslo, Norway	37.6%
Helsinki, Finland	44.0%	Ottawa, Canada	37.6%
Oxford, UK	42.8%	Bonn, Germany	37.3%
Munich, Germany	42.2%	Toronto, Canada	37.1%
Malmö-Lund, Sweden	41.4%	Copenhagen, Denmark	36.8%
Cambridge, UK	41.2%	Stuttgart, Germany	36.6%
London, UK	41.2%	Leicestershire, UK	36.2%
Berlin, Germany	39.3%	Leeds, UK	35.3%
Hamburg, Germany	38.2%	Paris, France	35.1%

Source: Florida (2012) based on data from Boschma and Fritsch 2009. Data for French metros provided by S. Chantelot.

2009; Asheim and Hansen 2009; Andersen et al. 2010; Florida 2012; Florida and Mellander 2013a, 2013b). Table 15.1, which is adapted from this project, shows shares of the creative class in a number of selected European and Canadian cities. We should emphasize that this list is illustrative only – there are many creative class regions, nations, and cities around the world where data are either unavailable, have not been organized or published, or of which we are not aware.

Amsterdam and Stockholm top the list, with 46% of their workforce in the creative class. The creative class makes up more than 40% of workers in Helsinki, Oxford, Munich, Malmö, London, and Cambridge. This is roughly the same as the top-ranked US metros at the time –Boulder and San Jose (Silicon Valley). The creative class made up 35–40% of the workforce in Paris, Toronto, Hamburg, Berlin, Oslo, Copenhagen, and several other metros – more than Boston, greater Washington, DC, Austin, or San Francisco at the time (Florida 2012).

The link between large metro areas and creative class growth are clear from a number of national cases. In Canada, for instance, economic prosperity has been linked to rising levels of national urbanization. Cities in that country, namely larger metros like Toronto or Vancouver, attract a disproportionate number of talented people, making them the engines of contemporary economic growth (Florida, Mellander, and Stolarick 2014). The same is true within other advanced, Anglophone countries, particularly Australia. There, the creative class is most likely to be found in the most amenity-rich cities, the economic engines of the Australian economy (Stolarick 2014).

A study of the United Kingdom (Clifton 2008) highlighted the uneven geography of the creative class, especially in and around London. Larger, denser cities with close links to the amenities and infrastructure of the capital have particularly high levels of the creative class (Clifton 2014). A study of Denmark (Andersen and Lorenzen 2009) found that the creative class is concentrated in Copenhagen and Denmark's larger cities, which are able to provide a diverse range of service and cultural offerings and tolerance to alternative lifestyles, though interviews indicate that creatives are also attracted to smaller cities because of the cost advantages, specialized employment offerings, attractive work/life balance, authenticity, and sense of community that they offer.

A study of the Netherlands (van Aalst et al. 2014) found a substantial relationship between the creative class and employment growth, and a smaller but still significant connection between the creative class and the growth of new businesses across Dutch regions. However it found only a weak association between regional "openness" and the creative class. Moreover, cities with greater shares of the population in the creative class in the Netherlands have been shown to do much better economically than other cities, thus underscoring the pivotal role of metros in national economic well-being (Marlet and van Woerkens 2014).

Several studies of Germany highlight the uneven geography of the creative class in that country. Fritsch and Stuetzer (2009) note that while Germany's creative class is especially concentrated in larger cities, considerable concentrations can also be found in smaller cities and rural places. In contrast with the findings for the Netherlands, the creative class weighs ethnic and cultural diversity more than employment opportunities in their location decisions. Education and public health care are strongly associated with the creative class as well.

A 2011 study (Falck, Fritsch, and Heblich) tracked economic growth in German cities that had built opera houses in the seventeenth and eighteenth centuries. "Proximity to a baroque opera house is a strong predictor of a region's equilibrium share of high-human-capital-employees," its authors found, even though the construction of the opera houses predated those jobs by centuries. "It is the local level of high-human-capital employees who value their proximity to a baroque opera house that shifts a location to a higher growth path."

A comparative study (Boschma and Fritsch 2009) examined the geography of the creative class across cities in Denmark, Finland, Germany, the Netherlands, Norway, Sweden, and the United Kingdom. It found the creative class to be over-represented in larger cities across Europe, especially in London. Indeed, all of the regions in Europe with substantial creative class populations tended to be around the main urban areas of their respective countries or regions (Boschma and Fritsch 2014). It also found that tolerance and openness have a positive effect on the regional share of the creative class, while the provision of public facilities in healthcare and education have only minor, if any, impacts.

A separate study (Andersen et al. 2010) examined the creative class across the Scandinavian and Nordic countries. It found the creative class to be highly concentrated in "capital regions," which possess the thickest labor markets. That city size is an especially important factor in fueling economic growth and attracting the creative class is borne out by the Swedish case, where one of the best indicators of creative class share has been shown to be city size (Hansen 2014). Exploring the relative effects of the "people climate" and "business climate" in creative class locations, it found that both play a role, though an attractive job market or business climate is especially important. This underlines the important point that people climates are heterogeneous, varying considerably from place to place. The authors also examine the impact of welfare state policies in the Nordic countries.

A study of China (Florida, Mellander, and Qian 2012) found that the distribution of the creative class and talent is considerably more concentrated there than in the United States or any other advanced economies. Universities are a key factor in shaping the distribution both of talent and of technological innovation in China; tolerance also plays an important role. All that said, neither talent nor technology are associated with the economic performance of Chinese regions, a finding that stands in sharp contrast to the pattern in more advanced economies, and casts doubt on China's transition from a traditional manufacturing to a creativity-driven economy.

Taken as a whole, this body of work confirms the uneven geography of the creative class and its tendency to concentrate in large cities or those with knowledge-based institutions like universities.

Innovation and Cities

In the industrial age, the key role in innovation was played by nation-states, which subsidized large-scale research, the R&D departments of major corporations, and also clusters of smaller, more entrepreneurial firms which spearheaded and commercialized breakthrough technologies. With the rise of knowledge-based capitalism, cities and locations themselves have come to play a more central role. Marshall (1890) initially noted the role that agglomerations or clusters of firms played in spurring innovation and productivity improvement. Agglomeration is a way of organizing the division of labor horizontally between firms as posed to vertically within large firms.

Later Porter (1996, 1998), Piore and Sabel (1984) and others noted the rise of clusters of firms and of industrial districts as key features of knowledge-based capitalism. Saxenian (1994) applied the theory of industrial districts to leading innovative centers like California's Silicon Valley. Jacobs (1969, 1984) identified the role of talent clustering in locations as the primary economic driver of activity. Her insights were later formalized by Lucas (1988, 2001) who identified the role of human capital externalities tied to location as the primary motor force or basic underlying "mechanism" of economic development. There is now a large body of literature on the higher rate of innovation and entrepreneurship and on the accumulation of talent and human capital in cities.

Studies of innovation identify the geographic concentration of innovation and university R&D and science and technology in cities, metro areas, and mega-regions. Florida notes that science and innovation are the spikiness of all economic activity, concentrated in an even smaller number of larger cities and metro areas than are population or economic activity (Florida 2005b).

A study of mega-regions (Florida, Gulden, and Mellander 2008), clusters of economically interconnected cities whose boundaries are delineated by night-time light emissions, identified 40 worldwide, including the Boston-Washington corridor, Greater London, Amsterdam-Brussels-Antwerp, Rome-Milan-Turin, Osaka-Nagoya, and Greater Tokyo. Those mega-regions house 23% of the world's population but account for 59% of its economic output and 77% of patents globally. Another study (Florida et al. 2010) identified approximately 680 metropolitan regions with half a million inhabitants or more. Together, these regions account for more than 60% of the world's economic activity, while housing just a quarter of the world's population.

Elite academic activity, particularly in science and technology, is also spiky, concentrated and clustered in a small number of cities globally. Figure 15.2 (via Florida 2010 and based on the Academic Ranking of World Universities) charts the concentration of academic activity in metros around the world. The leading centers are Boston-Cambridge, New York, San Francisco, and Los Angeles in the United States, and London, Paris, and Zurich in Europe. In general, there is a strong geographic concentration of top universities on the East and West Coasts of North America, Western Europe, and a few places in Asia and Australia/New Zealand.

Venture capital and startup activity is also extremely concentrated across the globe. Figure 15.3 shows the location of start-up companies across the world (via

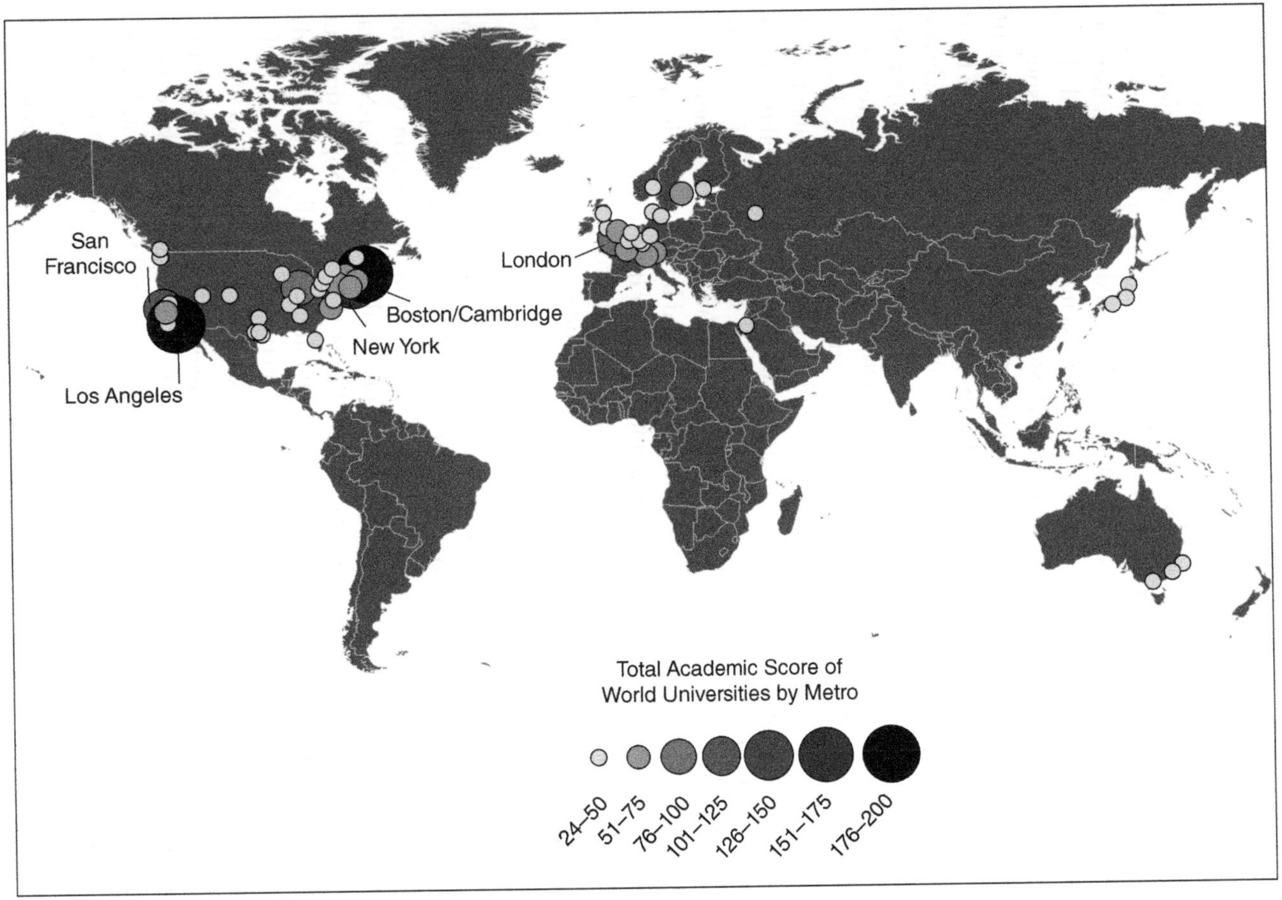

Figure 15.2 Total academic score of world universities by metro.
Source: Florida (2010).

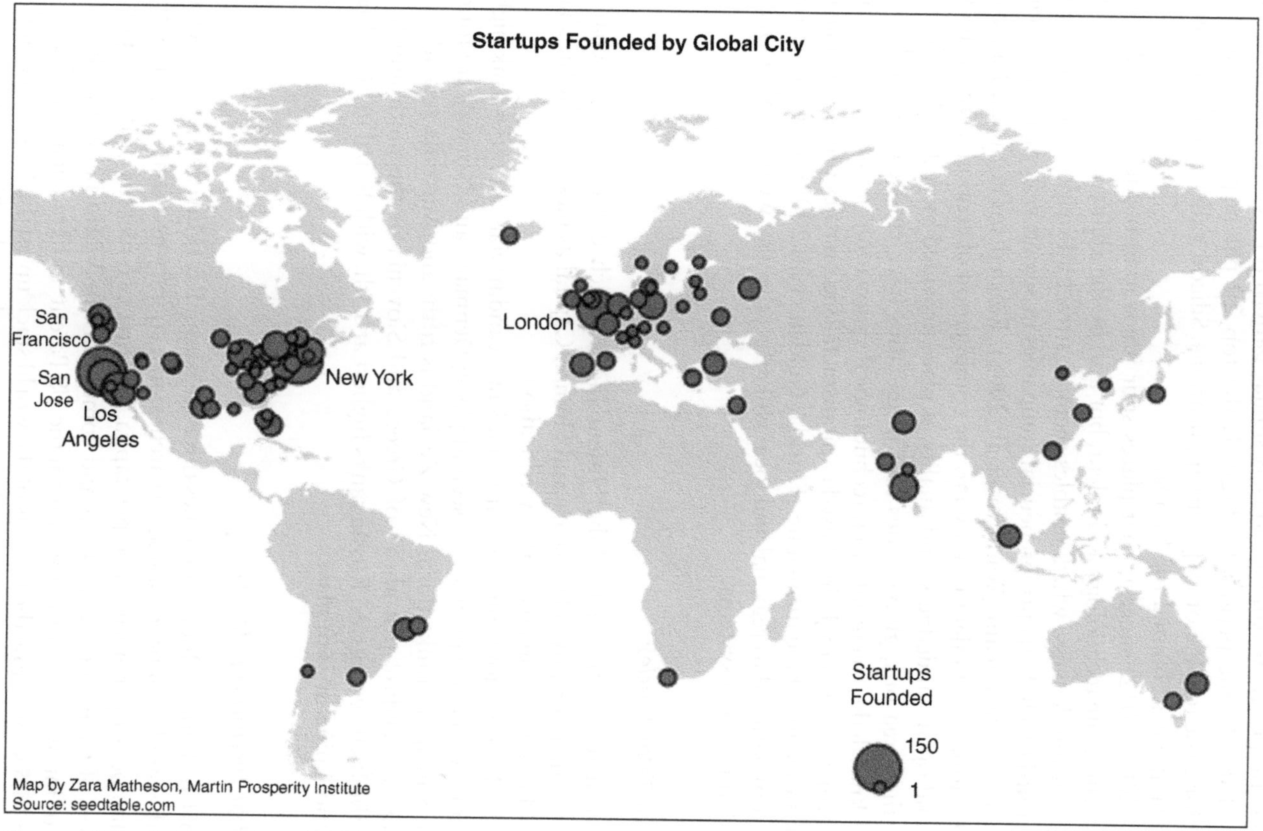

Figure 15.3 Start-ups founded by global metro.
Source: Florida (2013a).

Florida 2013a and SeedTable). The leading centers include the San Francisco Bay Area, New York, London, Los Angeles, Toronto, Boston-Cambridge, Chicago, Berlin, Bangalore, Austin, Seattle, and São Paulo. This is due to the fact that technology has a more complex set of inputs and is more closely tied to end users; also technology talent in general prefers urban locations. A similar pattern is revealed for innovation (Florida 2013b). In the period 2001–2011, the top 20 patent-producing metros accounted for close to 64% of all patent applications. Silicon Valley and the Bay area in California alone accounted for 13.6%.

Florida (2002, 2012) suggests that places spur innovation not just through their ability to accumulate technology and talent, but also by the ability to engender tolerance and open-mindedness, which enables them to attract a broader range of talent. Roughly a third to a half of entrepreneurial companies in Silicon Valley have a foreign-born founder on their team (Wadhwa et al. 2008). Related studies (e.g., Inglehart 1989, 1997) found that talent is drawn to the combination of economic opportunity, tolerance and open-mindedness, and quality of place provided by certain locations.

For all these reasons, global cities and mega-cities have emerged as the loci of talent and knowledge accumulation, technological innovation, and entrepreneurial business formation in knowledge-based capitalism, taking on roles previously played by corporations and the nation-state (Florida, Gulden et al. 2008). In the long run, this shift from national to regional and local will affect the way we think about innovation and innovation systems.

The Global Talent Index

We now turn to the broader Global Talent Index which combines two separate measures of talent: creative class share and Human Capital as measured by the share of adults who have completed tertiary education.

Figure 15.4 shows how countries fare on talent. Scandinavian countries rank high – Finland and Sweden are in first and second place; Denmark and Norway are fourth and sixth; Singapore ranks third; New Zealand is fifth; and Australia seventh. The United States is eighth, just ahead of Greece and Slovenia in the ninth and tenth spots. Of the BRIC countries, Russia ranks highest at 13th, with Brazil in 66th, India in 75th, and China in 76th place.

The Global Technology Index

The Global Technology Index combines three measures: R&D Investment as a percent of GDP; the number of professional researchers in a country as a share of population; and the number of patents granted per capita. The first two measure critical inputs to the process of technology generation and come from the World Bank; the third is a measure of innovative output and is from the US Patent and Trademark Office.

Figure 15.5 shows how the nations of the world stack up on the Global Technology Index. Finland, the home of Nokia, ranks first in researchers, third in R&D investment, and fourth in innovation. Japan ranks fourth in R&D investment, third in researchers, and second in innovation. Japanese companies have not only consistently pushed the technology envelope, they have been if anything even better at building reliable, follow-on generations of products, from high-quality cars to flat panel displays.

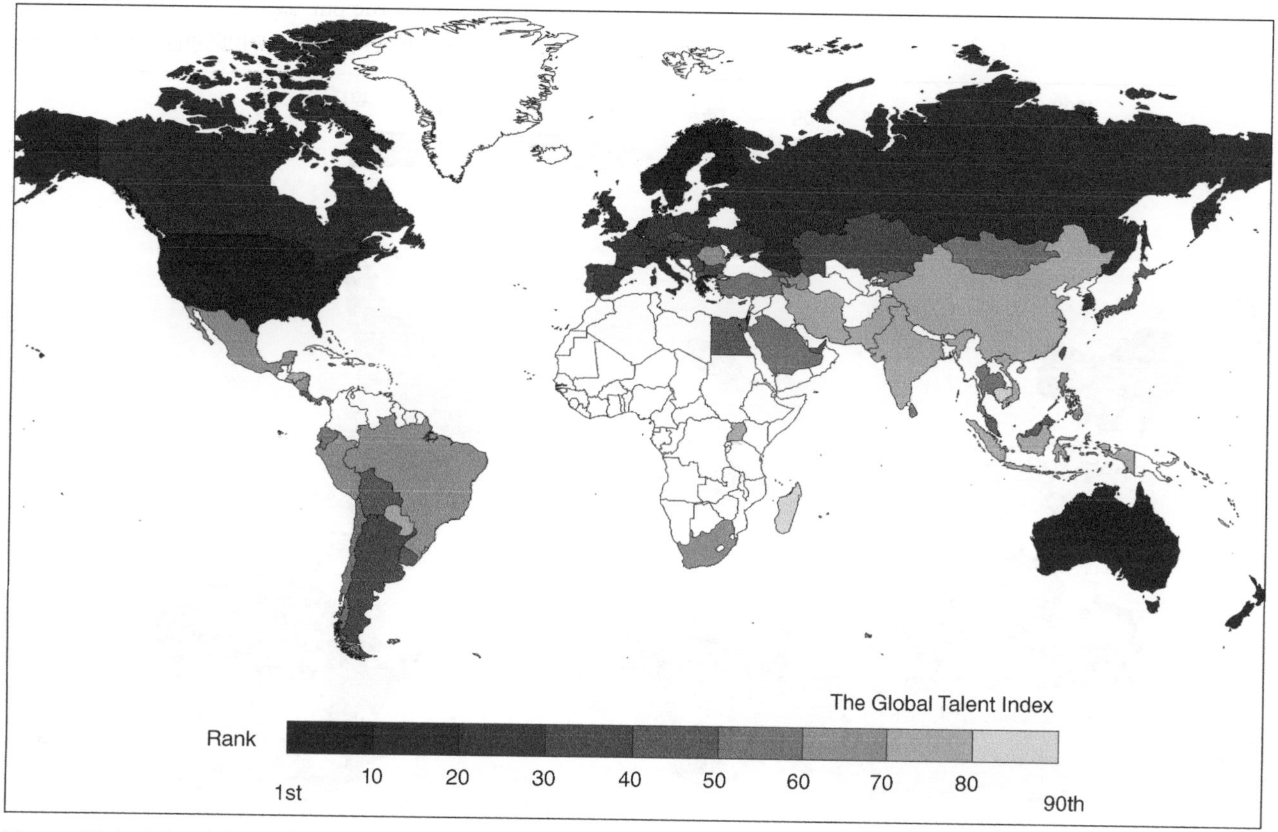

Figure 15.4 The global talent index map.
Source: Florida, Mellander, and Stolarick (2011).

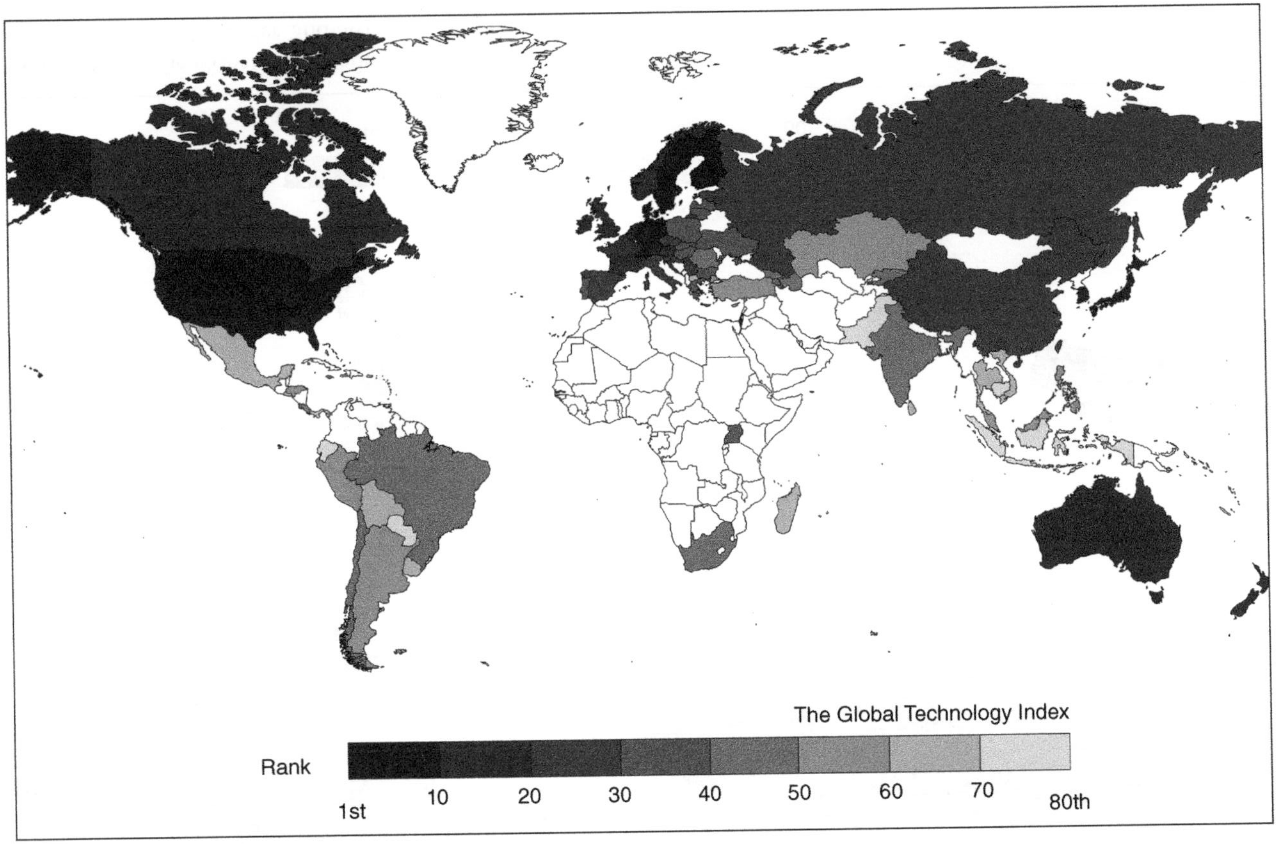

Figure 15.5 The global technology index map.
Source: Florida, Mellander, and Stolarick (2011).

The United States ranks third, finishing sixth in R&D investment and seventh in researchers, but solidly in first place for innovation. With its infrastructure for entrepreneurial venture capital finance in Silicon Valley and elsewhere, the United States has seen a long list of high-tech start-ups turn into global giants, such as Microsoft, Apple, Google, and Yahoo.

Fourth-ranked Israel has the highest concentration of engineers in the world – 135 per 10,000 people, compared to 85 per 10,000 people in the United States (Senor and Singer 2009). Sweden takes fifth place and Switzerland, Denmark, Korea, Germany, and Singapore round out the top ten. Canada ranks 11th. While much has been made of the ascendance of the BRIC countries, they do not rank highly on our technology measure. The highest ranking of them is Russia, in the 28th spot. China ranks 37th, about the same as Latvia and Bulgaria. Brazil takes 48th place and India 49th, just behind Serbia and Croatia.

The Global Tolerance Index

The Global Tolerance Index combines two measures of tolerance, both from the Gallup World Poll. The first captures a country's openness to ethnic and racial minorities; the second its openness to gay and lesbian people.

Figure 15.6 shows how the nations of the world stack up on global tolerance. Canada ranks first of the 82 countries and Ireland second. The Netherlands ranks third: it is the only country among the top five that is more open to gay and lesbian people (83%) than it is to racial and ethnic minorities (73%). New Zealand ranks fourth, followed by Australia in fifth place. Spain is in sixth place, followed by Sweden and the United States.

The Global Creativity Index

The 3Ts, technology, talent, and tolerance, work together in mutually reinforcing ways. Any one T is a necessary but in itself insufficient condition for economic success. For a nation or region to effectively compete in the creative economy, all three Ts have to work together. The GCI combines all three indexes into an integrated measure of a nation's overall creative economic potential.

Figure 15.7 charts how the nations of the world compare on the overall GCI. Sweden is first, topping the United States in second place. Finland takes third place, followed by Denmark in fourth, Australia in fifth, and New Zealand in sixth place. Canada and Norway take seventh and eighth place and Singapore and the Netherlands round out the top ten. Of the BRIC nations, Russia scores the best in 36th place, followed by Brazil in 46th, India in 50th, and China in 58th place.

Creativity and National Economic Performance

We now turn to the associations between creativity measured by the GCI and economic and social performance. Our analysis is structured around four key questions.

- Are more creative economies also more productive and competitive?
- Are more creative nations associated with higher levels of human development more generally?

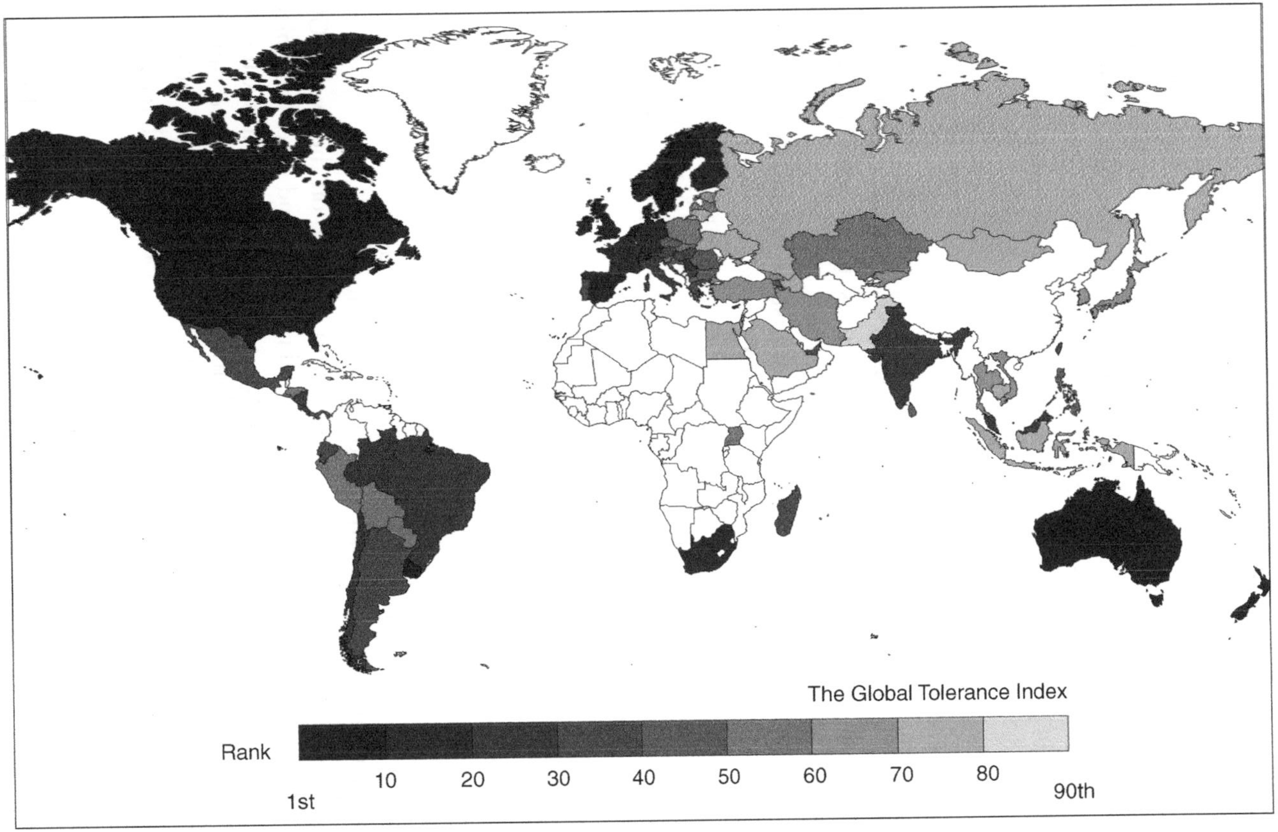

Figure 15.6 The global tolerance index map.
Source: Florida, Mellander, and Stolarick (2011).

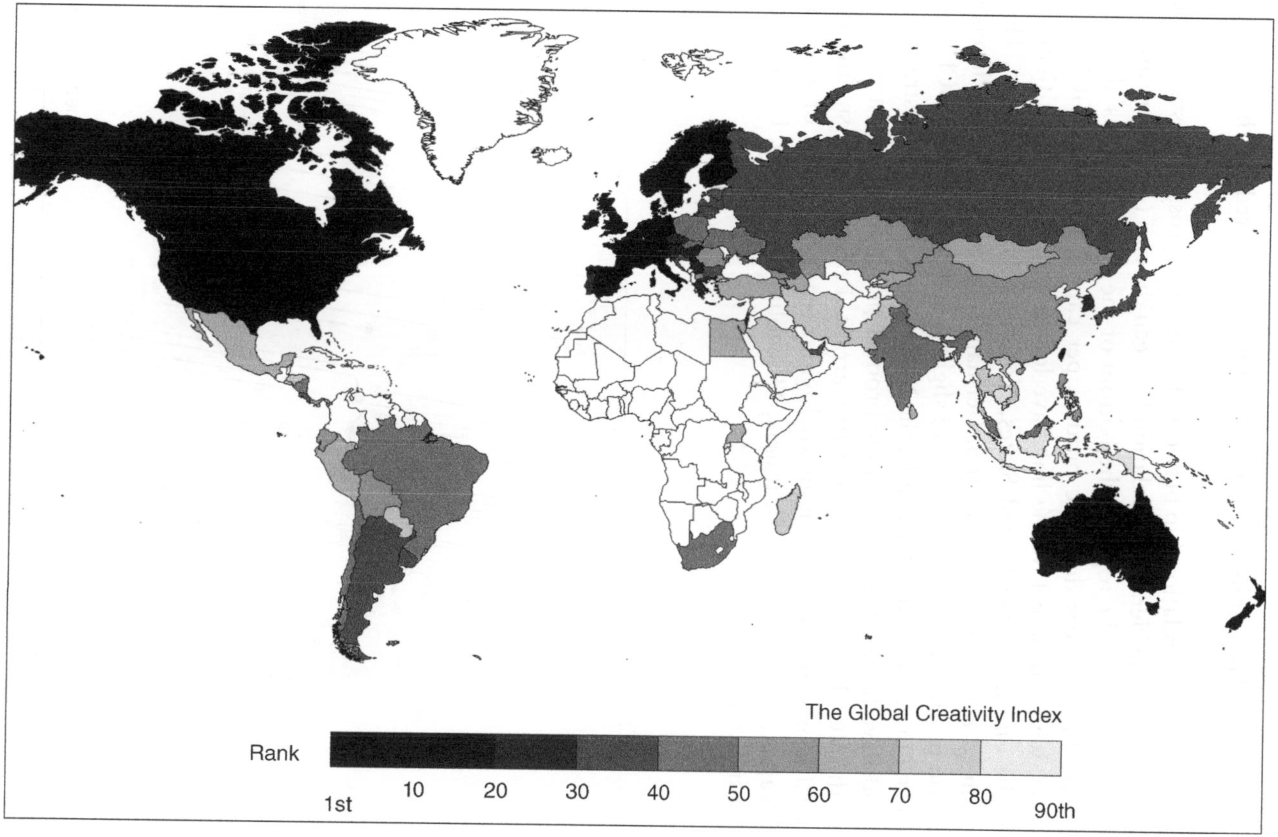

Figure 15.7 The global creativity index map.
Source: Florida, Mellander, and Stolarick (2011).

- Do more creative economies generate higher levels of happiness for their residents?
- Are creative economies more or less equal?

Creativity and Economic Output

We begin with the relationship between creativity as measured by the GCI and economic output per capita.

Figure 15.8 shows the relationship between the GCI and gross domestic product per capita. There is a considerable correlation between the two (.84); there are only a few extreme outliers at the top and the bottom of the chart. Nations above the fitted line have higher gross domestic product per capita than their GCI scores would predict, while those below the line have lower economic output than predicted. On the one hand, the United States, Norway, Switzerland, Japan, Hong Kong, the United Kingdom, Israel, Austria, Germany, and Korea all have levels of gross domestic product per capita which are slightly higher than their GCI scores would predict. On the other hand, Canada, the Netherlands, Finland, Australia, and New Zealand have levels of gross domestic product per capita which are slightly lower than their GCI scores would seem to warrant. Perhaps more significantly, very low GCI scores appear to be associated with even lower levels of economic output per capita, as the cases of Nicaragua, Mongolia, Kyrgyzstan, Uganda, and Madagascar indicate.

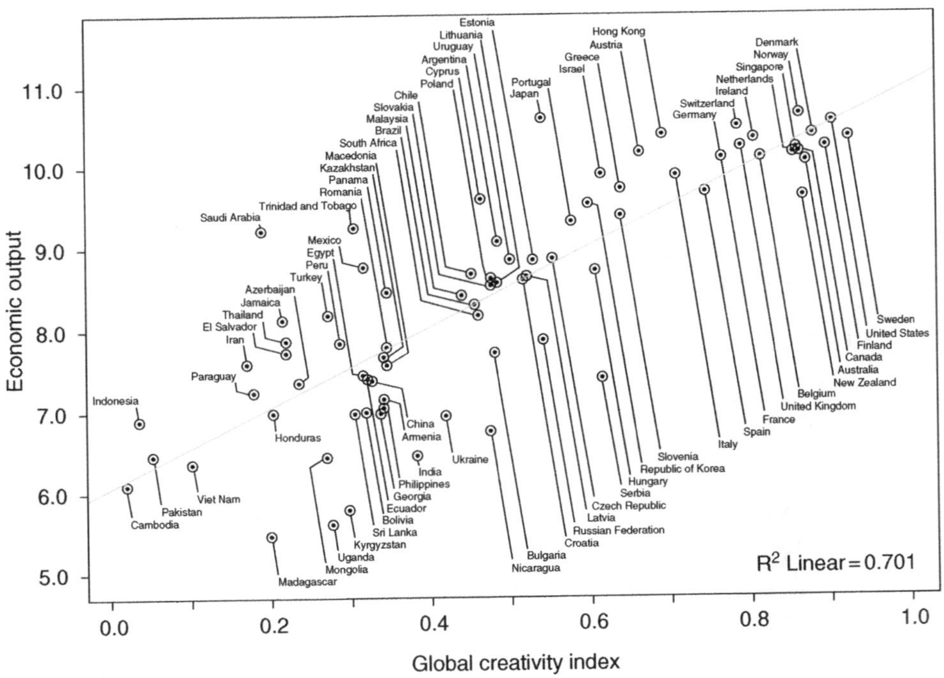

Figure 15.8 Global creativity and economic output.
Source: Florida, Mellander, and Stolarick (2011). Economic output measured as log of GDP per capita.

Creativity and Economic Competitiveness

We now turn to the relationship between the GCI and a well-established measure of global competitiveness: the Global Competitiveness Index developed by Michael Porter et al. (2000) for the World Economic Forum.

Figure 15.9 plots the relationship between the GCI and the Global Competitiveness Index. The correlation between the two is again substantial (.79), however there is more "scatter" about the line. The United States, Singapore, Switzerland, the United Kingdom, Japan, Hong Kong, Germany, and Denmark all perform better in terms of competitiveness than their GCI scores would lead one to expect. Canada performs just slightly better than its GCI scores predict. China and India both perform significantly better on competitiveness than their GCI scores suggest they should. On the other hand, New Zealand, Ireland, and Spain perform lower on competitiveness than their GCI scores would seem to predict.

Creativity and Entrepreneurship

Schumpeter (1934, 1942, 1947a, 1947b) long ago showed how innovation and entrepreneurship come together to set in motion the "creative destruction" that drives economies forward. Research in psychology finds close connections between creative and entrepreneurial people (Sternberg and Lubart 1999). We chart the

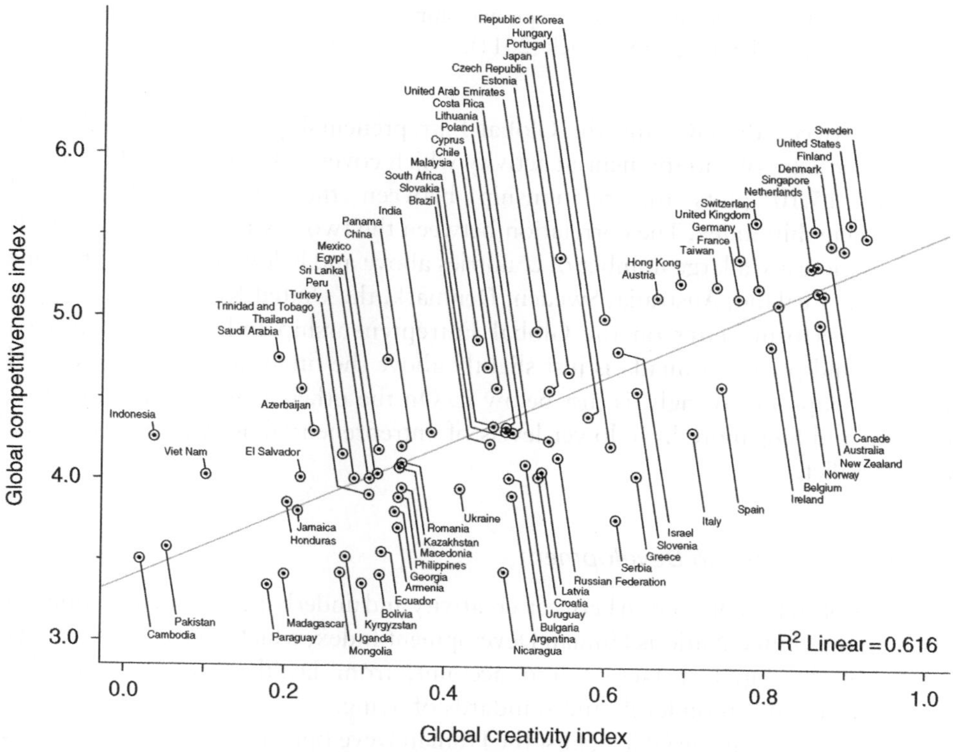

Figure 15.9 Global creativity and competitiveness.
Source: Florida, Mellander, and Stolarick (2011).

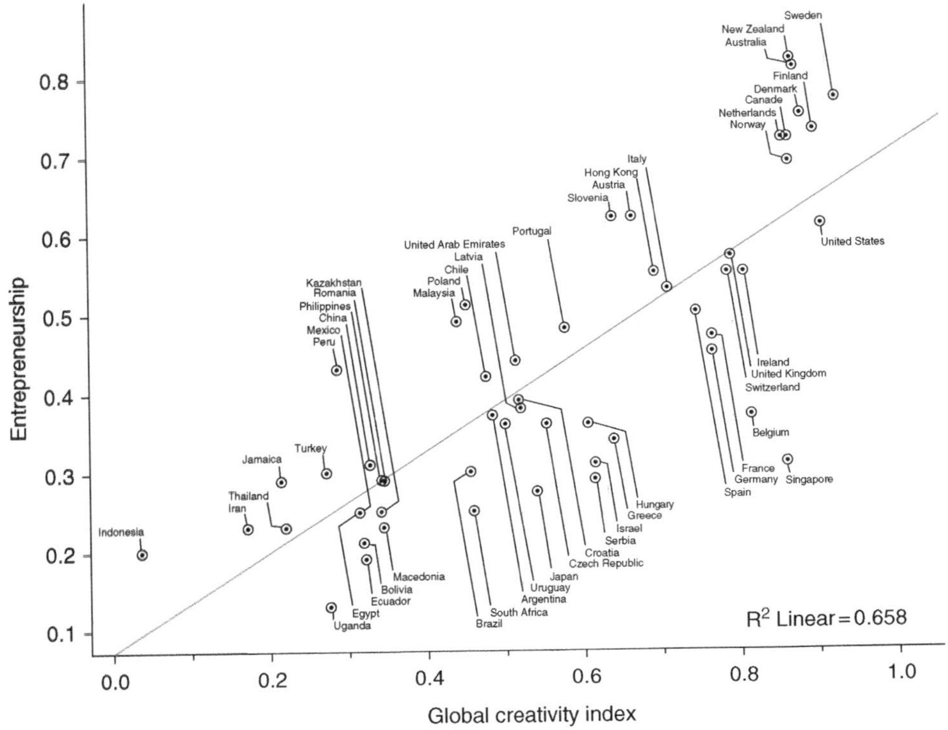

Figure 15.10 Global creativity and entrepreneurship.
Source: Florida, Mellander, and Stolarick (2011).

relation between the GCI and the Global Entrepreneurship Index (Acs and Szerb 2010), a measure of entrepreneurial activity which covers 54 nations worldwide.

Figure 15.10 plots the relationship between the GCI and the Global Entrepreneurship Index. The correlation between the two is strong (0.81). The fit is good but there are a large number of countries above and below the line. On the one hand, New Zealand, Australia, Sweden, Denmark, the United Kingdom, and Hong Kong all perform better on the Global Entrepreneurship Index than their GCI scores would predict. Canada is just slightly above the fitted line, while the United States, perhaps surprisingly, is just below it. On the other hand Germany, France, Belgium, and Singapore have lower levels of entrepreneurial activity than the GCI would predict.

Creativity and Human Development

We now look at the connection between creativity and underlying human development based on the United Nations Human Development Index, which takes a wide variety of human development factors into account, from health conditions and life expectancy to education levels and standards of living.

Figure 15.11 plots the GCI against the Human Development Index for the nations of the world. The fit is good, with outliers mainly at the lower left hand quadrant of the graph, where some of the least developed nations of the world are found. The

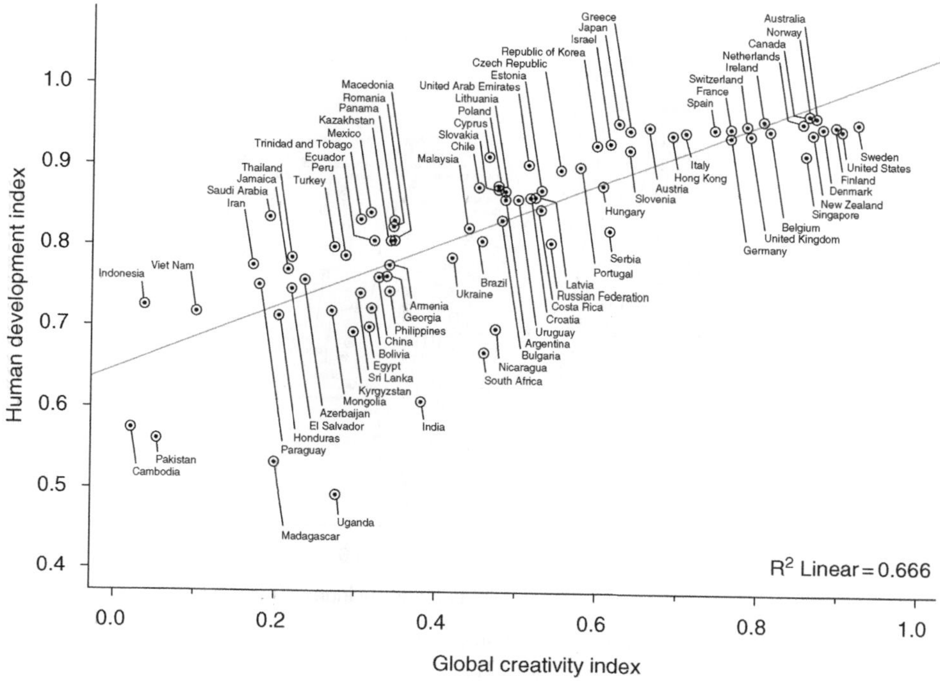

Figure 15.11 Global creativity and human development.
Source: Florida, Mellander, and Stolarick (2011).

overall correlation between GCI and the Human Development Index is again substantial (0.82). The United States performs considerably less well on the Human Development Index than its GCI score would predict; Canada performs slightly better. Of the BRIC countries, India performs significantly worse on Human Development Index than its GCI score would seem to warrant. Four less developed nations, Cambodia, Pakistan, Madagascar, and Uganda, lag significantly on Human Development when their GCI score is taken into account.

Creativity and Well-Being

There is a considerable ongoing debate concerning the relationship between economic development and subjective well-being, much of it revolving around the effects of money or material well-being on happiness. While it was initially found that the relationship between income and happiness only holds within and not across countries – the so-called "Easterlin effect" (Easterlin 1995) – more recent econometric studies by Deaton (2008) and Stevenson and Wolfers (2008) have challenged this view, finding that income exerts strong effects on happiness across the board.

We examine the relationship between the GCI and a comprehensive measure of happiness and life satisfaction collected by the Gallup Organization's World Poll (2010), using a standard set of core questions that asks individuals to rank their satisfaction with aspects of their lives.

Figure 15.12 plots the GCI against life satisfaction. The correlation between the two is substantial (0.74). The fit is reasonably good, with outliers mainly at the

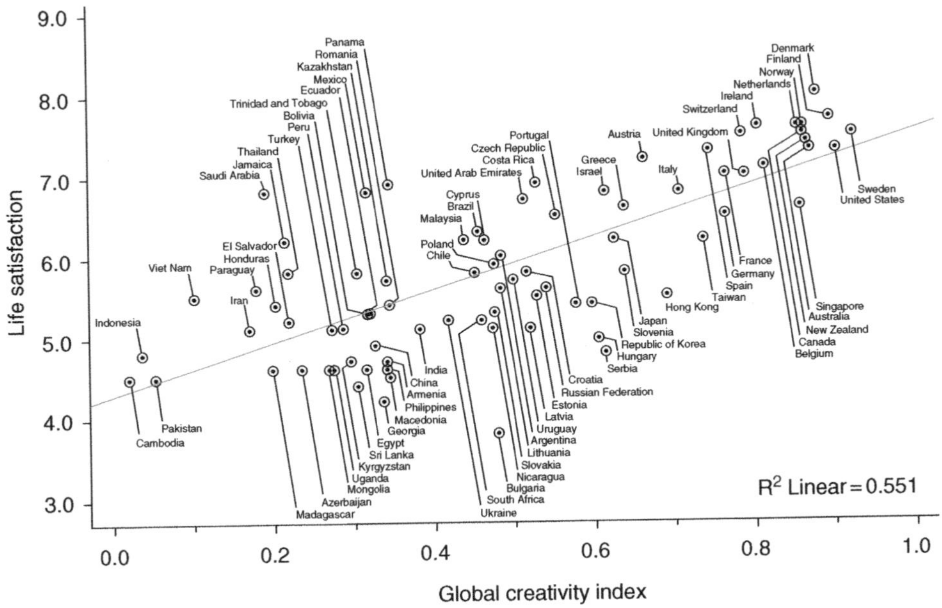

Figure 15.12 Global creativity and life satisfaction.
Source: Florida, Mellander, and Stolarick (2011).

bottom quadrants of the graph, that is, among the less developed nations. The relationship between the GCI and life satisfaction is strongest among the more advanced nations. Denmark, Finland, the Netherlands, Ireland, Switzerland, New Zealand, and Canada have higher levels of life satisfaction than their GCI scores would predict. The United States has a level of life satisfaction that is roughly in line with its GCI score. Singapore, the United Kingdom, Taiwan, Hong Kong, and Korea have lower levels of life satisfaction than their GCI scores would predict. Among the BRIC countries, Brazil has a significantly higher level of life satisfaction than its GCI score would predict, while Russia's is considerably lower. Both India and China have lower levels of life satisfaction than the GCI would predict.

Creativity and Inequality

It is widely believed that the shift from an industrial to an innovative, knowledge, and creativity-driven economy exacerbates inequality, as former high-paying, family-supporting manufacturing jobs decline and the labor market is bifurcated into higher-pay, higher-skill knowledge and professional jobs and lower-pay, lower-skill service jobs. A series of studies document the growth in income inequality across the advanced world (e.g., Sherman and Stone 2010). But do higher levels of creative performance and higher levels of income inequality necessarily go together?

Figure 15.13 plots the association between the GCI and the basic measure of income inequality, the Gini coefficient. The line slopes downward and the correlation between the two is negative (–0.43); in other words, the GCI is systematically associated with *lower* levels of income inequality – and hence greater equality – across the nations of the world.

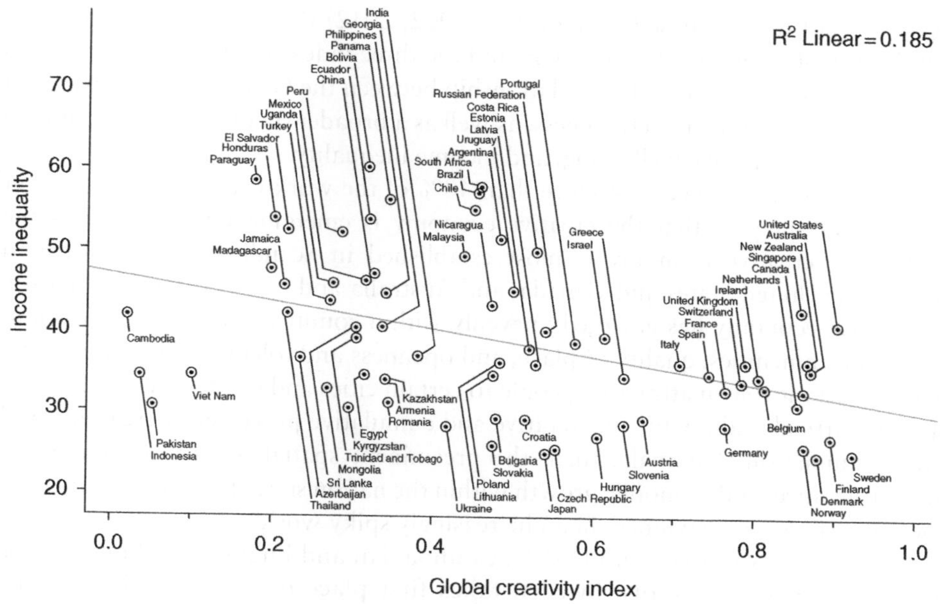

Figure 15.13 Global creativity and income inequality.
Source: Florida, Mellander, and Stolarick (2011). Income inequality measured as a Gini Index.

A closer look at the chart reveals two distinctive trajectories for the relationship between creativity and inequality. On the one hand, there are countries like the United States, the United Kingdom, Singapore, and to a lesser extent, Australia and New Zealand, in which high levels of creativity as well as productivity and economic competitiveness go hand in hand with higher levels of inequality. But there are also a substantial number of countries in which high levels of creativity, competitiveness, and productivity combine with much lower levels of inequality. These are largely Scandinavian and Northern European countries, including Sweden, Denmark, Finland, Norway, the Netherlands, and Germany. Japan is represented as well. Among the less developed nations, we find high levels of inequality in South American nations like Paraguay, Bolivia, Panama, Brazil, Honduras, Ecuador, and Argentina. Of the BRIC nations, China, Russia, and especially Brazil all exhibit much higher levels of inequality than their GCI scores would predict.

Each high-creativity, high-inequality nation has a high-creativity, low-inequality counterpart. This is a likely reflection of these countries' differing levels of social welfare. Though more systematic study is needed before we can draw any firm conclusions, this finding gives reason for optimism. At the same time, it suggests that sustainable, long-term prosperity requires a significant investment in education and skill development.

Conclusion

This chapter has charted the role of creativity in the economic and social performance of 82 nations across the advanced and developing worlds, using both the Global Creativity Index, a composite measure of the technology, talent, and tolerance capabilities of nations based on Florida's (2002, 2012) theory of the creative class and the 3Ts of economic development, and the three indexes that make it up. It has also systematically examined the relationship between the GCI and key measures of economic output and competitiveness, as well as a broader set of measures of human development, subjective well-being, and income inequality.

The creative class makes up more than 40% of the workforce in 40 nations, but our analysis suggests that the creative economy is emerging unevenly across the world. The creative economy is most established in Scandinavia and Northern Europe, the United States and Canada, and Australia and New Zealand. Moreover, the creative economy has emerged unevenly within countries as well as between it. Factors like amenities, quality of place, and openness and tolerance play an increasingly important role in attracting people to certain cities and creative centers. Larger cities in particular allow for the iterative and cumulative processes of creativity and skill-accumulation that make them the preeminent spatial homes of technology, talent, tolerance, and economic growth within the nation state. These larger cities are emerging as economic winners in an increasingly spiky world.

When it comes to the overall GCI, Scandinavian and English-speaking nations dominate the top ten spots. Sweden takes first place on the overall GCI. The United States is second, followed by Finland, Denmark, and Australia; Canada ranks eighth; New Zealand, Norway, Singapore, and the Netherlands round out the top ten. Despite their rapid economic rise, the BRIC nations still do not crack the upper tiers on the GCI: Russia ranks 31st, Brazil 46th, India 50th, and China 58th.

The GCI is closely associated with key measures of economic and social progress. Nations that score high on the GCI have higher levels of economic output, entrepreneurship, and overall economic competitiveness. Nations that invest in creativity and that achieve on the 3Ts of economic development also have higher levels of human development, life satisfaction, and happiness.

This chapter suggests that the economic, technological, and entrepreneurial performance of nations is due to a set of factors that extend beyond just science and technology. In line with other research at the national and subnational levels (see Barro 1991; Rauch 1993; Simon and Nardinelli 1996; Simon 1998; Glaeser 1999; Glaeser and Saiz 2003), we find that talent plays a substantial role in national economic performance. Our research further suggests that talent spans more than the conventional educational measure of human capital, but also includes occupational skill based on the kind of work people do. Overall, our research suggests that technology and talent are mutually reinforcing and complementary factors in cross-national economic performance.

This chapter further suggests that non-market factors, which we refer to as tolerance and openness to human capital, play an additional complementary role in the economic performance of nations. Nations that are more open to talent, which have lower barriers to entry for talent, benefit from the ability to attract and to mobilize a wider pool of talented individuals. This is in line with other research that finds that tolerance, diversity, and openness are associated with higher rates of innovation (Page 2007) and higher levels of national economic performance (Noland 2005; Noland and Pack 2004). Generally speaking, technology, talent, and tolerance operate as capabilities that reinforce each other.

This chapter also sheds light on the complex relationship between creativity, economic progress, and inequality. Generally speaking we find creativity (measured via the GCI) to be negatively associated with income inequality. While some countries, like the United States and the United Kingdom, achieve high GCI scores alongside relatively high levels of inequality, in general elevated levels of global creativity are associated with lower levels of inequality. We identify two paths here: a low road path where higher creativity is associated with greater levels of inequality, and a high road path where higher levels of creativity are associated with lower levels of inequality. This is at odds with the notion that it is large disparities in income that create the incentives and motivations that drive economic progress.

This provides some cause for optimism, even in the wake of the global economic crisis, as it suggests that economic growth and development can go along with lower levels of inequality – and that our future prosperity increasingly turns on the full development of each and every human being.

Note

1 This an updated, edited, and abridged edition of Florida, Mellander, and Stolarick (2011) (see also Florida and Tinagli 2004; Florida 2005a). Thanks are due to Karen King, Nick Lombardo, and Arthur Goldwag for support with research and editing of this chapter.

References

Acs, Zoltan J., and László Szerb. 2010. "A Global Entrepreneurship Index." *Foundations and Trends in Entrepreneurship* 5: 341–435.

Andersen, Kristina Vaarst, and Mark Lorenzen. 2009. *Den Danske Kreative Klasse: Hvem er den, hvordan ser den ud, hvor bor den – og hvad betyder det?* [The Danish Creative Class: Who, How, and Where – and So What?] Aarhus: Klim.

Andersen, Kristina Vaarst, Markus M. Bugge, Høgni Kalsø Hansen, Arne Isaksen, and Mika Raunio. 2010. "One Size Fits All? Applying the Creative Class Thesis onto a Nordic Context." *European Planning Studies* 18: 1591–1609.

Asheim, Bjørn. 2009. "Guest Editorial: Introduction to the Creative Class in European City Regions." *Economic Geography* 85: 425–442.

Asheim, Bjørn, and Høgni Kalsø Hansen. 2009. "Knowledge Bases, Talents, and Contexts: On the Usefulness of the Creative Class Approach in Sweden." *Economic Geography* 85: 425–442.

Barro, Robert J. 1991. "Economic Growth in a Cross Section of Countries." *Quarterly Journal of Economics* 106(2): 407–443.

Becker, Gary. 1964. *Human Capital.* New York: Columbia University Press for the National Bureau of Economic Research.

Boschma, Ron, and Michael Fritsch. 2009. "Creative Class and Regional Growth: Empirical Evidence from Seven European Countries." *Economic Geography* 85(4): 391–423.

Boschma, Ron, and Michael Fritsch. 2014. "The Location of the Creative Class in Seven European Countries." In *The Creative Class Goes Global*, ed. Charlotta Mellander, Richard Florida, Bjørn Asheim, and Meric Gertler, 227–240. Abingdon: Routledge.

Clifton, Nick. 2008. "The Creative Class in the UK: An Initial Analysis." *Geografiska Annaler, Series B, Human Geography* 90(1): 63–82.

Clifton, Nick. 2014. "Location, Quality of Place, and Outcomes: Applying the '3Ts' Model to the UK." In *The Creative Class Goes Global*, ed. Charlotta Mellander, Richard Florida, Bjørn Asheim, and Meric Gertler, 183–209. Abingdon: Routledge.

Currid-Halkett, Elizabeth, and Kevin Stolarick. 2013. "Baptism by Fire: An Occupational Analysis of Creative Sector Impact on Regional Economies." *Cambridge Journal of Regions, Economy and Society* 6(1): 55–69.

Deaton, Angus. 2008. "Income, Health and Wellbeing around the World: Evidence from the Gallup World Poll." *Journal of Economic Perspectives* 22(2): 53–72.

Drucker, Peter. 1969. *The Age of Discontinuity: Guidelines to Our Changing Society*. New York: Harper and Row.

Drucker, Peter. 1993. *Post-Capitalist Society*. New York: HarperCollins.

Easterlin, Richard. 1995. "Will Raising the Incomes of All Increase the Happiness of All?" *Journal of Economic Behavior & Organization* 27(1): 35–47.

Falck, Oliver, Michael Fritsch, and Stephan Heblich. 2011. "The Phantom of the Opera: Cultural Amenities, Human Capital, and Regional Economic Growth." *Labor Economics* 18(6): 755–766.

Florida, Richard. 2002. *The Rise of the Creative Class*. New York: Basic Books.

Florida, Richard. 2005a. *The Flight of the Creative Class: The New Global Competition for Talent*. New York: HarperBusiness.

Florida, Richard. 2005b. "The World is Spiky." *The Atlantic Monthly October*: 48–51.

Florida, Richard. 2010. *The Great Reset: How New Ways of Living and Working Drive Post-Crash Prosperity*. New York: Harper Collins.

Florida, Richard. 2012. *The Rise of the Creative Class, Revisited*. New York: Basic Books.

Florida, Richard. 2013a. "The New Global Start-Up Cities." *Citylab.com*, June 4, http://www.theatlanticcities.com/jobs-and-economy/2013/06/new-global-start-cities/5144 (accessed January 1, 2014).

Florida, Richard. 2013b. "Where America's Inventors Are." *Citylab.com*, October 9, http://www.theatlanticcities.com/jobs-and-economy/2013/10/where-americas-inventors-ara/7069/ (accessed January 1, 2014).

Florida, Richard, and Charlotta Mellander. 2013a. "The Creative Class Goes Global." In *The Creative Class Goes Global*, ed. Charlotta Mellander, Richard Florida, Bjørn Asheim, and Meric Gertler, 1–8. Abingdon: Routledge.

Florida, Richard, and Charlotta Mellander. 2013b. "The Creative Class around the World." In *The Creative Class Goes Global*, ed. Charlotta Mellander, Richard Florida, Bjørn Asheim, and Meric Gertler, 308–312. Abingdon: Routledge.

Florida, Richard, and Irene Tinagli. 2004. *Europe in the Creative Age*. London: Demos.

Florida, Richard, Tim Gulden, and Charlotta Mellander. 2008a. "The Rise of the Mega-Region." *Cambridge Journal of Regions, Economy and Society* 1: 459–476.

Florida, Richard, Charlotta Mellander, and Tim Gulden. 2010. "Global Metropolis: Assessing Economic Activity in Urban Centers Based on Night Time Satellite Images." *The Professional Geographer* 64: 178–187.

Florida, Richard, Charlotta Mellander, and Haifeng Qian. 2012. "China's Development Disconnect." *Environment and Planning A* 44: 628–648.

Florida, Richard, Charlotta Mellander, and Kevin Stolarick. 2008b. "Inside the Black Box of Regional Development." *Journal of Economic Geography* 8: 615–649.

Florida, Richard, Charlotta Mellander, and Kevin Stolarick. 2011. *The Global Creativity Index*. Toronto: University of Toronto Martin Prosperity Institute.

Florida, Richard, Charlotta Mellander, and Kevin Stolarick. 2014. "Talent, Technology and Tolerance in Canadian Regional Development." In *The Creative Class Goes Global*, ed. Charlotta Mellander, Richard Florida, Bjørn Asheim, and Meric Gertler, 50–84. Abingdon: Routledge.

Fritsch, Michael, and Michael Stuetzer. 2009. "The Geography of Creative People in Germany." *International Journal of Foresight and Innovation Policy* 5(1–3): 7–23.

Gabe, Todd. 2009. "Knowledge and Earnings." *Journal of Regional Science* 49(3): 439–457.

Gallup. 2010. Gallup World Poll, see http://www.gallup.com/services/170945/world-poll.aspx.

Glaeser, Edward L. 1999. "Learning in Cities." *Journal of Urban Economics* 46: 254–277.

Glaeser, Edward L. 2011. *The Triumph of the City: How Our Greatest Invention Makes Us Richer, Smarter, Greener, Healthier, and Happier*. London: Pan Macmillan.
Glaeser, Edward L., and Albert Saiz. 2003. "The Rise of the Skilled City." NBER Working Papers 10191, National Bureau of Economic Research.
Hansen, Høgni Kalsø. 2014. "Florida's Creative Class in a Swedish Context: The Problem of Measuring Tolerance and Amenity-Driven Growth." In *The Creative Class Goes Global*, ed. Charlotta Mellander, Richard Florida, Bjørn Asheim, and Meric Gertler, 87–102. Abingdon: Routledge.
Inglehart, Ronald. 1989. *Culture Shifts in Advanced Industrial Society*. Princeton, NJ: Princeton University Press.
Inglehart, Ronald. 1997. *Modernization and Post-Modernization*. Princeton, NJ: Princeton University Press.
Jacobs, Jane. 1969. *The Economies of Cities*. New York: Random House.
Jacobs, Jane. 1984. *Cities and the Wealth of Nations*. New York: Random House.
Lorenzen, Mark, and Kristina Vaarst Andersen. 2009. "Centrality and Creativity: Does Richard Florida's Creative Class Offer New Insights into Urban Hierarchy?" *Economic Geography* 85(4):363–390.
Lucas, Robert. 1988. "On the Mechanics of Economic Development." *Journal of Monetary Economics* 22: 3–42.
Lucas, Robert. 2001. "Externalities and Cities." *Review of Economic Dynamics* 4: 245–274.
Machlup, Fritz. 1962. *The Production and Distribution of Knowledge in the United States*. Princeton, NJ: Princeton University Press.
Marlet, Gerard, and Clemens van Woerkens. 2004. "*Skills and Creativity in a Cross-Section of Dutch Cities*." Utrecht: Tjalling C. Koopmans Institute, Discussion Paper Series 04–29.
Marlet, Gerard, and Clemens van Woerkens. 2014. "Tolerance, Aesthetics, Amenities or Jobs?: The Attraction of the Dutch City to the Creative Class." In *The Creative Class Goes Global*, edited by Charlotta Mellander, Richard Florida, Bjørn Asheim, and Meric Gertler, 138–163. Abingdon, UK: Routledge.
Marrocu, Emanuela, and Raffaele Paci. 2012. "Education or Creativity: What Matters Most for Economic Performance?" *Economic Geography* 88(4): 369–401.
Marshall, Alfred. 1890. *Principles of Economics*, 8th edn. London: Macmillan.
Marx, Karl. 1887. *Capital*, vol. 1. London: Sonnenschein.
McGranahan, David, and Timothy Wojan. 2007. "Recasting the Creative Class to Examine Growth Processes in Rural and Urban Counties." *Regional Studies* 41(2): 197–216.
Mellander, Charlotta. 2009. "Creative and Knowledge Industries: An Occupational Distribution Approach." *Economic Development Quarterly* 23(4): 294–305.
Mincer, Jacob. 1974. *Schooling, Experience and Earnings*. New York: Columbia University Press for the National Bureau of Economic Research.
Mowery, David C., and Nathan Rosenberg. 1989. *Technology and the Pursuit of Economic Growth*. Cambridge: Cambridge University Press.
Nelson, Richard. 1959. "The Simple Economics of Basic Scientific Research." *Journal of Political Economy* 67: 297–306.
Nelson, Richard. 1977. "In Search of Useful Theory of Innovation." *Research Policy* 6: 36–76.
Nelson, Richard, and Sidney Winter. 1982. *An Evolutionary Theory of Economic Change*. Cambridge, MA: Harvard University Press.
Noland, Marcus. 2005. "Popular Attitudes, Globalization and Risk." *International Finance* 8(2): 199–229.
Noland, Marcus, and Howard Pack. 2004. "*Islam, Globalization, and Economic Performance in the Middle East*." Washington, DC: Petersen Institute for International Economics.
Page, Scott E. 2007. *The Difference: How the Power of Diversity Creates Better Groups, Firms, Schools, and Societies*. Princeton, NJ: Princeton University Press.
Piore, Michael, and Charles Sabel. 1984. *The Second Industrial Divide: Possibilities for Prosperity*. New York: Basic Books.

Porter, Michael. 1996. "Competitive Advantage, Agglomeration Economies, and Regional Policy." *International Regional Science Review* 19(1): 85–94.

Porter, Michael. 1998. "Clusters and the New Economics of Competition." *Harvard Business Review*, November-December: 77–90.

Porter, Michael, Jeffrey Sachs, and Andrew Warner. 2000. *The Global Competitiveness Report*. New York: McGraw-Hill.

Rauch, James. 1993. "Productivity Gains from Geographic Concentration of Human Capital: Evidence from the Cities." *Journal of Urban Economics* 34: 380–400.

Romer, Paul M. 1986. "Increasing Returns and Long-run Growth." *Journal of Political Economy* 90: 1002–37.

Rosenberg, Nathan. 1983. *Inside the Black Box: Technology and Economics*. Cambridge: Cambridge University Press.

Rosenberg, Nathan, and Richard R. Nelson. 1994. "American Universities and Technical Advance in Industry." *Research Policy* 23(3): 323–348.

Saxenian, AnnaLee. 1994. *Regional Advantage: Culture and Competition in Silicon Valley and Route 128*. Cambridge, MA: Harvard University Press.

Schumpeter, Joseph A. 1934. *The Theory of Economic Development* (trans. Redvers Opie). Cambridge, MA: Harvard University Press.

Schumpeter, Joseph A. 1942. "The Process of Creative Destruction." In Schumpeter, *Capitalism, Socialism and Democracy*, 81–86. New York: Harper and Brothers.

Schumpeter, Joseph A. 1947a. "The Creative Response in Economic History." *Journal of Economic History* 7(2): 149–159.

Schumpeter, Joseph A. 1947b. "Theoretical Problems of Economic Growth." *Journal of Economic History* 7: 1–9.

Senor, Dan, and Saul Singer. 2009. *Start-Up Nation: The Story of Israel's Economic Miracle*. Toronto: McClelland and Stewart.

Sherman, Arloc, and Chad Stone. 2010. "Income Gaps Between Very Rich and Everyone Else More Than Tripled in Last Three Decades, New Data Show." Center on Budget and Policy Priorities, http://www.cbpp.org/cms/index.cfm?fa=view&id=3220 (accessed January 1, 2014).

Simon, Curtis J. 1998. "Human Capital and Metropolitan Employment Growth." *Journal of Urban Economics* 43: 223–243.

Simon, Curtis J., and Clark Nardinelli. 1996. "The Talk of the Town: Human Capital, Information and the Growth of English Cities, 1861–1961." *Explorations in Economic History* 33: 384–413.

Solow, Robert M. 1956. "A Contribution to the Theory of Economic Growth." *Quarterly Journal of Economics* 70: 65–94.

Spengler, Joseph J. 1977. "Adam Smith on Human Capital." *American Economic Review* 67: 32–36.

Sternberg, Robert J., and Todd I. Lubart. 1999. "The Concept of Creativity: Prospects and Paradigms." In *Handbook of Creativity*, ed. Robert J. Sternberg, 3–15. Cambridge: Cambridge University Press.

Stevenson, Betsey, and Justin Wolfers. 2008. "Economic Growth and Subjective Well-Being: Reassessing the Easterlin Paradox." *Brookings Papers on Economic Activity* 1: 1–87.

Stolarick, Kevin. 2014. "The Creative Class 'Down Under': Exploring the Creative Class Theory in Australia." In *The Creative Class Goes Global*, ed. Charlotta Mellander, Richard Florida, Bjørn Asheim, and Meric Gertler, 243–261. Abingdon: Routledge.

van Aalst, Irina, Oedzge Atzema, Ron Boschma, and Frank van Oort. 2014. "The Creative Class, Related Variety and Economic Growth in Dutch City-Regions." In *The Creative Class Goes Global*, ed. Charlotta Mellander, Richard Florida, Bjørn Asheim, and Meric Gertler, 164–182. Abingdon: Routledge.

Wadhwa, Vivek, AnnaLee Saxenian, Ben A. Rissing, and Gary Gereffi. 2007. "America's New Immigrant Entrepreneurs: Part I." Duke University, Duke Science, Technology & Innovation Paper 23.

Wadhwa, Vivek, AnnaLee Saxenian, Ben A. Rissing, and Gary Gereffi. 2008. "Skilled Immigration and Economic Growth." *Applied Research in Economic Development* 5(1): 6–14.

Chapter 16

Global Science Collaboration

Stefan Hennemann and Ingo Liefner

Introduction: The Organization of Science Collaboration

The internationalization or globalization of scientific collaboration is both a key feature of today's science and a political mandate. Several mechanisms are contributing to the internationalization of science. First, some of the most important institutions and frameworks for conducting science have become increasingly global. An obvious one is the use of English as the global science language in publications, conferences, and communication (Ammon 2001). Closely related to this is the use of peer reviews for assessing the quality of publications, grant proposals, and concepts (Smith 2006). Second, international collaboration means accessing and combining the scientific resources, ideas, and competencies available worldwide. This can raise the quality not only of research, but also of education, and may accelerate discovery processes (Carlsson 2006; de Wit 2002; Kafouros et al. 2008). During recent decades, reductions in travel and communication costs as well as advances in communication tools, including the Internet and social media, have made internationalization possible and helped these forces to shape the organization of science. Politicians and managers in higher education and science, however, have responded to these changes, putting internationalization high on their agendas. One example is the demand for international collaboration in research and teaching as a means of quality control and a method of broadening the scope of activities (Knight 2004). An existing international collaboration is a performance measurement which is easy to monitor and use as part of evaluation and performance-based funding procedures (cf. Geuna and Martin 2003).

However, the counter-hypothesis of primarily national or even regional science organization is not far away (Ponds 2009). Powerful mechanisms such as national languages, policies, and funding agencies implicitly favor domestic collaboration.

The Handbook of Global Science, Technology, and Innovation, First Edition.
Edited by Daniele Archibugi and Andrea Filippetti.

The advantages of close spatial proximity for frequent personal contacts and knowledge-sharing are obvious, and with respect to commercialization, local industry may seek scientific partners around the corner. These factors are at the core of some well-established theoretical concepts such as the National Innovation System (Lundvall 1992; Nelson 1993) and the Regional Innovation System (Cooke 1998). All these concepts provide the line of arguments for the hypothesis of a nationalism of technology systems (cf. Archibugi and Michie 1997), focusing on the complex interactions among the main groups within an innovation system, namely the industrial base, the governing state, and the science base, with scientific collaboration being one of the sub-dynamics within such complex innovation systems.

Network science allows us to analyze whether our images of how scientific collaboration is organized are accurate. For most scientific fields, network science shows that a national and regional collaboration bias is highly relevant and demonstrates features of a universally applicable law. While social networks have their own structural properties compared to other forms of networks (Newman and Park 2003), they represent a potentially large explaining factor for a universal function relating collaboration and space. A technical explanation for a decreasing probability of collaboration with distance is related to the fact that the actors and ties in networks are not randomly distributed. Large centers of knowledge creation are more likely to be found in urban agglomerations and their dominating organizations, that is, those which already have a large number of collaborations are more likely than small units to receive an additional one (cf. Yook, Jeong, and Barabási 2002). Still, this does not explain the spatial adjacency in which this linking activity is taking place.

Findings reported in the more recent literature on scientific collaboration in networks stem from three different perspectives: (1) looking at forces that increase internationalization; (2) regarding impediments to internationalization; (3) using network analysis to explain network structures and dynamics.

Many empirical studies have collected evidence for the internationalization of science and an increase in collaboration distances (Luukkonen, Persson and Sivertsen 1992: Archibugi and Iammarino 2002; Glänzel and Schubert 2005b; Leydesdorff and Wagner 2008; Tijssen, Waltman and van Eck 2011; Waltman, Tijssen and van Eck 2011; Ahlgren, Persson, and Tijssen 2013) that is sidelined by a general increase in research team sizes (cf. Hicks and Katz 1996). One important finding relates to an increase of the mean spatial distance between a paper's co-authors from the 1980s to today. This gives quantitative proof for the assumption that internationalization is a structural feature of today's science. However, Ahlgren et al. (2013: 782) also found that there is a quantitative counter-effect on the local scale, that is, an increase of local collaboration frequencies.

Quantitative evidence alone may say little about the true importance of internationalization and be of no help assessing whether an internationalization of scientific collaboration is indeed desirable. The quality of scientific research, however, may also rise with collaboration internationalization. This is the essence of findings presented by Jones, Wuchty, and Uzzi (2008) and Narin, Stevens and Whitlow (1991), who show that multi-collaborator papers spanning organizational boundaries receive more citations and thus seem to be more successful. This finding relates to the diversity argument for complex processes in Kafouros et al. (2008), who argue that innovation activity (as a complex process) is enhanced when firms internationalize, that is, absorb ideas from diverse sources. Diversity as a driver of opportunities and performance is also a prominent topic in complex network research (cf. Eagle, Macy, and Claxton 2010). These two lines of argument (network diversity, geographical distance) are likely to be interrelated, that is, the network-based diversity increases with geographical distance, simply because the physically distant collaborators have different local embeddings (e.g., institutional framework, cultural idiosyncrasies). Network-based approaches lead to the second large body of literature, which is concerned with the structural impediments to internationalization of scientific collaboration, coming up with equally convincing results. Examples of empirical studies that address limits to internationalization and stress the importance of national border effects have been presented by Hennemann, Rybski, and Liefner (2012) and Hoekman, Frenken, and Tijssen (2010). Arguments against growing internationalization have also been expressed from the perspective of economic theory. For example, Beckmann (1993, 1994) suggested classical gravity models as a perspective from which to understand collaboration patterns. According to this view, scientific collaboration is mainly the outcome of (rational) choices of individual researchers, who wish to maximize or (being more modest) to satisfy their own research output (Beckmann 1993: 8). Due to cognitive limits, the individual does not know all potential collaborators (such search restrictions can be extended by transaction costs, e.g., for traveling, meeting, while other constraints include language barriers: Beckmann 1994: 242), and collaboration thus tends to be more local.

However, when considering the evolution of collaboration networks in science, Barabási et al. (2002) collected interesting empirical evidence that may help to save the idea of productive science systems with strong local structures by showing that it is possible to find decreasing average shortest path lengths among the collaborators in growing scientific networks. This finding partly contradicts findings from other research on (social) networks that are governed in a similar way by hub-and-spoke structures. It seems that the hubs (e.g., the top league of researchers in a field) are perfectly optimizing their immediate network neighborhoods and contribute to a greater system-wide knowledge access. Having said this, it could be hypothesized that knowledge produced and disseminated in such a global system may well be leading toward more globalized scientific knowledge over time, with diversified ideas improving the overall outcome quality. Jansen, von Görtz, and Heidler (2010) attempt to dissemble relevant structural properties of science networks by testing the influence of different types of sub-structures within larger networks on the knowledge production performance. Neither clustered areas of networks (e.g., intensively cross-connected groups of researchers) nor strategic positioning alone (e.g., researchers who attempt to utilize structural holes, that is, provide non-redundant connections in networks) are per se beneficial for enhancing system-wide knowledge production. One potentially hampering factor for quality enhancement can be derived from the deeper analysis of the results of Barabási and colleagues (2002), who conclude that in growing scientific networks, the actors already present dominate the community to the detriment of the new entrants, that is, the preference of collaborating with researchers already present is much higher than that with new entrants (2002: 611, cf. also Zitt, Bassecoulard, and Okubo 2000). One might wonder whether this poses a potential threat for scientific progress.

This brief review of the related literature has shown two things: first, there is evidence of a quantitative and qualitative trend toward the internationalization of scientific collaboration, while convincing counter-evidence is also present. Second, from analyzing the structures of collaboration, a number of arguments emerge that indicate which of the structures seem to be more desirable from the point of view of science, society, or politics. However, there is no common understanding of whether science is becoming more global, and whether this should indeed be the case.

Against this background, this chapter aims to provide a compressed piece of evidence that clarifies the respects in which scientific collaboration is international (or global) and the respects in which there is a persistent national and regional bias. It will be shown that both features can be present at the same time. This chapter further summarizes and explores several arguments as to what future development of scientific collaboration might be considered beneficial. There are two major equitable views of how to explain collaboration at the individual level. On the one hand, if maximizing the outcome is important, the best partner available will be chosen no matter where this partner is located geographically. On the other hand, if resource constraints are present, the selection of geographically adjacent partners will be favored.

In order to provide a consistent research framework for the purpose of evaluating the structures and trends in global science collaboration, it is necessary to define comparable research units, because similarities in research communities are considered to be important proxies for space-independent collaboration, thus leveling out the described network-based diversity effects, at least in terms of the core topics. Therefore, the concept of epistemic communities serves as a reference notion for the empirical section. Epistemic communities are mainly referred to as being informal collaboration networks (Ynalvez and Shrum 2011). Members of epistemic communities are commonly characterized as being professionally similar, that is, they share the same or similar notions or beliefs about subject-specific applications and techniques (cf. Haas 1992; Weisberg and Muldoon 2009). This should enable all members to understand one another easily. With increasing improvements in technical infrastructure, it is the cognitive proximity (cf. Boschma 2005; Nooteboom et al. 2007) that enables members of the epistemic community to compensate for a possible lack of spatial proximity, also transcending organizational boundaries (cf. Gertler 2003) and forming global science networks of researchers who specialize in similar fields (Moodysson 2008). Especially at the frontiers of science, new knowledge can be expected to be created through joint efforts in international teams of excellence (Power and Malmberg 2008).

Concerning the interplay of the science sector with the (private) business sector, it is becoming increasingly prevalent for the overall economic development that comprehensive collaboration strategies, both national and international, are increasing competitiveness. This is particularly relevant in science-based industries and with respect to the technology and creative side of globalization, beyond production chains (cf. Archibugi and Michie 1997; Florida and Mellander, this volume, Chapter 15). Catchphrases such as "triple helix" (Etzkowitz and Leydesdorff 2000) or "national systems of innovation" (cf. Lundvall 1992; Nelson 1993) seek to explain this complex interplay of industries, governments, and the public research sector. It is not only advanced economies that are addressed by these concepts, but also rapidly growing latecomer economies, such as China, who are aware of the merits of intensive interaction, and the science sector can be seen to a certain extent as an integrator of global knowledge and local sources, even in the industrial sector (cf. Hennemann 2011; Barnard et al., this volume, Chapter 18). Therefore, the way science is organized geographically, as discussed in this chapter, relates directly to the ongoing debate concerning the strategic development paths of nations (e.g., the "techno-nationalism vs. techno-globalism" debate: Archibugi and Michie 1997; Ponds 2009).

How Researchers Connect Given a Set of Potential Partners at Various Distances

The measurement of scientific activity in quantitative terms is the major undertaking in *scientometrics*, a science field that dates back to the 1950s (cf. Garfield 1955; de Solla Price 1965). Scientometric research, as the science of science in a narrow sense, is commonly restricted to the measurement of pure (journal) publication activity. However, with the ubiquitous availability of other data sources such as research databases or the Internet, with its social communication platforms, this narrow view was relaxed to account for such new sources. This wider view is often termed *informetrics*, the measurement of information and knowledge production and dissemination. Consequently, three different sources of data are commonly used for the quantitative analysis, including academic publications, patents, and Internet resources. In recent years, the collaboration measurement in epistemic communities has gained importance in evaluating the knowledge production and dissemination in science, technology, and innovation systems at the system level. The evaluation and analysis of effects of such developments in the science sector is investigated within the methodological framework of the "science of team science" (Börner et al. 2010) or within the framework of "spatial scientometrics" (Frenken, Hardeman, and Hoekman 2009; Hennemann et al. 2012). The complex systems under investigation can be represented by networks (or mathematical graphs). Compared to simplistic frequency counting of the number of publications or patents in a given science field, the network approach to collaboration in science has the advantage of capturing a great variety of structural effects of the system itself that are present in most social systems, and allows for a direct translation of the network-oriented theoretical concepts into empirical research. The following empirical part of this chapter uses this systemic perspective and seeks to reveal the collaboration patterns in different epistemic communities in geographical space. The main question in this endeavor is to find out whether the geographical distance between any two collaborators in a given (and narrowly defined) science field is relevant for their probability of conducting joint research and eventually co-authoring a paper. Other questions of interest include the differences among science fields, for example, finding out whether analytical sciences such as biotechnology-related research is differently organized compared to more synthetically driven fields such as electrical engineering. One hypothesis often presented in recent decades relates to the impact of a shrinking geographical space due to the increasing capabilities of interacting through technical systems (e.g., the Internet) or through improvements in infrastructures that sharply reduce travel time (e.g., availability of high-speed trains, dramatically increased flight schedules). The basic rationale behind this is that knowledge production is made easier on a global scale, hence gaining in importance compared to local or national collaboration, and subliminally suggesting that this would improve the quality of the knowledge produced.

How Collaboration Activity in Science Systems Can Be Measured

In order to build co-authorship networks, publications from Thomson Reuters' Science Citation Index Expanded™ (SCI-E) are used, which is common in scientometric research (Glänzel and Schubert 2005a). Other international databases such as SCOPUS or Google Scholar offer similar quality with differences in the details. These data sources can be considered the best sources available. It needs to be discussed whether indicators such as journal publications can be a reliable source of information for the evaluation of science systems. Obviously, there are differences across disciplines – and perhaps also between national science systems – in terms of the publication culture. For example, the main publication outlets such as proceedings, journals, books, edited volumes, and the publication frequencies vary between humanities and natural sciences. The focus on publications may also vary between countries according to the importance that is attached to research assessments and performance-based funding. The limitations of bibliographic data are discussed in greater depth, for example, in Bornmann and Leydesdorff (2011). However, it is generally acknowledged that co-authorships from academic journal publications are a reliable indicator especially for the collaboration activity and knowledge production in academia. Academic papers in international journals represent the last step in a longer pre-publication phase during which the collaborators interact and share their knowledge on this certain topic, so that in most cases the publication is the visible outcome of joint research.

The concept of epistemic communities is used here as a rationale to define narrow scientific sub-fields. As described above, an epistemic community assembles professionals on the basis of similar contents, so that a scientific sub-field that is concerned with a specific technique or research object is an epistemic community. Since the content and technical language is a constituting element, the actual boundary of an epistemic community may be quite blurred. However, due to the proximity of the professional contents, scientists in a specific field can be assumed to be potential members of the community with the ability to collaborate principally with all other (potential) members of this community. This is the basic idea of the method proposed by Hennemann et al. (2012) that is adopted and extended here.

In this research, six distinct but internally narrow scientific sub-fields or techniques (bluetooth, image compression algorithms, heart-valve research, research on H5N1, tissue engineering, nanotube carbon-related research) were analyzed based on publications from 2004 to 2008 in order to build co-authorship networks (see Table 16.1). These fields were selected to represent various modes of knowledge production and community organization, which may influence the collaboration activity. Moreover, only cross-organizational co-authorships were included in order to eliminate university research team effects. In order to evaluate the development of collaboration structures in scientific sub-fields over time, the case of tissue engineering is analyzed in greater depth (see below).

Table 16.1 Characterization of the six epistemic communities 2004–2008.

Technique / research field	*Community character*	*Community size* (no. of organizations)*
Bluetooth	The wireless communication standard was defined by a company-driven research consortium. The underlying technology is based on electrical engineering and has little cross-disciplinary and enabling character. This community is engineering oriented.	455
Image compression algorithms	Software to reduce the amount of required storage capacity. As general purpose technology, this community is organized across discipline borders, ranging from medicine-related imaging processing to cineastic usage.	614
Heart valves	The related technologies combine fields such as human medicine, veterinary research, and engineering. For this community, moreover, it can be assumed that non-research-related stakeholders exert great influence on collaboration (e.g., companies, public health systems and policies, ethics commissions).	1,034
H5N1 (bird-flu virus)	This is a rather spontaneously emerging community where ad hoc solutions were needed to prevent a pandemic from spreading. This external pressure involves interaction between vaccine-producing companies, government agencies, and global research hot spots.	1,271
Tissue engineering	A cross-disciplinary medicine-related field, linked to material research and biology. The general issue is condensed in ethical debates. Like in all medicine-related communities, the stakeholder influence may have a strong impact on collaboration.	2,721
Nanotubes (carbon)	Highly cross-disciplinary and potentially enabling for other science and technology fields. This field uses costly equipment, which may put pressure on the collaboration activity.	4,483

Note: * as evident from the SCI-expanded sources.

Non-Linear Character of Complex Systems

Before such co-authorship networks can be explored and analyzed in greater depth, a method is required for capturing interdependencies among scientific collaborators that goes beyond simplistic counting of who collaborates with whom. This includes the need for a method that helps to assess whether an empirical observation in the network is significant or not. As already stated, graphs, as the mathematical form of a network, are capable of capturing the complexity that is inherent in collaboration and knowledge exchange. With such a network approach, it becomes easier to analyze the systemic effects and the dynamics. Moreover, this network approach simplifies the process of developing a randomized benchmark (or baseline) model in order to obtain an idea of the relevance of the empirical findings, that is, answering the question of whether the results are significant or simply the outcome of a random process.

In standard (inference) statistics where mainly linear phenomena are present and the independence of the observations is guaranteed, one of the common benchmark models is the Gaussian (or normal) distribution. If the empirically observed value/parameter is outside a pre-defined range (e.g., a range in which 95% of all values are encompassed), the empirical observation is termed *significantly different* from the theoretical expectation suggested by the model of Gauss. It is assumed that many empirically observable phenomena can be described by the Gaussian bell-shaped curves. However, in the world of networks and complex social systems, there are no such common and universal benchmarks that fulfill similar functions to the Gaussian model. This is related to the prevalent behavior of the system/network elements, which are dependent on each other, meaning that a change in one individual actor affects the behavior of all other actors in the network. While eliminating one random element from a Gaussian distribution does not change the character of the distribution, taking out a well-connected node from a network can fundamentally change the network structure, and with it the distribution of parameters. In addition to these interdependence and non-linear effects, the system's basic organizational principle affects the elements. It is easy to comprehend that in a very dense network where the nodes are almost completely cross-connected, the individual properties of the nodes do not differ much compared to nodes in a network that has a few heavily connected hubs and otherwise sparse areas with little cross-connection among the remaining nodes. Having said this, it is obvious that complex systems such as science collaboration networks need special treatment.

One strategy to produce a baseline model of a given empirical network, thus allowing for a "test" of empirically established network properties, is to randomize the present connections by picking any two edges randomly and cross-wiring (or swapping) them such that the original edges a-b and c-d become a-c and b-d (this idea was first proposed by Maslov and Sneppen 2002). This simple procedure repeated for a large number of swaps guarantees that the outcome network is structurally equivalent to the empirical network, that is, a dense network is still dense and the hubs remain hubs, etc. In addition, all empirical connections between the nodes are destroyed through the cross-wiring process, as required for significance testing. When the empirical observation is now compared to this random twin network, unusual and significant observations can be identified easily. If this randomization is repeated for a large number of configurations (i.e., a large number of random twin networks is produced), the confidence about the parameter estimate increases.

The main parameter of interest in this chapter is the collaboration probability with respect to the geographic distance between the collaborators: the empirical observation of collaboration among researchers at certain distances will be compared to a random collaboration process in a structurally equivalent theoretical world where researchers in an epistemic community collaborate with each other without paying attention to the geographical distance that separates them.

Here, the probability is defined as the number of empirically identified collaborations divided by the number of possible collaborations among the involved members of the epistemic community at a given distance. This approach essentially mimics the selection process of a given researcher A to connect or collaborate with another researcher B in the defined population of available counterparts that are located at different distances around A. It is important to note that this is a conditional probability (i.e., bound at a given distance), which is different from the measurement approach in most other cases when dealing with mean collaboration distances in kilometers (cf. Waltman et al. 2011). In a situation of two potential partners for researcher A, where one partner B is geographically nearby and one partner C is far away, without spatial condition, connecting to either of these partners would lead to equal probabilities (in absolute terms). Under the condition of available partners at a given distance, selecting B would have a greatly different meaning compared to collaborating with C, because the number of potential collaborators at the same distance is much lower in the case of selecting B than the much higher number of potential partners at large distances (in relative or conditional terms). In other words, a researcher who chooses 10 local partners from a population of 20 and also 10 distant partners from a population of 2000 obviously has a high propensity for choosing local collaboration partners.

The Geographical Patterns and Collaboration Probabilities in Six Scientific Communities

Figures 16.1 to 16.6 show the world maps for the collaboration activity on the left panel. While for the first two communities, bluetooth (Figure 16.1) and image compression (Figure 16.2), the Asian region is comparatively present, research on heart valves (Figure 16.3) in this part of the world is very weak. While Asian research activity can be neglected in heart-valve research, it is again very pronounced in the case of the bird flu virus research. African research steps in, showing contributions from a continent that is usually hardly visible in the global research system. Asia is the assumed origin of the bird flu virus (Figure 16.4). It is thus a reasonable result that these areas are included in the global research system, adding competence to the

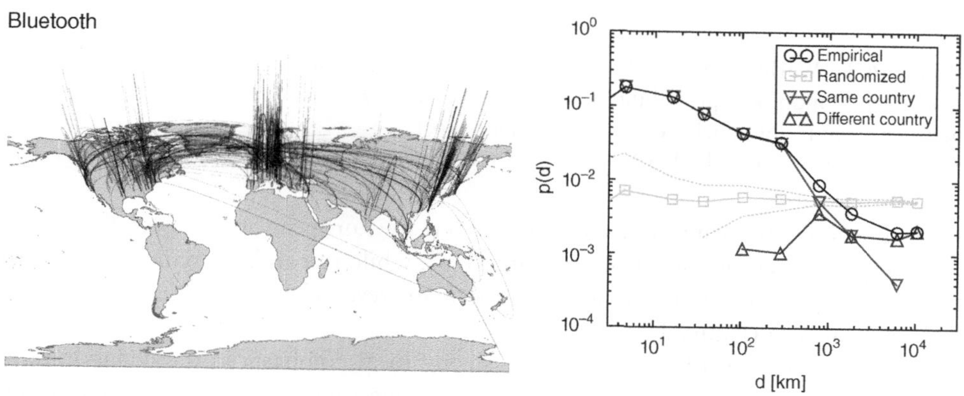

Figure 16.1 World map of collaborations in bluetooth research (left panel) and log–log plots of the conditional probability of co-authorships between organizations and the distance between them (right panel). The empirical probability of having a co-authored paper is indicated by the black circle line. The rewired random version is shown in squares (the dotted lines represent ±2 standard deviations of the configurations). The triangle down indicates the empirical probability of co-authorships within the same country, while the triangle up indicates cross-country co-authorships.
Source: Hennemann et al. (2012).

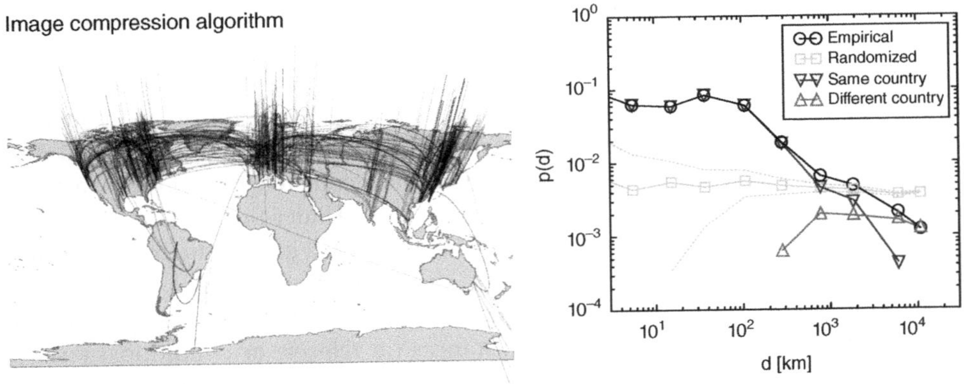

Figure 16.2 World map of collaborations in image compression research (left panel) and log–log plots of the conditional probability of co-authorships between organizations and the distance between them (right panel). Analogous to Figure 16.1.
Source: Hennemann et al. (2012).

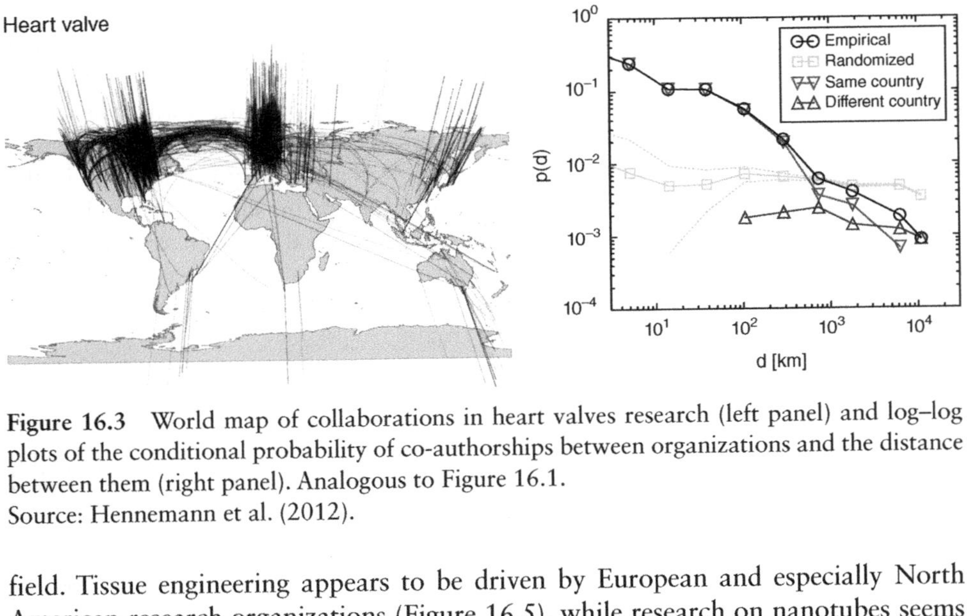

Figure 16.3 World map of collaborations in heart valves research (left panel) and log–log plots of the conditional probability of co-authorships between organizations and the distance between them (right panel). Analogous to Figure 16.1.
Source: Hennemann et al. (2012).

field. Tissue engineering appears to be driven by European and especially North American research organizations (Figure 16.5), while research on nanotubes seems to be equally distributed among the three main research regions North America, Europe, and Asia (Figure 16.6).

A striking element of all maps is the strong local emphasis in the collaboration activity, which can be derived from the way the connecting lines between collaborating organizations were drawn. The closer the partners are located, the higher the "spikes" are, whereas long-range connections can be identified by the smooth-curve character. This geographical collaboration pattern is also reflected in the probability plots on the right panels in Figures 16.1 to 16.6. All six plots show very similar curve shapes for all indicators presented. The empirical curve (circle line in the plots) shows the overall distance relation for all collaborations. This relation is subdivided into national (triangle-down) and international (triangle-up) collaboration. The collaboration

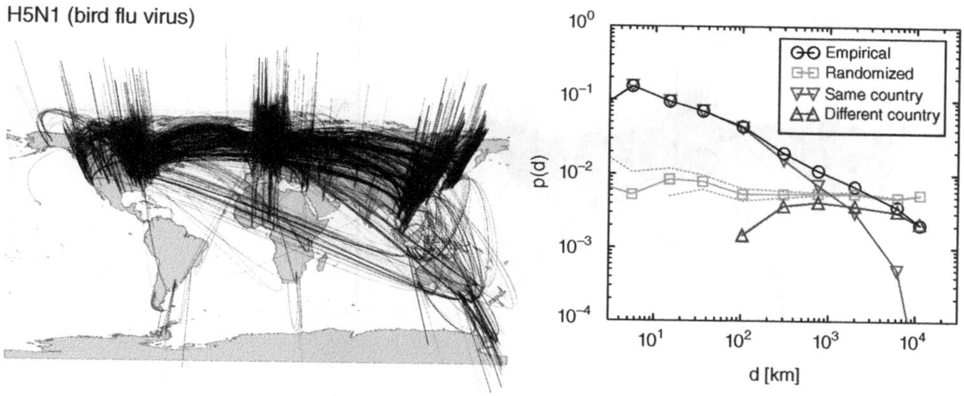

Figure 16.4 World map of collaborations in H5N1 (bird flu) research (left panel) and log–log plots of the conditional probability of co-authorships between organizations and the distance between them (right panel). Analogous to Figure 16.1.
Source: Hennemann et al. (2012).

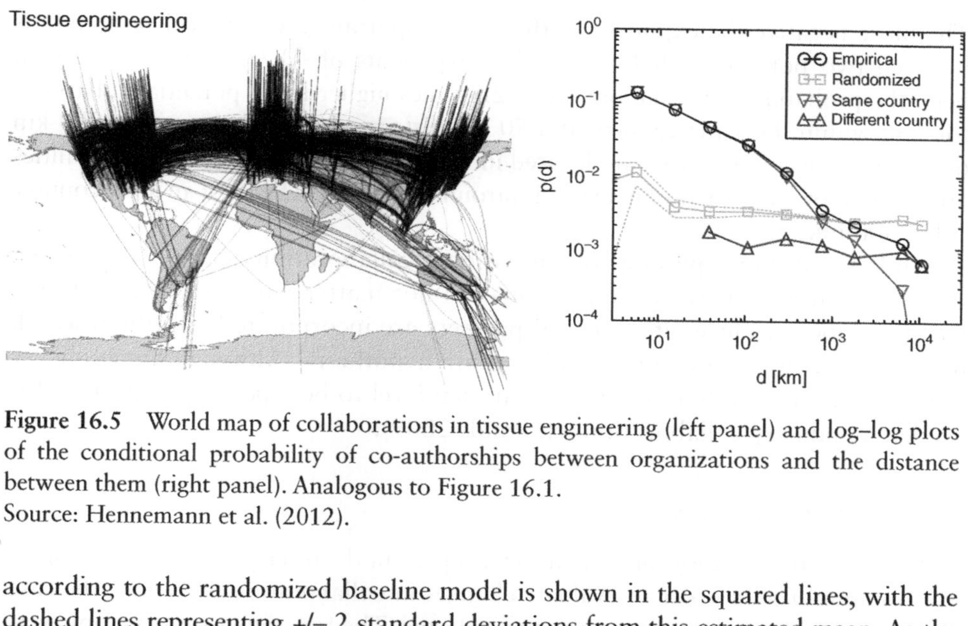

Figure 16.5 World map of collaborations in tissue engineering (left panel) and log–log plots of the conditional probability of co-authorships between organizations and the distance between them (right panel). Analogous to Figure 16.1.
Source: Hennemann et al. (2012).

according to the randomized baseline model is shown in the squared lines, with the dashed lines representing +/– 2 standard deviations from this estimated mean. As the plots are scaled in logarithm for both axes, straight lines would indicate a power-law relation. It is obvious that there is a very high propensity for local collaboration in all six communities. This finding is significantly different from the random model. In turn, international collaboration is significantly lower than expected. The mainly horizontal line for international collaboration shows that once a decision is made to select international partners, the location of that partner becomes irrelevant. Overall, the community structures as represented by the probability plots (right panels) show that the relation between distance and the chance of collaborating with a particular partner is indeed very similar in all fields analyzed.

In summary, the slope of the probability function can be interpreted as an indicator for the relevance of geographical proximity for collaboration in epistemic communities.

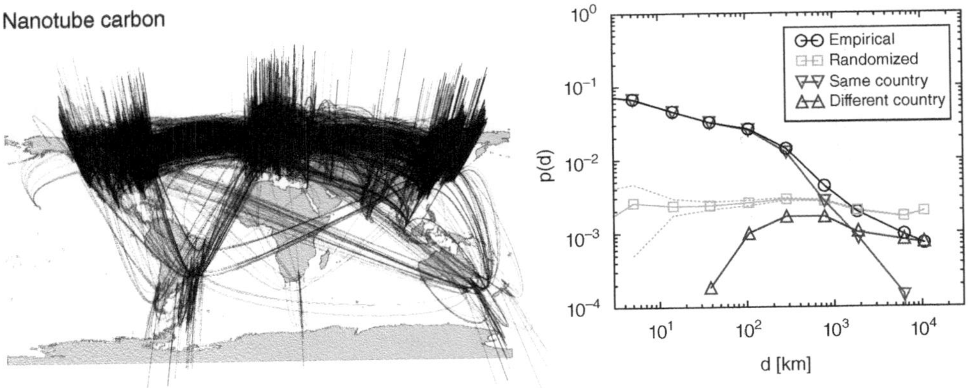

Figure 16.6 World map of collaborations in carbon nanotube research (left panel) and log–log plots of the conditional probability of co-authorships between organizations and the distance between them (right panel). Analogous to Figure 16.1.
Source: Hennemann et al. (2012).

The more negative the exponent is, the more important geographical proximity is between two potential collaborators. For exponents of –0.75, as typically found here, the probability of collaborating is 2.8 times higher if the potential collaborators are within the same country at a 50 km distance than if they are at a 200 km distance. If the collaborators are located in two different countries, the probabilities are different by two orders of magnitude (around 50 times) in favor of an intra-country collaboration.

While the distance plays a role in national collaboration activity (supporting the restricted resources hypothesis), it does not matter where the international partner is located once international partners are incorporated into the research team (supporting the outcome maximization hypothesis). Although not explicitly tested here, other authors found the national level to be especially important for academic-industry collaboration (cf. Ponds 2009).

Collaboration Over Time

The collaboration patterns presented so far represent the recent activity in the respective communities. In order to evaluate changes in collaboration over time, the case of the tissue engineering community was selected. The first paper on that topic was written in 1984. However, it was a single-author paper and therefore did not meet the requirements of this study. The fraction of co-authored papers increases over time. While two-thirds of the papers in tissue engineering were written by a single author in the 1980s and early 1990s in the initial phase of this science field, nowadays more than 80% are the joint work of two or more researchers. However, this trend is not exclusive to tissue engineering, but present in most disciplines (cf. Ponds 2009: 83, who found the same shares for a variety of disciplines). Reasons include increasing division of labor due to the increasing complexity in the scientific work process and the shortening of the knowledge life-cycle, which increases the pressure to publish, especially for younger researchers (cf. Bozeman and Corley 2004). The latter point is strongly linked to increasing evaluation and performance-based

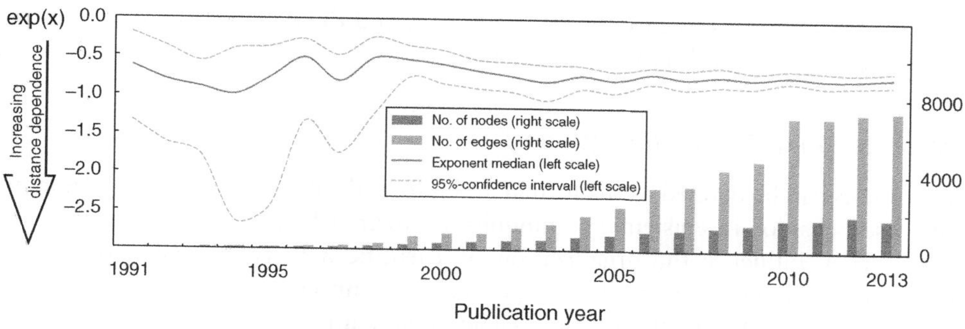

Figure 16.7 The development of the collaboration patterns in tissue engineering research 1991–2013. The exponent of the collaboration probability as function of the distance between the collaborators is an indicator for the overall dependency of collaboration and distance. If the exponent is 0.0, the collaboration is independent from distance. The more negative the exponent is, the more dependent is collaboration on the distance. The solid line represents the median exponent value of 2000 configurations (i.e., simulations of the network), the dashed lines represent the 95%-confidence interval around the median. The bars indicate the number of nodes (=institutions) in the network (dark gray) and the number of edges (=collaboration acts) between the nodes (light gray). While the networks of the first years are too small for sound interpretations, it is safe to say that the distance gains in importance from 1999 until 2003 and then fluctuates around a stable state, that is, although the network size increases significantly, there is no change in the collaboration propensity patterns, with exponents around –0.75.

funding (cf. Liefner 2003). However, what is striking in the dynamics of a science field is the remarkable stability of the geographical influence (Figure 16.7). The exponent of the collaboration probability function fluctuates at around –0.9 to –0.7. Although it appears that the role of space is even stronger at the beginning of the evolution of the science of tissue engineering, this should be interpreted with caution due to the large confidence intervals around the median in the first years when the epistemic community was small. We can conclude that given the dramatic technological (e.g., Internet) as well as organizational developments (e.g., political incentives to internationalize) during the period from the 1980s until today, there is almost no sign of a change concerning the collaboration activity in geographical space. This is supported by the findings of Ponds (2009), who also found a stable proportion of international collaboration compared to national collaboration over this period of time.

Explanatory Factors: Institutions, Network Dynamics and Attributes of Knowledge

As Figure 16.7 illustrates, science collaboration is both international (involving distant international collaboration partners) and shaped by a strong distance decay (thus a high propensity of choosing partners nearby). The following section aims to answer the question of why the collaboration pattern in science has this dual feature and which factors contribute to its emergence and permanence.

Of greater importance are several bundles of factors briefly touched upon in the introductory section, namely the institutions that shape the organization of science, the mechanisms of network evolution, and the content of collaboration regarding the types of knowledge involved.

Organization of Science (Institutions)

Institutions can be defined as the formal and informal rules that govern behavior, in this case behavior in epistemic communities (North 1990; Powell and DiMaggio 1991). They influence the organization of scientific activity, including the goals attached to collaboration and the ways of making connections, as well as the nature of collaboration. Most importantly, science policy and administration create the formal institutions on the national scale.

As noted in the introductory section, the institutions most relevant for science are present on the national scale. Many of these institutions – although implicitly or explicitly supporting mainly domestic collaboration – also affect international collaboration. Many national science policy programs, for example, include measures to encourage international collaboration, for example, in the form of the obligatory or voluntary inclusion of foreign partners in research consortia, or in the form of international expert review panels (e.g., Hoekman, Frenken, and van Oort 2009). A much more targeted impact on international collaboration can be attributed to programs in higher education that provide funding for individual mobility, such as the Fulbright Foundation, Alexander von Humboldt Foundation, China Scholarship Council (Altbach and Knight 2007). The alumni of these programs may maintain their connections over long time periods. Moreover, according to Knight (2004), institutions in the national higher education system exert considerable influence on efforts to internationalize at the level of higher education institutions. Internationalization may often be a part of branding one's university and building reputation (2004: 22). Similarly, valuable international connections may promote individual scientists' careers (van Rijnsoever, Hessels, and Vandeberg 2008).

However, national institutions still often induce a bias toward national collaboration. Ponds (2009) discusses the importance of techno-nationalism as a factor limiting the internationalization of science. Techno-nationalism involves aims to improve national competitiveness through specialization and a superior knowledge base. Unintended knowledge leakage through international collaboration may interfere with strategic decisions to improve the national science base.

Even stronger support for domestic collaboration may arise through non-intended effects of the institutions governing scientific activity. Among these institutions, the national standards for scientific merit review may have a most pronounced impact. The different stages of national development and large variation in the functions attributed to science in different countries further limit the scope for international collaboration (Suresh 2011).

Social and cultural factors are nationally, regionally, and sometimes locally set by institutions such as common ethics, regulatory frameworks, legal ground, or fiscal idiosyncrasies. These strong forces are reflected in national grant allocation systems, causing a potential mismatch between two funding bodies. This compromise is in turn producing disincentives to apply jointly for cross-national grants on the individual level, while tax-funded research grants are often exclusively accessible for national organizations and scientists.

Hence, strong institutions in science favor collaboration on the national scale over international and local collaboration. Institutions promoting international collaboration are positively related to reputation and brand-building and negatively affected by fears of knowledge leaking to partners with inferior technology. Global science collaboration can thus be expected to be most prominent when partners regard each other as equal with respect to their knowledge, or as adequate partners for the purpose of achieving certain tasks, for example acquiring grant money and writing publications. Such exceptions from the rule include the big sciences, scientific endeavors that are too complex and costly to be implemented by individual nations alone (e.g., sub-atomic particle research, space programs) (cf. Galison and Hevly 1992). Furthermore, there are increasing initiatives by the European Union to promote cross-country collaboration. Examples include the Human Brain Project, a large-scale big science initiative, and hundreds of smaller-scale research domains that are funded through the Framework Programmes or the recently established flagship initiative Horizon 2020. Consortia must typically involve partners from different EU countries to qualify for funding. As these programs are unique to the EU and not matched by similar systematically arranged programs at the supra-national level in other world regions, they do not influence the parameters of global science collaboration significantly.

Network-Inherent Factors

While institutions usually still give privilege to national collaboration, the basic network mechanisms that are present in social networks may favor international and local collaboration. For an efficient network structure under given resource constraints, only a few hubs are necessary and only a few long-range connections are present, while most collaboration is local. Such a network ensures a rapid distribution of knowledge through a short distance between all network partners, and at the same time keeps the costs of maintaining linkages low for most connections (cf. Watts and Strogatz 1998). As hypothesized, a top layer of wisely connected world leading scientists would then form the international backbone of the research system. Hence, if knowledge dissemination in the global science network was perfect, "the closer the better" would apply for most collaboration. Such a network would resemble the notion of a combination of global pipelines and local buzz (Bathelt, Malmberg, and Maskell 2004).

In real science networks, however, institutions come into play again, for example in the form of mutual trust. Individual researchers are facing a competition for grants and publication-centered reputation. Therefore, trust among collaborators is crucial. Local collaboration, coupled with frequent personal encounters between the individuals involved, who also share a common language and methods of communication, has a higher chance of creating trust and contributing to the emergence of local institutions that facilitate knowledge sharing (Cooke 1998).

Types of Knowledge Involved

Approaches to studying global science collaboration usually focus on the structures of science networks, on the institutions that influence collaborative behaviors, and on the incentives stemming from the institutions. The content of collaboration, an exchange of scientifically relevant knowledge, is seldom taken into account explicitly.

Scientifically relevant knowledge, however, may come in different forms and have very different characteristics. For example, the collaboration that eventually becomes manifest in a co-authored article may involve intensive discussion, joint learning and testing of tools and approaches, and the search for novel ways of handling a particular question. In such a case, the collaboration would include tacit knowledge and require intensive personal interaction. In other cases, a co-authored article may involve much less interaction and the sharing of codified knowledge only. Intensive personal interaction depends on time spent together, and thus requires long visits or a given spatial proximity of individuals. Therefore, international collaboration and local collaboration may at least to some degree reflect differences in the kinds of activity carried out.

However, a technical amplification of local communication patterns may occur in modern communication systems, in that researchers contact colleagues much more often via email, Skype, and other Web 2.0 applications whom they already know personally from their geographical neighborhoods. This behavior helps to reduce the "coordination dilemma" (Beckmann 1993) that is inherent in situations of complex negotiation, such as in research projects. Social media may in some cases provide a substitute for spatial adjacency and in other cases reinforce existing patterns of local collaboration.

Science Collaboration: Truly Global and Better?

From the perspective of national institutions, global science collaboration is tied to reputation and quality assessments, while the notions of network science and types of knowledge represent a view of global science collaboration as a means to ensure effective dissemination of mostly codified knowledge. So why should global science collaboration be worth the investment of time and money?

The core argument is centered on an increase of research productivity through international collaboration. For example, Ponomariov and Boardman (2010) find a positive relation between university research centers, collaboration, and improved productivity. Similarly, Barjak and Robinson (2008) claim that it is indeed beneficial to integrate researchers from another country in order to create well-balanced research teams. Scientometric analyses give support to these notions. At least for the level of inter-organizational collaboration, quality improvements could be empirically shown (cf. Jones et al. 2008). With its general early internationalization, and especially with the strong increase in international activity (as measured in internationally co-authored articles through the 1980s and 1990s), the academic world has somehow preempted trends in the corporate world (Archibugi and Iammarino 2002: 118). Referencing back, analogies from the business world help to rationalize the merits of global collaboration activities, since at least for companies, being present in many different markets is seen as a prerequisite for success (Kafouros et al. 2008, cf. also Carlsson 2006). Organizing science in such a way as to internalize the diversity of the world's many national science systems can be thought of as equally promising. Overall, one cannot eliminate the possibility that the few international collaborations have the greatest impact in terms of scientific progress. However, direct proof for this is hard to find and the greater success for international science teams remains ambiguous (Abbasi and Jaafari 2013, cf. also the discussion in Persson 2010).

So there is little doubt that science can benefit from global collaboration. The main counter-arguments, however, have been explored above. First, international collaboration is more expensive than national or local cooperation, and second, keeping science national may secure some advantages in international competition. Competition in science may also spur innovation as it does in business.

Independently of these considerations, recent decades have witnessed some new entrants in global science, namely East Asian countries and research organizations, including China (see, e.g., Figure 16.4). Overall, the gap between countries traditionally leading in economic terms and those that follow at a distance has narrowed with respect to innovative capabilities such as patents and article publications (Castellacci and Archibugi 2008).

Consequently, one must ask whether this new trend is more an opportunity or a burden for developing country entrants (Archibugi and Pietrobelli 2003). Drawing on similarities with business again, international collaboration may benefit both parties involved, allowing learning on both sides (Thorsteinsdóttir et al. 2010). Upgrading the scientific competence of developing countries, for example with the help of foreign knowledge through science collaboration or by means of domestic investments, may thus contribute to technological convergence across countries (cf. Archibugi and Iammarino 1999; Castellacci and Archibugi 2008). However, the task of converting a newly acquired scientific competence into upgrading and competitiveness of domestic firms requires additional political support and rising absorptive capacity among firms (Ernst and Kim 2002).

Conclusion

Science is global, but at the same time also a national and local activity, with a much higher propensity for choosing short-distance collaboration. From a network perspective, this is not an unusual pattern, as it combines the benefits from long-distance networking with efficiency and cost-effectiveness of short-distance networking. There is enough evidence to assume that international collaboration can provide a different quality of knowledge input and can thus give unique stimulus for scientific activity. But there is no reason to demand that international collaboration should be promoted beyond its current level at the expense of national or local collaboration. Money should be spent wisely, and only science organizations with strong internal capacities or individuals with unique expertise can engage in meaningful international collaboration (cf. Barnard et al., this volume, Chapter 18).

The nation turns out to be still the most important entity regarding institutions and infrastructures that shape science collaboration and technology systems (cf. Archibugi and Michie 1997; Pavitt 1998; Archibugi and Iammarino 1999). In this context, it should also be acknowledged that global scientific progress may come not only through international collaboration, but also through competition among groups of local scientists in the arena of international science (e.g., Marginson 2006, cf. Dasgupta and David 1994).

Hence, the spatial pattern of global science collaboration can be expected to remain unchanged. Relatively few global connections will stretch between science hubs, with the majority of connections being established at short distance around the hubs. However, it seems desirable to include all nations in the world science system and not to forego ideas that may evolve under the particular circumstance found in countries that do not yet have a visible role in global collaboration.

References

Abbasi, Alireza, and Ali Jaafari. 2013. "Research Impact and Scholars' Geographical Diversity." *Journal of Informetrics* 7(3): 683–692.

Ahlgren, Per, Olle Persson, and Robert J.W. Tijssen. 2013. "Geographical Distance in Bibliometric Relations within Epistemic Communities." *Scientometrics* 95(2): 771–784.

Altbach, Philip G., and Jane Knight. 2007. "The Internationalization of Higher Education: Motivations and Realities." *Journal of Studies in International Education* 11(3–4): 290–305.

Ammon, Ulrich (ed.). 2001. *The Dominance of English as a Language of Science*. Berlin: de Gruyter.

Archibugi, Daniele, and Simona Iammarino. 1999. "The Policy Implications of the Globalization of Innovation." *Research Policy* 28(2–3): 317–336.

Archibugi, Daniele, and Simona Iammarino. 2002. "The Globalization of Technological Innovation: Definition and Evidence." *Review of International Political Economy* 9(1): 98–122.

Archibugi, Daniele, and Jonathan Michie. 1997. "Technological Globalization or National Systems of Innovation?" *Futures* 29(2): 121–137.

Archibugi, Daniele, and Carlo Pietrobelli. 2003. "The Globalization of Technology and Its Implications for Developing Countries – Windows of Opportunity or Further Burden?" *Technological Forecasting & Social Change* 70(9): 861–883.

Barabási, Albert-Laszlo, Hawoong Jeong, Zoltan Néda, Erzsebet Ravasz, Andras Schubert, and Tamas Vicsek. 2002. "Evolution of the Social Network of Scientific Collaborations." *Physica A – Statistical Mechanics and Its Applications* 311(3): 590–614.

Barjak, Franz, and S. Robinson. 2008. "International Collaboration, Mobility and Team Diversity in the Life Sciences: Impact on Research Performance." *Social Geography* 3: 23–36.

Bathelt, Harald, Anders Malmberg, and Peter Maskell. 2004. "Clusters and Knowledge: Local Buzz, Global Pipelines and the Process of Knowledge Creation." *Progress in Human Geography* 28(1): 31–56.

Beckmann, Martin J. 1993. "Knowledge Networks – The Case of Scientific Interaction at a Distance." *The Annals of Regional Science* 27(1): 5–9.

Beckmann, Martin J. 1994. "On Knowledge Networks in Science: Collaboration Among Equals." *The Annals of Regional Science* 28(3): 233–242.

Börner, Kati, Noshir Contractor, Holly J. Falk-Krzesinski, et al. 2010. "A Multi-Level Systems Perspective for the Science of Team Science." *Science Translational Medicine* 2(49): 49cm24.

Bornmann, Lutz, and Loet Leydesdorff. 2011. "Which Cities Produce More Excellent Papers Than Can Be Expected? A New Mapping Approach, Using Google Maps, Based on Statistical Significance Testing." *Journal of the American Society for Information Science and Technology* 62(10): 1954–1962.

Boschma, Ron. 2005. "Proximity and Innovation: A Critical Assessment." *Regional Studies* 39(1): 61–74.

Bozeman, Barry, and Elizabeth Corley. 2004. "Scientists' Collaboration Strategies: Implications for Scientific and Technical Human Capital." *Research Policy* 33(4): 599–616.

Carlsson, Bo. 2006. "Internationalization of Innovation Systems: A Survey of the Literature." *Research Policy* 35(1): 56–67.

Castellacci, Fulvio, and Daniele Archibugi. 2008. "The Technology Clubs: The Distribution of Knowledge Across Nations." *Research Policy* 37(10): 1659–1673.

Cooke, Philip N. 1998. "Introduction: Origins of the Concept." In *Regional Innovation Systems: The Role of Governances in a Globalized World*, ed. Philip N. Cooke, Hans-Joachim Braczyk, and Martin Heidenreich, 2–25. London: Routledge.

Dasgupta, Partha, and Paul A. David. 1994. "Toward a New Economics of Science." *Research Policy* 23(5): 487–521.

de Solla Price, Derek J. 1965. "Networks of Scientific Papers." *Science* 149(3683): 510–515.
de Wit, Hans. 2002. *Internationalization of Higher Education in the United States of America and Europe: A Historical, Comparative, and Conceptual Analysis*. Westport, CT: Greenwood Press.
Eagle, Nathan, Michael Macy, and Rob Claxton. 2010. "Network Diversity and Economic Development." *Science* 328(5981): 1029–1031.
Ernst, Dieter, and Linsu Kim. 2002. "Global Production Networks, Knowledge Diffusion, and Local Capability Formation." *Research Policy* 31(8–9): 1417–1429.
Etzkowitz, Henry, and Loet Leydesdorff. 2000. "The Dynamics of Innovation: From National Systems and "Mode 2" to a Triple Helix of University–Industry–Government Relations." *Research Policy* 29(2): 109–123.
Frenken, Koen, Sjoerd Hardeman, and Jarno Hoekman. 2009. "Spatial Scientometrics: Towards a Cumulative Research Program." *Journal of Informetrics* 3(3): 222–232.
Galison, Peter, and Bruce William Hevly (eds.). 1992. *Big Science: The Growth of Large-Scale Research*. Stanford, CA: Stanford University Press.
Garfield, Eugene. 1955. "Citation Indexes for Science: A New Dimension in Documentation through Association of Ideas." *Science* 122(3159): 108–111.
Gertler, Meric S. 2003. "Tacit Knowledge and the Economic Geography of Context, or the Undefinable Tacitness of Being (There)." *Journal of Economic Geography* 3(1): 75–99.
Geuna, Aldo, and Ben R. Martin. 2003. "University Research Evaluation and Funding: An International Comparison." *Minerva* 41(4): 277–304.
Glänzel, Wolfgang, and András Schubert. 2005a. "Analysing Scientific Networks through Co-Authorship." In *Handbook of Quantitative Science and Technology Research*, ed. Henk F. Moed, Wolfgang Glänzel, and Ulrich Schmoch, 257–276. Heidelberg: Springer.
Glänzel, Wolfgang, and András Schubert. 2005b. "Domesticity and Internationality in Co-authorship, References and Citations." *Scientometrics* 65(3): 323–342.
Haas, Peter M. 1992. "Introduction: Epistemic Communities and International Policy Coordination." *International Organization* 46(1): 1–35.
Hennemann, Stefan. 2011. "The Role of Transnational Corporations in the Chinese Science and Technology Network." *Erdkunde* 65(1): 69–81.
Hennemann, Stefan. 2012. "Evaluating the Performance of Geographical Locations within Scientific Networks Using an Aggregation–Randomization–Re-Sampling Approach (ARR)." *Journal of the American Society for Information Science and Technology* 63(12): 2393–2404.
Hennemann, Stefan, Diego Rybski, and Ingo Liefner. 2012. "The Myth of Global Science Collaboration – Collaboration Patterns in Epistemic Communities." *Journal of Informetrics* 6(2): 217–225.
Hicks, Diana M., and J. Sylvan Katz. 1996. "Science Policy for a Highly Collaborative Science System." *Science and Public Policy* 23(1): 39–44.
Hoekman, Jarno, Koen Frenken, and Robert J.W. Tijssen. 2010. "Research Collaboration at a Distance: Changing Spatial Patterns of Scientific Collaboration within Europe." *Research Policy* 39(5): 662–673.
Hoekman, Jarno, Koen Frenken, and Frank van Oort. 2009. "The Geography of Collaborative Knowledge Production in Europe." *The Annals of Regional Science* 43(3): 721–738.
Jansen, Dorothea, Regina von Görtz, and Richard Heidler. 2010. "Knowledge Production and the Structure of Collaboration Networks in Two Scientific Fields." *Scientometrics* 83(1): 219–241.
Jones, Benjamin F., Stefan Wuchty, and Brian Uzzi. 2008. "Multi-University Research Teams: Shifting Impact, Geography, and Stratification in Science." *Science* 322(5905): 1259–1262.
Kafouros, Mario I., Peter J. Buckley, John A. Sharp, and Chengqi Wang. 2008. "The Role of Internationalization in Explaining Innovation Performance." *Technovation* 28(1): 63–74.
Knight, Jane. 2004. "Internationalization Remodeled: Definition, Approaches, and Rationales." *Journal of Studies in International Education* 8(1): 5–31.

Leydesdorff, Loet, and Caroline S. Wagner. 2008. "International Collaboration in Science and the Formation of a Core Group." *Journal of Informetrics* 2(4): 317–325.

Liefner, Ingo. 2003. "Funding, Resource Allocation, and Performance in Higher Education Systems." *Higher Education* 46(4): 469–490.

Lundvall, Bengt-Åke (ed.). 1992. "National Systems of Innovation: Towards a Theory of Innovation and Interactive Learning." London: Pinter Publishers.

Luukkonen, Terttu, Olle Persson, and Gunnar Sivertsen. 1992. "Understanding Patterns of International Scientific Collaboration." *Science, Technology & Human Values* 17(1): 101–126.

Marginson, Simon. 2006. "Dynamics of National and Global Competition in Higher Education." *Higher Education* 52(1): 1–39.

Maslov, Sergei, and Kim Sneppen. 2002. "Specificity and Stability in Topology of Protein Networks." *Science* 296(5569): 910–913.

Moodysson, Jerker. 2008. "Principles and Practices of Knowledge Creation: On the Organization of 'Buzz' and 'Pipelines' in Life Science Communities." *Economic Geography* 84(4): 449–469.

Narin, F., K. Stevens, and Edith S. Whitlow. 1991. "Scientific Co-operation in Europe and the Citation of Multinationally Authored Papers." *Scientometrics* 21(3): 313–323.

Nelson, Richard (ed.). 1993. *National Innovation Systems: A Comparative Analysis*. Oxford: Oxford University Press.

Newman, Mark E.J., and Juyong Park. 2003. "Why Social Networks Are Different from Other Types of Networks." *Physical Review E* 68: 036122-8.

North, Douglass C. 1990. *Institutions, Institutional Change, and Economic Performance*. Cambridge: Cambridge University Press.

Nooteboom, Bart, Wim Vanhaverbeke, Geert Duysters, Victor Gilsing, and A.J. van den Oord. 2007. "Optimal Cognitive Distance and Absorptive Capacity." *Research Policy* 36(7): 1016–1034.

Pavitt, Keith. 1998. "The Social Shaping of the National Science Base." *Research Policy* 27(8): 793–805.

Persson, Olle. 2010. "Are Highly Cited Papers More International?" *Scientometrics* 83(2): 397–401.

Ponds, Roderik. 2009. "The Limits to Internationalization of Scientific Research Collaboration." *Journal of Technology Transfer* 34(1): 76–94.

Ponomariov, Branco L., and P. Craig Boardman. 2010. "Influencing Scientists' Collaboration and Productivity Patterns Through New Institutions: University Research Centers and Scientific and Technical Human Capital." *Research Policy* 39(5): 613–624.

Powell, Walter W., and Paul J. DiMaggio. 1991. *The New Institutionalism in Organizational Analysis*. Chicago, IL: University of Chicago Press.

Power, Dominik, and Anders Malmberg. 2008. "The Contribution of Universities to Innovation and Economic Development: In What Sense a Regional Problem?" Cambridge Journal of Regions, *Economy and Society* 1(2): 233–245.

Smith, Richard. 2006. "Peer Review: A Flawed Process at the Heart of Science and Journals." *Journal of the Royal Society of Medicine* 99(4): 178–182.

Suresh, Subra. 2011. "Moving Toward Global Science." *Science* 333(6044): 802.

Thorsteinsdóttir, Hella, Christina C. Melon, Monali Ray, et al. 2010. "South-South Entrepreneurial Collaboration in Health Biotech." *Nature Biotechnology* 28(5): 407–416.

Tijssen, Robert J.W., Ludo Waltman, and Nees Jan van Eck. 2011. "Collaborations Span 1,553 Kilometres." *Nature* 473(7346): 154–154.

van Rijnsoever, Frank J., Laurens K. Hessels, and Rens L.J. Vandeberg. 2008. "A Resource-Based View on the Interactions of University Researchers." *Research Policy* 37(8): 1255–1266.

Waltman, Ludo, Robert J.W. Tijssen, and Nees Jan van Eck. 2011. "Globalisation of Science in Kilometres." *Journal of Informetrics* 5(4): 574–582.

Watts, Duncan J., and Steven H. Strogatz. 1998. "Collective Dynamics of 'Small-World' Networks." *Nature* 393: 440–442.

Weisberg, Michael, and Ryan Muldoon. 2009. "Epistemic Landscape and the Division of Cognitive Labor." *Philosophy of Science* 76(2): 225–252.

Ynalvez, Marcus Antonius, and Wesley M. Shrum. 2011. "Professional Networks, Scientific Collaboration, and Publication Productivity in Resource-Constrained Research Institutions in a Developing Country." *Research Policy* 40(2): 204–216.

Yook, Soon-Hyung, Hawoong Jeong, and Albert-Lászlo Barabási. 2002. "Modeling the Internet's Large-Scale Topology." *Proceedings of the National Academy of the USA* 99(21): 13382–13386.

Zitt, Michael, Elise Bassecoulard, and Yoshiko Okubo. 2000. "Shadows of the Past in International Cooperation: Collaboration Profiles of the Top Five Producers of Science." *Scientometrics* 47(3): 627–657.

Chapter 17

International Mobility of Scientists

Kieron Flanagan

Introduction

A strong aspiration toward universalism predates the government-funded, professionalized, and internationalized modern scientific enterprise by several centuries (Musselin 2004; Kim 2009). In this sense, international mobility is central to conceptions of knowledge production. However the nature and purposes of mobility have evolved as the sciences have professionalized. Over time, mobility behavior has become more purposeful and more clearly associated with research and career objectives rather than more general considerations of scholarship (Heffernan and Jöns 2013).

Recent decades have seen mobility actively constructed as a problem to be resolved by policy action (as in the fear of a "brain drain" of talented British scientists to the United States in the 1960s) or as a solution to the challenges faced by national research systems in a globalized world (harnessing the power of "brain circulation" and the transnational networks thus created). The current preoccupation with "brain circulation" as a means of raising the "excellence" of national research systems and boosting the exchange of knowledge is exemplified by the efforts of the European Commission to promote mobility of scientists between European Union member states.

Yet there remains surprisingly little consensus as regards how best to conceptualize scientific mobility. The many studies extant represent partial snapshots of a complex, entangled cluster of phenomena taken from a single perspective and usually focusing on a very specific population of scientists or engineers. In this chapter we examine how mobility has been conceptualized and explored, and what those studies tell us about the factors influencing mobility and the impacts mobility may have. Because the question of scientific mobility is politicized and because the

The Handbook of Global Science, Technology, and Innovation, First Edition.
Edited by Daniele Archibugi and Andrea Filippetti.

framing of policy debates influences measurement and analysis, we will also consider these policy framings and their strengths or weaknesses.

Definitional Challenges

Several definitional challenges confront those wishing to characterize and understand the international mobility of scientists, its drivers, and its impacts. Specifically, these concern the units of analysis: what constitutes a "scientist" and what constitutes "international mobility." We briefly explore these issues below.

Defining "Scientist"

Although "science and engineering professionals" appears in the international standard classification of occupations (ISCO-08) this category includes a wide range of occupations other than those where research is a major or the primary role. In addition, many of those engaged in medical and other health research would be covered under the category of "health professionals," while researchers in higher education institutions might find themselves categorized under "university and higher education teachers." Those engaged in research toward a higher degree (for instance doctoral researchers) would in many countries not be classed as being in an occupation at all but rather as students. Further, many studies of scientific mobility have focused only on those working in public or not-for-profit research organizations, excluding R&D workers in private enterprises. A few studies implicitly or explicitly address both populations (for instance, studies of inventors using patent data) and in some cases the two groups are simply conflated.

Given these definitional difficulties, studies of scientific mobility have tended either to rely upon data about doctoral candidates or doctorate holders or have taken advantage of information collected about populations of researchers for other purposes, especially CVs and biographical information provided as part of research funding applications or posted online, or information derived from bibliometric or patent databases. Primary data collection with a dedicated survey instrument is comparatively rare and has tended to focus on very specific populations, for instance as part of a program evaluation or investigation into the motivations and experiences of internationally mobile doctoral researchers or elite "star scientists." Many of these have been conducted by policy or professional organizations or interest groups rather than by social scientists. Relatively few could be described as comprehensive in terms of discipline, career state, institutional context, and country coverage, or rigorous in terms of conceptualization, design, and testing.

Defining "International Mobility"

A further definitional challenge is the nature of international scientific mobility itself. The term "mobility" tends to be used interchangeably to refer both to the international *migration* of scientists – moving from a position of employment in one country to one in another, whether for a fixed-term or open-ended period – and what we might call "non-job" international mobility, that is a time-limited visit (or repeated visits) to another site in a different country for the purposes of collaboration or scholarship. The former is a special case of a much broader socioeconomic phenomenon, the international migration of highly skilled professionals, and should be analyzed as such; while the latter is a phenomenon more specific to the world of research (MORE Consortium 2011; Børing et al. 2015).

There is debate about the extent to which there yet exists a truly international market for scientific labor: for instance, Musselin (2004) argues that differences in regulations, norms, and practices, as well as the obvious language differences, work to prevent the emergence of a functioning European academic labor market. However, some national research systems are undeniably more open to foreign scientific labor than others, and multiple labor markets coexist or overlap: there is an elite global scientific labor market, in which star scientists move between elite institutions; an emerging international scientific labor market in which young scientists take opportunities in more dynamic and open national systems; and finally, running alongside these, the continued existence of national labor markets. (A similar statement could be made about other highly skilled professions.)

Non-job mobility, while not entirely unique to scientific professionals, is more closely bound up with the practices of research and scholarship. Non-job mobility may be driven by the need to access research collaborators, acquire new skills and techniques, or gain access to materials, samples, specialized research equipment, or infrastructure. It may be supported by general research funding or dedicated initiatives such as the European Commission Marie Curie Programme (Ackers 2005b, 2005a) or through institutional support such as sabbatical or hosting policies (Heffernan and Jöns 2013). Such mobility is clearly intended to be time-limited in the sense that at some point a return to the normal place and country of employment will be expected (or formally required) by the parties involved.

Visits need not be short stays: a periodic one-year sabbatical is a common academic expectation and some research visits may be formalized through some kind of honorary or visiting position that does not involve a contract of employment but which may carry symbolic value to the institutions concerned and to the scientist in question, whose prestige and social capital is enhanced as a result (Ackers 2008). Further, norms and expectations about the need for such mobility may vary from discipline to discipline and from national system to national system (see, e.g., Ackers 2005a).

Conceptualising Mobility

In their excellent critical reviews of the economic literature on "brain drain" and mobility, Cañibano and Woolley (2010a, 2010b) highlight a conceptual confusion inherent in traditional approaches to considering "drain" and "gain": namely the conflation of the demographic categories of scientist or engineer (often termed human resources for science and technology, or HRST for short) with the economic concept of "human capital." Cañibano and Woolley argue that this tendency to treat human capital as if it were somehow encapsulated in the form of individual scientists results in a view in which movements of individuals from one system to another are considered as zero-sum additions to, and subtractions from, the respective national "balance sheets" of knowledge and talent (see also Meyer 2001). This focus on the *quantity* of human capital has latterly been supplemented by more attention to the *quality* of human capital (see, e.g., Jones 2008) but the way in which human capital is modeled still arguably conflates knowledge and skills and, even when emphasizing the role of complementary advanced skills in enabling the most productive use of the skills of another, the human capital perspective downplays the role of other complementary factors of scientific knowledge production.

Rather (and as noted by Cañibano and Woolley 2010b) human capital is only partly accounted for by the knowledge and skills embodied in individuals and is primarily a function of social structure and context (see also Davenport 2004; Bozeman, Dietz, and Gaughan 2001). The research potential of an individual scientist may be enhanced or constrained by their organizational and national contexts, not just in terms of access to potential collaborators and their knowledge and skills or access to advanced training, equipment, infrastructure, methods, samples, or data, but also in terms of the incentives and opportunities offered by the funding and research culture of the organizational or national context.

Mobility, then, is not simply a matter of the movement of discrete units of human capital and the consequent transfer or "recombination" of the knowledge embodied in those individuals (or through skill matching, unlocking the potential of one individual's skills by combination with another's). Context matters for the production of new knowledge, and we must therefore consider a wider range of contextual factors in conceptualizing mobility than would be suggested by simple human capital models. This is broadly in line with the so-called Stanford-Yale-Sussex synthesis (Dosi et al. 2006) and more specifically the "new economics of science" sparked by the contribution of Dasgupta and David (1994), although a search of the subsequent literature in this area finds few studies focusing specifically on the role played by the international, as opposed to intra- or inter-sectoral, movement of people. An honorable exception is the work of Edler, Fier, and Grimpe (2011) on the knowledge transfer activities of internationally mobile academic scientists.

The Context of Mobility: Drivers, Impacts and Barriers

As argued above, the international job mobility of scientists is best seen as a subset of the broader category of international migration of highly skilled professional workers. There is a substantial literature in work psychology and organizational sociology around job mobility and how decisions about changing job are made. Ng et al. (2007) offer a synthesis that identifies several levels of factors that are likely to be in operation in such decisions: structural factors affect the opportunities for job mobility and exist at the societal, regional, industry, and organizational levels of analysis; different individuals have different personal goals and motivations and thus preferences concerning mobility; and decisional factors (the norms to which the individual subscribes,[1] the individual's appetite for and readiness for change, and their calculation of the desirability of the mobility opportunity) ultimately determine whether the decision is made to change job, and also exist at the individual level of analysis.

The literature on scientific mobility rarely draws upon these potentially relevant work psychology or organizational sociology literatures. However, there is a loose conceptual framework implicit in much of the literature in which mobility is similarly seen in terms of *driving factors*. These can be conceptualized as operating on multiple levels, akin to those identified by Ng et al. (2007). Following the MORE Consortium (2009, 2011) we suggest that it is possible to identify three distinct but interrelated contextual levels, namely the *macro-level* of the national research or innovation system and its associated labor market; the *meso-level* – the research-performing organization, group, lab, or team in which a researcher works;[2] and the *micro-level* of factors relating to the individual researcher.

Further, the monolithic concept of "driver" is generally broken down into *push factors* that help to explain the decision to move away from one macro-, meso- or micro-context and *pull factors* that help to explain the decision to move into a different one (Thorn and Holm-Nielsen 2008). In this sense, mobility is "driven" by the interaction of personal preferences, local factors, institutional settings, and broader systemic contexts (Børing et al. 2015; Ackers and Gill 2008).

A further element which is often included in conceptualizations of mobility (especially in policy studies and documents) is the notion that there exist factors that act as *barriers* that suppress the level of mobility, with potentially negative effects on human capital formation, knowledge exchange, and the pursuit of "excellence." Assuming that such barriers exist, we can again conceptualize them as operating at the micro-, meso- and macro-levels.

We will explore below what the literature has to say about potential push and pull factors, and will also look at the evidence offered for the existence of barriers to mobility. First, we briefly discuss some measurement problems in the study of mobility. These challenges are not new: many were encountered by those designing the survey of UK university department heads on which the original 1963 Royal Society report that sparked the original "brain drain" panic was based (Balmer, Godwin, and Gregory 2009).

Measuring Mobility

Much of the literature on mobility tends to make a distinction between "temporary" and "permanent" migration (see, e.g., Cañibano, Otamendi, and Solís 2011; Baláz, Williams, and Kollár 2004), or between long and short stays (see, e.g., Mahroum 2000). The former distinction is problematic: studies almost always present a snapshot of the distribution or intentions of individuals at different points in their lives and careers. At what point should we consider the migration of an individual to be "permanent" while the possibility of return remains open in practice? It is also difficult to determine how best to treat duration: for instance, might the same or greater benefits be gained through repeated short visits rather than one longer visit? Perhaps the most practically useful distinctions we can make are between job migration and those forms of mobility not involving a change of employer, and between mobility with an intention or obligation to return versus mobility that can be seen as open-ended.

Further problems arise in seeking to understand "stocks" and "flows" of researchers. Specifically, it is not necessarily straightforward to determine the "home" or "sending" country. Some studies take the country of birth or of citizenship. Some individuals hold multiple citizenships and in any case the country of birth or citizenship will not necessarily be the country in which an individual has grown up and received their schooling or higher education. Given that the literature speculates that experience of international mobility earlier in the life course may be associated with a greater interest in mobility later in life, understanding what it means to be a British, a French, or an American scientist becomes even more important. More broadly the notion of measuring stocks and flows reflects the "national accounting" framing of mobility influenced by the dominance of human capital perspectives on mobility.

Levels of Scientific Mobility

The European IISER study (Moguérou and Pietrogiacomo 2007) used data on doctorate holders and doctoral candidates to examine the circulation of researchers across Europe as well as "flows" into and out of Europe. New data was collected to update this study in 2010 (MORE Consortium 2011). These show that 7% of doctoral candidates in the EU27 in 2007 held the citizenship of another member state, while Germany, Italy, France, Romania, Spain, the United Kingdom, Greece, and Bulgaria were among the top 30 countries of origin for holders of doctorates awarded in the United States in 2008. Finally, the update showed that, in 2007, China was the most important sender of doctoral candidates to the EU27 (with around 6500 doctoral candidates) followed by Mexico and the United States (sending 4000 and 3600 candidates respectively).

The same MORE study conducted two EU27 wide surveys of research active staff in universities and in non-university public research institutes, respectively ((MORE Consortium 2009, 2011; Børing et al. 2015). 54% of a sample of 4538 university researchers and 65% of a sample of 5050 non-university public researchers had experienced international mobility (defined as taking up an employed position or spending a research visit of three months or greater in a country other than the country in which they received their highest educational attainment). Across the two sectors, 76% of respondents who had experienced mobility as an exchange student subsequently experienced mobility as a researcher, a finding in line with the suggestion from Findlay et al. (2012) that mobility in early life and later career mobility may be linked.

In a study of patterns of academic mobility between the United Kingdom and the rest of the world, Sastry (2005) finds that it is relatively junior postdoctoral researchers who account for most movement. They also found that a majority of senior academics had spent time abroad in postdoctoral positions, suggesting that early career mobility experience is associated with career development. Børing et al. (2015) also find that propensity to be mobile seems to be greater in early career stages. Taking changes in author affiliation to or from an institution in a country other than the United Kingdom as a proxy for international mobility, Elsevier (2013) use Scopus data for 265,000 researchers over the period 1996–2012 to show the very high level of mobility into and out of the UK research system, identifying a lower level of "sedentary" researchers than for most other comparator countries and higher levels of outflowing, inflowing, and returning migrants. Inflowing, outflowing, and returning migrants are of high quality measured both in terms of productivity (relative to all UK researchers) and in terms of field-weighted citation impact.

Data from the MORE surveys confirms the internationalism of the UK scientific labor market and shows that, while the level of non-job mobility is high across

Europe, some national labor markets are more highly internationalized than others. Figure 17.1 shows the proportion of internationally mobile researchers with experience of international job mobility by country of affiliation for the European Union Higher Education sector. This share is highest for internationally mobile researchers from the United Kingdom (75%) and Austria (73%), and lowest for those from Slovakia (9%) and Romania (18%). Figure 17.2 shows the same data for the European Union non-university research institute sector. Again, this share is highest for internationally mobile researchers from the United Kingdom and Germany (both 79%), and lowest for those from Romania (21%) and Bulgaria (30%).

Jöns (2011) and Ackers and Gill (2008) suggest that different lengths of stay may be attractive or feasible for those with family or caring responsibilities and that this is very likely to have a gendered dimension. Børing et al. (2015) find that female respondents are less likely to have experienced international mobility than male respondents and, among those females who have experienced mobility, job mobility was less common but non-job mobility was no less common than for males. Interestingly, they also find no significant difference between the likelihood of female and male researchers having recent experience of international mobility (that is, in the three years running up to the survey). Børing et al. find no significant effect of marital/relationship status on experience of international mobility although parents have a relatively lower likelihood of having been internationally mobile. Nonetheless

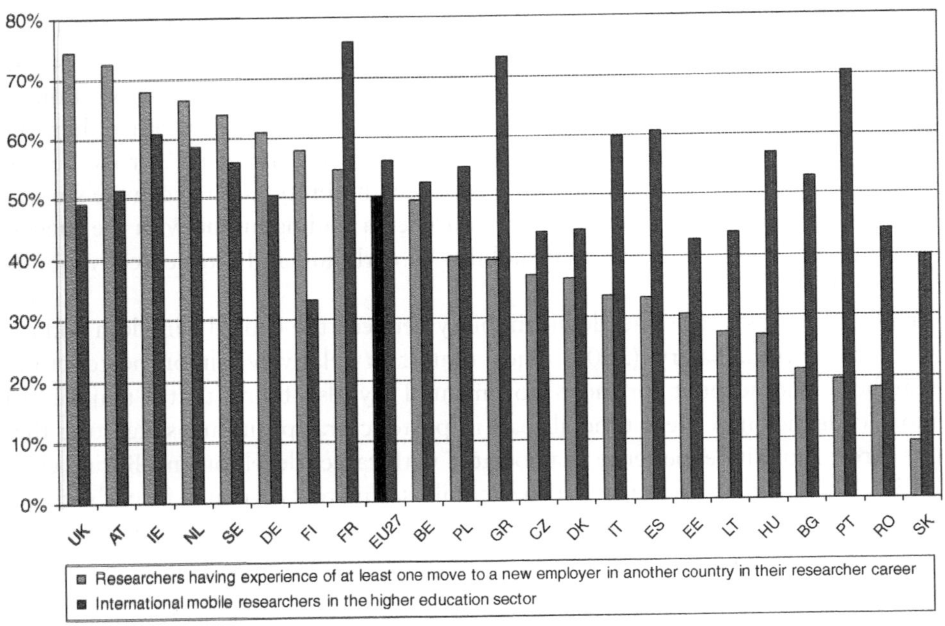

Figure 17.1 Proportion of researchers in the EU27 higher education sector with experience of at least one move to a new employer in another country by country of affiliation (left-hand columns) amongst all researchers with experience of international mobility, including non-job mobility. Proportion of all researchers in the EU27 higher education sector with experience of international mobility, including non-job mobility (right-hand columns). n = 2586.

Source: MORE Consortium (2010a): Cyprus, Latvia, Luxembourg, Malta, and Slovenia are excluded given the low response rate for those countries.

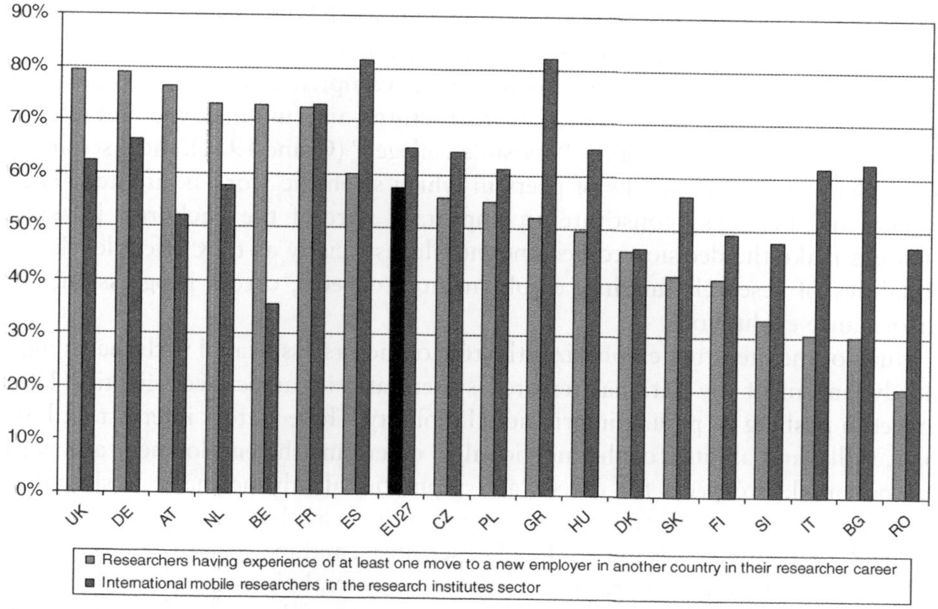

Figure 17.2 Proportion of researchers in the EU27 research institute sector with experience of at least one move to a new employer in another country by country of affiliation (left-hand columns) amongst all researchers with experience of international mobility, including non-job mobility. Proportion of all researchers in the EU27 research institute sector with experience of international mobility, including non-job mobility (right-hand columns). n = 3285.
Source: MORE Consortium (2010b):
Note: Cyprus, Estonia, Latvia, Lithuania, Luxembourg, Ireland, Malta, Portugal, and Sweden are excluded given the low response rate for those countries.

having children proves not to be a statistically significant predictor of whether respondents are likely to have previously experienced mobility.

At the same time Ackers has speculated that there may be a move away from traditional patterns of mobility characterized by long duration sabbaticals and job migration toward new forms of non-migratory internationalization such as holding positions simultaneously in two or more institutions in different countries, cross-border working, and new combinations of virtual collaboration and repeated very short research visits. The MORE surveys found some qualitative support for this hypothesis.

Push and Pull Factors

Macro- and Meso-Level Factors

Some conception of a "national system" is evident not only in accounts of the differential performance of national economies in terms of research, technological change, and economic growth (e.g., Lundvall 1992) but also in terms of broader varieties of capitalism (Hall and Soskice 2001). Other authors talk of national systems of research, higher education, and national systems of skills formation (Rip and van der Meulen 1996; Marginson and Rhoades 2002; Sung, Turbin, and

Ashton 2000). As already noted, many professional labor markets are at least partly internationalized, but scientists in particular work in overlapping national and internationalized systems, the former being comprised of national regulations and practices around employment, career structures, and funding dynamics referred to above, and the latter being the "invisible colleges" (Crane 1971), the disciplinary or problem-focused networks of peers in which scientific work is situated. These systems and networks constitute an important part of the backdrop in which scientists make the decision to become mobile, especially as they affect levels and modalities of research funding, employment prospects, career progression, and performance evaluation.

Much of the literature emphasizes the role of factors associated with the strength and dynamism of the national research system and its associated scientific labor market in pushing or pulling international mobility. These factors intersect with the goals, skills, and abilities of the individual to determine the employment and earnings potential, prospects for career, and reputational advancement. It should be noted once more that migration – the movement from a job in one country to a new job in another country – is not confined to scientists, and many of the factors identified in the literature are also relevant to the more general phenomenon of the migration of highly skilled professionals (for instance, issues of the availability and quality of job or training opportunities, level of remuneration, working conditions and broader issues around quality of working and personal/family life).

Ackers (2008) notes that some specific destinations may have intrinsic reputational and social capital over and above the intrinsic value of what may be done or what may be learnt there and that the international cosmopolitan flavor of some destination systems or research institutions may be an inherent attractor (see Florida and Mellander in this volume, Chapter 15).[3]

Differences between disciplines may lead to different observed mobility patterns. For instance Jöns (2007) argues that the degree of abstraction from place-specific realities inherent in the research work of different disciplines, the standardization of the practices involved, and their materiality (for instance in terms of equipment intensity) all imply different spatial relations, thus affecting the degree to which research is place specific or conducted in a variety of locations.

This approach has been elaborated further by Ackers (2013), who describes a continuum from more contextualized disciplines, such as anthropology or history, to highly standardized and abstract disciplines such as mathematics. Using data from the MORE surveys of EU27 university and non-university public researchers, Børing et al. (2015) found that those with social sciences and humanities qualifications were more likely to have been mobile in the last three years, suggesting that these fields have become more internationalized in recent years. However, across the whole of the research career, they found that respondents qualified in the broad natural sciences domain were most likely to have experienced international mobility.

Not surprisingly a lack of career opportunities in the sending system and better opportunities in the receiving system can operate as push and pull factors respectively (see, e.g., van de Sande, Ackers, and Gill 2005; Schiller and Diez 2009). However, unpacking what is really going on here can be difficult. Working conditions for scientists go beyond issues of remuneration and benefits, working hours and conditions, and the like to include issues such as freedom to pursue one's own agenda and access to funding and facilities.

A number of studies suggest that access to appropriate research facilities and collaborators, and the level of (and mode of access to) research funding are key push/pull factors (see, e.g., Franzoni, Scellato, and Stephan 2014; Kannankutty and Burrelli 2007; Moguérou and Pietrogiacomo 2007). Indeed, the MORE surveys (MORE Consortium 2011; Børing et al. 2015) found that factors relating to the quality and dynamism of the research system are more important in determining the attractiveness of a potential "target" country for international mobility than more general labor market and career structure factors such as salary and incentives, immigration rules, and so on. Interestingly, they found that broader innovation system factors, such as opportunities to build links with companies, are more important as pull factors making a destination system more attractive than they are as push factors making the home system less attractive. MORE respondents were also asked to nominate national research systems to which they would like to move in the future. There were some differences between disciplinary domains and between respondents with and without prior mobility experience, but overall the United States was the most commonly identified target for future mobility, with the United Kingdom, Germany, and France also very commonly identified.

Micro-Level Factors

MORE Consortium (2011) and Børing et al. (2015) find that personal or family factors are an explanatory factor for the lack of mobility of non-mobile researchers, while quality of life motives, career progression goals, personal research agenda goals, and training and development goals are all explanatory factors for mobility. Of these, all except quality of life factors seem to play a role in both kinds of mobility (it is not surprising that quality of life issues might be less determinant in relation to research visits not involving a change of job).

They also find, in line with the work of Ackers and others, that there are changes in perspective across the career and life-course of the researcher, with personal and family factors in general becoming more important to considerations of future mobility for previously mobile respondents than they had been in relation to past decisions about mobility. In a similar vein, Franzoni, Scellato, and Stephan (2012, 2014) find that personal or family factors seemed to become more important in influencing a decision to return to the home country.

Finally, Børing et al. find that cross-border workers – researchers who live in one EU27 member state but work in another – are more likely to have been internationally mobile in the past than those who are employed by an institution in their current country of residence. They also find that experience of international mobility as a student (for instance through a formal exchange program such as the European Union's ERASMUS scheme) increases the likelihood of having made at least one international research visit, though it has no significant effect on the probability of international job mobility. On the other hand they find that experience of an industrial work placement as a student increases the likelihood of job mobility during the research career, but has no significant effect on the likelihood that an individual will demonstrate non-job mobility.

Barriers to Mobility?

Cullen, Jones, and Hadjivassiliou (1998) note that "barriers" to labor movement may take two forms: namely factors militating against an impetus to move (usually associated with the "sending" system) and factors that might create obstacles to the individual (and their family) settling in the new "receiving" system. They also note that the barriers faced by those who exhibit cross-border working tend to be different and are more likely to revolve around technical issues relating to taxation and social security regulations and the risk of "double payments."

However, Børing et al. (2015) found that, while factors such as healthcare and pensions arrangements are of course experienced as difficulties by many mobile researchers, they seldom register as barriers to mobility, while in contrast caring and personal relationships and the challenges of obtaining funding and finding an appropriate position are often seen as barriers to mobility. How individuals calculate the trade-off between likely difficulties and potential benefits of mobility is likely to be very influenced by the culture and dynamism of the sending and receiving national research systems.

Impacts of Mobility

Macro- and Meso-Level Impacts

Mobility can have impacts on the content and direction of an individual's research, on their career, family, and social life and on the research performing organizations, disciplinary and problem-oriented networks, and national research systems in which they work. Much of the policy debate around mobility revolves around the desire to minimize negative systemic impacts and maximize positive ones. Debate has slowly moved away from a zero sum "national accounting" perspective of preventing "drains" and maximising "gains" and toward one emphasizing the potential benefits to both sending and receiving systems as well as to the individuals concerned, and this manifests itself in attempts to attract back or at least maintain productive links with expatriate scientists and to encourage mobility as part of early career development.

Edler et al. (2011) suggest that internationally mobile academic scientists engage in knowledge transfer activities in both their destination and their home systems, and that "frequent travelers" – scientists who engage in multiple instances of mobility – may be particularly active in transferring knowledge and technology back to their home system from their destinations (see also this volume, Barnard et al., Chapter 18and Hennemann and Liefner, Chapter 16). Trippl (2013) argues that internationally mobile elite "star scientists" may be particularly effective in transferring knowledge between the sending and receiving systems, especially – but not only – if they return to the former.

The famous study of Saxenian (2005) of the positive impacts of highly skilled engineers and entrepreneurs returning to Asia from time spent working in Silicon Valley popularized the term "brain circulation" and subsequently policymakers in many countries have put in place schemes to try to attract returnees. Studies focused on inventors and entrepreneurs echo Trippl's finding for "star scientists" that positive impacts may be felt by the sending system even without the migrant returning, due to the creation of productive social ties which serve to link the two systems allowing for knowledge flow and substantive collaboration (see, e.g., Meyer 2001; Agrawal, Cockburn, and McHale 2006; Oettl and Agrawal 2008). Such arguments have encouraged policymakers to look at how to promote the formation of productive "diaspora networks."

Van Heeringen and Dijkwel (1986) and Yano and Tomita (2006) argue that mobility is a characteristic of excellent scientists (rather than mobility effecting productivity) and that the performance of scientists is positively linked with achieving a position in "better" research organizations. Gibson and McKenzie (2014), using survey data to explore the migration experiences and scientific productivity of researchers from three small island countries, find that current migrants are more productive than either those who never migrate or those who have made a return migration, suggesting that productivity is conditioned by the environment in which they work and the opportunities it presents. However, Franzoni et al. (2014) find evidence that migrant scientists outperform non-migrant scientists in terms of publication impact, even after controlling for selection effects. They argue that this provides evidence to support the knowledge recombination theories of authors such as Agrawal and Saxenian, and the skill matching theory of Jones (2008), which they implicitly extend to include a wider range of complementary factors beyond an individual or a team's complement of potential or realized knowledge and skills.

Combining CV analysis and bibliometrics to explore the relationship between mobility and performance for a sample of 326 medical researchers, Sandström (2009) argues that two kinds of selection effect are in operation: on the one hand high performing institutions will select high performing researchers; while on the other hand researchers will also seek to improve their own performance by opting to become mobile. More broadly, the idea that excellence and mobility always go hand in hand neglects the reality that much mobility is effectively enforced by the perception or reality of poor prospects, poor working conditions, poor funding, or poor infrastructure in the home country (Ackers 2008).

Micro-Level Impacts

Previously mobile respondents to the MORE surveys of EU27 university and non-university public researchers overwhelmingly reported positive impacts on their career (MORE Consortium 2011; Børing et al. 2015). Interestingly, respondents without experience of mobility also strongly associated mobility with positive career impacts. Of course, an instance of mobility could have a positive impact on some aspects of an individual's life, such as career progression, but at the same time negative on others, such as personal and family life, and how this trade-off is calculated will depend on the specific context: for instance researchers from countries with poorly functioning research systems may be more inclined to accept negative impacts. Mobility is an event in the life course of an individual – personal impact is filtered through a complex set of characteristics of the sending and hosting country, the sending and hosting institution, the individual involved (career stage, age, family condition, research field, personal ambitions, etc.), and the features of the mobility experience itself.

The Construction of Mobility as a Policy Issue

The term "brain drain" was first coined in 1963, not in the context of the loss of scientific human capital from developing countries to developed ones but rather to describe a perceived increase in the rate at which UK scientists left the country for positions in Canada and especially the United States (Godwin et al. 2009; Balmer et al. 2009).[4] This debate, which ran on into the early 1970s, is characterized by Godwin, Gregory, and Balmer as tending to rely on patchy and often contradictory evidence; to conflate evidence of the migration of some elite scientists with the possibility of mass migration; and to conflate scientists and engineers, leaving for very different kinds of jobs in different sectors. Similar criticisms can be made of much of the more recent policy debate in this area.

Concerns about the extent, causes, and consequences of mobility were very much bound up with broader concerns about relative economic decline in the post-Second World War period, but were stoked by a landmark report from the Royal Society, the United Kingdom's national academy of science, in 1963 (Balmer et al. 2009).[5]

Intriguingly, the post-war brain drain debate in the United Kingdom tended to treat emigration to North America as of grave concern while the migration of scientists to less developed Commonwealth countries raised fewer concerns, reflecting the roots of this debate in concerns over what today would be called "national economic competitiveness." However, it also raised questions in Government about what level of scientific emigration was "natural" and to be expected, about whether the United Kingdom was simply training too many scientists, and about the immigration of scientists from other parts of the world into the United Kingdom. Finally, it shed light on differences between the dynamism and culture of the North American context as against the UK one and may thus have contributed to broader debates about the modernization of UK industry, science, and universities (Godwin et al. 2009).

At around the same time as "brain drain" emerged as a public and policy concern in the United Kingdom, economic theories of human capital were being developed and applied to the phenomenon and its consequences for the countries from which the brains are drained (Cañibano and Woolley 2010a, 2010b). As already noted, these authors argue that early accounts of mobility frequently conflated HRST with human capital. In their view this led to two dominant framings for mobility: a nationalist-pessimistic view ("brain drain") and an internationalist-optimistic view, what would now perhaps be characterized as "brain circulation" following the influential work of Saxenian (2005) and others.

Such framings influence the design of studies and the data that is collected, and both these perspectives remain very influential in science policy discourses. Davenport (2004) notes that "brain drain" debates have often had the character of "moral panics" and interprets the debate about "brain drain" from New Zealand during the early 2000s in those terms. In contrast, much of the European Union's recent policy discourse about scientific mobility has been framed along the "internationalist-optimistic" line.

Børing et al. (2015), Oliver (2012), and Ackers (2008) draw attention to the particular way in which mobility is framed in European Union science policy discourse (see, e.g., Morano-Foadi 2005; Fernández-Zubieta and Guy 2010) and in particular how it is increasingly constructed as intrinsically linked with "excellence."

At the same time mobility is seen as being suppressed by "barriers" and "obstacles", the removal of which should be a policy goal. Work in the internationalist-optimistic "circulationist" tradition (Ackers 2008), especially the contribution of Mahroum (2000), seems to have played a key role in encouraging this equation of "more mobility" with "more excellence" and in justifying a framing that there are "barriers" to mobility that can and should be removed.

Yet it is not clear why the "barriers" typically identified for policy action in the European science policy debate (mainly "technical" issues around health and social security provision, and pension portability) should be any more problematic and thus more worthy of intervention for scientists than any other group of potentially mobile highly skilled professional – save perhaps for an implicit admission that this professional group are typically paid less than other highly internationalized professions and therefore may have fewer personal resources to dedicate to the resolution of such problems than might professionals from other walks of life. While these factors do indeed create difficulties for those who are internationally mobile (and especially those who engage in cross-border working) the available evidence does not support the belief that such factors act as significant barriers preventing scientists from being mobile.

Conclusions

Mobility, whether job mobility or non-job mobility, is a complex event in the personal, family, and social life, as well as the professional life, of the individual scientist. The decision to move is influenced by conditions in both "sending" and "receiving" organizations and systems and, over time, may have a range of positive and negative impacts upon those organizations and systems as well as on the content and direction of the individual's research and on the progression of their career. While scientific mobility is clearly connected with the internationalization of science, the phenomenon of migration – the movement from a job in one country to a new job in another country – is not unique to scientists, and many of the factors identified in the literature on scientific mobility – issues such as the availability and quality of job or training opportunities, salaries and incentives, working conditions, and broader issues around quality of working and personal/family life – are also relevant to the more general phenomenon of the international migration of highly skilled professionals.

Nonetheless, scientific mobility is often framed in an exceptionalist way, and these framings matter because they shape how we problematize and measure mobility phenomena. Most of the available data on mobility is secondary data originally collected for some other purpose, data collected on behalf of a policy actor, or data collected under specific scholarly framings (such as human capital views or knowledge transfer views) which are themselves influenced by (and are influences on) policy framings. Mobility has become a hot topic, yet the evidence base is made up of a series of partial snapshots from different perspectives of a complex, entangled cluster of phenomena.

Furthermore, scholarly and policy understandings of scientific mobility are at risk of becoming disconnected from the reality of mobility as experienced by mobile scientists. Ackers (2008) argues that excellence and mobility are not "mutually constitutive" as is assumed in the circulationist framing of mobility in current science policy debates, and questions why policymakers and funding agencies seem so preoccupied with the barriers to mobility and with the quantity of mobility, rather than with drivers of mobility (a significant proportion of which is being "forced" by serious structural problems in national systems) and with the quality of mobility as experienced by those who undertake it. "The fact of mobility has been divorced from the objectives associated with it to become almost a rite of passage: a convenient, independent, indicator of excellence rather than a means to an end" (Ackers 2008: 418).

The act of mobility raises questions about the nature of the research of the individual but also about the sending and receiving organizations and systems and the way that they work. The framing of "brain circulation" has come to dominate thinking to the extent that there is a widespread commonsense understanding that the more mobility the better[6] – so long as at least some of those who leave return.

Yet the high levels of mobility of scientists in Europe observed by recent studies arguably tell us less about the growing excellence of the European Research Area than about, on the one hand, the emergence of an international labor market for elite researchers and elite institutions, and on the other, how dysfunctional some national research systems, labor markets, and career and reward structures still remain.

There is good evidence that international mobility can play an important role both in the transfer of knowledge and technology, and in the development of scientific human capital. The world's most successful national research systems undoubtedly owe much of their success to their ability to attract the brightest and the best scientists, wherever they come from. And these benefits are not won at the cost of equivalent losses in the "sending" systems in a zero-sum game: diaspora networks and migration/return can in principle benefit both systems.

However, it is lazy thinking to extrapolate from all this that mobility must always equate to excellence, and that more will always be better. The human and personal costs of enforced mobility, and the signals that mobility may send about dysfunctions in some national systems, deserve more scholarly and policy attention than they currently receive, especially since changes in the modes of production and dynamics of science such as the highly internationalized nature of some national research systems, the rise of cross-border working and dual positions, and the emergence of new patterns of ICT-enabled virtual collaboration, may be rendering traditional modes of mobility less attractive or necessary to scientists themselves.

Notes

1 Clearly, for scientists, a normative belief that international mobility is an important matter of personal, scientific, and career development may co-exist with a more general valuing of the universalism of science. Indeed (Ackers 2005a; Oliver 2012) argues that the former belief is increasingly widespread, at least in European researchers.

2 The macro- and meso-levels we identify here would collectively constitute the level of "structural" factors in the schema of Ng et al. (2007).

3 Conversely, scientists already working in highly internationalized systems or institutions will not be mobile in order to be highly internationalized.

4 The first recorded use of the term is by the London *Evening Standard* but, as Godwin et al. (2009) observe, was made famous by the response of the then UK Science Minister to a report from the Royal Society on the emigration of British scientists.

5 This can be seen as an incident in a larger struggle for authority between the venerable Society and the newer and increasingly assertive state advisory committees and funding bodies for science.

6 In the European Union in particular, this view of mobility may be underpinned by the more general belief that encouraging people to move around Europe somehow makes them more "European."

References

Ackers, Louise. 2005a. "Moving People and Knowledge: Scientific Mobility in the European Union." *International Migration* 43(5): 99–131.

Ackers, Louise. 2005b. "Promoting Scientific Mobility and Balances Growth in the European Research Area." *Innovation: The European Journal of Social Science Research* 18(3): 301–317.

Ackers, Louise. 2008. "Internationalisation, Mobility and Metrics: A New Form of Indirect Discrimination?" *Minerva* 46(4): 411–435.

Ackers, Louise. 2013. "Internet Mobility, Co-Presence and Purpose: Contextualising Internationalisation in Research Careers." *Sociología y tecnociencia / Sociology & Technoscience / Sociologia e Tecnociência* 3(3): 117–141.

Ackers, Louise, and Bryony Gill. 2008. *Moving People and Knowledge: Scientific Mobility in an Enlarging European Union*. Cheltenham: Edward Elgar Publishing.

Agrawal, Ajay, Iain Cockburn, and John McHale. 2006. "Gone But Not Forgotten: Knowledge Flows, Labor Mobility, and Enduring Social Relationships." *Journal of Economic Geography* 6(5): 571–591.

Baláz, Vladimír, M. Alan Williams, and Daniel Kollár. 2004. "Temporary Versus Permanent Youth Brain Drain: Economic Implications." *International Migration* 42(4): 3–34.

Balmer, Brian, Matthew Godwin, and Jane Gregory. 2009. "The Royal Society and the 'Brain Drain': Natural Scientists Meet Social Science." *Notes and Records of the Royal Society* 63(4); 339–353.

Børing, Pål, Kieron Flanagan, Dimitri Gagliardi, Aris Kaloudis, and Katerina Karakasidou. 2015. "International Mobility: Findings from a Survey of Researchers in the European Union." *Science and Public Policy*.

Bozeman, Barry, James Dietz, and Monica Gaughan. 2001. "Scientific and Technical Human Capital: An Alternative Model for Research Evaluation." *International Journal of Technology Management* 22(7/8): 716.

Cañibano, Carolina, and Richard Woolley. 2010a. "Beyond the Economics of the Brain Drain." Paper presented at Globelics 2010, Kuala Lumpur.

Cañibano, Carolina, and Richard Woolley. 2010b. "Towards a Socio-Economics of the Brain Drain and Distributed Human Capital." Paper presented at Globelics 2010, Kuala Lumpur.

Cañibano, Carolina, F. Javier Otamendi, and Francisco Solís. 2011. "International Temporary Mobility of Researchers: A Cross-Discipline Study." *Scientometrics* 89(2): 653–675.

Crane, Diana. 1971. "Transnational Networks in Basic Science." *International Organization* 25(3): 585–601.

Cullen, J., B. Jones, and K. Hadjivassiliou. 1998. *COMPETE: Competence Evaluation and Training in Europe. Final Report*. European Commission DG-XII, Targeted Socio-Economic Research Programme, Contract No. SOE2-08-2038.

Dasgupta, Partha, and Paul A. David. 1994. "Toward a New Economics of Science." *Research Policy* 23(5): 487–521.

Davenport, Sally. 2004. "Panic and Panacea: Brain Drain and Science and Technology Human Capital Policy." *Research Policy* 33(4): 617–630.

Dosi, Giovanni, Franco Malerba, Giovanni B. Ramello, and Francesco Silva. 2006. "Information, Appropriability, and the Generation of Innovative Knowledge Four Decades after Arrow and Nelson: An Introduction." *Industrial and Corporate Change* 15(6): 891–901.

Edler, Jakob, Heide Fier, and Christoph Grimpe. 2011. "International Scientist Mobility and the Locus of Knowledge and Technology Transfer." *Research Policy* 40(6): 791–805.

Elsevier. 2013. *International Comparative Performance of the UK Research Base – 2013*. Report prepared for the UK's Department of Business, Innovation and Skills.

Fernández-Zubieta, Ana, and Ken Guy. 2010. *Developing the European Research Area: Improving Knowledge Flows via Researcher Mobility*. European Commission, JRC-IPTS Scientific and Technical Report.

Findlay, Allan M., Russell King, Fiona M. Smith, Alistair Geddes, and Ronald Skeldon. 2012. "World Class? An Investigation of Globalisation, Difference and International Student Mobility." *Transactions of the Institute of British Geographers* 37(1): 118–131.

Franzoni, Chiara, Giuseppe Scellato, and Paula Stephan. 2012. "Foreign-Born Scientists: Mobility Patterns for 16 Countries." *Nature Biotechnology* 30(12): 1250–1253.

Franzoni, Chiara, Giuseppe Scellato, and Paula Stephan. 2014. "The Mover's Advantage: The Superior Performance of Migrant Scientists." *Economics Letters* 122(1): 89–93.

Gibson, John, and David McKenzie. 2014. "Scientific Mobility and Knowledge Networks in High Emigration Countries: Evidence from the Pacific." *Research Policy* 43(9): 1486–1495.

Godwin, Matthew, Jane Gregory, and Brian Balmer. 2009. "The Anatomy of the Brain Drain Debate, 1950–1970s: Witness Seminar." *Contemporary British History* 23(1): 35–60.

Hall, Peter A., and David Soskice (eds.). 2001. *Varieties of Capitalism: The Institutional Foundations of Comparative Advantage*. Oxford: Oxford University Press.

Heffernan, Michael, and Heike Jöns. 2013. "Research Travel and Disciplinary Identities in the University of Cambridge, 1885–1955." *The British Journal for the History of Science* 46(2): 255–286.

Jones, Benjamin F. 2008. *The Knowledge Trap: Human Capital and Development Reconsidered*. Washington, DC: National Bureau of Economic Research.

Jöns, Heike. 2007. "Transnational Mobility and the Spaces of Knowledge Production: A Comparison of Global Patterns, Motivations and Collaborations in Different Academic Fields." *Social Geography* 2(2): 97–114.

Jöns, Heike. 2011. "Transnational Academic Mobility and Gender." *Globalisation, Societies and Education* 9(2): 183–209.

Kannankutty, Nirmala, and Joan Burrelli. 2007. "Why Did They Come to the United States? A Profile of Immigrant Scientists and Engineers." NSF Infobrief 07-324.

Kim, Terri. 2009. "Shifting Patterns of Transnational Academic Mobility: A Comparative and Historical Approach." *Comparative Education* 45(3): 387–403.

Lundvall, B.-A. (ed.). 1992. *National Systems of Innovation: Towards a Theory of Innovation and Interactive Learning*. London: Pinter Publishers.

Mahroum, Sami. 2000. "Highly Skilled Globetrotters: Mapping the International Migration of Human Capital." *R&D Management* 30(1): 23–32.

Marginson, Simon, and Gary Rhoades. 2002. "Beyond National States, Markets, and Systems of Higher Education: A Glonacal Agency Heuristic." *Higher Education* 43(3): 281–309.

Meyer, Jean-Baptiste. 2001. "Network Approach Versus Brain Drain: Lessons from the Diaspora." *International Migration* 39(5): 91–110

Moguérou, P., and M.P.D. Pietrogiacomo. 2007. *Integrated Information System on European Researchers II: WP 1, 2 & 3: Indicators on Researchers' Stock, Career and Mobility*. Sevilla: Institute for Prospective Technological Studies, European Commission, Joint Research Centre.

Morano-Foadi, Sonia. 2005. "Scientific Mobility, Career Progression, and Excellence in the European Research Area." *International Migration* 43(5): 133–162.

MORE Consortium. 2009. *Technical Report 1: Scoping Paper*. Study on Mobility and Career Paths of EU Researchers.

MORE Consortium. 2010a. *Technical Report 2 (Part One): Report of the Higher Education Survey*. Study on Mobility and Career Paths of EU Researchers.

MORE Consortium. 2010b. *Technical Report 2 (Part Two): Report of the Non-University Research Institutes Survey*. Study on Mobility and Career Paths of EU Researchers.

MORE Consortium. 2011. *Final Report: Mobility and Career Paths of Researchers in Europe (MORE)*. Report to the European Commission Research DG.

Musselin, Christine. 2004. "Towards a European Academic Labour Market? Some Lessons Drawn from Empirical Studies on Academic Mobility." *Higher Education* 48: 55–78.

Ng, Thomas W.H., Kelly L. Sorensen, Lillian T. Eby, and Daniel C. Feldman. 2007. "Determinants of Job Mobility: A Theoretical Integration and Extension." *Journal of Occupational and Organizational Psychology* 80(3): 363–386.

Oettl, A., and A. Agrawal. 2008. "International Labor Mobility and Knowledge Flow Externalities." *Journal of International Business Studies* 39(8): 1242–1260.

Oliver, E.A. 2012. "Living Flexibly? How Europe's Science Researchers Manage Mobility, Fixed-Term Employment and Life Outside Work." *International Journal of Human Resource Management* 23(18): 3856–3571.

Rip, Arie, and Barend J.R. van der Meulen. 1996. "The Post-Modern Research System." *Science and Public Policy* 23(6): 343–352.

Sandström, Ulf. 2009. "Combining Curriculum Vitae and Bibliometric Analysis: Mobility, Gender and Research Performance." *Research Evaluation* 18(2): 135–142.

Sastry, T. 2005. "Migration of Academic Staff to and from the UK." Higher Education Policy Institute Report.

Saxenian, AnnaLee. 2005. "From Brain Drain to Brain Circulation: Transnational Communities and Regional Upgrading in India and China." *Studies in Comparative International Development* 40(2): 35–61.

Schiller, Daniel, and J. Revilla Diez. 2009. *Mobile Star Scientists as Regional Knowledge Spillover Agents*. IAREG Working Paper 2/7.

Sung, Johnny, Jill Turbin, and David Ashton. 2000. "Towards a Framework for the Comparative Analysis of National Systems of Skill Formation." *International Journal of Training and Development* 4(1): 8–25.

Thorn, K., and L.B. Holm-Nielsen. 2008. "International Mobility of Researchers and Scientists: Policy Options for Turning a Drain into a Gain." In *The International Mobility of Talent: Types, Causes, and Development Impact*, ed. A. Solimano, 145–167. Oxford: Oxford University Press.

Trippl, Michaela. 2013. "Scientific Mobility and Knowledge Transfer at the Interregional and Intraregional Level." *Regional Studies* 47(10): 1653–1667.

van de Sande, D., Helen Louise Ackers, and Bryony Gill. 2005. *Impact Assessment of the Marie Curie Fellowships under the 4th and 5th Framework Programmes of Research and Technological Development of the EU (1994–2002)*. Report of IMPAFEL 2, Contract No. IHP-D2-2003-01.

van Heeringen, A., and P.A. Dijkwel. 1986. "Mobility and Productivity of Academic Research Scientists." *Czechoslovak Journal of Physics B* 36: 58–61.

Yano, Masaharu, and Junichi Tomita. 2006. "Mobility Principle among Japanese Professors: Based on the Example of Professors in the Economics Field." *International Journal of Educational Management* 20(5): 338–347.

Chapter 18

The Role of Global Connectedness in the Development of Indigenous Science in Middle-Income Countries

Helena Barnard, Robin Cowan, Marta Fernandez de Arroyabe Arranz, and Moritz Müller

Introduction[1]

The study of globalization and internationalization is expanding beyond the traditional fields of economics and politics; literature on higher education and science describes the internationalization of universities and the creation of knowledge across borders (Altbach and Knight 2007), which suggests that science should be viewed as a global system. Derived from this internationalization, we observe an increasing international mobility of academics and scientist, resulting in the creation of numerous connections among different science centers and universities across the globe. This global science system comprises not only advanced economies but also academic systems from less developed countries (Altbach, Reisberg, and Rumbley 2009). But the way in which less developed countries are connected to this global system is not fully understood and needs further research.

There is extensive evidence that global connectedness matters, not only for the leading technological and economic powers, but also – and perhaps especially – for countries from behind the technology frontier (Fagerberg and Godinho 2005). Global connections provide access to technologies and know-how from a different context, and so generally accelerate technological and ultimately economic growth (Kemeny 2010). Furthermore, global connections present an opportunity for developing countries to take advantage of international knowledge diffusion through access to resources and knowledge flows that facilitate upgrading and catch-up (Ernst 2002; Lorenzen and Mudambi 2013). Global connectedness has been recognized in the literature as a key factor in the upgrading process of emerging economies (Lall and Narula 2004; Marin and Bell 2006; Fagerberg and Srholec 2008). For this reason, many countries behind the technology frontier are keen to increase their global connectedness.

The Handbook of Global Science, Technology, and Innovation, First Edition.
Edited by Daniele Archibugi and Andrea Filippetti.

But to accelerate economic growth, Nelson (2005) argues, indigenous public research has become especially important, and will become even more so, giving two reasons. The first is the role that indigenous public research plays in developing "social technologies," the practices, norms, and expectations that enable the ongoing development of new science and physical technologies. The second is the importance but also the challenge for lower-income countries of developing appropriate institutions in which those practices are embedded. Based on experiences in China, Fu (2008) and Fu, Pietrobelli, and Soete (2011) find that the benefits derived from global connectedness can only be delivered when sufficient indigenous R&D, human capital, and modern institutional and governance structures are present.

An emphasis on global connections is sometimes seen as somewhat at odds with an emphasis on developing local research capability. For example, Pouris and Ho (2014) express concern that the extensive collaborations of African scholars may occur at the cost of the development of local and regional research systems. The two options may well have different benefits: connectedness with the global community of science may help advance science, but only in small enclaves; while developing indigenous public research capability may contribute to a broad base of researchers, but operating largely behind the technological frontier. Moreover, developing countries may have research agendas that do not coincide with those of the leading developed community. Indeed, Shrum (1997) has found that a substantial proportion of indigenous research in lower-income countries is "invisible" to the developed world. Less developed countries are also by definition resource-constrained, which implies that some kind of trade-off may be required. It may seem that middle-income countries need to decide whether their economic and technological development is best served by investing scarce resources in seeking out global connections, or by investing in institutional reforms of public research entities.

This tension also matters in terms of policy, because the main criterion for allocating research funds is typically excellence, but this is often measured by looking at international peer recognition. Should internationally oriented researchers detach from the local scientific community, scientific catch-up might be hampered by such an incentive scheme.

However, we argue that such a question is itself flawed, and that the global/local tension in the public science institutions of developing countries is in fact a productive one with far more benefits than costs. There are two main reasons. First, scientific collaborations can be anchored around (small groups of) individuals (rather than entire institutions) and this dimension of scientific collaboration makes it possible to establish meaningful global connections even within underdeveloped institutions. Second, much of the perceived tension between local development and global connections originates from an overly simplified view of that relationship. Engaging globally does not preclude local upgrading; on the contrary, there is a potentially mutually beneficial relationship between the two. In the next section, we discuss those two aspects.

Understanding Global Connectedness in Science

Individual versus Institutional Collaboration

Research collaboration can be defined as "social processes whereby human beings pool their human capital for the objective of producing knowledge" (Bozeman, Fay, and Slade 2013). A recurring theme in work on research collaboration is the relative importance of organizational versus individually-driven collaborations, with studies generally concluding that both are important.

One body of literature trying to make sense of this duality relates the distinction between organizational versus individual drivers to the nature of the science conducted. Thus Chompalov and Shrum (1999) find that certain characteristics of research projects (namely project formation, magnitude, bureaucratic organization, interdependence, communication, participation, and technological practice) not only span scientific fields, but also result in a useful classification of how collaborative projects are managed. In particular, what they term "technological practice," the development of instrumentation specifically for the project, is useful in differentiating among four different types of approaches with different levels of institutional control. But it is noteworthy that of the four types, the formal institutional arrangement plays an important part only in the "managerial" approach. In a subsequent study, Chompalov, Genuth, and Shrum (2002) find that greater interdependence in data acquisition, analysis, and communication of results is correlated with less formal structures and greater success. They point out that if data are jointly obtained and verified by collaborators, a hierarchical arrangement becomes less important. Similarly, in a study of six fields of science, Wagner and Leydesdorff (2005) report that as ICTs and communication gain importance in collaboration, the selection of partners and the location of research is decided by the individuals rather than institutions or organizations.

Corley, Boardman, and Bozeman (2006) make a similar point using a different paradigm. They argue that collaborative research projects take place within the context of both epistemic norms and a certain type and level of development of organizational structure. In the case of epistemic norms, the definition of a scientific field could be seen as an indication of a high level of maturity; its organizational counterpart would be the establishment of an academic department. Collaborative research projects differ in terms of the level of development of the epistemic norms and/or organizational structure within which they operate, and Corley et al. (2006) argue that the two interact in a predictable way. When the epistemic framework is underdeveloped, there is a need for a stronger organizational structure, but when the epistemic dimension is already well-developed, a shared, common "language" reduces the need for strong organizational management.

As suggested by Nelson (2005), knowledge of such a language is an important "social technology." Scholars from less developed countries are likely to find it relatively easier to participate in projects in already defined fields where a shared scientific language is well-established and can act as a coordinating mechanism. This is particularly likely for scholars who were trained abroad and know those norms, but even when a scholar from an underdeveloped context is not entirely au fait with the norms of the field, s/he is likely to learn them more quickly to the extent that they are more strongly shared.

But where there are not strong epistemic norms, Nelson argues that coordination and governance may present a special challenge for scholars from institutionally underdeveloped contexts. He points out the complex array of tasks that support academic endeavor – establishing an appropriate division of labor; management, control, and coordination; hiring, rewarding, and occasionally releasing labor; financing and so forth. Developing those is both a prerequisite for and a desired outcome of effective indigenous research.

Relatively little work has been done on how individual versus institutional considerations vary for the establishment of research collaborations from more versus less developed research contexts. However, evidence from a related field, university–industry linkages, suggests that in contexts with weaker institutions, the role of individual drivers becomes more important. Thus a study of Brazil has shown that personal relationships are an important mechanism for university–industry linkages in emerging industries where public knowledge and support are limited, and less so in mature industries (Bodas Freitas, Marques, and Silva 2013). A similar pattern of relying on personal ties when institutions are weaker has been found in Italy where small firms rely on personal contacts to establish university–industry linkages, whereas large, vertically integrated firms use institutional arrangements (Bodas Freitas, Geuna, and Rossi 2013).

This suggests that individuals and individual relationships play an important role in establishing research collaborations in less developed countries. In the case of research collaborations with international partners, it is likely that useful practices from more developed contexts will be borrowed. In the process, scholars from a less developed context stand a chance of contributing not only to the local knowledge base, but also to its institutional development.

Learning from Local and Global Connections

Traditionally, global connections occurred among developed countries such as Japan, Europe, and the United States with a technology-seeking aim (Song and Shin 2008); recently, however, there has been an increase not only in the inflows of knowledge and FDI to developing economies (UNCTAD 2013) but also the emergence of outflows from those countries (Cuervo-Cazurra 2007; Gammeloft, Barnard and Madhok 2010; Álvarez and Marín 2013). Global connections in the literature on less developed countries can be categorized as either "inbound," where not only the desired new knowledge, but also much of the initiative for obtaining it comes from abroad, or "outbound," where the desired new knowledge lies elsewhere, but the initiative for seeking it is local. The case when advanced MNCs set up and nurture subsidiaries in less developed contexts (Narula and Dunning 2000, 2010) is perhaps the best documented example of inbound global connections. And when emerging MNCs (EMNCs) engage in FDI in especially more developed contexts (e.g., Nam and Li 2013), those global connections can be described as outbound.

There are benefits but also problems associated with both those paths. There is a growing body of evidence that outbound connections may not yield the desired benefits. The attempts of EMNCs to source knowledge and capabilities through establishing global connections do not always succeed. In particular, they seem to some extent excluded from the more tacit sources of knowledge, and better able to source market-based assets (such as more codified knowledge) (Barnard 2010) or operate in an incremental, exploitative way (Rabbiosi, Elia, and Bertoni 2012). Furthermore, EMNCs face the challenge of overcoming the liability of foreignness, that is, the cost of doing business abroad (Barnard 2010), and overcoming the disadvantage of weaker capabilities located behind the technological frontier (Ramamurti 2009; Álvarez and Marín 2013); for example, the average asset base of EMNCs is only 15% that of the MNCs of the developed world (UNCTAD 2013).

As for advanced MNCs, there is also considerable evidence that the host context does not always benefit from the entry of the MNC (Meyer and Sinani 2009). The reasons given that global connections fail to facilitate technological advance are often that the local context lacks the capability to take advantage of the knowledge, for example, because the social ties that would enable knowledge sharing to take place are limited (Eapen 2012), or because the host country lacks the sufficient absorption and adaptation capabilities (Kumaraswamy et al. 2012). Another common explanation is that the MNC does not extensively share knowledge with the subsidiary in the developing country, generally because the mandate of the subsidiary is limited, for example, when the MNC sought out the location for its low wages, and consequently located mainly low value-adding activities there (Sethi et al. 2003), but sometimes also because of political considerations inside the MNC (Becker-Ritterspach and Dörrenbächer 2009).

We argue that an overly simplified view where parties are seen as either "providers" or "receivers" of knowledge is problematic. Moreover, in such a view, less developed countries are typically seen as receivers rather than creators of knowledge. With the stark conceptual divide between receivers and providers of knowledge, it is hard to see the possibilities of feedback loops and mutual benefit. This analytic oversimplification makes it hard to understand the role of both local engagement and global connections in upgrading.

It is telling that this analytic oversimplification is less often found in studies of the global connections of advanced economies. In those studies, there is generally an appreciation for the fact that different countries have different specializations, and that entities in both contexts stand to benefit from sharing knowledge (Cantwell and Piscitello 2002, 2005). Studies have explicitly examined the benefits not only to the "recipient" (e.g., Chung and Alcácer 2002), but also to the "provider" (Singh 2007; Yang, Phelps, and Steensma 2010) of knowledge, and consistently find that knowledge does not flow only one way. Bartlett and Ghoshal (1998) conclude that the interactions between MNC and host countries are much more complex and interactive than simply establishing "replicas" abroad and in fact are close to the concept of an international network. Seen in this way, a more internationalized value chain (Kaplinsky 2004) ensures that host countries exploit competencies over the firm's network and also develop new competencies and knowledge resources (Rugman and Verbeke 2001; Yang, Mudambi, and Meyer 2008). Singh (2007) finds not only that the knowledge outflows from the host country back to the MNC in more developed countries (or more developed sectors) are greater than their knowledge inflows to the host country, but that even in the case of relatively less developed countries or sectors, outflows to the MNC almost equal the knowledge inflows from the MNC.

Although it is generally true that the capabilities of less developed countries are more limited and their institutions less developed than those of the leading economies, Marin and Bell (2006) provide evidence that the most beneficial global connections are those where the "inbound" and "outbound" distinction is less clear. In their study, they found that innovation in Argentinean subsidiaries of advanced MNCs was especially likely when the subsidiary was connected to the headquarters of the MNC but also to local providers of knowledge and expertise.

In contrast to the elusiveness of technological upgrading when the source of knowledge is distinct from the recipient, the evidence from Marin and Bell (2006) suggests that the best way for learning to take place is when both parties play an active role as partners and collaborators in the process of learning and knowledge production. For example, Rasiah (2003) reports how Malaysia and Thailand, key exporters in the electronics sector, have improved their technology capabilities as FDI moved from just the expansion of production to the development of technology. Indeed, evidence about the ongoing importance of absorptive capacity (Goode 1959; Cohen and Levinthal 1990) in its original conceptualization (i.e., with the emphasis on the considerable effort needed to be able to benefit from knowledge spillovers) suggests that the notion of a distinct "sender" versus "receiver" of knowledge is problematic. The process seems more robust when both parties are recognized as potential sources of knowledge. Thus, to advance research in a developing country, it is not enough for the role-players from the developed world to be highly competent; their counterparts from the developing world itself should be too (Barnard 2010).

The arguments put forward in the literature on knowledge flows within and across organizations apply as well to individuals. In fact most arguments, particularly those related to absorptive capacity, are rooted in theories of individual behavior, cognition, and psychology. Collaboration among scientists takes many forms, ranging from informal advice to joint research and co-authorship. Most scientific collaboration studies investigate co-authorship relations. A consistent finding is that scientists with higher reputation, essentially having more publications, attract more collaborators (Newman 2001). In other words, there is a strong tendency of young scholars to connect with scholars of high reputation and, at the same time, a strong tendency among highly visible scientists to collaborate (Verspagen and Werker 2004). Thus, many co-author relations are marked by a strong differential in terms of experience, and hence knowledge might flow predominantly in one direction. However, since by definition every author contributes to the joint scientific paper, knowledge flow will never be strictly one-directional; and even less so when we differentiate between generic knowledge (mainly from supervisor to student) and specific knowledge (mainly from student to supervisor). Knowledge differentials, averaged over specialties, between "world leading" scholars are going to be small in general, which strongly suggests bidirectional knowledge flows. When there is a "world leading" collaboration between scholars across more and less developed contexts, we may therefore expect both benefits in terms of the global production of knowledge, and also the development of indigenous knowledge and capacity.

This perspective suggests that scientific collaboration among scientists from developed and developing countries is essentially a win-win situation. How could the developing country then be negatively affected by numerous collaborations spanning the global and local science system? In the debate among scholars and policymakers two concerns are frequently raised.

The first concern is that excessive international collaborations might result in a local science system that fits the science systems of developed countries well but is maladapted to the local context. Developing countries may have research agendas that do not coincide with those of the leading developed communities because they are endowed by different resources and have different societal needs. This confrontation of research priorities has been extensively reported in the case of Asian business schools, where the research agendas, driven by developed country contexts, struggle to adapt to local circumstances, and thus fail to tackle issues that concern local managers (Meyer 2006; Leung 2007). This, however, is not unique to emerging countries, and has been also found in advanced economies such as the Netherlands, where the fruits of research bring prestige but do not bring profits to the region (Rip 2002). Based on similar arguments, Pouris and Ho (2014) express concern that the extensive collaborations of African scholars may occur at the cost of the development of local and regional research systems.

The second issue originates from the fact that developing countries are marked by strong heterogeneity. The local science system includes few scientists and institutions of research excellence. Most scientific institutions lag largely behind the scientific frontier as they lack the necessary resources. Only the few centers of excellence in the local science system are able to engage beneficially in international collaborations at the scientific frontier. Thus, supporting global connections means that one needs to invest more in centers of excellence rather than in the broader array of institutions. Individual scientists of excellence might then devote most of their collaboration time and resources to international collaborations and collaborate less with their less privileged local colleagues.

The concern then is that the centers of excellence integrate within the global science system but at the same time disconnect from the rest of the local science system. Instead of upgrading the overall indigenous science system, fostering international collaborations would increase heterogeneity, and weaken coherence, within the indigenous science system. Connectedness with the global community of science may help advance science, but only in small enclaves, while developing indigenous public research capability may contribute to a broad base of researchers, but operating largely behind the technological frontier.

The literature documents several reasons why globally connected but locally disconnected enclaves may develop around the most competent researchers in a developing country (see, e.g., Barnard, Cowan, and Müller 2012). Such enclaves have been previously documented in international business contexts in less developed countries, especially in those cases when quality is of particular importance (Akbar and McBride 2004; Feinberg and Majumdar 2001). In an earlier study of scientific collaboration, we found that the world-leading scientists in a developing country – rather than scholars whose emphasis is essentially local – tend to be well connected internationally, but at the same time remain strongly connected locally (Barnard et al. 2012). Thus, the concern of enclave formation could not be supported empirically in that study. On the contrary, "world-leading" scholars with a rich international collaboration network appeared to be important in the diffusion of global knowledge locally.

To this point our discussion has followed a stylized image where knowledge flows through scientific collaborations among scientists that are either from the developed or from the developing world. This picture is refined in the next section by taking into account that scientists are actually highly mobile between the local and global science system.

How Mobile Scientists Connect Global and Indigenous Science Systems

The analysis of the process of knowledge creation and diffusion cannot be separated from the study of individual scientists. On the one hand, Fischer (2001) points out that it is individuals, not organizations, who are the generators of knowledge; on the other hand, a large part of this knowledge is embodied in individuals as tacit knowledge (Polanyi 1983). The localization of knowledge flows is thus tied to and can be explained through local mobility of knowledge workers or inventors (Breschi and Lissoni 2009; Møen 2005).

Extensive literature suggests that knowledge diffusion occurs to a large extent through the movements of individuals (Saxenian 1996; Argote and Ingram 2000; Rosenkopf and Almeida 2003). In the context of inter-firm knowledge transfer, Malecki (1997) argues that individuals are an important channel of transfer. In the semiconductor industry, for example, studies have found that when inventors move from one firm to another, they carry knowledge from their old firm to the new one, affecting the subsequent patenting activity (Saxenian 1990; Almeida and Kogut 1990; Song, Almeida, and Wu 2001). Furthermore, organizational learning through employee turnover, or learning-by-hiring, is a significant driver of mobility in innovative firms (Cassiman and Veugelers 2006) and high-tech industries (Song, Almeida, and Wu 2003; Palomeras and Melero 2010); for example, the hiring of star scientists (Schiller and Diez 2009), and in the case of pharmaceutical industry, the adopting of biotechnology (Darby and Zucker 2001).

Similarly, in the context of science, PhD training, exchange of researchers, and professorship appointments are important ways of transmitting not only scientific knowledge but also social knowledge, as discussed for example by Nelson (2005). This has been pointed out at least since Polanyi (1958). More recent is the discussion of Scherer (2000) on how the emigration of many Austrian and German economists to the United States during the Third Reich affected American economic development. Similarly, Kogut and Macpherson (2011) provide evidence that during the 1990s economic paradigms (e.g., on privatization) traveled from the United States to Europe through mobile scientists.

Thus knowledge moves together with the individuals who hold that knowledge. This idea is at the origin of the brain drain discussion, which started in the 1960s when scientists and other knowledge workers trained in the United Kingdom migrated to the United States (Cervantes and Guellec 2002). The same label has also been used to describe the phenomenon in which untrained knowledge workers migrated from developing to developed countries, mainly the United States, in order to receive training. In the former case the loss is in actual human capital (the most skilled leave), while in the second it is the potential human capital (the brightest leave) (see, e.g., Gaillard and Gaillard 2001). Meyer (2001) reports that as the academic system and research network connections are more complete, the first kind of migration, the loss in actual human capital, is more common; for instance, having a more complete techno-economic network, South African expatriates tend to go abroad when they are more qualified, while Colombian expatriates, with short and limited networks at home, emigrate with only a bachelor degree, seeking to pursue further studies abroad.

There have been many policy initiatives of developing countries since the 1970s that aimed to repatriate skilled expatriates, through for example compensation or taxation schemes (Meyer 2001). Most of these policies failed, except perhaps for some cases in Far East (e.g., South Korea, Singapore, and Taiwan) (Gaillard and Gaillard 2001). This led policymakers in the 1990s to acknowledge (or hope?) that even when expatriates stay abroad they may bring some value for their country of origin, so that they should not be considered a loss but rather an asset that could be mobilized (Meyer 2001). As a consequence, network initiatives emerged to create networks among and with expatriates in order to create social capital (Gaillard and Gaillard 2001; Meyer 2001). Brown (2000) reported the creation of more than 40 networks in developing countries with the purpose of bringing closer expatriates' skills to their country of origin (e.g., SANSA initiative of South Africa or CALDAS in Colombia). In recent years, however, studies have reported a decrease in the level of permanent migration (Koser and Salt 1997; Cervantes and Guellec 2002), temporary mobility being far more common; especially among PhD students and post-docs with the aim of gaining knowledge, experience, and contacts that can be exploited upon their return to the home country (Edler, Fier and Grimpe 2011).

These migration patterns and the choice of location are determined by the social (Li et al. 1996), professional (Mitchell 2000; Beaverstock 2005), and co-ethnic (Meyer and Brown 1999; Saxenian 2002) relations that an individual has. Melin (2004) reports that when post-docs move they develop a richer network but often they move along established connections of their senior scientists at their home institution. Therefore, how scientists move is both a result and a driver of the collaboration network (Scellato, Franzoni, and Stephan 2012; Cowan, Feldman, and Kogler 2006).

Overall, the reason behind the temporary mobility of scientists is the perception that being part of the global network and collaborating with scientists abroad improves capabilities and research productivity (Edler 2007; Defazio, Lockett, and Wright 2009). Jonkers and Tijssen (2008), in an empirical study of China, and Scellato et al. (2012), in a study of 16 countries, show that mobility fosters collaboration networks and that scientists who experienced foreign stays have richer international collaboration networks and higher scientific standing. Both studies report a positive correlation between foreign experience and the number of international co-publications. Moreover, in a study of Swedish scientific mobility, Melin (2004) finds that a post-doc abroad has important integrating effects between the home and the host country.

Based on these ideas, we argue that scholars in developing countries who have had extensive experience working and training with scholars from the leading knowledge producers will play a special role in development of indigenous research. Ynalvez and Shrum (2011) point out in a Philippines-based study that benefits could occur because of training which would ensure that they are familiar with the norms for quality research in their field, but also because of the value of foreign socialization and developing contacts. However, developing relationships with scholars abroad can occur in many different ways. Research partners may be diasporans who are keen to establish collaborations with scholars at home, collaborators may have met at conferences, or a third party with extensive international experience may broker the relationship. Either way, research collaborations between local researchers from the developing country context and foreign scholars are likely to play an important role in the development of local science. In a study of the returning process to China of scholars and students overseas in the period 1997–2001, Zweig, Changgui, and Rosen (2004) found that returnees holding foreign, and to a lesser extent domestic, PhDs helped domestic PhDs to establish global networks; they also increased their human capital and attracted more grants. This phenomena of collaboration has also been observed in developed nations like Canada, where academic Canadians living abroad have been found to have an active role in Canadian knowledge creation, and thus, not constituting a pure brain drain (Cowan et al. 2006).

The extant literature on the mobility of scientists thus emphasizes the effect of mobility on the number and quality of scientists in a system (the human capital aspect) and how mobility relates to social connections within and among systems (the social capital aspect). Both international mobility and international collaborations foster global connectedness and affect the development of indigenous science in middle-income countries. To illustrate this hypothesis in one context, the next section presents an illustration of the importance of global connectedness in South Africa. South Africa is a useful setting and a representative example, because its level and patterns of scientific expertise are typical of middle-income countries generally. Thus, the results obtained in the empirical study, even though restricted to the case of South Africa, represent in general the issue of connectedness in developing countries and are intended to illustrate the theories presented in the first part of this chapter.

International Knowledge Flows and South African Science

Here, we show that the South African science system has indeed a position in the global science system which is comparable to that of other developing economies. After a short discussion of our data and sample, an analysis illustrates theoretical arguments in two steps.[2] First, we show that South African scientists contributing to frontier knowledge internationally act as gatekeepers in the South African science system (Barnard et al. 2012). Conceptually, gatekeepers of a system source knowledge from outside the system and diffuse it within the system (Allen 1977). In our study, gatekeepers are those scientists who strongly connect to the international science system and, at the same time, are central in the local, South African science system. This result shows that engaging globally does not preclude local engagement by individual scientists. Thus, local and global connectedness in research should not be considered as substitutes but rather as complements.

A second analytical step provides a closer look at how social and human capital relates to the ability to participate in scientific advancement at the global, international level. Here we observe that South African scientists having obtained a doctorate outside South Africa, rather than South African PhDs, independently expand their international contacts during their subsequent career, and are more likely to become internationally renowned scientists. The analysis suggests that international training and research within the international community is a paying investment for becoming a nationally as well as an internationally renowned scientist. By taking a central position in both worlds, researchers may function as gatekeepers for the development of indigenous science.

Relative Positioning of the South African Science System

A relative ISI citation ranking from 2013 reveals that most of South African science is behind the technological frontier. Table 18.1 indicates that the impact of most South African research is similar to the impact of work done in peer countries, other leading emerging economies like Argentina, Brazil, Saudi Arabia, and Mexico. Even in the best performing disciplines, the work of most South African researchers

Table 18.1 Relative impact in scientific field (number of papers multiplied by number of citations).

Field	*South Africa rank*	*Country ranked directly below South Africa*	*Country ranked directly above South Africa*
Agricultural Sciences	35	Austria	Czech Republic
Biology & Biochemistry	37	Mexico	Chile
Chemistry	39	Romania	Thailand
Clinical Medicine	31	Singapore	Czech Republic
Computer Science	37	Russia	Mexico
Economics & Business	32	Brazil	Lithuania
Engineering	44	Argentina	Saudi Arabia
Environment/Ecology	22	Norway	Portugal
Geosciences	26	Taiwan	Greece
Immunology	21	Israel	Finland
Materials Science	45	Bulgaria	Saudi Arabia
Mathematics	39	Serbia	Saudi Arabia
Microbiology	28	Mexico	Portugal
Molecular Biology & Genetics	41	Croatia	Thailand
Multidisciplinary	27	Brazil	New Zealand
Neuroscience & Behavior	39	Chile	Iran
Pharmacology & Toxicology	39	Argentina	Russia
Physics	46	New Zealand	Chile
Plant & Animal Science	22	New Zealand	Finland
Psychiatry/Psychology	28	Turkey	Greece
Social Sciences, General	19	Brazil	Denmark
Space Science	28	Finland	Hungary
All Fields	35	Iran	Mexico

Source: ISI Web of Knowledge, accessed December 8, 2013.

remains quite far from the frontier, and has a comparable impact to that of researchers in the non-core disciplines of small late-coming economies like Finland, Singapore, and Taiwan.

Citation rates are a commonly used measure of the impact of research and used to establish quality in a variety of knowledge production contexts (DuBois and Reeb 2000; Jaffe 1989; Starbuck 2005). Although there is evidence that citations reflect a social as much as a cognitive process (Leydesdorff and Amsterdamska 1990), it is a useful high-level indicator of the relative prominence, impact, and arguably also quality of research.

Because citation rates vary by discipline, we show the South African citation rate for each discipline separately (see Table 18.2.) There is evidence of a number of centers of excellence, such as Clinical Medicine and Immunology, related to the HIV epidemic in South Africa, Space Science, which reflects the extensive government support for South Africa's ultimately successful bid to host the Square Kilometer Array,[3] and the small but highly cited body of work on Computer Science. However, the overall average citation rate of publications by South African authors in ISI is almost exactly the average citation rate of ISI publications overall.

Table 18.2 Average citations per paper.

Field	*South Africa*	*Entire ISI database*
Agricultural Sciences	6.83	7.22
Biology & Biochemistry	**12.60**	**11.80**
Chemistry	7.81	9.13
Clinical Medicine	**14.26**	**12.09**
Computer Science	**6.43**	**3.98**
Economics & Business	2.36	4.76
Engineering	4.54	5.01
Environment/Ecology	**12.48**	**11.17**
Geosciences	8.84	9.65
Immunology	**22.76**	**17.42**
Materials Science	5.94	7.79
Mathematics	3.28	3.31
Microbiology	14.96	15.75
Molecular Biology & Genetics	14.66	19.26
Multidisciplinary	3.20	14.14
Neuroscience & Behavior	13.12	14.33
Pharmacology & Toxicology	9.93	10.62
Physics	7.77	8.87
Plant & Animal Science	6.98	7.18
Psychiatry/Psychology	6.22	9.10
Social Sciences, General	**4.24**	**4.17**
Space Science	**16.17**	**15.28**
All Fields	**9.20**	**9.15**

Note: Fields in bold exceed the ISI average.
Source: ISI Web of Knowledge, accessed December 8, 2013.

It is likely that other middle-income countries will exhibit a similar pattern: a few centers of excellence that relate mainly to local conditions, but generally scholarly work has "average" impact with a small community of researchers. Clearly, such countries will stand to benefit from being connected to scholars from the global technology leading countries.

Data Source

We employ a unique data set made available by the National Research Foundataion (NRF) of South Africa. The NRF is one of the main government research funding agencies, and has as its mission to promote research and develop research capacity. As part of the process by which it awards research grants, the NRF has instituted a "rating" process: before applying for funding, any researcher must apply to be rated. The process involves submission of evidence of all past research output in addition to the candidate's work history and other relevant details. The NRF assessment is based on roughly six referee reports, written by local and international scholars. The referee reports are used by an independent assessor and a specialist review panel to arrive at a rating for the applicant. A given rating is valid for a fixed period,[4] and when it lapses the researcher must reapply. Reapplication may result in a change in rating, either up or down, so ratings are a good reflection of very recent performance.

Table 18.3 defines the different possible rating categories according to experience and quality of work. Conveniently, NRF categories explicitly observe where researchers' work fits in the overall global or local context. A-rated and to a lesser extent B-rated researchers are seen as contributing to the international knowledge frontier, whereas the focus of C-rated researchers is primarily local. As the rating of scientists as world-leading scholars (A-rated), internationally recognized scholars (B-rated), or scholars acting within the local scientific community (C-rated) has strong implications for research funding, scholars, and their universities, are certainly motivated to achieve the highest possible rating. The data are therefore well suited to explore the local/global tension and potential in a technologically less developed context.

Sample

The study focuses on researchers working in science, engineering, and social science at South African universities. All NRF filings from the beginning of 2002 until the end of 2012 are used for the analysis. The data include socio-demographic

Table 18.3 Definitions of NRF research ratings.

Category	*Definition*
A	**Leading international researcher** Researchers who are unequivocally recognized by their peers as leading international scholars in their field for the high quality and impact of their recent research outputs.
B	**Internationally acclaimed researcher** Researchers who enjoy considerable international recognition by their peers for the high quality of their recent research outputs.
C	**Established researcher** Established researchers with a sustained recent record of productivity in the field who are recognized by their peers as having produced a body of quality work, the core of which has coherence and attests to ongoing engagement with the field as having demonstrated the ability to conceptualize problems and apply research methods to investigating them.
P	**NRF President's Awardee** Young researchers (normally younger than 35 years of age), who have held the doctorate or equivalent qualification for less than five years at the time of application and who, on the basis of exceptional potential demonstrated in their published doctoral work and/or their research outputs in their early post-doctoral careers, are considered likely to become future international leaders in their field.
Y	**Promising young researcher** Young researchers (below 40 years of age), who have held the doctorate or equivalent qualification for less than five years at the time of application, and who are recognized as having the potential to establish themselves as researchers within a five-year period after evaluation, based on their performance and productivity as researchers during their doctoral studies and/or early post-doctoral careers.

Source: http://www.nrf.ac.za/rating, accessed January 5, 2015.

information (age, sex, race, and marital status) as well as professional information (title, scientific field, and list of publications) on the researcher and the rating by the NRF. Our focal population for the analysis consists of the 2778 established scientists who received at least one valid rating (A, B, C, P, Y, or "Rating Unsuccessful") during the sample period.

We construct a dynamic co-authorship network using peer-reviewed publications, evolving from 2002 to 2012 using a lagged five-year moving window, and use this network to measure the network position of focal researchers at the time of rating. Because the rating might occur any time in a given year, we consider the co-authorship network five years prior to rating. If for example a focal rating was obtained in 2002, we measure the network position on the network that accumulated from 1997 to 2001.

For the analysis it is crucial to clearly identify foreign co-authors and rated co-authors. Yet, some co-authors in the publication list might be South African but not rated. We identify non-rated but South African researchers based on further data. The quality of the procedure for the identification of foreign researchers has been checked and found satisfactory with errors in the lower single-digit range, and conservative with respect to rated and foreign researcher categorization. After the whole cleaning procedure, the complete network accumulated from 2002 to 2012 includes 36,633 researchers (32,361 foreign and 4272 rated South African) connected through 211,207 single or multiple co-authorships.

Network positions, such as the gatekeeping score, are measured on the weighted network. We follow the approach of Newman (2001) and weight each link that originates from a paper with n co-authors by $1/(n-1)$. For example, a paper that is co-authored by three researchers adds three links of strength 1/2 to the network. The idea behind that weighting scheme is that each researcher invests a fixed amount of interaction time in each publication which is divided equally among all co-authors.

The Gatekeeping Function of Internationally Renowned Scientists

Our network measure of gatekeeping (Barnard et al. 2012) quantifies the extent to which an individual scientist occupies a position in the co-author network that is characterized by both strong research connections with international, non-South African scientists for knowledge acquisition and, at the same time, high centrality within the South African science system for knowledge diffusion.

The discussion on local and global connectedness in the literature argues that scientists may follow one of two strategies which tend to exclude each other. One strategy is to aim at international recognition by tackling research issues that are acknowledged within the international research community through extensive collaboration with international scientists. The other strategy is national. Here the researcher tackles research issues with more local relevance in collaboration with local scientists. Researchers that follow exclusively the international strategy would pursue international collaborations for the sake of international connectedness, while researchers that follow an exclusively "national" strategy tend to lose sight of international contacts. The result in both cases would be a poor performance in terms of our measure of gatekeeping. To capture these two key aspects – connection to the international knowledge frontier and connection within the South African community – we define the gatekeeping measure as follows:

$$Gatekeeping = external\,weight \times average\,weighted\,reach$$

The *external weight* of a researcher reflects his/her international connectivity and is captured by the (weighted) number of papers with international scholars. The knowledge diffusion capacity within the South African science system is measured through the *average weighted reach* of the focal researcher to all other researchers in South Africa. Average weighted reach is defined as the average of the inverse of the shortest weighted path from the focal agent to all other agents in the network. The higher the average weighted reach of the focal agent, the closer the focal agent is on average to other agents in the network. This is beneficial for knowledge diffusion because knowledge sent by a researcher spreads more easily and faster to recipients who are close in the network. Thus gatekeeping of a researcher increases as her external weight increases and as average weighted reach to other South African scientists increases.

Table 18.4 shows how the researchers' gatekeeping score relates to their ratings. The second column, "Observations," provides the number of observations in each category. A-ratings (world-leader) are relatively rare and C-ratings (local participant) are by far the most common. The roughly 20% of unsuccessful ratings includes cases where the track record of the applicant was either insufficient for rating or where a criteria for rating was not met (such as incomplete submission or main workplace outside South Africa).

The average gatekeeping scores, third column of Table 18.4, are decreasing from A- to B- to C-rating, and even lower for unsuccessful applicants. Since gatekeeping is composed of two factors, embeddedness in the local science system and access to the international science system, it is interesting to investigate the source of the gatekeeping advantage of higher rated researchers. Local embeddedness, that is, average weighted reach, is very similar across rating categories. Clearly, systematic differences in gatekeeping scores arise from differences in the strength of external co-authorship ties (see fourth and fifth columns "Average weighted reach" and "External weight").

A positive correlation of higher ratings on the one hand, and higher gatekeeping scores (higher external weight) on the other hand is also observed when we focus on individual time periods or individual scientific disciplines.[5] For example, the same pattern is observed for the sub-population of biologists. Biology is one of the fields where South African scientists are, according to ISI, recognized as just above the

Table 18.4 Gatekeeping score and its composition, mean (s.d. of the mean).

Rating	*Observations*	*Gatekeeping score*	*Average weighted reach*	*External weight*
A	99	6.70 (0.84)	0.57 (0.04)	8.98 (1.05)
B	620	3.17 (0.19)	0.54 (0.02)	4.94 (0.26)
C	1,673	1.46 (0.06)	0.50 (0.01)	2.44 (0.09)
Rating unsuccessful	386	0.55 (0.06)	0.38 (0.02)	1.42 (0.11)

mean. In this sense the biology discipline is representative for the whole, and we are able to enrich the global statistics by an image of the network among biologists.

Figure 18.1 presents the main connected component of the co-authorship network among focal (i.e., senior rated) biologists, taking into account all peer-reviewed scientific articles between 2006 and 2010. The size of the nodes increases with the external weight (in logs). Visual inspection of Figure 18.1 suggests that global patterns described above are also reflected in this relatively small sub-sample of the network: A-rated researchers (black in Figure 18.1) are relatively well embedded (central in Figure 18.1) in the South African co-author network, and have more external collaborations (large node size in Figure 18.1); thus, taking important gatekeeping positions in the network.

The example puts some light on two more general issues discussed above. Firstly, we observe that higher ranked researchers have more international collaborations.

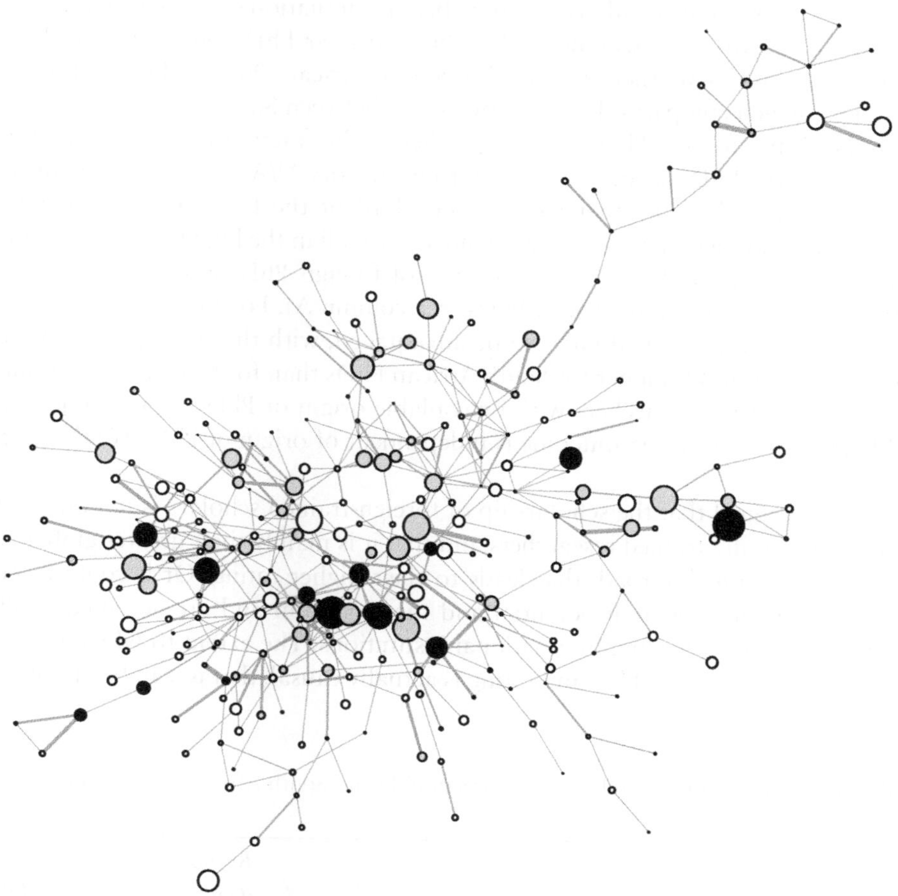

Figure 18.1 Co-author network among senior biology scientists accumulated from 2006 to 2010; main connected component including 278 researchers and 464 co-author ties. Each node is a rated South African biologist. Colour of nodes: A-rated in black (18), B-rated in grey (70), C-rated (162) and unsuccessful ratings (28) in white. Size of nodes increases with external weight. Edges are co-author ties. Edge width increases with the (summed) weight of the accumulated co-authorship links.

This suggests that knowledge flows not only one-way, from the global to the local science system, but both ways. In particular, A-rated South African scientists contribute as world-leaders (by definition) to global knowledge in international collaborations. The second observation is that researchers with many international connections tend to remain connected within the local science system. Thus, a strategy of international connectedness does not necessarily imply a cutting-off from the local science system but may well complement local upgrading.

The Role of Foreign PhDs in the South African Science System

The theoretical discussion above highlighted the relevance of individual mobility for international collaboration and knowledge flows. This section gives some intuition on this issue for the South African case. In particular, the following statistics suggest that exposure to global science during PhD helps scholars become better South Africa-based researchers and also become better internationally connected.

First, South African researchers who obtained their PhD abroad achieve higher rating categories than their peers with South African PhDs. This is shown in Table 18.5 which compares the rating outcomes between South African PhDs (1814 researchers) and foreign PhDs (840 researchers). The fraction for which we do not know where the PhD has been obtained, row-category N/A, is relatively small with 126 researchers. We note that roughly one third of the focal researchers have a foreign PhD and these have a stronger tendency to fall in the higher rating categories than South African PhDs. For example 5% of foreign PhDs obtained an A-rating compared to 3% of South African PhDs (see column A). For rating category B, we also observe foreign-trained PhDs to be advantaged, with the consequence that the fraction of C-ratings is higher for South African PhDs than for foreign PhDs. Fisher's exact test, ignoring researchers with unavailable origin of PhD, strongly rejects the null hypothesis that rating outcome is independent of origin of PhD (South African or foreign).

It can be argued that the scholarship of foreign-trained scholars is not any better than that of locally trained researchers and that it is mainly their greater global visibility, having trained abroad, that leads to their higher rating. Or it could be that they have a higher research potential and therefore aim and are accepted as PhD candidates at world-leading international institutions. The extent to which the relationship between foreign PhD and rating is actually causal, that is, can be interpreted

Table 18.5 Ratings of researchers with foreign PhD and South African PhD, frequencies (row-percentages).

	A	*B*	*C*	*Rating unsuccessful*	*Total*
South African PhD	54 (2.98)	341 (18.80)	1139 (62.79)	280 (15.44)	1814 (100)
Foreign PhD	41 (4.88)	243 (28.93)	450 (53.57)	106 (12.62)	840 (100)
N/A	4 (3.17)	36 (28.57)	84 (66.67)	2 (1.59)	126 (100)
Total	99 (3.56)	620 (22.30)	1673 (60.18)	388 (13.96)	2780 (100)

as a treatment effect, requires additional research. But even should foreign-trained scholars be nothing but the beneficiaries of social processes, it is likely that their status and visibility both locally and abroad allow them to play an important role in the domestic upgrading of their scientific field.

Further evidence suggests that the science system may benefit directly from having foreign-trained PhDs. One of the observable pathways to a higher rating is the publishing activity of the researcher. Table 18.6 (second column) shows the number of papers in the five-year period prior to rating of our focal researchers. Researchers with foreign PhDs publish on average about one paper more than South African PhDs. This translates to a difference of around three co-authors and four foreign co-authors on average (see third and fourth column in Table 18.6 respectively).

Unique foreign co-authors are those foreign researchers who collaborate only with the focal researcher and no other South African researchers. This statistic is of interest since one can consider each foreign collaborator as a unique source of knowledge for the South African science system. If the focal researcher were not part of the South African science system, the knowledge of her unique foreign co-authors would not be directly available to the system. Our evidence suggests that foreign-trained PhDs have on average three more unique foreign collaborators than locally trained scholars, which means that they tap into more unique knowledge than their local counterparts. A further finding not shown here is that, compared to locally trained PhDs, foreign-trained PhDs have not only more but also more often repeated, that is, stronger, international links.

In sum, statistical patterns of South African scientists who obtained their doctoral degrees abroad illustrate well two points stressed in the literature on mobility of scientists: First, one way of connecting the global and local science systems is the mobility of scientists. In the case of South Africa, scientists with foreign PhDs have a stronger standing in the international community than locally trained PhDs; they are more likely to become world participants (B-rated) or even world-leaders (A-rated). Second, individual, international mobility shapes international connections. In our example, South African scientists with foreign PhD have a richer international network than those with local PhD.

Overall, researchers in South Africa who obtained their PhDs abroad have a different career path than their peers with South African PhDs. They are more independent in expanding their international research network which yields more and stronger contacts with foreign co-authors, and in particular allows for accessing foreign co-authors who would otherwise not connect with the South African

Table 18.6 Papers and number of (foreign, unique foreign) co-authors within the five year period prior to rating, mean (s.d. of the mean).

	Papers	*Co-authors*	*Foreign co-authors*	*Unique foreign co-authors*
South African PhD	8.34 (0.17)	12.31 (0.40)	7.58 (0.32)	2.70 (0.12)
Foreign PhD	9.15 (0.27)	15.62 (0.81)	12.02 (0.72)	5.70 (0.38)
N/A	7.82 (0.71)	11.45 (2.36)	8.86 (2.03)	4.34 (0.86)

Note: 11 outliers with more than 50 papers removed.

community. In this sense, the doctorate is indeed found to be an early socialization phase that remains influential throughout the subsequent career of the scientist. Moreover, since it is the researcher's external contacts that mainly makes the difference in terms of gatekeeping potential, international experience and contacts of foreign PhDs are likely to be an important resource for upgrading the scientific system.

Discussion

The role of international partners in the building of research capacity in less developed countries is often a controversial matter, both in scholarship (e.g., Pouris and Ho 2014) and among officials who need to decide how to best allocate scarce resources. The local/foreign tension in particular permeates many decisions, for example whether to develop institutions so that doctorates can be trained locally, or to support gifted students in study abroad, with the associated risk that they may not return to their home country.

The evidence presented in this chapter suggests that the foreign connectedness of scientists generally results in accelerated domestic diffusion of frontier knowledge. This finding is in line with evidence from numerous other areas, for example research on international business (see Narula and Dunning 2000, 2010) and diasporas and returnees (e.g., Liu et al. 2010; Saxenian 2005).

We highlight some of the important mechanisms by which such upgrading in the science arena takes place. First, in order to achieve a strong international standing as a scientist one needs to be part of the international community. One way to become part of such a community is certainly through collaboration with members of the international community. However, our analysis of South African researchers with foreign PhDs shows that the international and local science system is connected not so much through formal, institutional collaborative ties. Instead, it is the individual person who spans the local and global worlds.

Doing an external PhD and coming back to South Africa for teaching and research is one strategy allowing scholars to work across both the developed and developing world. We observe that those scientists independently expand their international contacts during their subsequent career, and are more likely to become internationally renowned. However, many career trajectories are not as straightforward. In a qualitative assessment of the data, we also observe researchers going back and forth as well as researchers with multiple affiliations, one in South Africa and one in the most developed countries. Actually living in both worlds is possibly instrumental but certainly complementary to foreign as well as local collaborations.

But although close ties with technologically advanced countries are more often found among the more esteemed scholars, researchers do not disconnect from the local system when they achieve international standing. International connectedness is associated with the simultaneous strengthening of international collaborations and maintenance of national ones. In other words, at the level of the individual researcher there is not a trade-off between local and global connectedness. We showed that South African scientists contributing to frontier knowledge internationally act as gatekeepers in the South African science system (Barnard et al. 2012). Conceptually, gatekeepers of a system source knowledge from outside the system and diffuse this external knowledge within the system (Allen 1977). Therefore, because globally connected scholars are well connected with their internationally less connected colleagues, the latter have the opportunity to tap into the knowledge sources from elsewhere through the former. This result shows that engaging globally does not preclude local engagement by individual scientists. Thus, local and global connectedness in research should not be considered as substitutes but rather as complements.

Taken together, our evidence suggests that the concern about middle-income countries participating in global science networks is misplaced. International training and research within the international community is a paying investment for becoming a nationally as well as an internationally renowned scientist. From the point of view of the development of science, it makes sense. The advancement of science takes place through global engagement and participation, not simply locally. But it also makes sense for middle-income countries. The connections between scientifically and technologically more and less advanced countries take place through individual scientists. Stronger scientists not only contribute globally, but also play an important role in connecting their local peers to the global community, sometimes through their author networks, and sometimes through their institutions. Therefore, by taking a central position in both worlds, researchers may function as gatekeepers for the development of indigenous science. Indeed, our evidence suggests that global connectedness is key to the development of indigenous science in middle-income countries.

Notes

1 This research has greatly benefited from the encouragement of the National Research Foundation (NRF) of South Africa. Partial funding has been provided by the University of Strasbourg ProjEx "Creative, Sustainable Economies and Societies."

2 Additional econometric analyses can be found in Barnard et al. (2012) and in work in progress. By controlling for various factors as well as technical issues, these studies support and quantify the insights suggested by the basic statistics that follow.

3 See https://www.skatelescope.org/, accessed January 5, 2015.

4 The duration of validity for ratings changes from time to time over the history of the system, but in general they last for 4 or 5 years.

5 Researchers belong to one of 18 scientific disciplines, with 5 social science and 13 natural science disciplines.

References

Akbar, Yusaf H., and J. Brad McBride. 2004. "Multinational Enterprise Strategy, Foreign Direct Investment and Economic Development: The Case of the Hungarian Banking Industry." *Journal of World Business* 39(1): 89–105.

Allen, Thomas J. 1977. *Managing the Flow of Technology*. Cambridge, MA: The MIT Press.

Almeida, Paul, and Bruce Kogut. 1999. "Localization of Knowledge and the Mobility of Engineers in Regional Networks." *Management Science* 45(7): 905–917.

Altbach, Philip G., and Jane Knight. 2007. "The Internationalization of Higher Education: Motivations and Realities." *Journal of Studies in International Education* 11(3–4): 290–305.

Altbach, Philip G., Liz Reisberg, and Laura E. Rumbley. 2009. *Trends in Global Higher Education: Tracking an Academic Revolution*. Paris: UNESCO.

Álvarez, Isabel, and Raquel Marín. 2013. "FDI and Technology as Levering Factors of Competitiveness in Developing Countries." *Journal of International Management* 19(3): 232–246.

Argote, Linda, and Paul Ingram. 2000. "Knowledge Transfer: A Basis for Competitive Advantage in Firms." *Organizational Behavior and Human Decision Processes* 82(1): 150–169.

Barnard, Helena. 2010. "Overcoming the Liability of Foreignness without Strong Firm Capabilities – The Value of Market-Based Resources." *Journal of International Management* 16(2): 165–176.

Barnard, Helena, Robin Cowan, and Moritz Müller. 2012. "Global Excellence at the Expense of Local Diffusion, or a Bridge Between Two Worlds? Research in Science and Technology in the Developing World." *Research Policy* 41(4): 756–769.

Bartlett, Christopher A., and Sumantra Ghoshal. 1998. *Managing Across Borders: The Transnational Solution*, vol. 2. Boston, MA: Harvard Business School Press.

Beaverstock, Jonathan V. 2005. "Transnational Elites in the City: British Highly-Skilled Inter-Company Transferees in New York City's Financial District." *Journal of Ethnic and Migration Studies* 31(2): 245–268.

Becker-Ritterspach, Florian, and Christoph Dörrenbächer. 2009. "Intrafirm Competition in Multinational Corporations: Towards a Political Framework." *Competition & Change* 13(3): 199–213.

Bodas Freitas, Isabel Maria, Aldo Geuna, and Federica Rossi. 2013. "Finding the Right Partners: Institutional and Personal Modes of Governance of University–Industry Interactions." *Research Policy* 42(1): 50–62.

Bodas Freitas, Isabel Maria, Rosane Argou Marques, and Evando Mirra de Paula e Silva. 2013. "University–Industry Collaboration and Innovation in Emergent and Mature Industries in New Industrialized Countries." *Research Policy* 42(2): 443–453.

Bozeman, Barry, Daniel Fay, and Catherine P. Slade. 2013. "Research Collaboration in Universities and Academic Entrepreneurship: The-State-of-the-Art." *Journal of Technology Transfer* 38(1): 1–67.

Breschi, Stefano, and Francesco Lissoni. 2009. "Mobility of Skilled Workers and Co-Invention Networks: An Anatomy of Localized Knowledge Flows." *Journal of Economic Geography* 9(4): 439–468.

Brown, Mercy. 2000. "Using the Intellectual Diaspora to Reverse the Brain Drain: Some Useful Examples." Paper presented to Regional Conference on Brain Drain and Capacity Building in Africa, Addis Ababa, February 22–24.

Cantwell, John, and Lucia Piscitello. 2002. "The Location of Technological Activities of MNCs in European Regions: The Role of Spillovers and Local Competencies." *Journal of International Management* 8(1): 69–96.

Cantwell, John, and Lucia Piscitello. 2005. "Recent Location of Foreign-Owned Research and Development Activities by Large Multinational Corporations in the European regions: The Role of Spillovers and Externalities." *Regional Studies* 39(1): 1–16.

Cassiman, Bruno, and Reinhilde Veugelers. 2006. "In Search of Complementarity in Innovation Strategy: Internal R&D and External Knowledge Acquisition." *Management Science* 52(1): 68–82.

Cervantes, Mario, and Dominique Guellec. 2002. "The Brain Drain: Old Myths, New Realities." *OECD Observer* 230: 40–41.

Chompalov, Ivan, and Wesley Shrum. 1999. "Institutional Collaboration in Science: A Typology of Technological Practice." *Science, Technology & Human Values* 24(3): 338–372.

Chompalov, Ivan, Joel Genuth, and Wesley Shrum. 2002. "The Organization of Scientific Collaborations." *Research Policy* 31(5): 749–767.

Chung, Wilbur, and Juan Alcácer. 2002. "Knowledge Seeking and Location Choice of Foreign Direct Investment in the United States." *Management Science* 48(12): 1534–1554.

Cohen, Wesley M., and Daniel A. Levinthal. 1990. "Absorptive Capacity: A New Perspective on Learning and Innovation." *Administrative Science Quarterly* 35(1): 128–152.

Corley, Elizabeth A., P. Craig Boardman, and Barry Bozeman. 2006. "Design and the Management of Multi-Institutional Research Collaborations: Theoretical Implications from Two Case Studies." *Research Policy* 35(7): 975–993.

Cowan, Robin, Maryann P. Feldman, and Dieter F. Kogler. 2006. "Canadian Professional Networks: A Survey of Highly Skilled Canadian Workers." Skills Research Initiative working paper D-16.

Cuervo-Cazurra, Alvaro. 2007. "Sequence of Value-Added Activities in the Multinationalization of Developing Country Firms." *Journal of International Management* 13(3): 258–277.

Darby, Michael R., and Lynne G. Zucker. 2001. "Change or Die: The Adoption of Biotechnology in the Japanese and US Pharmaceutical Industries." *Research on Technological Innovation, Management and Policy* 7: 85–125.

Defazio, Daniela, Andy Lockett, and Mike Wright. 2009. "Funding Incentives, Collaborative Dynamics and Scientific Productivity: Evidence from the EU Framework Program." *Research Policy* 38(2): 293–305.

DuBois, Frank L., and David Reeb. 2000. "Ranking the International Business Journals." *Journal of International Business Studies* 31(4): 689–704.

Eapen, Alex. 2012. "Social Structure and Technology Spillovers from Foreign to Domestic Firms." *Journal of International Business Studies* 43(3): 244–263.

Edler, Jakob (ed.). 2007. *Internationalisierung in der deutschen Forschungs- und Wissenschaftslandschaft*. Stuttgart: Fraunhofer IRB.

Edler, Jakob, Heide Fier, and Christoph Grimpe. 2011. "International Scientist Mobility and the Locus of Knowledge and Technology Transfer." *Research Policy* 40(6): 791–805.

Ernst, Dieter. 2002. "Global Production Networks and the Changing Geography of Innovation Systems: Implications for Developing Countries." *Economics of Innovation and New Technology* 11(6): 497–523.

Fagerberg, Jan, and Manuel M. Godinho. 2005. "Innovation and Catching-Up." In *The Oxford Handbook of Innovation*, ed. Jan Fagerberg, David Mowery, and Richard R. Nelson, 514–542. Oxford: Oxford University Press.

Fagerberg, Jan, and Martin Srholec. 2008. "National Innovation Systems, Capabilities and Economic Development." *Research Policy* 37(9): 1417–1435.

Feinberg, Susan E., and Sumit K. Majumdar. 2001. "Technology Spillovers from Foreign Direct Investment in the Indian Pharmaceutical Industry." *Journal of International Business Studies* 32(3): 421–437.

Fischer, Manfred M. 2001. "Innovation, Knowledge Creation and Systems of Innovation." *Annals of Regional Science* 35(2): 199–216.

Fu, Xiaolan. 2008. "Foreign Direct Investment, Absorptive Capacity and Regional Innovation Capabilities: Evidence from China." *Oxford Development Studies* 36(1): 89–110.

Fu, Xiaolan, Carlo Pietrobelli, and Luc Soete. 2011. "The Role of Foreign Technology and Indigenous Innovation in the Emerging Economies: Technological Change and Catching-Up." *World Development* 39(7): 1204–1212.

Gaillard, Anne Marie, and Jacques Gaillard. 2001. "Science and Technology Policies in the Context of International Scientific Migration." In Science and Technology Policy, *Encyclopedia of Life Support Systems*. UNESCO-EOLSS.

Gammeltoft, Peter, Helena Barnard, and Anoop Madhok. 2010. "Emerging Multinationals, Emerging Theory: Macro- and Micro-Level Perspectives." *Journal of International Management* 16(2): 95–101.

Goode, Richard B. 1959. "Adding to the Stock of Physical and Human Capital." *American Economic Review* 49(2): 147–155.

Jaffe, Adam B. 1989. "Real Effects of Academic Research." *American Economic Review* 79(5): 957–970.

Jonkers, Koen, and Robert Tijssen. 2008. "Chinese Researchers Returning Home: Impacts of International Mobility on Research Collaboration and Scientific Productivity." *Scientometrics* 77(2): 309–333.

Kaplinsky, Raphael. 2004. "Spreading the Gains from Globalization: What Can Be Learned from Value-Chain Analysis?" *Problems of Economic Transition* 47(2): 74–115.

Kemeny, Thomas. 2010. "Does Foreign Direct Investment Drive Technological Upgrading?" *World Development* 38(11): 1543–1554.

Kogut, Bruce, and J. Muir Macpherson. 2011. "The Mobility of Economists and the Diffusion of Policy Ideas: The Influence of Economics on National Policies." *Research Policy* 40(10): 1307–1320.

Koser, Khalid, and John Salt. 1997. "The Geography of Highly Skilled International Migration." *International Journal of Population Geography* 3(4): 285–303.

Kumaraswamy, Arun, Ram Mudambi, Haritha Saranga, and Arindam Tripathy. 2012. "Catch-Up Strategies in the Indian Auto Components Industry: Domestic Firms' Responses to Market Liberalization." *Journal of International Business Studies* 43(4): 368–395.

Lall, Sanjaya, and Rajneesh Narula. 2004. "Foreign Direct Investment and Its Role in Economic Development: Do We Need a New Agenda?" *European Journal of Development Research* 16(3): 447–464.

Leung, Kwok. 2007. "The Glory and Tyranny of Citation Impact: An East Asian Perspective." *Academy of Management Journal* 50(3): 510–513.

Leydesdorff, Loet, and Olga Amsterdamska. 1990. "Dimensions of Citation Analysis." *Science, Technology & Human Values* 15(3): 305–335.

Li, F.L.N., A.M. Findlay, A.J. Jowett, and R. Skeldon. 1996. "Migrating to Learn and Learning to Migrate: A Study of the Experiences and Intentions of International Student Migrants." *International Journal of Population Geography* 2(1): 51–67.

Liu, Xiahui, Jiangyong Lu, Igor Filatotchev, Trevor Buck, and Mike Wright. 2010. "Returnee Entrepreneurs, Knowledge Spillovers and Innovation in High-Tech Firms in Emerging Economies." *Journal of International Business Studies* 41(7): 1183–1197.

Lorenzen, Mark, and Ram Mudambi. 2013. "Clusters, Connectivity and Catch-Up: Bollywood and Bangalore in the Global Economy." *Journal of Economic Geography* 13(3): 501–534.

Malecki, Edward J. 1997. "Technology and Economic Development: The Dynamics of Local, Regional, and National Change." Academy for Entrepreneurial Leadership Historical Research Reference in Entrepreneurship, University of Illinois at Urbana-Champaign.

Marin, Anabel, and Martin Bell. 2006. "Technology Spillovers from Foreign Direct Investment (FDI): The Active Role of MNC Subsidiaries in Argentina in the 1990s." *Journal of Development Studies* 42(4): 678–697.

Melin, Göran. 2004. "Postdoc Abroad: Inherited Scientific Contacts or Establishment of New Networks?" *Research Evaluation* 13(2): 95–102.

Meyer, Jean-Baptiste. 2001. "Network Approach versus Brain Drain: Lessons from the Diaspora." *International Migration* 39(5): 91–110.

Meyer, Jean-Baptiste, and Mercy Brown. 1999. "Scientific Diasporas: A New Approach to the Brain Drain." MOST Discussion Paper 41, World Science Conference, Budapest.

Meyer, Klaus E. 2006. "Asian Management Research Needs More Self-Confidence." *Asia Pacific Journal of Management* 23(2): 119–137.

Meyer, Klaus E., and Evis Sinani. 2009. "When and where Does Foreign Direct Investment Generate Positive Spillovers? A Meta-Analysis." *Journal of International Business Studies* 40(7): 1075–1094.

Mitchell, Katharyne. 2000. "Networks of Ethnicity." In *A Companion to Economic Geography*, ed. Eric S. Sheppard and Trevor J. Barnes, 392–407. Oxford: Blackwell.

Møen, Jarle. 2005. "Is Mobility of Technical Personnel a Source of R&D Spillovers?" *Journal of Labor Economics* 23(1): 81–114.

Nam, Kyung-Min, and Xin Li. 2013. "Out of Passivity: Potential Role of OFDI in IFDI-Based Learning Trajectory." *Industrial and Corporate Change* 22(3): 711–743.
Narula, Rajneesh, and John H. Dunning. 2000. "Industrial Development, Globalization and Multinational Enterprises: New Realities for Developing Countries." *Oxford Development Studies* 28(2): 141–167.
Narula, Rajneesh, and John H. Dunning. 2010. "Multinational Enterprises, Development and Globalization: Some Clarifications and a Research Agenda." *Oxford Development Studies* 38(3): 263–287.
Nelson, Richard R. 2005. "The Roles of Research in Universities and Public Labs in Economic Catch-Up." In *Technological Change and Economic Catch-Up*, ed. Grazia D. Santangelo, 19–33. Cheltenham: Edward Elgar.
Newman, Mark E.J. 2001. "Scientific Collaboration Networks. II. Shortest Paths, Weighted Networks, and Centrality." *Physical Review E* 64(1): 016132.
Palomeras, Neus, and Eduardo Melero. 2010. "Markets for Inventors: Learning-by-Hiring as a Driver of Mobility." *Management Science* 56(5): 881–895.
Polanyi, Michael. 1958. *Personal Knowledge: Towards a Post-Critical Philosophy*. London: Routledge & Kegan Paul.
Polanyi, Michael. 1983. *The Tacit Dimension*. Gloucester, MA: Smith.
Pouris, Anastassios, and Yuh-Shan Ho. 2014. "Research Emphasis and Collaboration in Africa." *Scientometrics* 98(3): 2169–2184.
Rabbiosi, Larissa, Stefano Elia, and Fabio Bertoni. 2012. "Acquisitions by EMNCs in Developed Markets." *Management International Review* 52(2): 193–212.
Ramamurti, Ravi. 2009. "What Have We Learned About Emerging Market MNEs?" In *Emerging Multinationals in Emerging Markets*, ed. Ravi Ramamurti and Jitendra V. Singh, 399–427. Cambridge: Cambridge University Press.
Rasiah, Rajah. 2003. "Foreign Ownership, Technology and Electronics Exports from Malaysia and Thailand." *Journal of Asian Economics* 14(5): 785–811.
Rip, Ariel. 2002. "Regional Innovation Systems and the Advent of Strategic Science." *Journal of Technology Transfer* 27(1): 123–131.
Rosenkopf, Lori, and Paul Almeida. 2003. "Overcoming Local Search Through Alliances and Mobility." *Management Science* 49(6): 751–766.
Rugman, Alan M., and Alain Verbeke. 2001. "Location, Competitiveness and the Multinational Enterprise." In *The Oxford Handbook of International Business*, ed. Alan M. Rugman and Thomas L. Brewer, 150–179. Oxford: Oxford University Press.
Saxenian, AnnaLee. 1990. "Regional Networks and the Resurgence of Silicon Valley." *California Management Review* 33(1): 89–112.
Saxenian, AnnaLee. 1996. *Regional Advantage: Culture and Competition in Silicon Valley and Route 128*. Cambridge, MA: Harvard University Press.
Saxenian, AnnaLee. 2002. *Local and Global Networks of Immigrant Professionals in Silicon Valley*. San Francisco, CA: Public Policy Institute of California.
Saxenian, AnnaLee. 2005. "From Brain Drain to Brain Circulation: Transnational Communities and Regional Upgrading in India and China." *Studies in Comparative International Development* 40(2): 35–61.
Scellato, Giuseppe, Chiara Franzoni, and Paula Stephan. 2012. "Mobile Scientists and International Networks." NBER Working Papers 18613.
Scherer, Frederick M. 2000. "The Emigration of German-Speaking Economists after 1933." *Journal of Economic Literature* 38(3): 614–626.
Schiller, Daniel, and J. Revilla Diez. 2009. "Mobile Star Scientists as Regional Knowledge Spillover Agents." IAREG Working Paper 2/7.
Sethi, Deepak, S.E. Guisinger, S.E. Phelan, and D.M. Berg. 2003. "Trends in Foreign Direct Investment Flows: A Theoretical and Empirical Analysis." *Journal of International Business Studies* 34(4): 315–326.
Shrum, Wesley. 1997. "View from Afar: 'Visible' Productivity of Scientists in the Developing World." *Scientometrics* 40(2): 215–235.

Singh, Jasjit. 2007. "Asymmetry of Knowledge Spillovers Between Mncs and Host Country Firms." *Journal of International Business Studies* 38(5): 764–786.

Song, Jaeyong, and Jongtae Shin. 2008. "The Paradox of Technological Capabilities: A Study of Knowledge Sourcing from Host Countries of Overseas R&D Operations." *Journal of International Business Studies* 39(2): 291–303.

Song, Jaeyong, Paul Almeida, and Geraldine Wu. 2001. "Mobility of Engineers and Cross-Border Knowledge Building: The Technological Catching-Up Case of Korean and Taiwanese Semiconductor Firms." *Research on Technological Innovation, Management and Policy* 7: 59–84.

Song, Jaeyong, Paul Almeida, and Geraldine Wu. 2003. "Learning-by-Hiring: When Is Mobility More Likely to Facilitate Interfirm Knowledge Transfer?" *Management Science* 49(4): 351–365.

Starbuck, William H. 2005. "How Much Better Are the Most-Prestigious Journals? The Statistics of Academic Publication." *Organization Science* 16(2): 180–200.

UNCTAD. 2013. *World Investment Report: Global Value Chains: Investment and Trade for Development*. Geneva: United Nations Publications.

Verspagen, Bart, and Claudia Werker. 2004. "Keith Pavitt and the Invisible College of the Economics of Technology and Innovation." *Research Policy* 33(9): 1419–1431.

Wagner, Caroline S., and Loet Leydesdorff. 2005. "Network Structure, Self-Organization, and the Growth of International Collaboration in Science." *Research Policy* 34(10): 1608–1618.

Yang, Hongyan, Corey Phelps, and H. Kevin Steensma. 2010. "Learning from What Others Have Learned From You: The Effects of Knowledge Spillovers on Originating Firms." *Academy of Management Journal* 53(2): 371–389.

Yang, Qin, Ram Mudambi, and Klaus E. Meyer. 2008. "Conventional and Reverse Knowledge Flows in Multinational Corporations." *Journal of Management* 34(5): 882–902.

Ynalvez, M.A., and W.M. Shrum. 2011. "Professional Networks, Scientific Collaboration, and Publication Productivity in Resource-Constrained Research Institutions in a Developing Country." *Research Policy* 40(12): 204–216.

Zweig, David, Chen Changgui, and Stanley Rosen. 2004. "Globalization and Transnational Human Capital: Overseas and Returnee Scholars to China." *China Quarterly* 179: 735–757.

Chapter 19

Global Trends in Brain Drain and Likely Scenario in the Coming Years

Alessio Terzi

Introduction

Over the course of the last decade, as part of a broader globalization trend, migration and international mobility at world level have been increasing. As evidenced by Arslan et al. (2014), the number of migrants (aged 15+) in OECD countries increased by 38% between 2000 and 2010, to 106 million. Of these, about 35 million have tertiary education: a number that has increased by 70% over the past decade. As evidenced in previous chapters of this manual, mobility of the highly educated[1] is not to be seen necessarily as a negative evolution (see Flanagan, Chapter 17 and Florida and Mellander, Chapter 15 in this volume). Several authors (Ackers 2008; Mahroum 2000) have highlighted the positive effects that mobility of researchers has on innovation and the circulation of ideas. This has been identified by the literature as (positive) "brain circulation" (Saxenian 2005). We can speak of "brain drain," on the other hand, when individuals with key skills and capacities leave a country and stop contributing to its development.

Even in the era of large detailed datasets, precisely measuring brain drain remains a daunting task. First, whereas it is possible to broadly measure the flows of highly educated individuals, tracking whether they return to their home countries at a later stage, bringing with them new skills and know-how, is still not possible on a large scale. Second, even were they not to return, in certain contexts, it is still possible that high-skilled emigrants retain ties and collaborations with their mother country. In such a way, the flow of knowledge would be ensured and we could not properly speak of a brain drain. Third, identifying the "home" country might not be as straightforward as one would think, as increasingly students spend parts of their early stages of education in different countries. Finally, as evidenced by Flanagan

The Handbook of Global Science, Technology, and Innovation, First Edition.
Edited by Daniele Archibugi and Andrea Filippetti.

(this volume, Chapter 17), measuring quantity (as in the flow of researchers, scientists, or highly educated individuals) might conceal important differences in quality of migration. With no pretence to address all these issues, in this chapter I will aim to give a sense of migration flows in selected (mostly OECD) countries, largely building on proxy measures of "brain drain." In doing so, I will build largely on the latest release[2] of the OECD/UN Database of Immigrants in OECD countries (DIOC) which contains detailed stock data on migration, broken down by country of birth, education attainment, age, gender, duration of stay, and labor force status.[3] The data was first released in 2008, based on 2000/01 data. Then updated using data from 2005/06 and the final round contains data for 2010/11. This allows estimating flows of migrants over this period of time.

Brain Drain: Migration Patterns of the Highly Educated

To get a sense of migration patterns of the highly educated we begin by looking at the geographical composition of migration to OECD countries. High-skilled migration originates mostly from Asia, which represents roughly one-third of all highly educated immigrants in the OECD in 2010 (Figure 19.1). Among all countries, India takes the lion's share with 2.2 million immigrants, followed by China (1.7m) and the Philippines (1.5m). Taken together, immigrants from these three countries represent one-fifth of all tertiary educated immigrants in OECD countries (Arslan et al. 2014). Interestingly, Europeans are also contributing highly to migration patterns worldwide.[4]

Moving on, it is interesting to see which countries are particularly successful at attracting or retaining talents. To this purpose, Figure 19.2 details the net flow (immigrants minus emigrants) of high-skilled workers for selected OECD countries in 2010, normalized by population. In the last year for which data is available, we note a few interesting patterns. First, four out of the six countries that are managing

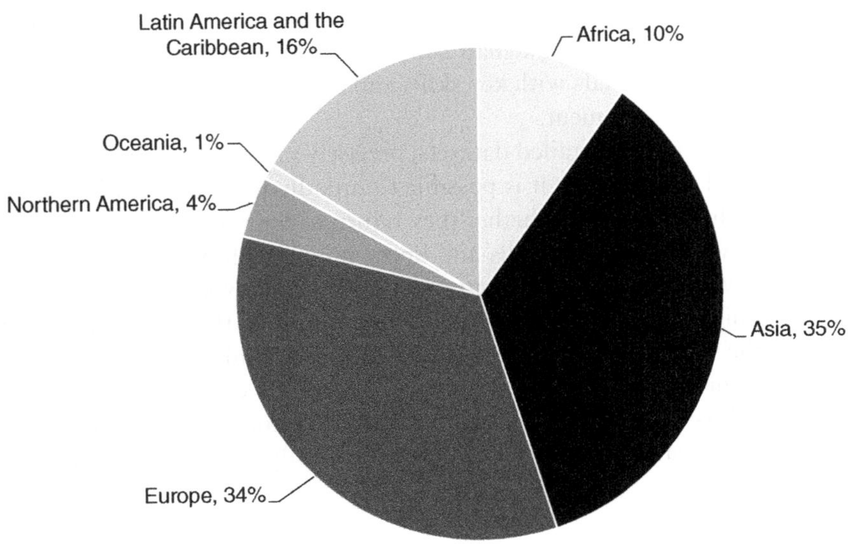

Figure 19.1 Share of highly educated migrants (15+) in the OECD by region of origin 2010/11. Source: DIOC 2010/11, Bruegel calculations.

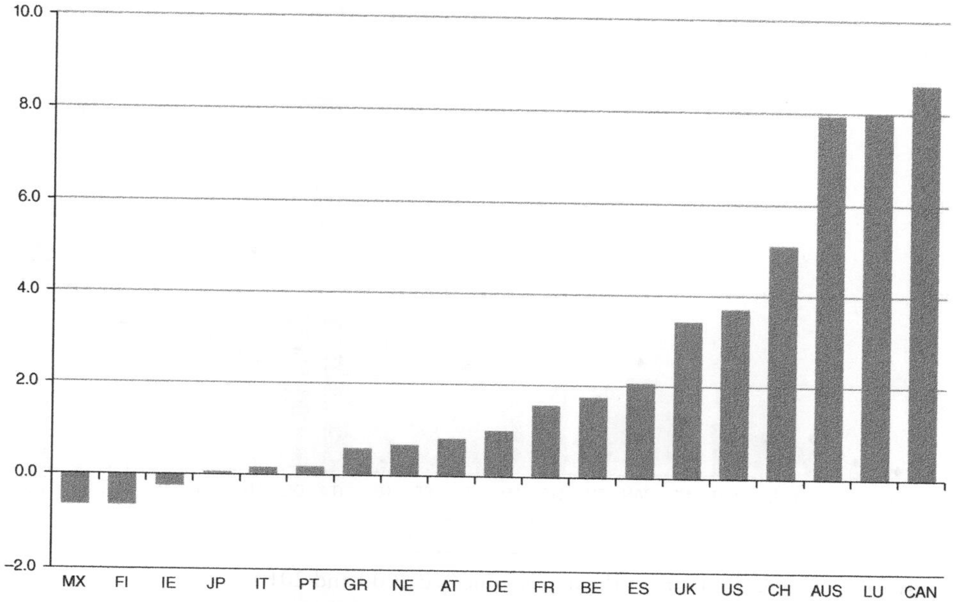

Figure 19.2 Net migration flow of highly educated normalized by population 2010. Source: DIOC 2010/11, Bruegel calculations.

to attract and retain top talents are English-speaking countries. Second, small countries at the centre of Europe (Switzerland, Luxembourg), have a particularly positive balance adjusted by population. Following these there is, a notch behind, a group of eurozone countries. Japan, with its particularly restrictive migration policy, is not taking much advantage of the creation of a global market for talents. Diametrically opposite is instead Canada which, on the other hand, was very successful at attracting and retaining highly educated individuals throughout the 2000s.

Aside from looking at countries' capacity to benefit from an international market for talents, it is interesting to see how effective they are at retaining highly educated people and how this has changed between 2000 and 2010. In doing so, however, we must account for the fact that education levels have been rising in time across the globe. To do this, we compute emigration rates of the highly skilled, which are constructed as the ratio of high-education emigrants over the number of people within their origin country with similar educational characteristics.[5] This is pictured in Figure 19.3.

First of all, it must be clarified that the number of highly qualified emigrants (the numerator of the ratio) for all the selected countries was increasing over the period analyzed. As such, changes between 2000 and 2010 are to be ascribed to a difference in the growth rate of highly educated emigrants with respect to the general growth of people with tertiary education. Interestingly, most of the selected countries analyzed have experienced an increase in emigration rates between 2000 and 2010. Exceptions to this are Canada and the United Kingdom, which had already been identified as potential "brain drainers" above.

Another interesting case is Greece: between 2000 and 2008 the country experienced a boom period, with real GDP growing cumulatively by 32.9%, compared to a mere 17.6% of the EU27. This resulted in a higher growth of people with university degrees who chose to stay within the country, with respect to those that were leaving.

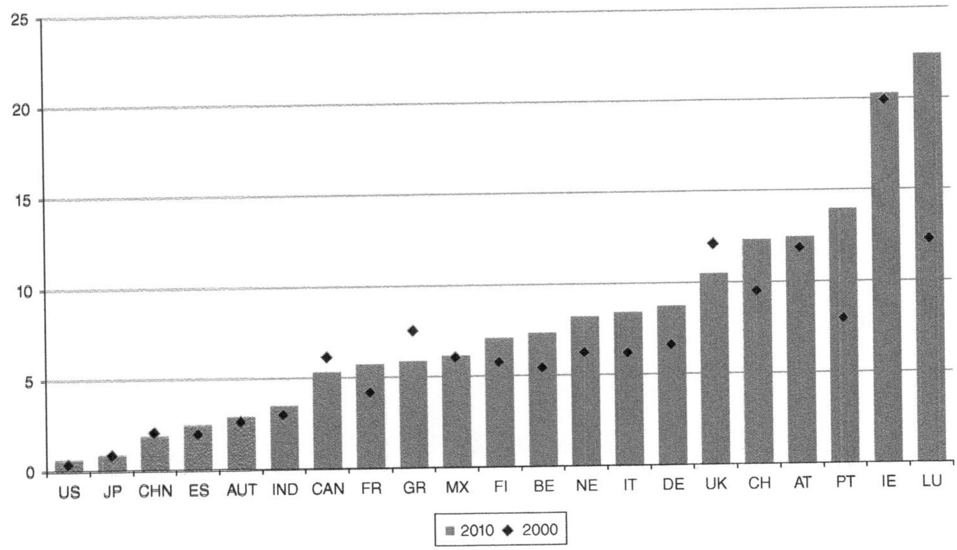

Figure 19.3 Emigration rate of the highly educated 2000 and 2010.
Source: DIOC, Bruegel calculations.

Two other countries are worth mentioning: Portugal and Luxembourg, which saw their emigration rates almost double over the period analyzed. While these results are less surprising for the former, which experienced sluggish growth over the whole decade (8% cumulative real GDP growth in 2000–2008), the latter is worth some reflection. When combined with the information contained in Figure 19.2, we have a picture of a country that attracts many highly talented people, but also lost one in five of its highly educated population. The small size of the country, however, makes this an outlier by definition.

Looking beyond changes, also levels are of interest. In line with the findings of Arslan et al. (2014) at global level, within the OECD small countries are those most susceptible to high emigration rates. Following these is the United Kingdom, and then a set of eurozone members. Finally, in the tail of the distribution, we find the United States and Japan, which have very small (and invariant) emigration rates of the highly qualified. Both China and India are relatively low in the distribution, suggesting that their dominance of the immigration stock in OECD countries is mostly due to their sheer size, rather than to a substantial loss of talent.

Clearly, as outlined in the introduction, it is not only the quantity that matters, but also the quality. It could be that India is "only" seeing 3% of its highly educated leave the country, but these could (and, to a certain extent, are actually likely to) be the top 3%. In order to get a sense of this, we would need a measure of "quality" of those leaving the country. To proxy for this, we build on another OECD dataset, namely the Science, Technology, and Innovation Scoreboard 2013. Figure 19.4 illustrates a normalized measure of the impact factor[6] of scientists who left a specific country over the period 1996–2011, and compares it to those who moved to the same country. Light and dark grey indicate whether a country was a net receiver or loser of scientists over the period analyzed, while the size of the bubbles is proportional to the sum of inflows and outflows.[7]

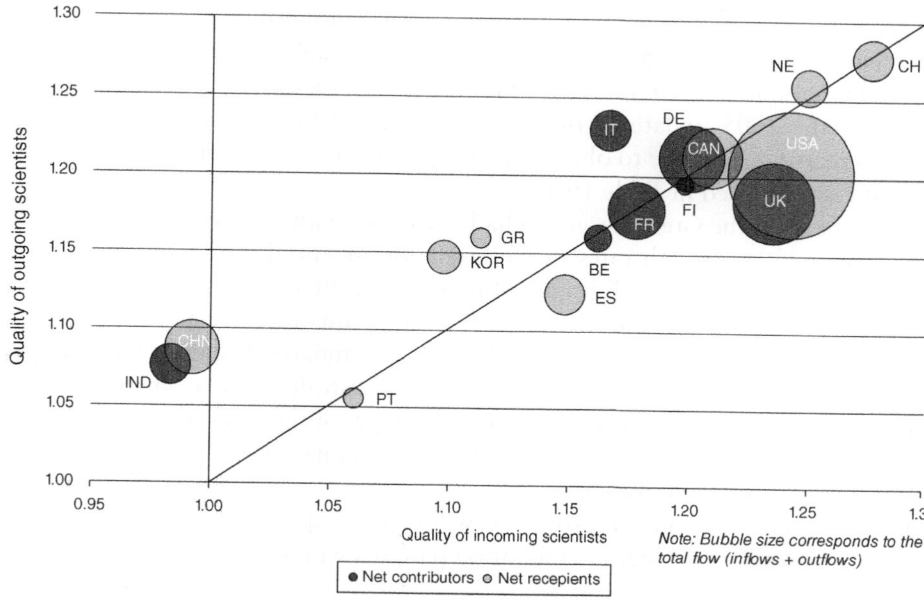

Figure 19.4 Quality of scientists 1996–2011.
Source: OECD (2013), Bruegel calculations.

Ideally, a country should want to be below or on the 45-degree line, indicating that the quality of the newcomers is just as high (or higher) as that of the leavers. Conditional on this, a country should also prefer a larger rather than smaller bubble, representing a sizeable flow of scientists and indicating a full exploitation of synergies gained from international cooperation. Finally, countries should aim to land in the top-right quadrant, indicating higher quality of both incoming and outgoing researchers.

Over the long-term period analyzed, we see that India, China, Korea, Greece, and Italy were faring relatively poorly, trading high-quality scientists with lower-quality ones. Italy and India were even in the negative mark area, indicating the number of those leaving was higher than those coming. Spain and the United Kingdom were placed relatively well in terms of quality, although the latter was strongly in the red.[8] The United States confirms itself as the greatest winner from the international market for scientists.

Aside from considerations of quality, this bubble chart also makes it immediately evident that few players alone dictate the direction of flows on a global scale. In particular, of the 33 countries analyzed by the OECD,[9] the United States and United Kingdom combined represented more than a third (34.8%) of total outflows and 38.0% of total inflows. By contrast, these values were merely 8.0% and 7.0% for China and India combined.

Recent Trends

As highlighted above, detailed standardized data on migration is highly cumbersome to generate. As such, the latest data available is from over four years ago. Although in normal times this would not be a problem, given that migration patterns tend to be relatively stable in time, at the current juncture 2010 sounds like a remote past.

The financial crisis, followed by the eurozone crisis, with its disruptive effects on the labor market and growth rates in several countries, is likely to have acted as a structural change also in terms of migration patterns. To get a sense of this effect, I compare data from the 2005 update of the DIOC with 2010 figures, and look at highly educated recent migrants,[10] to obtain a proxy of the direction of change in the flows. This data is illustrated in Table 19.1.

As can be seen, the Great Recession had a disruptive impact on trade flows of the highly qualified, as growth rates in the league of advanced economies plummeted into the red. However, a small league of nations benefited from this situation. This includes Norway, which jumped from 4000 highly qualified recent immigrants in 2005 to a whopping 38,000 just five years later. Similarly, Belgium, Finland, and the United Kingdpm, together with Austria and Australia, saw an increase in the share of highly qualified migrants[11] in the early stages of the crisis. On the other side of the distribution, Spain was the country where high-skilled migration dropped substantially as the economy stalled.

As most countries return to growth, it will be interesting in the years to come to observe whether these changes in migration patterns will prove temporary in nature

Table 19.1 Numbers (in 000s) and shares of highly educated migrants (aged 15+) 2010/11 and 2005/06.

	2010/11		*2005/06*		*Change*	
Country of residence	*number*	*share*	*number*	*share*	*number*	*share*
Norway	38.2	27.9	4.1	6.2	34.1	21.7
Belgium	140.4	40.1	57.1	25.0	83.3	15.1
Finland	12.6	25.9	7.6	22.2	5.0	3.7
United Kingdom	1335.5	39.2	793.9	37.0	541.6	2.2
Austria	52.0	23.8	40.7	23.2	11.3	0.6
Australia	405.3	23.3	278.3	23.1	127.0	0.2
Sweden	96.8	26.4	67.2	27.8	29.6	–1.4
Luxembourg	13.3	31.1	12.7	32.6	0.6	–1.5
Chile	15.3	31.9	14.8	33.7	0.5	–1.8
France	235.4	18.6	234.1	20.5	1.3	–1.9
Netherlands	42.7	10.8	39.9	13.2	2.8	–2.4
Canada	562.0	16.9	529.8	19.4	32.2	–2.5
Portugal	15.9	11.0	15.8	13.6	0.1	–2.6
Germany	269.3	13.7	238.0	16.6	31.3	–2.9
Denmark	9.4	13.7	9.1	16.9	0.3	–3.2
Greece	41.7	13.3	27.2	16.8	14.5	–3.5
OECD Total	**5684.1**	**19.5**	**5131.1**	**23.0**	**553.0**	–3.5
Switzerland	212.1	37.5	132.8	42.3	79.3	–4.8
United States	1268.2	11.2	1907.1	16.5	–638.9	–5.3
Italy	87.2	16.7	67.1	24.7	20.1	–8.0
Ireland	62.3	36.9	71.7	51.5	–9.4	–14.6
New Zealand	87.2	21.2	88.4	38.5	–1.2	–17.3
Spain	355.9	29.9	493.7	65.2	–137.8	–35.3

Source: DIOC 2010/11, Bruegel elaboration.

or more permanent. This will likely depend also on demographic factors, together with the growth prospects of Asia, Latin America, and Africa. The changes detailed above are likely to prove more permanent in particular for EU (and eurozone) countries, which have undergone a double-dip recession and are currently forecast to experience a protracted period of subdued growth. Moreover, in an effort to consolidate their public finances, several European governments have gone through a sharp and protracted period of "austerity." This is likely to have an effect on their capacity to attract and retain highly skilled workers going forward, an issue we now turn to before offering some concluding remarks.

Austerity and Brain Drain in the EU

For Europe, more than for any other area in the world, looking at 2010 data might be particularly outdated. As detailed hitherto, the continent has been broadly capable (with certain exceptions) to attract highly qualified individuals over the past decades. Not only are flows of highly educated positive for most EU countries, but also the quality of incoming scientists detailed in Figure 19.4 is quite high by global standards. With the financial and sovereign-debt crisis, however, unemployment rates have soared in several countries, and have now broadly stabilized at high levels. As job prospects for newcomers become grimmer, one can expect Europe's capacity to attract talented individuals to be negatively affected.

Moreover, within the EU itself there are large differences. Veugelers (2014) shows how fiscal consolidation has led countries that were already "innovation laggards" within the EU to cut disproportionately their Research and Innovation (R&I) expenditure with respect to other categories of public expenditure. Lower private and public spending on research is likely to have a significant impact on the capacity of countries to attract and retain talents in the longer term. With no pretence to trace a direct causality link, Figure 19.5 illustrates the strong correlation between percentage changes (approximated by the log function) in highly skilled migration flows and percentage changes in R&D expenditure (both public and private).

Survey evidence corroborates this hypothesis even when looking more specifically at researchers. The 2012 MORE2 survey on mobility patterns of researchers asks respondents, among other things, the reasons for moving in their post-PhD career. Figure 19.6 illustrates the main motives for moving to another EU country (left-hand side) or leaving the EU altogether (right-hand side). Career progression, research funding, facilities, and equipment (all of which are likely to be highly associated with R&D spending) appear among the top reasons for moving. This seems to be broadly corroborated also by the literature, as discussed in detail by Flanagan (this volume, Chapter 17).

A country that gives a high priority to R&D is one that is likely to generate growth in innovative sectors over the medium to long term (see Veugelers 2014). This is true both for the public sector (within universities' fields of research) and the private sector (in innovative business sectors). In turn, an economy where growth originates from innovative sectors is well placed to attract talents from abroad or create jobs for the highly qualified individuals it has trained. As such, we can expect

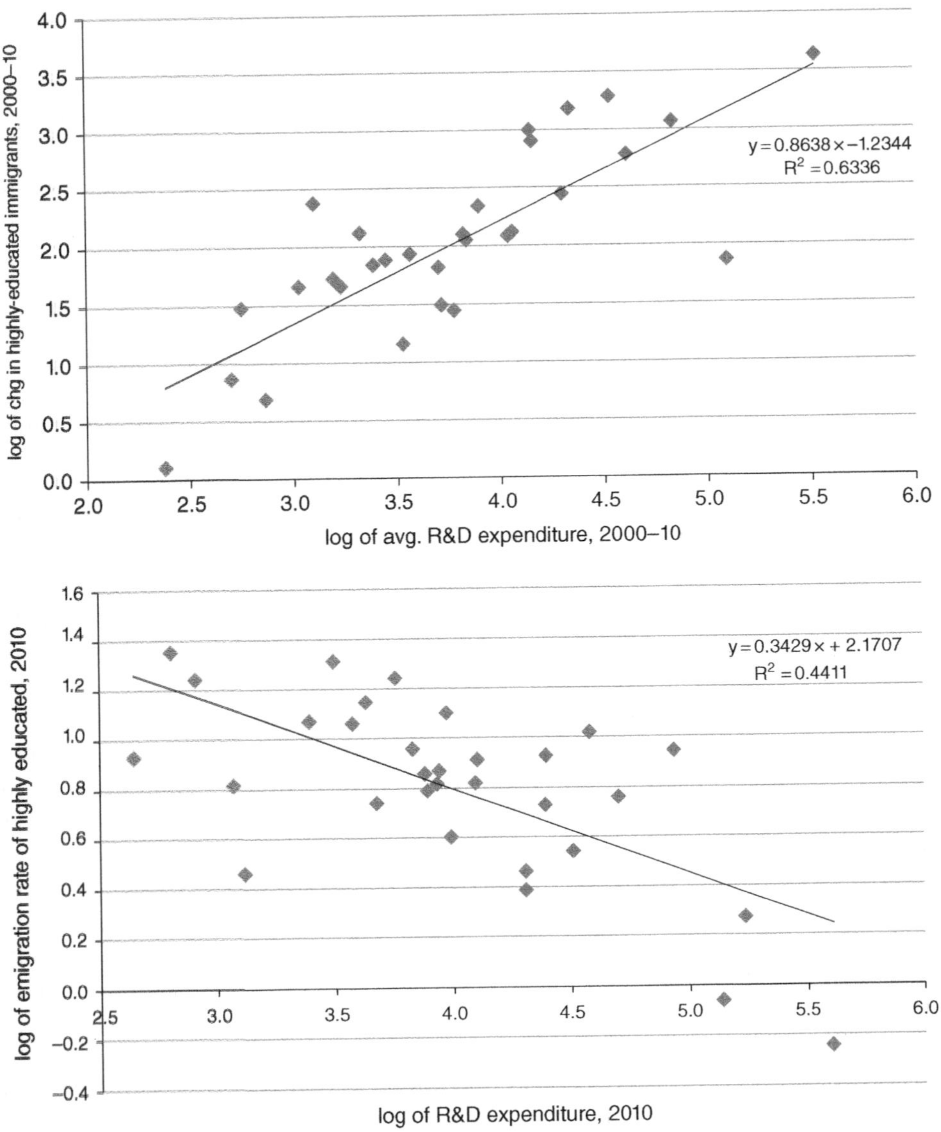

Figure 19.5 Correlations between R&D expenditure and highly educated immigration (top panel) or emigration rates (bottom panel).
Source: OECD (2013), Bruegel calculations.

that countries in the EU periphery where economic performance is sluggish and R&D spending has been slashed will see a higher incidence of brain drain in the years to come. Veugelers (2014) identifies these as Ireland, Spain, Italy, Portugal, and Greece.

It goes without saying that the impact of R&D spending on innovation, migration patterns, research facilities, and wages is likely to manifest only over long periods of time. As such, our analysis traces the likely scenario for these European countries only in the case in which the cuts to R&D are not rapidly reversed in the coming years.

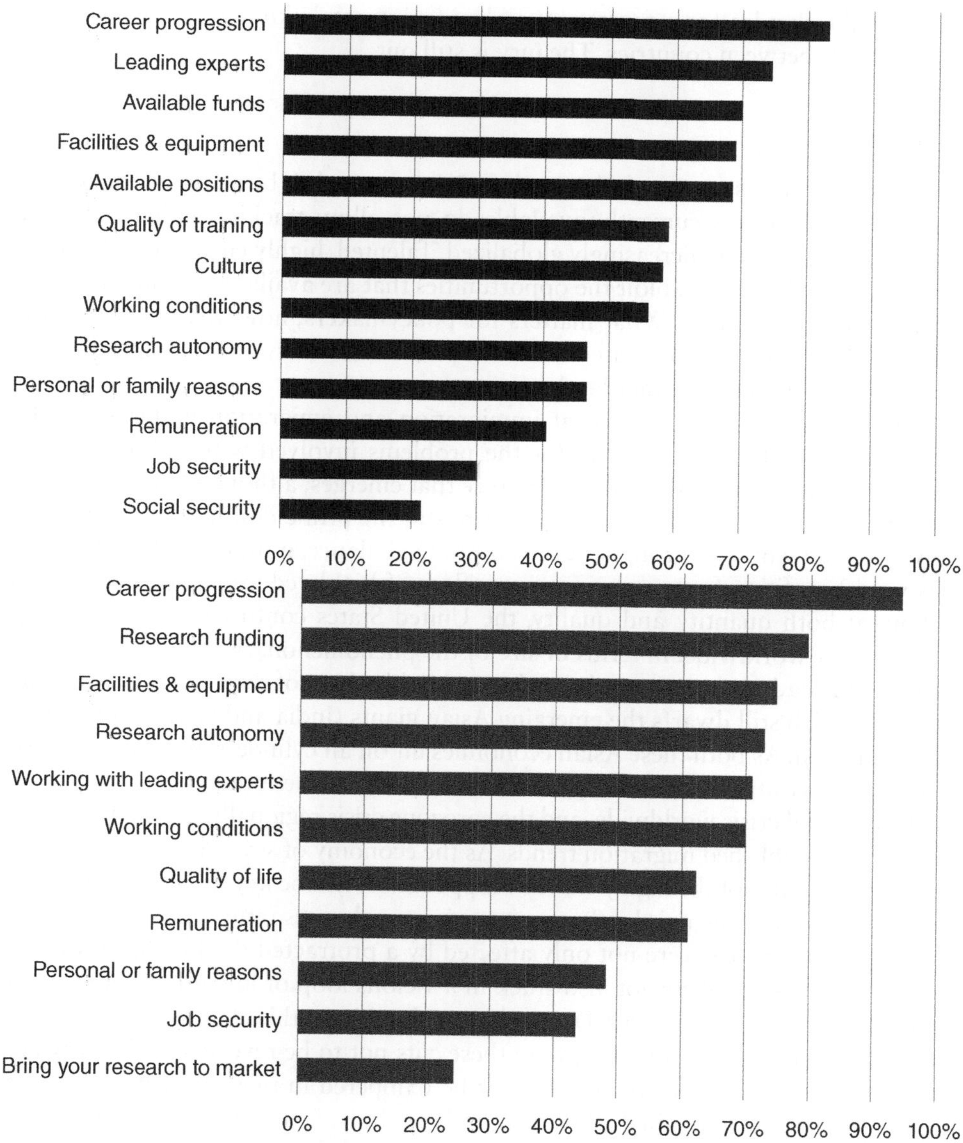

Figure 19.6 Reasons for moving to another EU country (top chart) or outside the EU (bottom chart).
Source: MORE2 Extra-EU Mobility Survey (2012). Used by permission of IDEA Consult.

Within the European context, migration, and with it therefore also "brain drain," are somewhat more problematic concepts. This is due to the fact that, in line with its mandate to create a single market for labor, the European Commission has been encouraging the free flow of workers and researchers in the EU.[12] As such, observing highly educated individuals leave countries where growth is sluggish and opportunities limited in favor of other (more economically successful) EU members is regarded as a positive development, indicating that the single market is at work. The question remains, however, whether these individuals will then continue contributing in some way to the prosperity of their origin country, or whether the single market will rather lead to a permanent and widening gap in knowledge, growth, and income levels between countries. The jury is still out.

Conclusions

Identifying clear instances of "brain drain" is a complex business. Comparative cross-country datasets currently available do not allow tracking individuals in a world that is becoming increasingly globalized. Talented, highly educated individuals are those best placed to exploit the opportunities that are available to those ready to move to another country. What matters for policymakers, however, is whether the individuals who have accumulated human capital in a country end up migrating and cutting all ties with the mother land.

In this chapter, I have looked at immigration and emigration patterns for the highly qualified, trying to correct for the problems involved with this necessarily approximate approach. All in all, the picture that emerges, although with significant differences, is that advanced economies were still the prime destination for increasingly educated workers throughout the 2000s. On balance, most of the OECD countries were benefiting from a positive net inflow of highly educated workers. When looking at both quantity and quality, the United States confirms itself as the top brain-drainer worldwide. In terms of size of the phenomenon, the United States and the United Kingdom commanded together a sizeable share of total flows of scientists (36.4%), which still dwarfs the emerging Asian giants (India and China represented 7.5% of flows). As both these Asian economies hit on all cylinders, however, it is not hard to see this situation reversing over the course of the next decade.

The financial crisis worldwide, and the eurozone crisis regionally, are likely to have affected the established migration trends. As the economy of several countries took a hit, migration rates of the highly skilled dropped sharply. The key question is whether they will return to pre-crisis levels as economic growth rates return into the black.

European countries were not only affected by a protracted double-dip recession, but several of its members got also stuck in a vicious loop of low growth and fiscal austerity. As a consequence, R&D spending took a hit. In this chapter, we offer preliminary evidence suggesting that, were these cuts not to be reversed as the pressure for consolidation recedes, countries might be hampered in their capacity to attract and retain talented individuals. Portugal, Ireland, Italy, Spain, and Greece are all at risk of being the victim, rather than the beneficiary, of "brain drain."

Notes

1 Throughout this chapter, we will use OECD definitions of educational attainment. As such, highly educated corresponds to tertiary education, medium education to upper secondary, and low to less than upper secondary.

2 November 2014.

3 For more information, refer to http://www.oecd.org/els/mig/methodology_DIOC_2010_11.pdf (accessed January 5, 2015).

4 Although beyond the scope of this short contribution, disentangling the effect of higher brain circulation within the European Union (also thanks to the policies implemented by the European Commission) and extra-EU migration would be of great interest and policy relevance.

5 To do this, Barro and Lee (2013) estimates of population by educational attainment levels are used.
6 For each author and mobility profile, the median across the relevant journals' Source-Normalized Impact per Paper (SNIP) over the entire period is calculated. A SNIP impact value that is higher than one means that the median attributed SNIP for authors of that country/category is above average.
7 International mobility of scientific researchers is inferred from authors listed in the Scopus Custom database of peer-reviewed scientific publications, with at least two documents during the reference period, based on changes in the location of their institutional affiliation. Outflows are defined on the basis of their first affiliation. Inflows are defined on the basis of the final affiliation and exclude individual authors who "return" to their original country of affiliation.
8 Elsevier (2013) confirm this finding and conclude however that the United Kingdom displays a very fluid research market, with low levels of "sedentary" researchers. Although the OECD data used tries to correct for returning scientists, it might not be appropriate to take this as evidence of "brain drain" in the UK scientific or research field.
9 Argentina, Australia, Austria, Belgium, Brazil, Canada, Switzerland, Chile, China, Czech Republic, Germany, Denmark, Spain, Finland, France, United Kingdom, Greece, Hong Kong, India, Israel, Italy, Japan, Korea, Mexico, Netherlands, Norway, New Zealand, Poland, Portugal, Russian Federation, Singapore Sweden, Turkey, Taiwan, United States, South Africa.
10 Recent migrants are defined by the OECD as those who have been living in the country of destination for less than five years.
11 Computed as short-term highly educated migrants over total highly educated migrants.
12 See European Commission initiatives geared toward the establishment of a European Research Area (ERA).

References

Ackers, L. 2008. "Internationalisation, Mobility and Metrics: A New Form of Indirect Discrimination?" *Minerva* 46(4): 411–435.

Arslan, C., J.-C. Dumont, Z. Kone, Y. Moulan, C. Ozden, C. Parsons, and T. Zenogiani. 2014. "A New Profile of Migrants in the Aftermath of the Recent Economic Crisis." OECD Social, Employment and Migration Working Papers.160.

Barro, R.J., and J.W. Lee. 2013. "A New Data Set of Educational Attainment in the World, 1950–2010." *Journal of Development Economics* 104: 184–198.

Elsevier. 2013. *International Comparative Performance of the UK Research Base – 2013.* Report prepared by Elsevier for the UK Department of Business, Innovation and Skills.

Mahroum, S. 2000. "Highly Skilled Globetrotters: Mapping the International Migration of Human Capital." *R&D Management* 30(1).

MORE2 Consortium. 2012. *Support for Continued Data Collection and Analysis Concerning Mobility Patterns and Career Paths of Researchers: Extra-EU Mobility Survey.* Report to the European Commission Research DG Directorate C.

MORE2 Consortium. 2013. *Final Report: Support for Continued Data Collection and Analysis Concerning Mobility Patterns and Career Paths of Researchers.* Report to the European Commission Research DG Directorate B.

OECD. 2013. *OECD Science, Technology and Industry Scoreboard 2013: Innovation for Growth.* Paris: OECD Publishing.

Saxenian, A. 2005. "From Brain Drain to Brain Circulation: Transnational Communities and Regional Upgrading in India and China." *Studies in Comparative International Development* 40(2): 35–61.

Veugelers, R. 2014. "Undercutting the Future? European Research Spending in Times of Fiscal Consolidation." Bruegel Policy Contribution, 2014/06.

Chapter 20

The Globalization of Intellectual Property Rights

Andrea Filippetti and Daniele Archibugi

Introduction[1]

The international economic landscape is periodically rattled by controversies concerning intellectual property rights (IPRs). To secure market shares, companies introduce new products and processes and this often leads to controversies for the real or supposed violations of patents, copyrights, and trademarks. Very often, the companies involved in these battles are based in different countries. The so-called smart phones war (Filippetti 2012; Graham and Vishnubhakat 2013; Wikipedia 2014) has recently captured the attention of public opinion because the product at stake is in the hands of top managers and leading politicians as well as in those of teenagers.

On some occasions, companies complain because supposedly some governments do not adequately protect their intellectual property, the most dramatic example being the coalition of large multinationals in the pharmaceutical industry, the so-called "big pharma," against the South African government for infringement of their IPRs on anti-HIV drugs (Fisher and Rigamonti 2005). In other occasions, the controversies do not directly involve competing companies, but rather the national governments themselves. This is, for example, the case with current controversies debated at the World Trade Organization (WTO) under the Trade Related Aspects of Intellectual Property Rights (TRIPS) agreement.

These controversies reflect the fact that in the current global economy, knowledge and intangibles have become increasingly important both as production factors and as consumption goods. While most of the international controversies of the past were associated to trade in physical merchandise, today there is an increase in litigation associated to IPRs.[2] Companies should bear greater investments in Research

The Handbook of Global Science, Technology, and Innovation, First Edition.
Edited by Daniele Archibugi and Andrea Filippetti.

and Development (R&D) and design in order to generate and bring to the market new products and services, but very often these core competitive assets can be imitated and replicated at costs that are substantially lower than the original cost. This is at the root of traditional tensions between innovators and imitators, a tension that for a long period mainly occurred within nations and that now has taken on a global dimension. On the other hand, the growth in international trade and foreign direct investment, associated to the rise of emerging markets, has thus increased the propensity of companies to search for profits related to their innovations at the global level. In addition, the rise of countries characterized by cheap labor costs has increased the number of potential imitators.

IPRs holders are highly concentrated in a restricted number of gigantic multinational corporations.[3] The same companies are also responsible for a corresponding amount of expenditure in R&D, industrial design, and investment in intangibles (Archibugi and Iammarino 2002; Archibugi and Pietrobelli 2003; Filippetti and Peyrache 2011). These companies have managed to assure that basic IPRs are protected in their home country as well as with their most important economic partners, namely in the areas where they concentrate the bulk of their sales. So far, however, they have not managed to get an equally effective protection in emerging and developing countries. These markets have grown in importance over the last decades and are more and more becoming part of their core business. This has led Western companies to demand greater international protection for their IPRs against unfair foreign imitators (Ryan 1998; Sell 2003).

The requests of multinational corporations to get stronger IPRs has generated passionate protests in developing countries since most inventions with commercial potential come from companies based in North America, Europe, and Japan. The full payment of IPRs would be, they argue, a net transfer of resources from developing to developed countries, which retards growth in developing countries along with pushing up prices. Critics of strong IPRs have also pointed out that, in some cases, large companies apply for IPRs to obtain exclusive rights on ancient forms of knowledge; in these cases the advantages of the legal protection are not provided to those that invest in fresh research, but rather to those who have the best attorneys and the most greedy managers (Shiva 2001; Farah and Tremolada, this volume, Chapter 22).

From the perspective of international political economy, this is an excellent case to explore the interplay of competition across companies *and* across states – an issue that Susan Strange and her followers address with great energy (see Strange 1988; Stopford and Sell 1991). There are two forms of adjudicating controversies on IPRs.

(1) When controversies relate to companies' interests they can be addressed in national courts. In spite of the presumption of impartiality, national courts are often suspected to be biased and to favor the interests of their own country and of home-based firms.
(2) When controversies relate to rules and legislation in force in countries, since 1995 the WTO is in charge. But since its members are governments only, it can examine and dispute controversies that are brought in by them only. Complaints of the business sector and of civil society can be discussed and taken into account at the WTO provided there is a government willing to receive them, under the assumption that their advocacies are part of their national interest.

This leads to the key question: what is the architecture of the emerging global IPR regime? Can IPR policy alone really regulate the generation and diffusion of knowledge and innovation across the world? And, ultimately, is a global regime of IPRs viable in practice? The aim of this chapter is to reframe the debate regarding the globalization of IPRs to place it into the right context. As students of innovation, we have learnt that the generation, transmission, and diffusion of knowledge is a complex phenomenon and that what IPRs can do and what they cannot do to reward inventors and innovators and to prevent imitators is limited. Both supporters and detractors of IPRs often tend to exaggerate the effects that they have in economy and society. The chapter is organized as follows. The next section outlines the general debate surrounding IPRs. We follow by describing the role of the United States in promoting IPRs and how this has gathered ideational strength. We then discuss the attempt to generate a global IPR regime through TRIPS. We then present two ideal typical models for and against IPRs and discuss their limitations. We continue discussing a few of the stages of a genuine global IPR regime, showing that it still lacks some important components. The last section reports the policy implications.

What Are Intellectual Property Rights?

IPRs are legal instruments designed to provide the holder of an IPR the exclusive use over certain creative activities (see Table 20.1). They include patents, copyrights, trademarks, utility models, geographic indications, and others. Each of these rights has a specific legislation and therefore covers a different domain. These IPRs have a distinct economic and social impact and can be more or less relevant according to the nature of the industry, of the technology, and of the geographical area. For example, patents are crucial for a few manufacturing industries, including pharmaceuticals, chemicals, and electronics; copyright is the core instrument to protect audiovisual and literary products; trademarks are relevant in the industries where brands represent an important competitive advantage; geographic indications are important for food and beverages (such as wine appellation). Often, companies use a combination of IPR instruments to increase the protection of their activities. For example, companies defend their software through a mixture of copyright and patents while drugs combine patents with trademarks.

The relevance of IPRs for science, technology, and innovation is also highly different. Patents are directly associated to invention and innovation since they are intended to protect ideas that, by requirement, should be both original and useful. A wealth of empirical literature has investigated the similarities and differences between scientific academic literature and patents, showing that the two are increasingly interconnected (Mayer 2000). Copyright was connected with the arts and less directly associated to the generation of new knowledge but it has become crucial for a leading knowledge-intensive industry of our age, namely software. Utility models are similar to patents, but are generally intended to protect inventions and innovations that have lower knowledge intensity, and where the design component is relevant. Trademarks and geographic indications are less directly associated to knowledge developments. This chapter will deal with IPRs that are relevant for innovation and scientific and technological knowledge, principally patents and copyright.

Table 20.1 Intellectual Property Rights.

Form of Intellectual Property Rights	*Description*	*Scientific, technological, and innovative content*
Patents	Exclusive rights granted for a limited period of time for the disclosure of an invention. The invention should be original and non-obvious	Very high. Used especially by high-tech manufacturing companies. Key instrument in industries such as drugs, chemicals, electronics, and mechanical engineering
Copyright	Legal right granted to the creator of an original work, usually for a limited period of time	Medium. Originally designed to protect literary work, music, and audio-visuals, is now used also for software
Utility model (or petty patent)	Very similar to the patent, but usually has a shorter term (often 6–15 years) and less stringent patentability requirements	Low-medium compared to patents
Industrial design rights	Property rights that protect visual designs. Easier to get and cheaper than patents	Scientific and technological content is rather low. It may have a greater impact in innovation, especially in products where the aesthetic dimension is important
Trademarks	A word, phrase, symbol, and/or design that identifies and distinguishes the source of the goods of one party from those of others	Low
Geographical indication	Name or sign associated to specific origin (town, region, or country)	Very low

The rules of the system evolved differently in each country. As shown by the country case studies reported in Odagiri et al. (2011), national institutions have evolved differently, and often they have tried to protect national industries through a variety of legal, institutional, and customary devices. Through complicated and long diplomatic negotiations – including the diffusion of experiences and attempts of companies and other economic agents to reach similar standards across countries – national IPR legislation and practices have converged (for an analysis of the patent offices in the new global context, see Drahos 2010). The number of countries that joined the international conventions on IPRs has steadily increased and the legal norms are certainly less different across countries than they were 100, 50, or even 25 years ago. If we look at the expansion of the IPR system, it is clear that in the beginning it involved countries at a comparable level of economic development. Progressively, the IPR regime has incorporated countries at the periphery and, consequently, the members of the world IPR system have become more heterogeneous.

One of the core principles of existing international conventions is to guarantee that public institutions in each country do not discriminate against foreigners. For many years, it has been acceptable that some countries had longer or shorter validity for patents or copyright, stronger or weaker protection, but the various conventions tried to establish the principle that home and foreign inventors, authors, and companies should be treated equally. The formal principle, however, is not easily enforced in practice. It is up to the national police to identify if counterfeited CDs and DVDs are sold in an open market, it is a national patent examiner who will decide if the application for an invention is genuinely original, and it is the remit of national courts to settle business controversies. In spite of all the principles stated in the various conventions, treaties, and national legislation related to non-discrimination, there have always been allegations that national institutions tend to favor national interests and that the police, patents and trademarks offices, and courts are inherently biased. The fact that recently Apple took Samsung to a Californian court, while Samsung responded by taking Apple to a Korean court, seems consistent with this view (Filippetti 2012).

As can be expected, countries that had more inventions and products to protect were keen to push the international system toward stronger protection. These countries saw IPRs as a method to increase their revenues and to strengthen the bargaining position of their national companies in host countries. Countries that relied more on knowledge and products generated elsewhere were keener to have a permissive IPR system. Catching-up countries interpreted a rigid IPR system as a further burden on their economy and their development strategy.[4]

This has continuously put developing countries in a dilemma: in order to acquire and benefit from the knowledge generated elsewhere, is it better to introduce IPR legislation and practice similar to the most developed economies or not? It can be questioned whether it is advantageous for developing and catching-up countries to implement the IPR system as designed by developed countries (see, e.g., the analysis of Drahos 2010). It has also been argued that weak economies have bad negotiators or have underestimated the costs associated to IPR agreements (May 2002; Sell 2003; Heller 2008). Others, on the contrary, have argued that countries with Western-like IPR systems have facilitated industrial development (Branstetter et al. 2010), induced foreign investment by multinational corporations (Dinopoulos and Segerstrom 2010), and induced technology transfers (Mansfield 1994, 1995). In the next sections, we will discuss how the situation has changed over the last two decades and if such a change has obtained its desired outcome.

The Rise of a Global IPR Regime

The Silent IPR Revolution in the United States

The United States has always been the country where violations of IPRs have been more vocally denounced. For the whole twentieth century the most spectacular controversies about intellectual property have occurred there and the twenty-first century looks largely the same. It is difficult to explain why the US economy, society, and legal system are so keen to consider somebody a villain when they use another's intellectual property without authorization. In Europe and Japan there are daily controversies on IPR infringements, but intellectual property is not surrounded by the aura that it has in the United States. The unauthorized download of a song, the cracking of software, the replication of a patented product are much less disapproved of in Europe than in the United States. Even more, in developing countries hackers and imitators are often admired as sorts of modern Robin Hoods who take knowledge from those who have it and provide it to those who do not have it (for a contemporary praise of the hacker, see Himanen 2001).

The reason why intellectual property in the United States is much more appreciated than in other countries has both material and ideological foundations. The United States is the largest R&D spender and patent generator in the world, its companies have invested massively in trademarks and brands, and it hosts the largest number of large multinationals with operations across the five continents. Old companies such as General Motors, Disney, and IBM, and new companies such as Microsoft, Google, and Amazon have business lines heavily dependent on IPR protection. Moreover, the US government has also the economic, political, and diplomatic muscle to protect the property rights of its companies both at home and abroad, instruments that are often lacking for other countries.

But there are perhaps also other cultural reasons that make the US society praise and reward individual creativity rather than the societal context in which knowledge is generated. The Anglo-American ideology strongly supports individual values and freedom even when they are at the expense of the public interest. In continental Europe, on the contrary, the public interest generally prevails over individual rights. This US vision spans across a variety of property rights and it extends to *intellectual* property rights. European societies are generally keen to consider limits to property and to intellectual property especially. It is therefore understandable that obtaining strong protection of IPRs has been a lesser priority in Europe than in the United States. The Asian vision is even more likely to praise collective effort over individual enterprise. In most of the Asian countries, the introduction of IPR legislation has been directed more by the need to please the United States and the other Western parties rather than by a genuine endogenous sentiment to reward individual creativity and company investment in immaterial goods.

It is true that Asian countries such as Japan and China have also been *catching-up* countries, the former in the 1950s and 1960s and the latter since the 1990s. As catching-up countries, they had a clear interest to adopt technologies developed abroad and a weak IPR system could serve such a purpose. But it is also true that since the 1980s Japanese companies started to be major producers of innovation and yet they have still been reluctant to use IPRs as their principal instrument for getting a return from their innovations.

The attempt to strengthen the IPR regime and to obtain greater enforcement started in the United States more than 30 years ago. A "silent revolution" (Jaffe and Lerner 2004) took place in the form of four interconnected changes. The first was the Bayh–Dole Act, which allowed Universities and other research centers to commercialize and profit from the innovations generated with public money – a legal transformation that has been later imitated by several OECD countries (Grimaldi et al. 2011). Second, the scope of the patent system has progressively grown, allowing patenting in new areas, such as software, which were previously covered by other forms of intellectual property or not covered at all. Third, the US Patent and

Trademark Office (USPTO) started to be funded through a fee charged on applicants rather than by the government. This apparently innocuous change made the USPTO keen to grant patents to as many applications as possible, even when the novelty and utility is not self-evident and when the invention is in areas traditionally excluded from the range of patents. Fourth, the courts have become increasingly tougher with violators of IPRs (see also Coriat and Orsi 2002).

The effects of the silent revolution in the United States are controversial and many critics argue that this has been harmful to the US economy since it has reduced the rate of innovation (Heller 2008; Heller and Eisenberg 1998; Jaffe and Lerner 2004), it has generated excessive litigation and increased costs (Bessen and Meurer 2008), and it has increased monopoly power (Boldrin and Levine 2008). So, within the United States there have been many voices that have argued that the "silent revolution" has discouraged innovation and distorted resources from innovation to patent protection. On July 4, 2013, President Barack Obama himself expressed concern that there are too many US patents that do not reflect genuine inventive activity, introducing appropriate legislation to reduce "frivolous litigations" (White House 2013). But the hegemonic country moved in the opposite direction, after having urged and pressed other countries to create a stronger regime of IPRs, which has already had significant international impact.

From the Silent Revolution to the Vociferous Global Scene

We have described above the "silent revolution" that considerably increased the strength of IPRs within the United States, and has been able to better protect US companies within their own internal market against foreign competitors. However, in a globalizing economy, this was only part of what US corporations desired, because supposed and real IPR infringements continued to occur outside the United States. The problem was that US companies had the possibility to use legal devices to block violations in their own country but had little possibility to retaliate when violations occurred elsewhere. The traditional legislation on intellectual property inaugurated in the nineteenth century required individual states to prevent discrimination and to protect foreign IPRs through legislation and, above all, enforcement. As with many other international covenants, there was no guarantee that states actually introduced homogeneous legislation and, even when they did, that national institutions were obliged to enforce it. Many governments had little interest to use their authority to protect the IPRs of foreign companies.

Since the end of the Second World War, the United States has been a generous supplier of knowledge, technology, and technical assistance to its allies. But at the end of the 1970s, US corporations and their government started to take a different turn. The undisputed technological leadership of the United States started to be eroded by its economic partners, productivity gaps narrowed, and competing countries continuously improved their innovative potential (see Nelson and Wright 1992; Pianta 1988). Japan constantly increased its export share in high-technology products and Europe was progressively performing better. The US trade deficit was no longer associated only with the import of raw materials and traditional products, but with high-technology products, and this was enough to shock a public which for most of the twentieth century was proud of its leadership in science, technology, and innovation. A growing concern emerged in the United States and the culprit was easy to be found: Japanese companies (and, to a lesser extent, European companies) had better performance because they were exploiting commercially the knowledge generated in the United States, often infringing IPRs. Less attention was devoted to looking at how companies based in Japan, Germany, Switzerland, Sweden, and other countries were investing massively in R&D and design, and even less to the fact that these companies generated an increasing number of inventions and innovations for which they claimed patent protection not only in their national patent offices, but also in the United States.

The stories heard in the 1980s soon re-emerged with reference to the Asian tigers. These countries followed patterns very similar to Japan. Comparable catching-up processes have occurred for more than a decade in China, a country with a population size much higher than Japan and the other East Asian tigers. The widespread feeling in the United States continued and still continues to be that the national investment in knowledge and in other intangible assets is not adequately rewarded and that its competitors in the international market make unauthorized use of technologies, design, and other intangible assets generated by US corporations. This position holds that, as long as the international market does not provide adequate remuneration for US innovations and intangibles, and institutions do not punish infringements, the United States will continue to be penalized.

It took a while for the US corporations to find a method that was able to better guarantee their IPRs outside the United States. The retaliations available ranged from diplomatic reproach to sanctions; none of them was particularly effective to protect companies' economic interests. As is well documented by Susan Sells (2003) and Michael Ryan (1998), a group of Chief Executive Officers of leading corporations joined forces through the creation of the Intellectual Property Committee (IPC) and the International Intellectual Property Alliance (IIPA). The IPC explored the available opportunities and eventually agreed that the most effective and perhaps sole way to penalize countries not adequately protecting IPRs was by retaliating on trade. Since negotiations to reduce trade barriers were already on the agenda, the US government effectively pushed with all its political, diplomatic, and economic muscle to make IPRs a crucial pillar of the GATT's replacement, the WTO – which is where TRIPS was born.

Some commentators were surprised that so many countries signed TRIPS: after all, if these countries were using knowledge and intangibles without paying a proper remuneration they should not have an interest to sign an agreement that implied greater transfer of resources from the imitators to the innovators (Wade 2003). Developing countries had an interest to liberalize the international trade in textile apparel – the Multifiber Agenda – and agricultural products. And the US market, along with the European market, was the most attractive for them. If the price to be paid to access the American and European markets was to promise to introduce a tougher IPR legislation, it seemed that this was an acceptable deal. TRIPS was born as a bargain: developing countries accepted the risk of retaliation if they were not able to introduce adequate legislation for IPRs in their own country in exchange for the opportunity to enter into the American and European markets. The powerless conventions of the twentieth century were therefore integrated with a potentially vigorous tool: trade retaliations.

Two Models on IPRs

Since the beginning, IPRs had enthusiastic supporters and fierce enemies (May and Sell 2006 trace the genealogy of these arguments). Much has changed in the generation, imitation, and diffusion of invention and creativity over the last few centuries, but the arguments in favor and against IPRs have somehow been repeated. It is possible to identify two opposite models in the contemporary debate, the first that recommends a strong regime of IPRs for economic development, and a second that argues that a weak regime or even the absence of IPRs is conducive to the diffusion of knowledge.

In Praise of Strong IPRs

The model that supports strong protection of IPRs at the national level argues that in the long term IPRs will not only be beneficial for the producers of knowledge but also for users and therefore for society at large. This model applies the same logic at the international level, assuming that IPRs will benefit developing as much as developed countries. The model stresses that incentives and rewards to inventive and innovative activity are crucial to generate further investment. It is not denied that *statically* the existence of IPRs may reduce the diffusion of innovations to those who cannot afford to pay the price, but *dynamically* this will be an incentive to invest more in the future. This is supported by four main arguments:

- Invention, innovation, and more generally creativity are all costly activities, especially since they involve a great degree of uncertainty about the outcome. But the outcome of these activities, when valuable, can be imitated or replicated at substantially lower costs. The absence of protection discourages profit-seeking agents to invest in these activities.
- IPR protection encourages full disclosure of inventions that eventually became part of the public domain. In absence of disclosure, knowledge may be kept secret and this reduces its dissemination.
- The revenues generated by inventions are one of the core resources to finance further projects. It is therefore relevant that the current inventive activities are able to generate profits to keep the system going.
- New entrants have to face the alternative of imitating existing devices or investing in generating fresh solutions and strong IPRs induce them to opt for the latter strategy that sparks inventiveness.

These reasons are used to justify IPRs within nations with the awareness that they provide advantages for individuals and companies that generate knowledge and disadvantages for consumers or supplier-dominated companies. But it is assumed that the advantages for the economic system are greater than the disadvantages and that well-tuned legislation could balance the interests of both producers and users of knowledge (for example by regulating the length and scope of IPRs and even by using compulsory licensing in cases of palpable public interest).

The pro-IPR model advances additional reasons to suggest that developing countries also benefit from a global IPR regime. It is a fact that the distribution of scientific and technological capabilities is extremely polarized across the globe (see Castellacci and Natera, this volume, Chapter 1). This means that the net recipients of IPR royalties and fees are based in the North and the net payers are based in the South. The advocates of IPRs argue that Southern countries with a strong regime of IPRs benefit from greater inflows of technology transfer. Companies in the North may be reluctant to establish production facilities, to build R&D labs, to license know-how, and to engage in strategic technology agreements in countries that do not properly guarantee IPRs. On the contrary, if IPRs are secure, companies may be willing not only to license the knowledge generated in their home country, but they may also decentralize in emerging and developing countries by moving some of their R&D and innovative facilities and to collaborate with local companies on common projects. In a world where developing and emerging countries still need to acquire the knowledge generated in the North, a well-functioning IPR system is the best guarantee that Southern companies would not be excluded.

This view has particularly emphasized the role of multinational firms and foreign direct investment. For instance, Dinopoulos and Segerstrom (2010) developed a model of North–South trade and find that stronger IPR protection in the South leads to: (a) a permanent increase in the rate of technology transfer to the South within multinational firms and (b) a permanent increase in R&D employment by Southern affiliates of Northern multinationals. Along similar lines it has been shown that improvements in IPR protection has led US-based multinational firms to increase technology transfer to their affiliates and to shift toward more technologically-based products abroad (Branstetter, Fisman, and Foley 2004; Branstetter et al. 2010). Other scholars point to the indirect benefits for the countries hosting multinational firms, in terms of growth of local suppliers (Javorcik 2004a), and transfer of advanced knowledge and skills to the local workforce (Görg and Strobl 2005; Poole 2013). Also, the establishment of reliable and harmonized IPR systems lead to the creation of "markets for technology" that facilitate and encourage knowledge diffusion trough formal transactions of technology (Arora, Fosfuri, and Gambardella 2001; Athreye and Cantwell 2007).

Against IPRs

The pro-IPR model has been opposed with fierce arguments. The first argument against IPRs is that they are harmful because, by creating a legal monopoly, they obstruct and reduce the diffusion of knowledge (Andersen 2006; Boldrin and Levine 2008). Since knowledge drives the generation of further knowledge, IPRs may create a vicious circle that stops inventive activity. This is particularly true within industries in which innovation activity is based on sequential inventions and complementary technologies. In these cases, imitation may promote further innovations, while strong patents might actually inhibit it (Bessen and Maskin 2009; Merges and Nelson 1990). By becoming tighter, IPRs increase legal costs more than investment in innovation, leading to a scientific and technological system based on litigation rather than on research (Bessen and Meurer 2008). Eventually, lawyers are the main beneficiaries of a strong system of IPRs. Second, IPRs may also distort the investment for knowledge since this is likely to be directed toward the areas that promise greater profits or that can be better protected rather than toward those that are more likely to generate socially useful results or where there are more technological opportunities. For instance, it has been shown that the type and strength of the patent regime influences not only the rate of innovation activity but also the direction of technical change (Moser 2005). It is therefore not in the public interest to get a strong IPR system (Macmillan 2006). Thirdly, it has been argued that too much patenting in some industries is hampering innovation rather than encouraging it (Heller 2008).

These beliefs are reinforced when the needs of developing countries are taken into account. In these nations, knowledge-generating institutions are still in their infancy and they are the most affected by a strong international regime where there is a price to be paid for any technology transfer. In a North–South perspective, the IPR regime may hamper or impede catching-up (Chang 2002). It has been claimed that transferring the IPR practices of the developed world to developing countries leads to reduced knowledge flows, lower imitation, and increased prices (Helpman 1993; Lai 1998; Parello 2008). Thus, a system prior to TRIPS, with a strong IPR regime in developed countries and a weaker one in developing countries, is more congenial so as to allow the lower- to middle-income countries to catch up by benefiting from the knowledge developed elsewhere. The consequence of TRIPS is therefore damaging development and it will make it more difficult to allow emerging and developing countries to build solid innovation systems. Moreover, it will extract resources from developing to developed economies and this will delay further catching-up.

From a historical perspective, it is often pointed out that most countries managed to catch up through copying and imitating from more developed countries (Boldrin and Levine 2008). Only at a later stage were IPR regimes established *as an effect* of the development of the country (Lerner 2002; Mokyr 2002: for a detailed list of national case studies, see Odagiri et al. 2011). By making international technology transfers more onerous through a strong regime of IPRs, developed countries "kick the ladder away" and make it more difficult for laggard countries to catch up (Chang 2002).

Some critics of IPRs also focus on selected key products protected by IPRs (see, e.g., Correa 2000; Shiva 2001). A paradigmatic case concerns the anti-retroviral drugs patented by leading US and European pharmaceutical corporations (the so-called "Big Pharma"), but marketed in South Africa through local generic and unauthorized manufacturers. This led to a legal battle between Big Pharma and the South African government. The government argued that the vast majority of South Africans affected by HIV could not afford to pay the price charged by Big Pharma (Fisher and Rigamonti 2005). The campaign was sufficiently powerful to induce Big Pharma to withdraw from the case. In this domain, intellectual criticism overlaps with social and political activism. Due to the public good related to certain products, new social movements have started to contest the pro-IPR policies carried out by multinational corporations, national governments, and the WTO. Big Pharma, Microsoft, and other companies have become the frequent targets of many campaigns against intellectual monopoly.

A Different View: How Powerful Are IPRs?

The arguments of those who are for and against IPRs describe real aspects of the generation, transmission, and diffusion of innovation. But, surprisingly, both models are based on a common textbook assumption: *they assume that the legal system of IPRs is much more powerful than it actually is*. Both of them take for granted that strong IPRs can guarantee the protection of invention, innovation, and intangibles and that weak IPRs, on the contrary, allow imitators to acquire the related knowledge. This is not the case and there is old and new empirical evidence that illustrates that IPRs are much less effective than generally assumed by both models.

From the perspective of the producers of invention, innovation, and intangibles, it should be clarified that their main economic interest is not to secure IPRs per se but rather to profit from them. In order to appropriate the returns from their inventions, innovations, and intangibles, companies have to develop complex strategies that include R&D, design, lead-time, ability to deliver products to market, and to combine effectively industrial secrecy with IPRs. IPRs are just one element in this strategy and certainly not the most important one. Surveys carried out for US and European manufacturing companies have consistently indicated that patents and other legal methods are, in fact, the two least important appropriability factors, while companies have ranked as more important lead-time, industrial secrecy, complementary manufacturing, and complementarily sales and services (Cohen, Nelson, and Walsh 2000; Arundel 2001).

From the perspective of potential users, would-be imitators cannot manage to acquire knowledge for production facilities just by getting knowledge unprotected by IPRs. The use of knowledge for production is associated with a much larger variety of factors. IPRs may, at most, report and protect some codified knowledge, but there is an equally important component represented by tacit knowledge that is not, nor can be, reported in patents, handbooks, software, blueprints, and other codes (Pavitt 1987; Nelson 1992).[5] Imitators will need to acquire this knowledge to properly use it. A musical score and a violin are not sufficient to play Beethoven's violin concerto, likewise the free use of relevant patents does not immediately result in a company manufacturing a good car. Patent rights last for not more than 20 years: after this period, the knowledge protected by patents is publicly available. It is, however, difficult to imagine that, if the protection accorded by patents was immediately abolished worldwide, developing countries would be able to assimilate and put into practice the knowledge of the most developed countries. The reason is very simple: the acquisition of knowledge is not mainly blocked by legal devices but rather by the lack of competences of prospective imitators and complementary assets. In a nutshell, IPRs are less important than assumed for both generators and users of knowledge.

Of course, it is difficult to make generalizations regarding a complex economy where there are many products and industries with radically different characteristics. In fact, the available empirical research has shown that industries and products are affected differently by IPRs. Within the manufacturing industry, pharmaceuticals and, to a much smaller degree, chemicals are heavily dependent on patents. According to companies' respondents, as many as two thirds of innovation in pharmaceuticals, and one third in chemicals, would not have been introduced in the absence of patent protection (Mansfield 1986). Likewise, copyright is a crucial factor for the audiovisual and software industries where the final consumers have the possibility to copy the products directly.[6] In most other high-tech industries, including computers, electronics, aerospace, automobiles, mechanical engineering, IPRs are overwhelmed and/or complemented by other methods of appropriation of innovation.

The fact that IPRs are effective only in a few industries is reflected in the composition of foreign direct investment. It has been shown that the importance of IPR protection varies between industries (Mansfield 1994, 1995) and that weak IPR discourages investors only in sensitive sectors (Javorcik 2004b). In a study examining the drivers of the surge of patents in China, it is shown that foreign direct investment is mostly limited to electric machinery, transportation equipment, and chemical industries (Hu and Jefferson 2009). Also, it has been argued that far from being automatic, adoption of foreign technologies by developing countries is contingent on the development of an adequate level of skills and technological capabilities (Benhabib and Spiegel 2005; Parello 2008). Other research that has tried to explain how Western companies have increased their patent applications in countries with weak appropriability regimes, has found that preventing imitation or securing royalties are only partial reasons (Keupp, Friesike, and von Zedtwitz 2012).

Innovating companies know well that in developing countries they cannot sell their products for the same price as they sell it for in developed countries. Many products have a substantial difference between the average and the marginal cost, and in products such as drugs, software, and audiovisuals the difference is enormous. Companies apply price discrimination in order to maximize the revenues from the same product innovation across different markets. What these companies are most worried about is the possibility that the same products are reimported into the Western markets. For example, one main concern of Big Pharma when they sued the South African government for its unwillingness to stop the diffusion of generic versions of the anti-retroviral drugs against HIV/AIDs was the concern that the generic version could also reach the much more lucrative Western markets (Muzaka 2011).

Most of the ancient and contemporary debate has focused on the pharmaceutical industry (see Scherer, this volume, Chapter 21). This confirms the importance of IPRs, and patents especially, to the pharmaceutical industry. Media, social movements, and public officers have also been keen to take positions when crucial drugs are fenced by IPRs, given their substantial and often immediate impact in health. On a moral level, there have been convincing claims that every citizen should have a right to medications regardless of income and in order to live a minimally decent life (Brown and Paremoer 2014). However, it is difficult to believe that, should TRIPS be abolished tomorrow, the pharmaceutical industry would suddenly become the engine of technological advances and growth in developing countries. Thus one has to make a distinction between the two problems related to IPRs in the pharmaceutical sector. The first concerns the right of access to drugs, while the second regards the capacity of economic and technological development in low- and middle-income countries. Our conviction is that TRIPS is affecting the delivery of drugs in low-income countries, but it is not impeding their economic and technological development.

To sum up, IPRs on their own cannot guarantee or reduce returns to invention, innovation, and intangibles. Profits for innovations are obtained through a variety of channels and, if companies are asked, most of them rank patents to prevent imitators and patents to secure royalties as subordinate methods to appropriate returns from innovation. This leads to a logical question: *if IPRs have so little relevance, how come companies, governments, lobbies, and social movements are so concerned about them?* A first tentative answer is to suggest that IPRs are somehow readily visible and can be more easily modified by institutions. But what they represent – the system of incentives to generate, transmit, and diffuse knowledge and creativity – are much more complex and often less visible. As a result the debate is seemingly concentrating on the finger rather than on the moon it is pointing to. A second tentative answer is that the current situation in patenting is the (inefficient) result of a game in which for each actor the optimal strategy is that of patenting as much as possible in order to prevent people being treated by competitors. In this context, companies are encouraged to build a large patent portfolio, regardless of the value they attribute to patents, not to protect themselves from potential imitators, but from real competitors.

Is a Global IPR Regime Possible?

In a planet of independent and interconnected states, it is not an easy task to build a global regime of IPRs. We can outline four different stages for its construction:

- *Harmonization.* Countries agree to have comparable, if not identical, legislation guaranteeing that there is no discrimination against foreigners. Harmonization allows differences in the regime of each country, but within each national system domestic and foreign citizens and companies should be on par.
- *Standardization.* When harmonization evolves into standardization, individual countries are no longer autonomous to decide which rules they can apply domestically. Through appropriate discussions, agreements, and treaties, rules and standards are collectively established. For example, treaties may require to members to introduce legislation protecting IPRs, that patents should be of a certain length, and that the legal system should contemplate penalties for infringements.
- *Control and dispute settlement.* The standards in force in each country are not exclusively under national sovereignty, but they are also assessed by multilateral institutions. This assessment should guarantee standardization. It also contemplates that countries that do not comply are requested to remedy or are exposed to sanctions.
- *Implementation with enforcement.* Procedures applied within countries are not left to national authorities only, but are submitted for evaluation to global institutions that have the mandate and the authority to assess them. Enforcement is not left to national authorities only, but is also provided by international authorities.

The existence of all the four stages distinguishes a strong global regime. Where are we in the case of IPRs? Table 20.2 reports the paradigmatic sources for each of these four stages.

The *harmonization* process started more than one century ago, with the Paris Convention for the Protection of Industrial Property, signed in 1883 and the Berne Convention for the Protection of Literary and Artistic Works, signed in 1886. The Conventions established the non-discrimination principle as it stated that the application for a patent, trademark, or copyright by a foreign citizen should receive the same treatment as for a national citizen. These were important attempts to reach a uniform legislation, although they coexisted with the desire of each nation to get the IPR system congenial to its own economic and social advantage (Odagiri et al. 2011 shows how this has worked in practice in several countries). If we read the evolution of IPRs in each country, we can appreciate that the process of harmonization has not prevented governments from shaping the system according to national economic

Table 20.2 The four stages of a global IPR regime.

Harmonization	Paris Convention for the Protection of Industrial Property (1883) Berne Convention for the Protection of Literary and Artistic Works (1886) Institution of the World Intellectual Property Organization (1967)
Standardization	World Trade Organization through TRIPS Agreement (1994)
Control and dispute settlement	TRIPS Agreement (1994). Coercion works through the WTO Dispute Settlement Process
Implementation and enforcement	National legislative institutions to implement remedies requested by WTO Dispute Settlement Body National police enforcement National courts in adjudicating IPRs controversies

and social needs. Some industries, and specifically drugs, have been excluded from patent protection, the life-span of patents has sometimes been reduced, and transition periods have been granted to facilitate the introduction of new rules. In spite of countries' patterns and strategies, national IPR regimes have strongly converged toward similar models over more than a century. Today, IPR systems are comparable across the largest number of countries.

When, with the institution of the WTO, IPRs started to be one of the fundamental pillars of the organization through TRIPS, a very important change was introduced (Drahos 2010). In order to be part of the WTO, and to benefit from the advantages of free trade, countries were requested to have not only an IPR system, but also to conform it to specific standards already operating in Western countries. As said above, the deal offered by developed to developing countries was clear: we open up our markets to your merchandise, but you should guarantee our IPRs in your countries. TRIPS moved an important step from harmonization to standardization.

But this was not the only change introduced by TRIPS. Perhaps more importantly, for the first time it introduced a form of multilateral control and dispute settlement previously absent. Control was, in the intention of the TRIPS architects, meant to be particularly effective because trade retaliations against those governments who did not comply with the agreement were a credible threat. Arguably, this is one of the main reasons why the global regime of IPRs has been brought into the WTO instead into the World Intellectual Property Organization (WIPO). The newly instituted Dispute Settlement Body was precisely designed as a procedure for trade quarrels and enlarging it to IPR quarrels was a major step toward the generation of a genuine global IPR regime. The institution of WTO worked at the same time for standardization (as a precondition to WTO membership) and control and dispute settlement. The fourth stage, namely implementation and enforcement, is left totally in the hands of national authorities. It would have been difficult to do otherwise: there are no cases of international organizations that manage to override governments in policing and judicial power.

More than 15 years have passed since TRIPS was introduced. It is perhaps possible to make an assessment of its role. The WTO, through TRIPS, has been rather successful in achieving standardization and now all WTO member countries have IPR legislation in line with Western tradition. But if we look at the effectiveness of the DSB, there are few reasons to assume that control and dispute settlement has been effective. In practice, the disputes brought at the WTO concerning IPRs are not many. From 1995 to 2011, the WTO Dispute Settlement Process (DSP) has been activated 29 times for IPR-related issues (Lee 2010–2011). In 17 of these cases the United States – the economically dominant country – is the complainant country. But only 7 of these complaints are directed toward developing countries, while 10 are directed toward other OECD countries. Of course, this does not imply that the most serious infringements of IPR occur in the OECD area, but rather that real and perceivable economic damages largely occur in OECD countries. The US government, the most active in using the WTO machinery, did not bother to use the DSP when the markets in which violations take place are not particularly attractive. Moreover, the DSP has some clear limitations:

- The DSP can be activated against unfair legislation but much less against the lack of effective enforcement of IPRs, which continue to be the prerogative of national governments. So far, the stage "Implementation and enforcement" continues to be firmly in the hands of national authorities.
- The DSP process is lengthy (it takes up to three years) and the remedies that a country agrees to implement may take up to a couple of years more. In areas of rapid technological change, a DSP decision may be taken when it is no longer relevant.
- The DSP may require individual countries to put into practice "remedies" and to change legislation, but cannot dictate their specific aspects. Often, losing countries do introduce changes, but not necessarily to the extent that the winning party is satisfied.
- The parties to the WTO and its DSP are states and not companies. Governments often act to pursue the interests of the companies based in their own country, and governments are often solicited by the companies themselves. But when dealing with multinational corporations, the national interests are more difficult to assess and governmental take-up will be weighed alongside other strategic interests.
- Trade retaliations have so far seldom been authorized.

The achievement of a global IPR regime is not limited to legislation only. It should also be based on implementation and enforcement. When the globalization of IPRs is discussed, it is crucial to keep in mind that the fourth stage is not under the control of an international organization, since we are dealing with legal and policing practices that are predominantly national in scope. While the legislation on IPRs can be made accountable to the WTO through the DSP, national enforcement is under the exclusive responsibility of national authorities. National police can be tougher or softer against violators and national courts take the final decision in controversies, often with different rulings according to the country where they are based and the nationality of the companies they are assessing. It is therefore not surprising that at least with reference to patents a clear trend toward stronger protection in developing countries is not visible (Allred and Park 2007; Park 2008).

There is huge anecdotal evidence that in most emerging and developing countries, IPR is far from being strongly prosecuted and enforced by national authorities. The reasons can be different. In some occasions, it could be the results of a specific political orientation of governments that are trying to favor domestic rather than foreign companies. The governments could allow a weak IPR enforcement even when there is more rigid legislation in place. On other occasions, it could be simply a matter of a lack of resources and institutions. An effective enforcement of intellectual property, in fact, requires a considerable amount of resources, including policing and courts. Despite the widespread adoption of the TRIPS, IPR enforcement in several countries is currently occurring more on paper than in reality.

In sum, there is not yet a global regime of IPRs in practice. This is something well known to practitioners who reported to us that it is basically pointless to sue Chinese companies for patent infringement in China, simply because it would be a waste of time and money. The main reason why currently a global IPR regime is not viable is because of the four stages necessary to make it work, the last one is still under the control of national governments and institutions, and often they have either a strong incentive in maintaining a weak IPR system, or a lack of resources to make it stronger. While states might retaliate against each other at the WTO, there is neither a global police nor a global court that can enforce the implementation of the TRIPS provisions within each nation-state. It is certainly telling that Apple and Samsung, two companies that with their smart phone war inflamed the IPRs debate, eventually agreed to give up legal cases against each other outside the United States (Jin and Levine 2014). The unpredictability of rulings and soaring legal costs have perhaps convinced these corporations that it is better investing in R&D than in attorneys.

What Is the Future of IPRs in a Global Economy?

There has been a consistent attempt to create a stronger global IPR regime led by top US corporations. TRIPS has been the main outcome of this attempt. This has generated heated debates in academia and outside it. Supporters of IPRs have argued that without them there is the risk that companies and other private investors will not find sufficient incentives to invest in R&D and innovation. Critics of IPRs, on the contrary, have argued that making IPRs stronger will make it more onerous and more difficult for developing countries to catch up.

Both these views, in spite of the very opposite policy conclusions they reach, share a similar view, namely that IPRs can really make a difference in the economy and in the society. We have, on the contrary, argued that IPRs per se do not change significantly the process of technology transfer and acquisition. The transfer of knowledge is hampered by other things and, in particular, by the absorptive capacity of recipient subjects. In the absence of investment in R&D, education, and infrastructures, developing countries are unlikely to benefit from the knowledge developed elsewhere. IPRs, either strong or weak, would be basically irrelevant.

In a changing economic environment, economic players are also able to adjust their behavior according to the price they have to pay for innovations. If the costs to acquire innovations and knowledge-intensive products rise, new players are ready to take it as an opportunity to enter into profitable markets. Strong IPR enforcement may induce their holders to raise prices rather than to upgrade products: when software started to be policed more seriously, many companies were forced to get rid of cracked programs, and this was certainly beneficial for Microsoft and other software companies. But when prices started to rise, consumers searched for other viable alternatives. This led to an unexpected diffusion of open source software and alternative operating systems such as Linux. In turn, this reduced the sales of copyrighted programs. When the United States complained against the illegal distribution of Hollywood films in India and obtained greater protection, this provided an impetus to scale-up new Bollywood productions into an industry that is now a potential threat to Hollywood itself not only in India, but in many other markets (Sunder 2011). Companies are often aware that the best way to profit from their inventions and innovations is to leave IPRs relatively flexible since this helps the diffusion of their products. This was the policy carried out by Microsoft in the 1980s.

It is often argued that Western countries should make a net profit from stronger IPRs. But, again, this conclusion is not substantiated by evidence. Or, more precisely, greater profits for the IPR-holding companies are not necessarily a gain for these countries' home country. As already indicated by the Lieberman Report (Lieberman 2004), the main aim of multinational corporations in obtaining a stronger global IPR regime was to offshore production facilities and knowledge-intensive jobs to countries with lower wages. This can hardly be called an advantage for the United States and other advanced economies. In fact, it opens up questions about whether the US government has actually pursued the interests of its citizens or whether it pursued the interests of multinational corporations in pushing for a harmonized global IPR regime.

Finally, we have argued that we are far from having an effective global IPR regime. While Western countries have, through their power in international organizations, successfully managed to reach an unprecedented level of harmonization of intellectual property rules, states still exert a great deal of control on the implementation and enforcement of TRIPS provisions within their borders. We have also shown that the DSP has some clear limitations and has been used in a rather limited way. A strong global IPR regime does not exist and it does not seem that it will be available in the short run.

Should the IPR regime be fixed? The economic practice is very different from what is often suggested as the institutional ideal. So far, even the advocates of TRIPS recognize that the agreement did not have much impact, either positive or negative, on the least developed countries (Hold and Mercurio 2012). There are clearly contesting forces at work, with net knowledge producers trying to appropriate the returns of their investments. But neither strong nor weak IPRs can guarantee technology transfer. What is needed for a successful strategy of technology transfer and innovation-based growth is an overall strategy of developing countries to assimilate, apply, transform, and improve the knowledge generated elsewhere. IPRs may block imitators in some specific areas, or they can make it more onerous, but in general IPRs cannot impede the use of knowledge to those who have the competences to use it. Once again, we are in the classical situation in which the main policy implication is "if it ain't broke, don't fix it" (Winter 1989).

Notes

1 This research has been supported by a Marie Curie Intra-European Fellowship project FP7-PEOPLE-2011-IEF-298167-REGIO_SPIN, under the EC Grant Agreement No: PIEF-GA-2011-298167 and from a Grant of the School of Business, Economics and Informatics, Birkbeck College, University of London. Parts of the chapter are based on a note in Mayer, Carpes, and Knoblich (2014).

2 See among others: Archibugi and Filippetti (2010); Drahos and Mayne (2002); Henry and Stiglitz (2010); Maskus (2000).

3 The concentration of patents in a few hundred multinational corporations has been analyzed by Patel and Pavitt (1997) and Cantwell and Iammarino (2003).

4 The issue is certainly not new and it was already addressed by Edith Penrose (1950, 1973).

5 The distinction between codified and tacit knowledge is, of course, due to Michael Polanyi (1967).

6 If products such as CDs, DVDs, and software could not be technically copied, they would not even need to rely on *intellectual* property rights since the standard property rights will be more than sufficient to protect the producer.

References

Allred, Brent B., and Walter G. Park. 2007. "Patent Rights and Innovative Activity: Evidence from National and Firm-Level Data." *Journal of International Business Studies* 38: 878–900.

Andersen, Birgitte (ed.). 2006. *Intellectual Property Rights: Innovation, Governance and the Institutional Environment*. Cheltenham: Edward Elgar.

Archibugi, Daniele, and Andrea Filippetti. 2010. "The Globalization of Intellectual Property Rights: Four Learned Lessons and Four Theses." *Journal of Global Policy* 1: 137–149.

Archibugi, Daniele, and Simona Iammarino. 2002. "The Globalization of Technological Innovation: Definition and Evidence." *Review of International Political Economy* 9: 98–122.

Archibugi, Daniele, and Carlo Pietrobelli. 2003. "The Globalisation of Technology and Its Implications for Developing Countries: Windows of Opportunity or Further Burden?" *Technological Forecasting and Social Change* 70: 861–884.

Arora, Ashish, Andrea Fosfuri, and Alfonso Gambardella. 2001. *Markets for Technology: The Economics of Innovation and Corporate Strategy*. Cambridge, MA: The MIT Press.

Arundel, Anthony. 2001. "The Relative Effectiveness of Patents and Secrecy for Appropriation." *Research Policy* 30: 611–624.

Athreye, Suma, and John Cantwell. 2007. "Creating Competition? Globalisation and the Emergence of New Technology Producers." *Research Policy* 36: 209–226.

Benhabib, Jess, and Mark Spiegel. 2005. "Human Capital and Technology Diffusion." In *Handbook of Economic Growth*, ed. Philippe Aghion and Steven N. Durlauf, 935–966. Amsterdam: North-Holland.

Bessen, James, and Eric Maskin. 2009. "Sequential Innovation, Patents, and Imitation." *The RAND Journal of Economics* 40: 611–635.

Bessen, James, and Michael J. Meurer. 2008. *Patent Failure: How Judges, Bureaucrats, and Lawyers Put Innovators at Risk*. Princeton, NJ: Princeton University Press.

Boldrin, Michele, and David K. Levine. 2008. *Against Intellectual Monopoly*. Cambridge: Cambridge University Press.

Branstetter, Lee, Ray Fisman, and C. Fritz Foley. 2004. "Do Stronger Intellectual Property Rights Increase International Technology Transfer? Empirical Evidence from U.S. Firm-Level Panel Data." *Quarterly Journal of Economics* 121: 321–349.

Branstetter, Lee, Ray Fisman, C. Fritz Foley, and Karmal Saggi. 2010. "Does Intellectual Property Rights Reform Spur Industrial Development?" *Journal of International Economics* 83: 27–36.

Brown, Garrett W., and Lauren Paremoer. 2014. "Global Justice and the Right to Health." In *Global Health Policy*, ed. Garrett W. Brown, Gavin Yamey, and Sarah Wamala. Oxford: Wiley-Blackwell.

Cantwell, John, and Simona Iammarino. 2003. *Multinational Enterprises and European Regional Systems of Innovation*. London: Routledge.

Chang, Ha-Joon. 2002. *Kicking Away the Ladder: How the Economic and Intellectual Histories of Capitalism Have Been Re-Written to Justify Neo-Liberal Capitalism*. London: Anthem Press.

Cohen, Wesley M., Richard R. Nelson, and John P. Walsh. 2000. *Protecting Their Intellectual Assets: Appropriability Conditions and Why US Manufacturing Firms Patent (or Not)*. Cambridge, MA: NBER Working Paper 7552.

Coriat, Benjamin, and Fabienne Orsi. 2002. "Establishing a New Intellectual Property Rights Regime in the United States: Origins, Content and Problems." *Research Policy* 31: 1491–1507.

Correa, Carlos M. 2000. *Intellectual Property Rights, the WTO and Developing Countries: The TRIPS Agreement and Policy Options*. London: Zed Books.

Dinopoulos, Elias, and Segerstrom, Paul. 2010. "Intellectual Property Rights, Multinational Firms and Economic Growth." *Journal of Development Economics* 92: 13–27.

Drahos, Peter. 2010. *The Global Governance of Knowledge: Patent Offices and Their Clients*. Cambridge: Cambridge University Press.

Drahos, Peter, and Ruth Mayne. 2002. *Global Intellectual Property Rights: Knowledge, Access and Development*. New York: Palgrave Macmillan.

Filippetti, Andrea. 2012. "Apple vs. Samsung: Something Doesn't Quite Fit!" *Opendemocracy.net*, September 18, http://www.opendemocracy.net/openeconomy/andrea-filippetti/apple-vs-samsung-something-doesn%E2%80%99t-quite-fit (accessed January 6, 2015).

Filippetti, Andrea, and Antonio Peyrache. 2011. "The Patterns of Technological Capabilities of Countries: A Dual Approach Using Composite Indicator and Data Envelopment Analysis." *World Development* 39: 1108–1121.

Fisher, William W., and Cyrill P. Rigamonti. 2005. *The South Africa AIDS Controversy: A Case Study in Patent Law and Policy*. Cambridge, MA: Harvard Law School.

Görg, Holger, and Eric Strobl. 2005. "Spillovers from Foreign Firms through Worker Mobility: An Empirical Investigation." *Scandinavian Journal of Economics* 107: 693–709.

Grimaldi, Rosa, Martin Kenney, Donald S. Siegel, and Mike Wright. 2011. "30 Years After Bayh–Dole: Reassessing Academic Entrepreneurship." *Research Policy* 40: 1045–1057.

Graham, Stuart, and Saurabh Vishnubhakat. 2013. "Of Smart Phone Wars and Software Patents." *Journal of Economic Perspectives* 27: 67–86.

Heller, Michael M. 2008. *The Gridlock Economy: How Too Much Ownership Wreck Markets, Stops Innovation, and Costs Lives*. New York: Basic Books.

Heller, Michael A., and Rebecca S. Eisenberg. 1998. "Can Patents Deter Innovation? The Anticommons in Biomedical Research." *Science* 280: 698–701.

Helpman, Elhanan. 1993. "Innovation, Imitation, and Intellectual Property Rights." *Econometrica* 61: 1247–1280.

Henry, Claude, and Joseph Stiglitz. 2010. "Intellectual Property, Dissemination of Innovation and Sustainable Development." *Global Policy* 1(3): 237–251.

Himanen, Pekka. 2001. *The Hacker Ethic and the Spirit of the Information Age*. New York: Random House.

Hold, Arno, and Bryan C. Mercurio. 2012. *Transitioning to Intellectual Property: How Can the WTO Integrate Least-Developed Countries into TRIPS?* NCCR Working Paper 2012/37.

Hu, Albert G., and Gary H. Jefferson. 2009. "A Great Wall of Patents: What Is Behind China's Recent Patent Explosion?" *Journal of Development Economics* 90: 57–68.

Jaffe, Adam B., and Josh Lerner. 2004. *Innovation and Its Discontents: How Our Broken Patent System Is Endangering Innovation and Progress, and What to Do About It*. Princeton, NJ: Princeton University Press.

Javorcik, Beata S. 2004a. "Does Foreign Direct Investment Increase the Productivity of Domestic Firms? In Search of Spillovers through Backward Linkages." *American Economic Review* 94: 605–627.

Javorcik, Beata S. 2004b. "The Composition of Foreign Direct Investment and Protection of Intellectual Property Rights: Evidence from Transition Economies." *European Economic Review* 48: 39–62.

Jin, Hyunjoo, and Dan Levine. 2014. "Samsung Electronics, Apple Call End to Patents War Outside U.S." *Reuters*, August 6, http://uk.reuters.com/article/2014/08/06/us-samsung-elec-apple-idUKKBN0G605920140806 (accessed January 6, 2015).

Keupp, Marcus M., Sascha Friesike, and Maximilian von Zedtwitz. 2012. "How Do Foreign Firms Patent in Emerging Economies with Weak Appropriability Regimes? Archetypes and Motives." *Research Policy* 41: 1422–1439.

Lai, Edwin. 1998. "International Property Right Protection and the Rate of Product Innovation." *Journal of Development Economics* 55: 133–153.

Lee, Edward. 2010–2011. "Measuring TRIPS Compliance and Defiance: The WTO Compliance Scorecard." *Journal of Intellectual Property Law* 18: 414–445.

Lerner, Josh. 2002. "150 Years of Patent Protection." *American Economic Review* 92: 221–225.

Lieberman, Joseph I. 2004. *Offshore Outsourcing and America's Competitive Edge*. Washington, DC: Office of Senator Joseph I. Lieberman.

Macmillan, Fiona. 2006. "Public Interest and the Public Domain in an Era of Corporate Dominance." In *Intellectual Property Rights: Innovation, Governance and the Institutional Environment*, ed. Birgitte Andersen, 46–69. Cheltenham: Edward Elgar.

Mansfield, Edwin. 1986. "Patents and Innovation: An Empirical Study." *Management Science* 32: 173–181.

Mansfield, Edwin. 1994. *Intellectual Property Protection, Foreign Direct Investment, and Technology Transfer*. Washington, DC: World Bank Discussion Paper 19.

Mansfield, Edwin. 1995. *Intellectual Property Protection, Foreign Direct Investment, and Technology Transfer: Germany, Japan and the United States*. Washington, DC: World Bank Discussion Paper 27.

Maskus, Keith E. 2000. *Intellectual Property Rights in the Global Economy*. Washington, DC: Institute for International Economics.

May, C. 2002. *A Global Political Economy of Intellectual Property Rights*. London: Routledge.

May, Christopher, and Susan Sell. 2006. *Intellectual Property Rights: A Critical History*. London: Lynne Rienner Publisher.

Mayer, Martin. 2000. "Does Science Push Technology? Patents Citing Scientific Literature." *Research Policy* 29: 409–434.

Mayer, Maximilian, Mariana Carpes, and Ruth Knoblich (eds.). 2014. *The Global Politics of Science and Technology*. Berlin and Heidelberg: Springer-Verlag.

Merges, Robert P., and Richard R. Nelson. 1990. "On the Complex Economy of Patent Scope." *Columbia Law Review* 90: 839–916.

Mokyr, Joel. 2002. *The Gifts of Athena: Historical Origins of the Knowledge Economy*. Princeton, NJ: Princeton University Press.

Moser, Petra. 2005. "How Do Patent Laws Influence Innovation? Evidence from Nineteenth-Century World's Fairs." *American Economic Review* 95: 1214–1236.

Muzaka, Valbona. 2011. *The Politics of Intellectual Property Rights and Access to Medicines*. Basingstoke: Palgrave Macmillan.

Nelson, Richard R. 1992. "What Is 'Commercial' and What Is 'Public'" About Technology, and What Should Be?" In *Technology and the Wealth of Nations*, ed. Nathan Rosenberg, Ralph Landau, and David C. Mowery, 57–71. Stanford, CA: Stanford University Press.

Nelson, Richard R., and Gavin Wright. 1992. "The Rise and Fall of American Technological Leadership: The Post-War Era in an Historical Perspective." *Journal of Economic Literature* 30: 1931–1964.

Odagiri, Hiroyuki, Akira Goto, Atsushi Sunami, and Richard R. Nelson (eds.). 2011. *Intellectual Property Rights, Development and Catch-Up: An International Comparative Study*. Oxford: Oxford University Press.

Parello, Carmelo P. 2008. "A North–South Model of Intellectual Property Rights Protection and Skill Accumulation." *Journal of Development Economics* 85: 253–281.

Park, Walter G. 2008. "International Patent Protection: 1960–2005." *Research Policy* 37: 761–766.

Patel, Pari, and Keith Pavitt. 1997. "The Technological Competences of the World's Largest Firms: Complex and Path-Dependent, But Not Much Variety." *Research Policy* 26: 141–156.

Pavitt, Keith. 1987. "On the Nature of Technology." Reprinted in Pavitt, *Technology, Management, and Systems of Innovation*. Cheltenham: Edward Elgar, 1999.

Penrose, Edith. 1950. *The Economics of the International Patent System*. Baltimore, MD: Johns Hopkins Press.

Penrose, Edith. 1973. "Patenting and the Less-Developed Countries." *Economic Journal* 83: 768–786.

Pianta, Mario. 1988. *New Technologies Across the Atlantic*. Hemel Hempstead: Harvester Wheatsheaf.

Polanyi, Michael. 1967. *The Tacit Dimension*. London: Routledge & Kegan Paul.

Poole, Jennifer P. 2013. "Knowledge Transfer from Multinational to Domestic Firms: Evidence from Worker Mobility." *Review of Economics and Statistics* 95: 393–406.

Ryan, Michael R. 1998. *Knowledge Diplomacy: Global Competition and the Politics of Intellectual Property*. Washington, DC: Brookings Institution Press.

Sell, Susan K. 2003. *Private Power, Public Law: The Globalization of Intellectual Property Rights*. Cambridge: Cambridge University Press.

Shiva, Vandana. 2001. *Protect or Plunder? Understanding Intellectual Property Right*. London: Zed Books.

Stopford, John, and Susan Strange. 1991. *Rival States, Rival Firms*. Cambridge: Cambridge University Press.

Strange, Susan. 1988. *States and Markets*. London: Pinter Publishers.

Sunder, Madhavi. 2011. "Bollywood/Hollywood." *Theoretical Inquiries in Law* 12: 275–308.

Wade, Robert. 2003. "What Strategies Are Viable for Developing Countries Today? The World Trade Organization and the Shrinking of Development Space." *Review of International Political Economy* 10: 621–644.

Wikipedia. 2014. Entries *Smartphone Wars* and *Apple Inc. v. Samsung Electronics Co., Ltd.*, consulted on August 27.

Winter, Sydney. 1989. "Patents in a Complex Context." In *Owning Scientific and Technical Information*, ed. Vivian Weil and John Snapper, 41–60. New Brunswick, NJ: Rutgers University Press.

White House. 2013. "Fact Sheet: White House Task Force on High-Tech Patent Issues." Released June 4, http://www.whitehouse.gov/the-press-office/2013/06/04/fact-sheet-white-house-task-force-high-tech-patent-issues (accessed January 6, 2015).

Part IV Global Institutions and Intellectual Property Rights

Chapter 21

Patents, Monopoly Power, and the Pricing of Pharmaceuticals in Low-Income Nations

F.M. Scherer

Introduction[1]

Invention patents are accorded particular importance in the pharmaceutical industry – a role both complicated and enhanced by the adoption of the TRIPS (Trade-Related Intellectual Property Rights) agreement under the Treaty of Marrakesh in 1995. Before TRIPS was adopted, many nations did not grant patents on pharmaceuticals, and especially on pharmaceutical products. TRIPS established deadlines for the universal granting of pharmaceutical product patents. Among 32 mostly high- and medium-income nations surveyed by Edson Kondo (1994: 62), 13 – Argentina, Brazil, Chile, Colombia, Ecuador, Greece, India, Mexico, Peru, Portugal, Spain, Thailand, and Venezuela – did not grant pharmaceutical product patents as of 1990. By 1996, the most highly developed World Trade Organization (WTO) members were expected to begin granting such patents, and the least developed nations were required to comply by 2006 (later extended to 2016). (International agency terminology favors using "least developed" to characterize "low-income" nations. In the context used here, "low-income" is more descriptive, even if not always politically correct.)

Patents are significant in pharmaceutical research and development for two main reasons.[2] First, focusing usually on a precisely defined chemical or biological molecule and its uses, they provide particularly clear and unambiguous property rights, less muddled by the ambiguities that plague mechanical, electrical, and communications apparatus patents. But second, numerous surveys have shown that having patent protection is particularly important in decisions to invest R&D resources to discover and prove the therapeutic efficacy of pharmaceutical molecules. In the first of several such studies, Taylor and Silberston (1973) asked industrial interviewees in England

The Handbook of Global Science, Technology, and Innovation, First Edition.
Edited by Daniele Archibugi and Andrea Filippetti.

what negative impact on their R&D expenditures a legal regime of requiring world-wide compulsory patent licensing at "reasonable" royalties would have.[3] The weighted average reduction for all industries was projected to be 8%, but pharmaceuticals was the outlier, with expected R&D cuts of 65%. There are two main reasons for this difference. For one, pharmaceutical R&D is particularly expensive, requiring discovery and testing outlays measured in the hundreds of millions of dollars to achieve a single new marketable molecule during the early years of the twenty-first century (see Scherer 2014). But second, much of the relevant expenditure is for what is best described as information, assumed to approximate unusually closely to a pure public good. Once a promising new molecule is synthesized and patented, its structure is there for all to see once patent applications are published. Moreover, much of the R&D that follows discovery is devoted to clinical trials that show whether the molecule is effective or not in alleviating disease. When regulatory authorities have validated that the molecule is indeed considered safe and efficacious, others could, absent patent or other regulatory protection, synthesize the identical drug for a few millions of dollars and enter competitive (generic) production. Although the original R&D pioneer would still enjoy reputational "first mover" advantages for its efforts (see Bond and Lean 1979), these might be insufficient to deter entry and allow recovery on average of sunk research, development, and testing costs. If so, incentives for original discovery could fail.

Thus, patents are unusually important in pharmaceuticals. The remaining questions are, how widespread geographically must the scope of patents be to maintain the incentives required to sustain a vigorous pace of discovery and testing? What tradeoffs are embraced when only some nations offer pharmaceutical patents? And since historically the nations with low to modest per capita incomes have also been among those not granting pharmaceutical product patent rights, how might the extension of those rights affect the availability of new medicines?

These questions have wide applicability, but they were particularly pertinent as the TRIPS agreement was being negotiated and then adopted during the early 1990s. Two phenomena gave the question unusual saliency then. For one, multinational pharmaceutical companies took a leading position in the advocacy, first in the United States and then on a united US-Japanese-European front, to secure a TRIPS requirement that patent protection no longer be denied on pharmaceutical products (see Ryan 1998). But second, these demands for increased patent protection coincided with a time when the world fully recognized a deadly new health scourge: HIV/AIDS. After the nature of the disease was first understood in the early 1980s, numerous new drugs were developed to combat it. Nearly all were protected by patents and sold at prices far in excess of their production costs. The key question arose, how should a firm that has a patent-protected monopoly on some life-saving therapy price that therapy across diverse world jurisdictions, some rich like the United States and Europe, some desperately poor (as in Africa, from which HIV/AIDS spread)?

To be sure, until the requirements of TRIPS are implemented by all nations in 2016 (or later, if there are further extensions), world-wide patent protection of new pharmaceuticals will be absent. But with some important exceptions that will be addressed later, the reach of pharmaceutical patents is extensive, in part because many nations offering no local patent protection lack the internal capabilities to produce new drugs and therefore are at least partly dependent upon sources in patent-granting jurisdictions.

Pricing New Drugs Under Monopoly Conditions

We begin with a normative question: What pricing strategies would one expect for a multinational drug company selling its products in a cross-section of markets directly or indirectly protected from competition by patents?

Figures 21.1a and 21.1b provide an introductory perspective. They assume that a particular patented drug is sold under essentially monopolistic conditions in two nations: wealthy nation A (e.g., the firm's home nation or similar nations) and low-income nation B.[4] Assuming for convenience similar population counts between the two nations, the demand curves for the drug nevertheless differ significantly because of an income effect that shifts the less affluent nation's demand curve downward and toward the origin. Only at very low prices is the quantity demanded in the low-income nation similar to that of the high-income nation. How should the drug provider set its prices? A well-established tradition in economics suggests that it should engage in what is often called Ramsey-Baumol-Bradford price discrimination (Baumol and Bradford 1970). Assuming identical marginal costs in both nations of $18 per prescription delivered and no leakage of product from the low-price to the high-price nation (i.e., no arbitrage or parallel trade), the firm maximizes its profits decentrally by setting a high price of say $58 per standard prescription unit in the rich nation and a low price of $26 per unit in the low-income nation. In this way it realizes a surplus of revenues over marginal costs of approximately $17.6 million per month from sales in the high-income nation and $1.6 million in the low-income nation. If on the other hand it were to levy the profit-maximizing rich-nation price of $58 per Rx in the low-income nation, it would under the assumed demand conditions sell nothing at all there, foregoing a contribution to profits (and to the recovery of its research and development investments) of $1.6 million per month. The logic is compelling.[5] Successful price discrimination (also called differential pricing, tiered pricing, or equity pricing) is more profitable when feasible than a single-price strategy. It also has another compelling feature reflecting other well-recognized advantages of price discrimination. When

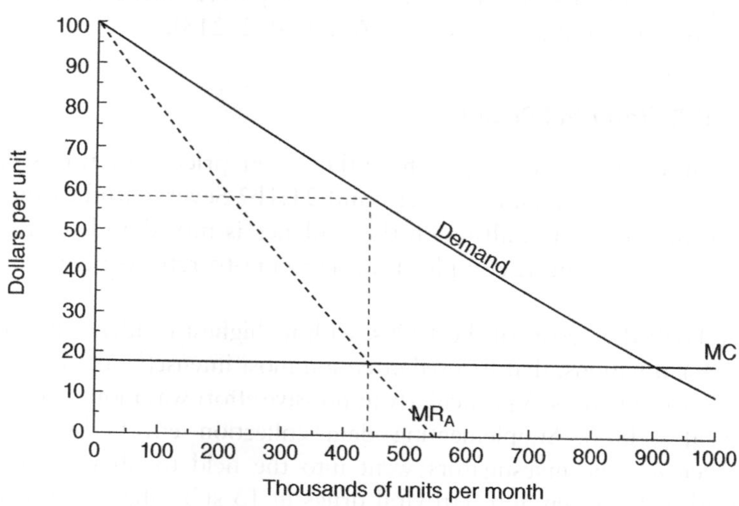

Figure 21.1a Price discrimination between markets of differing wealth. Wealthy nation A.

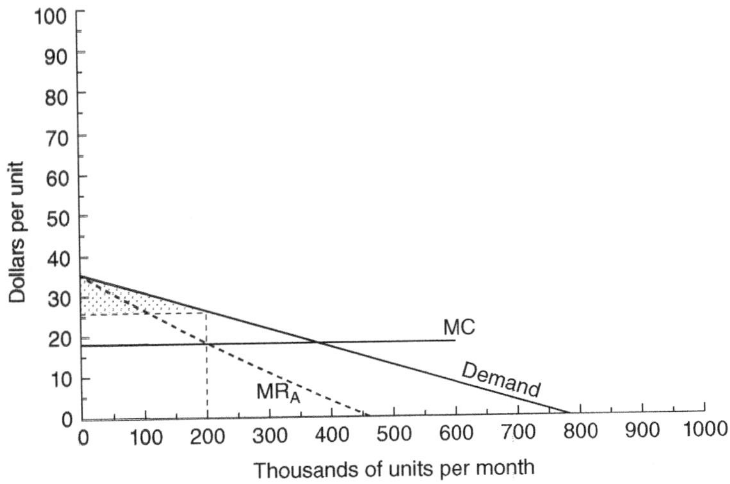

Figure 21.1b Price discrimination between markets of differing wealth. Low-income nation B.

a $26 price is set in the low-income nation, consumers with ability and willingness to pay more than that price realize a "consumers' surplus," shown in Figure 21.1b by the dot-shaded triangle above the $26 price line and below the demand curve. Under the near-linearity assumptions made, which probably underestimate the willingness to pay of especially affluent and needy consumers, setting prices well above the low-income nation's profit-maximizing discriminatory price causes consumers to forego a consumers' surplus of approximately $1.1 million per month.

Recognizing these advantages, a conference convened in Høsbjør, Norway during April 2001 by important but unusual partners – the World Health Organization (WHO) and the WTO – concluded: "Differential pricing can and should play an important role in ensuring access to existing essential drugs at affordable prices, especially in poor countries. In doing so, it could help to reconcile affordability with incentives for research and development" (Watal 2002: 218).

The Extent of Differential Pricing

Do pharmaceutical firms actually differentiate their prices in nations of diverse affluence, as suggested by Figures 21.1a and 21.1b? Several studies suggest that there is some such tendency, although the evidence is mixed and the reasons for price variation are much more complex than a pricing-to-reflect-income hypothesis can support.

It was the HIV/AIDS crisis of the 1990s, with its highest incidence in low-income African and Asian nations, that focused attention most intensely on the determinants of national price differences. A particularly impressive effort was mounted by Stéphane Lucchini et al. (2003). Supplementing data collection efforts by international healthcare agencies, the investigators went into the field to obtain point-of-entry transaction prices for seven anti-retroviral drugs in 13 sub-Saharan African nations plus Brazil. The most striking result of their study was a series of charts tracking the dramatic price decreases observed between 1997 and 2001 – on average from their regression equations (in their Table 3), by 82%, with an especially sharp decline in 2001. Among the variables they tested for explanatory power was GDP per capita in the target nations. Consistent with the differential pricing hypothesis, they found a positive relationship between prices and GDP per capita, but the relevant coefficient was not statistically significant. Its role, however, might have been blurred by the fact that their sample focused almost exclusively on low-income nations. Among other findings, prices tended to be systematically higher in sample nations with drug product patent protection and lower for drugs allowing significant generic competition.

Another study analyzing the prices of AIDS drugs in 18 intermediate and low-income nations, with benchmarking to prices in the much richer United States, was conducted by Scherer and Watal (2002a). Data were obtained from IMS, a pharmaceutical industry data compiler, on wholesale prices, usually charged to retailers, for 15 AIDS anti-retrovirals across the years 1995–1999. The sample differed from that of Lucchini et al. by emphasizing South and Central American nations, mostly with low to intermediate average GNP per capita, and only two sub-Saharan African nations. The main analyses focused on the ratio of observed prices in sample nations as compared to the wholesale prices of the same drugs in the United States. The central question was, were prices in the low- and medium-income nations systematically lower than those in the wealthy United States? The evidence of Ramsey-like discriminatory pricing was weak. On average, prices in the sample nations were 85% of reported US wholesale prices, which tended in turn to be 15–25% higher than actual US transaction prices. There was a weak positive correlation (+0.127) between the ratio of sample to US prices and GNP per capita in the surveyed nations, declining to essentially zero in the last covered year. In an unreported analysis of the original data, the US prices to which the sample nations' prices were compared were found in most (but not all) cases to have fallen substantially between 1995 and 1999, mirroring the similar price decreases observed by Lucchini et al. (2003) in sub-Saharan African nations.

Keith Maskus and Matthias Ganslandt (2002) approached the price discrimination question from a different perspective.[6] See also Danzon and Cho (2000), analyzing mainly high-income nations. The focus of Maskus and Ganslandt was on 20 typically well-known brand-name drugs, explicitly excluding AIDS anti-retrovirals, in 14 nations, of which only five fell into the low- or medium-income per capita category. They found only weak support for the income-correlated differential pricing hypothesis. Some lower-income nations did benefit from reduced average prices, but others paid even more than in the wealthier nations. Variables influencing the prices at which nations were supplied included the existence of national price controls, drug makers' strategies in targeting diverse income groups within national markets, the extent of competitive pressure from imports of the same drugs from lower-price jurisdictions, the way procurement efforts were organized, and the competitive efficiency of nations' wholesale distribution channels.

Reasons for the Paucity of Differential Pricing

There are several reasons why pharmaceutical prices have not been systematically adjusted to different nations' income per capita, reflecting ability to pay, under Ramsey-type pricing strategies.

One that has received particularly intense attention is so-called "parallel trade," that is, arbitrage transactions by middlemen who acquire pharmaceuticals (or other products) at sharply discounted prices in low-income nations and divert them from their intended recipients to the consumers of rich nations, reaping, needless to say, an arbitrageur's profit by doing so. During the 1990s, European Community authorities actively encouraged parallel trade in pharmaceuticals, hoping that the redirection of supplies from low-price Community member nations (most notably, Spain) to nations in which prices were higher would help perfect the Common Market (see Brittan 1992). A decade later, however, it was recognized that a policy suitable for encouraging parallel trade among rich nations was inappropriate to meeting the needs of consumers in low-income nations. The consensus report of the April 2001 WHO-WTO conference concluded that: "Markets for differentially priced drugs need to be tightly segmented to prevent leakage of differentially priced drugs to high-income markets. ... High-income countries may need additional legal authority in order to prevent the import of products marketed elsewhere at differential prices" (Watal 2002: 226–227).

In an apparent response to the conference sponsors' urging, European national customs officers soon thereafter executed a well-publicized confiscation of drugs reimported from low-income nations back into European markets.[7] More systematic policies were subsequently adopted to encourage "tiered pricing" and to discourage the parallel importation of low-priced drugs into Europe from less developed nations (see European Commission 2002). Among other things, it was recognized that low-price shipments to the least developed nations should have distinctive packaging.

A possibly more serious obstacle to the differential pricing of drugs comes from the formal price control systems some wealthy nations use to prevent the makers of patented drugs from exploiting their full pricing power. This is the so-called system of "external reference pricing," under which national price regulators set the maximum allowable price for a drug to approximate the level of prices observed in several "reference" nations. Intricate international reference price networks have been mapped (see Watal 2002: 229, reproducing a chart by Ed Schoonveld). The available evidence suggests that the least developed nations have seldom been included among the reference nations. However, this could change, and if the low prices offered under income-correlated differential pricing were incorporated in reference-based price control systems, the rich-nation price-reducing effect could substantially impair multinational pharmaceutical companies' incentives to offer drugs in low-income nations at prices approaching marginal cost.

This disincentive could also work without formal reference pricing control systems. If drug X is sold at $65 per 30 capsules in a rich nation and only $15 in a low-income nation, procurement authorities in the rich nation are likely to say to the supplier, "See here. You're only charging $15 for this drug elsewhere. You're discriminating against us. We demand that you reduce your prices here too." The multinational pharmaceutical manufacturer may be forced to concede, in which case, its incentive to offer the drug at low prices in less developed nations is undermined.

Apart from the linkages among rich nations in systems of pharmaceutical pricing, the segmentation of consumers within a particular low-income nation market can also lead to conscious drug manufacturer strategies favoring relatively high prices. Income inequality is pervasive, within rich nations and poor. Many low-income nations (South Africa comes to mind) can with some oversimplification be divided into two groups: a rich minority often employed by multinational enterprises at high wages and, very importantly, enjoying generous health insurance; and the vast majority who toil at low wages and lack formal health insurance plans. Figure 21.2 illustrates the elemental economic theory. It shows a national market in which consumers can be segmented into two groups: sub-market 1, comprising the rich, well-insured consumers (corresponding to the aggregate of consumers in Figure 21.1a) and another sub-market 2 in which are clustered the less affluent and poorly-insured potential consumers.[8]

In this case, the combined demand curve for both classes of consumers is kinked, with demand in the less affluent sub-market D_2 being added to demand in sub-market 1 only below the "choke price" of $35 per unit in the lower-income sub-market. If the supplier is able to engage in price discrimination across the two sub-markets, its profit-maximizing strategy must include ensuring that the marginal revenues in the two sub-markets are equalized. Otherwise a reallocation of output from the lower-MR to the higher-MR sub-market is profitable. Under the assumed circumstances, the supplier will compute its combined marginal revenue function (with a kink where the two sub-market MR functions intersect). It then sets a price of $58 in the rich segment and $32 in the low-income segment. However, within a single national entity, arbitrage is much more likely than across geographically separated nations, and so the effort at differential pricing may fail. In that case, the pharmaceutical supplier faces two alternative single-price options. It can try to serve all customers at a price of $32, or it can confine its sales to the rich consumers comprising sub-market 1 at $58. The surplus of revenues over marginal costs is approximately $4.6 million per month if both consumer classes are served at a uniform price. If sales are directed only toward affluent consumers, profits are $6.6 million per month. If arbitrage is probable and successful market segmentation unlikely, the producer will be inclined to charge a high price and exclude low-income consumers, whose "choke price" of $35 precludes their participation in the high-price sub-market. Among other things, the high-price strategy sacrifices substantial consumer surpluses for both rich and poor consumers even as it maximizes producer profits.

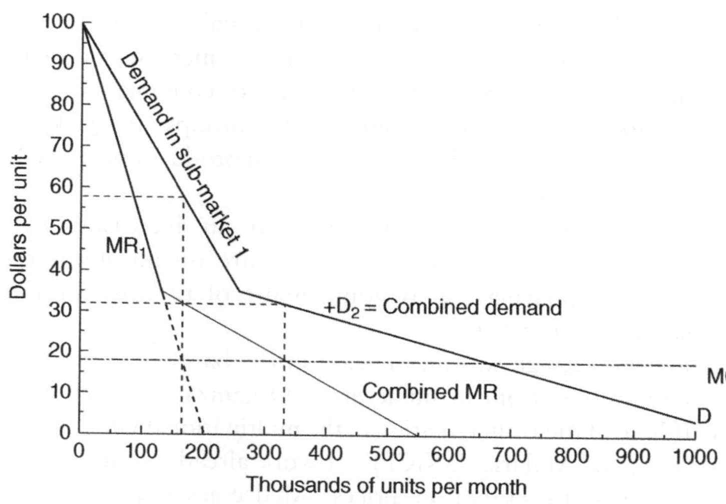

Figure 21.2 Market segmentation within a national market.

A complication overlooked in the analysis by Scherer and Watal (2002a) was articulated by Maskus and Ganslandt (2002: 68–69 and 78–79). Especially in the least developed nations, but also in China,[9] distribution channels linking pharmaceutical manufacturers to eventual consumers are often inefficient and infused with monopoly elements. In many sub-Saharan African nations, exclusive franchises are awarded, often as the result of political corruption, to drug wholesalers and sometimes to private companies that receive drugs at the initial importation node and then relay them to wholesalers. Competition at the drug retailing level is also limited. Given their monopoly positions, middlemen charge extraordinarily high prices for their services, raising the prices ultimately paid by retail consumers. The more monopolistic stages pyramided between the original product source and the consumer, the more the ultimate quantity supplied is likely to be restricted, all else being equal.[10] One consequence is that price concessions offered by manufacturers can be dissipated and hence blurred in the analysis of prices paid by retailers and consumers. This blurring is more likely in analyses that use data sources such as IMS, which focuses its efforts on the prices paid to wholesalers by retailers, than in studies such as that of Lucchini et al. (2003), who ventured into the field to obtain original source pricing data.

The Ultimate Triumph of Differential Pricing

Despite these problems, needy consumers in low-income nations ultimately became the beneficiaries of greatly reduced prices for life-saving drugs and vaccines, especially during the 1990s and the first decade of the twenty-first century. For example, the cost of providing three-drug therapy to persons afflicted with HIV/AIDS in the poorest nations fell from approximately $15,000 for a year's treatment in 2001 to an average of $127 in 2012.[11] These price reductions made possible an increase in the number of low- and middle-income nation citizens treated with anti-retrovirals from the low hundreds of thousands in 2001 to an estimated 10 million in 2012, along with an appreciable decrease in the number of persons dying from AIDS following peak mortality levels in 2005.[12] Similar price reduction trends can be observed for vaccines (beginning in an earlier decade) and drugs targeted inter alia against malaria and tuberculosis.

Several developments led to these changes. For one, increases in the cumulative and ongoing quantity of drugs procured probably led to economies of scale, although detailed recent information on cost-volume relationships are lacking. On the substantial cost reductions achieved in early penicillin production, see Federal Trade Commission (1958: 162–163).

Second, life-saving drugs became available to an increasing extent from government clinics and from international healthcare organizations rather than through private-sector pharmacies, avoiding many of the distribution channel inefficiencies characterized earlier.

Accompanying this change was an increase in the bargaining power of entities purchasing essential drugs from manufacturers. Organizations such as the WHO, UNAIDS (an affiliate of the United Nations), the multinational Global Fund to Fight AIDS, Tuberculosis, and Malaria, UNICEF (at work already in the 1990s to obtain and distribute vaccines at rock-bottom prices), Médecines Sans Frontières, Oxfam, the Gates Foundation, the Clinton Foundation, and PEPFAR (the US President's

Emergency Plan for AIDS Relief) brought to bear both moral and large-purchaser pressure on pharmaceutical manufacturers and then ensured that the purchased drugs reached utilization points expeditiously (see Kapstein and Busby 2013).[13]

By no means all of the dramatic reductions in AIDS drug prices came from price discrimination favoring low-income nations. Moral suasion and large-buyer pressure also had an impact in the rich nations that were home to research-based pharmaceutical companies. In the United States, for example, AZT (original brand name Retrovir), the first drug shown to be effective against AIDS, was sold in 1987 at prices amounting to approximately $10,000 for a year's treatment of a single patient. A patient advocacy group, the ACT UP (AIDS) coalition, disseminated information on the large disparity between estimated production costs and prices and mounted highly visible protests. Its activity stimulated investigatory hearings in the US Congress, eventually leading to substantial domestic market AZT price reductions. Similar pressures were exerted with respect to the numerous AIDS drugs reaching the market during the ensuing decade and a half.

Differentially, however, the producers of drugs with patent protection were induced to reduce prices dramatically in less affluent markets not only through appeal to the logic of Ramsey-type price discrimination, but also by the stern discipline of competition. Some nations, we have seen earlier, did not issue patents for pharmaceutical products at the time the TRIPS agreement was signed in 1995. Among these were India and Brazil. Although it was not required to do so at an early date, Brazil extended patent protection for pharmaceutical products shortly after the ratification of TRIPS. India chose not to do so until 2005. Even for Brazil, however, only the newest pharmaceutical products were covered by patents. Older molecules remained patent-free while protection continued in the nations that had earlier coverage. This made it possible for Brazilian companies not bound by patent rights to begin competitive generic supply and (more importantly) for Brazilian health authorities to draw upon Indian suppliers at greatly reduced prices. The multinationals were required either to follow suit or confine themselves to severely reduced market shares when they tried to maintain high prices in no-patent jurisdictions.

The history of India's role in these developments is particularly interesting. The chronicle begins much earlier in Europe, however.

Until 1978, Italy's laws expressly excluded pharmaceutical products from patent protection. In 1978, however, the Italian Supreme Court ruled in a law suit brought by both multinational and some Italian pharmaceutical firms that this exclusion was unconstitutional. The Italian parliament passed a law implementing the court's decision in 1982, and from that time on, new pharmaceutical products tended to receive Italian patent protection. Before that change, Italy was home to the world's leading export-oriented generic pharmaceutical industry. With no domestic patent barriers, Italian firms were able to supply new drugs competitively in their home market. But in so doing they also enjoyed a first-mover advantage in supplying those products to other nations (including Greece, Spain, and Argentina) that did not grant product patents. As the almost certain consequence of these 1978–1982 legal changes, Italy's balance of trade in pharmaceuticals turned sharply negative beginning in 1982 as the export advantage of its domestic producers faded, while pharmaceutical R&D in Italy rose at an annual rate below multinational company trends (see Weisburst and Scherer 1995; Challu 1995).

India's industry rose to fill the gap. It became the world's leading generic drug supplier. The AIDS crisis amplified that role. As the twenty-first century dawned, anti-retroviral prices in most of the world, including severely impacted African nations, were so high as to be unaffordable for most AIDS sufferers. With the encouragement of William Haddad, president of the US Generic Pharmaceutical Manufacturers Association, Médecins Sans Frontières, and an organization created by US consumer activist Ralph Nader, Indian firms, led by generic specialists Cipla and Ranbaxy, began exporting substantial quantities of anti-retrovirals and other important drugs to Africa and some southeast Asian nations at prices as low as two percent of those quoted by multinational patent holders.[14] Again, the consequence was a dramatic decrease in drug prices for low-income nations and indeed the emergence of even greater cross-nation price differences than those implied by the theory of differential pricing.

Compulsory Patent Licensing

The TRIPS treaty requirement that recalcitrant nations begin issuing patents on pharmaceutical products and other previously excluded subject matter, immediately for so-called developed nations and with delays for less developed nations, was not the sole determinant as to whether generic competition could emerge. The original TRIPS agreement included language permitting national governments to "march in" and negate otherwise binding patent rights through so-called compulsory licensing decrees under certain specified circumstances. Under Article 31, compulsory licensing was permitted when a would-be patent user is unsuccessful within a reasonable period of time in obtaining from a patent holder the right (i.e., a license) to supply competitively a patented invention "on reasonable commercial terms and conditions" – for example, with the payment of what are called (but hard to quantify) "reasonable royalties." The failed negotiations condition could be waived in cases of national emergency or extreme urgency or for non-commercial public use.[15] Subparagraph 31(k) also allowed compulsory licensing to correct "anti-competitive practices," which were spelled out more fully in Article 40. The combination of subparagraph 31(k) and Article 40 appears to track in a general way the history of the United States, where, especially between 1939 and 1956 and more sporadically thereafter, tens of thousands of US patents were subjected to compulsory licensing in full or partial settlement of antitrust complaints (see Hollabaugh and Wright 1960).

The exact legal procedures through which compulsory licenses can be issued under TRIPS were left unspecified, and publicity on their actual issuance is sparse. It is known that licenses were issued on at least seven pharmaceutical patents by Thailand, one by Brazil (with other threats settled by voluntary licensing), and one (with others under consideration) by India. The more important function of Article 31 may be to induce price reductions or the issue of voluntary patent licenses without the formal declaration of a compulsory license. In 2001, for example, both the United States and Canada threatened compulsory licensing of Bayer AG's patent on Ciproflaxin when, it was feared, terrorist activity might trigger an epidemic of otherwise untreatable anthrax. In that case, however, substantial price reductions were forthcoming and formal licensing was averted.

South Africa, with some of the most severe AIDS incidence, implicitly used the provisions of Articles 31 and 40 (plus its own competition laws) to secure licenses without carrying the action all the way to a formal declaration. By 2003, it had become known that a so-called "triple therapy," including three different anti-retroviral drugs, was the most effective way to treat AIDS. Three-drug therapy was more effective in abetting the frequent mutations that could render individual therapeutic molecules impotent, and combining three molecules in one twice-daily pill was a superior way to ensure daily compliance, again reducing the danger of mutation. But patents covering one of the drugs (AZT) were held by GlaxoSmithKline (successor to Burroughs-Wellcome) and those for two other key ingredients were held by the German firm Boehringer-Ingelheim. The two firms declined to cross-license each other, so no single-pill therapy was available. An action by the South African Competition Commission aided by CPTECH, an offspring of Ralph Nader's US consumer advocacy organization, induced the firms to offer what were in effect compulsory licenses, first to a South African generic supplier, Aspen Pharmacare, and then to foreign (e.g., Indian) suppliers (see Competition Commission of South Africa 2004).[16] New triple therapies became available at unprecedentedly low prices.

The original text of TRIPS Article 31 specified that the compulsory licenses it authorized had to be "predominantly for the supply of the domestic market."[17] This posed a difficulty. Many less developed nations lack the technological capability to produce advanced organic pharmaceutical molecules. A license issued only domestically for them would be of little use. For a supplier in India, on the other hand, the "domestic market" language could be a bar to licensing and then exporting in majority quantities to other needy markets. This issue was addressed at the 2001 WTO conference in Doha, leading to a declaration that: "We recognize that WTO members with insufficient or no manufacturing capacities in the pharmaceutical sector could face difficulties in making effective use of compulsory licensing under the TRIPS Agreement."

A definitive solution was postponed, but in 2003 the WTO formally declared that border-hopping licenses could be issued "in good faith to protect public health."[18] The WTO's Doha declaration also stated that "Each Member has the right to grant compulsory licenses and the freedom to determine the grounds upon which such licenses are granted" and that "Each member has the right to determine what constitutes a national emergency or other circumstances of extreme urgency, it being understood that public health crises, including those relating to HIV/AIDS, tuberculosis, malaria and other epidemics, can represent a national emergency or other circumstances of extreme urgency."

Many of the most advanced pharmaceuticals originate not in company laboratories but through basic research pursued by university scientists. Since universities lack the capabilities needed to manage large-scale clinical tests, manufacture new drugs, and distribute them to users, universities in the United States since at least passage of the Bayh–Dole Act (94 Stat. 3019, 1980) have granted typically exclusive licenses for relevant new drug entity patents to commercial firms. The practice attracted public scrutiny in 2001. Yale University licensed a patent on the anti-retroviral molecule stavudine (d4T) to Bristol-Myers-Squibb, whose high prices on third-world sales of its branded product Zerit stimulated student protests. Yale officials thereupon brought pressure to bear on Bristol-Myers to reduce its third-world prices drastically and to allow other companies to enter into competitive manufacturing overseas. This experience in turn led to a multi-university concord encouraging patent licensing strategies that "make affordable versions [of drugs, vaccines, and medical diagnostic tools] available in resource-limited countries" (see Universities Allied for Essential Medicines 2007).

Consequences for Research and Development

Dramatic changes have occurred in the supply of advanced pharmaceutical products since the TRIPS agreement was adopted in 1995. Nations that once chose not to do so are gradually being required to issue pharmaceutical product patents covering bona fide technological achievements. Numerous national and international organizations have been reoriented or formed de novo to bargain for low prices in procuring patented pharmaceuticals directed to low-income nations when suppliers were disinclined for diverse reasons not to engage in differential pricing voluntarily. The TRIPS agreement initially carved out exceptions to its strong pro-patent requirements with compulsory license clauses and then adopted further extensions. All this has led to huge decreases in the prices paid for life-saving drugs (if payment has been required at all) to the world's least advantaged citizens.

The key remaining question is, how has downward pressure on third-world pharmaceutical product prices affected company incentives to invest in research and the development of new drugs? As a first approximation, measures that limit the profits obtainable through pharmaceutical innovations also reduce incentives to invest in bringing new drugs onto the market (although the advent of generic competition against already marketed drugs also spurs R&D on more advanced replacements). Difficult tradeoffs must be faced. For drugs targeted toward diseases at least as prevalent if not more prevalent in rich nations, providing weaker patent protection and facilitating tougher competition in low-income nations undoubtedly has a small, even if not vanishing, negative impact. A deeper analysis (Scherer 2004) of the tradeoff suggests the following generalizations:

- The smaller is the ratio of attainable potential quasi-rents (roughly, profits before deduction of fixed costs) in low-income nations relative to those in wealthy nations, the stronger is the case for weak patents in low-income nations.
- The more the marginal utility of income diminishes with increased affluence, the stronger is the case on total economic welfare grounds for weak patents in low-income nations.
- The case for weak patents in low-income nations is stronger when R&D is a virtuous rent-seeking process, with R&D expenditures rising competitively to dissipate supra-normal profit opportunities, than when a prospect theory of R&D investment applies (i.e., when some companies hold preferred and relatively imitation-proof ex ante positions in developing a new product).[19]
- The case for weak patents in low-income nations is strengthened when indigenous firms build their technological capabilities through imitative (i.e., generic) production that can then serve as a platform for advancing to higher levels of productivity and innovativeness (mirroring the Japanese, South Korean, and most recently, Chinese economic development histories).

A special case of the first condition exists for so-called "tropical diseases" – that is, diseases endemic preponderantly in low-income nations and rare in highly developed nations. Malaria and dengue fever are examples. Then nearly all of the quasi-rents attainable through successful R&D come from low-income nations. The expectation of such rents may be so meager that R&D investments will not be made even under the most favorable market conditions. See Médecins Sans Frontières (2001), which found that among 1393 new drug chemical entities introduced into world markets between 1975 and 1999, only 13–15 were for so-called "tropical diseases." But in borderline cases, limiting the strength of patent rights is likely to weaken incentives even more. If so, some form of public-spirited philanthropy by private organizations or governments must fill the gap. The gap-filling problem can be addressed through subsidies on either the supply side or the demand side. On the supply side, governments (or even pharmaceutical companies, in their philanthropic modes) can support research, development, and testing of therapies for diseases neglected under normal market incentives. Or on the demand side, advance commitments can be made to purchase at subsidized prices new drugs and vaccines effective against neglected tropical diseases. Levine et al. (2005) propose in detail an "advance market commitment" initiative targeting HIV/AIDS, malaria, and tuberculosis vaccines that was considered favorably by G-8 government leaders at plenary meetings between 2005 and 2008 but never fully implemented. If such philanthropy could be successful, the conflict between technological progress and maximum consumer access to its fruits would be reduced.

Notes

1 The author is indebted to Jayashree Watal, Jamie Love, Daniele Archibugi, and Calestous Juma for helpful comments.

2 Some national laws also attain patent-like results by specifying minimum periods of market exclusivity before generic competition is permitted, e.g., 12 years for so-called "biosimilars" in the United States. Limits on the uses of clinical test data by would-be imitators have similar effects.

3 For a survey of later studies yielding similar insights, see Scherer (2009a: 171–175).

4 This analysis follows Scherer and Watal (2002a). The demand function for wealthy nation A is $P = 100 - .1\ Q + .00001\ Q^2$; for the low-income nation $P = 35 - .045\ Q + .00001\ Q^2$.

5 For extensions to differently shaped demand curves but retaining the assumption of an "income effect," see Scherer and Watal (2002b).

6 See also Danzon and Chao (2000), which focuses mainly (with the exception of India) on high-income nations and has no test of how income per capita differences affected prices.

7 See "Nearly $18M in Discounted AIDS Drugs Allocated for Africa Diverted by Wholesalers and Sold on European Markets," *Kaiser Health News*, October 3, 2002, 1.

8 The demand function for Market 1 is $P = 100 - .25\ Q$; the demand function for the low-income is identical to the low-income nation demand function assumed in Figure 21.1b.

9 See Li and Reimers (2009), Huang and Yang (2009), and Scherer (2009b).

10 These propositions were initially proved mathematically in a famous work by Augustin Cournot in 1838 and a much less well-known monograph by Charles Ellet Jr. in 1839. The phenomenon was recognized in less rigorous form by US Treasury Secretary Alexander Hamilton in *Federalist Paper* No. 22 (December 1787). High tariffs on drugs imported into the least developed nations also had a negative impact, albeit without a pyramiding effect.

11 "A New Approach to Solicitations for a Troubled AIDS Charity," *New York Times*, July 10, 2012, D6.

12 "The 30 Years War," *The Economist*, June 4, 2011, 89–91; and "A Dispatch from the Front," *The Economist*, September 28, 2013, 79.

13 Noticeably absent from the history is collaboration among the least developed nations to leverage their purchasing power collectively.

14 See "Selling Cheap 'Generic' Drugs, India's Copycats Irk Industry," *New York Times*, December 1, 2000, 1 and A12. The article includes a photo of Yusuf Hamied, chairman of Cipla and a key participant in the efforts to involve Indian generic suppliers in world supply networks.

15 Under US law, the US government was since at least 1952 permitted to infringe valid patents for government use. In the early 1960s, the US Department of Defense was procuring patented tetracycline from Italian generic suppliers. The practice was prohibited by a special act of Congress.

16 See also "Agreement Expands Generic Drugs in South Africa to Fight AIDS," *New York Times*, December 11, 2003, A24.

17 "Predominantly" was apparently interpreted informally to mean that a majority of sales were domestic.

18 Decision of the WTO General Counsel, August 30, 2003 (WT/L/540), amended in WT/L/641; and "WTO to Let Poor Nations Import Generic Drugs," *Philadelphia Inquirer*, August 31, 2003. In 2005 the compromise was adopted as a permanent amendment to TRIPS.

19 On the prospect theory, see Kitch (1977) and Nordhaus (1969). On rent-seeking, see McFetridge and Rafiquzzaman (1986). For a synthesis see Scherer (2010: 560–571).

References

Baumol, William J., and David Bradford. 1970. "Optimal Departures from Marginal Cost Pricing." *American Economic Review* 60: 265–283.

Bond, Ronald, and David Lean. 1979. *Sales, Promotion, and Product Differentiation in Two Prescription Drug Markets*. Washington, DC: US Federal Trade Commission.

Brittan, Sir Leon. 1982. "Making a Reality of the Single Market: Pharmaceutical Pricing." Speech before the IEA Health and Welfare Unit, London.

Challu, Pablo. 1995. "Effects of the Monopolistic Pricing of Medicines in Italy Since 1978." *International Journal of Technology Management* 10: 237–250.

Competition Commission of South Africa. 2004. "GSK and BI Issue Anti-Retroviral Licenses." *Competition Commission Newsletter* 15: 1–2.

Danzon, Patricia, and Li-Wei Cho. 2000. "Cross-National Price Differences for Pharmaceuticals: How Large, and Why?" *Journal of Health Economics* 19: 159–195.

European Commission. 2002. "Tiered Pricing for Medicines Exported to Developing Countries: Measures to Prevent Their Re-importation into the EC Market." DG Trade, working document, April 22.

Federal Trade Commission. 1958. *Economic Report on Antibiotics Manufacture*. Washington, DC: Federal Trade Commission.

Hollabaugh, Marcus A., and Robert Wright. 1960. *Compulsory Licensing under Antitrust Decrees*. Washington, DC: US Senate Committee on the Judiciary, Subcommittee on Patents, Trademarks, and Copyrights.

Huang, Yanfen, and Yiyong Yang. 2009. "Pharmaceutical Pricing in China." In *Prescribing Cultures and Pharmaceutical Policy in the Asia-Pacific*, ed. Karen Eggleston, 205–222. Stanford, CA: Shorenstein Asia-Pacific Research Center.

Kapstein, Ethan B., and Joshua W. Busby. 2013. *AIDS Drugs for All: Social Movements and Market Transformations*. Cambridge: Cambridge University Press.

Kitch, Edmund W. 1977. "The Nature and Function of the Patent System." *Journal of Law & Economics* 20: 265–290.

Kondo, Edson. 1994. "*Patent Laws and Foreign Direct Investment: An Empirical Investigation.*" PhD dissertation, Harvard University.

Levine, Ruth et al. 2005. *Making Markets for Vaccines: Ideas to Action*. Washington, DC: Center for Global Development.

Li, Mingzhi, and Kai Reimers. 2009. "Streamlining China's Pharmaceutical Distribution Channels." In *Prescribing Cultures and Pharmaceutical Policy in the Asia-Pacific*, ed. Karen Eggleston, 183–196. Stanford, CA: Shorenstein Asia-Pacific Research Center.

Lucchini, Stéphane, Boubou Cisse, Ségolène Duran, Marie de Cenival, Caroline Comite, Marion Gaudry, and Jean-Paul Moatti. 2003. "Decrease in Prices of Antiretroviral Drugs for Developing Nations." In *Economics of AIDS and Access to HIV/AIDS Care in Developing Countries*, ed. Jean-Paul Moatti et al., 170–211. Paris: International AIDS Economics Network.

Maskus, Keith, and Matthias Ganslandt. 2002. "Parallel Trade in Pharmaceutical Products: Implications for Procuring Medicines in Poor Countries." In *The Economics of Essential Medicines*, ed. Brigitte Granville, 57–80. London: Royal Institute of International Affairs.

McFetridge, Donald G., and M. Rafiquzzaman. 1986. "The Scope and Duration of the Patent Right and the Nature of Research Rivalry." *Research in Law and Economics* 8: 91–129.

Médecins Sans Frontières. 2001. *Fatal Imbalance: The Crisis in Research and Development for Neglected Diseases*. Geneva: MSF Access to Essential Medicines Campaign.

Nordhaus, William D. 1969. *Invention, Growth and Welfare*. Cambridge, MA: MIT Press.

Ryan, Michael P. 1998. *Knowledge Diplomacy: Global Competition and the Politics of Intellectual Property*. Washington, DC: Brookings Institution.

Scherer, F.M. 2004. "A Note on Global Welfare in Pharmaceutical Patenting." *The World Economy*, 27: 1127–1142.

Scherer, F.M. 2009a. "The Political Economy of Patent Policy Reform in the United States." *Journal on Telecommunications & High-Technology Law* 7: 167–216.

Scherer, F.M. 2009b. "Conclusion: A Critical Review of the Dynamics and Regulation of Asian Pharmaceutical Industries." In *Prescribing Cultures and Pharmaceutical Policy in the Asia-Pacific*, ed. Karen Eggleston, 367–386. Stanford, CA: Shorenstein Asia-Pacific Research Center.

Scherer, F.M. 2010. "Pharmaceutical Innovation." In *Handbook: Economics of Innovation*, ed. Bronwyn Hall and Nathan Rosenberg, 539–576. Amsterdam: North-Holland.

Scherer, F.M. 2014. "R&D Costs and Productivity in Biopharmaceuticals." In *Encyclopedia of Health Economics*, ed. Anthony Culyer and Patricia Danzon. Oxford: Elsevier Science.

Scherer, F.M., and Jayashree Watal. 2002a. "Post TRIPS Options for Access to Patented Medicines in Developing Nations." *Journal of International Economic Law* 5: 913–939.

Scherer, F.M., and Jayashree Watal. 2002b. "The Economics of TRIPS Options for Access to Essential Medicines." In *The Economics of Essential Medicines*, ed. Brigitte Granville, 32–56. London: Royal Institute of International Affairs.

Taylor, C.T., and Z.A. Silberston. 1973. *The Economic Impact of the Patent System: A Study of the British Experience*. Cambridge: Cambridge University Press.

Universities Allied for Essential Medicines. 2007. "Global Access Licensing Framework." http://uaem.org/global-access-licensing-framework/ (accessed January 6, 2015).

Watal, Jayashree. 2002. "Differential Pricing and the Financing of Essential Drugs." In *The Economics of Essential Medicines*, ed. Brigitte Granville, 209–231. London: Royal Institute of International Affairs.

Weisburst, Sandy, and F.M. Scherer. 1995. "Economic Effects of Strengthening Pharmaceutical Patent Protection in Italy." *International Review of Industrial Property and Copyright Law* 26: 1009–1024.

Chapter 22

Global Governance and Intangible Cultural Heritage in the Information Society: At the Crossroads of IPRs and Innovation

Paolo Davide Farah and Riccardo Tremolada

Introduction[1]

Globalization is the common word we use to describe the phenomenon of a contemporary world which is more and more interconnected and has no borders. Globalization is the result of a combination of factors. The role of technology is a fundamental feature in shaping what globalization is and what it means for society as a whole. The improvement of telecommunications, in particular Internet and the fast-growing development of transportation, has reduced production costs related to the circulation of goods and services. This allows multinational corporations and their supply chains to integrate assets located in different countries and centralize their production in low-cost regions. However, globalization is not only trade in goods. The world of the nation-states has to tackle problems which are no longer internal within the borders. Everything has become global, not only trade.

Challenges and risks for society such as environmental crisis, energy security, or terrorism are not local; private companies are no longer national, they are multinational, sometimes with annual turnovers higher than the GDP of certain developing countries, and they influence the life of the population of the states in which they operate and beyond.

At the same time, it has to be taken into account that there are also important societal values which are directly or indirectly affected by the global expansion of world trade and should be upheld to balance the excesses of globalization. Globalization without control risks becoming only standardization and unification; globalization without local concerns can endanger relevant issues such as good governance, human rights, right to water, rights to food, social, economic, and cultural rights, labor rights, access to knowledge, public health, social welfare,

The Handbook of Global Science, Technology, and Innovation, First Edition.
Edited by Daniele Archibugi and Andrea Filippetti.

consumer interests and animal welfare, climate change, energy, environmental protection and sustainable development, product safety, food safety and security.

There are several sets of rules and regulations implemented at the global level that are relevant to boosting certain objectives for society as a whole, like rewarding research and development, supporting creativity and innovation, but that do not always consider the needs of the developing countries which often did not have the opportunity to contribute to the discussion which led to the adoption of those international rules. Those regulations do not always fit with a developing country's societal needs and local characteristics, as will be described when we utilize intangible cultural heritage as a case study for our analysis.

Endeavors to protect and preserve intangible cultural heritage, against the risks of globalization, have been influenced by the emergence of the Information Society. Consequentially there have been various attempts to develop schemes recommending that heritage be archived, withdrawn from the public domain, and handed back to the exclusive control of its alleged inventors. This paper aims to investigate the relationship between intangible cultural heritage protection, intellectual property rights (IPRs) regime, and the Information Society, assessing some of the major issues at stake resulting from the interaction of, and conflict between, different evolving regimes.

Global Governance and the Excesses of Globalization: Preliminary Considerations

From Seattle in 1999 to "Occupy Wall Street" in 2011, civil society has continued to express a strong dismay and anger toward human, social, and environmental consequences of the global expansion of world trade and of the monetary and commercial translation of all interpersonal transactions. However, since the beginning of the industrial revolution, the success of the capitalist mode of production seems to be marked by its results in terms of demographic, economic, and technological development. Between 1810 and 2010, if the total income per capita is multiplied by nine, and the world population by six, and then the pace of technological innovation has grown exponentially.

However, we need to bear in mind two important caveats regarding the effectiveness and legitimacy of the industrial mode of production of goods and the commercial mode of the distribution of wealth in their global expansion. The first caveat is that the global development of the industrial mode of production of goods caused a disruption in the balance that regulates the interaction between man and the environment: climate change (Cossy and Marceau 2009), depletion of fisheries, soil erosion, and pollution are the flip side of the global triumph of the industrial revolution. As stated in the report *Limits to Growth* prepared by the Club of Rome in 1972, and as reformulated more recently in the UN Millennium Ecosystem Assessment (2005), "Human activity is putting such strain on the natural functions of Earth that the ability of the planet's ecosystems to sustain future generations can no longer be taken for granted." The fact that the current mode of development is not sustainable means that its continuation will soon result in its own end: human development has exhausted the Earth's production, absorption, and recycling capacity. The second caveat is that the global development of commercial wealth distribution has created a paradoxical increase in inequalities: "A steady trend of increasing inequality in the world economy between rich and poor economies has taken place since 1820 and throughout the 20th century, the exception being the 'golden age' period of 1950–1973" (Solimano 2001).

Therefore, the idea of limiting the excesses of globalization while controlling the international institutions that create them, such as the International Monetary Fund (IMF) or the World Trade Organization (WTO), may be, to a certain extent, justified. As the Director-General of the WTO Pascal Lamy once stated: "for better and for worse, globalization of the issues increases, on a daily basis, the need to organize democratic, global forms of governance that are both legitimate and efficient. In other words: democratic" (Lamy 2004). In fact, for many years, the General Agreement on Tariffs and Trade (GATT) and the WTO have been highly criticized and portrayed as the least transparent and democratic of all international organizations. Indeed, most of the meetings and debates are held privately with no public access and agreements are reached at the intergovernmental level without any public participation. The lack of participation of civil society in the ongoing negotiation rounds at the WTO affects the rights of all parties to both information and participation, as was made clear by the 1998 Aarhus Convention (UNECE 1998), Article 21 of the Universal Declaration of Human Rights (UN 1948), as well as Principle 10 of the Rio Declaration on Environment and Development (UNEP 1992). However, the WTO has recently begun to open up to civil society, both through the participation of Non-Governmental Organizations (NGOs) in the plenary sessions of the most important ministerial conferences, and through the establishment of a contact group dedicated to NGOs within the WTO secretariat. Additionally, the WTO website allows one to access a wide range of highly detailed information on trade issues and relevant committees. Moreover, the Appellate Body of the WTO has recently authorized the submission of *amicus curiae* briefs as well as the opening of public hearings (Marceau and Stilwell 2001). Thus, as responsibilities of trade institutions grow, the role of civil society in the institutions that govern international trade such as the WTO is ever increasing (Ullrich 2002; Tuerk 2003).

Hence, as noted by Thomas Zweifel, the right approach is not to oppose global sovereignty of certain international organizations, but rather to propose concrete reforms that would make them purely liberally oriented:

> There are three possible reactions to this democratic deficit of international institutions. First we can decry how undemocratic international institutions are and urge caution in their use. Second, we can reject international organizations on the grounds that democracy and international relations are two incompatible concepts ... Hence the third response: we can decry the lack of democracy of international institutions, but push for their reform and democratization. (Zweifel 2006).

To provide such institutions with the democratic legitimacy they currently lack, in addition to the increasing democratic transparency of their decisions, it is mostly the outer frame of analysis which should be amended (Hilf and Goettsche 2003; Howse 2003).

According to Ernst-Ulrich Petersmann, human rights should be used as a regulatory framework of the global expansion of WTO rules:

> In order to remain democratically acceptable, global integration law (e.g. in the WTO) must pursue not only "economic efficiency" but also "democratic legitimacy" and "social justice" as defined by human rights [Marceau and Doussin 2009]. Otherwise citizens will rightly challenge the democratic and social legitimacy of integration law if it pursues economic welfare without regard to human rights ... (Petersmann 2002)

However, the difficulty and the limits of this approach lie in the fact that it affirms both that human rights should guide the process of global legal integration and that such process should be implemented by the WTO itself: "As freedom for hunger and economic welfare are preconditions for the enjoyment of many other human rights, the WTO guarantees of economic liberties and of welfare-increasing cooperation across frontiers serve important human rights functions" (Petersmann 2000). As noted by Philip Alston (2002), saying that the WTO law itself guarantees the respect of fundamental human rights, implies a refusal to evaluate the business practices of organizations such as the WTO or the IMF in terms of these same rights.

Indeed, WTO rules and principles not only prevent Members from discriminating against other Members in general, but also severely restrain their ability to adopt trade measures against another Member whose practices do not respect human rights. Likewise, the accession to the WTO is not subject to any criterion related to the respect of human rights and its rules require that each Member, once part of the Organization, can enjoy the same commercial benefits as the others, thus preventing Member States from discrimination based on the protection of human rights. Nevertheless, the GATT has an exception, which allows Members to adopt measures that restrict trade in goods manufactured by prisoners. Another interesting example is the waiver recognized for all those countries which are parties to the Kimberley Process. The Kimberly Process is designed to certify the origin of "rough diamonds" (diamonds which are unworked or simply sawn, cleaved, or bruted: Kaplan 2003) from sources which are free from conflict fueled by diamond production (i.e., are not "blood diamonds": Feldman 2003). Although, thus far, there has been no discussion on human rights within the WTO, some of its provisions might be interpreted so as to give Members the opportunity to pursue such objectives. For example, the exceptions set out in GATT Article XX, in particular the exception on public morals, could provide a viable framework for action. On the other hand, the recent trend of signing regional trade agreements where states often choose restrictively with whom they wish to engage in such a partnership can render human rights more important in trade policy decisions and have consequences in terms of legal empowerment.

Part of this debate also includes the idea of replacing the notion of Non-Trade Concerns (NTCs). As noted by James R. Simpson and Thomas J. Schoenbaum, the integration of NTCs within the WTO decision-making process should add a regulatory reference to certain non-economic values and fundamental rights to the economic criteria that have been defining globalization so far:

> The excesses of globalization can be managed only by paying adequate attention to what are termed "Non-Trade Concerns" ... A balance must be struck in the WTO between liberalization on the one hand, and NTCs on the other. The economic dimension

> of trade must be balanced with non-economic values. Simply ignoring NTCs is not an option. In our considered opinion, if this is done, the Doha negotiations will fail. (Simpson and Schoenbaum 2003)

From this perspective, the success of the Doha Round will depend on the ability to bring together all the different partners in joint discussions on non-economic parameters.

Indeed, the right to development and other associated rights, such as the right to food, shelter, water, and so on, are all to some extent directly affected by international trade. International trade law, both within and outside the WTO, determines how global trade evolves and limits the ways in which trade policy can be used to encourage domestic policies, which, in third-world countries, allow greater legal empowerment (Scott 2002; Breining-Kaufmann 2005; Tamiotti 2009; Baxi 2009; Bates 2010). Allowing a more substantial participation of developing countries within the WTO and promoting the opening up of trade and custom barriers, for example for agricultural products, might have an extremely important impact on the issues connected to NTCs. Moreover, the rights of disadvantaged farmers in developing countries could be positively – though indirectly – affected. This is precisely why Simpson and Schoenbaum call for the expansion of the concept of NTCs to the dissemination and production of agricultural products, qualifying them global public goods:

> It seems reasonable that all citizens can, and should, have an interest in assuring that they have an input into deciding how, and in what form, the public goods related to agriculture are used and maintained for future generations. That is one of every person's fundamental rights. (Simpson and Schoenbaum 2003).

With regard to NTCs, the main issue is that developing countries do not have the same concerns as developed ones. When a country is still facing problems related to basic healthcare and hygiene, less resources will be left to address issues such as animal welfare or food safety. Developing countries will try to obtain a high level of protection of NTCs primarily in those areas where the same level of protection is granted by industrialized countries whose market they wish to access. The key challenge is how to satisfy the rights of developed nations to grant social values the degree of protection they consider appropriate, while minimizing the negative effects in terms of market distortion for their trading partners. This is exactly where the WTO can prove very useful. For example, the regulation of investments and protection of IPRs on products and production processes, as well as issues related to patents derived from traditional knowledge or Chinese traditional medicine, fall within both the international legal system (Agreement on Trade-Related Aspects of Intellectual Property Rights (TRIPS), Patent Cooperation Treaty) and the European one (European Patent Convention).

Definitional Issues: Information Society, Culture, and Social Involvement

The "intangible cultural heritage" (e.g., stories, designs, musical forms, and information) suggests a further inclusive management of traditional cultural expressions, while on the other hand it also generates crucial difficulties for policymakers. The studies on intangible cultural heritage not only have led commentators to reshape the concept of property, but have also underlined the relevance of the current wide notion of *information*, encompassing the practical, political, and moral impact of its proposed regulation in the context of the current Information Society, stemming from the new information technologies' pervasive presence and role in our society (Brown 2005). Information presents some peculiar features, such as its capacity to ubiquity, which carries the danger of erasing the line between real or authentic and counterfeit, just as it weakens the links that connect units of information together in meaningful regimes.

This study aims at assessing the relationship between intangible cultural heritage and Information Society, identifying potential strategies for the protection of this peculiar type of cultural expression. As we approach this topic, it is indeed necessary to perceive "information" in an holistic way, framing it as a system of mutually influencing ties and intertwined forces. The current Information Society comprehensively manages, or at least seeks to manage, complex systems involving different variables, whose modifications uncertainly reflect and affect the whole (Davenport 1997; Samuelson and Davis 2000). This extremely interconnected and multifaceted nature entails that intangible aspects of cultural heritage require a wide range of options of protection, going well beyond the typical IPRs instrument.

As for intangible cultural heritage, it is challenging to find a widely accepted definition. From an international law perspective, the definition in the UNESCO Convention for the Safeguarding of the Intangible Cultural Heritage adopted in 2003 (hereafter, ICH Convention) is descriptive and very broad. According to Article 2(1), intangible cultural heritage "means the practices, representations, expressions, knowledge, skills – as well as the instruments, objects, artifacts and cultural spaces associated therewith – that communities, groups and, in some cases, individuals recognize as part of their cultural heritage" (UNESCO 2003). Moreover, Article 2(2) of the ICH Convention sets forth a non-exhaustive list of the domains in which intangible cultural heritage is manifested: oral traditions; performing arts; social practices, rituals, and festive events; knowledge and practices concerning nature and the universe; and traditional craftsmanship.

In this regard, it is also crucial to define the multifaceted term "culture." From an anthropological perspective, this term has a polysemantic value, that is, the fact that it may mean many things depending on different backgrounds and points of view (Arizpe 2002). For example, according to the definition adopted in the UNESCO Universal Declaration on Cultural Diversity, culture is defined as "the set of distinctive spiritual, material, intellectual and emotional features of society or a social group, and all it encompasses, in addition to art and literature, lifestyles, ways of living together, value systems, traditions and beliefs" (Arizpe 2002). In relation to intangible cultural heritage, two classes of culture should be taken into consideration. On the one side, traditional culture, describing the cultural practices which a social group acquired from the past through intergenerational passages (even if these are recent inventions) and to which the group assigns a clear status; on the other side, popular culture, alluding to those cultural performances, often commercialized, through which a sub-group of a society manifests its unique identity. As Lourdes Arizpe observes, since every member of a sub-cultural group can have an impact on the practices of the groups in which they belong, the origin of all intangible cultural elements is to be found in the intrinsic capacity of human beings able "to create original meanings and imaginaries that build social practices and representations"

(2002). Intangible cultural heritage relates primarily to processes implying the gathering of people belonging to a cultural group (a village, a group, or a nation) who engender an intangible cultural performance in a cultural happening which may be more relevant than the event itself. What really matters is therefore the repetition of a continuous process of social involvement through its remarkable expressions such as "story-telling, myths or songs or other such time-capturing inventions" (2002).

The manifestations of "intangible cultural heritage," as defined in the ICH Convention, allow one to take into account also relevant economic assets of the communities or individuals who generate, implement, and preserve them. Hence, intangible cultural heritage and specific objects deriving from such heritage can be the contents of trade. Indeed, under Article 2, what is to be protected are not merely intangible aspects as such but also tangible items originating from intangible cultural heritage. From an anthropological point of view, the participation of people of the community to such events activates societal bonds. As underlined by Arizpe:

> this "activation" of bonds has a crucial role in updating the representation of the community in the eyes of all its members even if they are living elsewhere. Intangible heritage, then, in terms of the processes it involves, helps keep otherwise invisible bonds alive and updated among the members of a community. (2002)

Consequentially the distinction between tangible and intangible cultural heritage is not clear-cut and the two concepts may sometime overlap. According to the terminology used by the World Intellectual Property Organization (WIPO), intangible cultural heritage embraces traditional knowledge and traditional cultural expressions or expressions of folklore, thus including knowledge itself, which embraces know-how, skills, innovations, practices, and knowledge comprising traditional lifestyles as well as distinctive signs and symbols related to traditional knowledge. Traditional cultural expressions encompass phonetic or verbal expressions, such as stories, narratives, signs, and names; musical or sound expressions; expressions by action, such as dances, plays, ceremonies, rituals, and performances, whether fixed or unfixed; and material expressions of art, such as handicrafts (WIPO 2011). In fact, it is crucial to prevent commercial misappropriation and to guarantee that the commercial use of intangible cultural heritage does not misconstrue its connotation, significance, and aim for the group concerned, as also highlighted in the Operational Directives for the Implementation of the ICH Convention adopted by the General Assembly of the States Parties to the Convention at its second ordinary session in 2008 and amended at its third session in 2010 (para. 117).

Development of the Information Society

Commentators (Brown 2005) assessing the necessity of protection for intangible cultural heritage have highlighted the influence of various social forces, in particular, starting from the 1990s, the ascent of what has been variously defined the "Information Society," "Information Age," "Knowledge Economy," or, most recently, "Network Society." These terms refer to new information technologies which have deeply changed global society just as thoroughly as the industrial revolution transformed the nineteenth-century world. The Information Society is characterized by various factors: the rise of new economies relying on technology-based diffused knowledge; globalization of multinational companies and public institutions, leading to a reduction of the power stemming from traditional nation-states; and the shaping of new networks that refashion labor relations, politics, and economic activity (Castells 2000).

Among the positive effects of the modern Information Society deriving from extreme and multifaceted globalization, there is the rapid emergence of IPRs as a worldwide concern, as well as the ability of advocacy groups to gather international support for heritage protection. Local cultural integrity, languages, values, and traditional expressions of culture are being superseded by cultural influences originating principally in the Western developed world. In this vein, Brown observes that the local perspective is abandoned and a global approach is taken, demanding regulation of movement of alien images and information, in the interests of "local cultural integrity." The drawbacks of the Information Society in relation to cultural heritage are nevertheless numerous. In particular, the Information Society tends to menace the current outset of norms and institutions, while emphasizing the relevance of values and moral commitments, that is, culture. Simon Harrison (1999) remarks that in this context culture may become a "scarce resource to be defended as another form of property, either personal or collective." Consequentially cultural heritage and expressions of culture are remodeled into a highly politicized commodity (Harrison 1999; Lash 2002).

All the while, the growing economic value of information has amplified the value of IPRs as the global market economy greatly relies on IPR protection regimes. IPRs have benefited from the Information Society in the sense that globalization and the global market are continually trying to extend IPRs' scope in time, through industry's constant lobby to stretch the terms of copyright and patent protection, and in space, through widely accepted agreements, such as the TRIPS Agreement within the WTO framework (WTO 1994; see Filippetti and Archibugi, Chapter 20 and Scherer, Chapter 21 in this volume). Conversely, new challenges and complexity derive from the peculiarities of the Information Society and related technology. For instance, the fields of software engineering and biotechnology have enlarged the application of IPRs beyond the traditional areas, asserting IPRs in domains which encompass gene sequences, life forms, and manipulation of information in databases. These social changes and forces provoking these developments must be assessed vis-à-vis the role of the Information Society with respect to intangible cultural heritage and its protection, taking into account political, cultural, and economic aspects of the matter and adopting a holistic approach.

IPRs and Intangible Cultural Heritage: Options for Protection and the Commodification of Intangible Cultural Property

The protection of intangible cultural heritage must take into account the inherent dynamic nature of the subject matter. Although different forms of protection for cultural expression and knowledge have been envisaged since the late 1800s, the first international multilateral framework on this issue was not attained until October 2003, with the adoption by the UNESCO General Conference of the Convention for the Safeguarding of the Intangible Cultural Heritage. In the effort to evaluate the desirability of using IPRs for the protection of intangible cultural heritage, significant conflicts arise between these two domains (Le Goff 1997; Grosheide and Brinkhof 2002; von Lewinski 2004; Farah and Tremolada 2014a, 2014b). The clash between IPRs and intangible cultural heritage derives from the fact that the former aims at protecting the proprietary rights and economic interests of individuals (human or corporate), whereas the latter relies on preservation of the common heritage of a specific community or group. In this vein, scholars, indigenous communities, minority groups, NGOs, and UNESCO have fought for achieving a reform of the current intellectual property regime, stressing the different rationale for protection of intangible cultural heritage. Hence, two compelling solicitations on intangible cultural heritage protection have arisen: first, the solicitation for acknowledgement of the rights of intangible cultural heritage holders associated with their traditional knowledge, and, second, concerns about the unauthorized acquisition and access by third parties of IPRs over intangible cultural heritage. On this point, two kinds of protection have been complementarily elaborated and implemented:

- Positive protection, which empowers the intangible cultural heritage holders to take action or seek remedies against certain forms of misuse by obtaining and asserting rights in the protected material; and
- Defensive protection, aiming at preventing illegitimate and adverse IPRs being claimed by others over intangible cultural heritage matters.

As pointed out by various scholars (e.g., Karjala and Kirkwood Paterson 2003), one of the side effects of affirmation of traditional IPRs in this context is the consequent commodification of intangible cultural property, that is, the translation of intangible cultural property into articles of economic worth that can be exchanged for commercial profit by such means as licenses, rental, or sale. It is crucial to assess whether this resultant reification of intangible cultural heritage is desirable vis-à-vis the aim of granting protection to collective cultural expressions. Their conversion into property has been considered exploitation and commercialization, and consequentially misappropriation, of reified intangible cultural heritage. Meanwhile, the commodification of intangible cultural property by third parties other than their legitimate custodians has been exacerbated by the dissemination of intangible cultural property through digital technology and the Internet. Generally referring to traditional cultural expressions, Mira Burri Nenova (2007) points out: "With the digitization of content and existence of the Internet as a tool for dissemination, the risk of misappropriation has increased dramatically now that any tourist can use his or her phone to photograph and record a traditional cultural expression." She further observes: "exploiting the captured traditional cultural expression commercially can be accomplished quickly [by third parties], thereby usurping the chance to use traditional cultural expression as an asset of economic development."

Analyzing this issue, it seems more appropriate that, in order to provide protection for intangible cultural heritage, it is necessary not to recognize new IPRs. Rather, it is crucial to campaign for a reinterpretation of existing legal regimes concerning specific aspects of the existing IPR framework, such as privacy and unfair competition law, aiming at leveling what might differently be perceived as an unfair playing field (Farah and Tremolada 2014a, 2014b). This approach allows the claimants not to frame their legal rights in relation to existing classes of IPRs, thus eluding the charges of misuse and reification that have clung to such claims in the past. In fact, there have been several calls for differentiated and more coherent legal protection of the cultural heritage. At the same time, it has been stressed how the extension of IPRs over cultural heritage would generate a number of problems related to the very core of democratic conception of free speech and free expression, as carried out in both patent and copyright notions of public domain.

The perils affecting intangible cultural heritage and its expressions cannot be underestimated. It is sufficient to recall issues such as the employment of indigenous knowledge to generate a commercial product, such as a valuable drug, or the use by outsiders of tribal names or other identifiers, sacred symbols or images, or artistic designs generally (Burri Nenova 2007). If the community where the heritage generates intends to maintain secrecy in its rituals, employing unfair means to gain information about them breaches the underlying concept of privacy. Another hurdle pertains to the use of these names, symbols, or designs by authentic members of the community but in ways that are manifestly in conflict with the community's traditional use. Additionally, individuals external to the community may be confused as to whether an item is authentic, negatively affecting the community's ability to profit from commercial sales. Finally, a related issue is the "disturbance of an embedded landscape" (Karjala and Kirkwood Paterson 2003) in which cultural heritage is so intrinsically connected to nature that it cannot be used without either severing it from its authentic environment or reducing the usefulness of the heritage itself. In this vein, Karjala and Kirkwood Paterson note:

> from this notion it follows that local ecological knowledge should belong to the community as a whole and be considered inalienable. The sensitivity of these embedded landscapes argues for the right of indigenous people to restrict research on indigenous knowledge or biological resources where integrity of natural or cultural patrimony is threatened. (2003)

Experience in the context of protection of intangible cultural heritage has revealed that no single all-embracing, umbrella solution which fits with all the needs of communities in all countries is likely to satisfy all issues that such protection targets. Conversely, valid and serviceable protection should imply a plethora of varied and manifold options for protection, reinforced by an internationally recognized range of core principles. The outlining of these objectives in the international arena would frame the protection regime at the domestic level as well, allowing for ad hoc implementation and greater flexibility vis-à-vis the diverse needs of intangible cultural heritage holders from different countries.

The protection of intangible cultural heritage and related expression is today a multifaceted matter, whose complexity is exacerbated by lack of consensus on the definition of the subject matter, the *raison d'être* for protection, and the methods for achieving it. Nevertheless, this complexity should not lead to an impasse in trying to frame and outline some form of protection. Evaluating the existing IPR regimes, one can note their deficiencies as well as their general inappropriateness in relation to the claims for greater legal protection in respect of intangible cultural heritage. IPRs seem to be an unsatisfactory foundation on which to build a viable cultural heritage legal edifice. Rather than try to fit the justifiable claims of traditional communities into legal property rights categories that were not designed to accommodate their essential characteristics, our proposal is to focus on those aspects of communities' claims that can be addressed outside the IPR regimes of patent and copyright. We have found that traditional concepts of Western law, privacy, trade secrets, trademark, and contracts can take us a long way in the desired direction. At the same time, it is undeniable that the current IPR system can be relevant for the protection of some expressions of intangible cultural heritage, although this alone is not sufficient. It is thus crucial to elaborate a sui generis system of protection capable of meeting the heterogenic needs of respective communities, each with their own peculiarities, cultures, and resources, and to create legal rights that not only protect against the perceived abuse of cultural heritages, but that also protect, at least in principle, anyone who can satisfy its requirements. Cultural heritage rights developed on this basis will have enhanced credibility and compatibility with existing property rights and liability systems.

Moreover, some communities may oppose any commodification (or reification) of their cultural heritage, and thus perceive the application of IPRs over their intangible cultural heritage as inferring such commodification. Some commentators warn that the assertion of property concepts in intellectual property to intangible cultural heritage may prompt a transformation even of the social relations underlying the creativity processes in traditional communities where the heritage is generated. The extreme cases are arguments such as that IPR claims by individuals could deeply modify the relations of exchange and reciprocity vital to the communal genesis of intangible cultural heritage (Anderson 2009).

Taking a middle course, an alternative future framework could be expected, where specific types of IPR or sui generis protection might be shaped to furnish some means of legal protection for cultural heritage and related expressions, while also strengthening underlying traditional practices and social relations. In order to pursue this scenario, the International Institute for Environment and Development remarks that the protection of intangible cultural heritage would need to embrace various aspects of the communities where the heritage is created, and be grounded in a more integrated comprehensive approach, also taking into consideration customary principles and practices of those communities (Swiderska et al. 2009). Intangible cultural heritage results from social relations, and its protection must take into account that it represents a process which is subject to continuing evolution, stemming from heterogenic considerations and aspirations in human development, and pursuing a plethora of societal functions.

Tort of Cultural Misappropriation in the Context of the Information Society

The application, interpretation, and enforcement of protection against misappropriation of traditional knowledge, along with determination of equitable sharing and distribution of benefits, should be conducted with appreciation for the customary practices, laws, norms, and understandings of the holder of the heritage, comprising the sacred, spiritual, or ceremonial features of the traditional genesis of the heritage as influenced and modified by the inner peculiarities of the current Information Society.

A typical feature of the Information Society, as Michael F. Brown remarks, is its tendency to carve out the performative content, that is, images, sounds, practices, and information, from their original context and relocate them elsewhere. This phenomenon, generally referred to as "cultural appropriation," has triggered a debate addressing the movement of cultural elements from the politically weak to the politically strong actor on the global stage. Conversely, appropriation of cultural expressions by the weak (e.g., the manifest violation of copyright on popular music and film widespread in the developing countries) is either overlooked altogether or praised as an act of cultural resistance (Coombe 1998). Cultural appropriation is detrimental and disrespectful of the cultural values of the source community where the heritage originates, which rarely has sanctioned the imitation of its cultural expressions by outsiders. Moreover, it exposes that social group to material harm, not granting its legitimate economic gain, or by menacing shared understandings crucial to its social health.

Cultural expression has been targeted by misappropriation, modern-day cases of which encompass, inter alia, the employment by non-indigenous people of native symbols, songs, dances, words, and other forms of cultural expression (Coombe 1998). Cultural misappropriation is thus far beyond deprivation of mere economic gain, representing rather a sort of human rights abuse or, at least, an offense to the community's self-respect and identity. The tort of misappropriation is centered on economic rights and losses, whereas cultural appropriation claims go beyond these objectives and cannot be contended with merely economic remedies. In this vein Carlos Correa proposed a misappropriation regime which would allow national laws to determine appropriate measures to avoid misappropriation (including the obligation to stop using knowledge or to pay compensation for such use). In his view, "this regime should have three important points: documentation of traditional knowledge, proof of origin or materials, and prior informed consent" (Correa 2002).

Implicit support to protect against misappropriation is found in two United Nations documents: Decision V/16 of the CBD's Conference of the Parties, and the Principles and Guidelines for the Protection of the Heritage of Indigenous Peoples. Specifically, Paragraph 17 of Decision V/16 of the CBD's Conference of the Parties states:

> Request[ed] Parties to support the development of registers of traditional knowledge, innovations and practices of indigenous and local communities embodying traditional lifestyles relevant for the conservation and sustainable use of biological diversity through participatory programmes and consultations with indigenous and local communities, taking into account strengthening legislation, customary practices and traditional systems of resource management, such as the protection of traditional knowledge against unauthorized use.

On the other hand, the Principles and Guidelines for the Protection of the Heritage of Indigenous Peoples, elaborated in 1995 by the Special Rapporteur of the former UN Sub-Commission on the Prevention of Discrimination and Protection of Minorities, provides in paragraphs 26 and 27 the following:

> National laws should deny to any person or corporation the right to obtain patent, copyright or other legal protection for any element of indigenous peoples' heritage without adequate documentation of the free and informed consent of the traditional owners to an arrangement for the sharing of ownership, control, use and benefits. National laws should ensure the labelling and correct attribution of indigenous peoples'

artistic, literary and cultural works whenever they are offered for public display or sale. Attribution should be in the form of a trademark or an appellation of origin, authorized by the peoples or communities concerned.

Similarly, the WIPO (2011) draft provisions for the protection of traditional knowledge contain a provision on protection against misappropriation. Article 1 (2) (Protection against Misappropriation) states:

> Traditional knowledge shall be protected against misappropriation. Any acquisition, appropriation or utilization of traditional knowledge by unfair or illicit means constitutes an act of misappropriation. Misappropriation may also include deriving commercial benefit from the acquisition, appropriation or utilization of traditional knowledge when the person using that knowledge knows, or is negligent in failing to know, that it was acquired or appropriated by unfair means; and other commercial activities contrary to honest practices that gain inequitable benefit from traditional knowledge.

Intangible cultural heritage holders should also be safeguarded vis-à-vis other acts of unfair competition, inter alia, those named in Article 10-*bis* of the Paris Convention for the Protection of Industrial Property, such as false or misleading representations that a product, service, or expression is supplied with the approval of traditional cultural heritage holders, or that the commercial exploitation of products, services, or cultural expressions profits holders of cultural heritage (Bodenhausen 1968). Proper attention must be given to acts susceptible of creating confusion with a product or service of intangible cultural heritage holders, as well as to false allegations in the course of trade which discredit the products or services of traditional knowledge holders.

In the light of the above, while respect is particularly hard to impose by legislative measures, legal safeguards can be developed to foster protection for minorities from evident discrimination. Hence, analyses of cultural appropriation focus on the issue of material damages and promotion of legal reforms aiming at compensating communities for such offences and, when possible, at preventing them in the first place. Redressing civil wrongs related to intangible cultural heritage raises further issues concerning intellectual property, tort law, and debate on the notion of cultural ownership (Wüger 2004).

The Information Society Conundrum and the Convention for the Safeguarding of Intangible Cultural Heritage

The ICH Convention represents a fundamental effort toward the protection of intangible cultural heritage and related expressions. The Convention envisages a wide range of legal tools "aimed at ensuring the viability of intangible cultural heritage, including the identification, documentation, research, preservation, protection, promotion, enhancement, and transmission, particularly through formal and non-formal education, as well as the revitalization of the various aspects of such heritage" (Scovazzi 2010, 2012; Zagato 2008; Blake 2002; Lankarani 202). The ICH Convention is to be evaluated in the context of UNESCO's endeavors to promote global information democracy and to assure the right of sovereign nations to defend their cultures against unwanted external, and possibly harmful, influences.

One of the main flaws of this legal instrument is that its terminology, language, and envisaged measures are shaped on the very Information Society practices they are manifestly tackling and challenging. As Michael F. Brown observes, the convention "portrays intangible heritage as an objectified resource amenable to modern management techniques" (Brown 2005). He further argues that, in such a legalistic vision, heritage cannot be protected until it is thoroughly documented. Accordingly, the ICH Convention promotes the preparation of "one or more inventories of the intangible cultural heritage present in [each nation's] territory." This nevertheless requires a monumental documentation effort which seems unlikely to be feasible, in particular considering vast or outstandingly multi-ethnic states such as Peru, China, India, and Papua New Guinea. Assessing the efficacy of documentation in the preservation of intangible expression of cultural heritage, commentators agree on the fact that, in spite of the several national initiatives aiming at using electronic resources to address the loss of local cultural expression, numerous doubts arise on the desirability and adequacy of such "technocratic and top-down approaches." In particular, one shortcoming is the frailty or ephemeral nature of digital media forms, which continuously undergo change (Stille 2002). Furthermore, the Information Society and its major social force and tool, the Internet, advocate for a concept of knowledge that must be recorded by outsider individuals in order to preserve it from loss, in contrast with the protective and secrecy-oriented attitude of many traditional communities (Harding 2000; Brown 2003), who fear that documentation of their non-physical intangible heritage merely facilitates its exploitation, rather than protects it.

On the other hand, the WIPO has supported the technique of online "prior art databases" to counteract the drawbacks of the Information Society vis-à-vis the necessity to provide protection for intangible cultural heritage through IPRs (WIPO n.d.). The underlying rationale of the databases is to disclose and publish traditional pharmacopoeias in order to prevent patent applicants from claiming they have discovered something novel.

Concluding Remarks

Numerous doubts arise from our assessment of the relationship between the Information Society, intangible cultural heritage, and its potential protection through IPR regimes, either as currently constituted or as modified to create a sui generis system providing effective protection to this peculiar type of cultural expression. Commentators have striven to strike a balance between heritage as a resource for all of humanity and as something that properly belongs to, and remains controlled by, its communities of origin (e.g., Brown 2005). Modern policies generally consider cultural heritage to be common heritage of all humankind, while contrastingly attempting to subtract much of this heritage from the global commons and privatize it in the communities where it is generated, and where it may be regarded as communal property. Issues regarding the role of IPRs and their rationale come into play at this stage in the matter. The underlying policy grounds for conceding exclusive IPRs stems from the public goods issue, that is, that the unsuccessful attempt to safeguard intellectual creativity would result in fewer inventive and socially advisable works being created and disclosed to the public, since the initial effort required for producing cultural expressions is much greater than that for merely copying cultural expressions already existing. On the other hand, an overly strong or long protection would have a detrimental effect on the social availability of desirable works, because later authors and inventors depend and build upon works that have come before them (Arizpe 2002). The final objective of the IPR regime is to equilibrate these two forces, in order to take full advantage of the works made available to the public.

A public domain characterized by cultural internationalism can only thrive if minority cultures withstand cultural globalization and a finer equilibrium of openness and discretion is pursued. In this vein, Tyler Cowen defined the notion of the "paradox of diversity" thus: "[T]he world as a whole may be more diverse if some societies refuse to accept diversity as a value. Those cultures will continue to generate highly unique creations, given their status as cultural outliers" (Cowen 2002). Vice versa, generalized diversity may lead to further uniformity because of the cultural mixture that it ineluctably brings forth (Coombe 2003).

As we have shown in our analysis, the protection of intangible cultural heritage has mainly been centered on the role of IPRs, in particular focusing on the expansion of IPRs and sui generis regimes to traditional communities. The desirability and adequacy of the proposed solutions vary, and the complexity of this matter is increased by the difficulty faced when attempting to specify what amounts to intangible culture in the first place, especially in light of the continually changing content of heritage. Heritage protection and preservation should constitute a measure to promote societies in which minority communities have a voice in decisions about their future, and where they can attain the same prosperity available to everyone else. To this end, cultural heritage is as crucial to their well-being as public health, education, human rights, and self-government. The complex modern Information Society cannot avoid adopting a comprehensive approach vis-à-vis the multifaceted issues which characterize the matter of intangible cultural heritage, taking into account that the fragile equilibrium between technology innovation and cultural expression can be achieved by legal instruments, provided that the balance is continuously fine-tuned.

Note

1 This chapter is part of the results of the Research Team at University Institute of European Studies (IUSE) in Turin (Italy). The research leading to these results has received funding from the People Programme (Marie Curie Actions) of the European Union's Seventh Framework Programme (FP7/2007-2013) under REA grant agreement no. 317 767, acronym LIBEAC (2013–2016), entitled "Liberalism in Between Europe and China," coordinated by Université d'Aix-Marseille, CEPERC-UMR 7304 Epistémologie et Ergologie Comparatives (France), Work-package 2 and 4. The authors are grateful for the comments received on the draft from Daniele Archibugi, Andrea Filippetti, James Nafziger, Tullio Scovazzi, and Valentina Vadi. The article reflects the authors' views only and the usual disclaimer applies.

The article is the result of the joint research of the authors, as shown in the introduction and the concluding remarks. Paolo Davide Farah focused on the following sections: Global Governance and the Excesses of Globalization: Preliminary Considerations; Development of the Information Society; Tort of Cultural Misappropriation in the Context of the Information Society; The Information Society Conundrum and the Convention for the Safeguarding of Intangible Cultural Heritage; and Riccardo Tremolada focused on the following sections: Definitional Issues: Information Society, Culture, and Social Involvement; IPRs and Intangible Cultural Heritage: Options for Protection and the Commodification of Intangible Cultural Property. The respective comments and inputs are gratefully acknowledged. An extended version of this article is published in *West Virginia Law Review*, Vol. 118, Issue 1 (2015).

References

Alston, Philip. 2002. "Resisting the Merger and Acquisition of Human Rights by Trade Law: A Reply to Petersmann." *European Journal of International Law* 13(4): 815–884.

Anderson, Jane E. 2009. *Law, Knowledge, Culture: The Production of Indigenous Knowledge in Intellectual Property Law*. Cheltenham: Edward Elgar.

Arizpe, Lourdes. 2002. "Brief Analysis of the Concept of Intangible Cultural Heritage." In IIIrd Round Table of Ministers of Culture, "Intangible Cultural Heritage – A Mirror of Cultural Diversity," Istanbul, discussion guidelines. Geneva: UNESCO.

Bates, Rebecca. 2010. "The Road to the Well: An Evaluation of the Customary Right to Water." *Review of European Community and International Environmental Law* 19(3): 282–293.

Baxi, Upendra. 2009. "The Human Right to Water: Policies and Rights." In *Water and the Laws in India*, ed. Ramaswamy R. Iyer, 149–166. New Delhi: Sage.

Blake, Janer. 2002. *Developing a New Standard-Setting Instrument for the Safeguarding of Intangible Cultural Heritage: Elements for consideration*. Paris: UNESCO.

Bodenhausen, Georg Hendrik Christiaan. 1968. *Guide to the Application of the Paris Convention for the Protection of Industrial Property as Revised at Stockholm in 1967*. Geneva: World Intellectual Property Organization.

Breining-Kaufmann, Christine. 2005. "The Right to Food and Trade in Agriculture." In *Human Rights and International Trade*, ed. Thomas Cottier, Joost Pauwelyn, and Elisabeth Bürgi Bonanomi, 341–381. Oxford: Oxford University Press.

Brown, Michael F. 2003. *Who Owns Native Culture?* Cambridge, MA: Harvard University Press.

Brown, Michael F. 2005. "Heritage Trouble: Recent Work on the Protection of Intangible Cultural Property." *International Journal of Cultural Property* 12(1): 40–61.

Burri Nenova, Mira. 2007. *EC Electronic Communications and Competition Law*. London: Cameron May.

Castells, Manuel. 2000. "Materials for an Exploratory Theory of the Network Society." *British Journal of Sociology* 51(1): 5–24.

Coombe, Rosemary J. 1998. *The Cultural Life of Intellectual Properties: Authorship, Appropriation, and the Law*. Durham, NC: Duke University Press.

Coombe, Rosemary J. 2003. "Fear, Hope, and Longing for the Future of Authorship and a Revitalized Public Domain in Global Regimes of Intellectual Property." *De Paul Law Review* 53: 1171–1191.

Correa, Carlos M. 2002. *Protection and Promotion of Traditional Medicine: Implications for Public Health in Developing Countries*. Available at http://apps.who.int/medicinedocs/pdf/s4917e/s4917e.pdf (accessed January 7, 2015).

Cossy, Mireille, and Gabrielle Marceau. 2009. "Institutional Challenges to Enhance Policy Co-ordination – How WTO Rules Could Be Utilised to Meet Climate Objectives." In *International Trade Regulation and the Mitigation of Climate Change*, ed. Thomas Cottier, Olga Nartova, and Sadeq Z. Bigdeli, 371–394. Cambridge: Cambridge University Press.

Cowen, Tyler. 2002. *Creative Destruction: How Globalization Is Changing the World's Cultures*. Princeton, NJ: Princeton University Press.

Davenport, Thomas H. 1997. *Information Ecology: Mastering the Information and Knowledge Environment*. Oxford: Oxford University Press.

Farah, Paolo Davide, and Riccardo Tremolada. 2014a. "Desirability of Commodification of Intangible Cultural Heritage: The Unsatisfying Role of IPRs." *Transnational Dispute Management* 11(2).

Farah, Paolo Davide, and Riccardo Tremolada. 2014b. "Diritti di proprietà intellettuale, diritti umani e patrimonio culturale immateriale." *Rivista di diritto industriale* 2(1): 21–47.

Feldman, Daniel L. 2003. "Conflict Diamonds, International Trade Regulation, and the Nature of Law." *University of Pennsylvania Journal of International Economic Law* 24: 835–874.

Grosheide, Frederik Willem, and Jan J. Brinkhof. 2002. *Intellectual Property Law: Articles on Cultural Expressions and Indigenous Knowledge*. Antwerp: Intersentia.

Harding, Sarah K. 2000. "Cultural Secrecy and the Protection of Cultural Property." In *Topics in Cultural Resource Law*, 69–78. Washington, DC: Society for American Archaeology.

Harrison, Simon. 1999. "Identity as a Scarce Resource." *Social Anthropology, Marketing of Destinations Journal of Travel Research* 7(3): 3–14.

Hilf, Meinhard, and Goetz Goettsche. 2003. "The Relation of Economic and Non-Economic Principles in International Law." In *International Economic Governance and Non-Economic Concerns: New Challenges for the International Legal Order*, ed. Stefan Griller, 18–28. Vienna: Springer.

Howse, Robert. 2003. "How to Begin to Think about the 'Democratic Deficit' at the WTO." In *International Economic Governance and Non-Economic Concerns: New Challenges for the International Legal Order*, ed. Stefan Griller, 79–102. Vienna: Springer.

Kaplan, Marco. 2003. "Carats and Sticks: Pursuing War and Peace through the Diamond Trade." *New York University Journal International Law & Policy* 35: 559–617.

Karjala, Dennis S., and Robert Kirkwood Paterson. 2003. "Looking Beyond Intellectual Property in Resolving Protection of Intangible Cultural Heritage of Indigenous Peoples." *Cardozo Journal of International and Comparative Law* 11: 633.

Lamy, Pascal. 2004. *La démocratie-monde. Pour une autre gouvernance globale*. Paris, France: République des Idées. Editions du Seuil.

Lankarani, Leila. 2002. "L'avant-projet de Convention de l'UNESCO pour la sauvegarde du patrimoine culturel immatériel: évolution et interrogations." *Annuaire Français de Droit International* 48(1): 624–656.

Lash, Scott M. 2002. *Critique of Information*. London: Sage.

Le Goff, Jacques. 1997. *Patrimoine et passions identitaires: Actes des Entretiens du Patrimoine*. Paris: Fayard/Éditions du Patrimoine.

Marceau, Gabrielle, and Aline Doussin. 2009. "Le droit du commerce internationale et les droits fondamentaux et considérations sociales." *L'Observateur des Nations Unies* 27(2): 241–247.

Marceau, Gabriele, and Matthew Stilwell. 2001. "Practical Suggestions for the Administration of Amicus Curiae Briefs by WTO Adjudicating Bodies." *Journal of International Economic Law* 4(1): 155–187.

Paris Convention for the Protection of Industrial Property. 21 UST 1583; 828 UNTS 305.

Patent Cooperation Treaty. TIAS 8733; 28 UST 7645; 9 ILM 978 (1970).

Petersmann, Ernst-Ulrich. 2000. "From 'Negative' to 'Positive' Integration in the WTO: Time for 'Mainstreaming Human Rights' into WTO Law?" *Common Market Law Review* 37(6): 1363–1382.

Petersmann, Ernst-Ulrich. 2002. "Time for a United Nations 'Global Compact' for Integrating Human Rights into the Law of Worldwide Organizations: Lessons from European Integration." *European Journal of International Law* 13(3): 621–650.

Samuelson, P., and R. Davis. 2000. "The Digital Dilemma: A Perspective on Intellectual Property in the Information Age." Paper presented at the 28th Annual Telecommunications Policy Research Conference, Arlington, VA.

Scott, Joanne. 2002. "Integrating Environmental Concerns into International Economic Law." In *International Economic Governance and Non-Economic Concerns: New Challenges for the International Legal Order*, ed. Stefan Griller, 371–387. Vienna: Springer.

Scovazzi, Tullio. 2010. "La Convention pour la sauvegarde du patrimoine culturel immatériel." In *Democracy, Ecological Integrity and International Law*, ed. J. Ronald Engel, Laura Westra, and Klaus Bosselmann, 409–428. Cambridge: Cambridge Scholars Publishing.

Scovazzi, Tullio. 2012. "La Convenzione per la salvaguardia del patrimonio culturale intangibile." In *Il patrimonio culturale intangibile nelle sue diverse dimensioni*, ed. Tullio Scovazzi, Benedetta Ubertazzi, and Lauso Zagato, 3–27. Milan: Giuffré.

Simpson, James R., and Thomas J. Schoenbaum. 2003. "Non-Trade Concerns in WTO Trade Negotiations: Legal and Legitimate Reasons for Revising the 'Box' System?" *International Journal of Agricultural Resources, Governance and Ecology* 2(3/4): 399–410.

Solimano, Andrés. 2001. *The Evolution of World Income Inequality: Assessing the Impact of Globalization*. Santiago, Chile: United Nations Publications.
Stille, Alexander. 2002. "Are We Losing Our Memory? Or the Museum of Obsolete Technology." In *The Future of the Past*, ed. Alexander Stille, 299. New York: Farrar, Straus and Giroux.
Swiderska, Krystyna, et al. 2009. *Protecting Community Rights over Traditional Knowledge: Implications of Customary Laws and Practices*. London: International Institute for Environment and Development.
Tamiotti, Ludivine. 2009. *Trade and Climate Change: A Report by the United Nations Environment Programme and the World Trade Organization*. Geneva: WTO Publications.
Tosato, Andrea. 2011. "Security Interests Over Intellectual Property." *Journal of Intellectual Property Law and Practice* 6(2): 93–104.
Tuerk, Elisabeth. 2003. "The Role of NGOs in International Governance: NGOs and Developing Country WTO Members: Is There Potential for Alliance?" In *International Economic Governance and Non-Economic Concerns: New Challenges for the International Legal Order*, ed. Stefan Griller, 169–209. Vienna: Springer.
Ullrich, Heidi. 2002. "Expanding the Trade Debate: The Role of Information in WTO and Civil Society Interaction, Information and Communication Technologies and Human Rights Advocacy: The Case of Amnesty International." In *Civil Society in the Information Age*, ed. Peter I. Hajnal, 175–200. London: Ashgate.
UN. 1948. Universal Declaration of Human Rights. UN General Assembly.
UNECE. 1998. Convention on Access to Information, Public Participation in Decision-Making and Access to Justice in Environmental Matters. 2161 UNTS 447; 38 ILM 517 (1999).
UNEP. 1992. Rio Declaration on Environment and Development. UN Environment Programme.
UNESCO. 2003. Convention for the Safeguarding of the Intangible Cultural Heritage. 2368 UNTS 1.
UN Millennium Ecosystem Assessment. 2005. *Living Beyond Our Means: Natural Assets and Human Well-Being*. New York: United Nations.
von Lewinski, Silke. 2004. "Final Considerations." In *Indigenous Heritage and Intellectual Property: Genetic Resources, Traditional Knowledge and Folklore*, ed. Silke von Lewinski, 379–397. The Hague: Kluwer Law International.
WIPO. n.d. Portal of Online Databases and Registries of Traditional Knowledge and Genetic Resources. World Intellectual Property Organization, http://www.wipo.int/tk/en/resources/db_registry.html (accessed January 7, 2015).
WIPO. 2011. "The Protection of Traditional Cultural Expressions: Draft Articles." WIPO Intergovernmental Committee on Intellectual Property and Genetic Resources, http://www.wipo.int/edocs/mdocs/tk/en/wipo_grtkf_ic_18/wipo_grtkf_ic_18_4_rev.pdf (accessed January 14, 2015).
WTO. 1994. Agreement on Trade-Related Aspects of Intellectual Property Rights. World Trade Organization. 1869 UNTS 299; 33 ILM 1197 (1994).
Wüger, Daniel. 2004. "Prevention of Misappropriation of Intangible Cultural Heritage through Intellectual Property Laws." In *Poor People's Knowledge: Promoting Intellectual Property in Developing Countries*, ed. J. Michael Finger and Philip Schuler, 183–206. Washington, DC: World Bank.
Zagato, Lauso. 2008. "La Convenzione sulla protezione del patrimonio culturale intangibile." In *Le identità culturali nei recenti strumenti UNESCO: un approccio nuovo alla costruzione della pace*, ed. Lauso Zagato, 27–70. Padova, Italy: CEDAM.
Zweifel, Thomas D. 2006. *International Organizations and Democracy: Accountability, Politics, and Power*. London: Lynne Rienner Publishers.

Part V The Global Governance of Science and Technology

Chapter 23

Knowledge as Global Public Good

Daniele Archibugi and Andrea Filippetti

Introduction

Economic and social development is strictly associated to the capacity to generate, absorb, and diffuse knowledge. The economic characteristics of knowledge have for long intrigued thinkers. Already Richard Nelson (1959), Fritz Machlup (1962), and Kenneth Arrow (1962), among others, noted that knowledge is a very peculiar commodity. On the one hand, it is often generated for competitive purposes: armies and companies do invest time and money to develop new and superior knowledge to be used against their rivals. On the other hand, it happens seldom that those who generate knowledge manage to keep it for their own private benefit. Neither industrial and military secrecy nor intellectual property rights can, in the long run, impede the dissemination of knowledge. The knowledge about nuclear fission, in spite of all the efforts made by the US government to keep it confidential, was soon assimilated and put in practice by the Soviet rivals. And business corporations are certainly no better than the military in preserving for a long time the secrets of their innovations. But this is part of the story only. A large portion of those generating knowledge, including academics and their universities, diffuse the results *urbi et orbi* for the sole satisfaction of seeing their achievements acknowledged and their reputation enhanced. In fact, the dissemination of knowledge is the ultimate goal of their activity.

Related problems can be encountered when the spatial benefits provided by knowledge are taken into account. We know from the past that only craftsmen from Murano were able to produce mirrors above a certain size and that the best violins were produced in Cremona. Even today knowledge tends to cluster in specific districts, for example, some of the best software is generated in Silicon Valley. But these

The Handbook of Global Science, Technology, and Innovation, First Edition.
Edited by Daniele Archibugi and Andrea Filippetti.

clusters are both increasingly sparse across the globe as well as deeply integrated in the global economy (see this volume, Lorenzen and Mudambi, Chapter 10 and Iammarino and McCann, Chapter 14). These days, along with Silicon Valley and Cambridge, Bangalore and Shanghai have become global centers of knowledge production. Inventions and innovations tend to be disseminated across cities, regions, and countries, and they often involve brains, instruments, and equipment originated from worlds apart. In our own epoch dominated by information and communication technologies (ICT), the channels to transmit ideas and artifacts have increased at an exponential rate, and their costs have considerably decreased. Knowledge is generated globally through an increased variety of devices, and it is also transmitted globally, as several chapters in this handbook clearly indicate.

It is not easy to identify the components of knowledge that are "private" or "public," "national" or "global" (Nelson 1992). The debates on these issues are often passionate, especially when they have normative implications. Should knowledge be generated for the benefit of everybody? Is it right to keep secret or proprietary some knowledge that may have crucial implications for health or security? Should the outcome of knowledge generated with the money of the taxpayers of a country be disseminated globally? These discussions are not only theoretical but have important policy implications leading to very practical issues.

- To what extent should public institutions, regulations, and norms protect the intellectual property rights of inventors and innovators?
- Should universities make a profit from the ideas they generate?
- Should national institutions provide free access to scholars and students of rival countries?

One way to explore these issues is through the well-established framework that economists have developed for public goods. This chapter aims to investigate what are the characteristics of knowledge as a public or private good and to look at the implications for national and global actions. A better understanding of its nature will hopefully help to design better policies for its generation and diffusion. The next section recaps the basics of a public good, while the following section applies it to the case of knowledge. We continue to examine a key instrument to make knowledge appropriable, namely intellectual property rights. The global nature of knowledge as a public good is then discussed, illustrated by the case of research on vaccines. The last section discusses the implications for global governance and policy.

What Is a Public Good?

Sixty years ago, a young and brilliant economist, Paul Samuelson (1954), made an attempt to define public goods. He identified these goods as those "which all enjoy in common in the sense that each individual's consumption of such a good leads to no subtractions from any other individual's consumption of that good," in other words goods that are non-rival. A pen, for example, is rivalrous since either Laurel or Hardy can use it. Moonlight, on the other hand, is non-rival since both Laurel and Hardy can enjoy it. In addition, it should be considered if it is possible to exclude some potential consumers from the use of a good. Exclusivity applies when an owner

Table 23.1 Private and public goods.

	Rivalrous	*Non-rivalrous*
Excludable	*Pure private goods* (pen, bread, car)	*Network or club goods* (cinema, cable TV, canals, waterways)
Non-excludable	*Common goods* (ocean fisheries, free access pasture) *Congestible goods* (pest control, open pathways)	*Pure public goods* (clean air, defense, street lighting) *Public bads* (pollution, financial instability)

Source: Adapted from Ostrom and Ostrom (1977) and Kaul, Grunberg, and Stern (1999).

can effectively prevent others from using a good. Laurel can exclude Hardy from using his pen but he cannot impede him from enjoying the moonlight (Ostrom and Ostrom 1977). When a good is both rival in use and excludable it is very likely to be a typical private good. When, on the contrary, it is non-rival in consumption and it is non-excludable it is very likely to be a pure public good.

Public goods are generally considered those that benefit the whole community, and this in turn justifies the assumption that the group of beneficiaries should contribute to their provision and maintenance. While moonlight and sunshine are not produced by humans, and therefore do not need a pro-active role of the government, roads and street lighting are typical examples of goods that need somebody to be in charge of their production and/or maintenance. In the absence of such an intervention, it is unlikely that these goods will be produced at all.

The rise of adverse conditions that affect everybody and that generate negative externalities has also led to the development of the concept of *public bad*. Actions that generate negative externalities, such as littering in a public space, are public bads. Pollution and climate change are public bads that affect everybody and cannot be prevented through individual agency only. This category includes human-related phenomena (such as pollution, climate change, financial instability) but also natural phenomena (such as earthquakes and other natural disasters).

The concept of public goods implicitly indicates that supply is feasible when it is provided by collective agents such as the government. Profit-seeking agents have little interest to provide and to pay for public goods that everyone can use without paying: who will build a road if a toll cannot be introduced? Since each individual has an incentive to be free rider, it is not possible to provide roads and street lighting without coordination and voluntary or binding agreements. These types of goods require some forms of collective actions. In the absence of agreements about the production of these goods, the quantity produced of these goods could be lower than the level desirable for the society. In this sense, the theory of public goods can also be seen as a justification for the economic role of the government.

As reported in Table 23.1, it is also interesting to consider intermediate cases. Some goods can be rivalrous but non-excludable, and others can be excludable but non-rivalrous. The former are called *common goods*, the latter *network or club goods*.

The network or club goods are those that are not necessarily rivalrous in consumption, but whose access can be made excludable. Cinema is a typical example: there is no pleasure to watch a film in an empty theatre, indicating that enjoying a film is non-rivalrous, but those who do not pay the ticket can be excluded. Common goods are potentially rivalrous but there are not (yet?) the conditions to exclude consumers from access. Ocean fishery is rivalrous because fish is scarce, but it is not excludable since there is virtually no possibility to regulate the access of fishermen to the open sea. These are the goods more at risk of being depleted through over-exploitation, that is, those for which Hardin's "tragedy of commons" could happen (Hardin 1968).

The conditions of rivalry and excludability are not absolute. It is more precise to define goods on a scale and attribute to them a level of low, medium, or high rivalry and excludability. These conditions are different across locations (e.g., wood may be scarce and therefore rival in some dry country but may be abundant and therefore non-rival in Canada) and across the existing institutions (e.g., access to beach can be made excludable in some countries while it is open in other countries).

These conditions also change over time. Fish has become more and more scarce and, therefore, it has become more and more rivalrous. ICTs have also changed access to many services and they have allowed making excludable many products that a few years ago were not excludable. When TV programs were broadcast through aerial only it was not technically possible to exclude users and TV programs had the typical characteristics of pure public goods. Anybody who was not paying the TV license acted as a free rider. Cable TV programs continue to be non-rival but they have become excludable. Similarly, until a few years ago it was not technically possible to exclude only certain automobiles from access to the center of cities: if private traffic was banned, this applied to everybody. Through CCTV, the center of London and of other cities is selectively accessible to those who pay the access fee.

We have also learnt that public goods depend on a variety of social and institutional aspects. Fred Hirsch (1976) introduced the concept of "positional goods" to describe those goods that are not necessarily rivalrous, but to which access is associated to a peculiar social status. The premiere of a TV program in a theatre accessible to invitees indicates that what is rival is not the TV program itself (everybody can watch it at home), but the VIP status associated with the invitation. Similar cases can be found in several goods that are potentially non-rival but become rivalrous as status symbols (access to news, social clubs, and tourist resorts).

The concept of rivalry is also associated with the community served. Defense has for long been considered the typical public good and, on that ground, all citizens of a given state were asked to contribute to its costs. Since the foundation of the United States of America it has not asked the inhabitants near the frontier to pay more taxes to support the army than those farther away.[1] The basic assumption is that the benefits of defense are indivisibly provided to all inhabitants of a community. But since the frontiers of states have become blurred, the "public good" defense of state A might automatically increase the "public bad" of insecurity of state B. If both states A and B increase defense expenditure, the outcome might be more insecurity rather than the opposite, while the public good of security could be guaranteed by mutual security agreements such as non-aggression treaties. According to the community to which you apply the concept, defense can be either a pure public good or a club good (e.g., for states that belong to a defense union), or a public bad (e.g., for neighboring states that may feel threatened by next-door military capacity).

Is Knowledge a Public Good?

For many years, knowledge has been considered a public good. Economics Nobel Prize winner Kenneth Arrow (1962) contributed to the dissemination of this view, arguing that knowledge is costly to produce (as many other public goods) but could be disseminated as information at zero or very low costs. This rather persistent view was restated by other authoritative Nobel Prize winners such as Joseph Stiglitz (1999). The modern economics of science and innovation, on the contrary, has argued that knowledge is neither a totally private nor public good. It is, in fact, a combination of the two (Pavitt 1987; Nelson 1992; Callon 1994). Each single aspect of knowledge has a different balance of "private" and "public" components. Seen through the public goods lens, knowledge is only partially non-rivalrous and non-excludable. Let us examine in detail the two components.

Knowledge and Rivalry

If Laurel develops a new method to get rid of headaches, let us call it NoheadachesA, he will be able to use it and to get a direct benefit. If Hardy, a generic member of the public, buys, rents, imitates, or steals the new method, Laurel will not lose its control; he will continue to be able to use NoheadachesA. Two people will benefit from the same method and this will double the social advantages provided by the new knowledge. While the transfer of a physical commodity (a pen, a car, a machine) implies that the original proprietor loses control of the object, nothing like that happens when knowledge is acquired by or transferred to others. In this sense, knowledge is non-rivalrous in consumption.

If Laurel has invested massively to develop NoheadachesA, he might be willing to sell or license the knowledge rather than disseminate it for free. Laurel might be eager to find a market for his knowledge, especially since the sale will not imply that he will forget about it, and he can continue to use it. As with mass production, there are no limits to duplications, but variable costs are zero or very low. Even if Laurel sells the exclusive right of NoheadachesA to a client, he might continue to use the knowledge acquired to generate new one. If Hardy acquires the license for NoheadachesA for a limited period, at the end of the lease he will still possess it and it will be difficult to prevent him from using it again.

Both buyers and sellers of knowledge have to face a market that is highly imperfect.[2] Hardy might not be willing to pay a price for knowledge that he has not yet seen in operation (we do not often buy an item unless we have a degree of certainty that it will serve our purposes). But if Laurel discloses his knowledge to Hardy, up to the point that he can be sure that it serves his purposes, he will not need to pay a price for something that he already knows.

If another inventor, Bernard, also develops a competing device, a fierce competitive race may occur between Laurel and Bernard since both of them potentially aim to sell the product to the same consumers. Laurel and Bernard will compete for market shares and this will imply their attempts both to upgrade the knowledge to make it more appealing to prospective customers and to protect it according to the available systems. This process has all the characteristics of economic rivalry. Although there is no rivalry in consumption, there might be a substantial rivalry in the generation and upgrade of knowledge.

Knowledge and Excludability

The motivations that induce individuals to generate new knowledge are diverse. Some may be willing to invent for their own personal pleasure, and may not necessarily be looking for material rewards. For a large part of the academic community, fame and reputation are already sufficient rewards for their commitment. For this community, the diffusion of research results is a priority, and it is up to that community itself to undertake active policies, including science communication, to reduce the number of those excluded from access.

But if prospective gains induce Laurel to devote time, energy, and resources to generate new knowledge, he will try to do his best to profit from it. Therefore, although Laurel does not lose the control over his knowledge when Hardy is using it, he will not look for its personal use only but will look to commercialize and to profit from it. In order to get some form of remuneration, Laurel will search for methods to exclude individuals who do not pay a price to use the method (excludability). If Laurel succeeds in doing that, the knowledge generated is still non-rivalrous in consumption but it will become excludable, or accessible to members that have been granted access, that is, it will become a network or a club good.

Since knowledge is non-rivalrous in consumption, excludability needs to be artificially constructed. There are several methods that inventors and innovators can use to exclude others from the benefits of knowledge they generated:

- **Secrecy**. Keeping a discovery secret allows the inventor to prevent others using it without charge. The practice is rather diffused not only in the military sector but also in business companies. Secrecy offers only partial protection since it can be broken: military and industrial espionage, headhunting, reverse engineering, and other practices do not guarantee that the original inventor retains the secret. And, in parallel, the strategy of secrecy may hamper the marketing and commercial diffusion of an innovation.
- **Access codes**. Developing technical methods such as access codes, passwords, and software protection makes it more difficult to use the knowledge without authorization. Access codes are the paradox of the digital society. On the one hand, the technical reproducibility of artifacts increases the non-rival part of knowledge. On the other hand, the creation of devices such as access codes can prevent open access (see Hess and Ostrom 2007). But access codes, passwords, and other devices do not guarantee absolute protection, as hackers and their victims well know.
- **Intellectual property rights** (IPRs). The two economic methods above are complemented by IPRs, a family of legal devices (patents, copyrights, trademarks) that should guarantee the inventors and the innovators the exclusive right over the product of their activities, although for a limited span of time. These tools are an institutional solution provided by the government in the hope of solving the problem of underproduction of public goods. The rationale in setting IPRs is the comparison of static versus dynamic advantages: in the short run, there is a loss of welfare associated with the fact that the dissemination of knowledge is constrained, but in the long run they may induce profit-seeking agents to invest in

knowledge and/or to make it public rather than keep it secret. Even more than in other areas of economic life, institutions are needed to enforce the excludable nature of knowledge. Also IPRs are often infringed and they do not manage to guarantee full protection.

Companies generally use a combination of the three methods, integrating industrial secrecy, codes, and other technical devices and IPRs. The economic literature on technological appropriability (Levin et al. Levin, Klevorick, Nelson and Winter, 1987; Cohen, Nelson, and Walsh 2000; Arundel 2001; Odagiri et al. Odagiri, Goto, Sunami and Nelson, 2010) has shown that each technical field, industry, company, and country has specific ways to appropriate its knowledge using different excludability devices. If industrial secrecy cannot be guaranteed because the knowledge is visible and easy to imitate, inventors and innovators may devote a considerable amount of the resources to produce devices which will obstruct others from using the knowledge without permission. Software companies invest substantial resources not only to improve software, but also to prevent its unauthorized duplication.

Not all fields can be equally protected, and profit-seeking inventors will take that into account before devoting time and energy in one field rather than another. If we assume that some knowledge is more likely to be made excludable and it is easier for their inventors to generate profits, excludability may steer agents to work in an area rather than in another. If appropriability is not possible, there is the danger of a "tragedy of knowledge" similar to the tragedy of commons (Harding 1968), since the *economic incentive* to generate knowledge will be reduced. Since institutions have an impact, for example, making IPR protection weaker or stronger, excludability raises an important policy issue: What is the suitable level of excludability that governments should guarantee to inventors and innovators? The more institutional devices that contribute to make knowledge excludable the less it will have the characteristic of a pure public good.

Knowledge Is Not Freely Available

The sections above have looked at the generation of knowledge through the lens of non-rivalry and non-excludability. But the standard framework of public goods, when applied to knowledge, has also some limits. The typical cases of public good, such as defense and clean air, assume that economic agents receiving the "good" do not have to make an additional effort to reap the benefits. If a citizen lives in a well-defended country, or breaths clean air, he or she is not even supposed to notice it. The story is entirely different with knowledge: even when the producers of knowledge have the best intentions of transferring their expertise, the economic agent has to invest his or her time, efforts, and resources to acquire the expertise and to be able to benefit from it. And not even those efforts will guarantee that he or she will in the end manage to acquire and master the expertise. Stiglitz (1999) has argued that

> Knowledge of a mathematical theorem clearly satisfies both [non-rivalry and non-excludability]: if I teach you the theorem, I continue to enjoy the knowledge of the

theorem at the same time that you do. By the same token, once I publish the theorem, anyone can enjoy the theorem. No one can be excluded.

This statement does not take into account the costs and efforts needed to learn and to exploit the economic benefits of the theorem. To be sure, Stiglitz is not silent about the costs of acquiring and using knowledge but they are conceived basically in terms of transaction costs associated with the transmission of knowledge. However, he argues that this does not affect the public good nature of knowledge, since private owners will simply charge a fee reflecting the marginal cost of transmission. So overall the problem of transferring knowledge is largely underplayed.

The art of playing the violin is well codified in hundreds of books, scores, records, and videos, and all this wealth of information is publicly available at zero or very low costs. But only a few of those who make an attempt manage to become professional violin-players. Companies selling mobile phones or cameras do their best to transfer the knowledge on how to use these products to their clients up to the point that they invest resources to hire writers, designers, and cognitive psychologists to make instructions easily understandable. But only a few customers devote enough time and energy to use these appliances at their best. To play the violin decently and to use a mobile phone properly both requires that the would-be imitator devote time and resources to learning. This is even more evident in the case of technical knowledge in which the recipient needs to have basic knowledge to scan, understand, and use it. Economic development studies have convincingly shown that North–South technology transfer is unlikely to be successful unless the recipients of knowledge invest time and resources to absorb it (Bell and Pavitt 1997; Athreye and Kapur in this volume, Chapter 9). Technology transfer, like the generation of new knowledge, is an uncertain activity with successes and failures.

This is why it is wrong to equate knowledge and, above all, its technological component, to information (Pavitt 1987). Most knowledge, and certainly scientific and technological knowledge, is not available without costs since those willing to learn should make their own efforts. There are a few cases in which the benefit of knowledge is fully incorporated into products and it does not require an active role of users: for example, to benefit from a new drug it is sufficient to swallow the pill. These cases can be labeled *turnkey knowledge*. But turnkey cases are rather exceptional.

Callon (1994) has introduced the difference between knowledge and technology that is *freely available* (not protected by legal or technical devices) and knowledge and technology that *can be used without incurring costs* (i.e., can be directly applied without additional investment by the prospective user). Freely available knowledge is a rather large basin: all knowledge in textbooks and in the scientific literature is freely available. Also technology that in its origin is proprietary is often freely available: patents, for example, have a legal validity of not more than 20 years, and the majority of them lapse earlier than that since companies do not even bother to pay their renewal fees (Hikkerova, Kammoun, and Lantz 2013). But this does not mean that everybody can have a direct benefit from them. In order to exploit this basin of knowledge, a lot of additional effort in learning, tooling up, and development is needed. In this sense, the amount of knowledge that can be used without incurring costs seems to be very limited. As we have seen, not even the knowledge embodied in some products, such as computers, cameras, mobile phones, can be used without incurring costs, since the effort undertaken by consumers could considerably change the way in which they manage to exploit the products.

The economics of innovation has also shown that there are huge inter-industry differences. There are areas in which the transfer of knowledge is easier, and the infrastructures needed by would-be imitators are rather basic. In other areas it is the opposite. Drugs can be imitated and replicated rather easily while in nuclear physics entry barriers are much stronger, and only a few organizations are able to manufacture a nuclear reactor. There are very significant differences across typologies of knowledge, products, and industries. According to the product and the industry, excludability can change considerably. This recalls a basic fact, namely that knowledge is a highly heterogeneous commodity and in many cases it is not a commodity at all.

Summing up, knowledge has some characteristics only of a public good especially since it is non-rivalrous in consumption. There are economic and institutional methods that would potentially allow making knowledge excludable, but they are never totally effective. Knowledge is very close to being a pure public good when it can be used turnkey, namely when it is not required for users to properly understand how it works and how it is developed. There are, no doubt, cases of turnkey knowledge, but they are not very frequent. In most cases, users have to learn to use knowledge, and the more it is sophisticated and complex, the more it will require investment of time and resources. In these cases, even when knowledge is free to use, it can be used only if the relative costs are affordable. Therefore, what makes knowledge different from public goods is not the related production process, rather its process of diffusion, which has been scarcely addressed in standard economic theory.

Governments in the Generation and Distribution of Knowledge

Knowledge is so crucial for welfare and its characteristics as a commodity so peculiar, that public players have always taken a very active role in its promotion and distribution. Public policies and regulations aim in all countries to reward the producers of good ideas, to increase the investment in knowledge, and to induce inventors to disclose their discoveries. Governments have promoted knowledge to win wars, to increase security, to safeguard public health, to explore the sky, to improve communications, to advance education and learning. Policymakers may not perceive totally the public good characteristics of knowledge, but for sure they appreciate that generating and disseminating knowledge has strong positive externalities that do justify their concern. Public intervention to foster knowledge can be subdivided into four broad areas.

In-House Investment

The government develops the knowledge through publicly funded institutions such as research centers (e.g., NASA or the Max Planck Institutes) and universities. This includes the training of qualified people under the assumption that they will become a knowledgeable and technically competent class. In principle, the results of government-funded research should be in the public domain and freely available.

Procurement

The government, through its ministries and agencies, contracts to the business sector the development of the knowledge needed for its purposes (as it is often the case with military, space, and health programs). Procurement can be in the form of knowledge embodied in final products or entirely disembodied. In the first case, the government purchases products (e.g., an aircraft with given specifications) and the executing firm is required to develop the necessary knowledge. In the second case, the government can directly ask the business sector to develop new disembodied knowledge (for example, the prototype of a new vaccine). In the case of procurement, the government generally holds the IPRs, although contracting firms are likely to retain them de facto. Since the government holds the property rights, it may also distribute the knowledge to other fields and to other companies (as is often the case of procurement associated with defense and space programs). In both cases, since the contracting firm develops the knowledge, it will retain the expertise. Even if contracts require the contracting firm to erase all associated information, it will not be possible to erase the knowledge acquired.

Beauty Contests

The government rewards the individuals and the organizations that have produced socially relevant knowledge through "prizes." These prizes should be an incentive for individuals and companies to carry out scientific and technological investigation. The contest implies that private players disclose their knowledge, and the government can acquire it as a consequence of delivering the prize. But this is more likely to be effective for codifiable rather than tacit knowledge.[3]

Intellectual Property Rights (IPRs)

Finally, the government guarantees private inventors the fruits of their discoveries, and IPRs are the main way this is ensured. They provide an incentive to private agents to invest in the generation of knowledge and to disclose the knowledge developed. By making knowledge property, IPRs are designed to make it temporarily excludable and generate a market for it. As we have seen, the effectiveness of IPRs changes not only according to the institutional design in each country (Odagiri et al. 2010), but also according to the invention, the product, and the industry (Levin et al. 1987; Cohen et al. 2000; Arundel 2001).

In contemporary capitalist economies governments use all four methods, using excludability differently. In the first three cases, government intervention is designed to minimize excludability and often the interventions are combined with deliberate policies to promote the diffusion of knowledge. In the case of IPRs, on the contrary, the government guarantees excludability, although for a limited period of time. The basic rationale to explain the difference is that public sources generally finance in-house investment, while profit-seeking agents invest in the knowledge generation that IPRs should reward.

It is often stated that public intervention in the market for knowledge is the result of "market imperfections." However, this should not induce us to believe that, once upon a time, there was a market for knowledge, and that at a later stage governments intervened to regulate the imperfections. Historically, direct forms of government intervention have existed long before the creation of a formal market for knowledge. The first introduction of the modern patent system, the milestone for IPRs, was introduced in the Venetian Republic in 1474 (May 2002), while public investment in knowledge and education started a few thousand years ago.

Intellectual Property Rights as a Tool for Exclusion

IPRs explicitly aim to make knowledge excludable and therefore to prevent imitation and to make it possible to sell and hire it. It is worth pointing out that *intellectual* property rights are not needed if there are technical devices that impede the forgery of technology: for instance, if an effective device prevented the illegal copies of DVDs, the film industry would no longer need recourse to IPRs, since normal property rights are sufficient guarantee. The rationale for IPRs therefore resides precisely in the fact that knowledge in itself is often non-excludable.

Since the explicit aim of IPRs is to make knowledge excludable, it is worth asking if the government is contradicting itself: on the one hand, it encourages the generation and diffusion of knowledge on the ground that it has the characters of a pure public good; on the other hand, it makes it a private good by allowing making it excludable through its IPR legislation. This goes back to the different channels that the government can use to promote knowledge: when it uses in-house investment, there is the assumption that it does not only generate knowledge but also fosters its dissemination through policies for scientific and technological transfer.[4] But when the government is willing to foster the investment of profit-seeking agents, it uses IPRs. The underlying assumption is that, in absence of such a guarantee, individuals would either not disclose the knowledge or not produce it at all.

As mentioned above, IPRs' effectiveness in making knowledge excludable is not limited by government legislation only, but also by the nature of a competitive economy. In the real world, IPRs provide only a partial protection: competitors and would-be imitators are often able to by-pass the existing legal limitations by "inventing around" and many other ways. Consumers also often manage to acquire the knowledge associated with IPRs without paying its costs (or paying it only to a limited degree). As happens with pop songs, celebrity photos, and software, knowledge is used, copied, and imitated and the existing legal devices are far from able to provide full protection to the singer, photographer, software engineer, and inventor. Legal devices to guarantee excludability are, in the case of knowledge, largely ineffective.

All property rights can be infringed, as each person with a wallet knows. But governments are less willing to enforce *intellectual* property rights over knowledge than they are on rights over our wallets. Software thieves are more numerous than wallet thieves, but are much less likely to be found in jail. Why? Probably, because there is the tacit awareness that knowledge is non-rival in use and, therefore, it is in the social interest that excludability is not absolute. When physical property is stolen, there is a zero sum (or even a negative sum) outcome: if Hardy steals the car of Laurel, the number of cars in the community is still the same, and Laurel got a loss equal or greater to the gain of Hardy. But if Hardy steals Laurel's NoHeadachesA, society has two citizens with clear heads. This also applies in the case of two countries, where a company from country A imitates a technology of a company in country B.

In this context, public policies have tried to balance the private and the public interest. Without any enforcement, IPRs would lose their meaning and other methods would prevail. To enforce IPRs strictly could imply a fee be paid every time that "Happy Birthday to You" is sung, although it is difficult not only to enforce it, but even more to consider a welfare advantage to enforce it.[5] Public policy has therefore been rather keen to enforce IPRs with a certain relaxation, possibly on the assumption that a less than perfect protection would increase welfare and reduce the monopoly power associated with IPRs. This also depends on the stage of the development of the country as well as its industrial specialization. In general, governments in less developed countries tend to have low incentives for strong enforcement of IPRs. As long as domestic innovators come to the fore, the pressure for a stronger enforcement of intellectual property also rises (see Odagiri et al. 2010). As we have illustrated elsewhere (see Filippetti and Archibugi on the globalization of IPRs in this volume, Chapter 20), the constituency advocating a stronger enforcement of intellectual property rights in developing countries came from outside, that is, multinational firms in developed countries.

The somewhat relaxed attitude of governments in the face of IPR infringements is also associated with the idea that the generation of knowledge is seldom down to a single owner. Inventions and innovations in contemporary society often have a number of fathers and mothers (what is known as multiple independent discoveries), each of which have been able to exploit previous knowledge, to exchange information with colleagues, to absorb what is generated in public research centers and universities. It is impossible to be a successful producer of knowledge without being an even more successful absorber of knowledge. In the world of ideas, the difference between the "pupil," the "absorber," the "plagiarist," and the "robber" is very thin. To give full reward to the person who owns the IPRs does not necessarily mean to reward the person who does deserve it more.

The discussion on knowledge as a public good has a direct implication for the design of IPRs. Governments have a large number of instruments to make IPRs strong and weak: they can decide: (a) what is the length of protection provided to inventors; (b) the requirements of novelty to qualify for the granting of a patent; (c) the way in which Courts rule on controversies about infringements; (d) the level of enforcement guaranteed by the police against forgeries; (e) the industries involved (e.g., several countries in the past have prohibited patents in the pharmaceutical sector). The lens of global public goods indicates that the more the government decides to make IPRs strong, and therefore to promote exclusivity, the more it will need to use other channels to promote knowledge in its priority areas (such as in-house investment).

Knowledge as a Global Public Good

Health, environment, communications, mobility, and security all require the development of new knowledge. New scientific and technological competences are far from being confined to one country only. On the one hand, any significant breakthrough has an impact beyond the borders of the state that has actually produced the knowledge. On the other hand, it is very likely that most significant knowledge augmenting is the outcome of developments in different locations. In spite of this, many national governments develop their own science and technology policy agenda on the implicit assumption that soon or later they will manage to benefit from basic research funded and performed elsewhere. But when such an attitude becomes general, the free-riding syndrome will prevail: each country might be tempted to wait until others will invest in finding out a solution. This in turn will lead to an underproduction of the good.

In spite of the strong geographical interrelatedness of knowledge, its global governance is still very vague: How should it be organized? Who should provide the resources? The global public goods framework, developed by the UNDP (Kaul et al. 1999; Kaul et al. 2003), can perhaps provide some answers.

Public goods can be considered global when they cover a rather large group of countries and when it is difficult or impossible to identify a geographically restricted community of beneficiaries. Financial stability, peace, combating climate change and transmittable diseases, are all cases of public goods that do not have a clear geographical space: directly or indirectly they affect everybody. Besides this geographical criterion, Kaul et al. (1999) also stress the existence of a temporal dimension, goods (or bads) that could benefit (or damage) not only the current generation, but also future generations.

Knowledge might not totally fit into the *public* goods framework, as stated above, but it is certainly a *global* good: only in a few occasions, and for short periods of time, do institutions and companies manage to maintain their knowledge within their national borders. Even in the case of top secret investigations, such as those associated with the military sector, knowledge is likely to spill over to rival countries. Excludability may be obtained in the short run, but less and less in the long run. If the impact on future generations is also included, it is hard to imagine how some knowledge can provide benefits to a specific local or national community and not to all the others.

Knowledge as a global public good also challenges the idea that it could be provided by the business sector alone. Private incentives do not work very well in basic research project whose results are extremely uncertain and fixed costs very high, such as the case of research in particle physics whose benefit would span the entire scientific community.[6] Moreover, the notion of knowledge benefits (or damages) for future generations challenges the idea that market forces can provide them, since markets for future generations do not exist. The existence of appropriability mechanisms like IPRs does not guarantee a sufficient production of these goods. In these cases, single countries would tend to be free riders since the benefits of such a type of research are likely to span the globe and future generations. This poses a fundamental problem of *governance* related to knowledge as a global public good.

The distribution of competences is not uniformly distributed across countries, as other chapters in this volume have documented (see Castellacci and Natera, Chapter 1 and Zhou and Li, Chapter 5). Each country finances its own R&D and promotes knowledge-related activities. There is a huge discrepancy between the nature of the inputs devoted to promote knowledge, which continues to be national in scope, and the benefits, that are more likely to be global. An individual country can be compared to the single consumer/producer in the context of national public goods. Since each country cannot fully appropriate the returns of the knowledge produced at home, a free-rider strategy might arise leading to underproduction of knowledge. Greenhouse gas is a typical case of a global public bad: countries are trying to contain overproduction by international regulations within international organizations (e.g., the Kyoto Protocol at the United Nations Framework Convention on Climate Change), but many of them act as free riders. What are the implications of the public good analysis? If there are goods that provide widespread global benefits, who should be in charge of producing them? In particular, how will the free-rider syndrome, that has such an important role in many global public goods, work in the case of knowledge?

Public Institutions

Members of national academic communities and of other publicly funded institutions always had a strong propensity to exchange their wisdom, insights, and perceptions with foreign colleagues. Some of the instruments used to guarantee the international dissemination of ideas included academic societies, international journals, conferences, sabbatical years, and mobility grants. National governments have encouraged the academic community to be open to cross-border collaborations, probably because there is awareness that the outcomes are non-rivalrous and therefore they should also be made non-excludable, within borders as well as across borders. Moreover, there is a clear self-interest: collaboration implies not only knowledge outflows, but also inflows.

Citizens may potentially benefit from knowledge that has been generated elsewhere, and often paid by taxpayers of other countries. The propensity to act as a free rider, however, is constrained by the characteristics of knowledge: on the one hand, countries that do not invest enough have lower absorptive capacity and may be slower or inefficient in putting into practice what has been generated elsewhere. On the other hand, countries which invest more in knowledge are also those able to learn and assimilate the knowledge generated elsewhere. Countries manage to capitalize what they pay for.

The transfer of scientific and technological expertise to catching-up countries is constrained by the level of capabilities of the absorbing countries. This is a typical case where the difference between *freely available knowledge* and *knowledge that can be used without incurring costs* becomes relevant (Callon 1994). Even if significant portions of knowledge are freely available, this does not mean that other countries are able to collect the benefit without the necessary infrastructures and skills. The countries that have managed to absorb knowledge generated are also those that have invested massively in endogenous infrastructures, R&D, and education. Japan in the 1950s and 1960s, South Korea and Taiwan in the 1970s and 1980s, China in the 2000s are all cases of countries that have taken advantage of knowledge generated elsewhere because they have made an enormous endogenous effort to acquire it. To consider knowledge as a pure public good (Stiglitz 1999) risks diffusing the view that developing countries could benefit from the competences of developed countries if the latter are prepared to remove barriers to the transfer. But this is inaccurate since institutional (such as IPRs) or economic (such as industrial secrecy) barriers are not the main obstacles to the use of knowledge. The main obstacle faced by developing countries is the lack of endogenous absorptive capabilities.

The global public goods framework suggests also another important dimension: it is in the interest of all countries, including those that are already active in the generation of knowledge, to augment the partners involved in scientific and technological activities since this will enlarge the overall stock of knowledge. There is a potential mutual incentive for developed and developing countries to carry out and exchange non-proprietary knowledge. Public players have, in fact, pursued active policies to induce other countries to increase the pool of knowledge. Funded joint research programs, international conferences, international disciplinary academic associations, sabbatical years, and student exchanges are also methods to foster emerging and developing countries in the knowledge society. In these areas, governments are likely to have a healthy agonistic spirit similar to that encountered in the Olympic Games: each country is fostering its own academic community in order to acquire a better performance, but the methods and results are generally shared.

One way to gather empirical evidence on this form of cross-border collaboration is by looking at the scientific papers that are co-authored by scientists of different countries (see Hennemann and Liefner in this volume, Chapter 16),[7] an indicator of mutual collaboration in the generation of new knowledge without a hierarchy among participants. International science and engineering co-authored papers have substantially increased, thanks to the Internet and ICT. Internationally co-authored papers have grown from 16% of the total in 1997 to 25% in 2012 (National Science Foundation 2014: ch. 5). Already one fourth of academic papers are the outcome of direct cross-border collaboration. This form of collaboration has affected all countries, but it has become more and more important in small countries. Regions of the world with lower scientific capabilities are, in proportion, relying more on cross-border collaborations (National Science Foundation 2014: Appendix, Table 5-41).

The Business Sector

The business sector is increasing its role in the development of knowledge not only within borders, but also internationally. Companies contribute to perform R&D, to upgrade skills, to disseminate technical and engineering capacities at home and abroad. Firms can less and less be associated with a national territory. The activities they carry out outside their nation, including R&D, have increased substantially (see Ietto-Gillies in this volume, Chapter 6). Several leading multinational companies (MNCs) have built their own *intra*-firm and *inter*-national innovation centers. New products introduced by firms are traded in the international markets, new processes are scrutinized and diffused by competitors at home and abroad, and ultimately the externalities associated with companies' knowledge generation are less and less restricted to a specific nation. It might be discussed to what extent the knowledge generated by MNCs is private or public, but their investment in R&D and innovation is by definition multinational rather than national, and it generates substantial externalities across national borders.

MNCs are important vehicles for the international spread of knowledge. They do not necessarily manage to keep their ownership of knowledge, and very often they act as fertilizers for skills that are picked up and further developed in host countries. A significant case is represented by the software district of Bangalore: from an original foreign direct investment by Texas Instruments in the mid 1980s, a hub of excellence in ICTs and software has been developed, thanks also to education policies that have managed to train computer engineers at local universities (Chaminade and Vang 2008). This led to the birth and growth of a cluster of dynamic local firms, which has further induced other MNCs to invest in the place. A deliberate policy to build competences has successfully upgraded technological capabilities.

In recent years, firms have become more willing to collaborate with other firms to develop their technological knowledge. A vast literature has started to collect evidence and data on inter-firm technology agreements, defined as something more than an occasional collaboration, involving two or more independent firms and where the generation and application of knowledge is a key component (see Narula and Martínez-Noya in this volume, Chapter 7). Why do firms share a key competitive resource such as technological capability with their rivals? It has emerged that the need to share costs and risks with others is often a more important factor than the need to keep research projects confidential. The implication is that even profit-seeking agents are partially abandoning the search for exclusivity on their knowledge and are prepared to share it with actual or potential rivals.

International Organizations

International organizations play an important role in generating and disseminating knowledge. Since inter-governmental organizations are established by states to address common problems, it seems natural that they are also in charge of the improvement of general-purpose knowledge. Can international organizations be the genuine providers of knowledge as a global public good?

First, international organizations set common standards that allow all countries to benefit from best-practice knowledge. Transport and ICTs can operate internationally only if there are joint standards. The value of the standards increases with the number of players able to use them: by definition, a telephone that does not allow communication with anybody has no value. Producers have an incentive to transfer the relevant knowledge and expertise to the largest number of potential users. International organizations devoted to establish, disseminate, and upgrade standards therefore do play an important role in the transmission of knowledge and technical competence. Standards themselves have many of the attributes of pure public goods. While the implementation of standards may seem to be just a "passive" acquisition of knowledge, often they require also the making of endogenous competences in the countries that join the standards, then enabling also "active" knowledge developments.

Second, international organizations are also in charge of the transmission of scientific and technological expertise. Specialized agencies such as FAO, UNICEF, UNIDO, WHO, and the World Bank do play a crucial role in allowing countries to acquire competencies. This does not happen in development projects only. When international organizations allow people from different countries to sit together, they provide an important learning opportunity. The role of international organizations should also be stressed to counterbalance the typical and dominant direction of technology transfer, which traditionally implies the transfer of knowledge from the North to the South. In some occasions, the knowledge transferred may not necessarily be the most appropriate for the needs of the South. International organizations provide the opportunity to exchange also South–South knowledge. In many cases, knowledge developed in some countries of the South is more effective for other countries in the South than what has been cultivated for different purposes in the North, empowering what has been labeled "reverse innovation" (Govindarajan and Trimble 2012).[8]

Third, there are also a few research centers promoted and funded under the auspices of international organizations where innovation and learning work as two sides of the same coin. In fields with high fixed costs, governments have promoted joint research centers, as in the case of CERN. The European Union has established several international centers in high-cost and high-risk scientific domains, under the assumption that the outcomes will equally benefit all member countries. Other research centers have been established by the United Nations University or under the auspices of UN specialized agencies (as in the case of the UNIDO International Centre for Science and High Technology). These are pioneering cases of a genuine transnational public financing of public goods. The research centers and learning institutions are just a few if compared with the total resources provided to the national ones. Still, they provide significant models that point to a new way to promote knowledge for development.

If there is the genuine political interest of making knowledge a pure global public good in some areas, it might be wondered why the financing of cooperative international research is not larger, while the bulk of government spending for R&D is still directed toward national institutions. There are many reasons that can explain why. First, governments are well aware that R&D generates a lot of externalities, and that the local dimension is equally important. Second, government R&D spending often occurs in an area of inter-state rivalry, either because it is associated with national power (as in the case of military programs) or because it is devoted to supporting the competitiveness of national firms. But if there is the political willingness to disseminate the benefits of knowledge, transnational and collaborative R&D centers seem to be the appropriate instrument.

Vaccine Research as a Case of a Global Public Good Management

Human necessities are different across geographical areas, and even then the needs of inhabitants are not equal. Take the typical case of health research: poor countries undertake a tiny fraction of the world's R&D health expenditure, although they have 90% of the total disease burden (what has been labeled the 10/90 gap, see Global Forum for Health Research 2004). The countries that perform the other 90% of health R&D might have very different priorities, since many of the diseases affecting the South have already been eradicated in the North, for example, through improved hygienic conditions. For the South, the priority might be to find workable drugs to prevent death from diarrhea, while the problem is so rare in the North that the priority is often to find drugs for hair re-growth (for men) or hair removal (for women).

A typical case is represented by scientific research on vaccines. Knowledge leading to the identification of a successful vaccine is very close to being deemed a pure global public good: first, effective vaccines are costly to invent, but can be produced and transmitted at rather low cost.[9] Second, they can provide benefits globally and across generations. There is an inter-generational advantage both if future generations use the vaccine or, even more so, if the disease is eradicated. The transfer of knowledge requires that recipient countries should development the local competences and capabilities required to administer it (Yaqub and Nightingale 2012). The case of smallpox eradication, as a consequence of the World Health Organization program, is a successful attempt that has generated advantages for both the North and the South (see Fenner et al. 1988). For example, it is estimated that the United States recovers its own share of investment for smallpox eradication every 26 days (Arhin-Tenkorang and Conceição 2003).

With the appropriate investment, vaccines can be found for some of the major fatal illnesses of our age, such as malaria, tuberculosis, and HIV.[10] These three diseases are responsible for several million deaths a year (Archibugi and Bizzarri 2004). Major long-term investments could lead to an effective solution to this problem. In spite of that, investment is still negligible. Business companies apparently find it a risky area to invest. To the scientific and technological uncertainty, they add the social risk: if they develop a well-functioning vaccine, it is hard for them to make it excludable through patents since the social pressure for diffusion will be very high. Surprisingly, the public investment devoted to R&D in this field is also negligible (Archibugi and Bizzarri 2004). The relatively higher level of investment in vaccines for HIV is due to the fact that this disease is a top priority for the North, while malaria and tuberculosis are not a major concern for the North since improved environmental and hygienic conditions have managed to almost eradicate the disease.[11]

It is, however, significant how globalization is changing the perspective of the Northern countries as well: inward migration is bringing tuberculosis back and outward tourism is exposing Westerners to malaria. Vaccines are therefore a typical area where both the North and the South will have an advantage, as already shown by the smallpox case. The relative advantage of the North will be comparatively smaller than the advantage of the South, but this should not hide the fact that countries of the North will get from this investment much higher returns than for many other fields.

A new impetus for research, which has already achieved some important breakthroughs for malaria, was possible thanks to resources coming from private sources such as the Bill and Melinda Gates Foundation (White 2011). It is, however, significant to compare the different strategies pursued in the 1970s to eradicate smallpox with what is currently done to combat malaria and tuberculosis: in the 1970s, the program was promoted and carried out by an international organization, WHO, in collaboration with national governments. Today, it was promoted and mostly funded by the resources provided by the richest businessman in the world: the shift from public responsibility to private engagement is palpable. It may be wondered why such an important priority is funded through business resources rather than through public funding. The global public goods framework does indicate that the latter is a more convincing strategy.

The Governance of Knowledge as Global Public Good

The public goods concept has strong normative implications. In a pure market economy, the goods and services that have these characteristics risk being underprovided for, since profit-seeking agents have low interest in producing them and potential consumers have a tendency to act as free riders. If they need to be produced and maintained, a collective agent needs to be in charge. Within a national context, the public goods concept provides: (a) justification for the economic activity of the government; (b) legitimacy to the collection of taxes; and (c) guidance for indicating the areas on which the public sector should operate.

When we deal with *global* public goods, similar problems emerge. Individual governments might also be affected by the free-riding syndrome, expecting that other governments are going to invest to address and solve common problems. In the case of knowledge, we also have to take into account that scientific and technological capabilities are unevenly distributed across countries (as indicated in this volume by Castellacci and Natera, Chapter 1 and Le and Tang, Chapter 2), and therefore there is not only a problem of nations that might behave as free riders, but also the fact that many countries do not have the resources and the expertise required to address societal needs.

In this context, it becomes crucial to operate a distinction between the *production* of knowledge on the one hand and its *absorption and dissemination* on the other hand. We wish to point out here three important policy implications of the public good discourse applied to global knowledge:

- The tendency toward privatization of knowledge.
- The relevance of absorptive capability for disseminating knowledge across countries, particularly in developing countries.
- The importance of international collaboration, between countries and also through knowledge-generating institutions, to target the areas with greatest scientific and technological opportunities and of greater social relevance.

The Tendency Toward Privatization of Knowledge Commons

The public sector has various methods to achieve the proper generation and dissemination of knowledge. The first is to promote, finance, and perform work in scientific and technological institutions. Such a portion of knowledge is generated through public money and is in the position to be fully disseminated in the social and economic fabric. The second is to guarantee sufficient incentives to profit-seeking agents willing to invest in knowledge, providing the institutional context that would allow them to profit from it. IPRs and other institutional devices are precisely designed to provide these incentives by making knowledge excludable and tradable. We have noted that there is an intrinsic tension in government's behavior: on the one hand, it uses public resources to promote knowledge because it has the characteristics of a pure public good, on the other hand, it provides norms such as those of the IPRs that allow the business sector and, more and more, also the public sector,[12] to privatize the knowledge generated.

Governments should make an attempt to manage this tension between the proprietary and the public components of knowledge. Any actions aiming at reinforcing

Table 23.2 R&D by source of financing and performance, selected countries over selected years. GERD = Government Expenditure on Research and Development.

Country	*1981*	*1985*	*1990*	*1995*	*2000*	*2011*
	Percentage of GERD financed by industry					
Germany	56.85	61.09	63.54	60.03	66.04	65.63
Japan	62.28	68.87	73.11	67.11	72.42	76.52
United Kingdom	42.05	45.89	49.60	48.21	48.31	45.86
United States	49.41	50.31	54.60	60.23	69.03	58.58
	Percentage of GERD financed by government					
Germany	41.79	37.48	33.81	37.88	31.40	29.83
Japan	26.95	21.00	18.05	22.84	19.58	16.41
United Kingdom	48.10	43.50	35.54	32.84	30.23	30.45
United States	47.80	46.89	41.61	35.41	26.24	31.17
	Percentage of GERD performed by the business enterprise sector					
Germany	68.97	72.24	72.08	66.28	70.33	67.65
Japan	60.67	66.81	70.86	65.21	70.96	76.96
United Kingdom	62.96	64.35	69.37	64.96	64.96	63.58
United States	69.31	71.50	70.48	70.53	74.19	68.53

Source: OECD Main Science and Technology Indicators 2014.

IPRs should be matched with policies devoted to foster the public dissemination of knowledge. If IPRs are strengthened, the public space for knowledge dissemination should be compensated through other policies such as: (a) expanding government-performed R&D to generate a pool of public knowledge to which both public and business players could draw; (b) financing of pre-competitive and cooperative research in the business sector to increase flows; (c) rewarding profit-seeking inventors and innovators that are prepared to disseminate the results they have achieved through open source strategies.

Unfortunately, in the last three decades, we have seen the adoption of dangerous strategies to privatize knowledge. On the hand, we have witnessed a substantial reinforcement of IPRs, enlarging the scope of the Western system also to developing countries and privatizing knowledge that was previously public (Maskus and Reichman 2004). On the other hand, the public sector has also reduced its commitment to the promotion of knowledge. For example, the share of R&D of the business sector has substantially grown in most advanced countries while the share of government-funded R&D has declined correspondingly. Table 23.2 reports some data for four of the largest R&D spending nations: Germany, Japan, the United Kingdom, and the United States. From 1981 to 2011 the share of R&D financed by the government has been reduced in all countries. The dominant R&D spender is now the business sector. Where the government still finances a substantial share of R&D, as in the case of the United Kingdom and the United States, a growing portion is performed in the business sector: in the United Kingdom, the business sector finances 48.9% of the national R&D but it performs as much as 63.6%. Similarly, in the United States the business sector finances 58.6% of the national R&D but it performs 68.5%. The difference is associated mostly with public contracts provided to companies.

Finally, there has even been an attempt to allow private profits from R&D funded by the government and performed in the public sector. We have already highlighted the tendency to make such knowledge proprietary allowing the executing players (including universities, public research centers, and companies) to use IPRs and therefore to appropriate privately what has been funded by taxpayers.

If knowledge should really be sustained as a global public good, the first step is to reverse the privatizing trend that has dominated over the last 30 years.

Absorptive Capability for Disseminating Knowledge Across Countries, Particularly in Developing Countries

Although knowledge can be assimilated to a public good in the production phase, we have strongly argued that the successful diffusion and use of knowledge requires that recipient agents invest time and resources in learning and applying it. The distinction between production and diffusion is already relevant at the country level, but we can still assume that, within the borders of an individual country, the capabilities and the institutions generating knowledge manage to distribute it to users. At the global level, the transmission of knowledge between producers and users is much less likely to be automatic since there is less proximity. In particular, in the absence of absorptive capability by the recipient country, the availability of knowledge is of no use.

This has major implications for an international technology transfer strategy. To build absorptive capacity requires time, effort, and investment: infrastructures, education, training, and even R&D labs are needed to grasp, learn, and take advantage of the existing stock of knowledge. To focus on the supply from developed countries without taking into account the absorptive capacity of developing countries could lead to a waste of resources. It is a waste of time to bring the horse to the water if nobody has taught to the horse how to drink.

For this reason, the notion that knowledge is be a public good (as argued by Arrow 1962 and Stiglitz 1999), which we claimed to be theoretically inaccurate, is also politically harmful since it may diffuse the view that would-be users could reap its advantages freely. We maintain, on the contrary, that individuals, companies, and countries do not manage to take advantage of knowledge developed elsewhere unless they also make a substantial investment to effectively absorb it. When we consider the global public dimension of knowledge, the diffusion problem should be addressed, to guarantee that individuals or countries which should benefit from it have the necessary capabilities to absorb it. In brief, public intervention is required not only to make global knowledge *producible*, but also *absorbable*.

The Importance of International Collaboration

Since knowledge is non-rival in consumption, the larger the pool of resources devoted to it the more positive externalities will be generated. There is a direct interest of national governments to exchange results and to foster scientific and technological institutions in countries where they are weak. An effective global governance of knowledge, based on cooperation among governments and across national academic institutions, is therefore desirable. This, in turn, requires an increased coordination among national efforts, but also an enhanced role for international organizations dedicated to knowledge generation and dissemination.

The current distribution of scientific and technological capabilities affects the priorities of scientific and technological investigation. Capabilities are strongly concentrated in one part of the world only, the North. Not surprisingly, the public expenditure and other government interventions for the advancement of knowledge are generally directed toward the problems of the North rather than those of humanity. As illustrated in the case of vaccine research, companies in developed countries do not have enough incentives to address these priorities when the potential market is uncertain and IPRs do not guarantee the appropriation of possible returns. Companies and public institutions in developing countries may lack the required capabilities, at least in the short term. As a result, neither the public nor the business expenditure is successfully addressing some of the most important world human necessities.

In the long term, it is desirable and feasible that countries in the South will expand their research capabilities that will allow them to address their own societal needs by developing the required new knowledge. Some emerging countries, especially in East Asia, are expanding their knowledge-base, although their focus is more on economic development than on human necessities. But there is an urgent need of a policy shift, replicating the North–South forms of cooperation that have led, for example, to the eradication of smallpox.

It was possible to pull resources and competences to eradicate smallpox when scientific targets and societal benefits were clearly identified and aligned. If this example should be replicated, transparent and accountable R&D targets, addressing basic human necessity, must be identified. This in turn will require inclusive governance systems. It is unlikely that more public resources will be steered to knowledge if stakeholders are unable to assess what the resources are going to be used for.

A crucial aspect will be to identify in which fields investment in the generation of knowledge should be directed. Fields considered important by general consensus are those that are mutually advantageous for current and future generations and less likely to be associated with trade rivalry. Astronomy, space, and theoretical physics are already some of the areas where strong international collaboration has emerged, and where permanent institutions have also been established. The public goods framework, however, has also indicated that global health and other basic human needs have, so far, been seriously denied by the research agenda.

Notes

1 One of the strong arguments used by the authors of *The Federalist* (Hamilton, Jay, and Madison 1788) to create the new Federal state with a single army paid by taxes collected in all the states was precisely that it was unfair to put the burden of defense on the shoulders of the states more at risk to be taken back by Britain.

2 The imperfections of the market for technology are explored in Arora, Fosfuri, and Gambardella (2001).

3 In a brilliant book, Dava Sobel (1995) reports how in 1714 the British government made use of a contest to solve one of the fundamental problems in marine navigation, that of determining longitude at sea. The prize was won by a clockmaker who created the first clock (chronometer) sufficiently accurate to be used to determine longitude at sea – an important development in navigation.

4 Over the last 30 years a fundamental change has occurred in many countries: through the Bayh–Dole act of 1980 and similar legislation in other countries, governments have allowed universities to commercialize the results of their inventions even when funded from public sources. For an assessment, see Grimaldi et al. (2011). For the changes in European countries, see Geuna and Rossi (2011).

5 The legal controversies about "Happy Birthday to You" are summarized in the Wikipedia page with the song's title.

6 It is worth noting that benefits for industry from publicly funded projects even in basic research can be considerable. A case in point is represented by research at CERN, whose results have had a big impact in terms of technological transfer to European industry (see http://knowledgetransfer.web.cern.ch/, accessed January 8, 2015).

7 In spite of the increasing amount of publications generated by researchers in the business sector, the bulk of scientific papers are still authored by members of academic institutions. Researchers and engineers in industry are contributing more and more to academic journals, showing that even the results of their research are sometimes publicly disclosed.

8 UNDP (2001: 28) provides the case of the oral rehydration therapy developed in Bangladesh to fight diarrhea. Developed countries have sufficient medical infrastructure to combat the same disease by providing sterilized liquid through an intravenous drip. But when these medical infrastructures are not available, a second best is a simple tablet with the appropriate combination of salt and sugar. The therapy has often proved effective to prevent children's death. Since the same problem affects the majority of developing countries, it is a typical case where South–South technology transfer is needed.

9 Certainly it is much easier to transfer between countries the knowledge to successfully administer vaccines than the knowledge related to nuclear programs. However, not even in the case of vaccines is it possible to ignore absorptive capacity. Woodle (2000) shows the difficulty encountered by developing countries when they have to rely on external supply sources only.

10 The World Health Organization periodically assesses the number of infections and casualties of these diseases. See, e.g., http://www.who.int/topics/millennium_development_goals/diseases/en/ (accessed January 8, 2015).

11 The only high-income country with reported malaria mortality is Barbados (UNDP 2013: 168, table 7).

12 We refer here to the propensity of an increasing number of governments to allow publicly funded R&D to be protected through patents and other IPRs. See note 4.

References

Archibugi, Daniele, and Kim Bizzarri. 2004. "Committing to Vaccine R&D: A Global Science Policy Priority." *Research Policy* 33: 1657–1671.

Arhin-Tenkorang, Dyna, and Pedro Conceição. 2003. "Beyond Communicable Diseas Control: Health in the Age of Globalization." In *Providing Global Public Goods*, ed. Inge Kaul, Pedro Conceição, Katell Le Goulven, and Ronald U. Mendoza, 484–515. Oxford: Oxford University Press.

Arora, Ashish, Andrea Fosfuri, and Alfonso Gambardella. 2001a. *Markets for Technology: The Economics of Innovation and Corporate Strategy*. Cambridge, MA: MIT Press.

Arora, Ashis, V.S. Arunachalam, Jai Asundi, and Ronald Fernandes. 2001b. "The Indian Software Services Industry." *Research Policy*,30: 1267–1287.

Arrow, Kenneth. 1962. "Economic Welfare and the Allocation of Resources for Invention." In *The Rate and Direction of Inventive Activity: Economic and Social Factors*, ed. Richard Nelson, 609–626. Princeton, NJ: Princeton University Press.

Arundel, Anthony. 2001. "The Relative Effectiveness of Patents and Secrecy for Appropriation." *Research Policy* 30: 611–624.

Bell, Martin, and Keith Pavitt. 1997. "Technological Accumulation and Industrial Growth: Contrasts Between Developed and Developing Countries." In *Technology, Globalisation and Economic Performance*, ed. Daniele Archibugi and Jonathan Michie, 83–137. Cambridge: Cambridge University Press.

Callon, Michel. 1994. "Is Science a Public Good?" *Science, Technology, & Human Values* 19: 395–424.

Chaminade, Cristina, and Jan Vang. 2008. "Globalisation of Knowledge Production and Regional Innovation Policy: Supporting Specialized Hubs in the Bangalore software Industry." *Research Policy* 37: 1684–1696.

Cohen, Wesley M., Richard R. Nelson, and John P. Walsh. 2000. *Protecting Their Intellectual Assets: Appropriability Conditions and Why US Manufacturing Firms Patent (or Not)*. Cambridge, MA: NBER Working Paper 7552.

Fenner, Frank, Donald A. Henderson, Isao Arita, Zdenek Jezek, and Ivan Danilovich Ladny. 1988. *Smallpox and Its Eradication*. Geneva: WHO.

Geuna, Aldo, and Federica Rossi. 2011. "Changes to University IPR Regulations in Europe and the Impact on Academic Patenting." *Research Policy* 40: 1068–1076.

Global Forum for Health Research. 2004. *The 10/90 Report on Health Research*. Geneva: Global Forum for Health Research.

Govindarajan, Vijay, and Chris Trimble. 2012. *Reverse Innovation: Create Far From Home, Win Everywhere*. Cambridge, MA: Harvard Business Review Press.

Grimaldi, Rosa, Martin Kenney, Donald S. Siegel, and Mike Wright. 2011. "30 Years After Bayh–Dole: Reassessing Academic Entrepreneurship." *Research Policy* 40: 1045–1057.

Hamilton, Alexander, John Jay, and James Madison. 1788. *The Federalist: A Collection of Essays Written in Favor of the New Constitution*. New York: John and Andrew McLean.

Hardin, Garrett. 1968. "The Tragedy of the Commons." *Science* 162: 1243–1248.

Hess, Charlotte, and Elinor Ostrom (eds.). 2007. *Understanding Knowledge as a Commons*. Cambridge, MA: The MIT Press.

Hikkerova, Lubica, Niaz Kammoun, and Jean-Sébastien Lantz. 2013. "Patent Life Cycle: New Evidence." *Technological Forecasting and Social Change*, 88: 313–324.

Hirsch, Fred. 1976. *Social Limits to Growth*. London: Routledge and Kegan Paul.

Kaul, Inge, Isabelle Grunberg, and Marc A. Stern (eds.). 1999. *Global Public Goods: International Cooperation in the 21st Century*. Oxford: Oxford University Press.

Kaul, Inge, Pedro Conceição, Katell Le Goulven, and Ronald U. Mendoza (eds.). 2003. *Providing Global Public Goods*. Oxford: Oxford University Press.

Levin, Richard, Alvin Klevorick, Richard Nelson, and Sidney Winter. 1987. "Appropriating the Returns from Industrial Research and Development." *Brookings Papers on Economic Activity* 3: 783–831.

Machlup, Fritz. 1962. *The Production and Distribution of Knowledge in the United States*. Princeton, NJ: Princeton University Press.

Maskus, Keith E., and Jerome H. Reichman. 2004. "The Globalization of Private Knowledge Goods and the Privatization of Global Public Goods." *Journal of International Economic Law* 7: 279–320.

May, Christopher. 2002. "The Venetian Moment: New Technologies, Legal Innovation and the Institutional Origins of Intellectual Property." *Prometheus: Critical Studies in Innovation* 22: 159–179.

National Science Foundation. 2014. *Science and Engineering Indicators*. Washington, DC: NSF.

Nelson, Richard R. 1959. "The Simple Economics of Basic Research." *Journal of Political Economy* 30: 297–306.

Nelson, Richard R. 1992. "What Is Commercial and What Is Public About Technology, and What Should Be Done?" In *Technology and the Wealth of Nations*, ed. Nathan N. Rosenberg, Ralph Landau, and David Mowery, 57–71. Stanford, CA: Stanford University Press.

Odagiri, Hiroyuki, Akira Goto, Atsushi Sunami, and Richard R. Nelson (eds.). 2010. *Intellectual Property Rights, Development, and Catch-Up: An International Comparative Study*. Oxford: Oxford University Press.

Ostrom, Vincent, and Elinor Ostrom. 1977. "Public Goods and Public Choices." In *Alternatives for Delivering Public Services: Towards Improved Performance*, ed. E.S. Savas, 7–49. Boulder, CO: Westview Press.

Pavitt, Keith. 1987. "On the Nature of Technology." Reprinted in Pavitt, *Technology, Management, and Systems of Innovation*, 3–14. Cheltenham: Edward Elgar, 1999.

Samuelson, Paul. 1954. "The Pure Theory of Public Expenditure." *Review of Economics and Statistics* 36: 387–389.

Sobel, Dava. 1995. *Longitude: The True Story of a Lone Genius Who Solved the Greatest Scientific Problem of His Time*. New York: Walker and Company.

Stiglitz, Joseph. 1999. "Knowledge as a Global Public Good." In *Global Public Goods: International Cooperation in the 21st Century*, ed. Inge Kaul, Isabelle Grunberg, and Marc A. Stern, 308–325. Oxford: Oxford University Press.

UNDP. 2001. *Human Development Report 2001: Making New Technologies Work for Human Development*. New York: UNDP.

UNDP. 2013. *Human Development Report 2013: The Rise of the South: Human Progress in a Diverse World*. New York: UNDP.

White, Nicholas J. 2011. "A Vaccine for Malaria." *New England Journal of Medicine* 365: 1926–1927.

Woodle, Dian. 2000. "Vaccine Procurement and Self-sufficiency in Developing Countries." *Health Policy and Planning* 15: 121–129.

Yaqub, Ohid, and Paul Nightingale. 2012. "Vaccine Innovation, Translational Research and the Management of Knowledge Accumulation." *Social Science & Medicine* 75: 2143–2150.

Chapter 24

From Governmental Open Data Toward Governmental Open Innovation (GOI): A Global Perspective

Sabine Brunswicker and Jeremiah Johnson

Introduction

Over the last decade, the burgeoning discourse on "open innovation" highlights the shift from an innovation model of private investment and hierarchical control toward an increasingly "open" and boundary-spanning notion of innovation (Chesbrough 2003b, 2006). Open innovation represents a distributed innovation process based on purposive managed knowledge flows across organizational boundaries (Chesbrough and Bogers 2014). The growing body of literature on open innovation reports that some firms waive control of some internal innovation-related knowledge, and make such knowledge freely available to organizational actors and individuals that participate in the distributed process of innovation (Alexy, Goerge, and Salter 2013). The application of such "selective" revealing is relatively rare among profit-seeking organizations, but there is evidence that innovation-related knowledge and alleviating control over knowledge flows may create innovation benefits (Henkel, Schöberl, and Alexy 2014; Chesbrough and Brunswicker 2014).

While the early discourse about open innovation and selective revealing, both in practice and among scholars, was primarily concerned with profit-oriented, R&D-intensive firms, it is recently expanding to new levels of analysis and new innovation contexts. One of these levels is the governmental organization that is mandated by law to secure economic, social, and ecological welfare. Governmental actors increasingly engage more purposively in "open" forms of governmental innovation (Chesbrough and Bogers 2014). In this chapter, we introduce the concept of Governmental Open Innovation (GOI), which describes how governmental actors *purposively* manage the knowledge flows that span organizational boundaries and *selectively* reveal innovation-related knowledge and information to the public with

The Handbook of Global Science, Technology, and Innovation, First Edition.
Edited by Daniele Archibugi and Andrea Filippetti.

the aim to spur innovation for a higher economic and social welfare at regional, national, or global scale (Almirall, Lee, and Majchrzak 2014; Alexy et al. 2013). Selective revealing describes the process by which governmental actors make some innovation-related knowledge available as a "public" good, while at the same time supporting the privatization of innovation through intellectual property rights (IPR) and other protection mechanisms (Henkel et al. 2014). We explicate three revealing strategies in GOI, namely problem revealing, solution revealing, and governmental Open Data (OD). Governmental OD efforts have been the trigger for the emergence of GOI. Since 2008, governments around the world have released machine-readable datasets that (potentially) are of general civic interest, which were previously inaccessible by the general public and tightly controlled by governmental rules and regulations. One objective of governmental OD policies is to fuel innovation and to equip citizens, civic entrepreneurs, and also private businesses with the opportunity to turn this data into novel OD applications and solutions (Veit and Huntgeburth 2014; OSTP 2013). For example, in 2008 the federal US government created Data.gov, a repository of governmental open data of more than 150,000 machine-readable datasets that are freely available to the public and provide them with the opportunity to turn this data into new tools, applications, and governmental services (Obama 2009). Beyond governmental OD, problem revealing is another central strategy of GOI (Almirall et al. 2014; Nam 2012). Governments have started to launch so-called crowdsourcing initiatives, in which they publicly release governmental innovation problems and invite a large number of citizens to find a solution for such problems (Almirall et al. 2014; Afuah and Tucci 2012). We observe that "problem revealing" goes hand in hand with governmental OD revealing. For example, on January 4, 2011 President Obama signed the America Competes Reauthorization Act, which led to the creation of Challenge.Gov, and allowed all Federal agencies the authority to award prizes for agency-sponsored OD contests, which invite citizens via an open call to solve governmental innovation problems (OSTP 2013).

In sum, GOI describes a novel concept of governmental innovation that departs from "traditional" assumptions about the role of governments in regional, national, and societal innovation and theoretical explanations of knowledge flows and their effects on system-level innovation. We are aware that prior innovation literature and theory have already pointed out the critical role of governmental actors in fostering a region's or a country's innovation performance and enabling critical knowledge spillovers. Scholars on national systems of innovation, evolutionary economics, and innovation learning have pointed out the critical role of the network of institutions and in particular governments to create, store, and transfer the knowledge, skills, and artifacts which foster innovation learning and the diffusion of new technologies (Nelson and Winter 1977; Nelson and Sampat 2001). In addition, governments have been some of the first institutions that relied on theoretically grounded practices and institutional mechanisms like contests and tournaments that are also critical governance modes for solving innovation problems and fueling governmental innovation in the context of GOI. As far back as 1714, the British Government offered a cash prize – the Longitudinal Prize – to anyone who could come up with a solution to identifying the position of ships at sea (Afuah and Tucci 2012; OSTP 2013). GOI was triggered by the release of governmental OD. Indeed, waiving control over different types of innovation-related knowledge and information, problems, and solutions, is a central characteristic of GOI which is at odds with traditional strategies of how governments engage with external solution providers following rigid rules and regulations. GOI puts a large emphasis on non-expert citizens and the "citizen" crowd as a key actor in solving governmental innovation problems, and turning OD into innovative solutions that increase societal welfare. This particular nature of GOI results from the global policy movement of "Open Government" and is afforded through the digitization of governmental activities, the pervasiveness of digital technologies and the Internet at a global scale, and the continuously sinking costs for collecting, storing, and disseminating information (Meijer, Curtin, and Hillebrandt 2012; Lathrop and Ruma 2010). Following Obama's 2009 Memorandum on Transparency and Open Government, Open Government efforts in the United States are guided by the principles of *transparency*, *participation*, and *collaboration* (Tauberer 2014; Obama 2009; OSTP 2013). Governmental OD efforts are a core part of the Open Government movement, and recently, transparency-driven *innovation* is a central policy objective of OD efforts (Chopra 2011).

Indeed, the release of governmental OD is a global trend, which has evolved into GOI as new global model of governmental innovation. GOI implies that governments purposively reveal different types of innovation-related knowledge and information – governmental data, problems, and solutions – to fuel innovations that enhance economic and social welfare. In addition, GOI relates to a governmental innovation ecosystem that spans four major actors: governments, firms (large and small entrepreneurial ones), universities and research organizations, and the citizen crowd. Despite the fact that GOI is a global phenomenon, not all governments have evolved from governmental OD toward GOI to the same degree and the same way. There are differences in the objectives, strategies, the relative importance of different actors, and the organizational mode used to fuel innovation through revealing strategies. To introduce GOI as a global concept and at the same time acknowledge the local differences, this chapter presents a historical analysis of the emergence of GOI in different geographic areas around the world. First, we introduce GOI by providing a brief overview of Open Government as a global policy movement and introducing revealing strategies in a governmental innovation context and their differences from revealing strategies pursued by rent-seeking firms. Following that we present a historical analysis of GOI on four different continents to highlight the emergence of GOI at a global scale.

Governmental Open Innovation: "Openness" of Governments for Innovation

Open Government and Governmental Open Data: What's Behind This Policy Movement?

Open Government has become a popular term in discussion among policymakers at a global scale. The European Union states the term in its foundation Treaty, and President Obama signed the Open Government Directive on January 21, 2009, his first day in office (Meijer et al. 2012; Obama 2009). Open government is regularly associated with three principles of governmental activities, namely *transparency*, *participation*, and *collaboration* (Jaeger and Bertot 2010; Obama 2009).

To achieve these principles, and in particular the principle of *transparency*, governments around the world have recently revisited their information policies and implemented governmental OD efforts as information policy for greater governmental transparency (Jaeger and Bertot 2010). They make use of Internet-based technologies and facilitate access to governmental documents and machine-readable datasets, which were previously hidden from the public. For example, the European Commission has launched a European Open Data Portal (open-data.europe.eu) to create a single point of access to data of institutions and other bodies of the European Commission and its member states, which are free of use and reuse. In a similar vein, the US Federal Government offers Data.gov that provides access to a diverse set of machine-readable datasets that were previously not accessible.

The concept of transparency is not new in the scholarly discussion in political science and public administration. However, since 2008 it has been revitalized in the emerging discussion on Open Government and digitalization of governmental activities. Information technologies have changed the nature of how governments realize "transparency" (Brunswicker, Almirall, and Lee 2013; Medaglia 2012). Broadly speaking, information transparency can be described as the process of making information visible and inferable, and the degree to which information is visible and accessible (Bertot, Jaeger, and Grimes 2010). Thus, researchers and policymakers often use the term "open data" as a shorthand reference for the ability to "see" the activities of governmental agencies through the data that they generate and make available to the public. Scholars regularly consider transparency as a generative mechanism for deliberation, democracy, and thus, – often only implicitly – build upon principles of established democracy theories (e.g., Habermas's theory of democracy) (Susha and Grönlund 2012).

To be generative for these deliberation and democracy effects, information transparency implies both a "technical" and "social/organizational" dimension (Jaeger and Bertot 2010). Over the last few years OD efforts have focused intensively on the technical dimension of transparency. For example, governmental agencies developed open data standards, open data metadata, open data interoperability, and took measures to create "linked" governmental data (Huijboom and van den Broek 2011; Fioretti 2011; Guijarro 2007). However, scholars point out that for true, citizen-centered transparency to be realized, participants must have an understanding of what that data means, and the ability to share that content and interact with others. Unfortunately, many issues like accessibility, usability of open data portals, and consideration of the social and interactive dimension of information search have not been resolved yet, and lack of participation of governmental agencies in open data programs remains a major barrier for open data efforts to be generative for democracy effects (Burnett, Jaeger, and Thompson 2008; Jaeger and Bertot 2010). In sum, a large proportion of the scholarly dialogue on open data policies is concerned with transparency as a central generative mechanism for deliberation and democracy and points out that governmental OD efforts may realize a new kind of citizen-centered transparency if the vision of governmental OD becomes reality (Meijer et al. 2012; Jaeger and Bertot 2010).

However, information transparency and democracy effects are not the only rationale for governmental OD efforts. Another reason for revealing OD to the public is *economic* in nature, and relates to the creation of an information market (Janssen 2011;

Janssen, Charalabidis, and Zuiderwijk 2012; Chan and Pan 2008). The aim is to increase the amount of information available in order to inspire innovation, creation of applications and companies, and derive a greater value from the data that is being collected. Governments collect large amounts of data, and that data has value beyond the governmental context in which it was collected (Chan 2013). Releasing this data to the public allows businesses and individuals to access previously hidden data and create new products and services that can create value for both public and private interests. As a result, OD efforts have become relevant for *innovation*. For example, in the progress report on the US Open Government Initiative of June 11, 2011, the Executive Office of the President celebrates the *Democratization of Government Data* and the Data.gov platform launched in 2009 as leading a policy practice for innovation that is intended to be scaled across federal agencies (Chopra 2011). Thus, governmental OD efforts create a potential generative mechanism for transparency and at the same time may also act as source of a new mode of governmental innovation. With the release of governmental OD, innovations at the governmental level have received a new dimension of "openness," which creates opportunities for innovations facilitated by governmental actors. Governmental agencies release data that are of public interest and hold some public properties but were previously not accessible by individual citizens and industry actors that were not in a contractual relationship with governments. This data has potential value for innovations that address societal problems or may motivate private and rent-seeking investors to turn this data in a novel product or service. This activity points out that "openness" of governmental activities is not just to do with democracy effects, deliberation, and inclusion. In a governmental context, "openness" and revealing (or disclosure) may also become a generative mechanism for innovation.

Revealing Mechanisms in a Governmental Open Innovation Context

Sharing and revealing of know-how and information to spur innovation are inherent characteristics of open innovation activities, which have predominantly been associated with firm-level innovation activities (Alexy et al. 2013; Henkel et al. 2014; Henkel 2006; Chesbrough 2003a; Chesbrough 2014; West and Bogers 2014). To understand the notion of revealing in a governmental innovation context and the particular nature of governmental OD revealing, we briefly review the concept of open innovation and nature of revealing in it. Open innovation describes an innovation process, in which firms *purposively* manage knowledge flows across organizational boundaries (Chesbrough and Bogers 2014). Following the logic of open innovation, the focal organizations need to transcend their boundaries and make use of external sources of innovation in order to respond to the increasing environmental uncertainty and complexity of the innovation problems they are exposed to (Felin and Zenger 2014; West and Bogers 2014). While the early discussion on open innovation conceptualized open innovation as a linear process, today scholars acknowledge that open innovation is interactive in nature and implies mutual inflows and outflows of knowledge (West and Bogers 2014; Henkel et al. 2014). In open innovation firms purposively manage these boundary spanning inflows (or inbound) and outflows (or outbound) of knowledge in order to appropriate value from them.

There are different mechanisms for managing the "outflow" of knowledge and information. Some scholars argue that strong control over the intangible assets and legal ownership and immediate financial compensation for the outflow are essential in order to create and capture value from sharing intangible assets and innovation-related knowledge (Chesbrough 2003b, 2003a; West and Gallagher 2006). Examples range from biotech alliances, to contributions to standards (Simcoe, Graham, and Feldman 2009), or transactional relationships in technology markets (Arora, Fosfuri, and Gambardella 2001; Arora and Gambardella 2010), that are regularly offered by open innovation intermediaries (Howells 2006).

In some instances, innovators purposively waive control rights and *reveal* some innovation-related knowledge for free and without any contractual obligation for the receiver (Henkel et al. 2014; Alexy et al. 2013). This move is contrary to established "wisdom" and assumptions of theory and practice of strategy and innovation. Control over valuable knowledge has widely been considered as one of the most critical sources of competitive advantage and innovation (Teece, Pisano, and Shuen 1997). The phenomenon of revealing can be partly explained with the theory of private-collective innovation (Hippel and Krogh 2003; Alexy and Reitzig 2013) which describes a mode of value creation that lies between two opposing ends, the private-investment model on one end and the collective-action innovation model on the other. In the private-investment model, the innovator has control over the intellectual assets in order to secure the appropriation of innovation rents. In the collective-action innovation, the objective is to create "public goods" in which the innovative solution is non-exclusive and non-rival (Hippel and Krogh 2003). In the private-collective innovation, firms are willing to selectively contribute to the "collective" good as they may directly and indirectly benefit from it, for example, through the use of complementary assets resulting from the collective good or innovation learning from self-organizing communities (Alexy et al. 2013). In an open innovation mode, firms may very purposively and strategically make use of revealing to ensure that they derive direct and indirect benefits.

Open innovation literature furnishes two archetypes of "revealing" strategies at the firm level. There are two critical types of knowledge that innovating organizations usually try to "control," namely *problem-related* knowledge, and *solution-related* knowledge (Hippel 1994; Felin and Zenger 2014; Alexy et al. 2013). Problem-related knowledge relates to market needs that the firm aims to address successfully. Solution-related knowledge relates to technological solutions that address these problems, and also knowledge to develop these solutions. As a result, we can distinguish two *revealing strategies*, namely *problem revealing*, and *solution revealing*. The open innovation literature has documented a wide adoption of problem revealing. For example, the large number of crowdsourcing studies document that firms engage in problem revealing (see Cordova, Dolci, and Gianfrate in this volume, Chapter 12). They furnish empirical evidence that problem revealing allows them to identify solutions of high novelty and helps them to overcome the problem of local search (Afuah and Tucci 2012). Firms engage in problem-related revealing to send signals to the external environment that they are unable to solve a particular technological problem on their own. This provides them with the opportunity to tap into the generative mechanism of "diversity" as the participation of a large public crowd increases the probability that they identify a novel "outlier" solution by technological and socially marginal problem solvers (Jeppesen and Lakhani 2010). Beyond that, problem revealing allows them to invite external actors to collude on existing technological paths they had started to developed internally but where multiple problems are not yet solved, and creates opportunities for detailing and expanding a technology path they have already invested in (Alexy et al. 2013). Thus, they can either spread a problem (or issue) to find novel solutions they would not identify internally, or make use of problem revealing to shape the agenda of innovators who learn about the problem. However, problem revealing is not confined to profit-seeking organizations only and has been adopted by governmental actors long before the governmental OD movement. In the eighteenth century, the British Government already spread the problem of "locating ships in the sea" when running the widely cited Longitudinal Prize mentioned above in the introduction (Afuah and Tucci 2012; OSTP 2013). In this case, the government directly benefited from a novel solution but also shaped the agenda for others that developed novel technological solutions for location identification.

Open innovation literature also provides empirical evidence of the adoption of *solution revealing*. For example, open source software companies disclose the source code to the general public, who can use and modify the code as they wish (Hippel and Krogh 2003). Solution revealing is also emerging in other non-ICT sectors. For example, pharmaceutical firms like Novartis and GlaxoSmithKline recently revealed some of their solution knowledge, their patents, for free usage and reuse in order to engage a larger community of researchers to solve a complex disease problem such as diabetes (Brunswicker and van de Vrande 2014). Solution revealing creates network externalities, enables the adoption of a new technology, and drives development of complementary solutions for a particular technology (Teece et al. 1997). Because of such benefits, solution revealing is also a relevant strategy pursued by governmental actors. For example, R&D grant policies and guidelines issued by governmental agencies increasingly make it mandatory to use creative commons or open source software licensing in order to increase the adoption and refinement of novel technological knowledge.

Against this background and the distinction of two revealing strategies, we observe that governmental OD represents a very particular form of revealing of innovation-related knowledge and information that contains public properties. Governmental data, such as data about crime rates, water quality, and other public matters, hold public properties and are of public interest. They furnish information that can turn into an innovative solution for a particular innovation problem. This data is regularly, though not always, released without the encumbrance of restrictive IPR that would confine their use to specific non-commercial solutions. In this case, OD represent public goods, which are non-rival, non-exclusive, and available to the society (Hippel and Krogh 2003; Ostrom 1990; Archibugi and Filippetti in this volume, Chapter 23). However, some countries have opted to restrict the use of their OD to the commons, believing that private returns to innovation are not well aligned with social returns (Dosi and Stiglitz 2014). Other countries have opted to allow private citizens and industry to develop solutions utilizing OD while maintaining IPR, and to extract rents from those solutions beyond contest awards. Many, including the United States, utilize multiple levels of rulemaking to adapt the IPR to best fit the local needs (Hess and Ostrom 2005) by allowing agencies to set their own policies with respect to the ownership of solutions that are created with their data.

In cases where governments release their data without restrictions on the IPR that govern their use, the revealing of governmental OD does not specify a particular problem that should be tackled by the public, nor does it preclude the innovator who makes use of this data to control the solution. Both private citizens and industry are able to appropriate the data for their own innovative uses and goals, and can freely choose to work alone or in concert with others to develop a solution. Revealing OD triggers innovation in a completely self-organizing way, following the idea of a collective innovation model (Alexy and Reitzig 2013; Ostrom 1990). OD are public goods, which may create potential societal benefits.

With the emergence of governmental OD as a revealing strategy of governments that should fuel innovations, we see that governments have taken measures to combine governmental OD with the two revealing strategies discussed above. In a sense, they move from a "collective-action" innovation model, toward a GOI model, in which they combine three revealing strategies. They selectively reveal data, problems, and solutions. They aim to turn innovation-related knowledge (though not in all cases) into a public non-rival good, but they also support some level of control over the intellectual asset. National governments and city authorities around the world have started to publicly release governmental innovation problems and invite a large number of citizens to find a solution for such problems (Almirall et al. 2014; Afuah and Tucci 2012). One of the most widely known "problem revealing" initiatives in the United States is Challenge.gov. The website publishes innovation problems that are of interest to the federal government, and regularly these problem statements explicitly point to the usage of governmental OD to develop an innovation solution. Thus, "problem revealing" goes hand in hand with governmental OD revealing. In some cases, governments allow innovators to maintain ownership over the solutions they create from OD. However, some governments also combine governmental OD with solution revealing when insisting on making the solutions developed by the innovator free of use (or reuse). For example, in Australia the governmental OD efforts are tied to a creative commons license, which limits the use of this data for solutions with tight control over technologies developed from OD (Jeppesen and Lakhani 2010). Table 24.1 summarizes the three revealing strategies available in a governmental innovation context.

Objectives of Governmental Open Innovation

Governmental OD furnishes a novel revealing strategy and creates a novel dimension of "openness" of governmental activities. OD efforts are potentially generative for democracy effects. At the same time, they create a new "information assets" that may spur self-organizing innovations. Since the emergence of the open data movement in 2008, we observe that governments combine governmental OD efforts with other revealing strategies, namely problem revealing and solution revealing. GOI may have multiple innovation objectives. They range from novel processes or policy solutions that make governments more efficient and cost-effective, toward new products and services that create value for a region, a state, a nation and increase its ecological, economic, and social welfare (Veit and Huntgeburth 2014; Almirall et al. 2014; Mergel and Desouza 2013). For example, OD innovations may help the society to manage our natural resources, like energy, in a more sustainable way by having

Table 24.1 Free revealing in Governmental Open Innovation.

	Revealing Mode	*Description*
Open data revealing	Innovation opportunity creation (self-organized)	Governmental actors reveal governmental open data to motivate innovators to develop innovative solutions in a completely self-organizing way
Problem-oriented revealing	Issue spreading and agenda shaping	Governmental agencies reveal innovation problems that they consider critical in order to spread a particular "innovation" issue and shape the agenda of OD innovators
Solution-oriented revealing	Solution adoption (enhancement) and collective good creation	Governmental agencies make solutions freely available (e.g., via creative commons license or OSS license) to support the adoption of an OD solution and to create a truly collective good

Source: Authors' elaboration based on Alexy et al. (2013).

better access to energy usage behavior through smart metering solutions (Chesbrough 2014). However, OD innovations may also result in new services that allow citizens to save costs on public administration and help them to better adhere to regulations and laws (such as mandatory inspections of vehicles, etc.), or a new visualization tool to learn about the best schools in a particular district. Thus, GOI may address incremental as well as more radical innovations that relate to new processes, products, and services.

Governance Modes of Governmental Open Innovation

There at least four stakeholder groups that are essential for GOI to flourish, namely (1) governments, (2) firms and start-ups, (3) universities and research organizations, and (4) the civic crowd. Depending on the objectives of the GOI activities, the relative importance of different actors may vary. However, in general GOI moves beyond the triple helix model and considers the citizens and the civic crowd as an essential actor in GOI (Almirall et al. 2014).

Open innovation literature discusses a range of governance modes to coordinate innovation among these stakeholders. They range from market-based transactions like licensing, alliances, innovation tournaments and contests, to self-organizing innovation communities (Felin and Zenger 2014). While the former are more suited to organize "designated" agents, the latter allow involving a large number of unknown innovation actors, including the civic crowd. In general, one can differentiate between two ways of organizing the external "crowd": competitive contests, in which participants compete with each other, or collaborative communities (Boudreau and Lakhani 2013; Almirall et al. 2014). In GOI, the most widely used organizational forms are innovation contests (often referred to as challenges), in which they release a problem and motivate innovators via competitive contests and prize awards to develop novel solutions. They make use of rivalry as an incentive driving efforts and increasing the probability for novel solutions. Hackathons are a particular form of a one- to two-day innovation contest, in which governments invite software developers to a physical location in order to develop a solution in response to a revealed "problem." Application developer contests are a virtual form of such hackathons, which invite developers to develop OD solutions and win a prize award independent from their physical location. While contests are the dominant mode of organizing GOI, some GOI activities are formed in a more collaborative manner, and allow the formation of OD communities that self-organize to respond to a particular governmental OD problem that is particularly complex. We also see forms of "collaborative" contests that address the overall ecosystem, and not just a particular subgroup of actors, to co-create OD innovations in a collaborative manner (Almirall et al. 2014; Afuah and Tucci 2012).

The Emergence of Governmental Open Innovation: A Global Perspective

GOI has evolved in ways that are unique to each country, and this evolution has taken place at least along four dimensions, namely (1) objectives, (2) revealing strategies, (3) actors involved, and (4) governance modes (see Table 24.2). The following historical section examines the rise of governmental OD and transition to GOI in selected countries across several continents, and highlights the initial objectives and the actual outcomes of GOI programs. In the following examples, we illustrate how governments have implemented open data while stating that the motivation is transparency, and how the results from these efforts have not fully satisfied government needs. We will also examine the motivation to create value from this data, and how it has helped to push governmental open data toward open innovation.

Europe

As an early adopter of the Open Data (OD) movement, the European Union currently boasts the largest number of OD portals of any region. The high density of countries, combined with an official encouragement to share public sector information (PSI Directive 2003), has created a vibrant area for governmental OD development to be compared between countries. Current OD efforts in the European Union are characterized by strong national efforts, and less mature umbrella programs, such as OpenData.europa.eu and PublicData.eu, which seek to catalog data across governments. Several countries, including Ireland, Italy, France, and Spain, have also joined the Open Government Partnership (OGP), which is a worldwide multinational partnership dedicated to helping countries effectively create and release their data.

For our historical analysis, we focus on the United Kingdom and Denmark, which offer excellent examples of policy changes leading to new OD opportunities. Both countries implemented their OD policy in 2009, riding the first wave of global interest in making data more available to the public. Though they started at the same time, their stated objectives for releasing data were particular to the national context that they were developed in. Denmark pushed for "new services and insights" from open data (Danish Agency for Digitisation 2013) while the United Kingdom proposed "Putting the frontline first: smarter government" (Byrne 2009). The difference between these

Table 24.2 Overview of Governmental Open Innovation across different countries.

	Objectives	*Revealing Mechanism*	*Governance Modes*
United Kingdom	Put the frontline first: Smarter Government	Open Data through a large central site and multiple agency repositories Problems actively revealed Solutions often shared with public	Large contests and awards offered by agencies
Denmark	New services and insights	Open Data through multiple data repositories Problems actively revealed Solutions often shared with public	Large contests and awards offered by agencies
United States	Government should be transparent, participatory, and collaborative	Open Data through Central Repository Problems actively revealed Solutions often shared with public	Large contests and awards offered by agencies
Mexico	Transparency in resource management, fiscal matters, and citizen interests	Open Data through small centralized repository and diverse local data Solutions often shared with public	Contests, local communities
Australia	Realize the economic, social, and environmental potential of releasing Government data	Open Data through small central repository and territorial repositories Solutions often shared with public	Annual contests small prizes, citizen centered
China	Big data is a source of innovation, competition, and productivity	City Open Data, No Federal Declaration Solutions may require approval	City contests, small awards, citizen organized

Source: Authors' elaboration.

objectives may appear to be slight, but the ramifications of their vision were revealed throughout the development of their OD policy. Denmark placed a premium on digital solutions and "waves of digital self-service" which would allow citizens to use e-government sites and data on their own. In comparison, the United Kingdom viewed OD as a way to reduce disparities between the "frontline" of their government services and the "center" where decisions are made.

Both Denmark and the United Kingdom observed the need to encourage particular innovations using OD, and they choose different paths to achieve that goal. In Denmark, the launch of the Open Digital Innovation Strategy (ODIS) in 2013 created a Department of Digitalization to speed the transition toward recording all government activities in a digital form. Denmark also embraced OD contests at the onset of ODIS, and they are planning an annual contest (Smart Aarhus 2014). In the United Kingdom, a "non-departmental public body" called Innovate-UK was formed, under the guidance of the department for Business, Innovation and Skills, to promote OD and drive innovation through innovation contests. In both countries, there is a tendency toward governing the use of OD with large annual contests worth hundreds of thousands of kroner/pounds, in addition to small hackathons and other OD contests. Solutions developed from these contests are often released to the public once the contests conclude (Opening Up Government 2014).

The United Kingdom and Denmark place a premium on local and regional governance of data, tacitly recognizing that citizen participation is needed to identify problems with the current system, but rarely looking to individual citizens for solutions. Each government is also keenly aware of the impact that OD can have on industry, and action plans from Denmark's Open Government Partnership affirm the strong interest of the Danish government in promoting innovation (Open Government Partnership 2013). In the United Kingdom, several contests are specifically targeted to help particular sectors, such as advanced materials (Innovate UK 2014b). Academia is often mentioned as a partner along with industry, however there are few examples where academia is specifically targeted, and fewer where they participate.

Finally, both the United Kingdom and Denmark conduct their OD in a widely distributed fashion common to most of the European Union. They offer powerful sites where they conduct OD contests, separate sites where they host data, and additional sites for departmental, national, and international repositories. Both also offer networking tools and events to bring participants together (Innovate UK 2014a), and conferences on OD have been held in both countries. There is no unified virtual space or forum where OD participants, industry, and academics can go to interact, nor is there a central collection of resources. These similarities, particularly in the large number of sites and lack of a centralized social infrastructure, echo the challenges that exist with OD in Europe: many cultures, many languages, and many problems that are embedded in their local context.

North America

Many nations mention the 2009 Memorandum on Transparency and Open Government, which was signed by US President Barack Obama on his first day in office, as a point of inspiration for their own Open Government efforts. While many states and municipalities around the world had already embarked on a journey toward OD, this simple document and the subsequent Open Government Directive (Obama 2009) produced intense international attention to OD. The memorandum proposed three simple principles that government should follow: transparency, participation, and collaboration. The Open Government Directive that followed in December 2009 outlined a method to publish government information online, improve the quality of government information, create and institutionalize a culture of Open Government, and create an enabling policy framework for Open Government. The United States also became a founding member of the OGP in 2011, along with Mexico, Brazil, Indonesia, Norway, Philippines, South Africa, and the United Kingdom (US Department of State 2014).

Mexico has rapidly adopted the OGP platform, becoming the first Latin American nation to chair the OGP, and launching their own National Digital Strategy in 2013

(Mexico 2014) that includes government transformation with OD. The desire for transparency is not a new issue in Mexico, which has had historical issues with state secrecy throughout the 71 years when the PRI (Institutional Revolutionary Party) controlled politics in the country. In 2000, Vicente Fox ousted the PRI from the Presidency, gaining popular support with promises of transparency and accountability that would fight state corruption and secrecy (Vega 2012). In 2003, Mexico enacted their first federal Transparency and Freedom Act, which released government information to academics, journalists, and activists through a website. In many ways, this was one of the earliest digital Open Government efforts, but the implementation soon faltered, due to conflict between states and local governments and the federal government. In their latest Open Government efforts, special attention has been paid to this failure, and their strategy for open government has taken a more regional approach (Vega 2012).

In the United States, the creation of Data.gov ushered in a new era in data access for citizens. Starting with 47 datasets in 2009, a directive for each government agency to add at least two high-value datasets led to the platform amassing over 300,000 by the end of 2013. Data.gov has become a central repository for national, state, municipality, and some academic data. The migration to an open source platform, Comprehensive Knowledge Archive Network (CKAN), reduced the duplicates and errors, and currently hosts over 130,000 datasets from hundreds of government entities. On January 4, 2011 President Obama signed an expanded America Competes Act, which allowed agencies to fund prize competitions to spur innovation (US Department of Commerce 2014). The creation of Challenge.gov allowed agencies to expose their problems to the public and to govern the solutions through large prizes for innovative solutions. To date, there have been over \$63 million in prizes awarded by over 60 agencies, which engaged 42,000 citizen solvers (Dorgelo 2014).

Mexico has also released a new OD transparency portal (Federal Institute for Access to Information and Data Protection 2014) that offers a variety of governmental datasets to the public. Though the number of datasets is currently rather small, the Mexican government has placed governing emphasis on local action through local civic innovation and hackathons. This local focus has been driven by Mexico's history of OD failure, and a wellspring of civic-minded hackers who have helped change the technology landscape. In March 2013, the Mexican government had voted to spend 115 million pesos (9.3 million USD) for a mobile app for approximately 500 lawmakers to track their legislation. A group of civic hackers quickly created a competing app for 11,500 peso prize (\$930 USD) (Codeando México 2013; Vega 2012). These hackers went on to work with the government to create a series of federal civic innovation offices, where they work with local policy experts and citizens to leverage technology and OD to solve civic problems.

Comparing the Open Government strategies between these two founding members of OGP reveals the national differences that drive adoption and use of OD. The United States and the Mexican government have both adopted a central repository for their OD, however the level of centralization is markedly different. In the United States, all federal programs are centralized on Data.gov, and a large number of states and municipalities have also uploaded their data there. This has led to criticisms of Data.gov as a dumping ground for data, where high-value datasets are lost amid the clutter of maps and GIS data that dominate the platform (Peled 2011). In Mexico, the central repository contains data that is purposefully selected due to its national importance, and regional data is incorporated into the Federal Innovation Offices.

In terms of participation, both the United States and Mexico refer to transparency, efficiency, and innovation as key benefits of citizen involvement in Open Government. Both countries also use contest awards and hackathons to encourage solution development, and those solutions are shared with the public, but implementation of contest outcomes appears to have more impact at the highest levels of government in Mexico. Industry participation with Governmental OD is also commonplace in both countries, with the United States listing dozens of companies working with Governmental data (US General Services Administration 2014), and Mexico has launched the Open Data 100 Mexico (2014) digital innovation team to identify and support companies working on their data. Academic participation is often mentioned as a key component of OD success, however academics currently play a small role in the governmental OD landscape. Both countries host OD conferences that attract academic participation, and both have datasets that have been supplied by institutions of higher education.

Again, we have found no unified virtual space or forum where OD participants, industry, and academics can go to interact, though there are many targeted communities and there is a central collection of resources. The differences between Mexico and the United States, particularly in the centralization of power and the diffusion of local innovation centers, highlight the challenges that face a large federal government and the innovativeness that can be gained by supporting local "hacktivisim."

Australia

Following the recommendations of their Government 2.0 Taskforce, the Australian government adopted an Open Government Declaration in July 2010, citing transparency and innovation as key reasons to open up government data. The Open Government Declaration recognized the "economic, social, and environmental potential of releasing Government data," and divided the effort into three key principles: informing through right to access, engaging to collaborate on policy, and participating to make government more consultative. In concert with these goals, the government launched Data.gov.au as a central repository for governmental OD (AGIMO 2014). The focus on innovation from governmental OD is in stride with previous countries that have been discussed above, however the emphasis on environmental impact is particularly poignant in a country that was home to the world's first Green Party (Miragliotta 2012).

The Australian implementation of governmental OD has been touted as an invitation for citizens to innovate with OD, but there have been several setbacks on the journey toward this goal. In 2013, the Australian government acknowledged that 700 of their 1200 available datasets were just links to webpages or missing files (Sheridan 2013) and that there were very few users of this data. In 2014, the government bolstered the number of datasets by adding 2500 maps from the Department of Geoscience, and the repository contained over 5000 datasets at the end of the year. This growth has been hampered by a reticence on the part of territorial and local governments to share data (Hilvert 2013), and each territory continues to maintain its own OD site that is not searchable from the main Data.gov.au. Licensing of datasets is also an issue for Australian OD efforts, because two thirds of the data released by agencies is not licensed for open use (Hilvert 2013). Finally, data has been slowly released because it is difficult to convince agency leaders to spend money to prepare their data for consumption, leading to low participation among agencies.

While the state of Australian OD is in flux, the use of data to create innovative solutions is increasing rapidly. Utilizing a mixed cooperative and competitive hackathon approach, GovHack.org hosts a yearly competition that has grown from a small two-city event in 2012 to an 11-city event with 1300 participants in 2014 (GovHack.org 2014). Utilizing hackathons, Australians have developed a model for future civic innovation with governmental OD, which incorporates a diverse field of participants and supporters. GovHack organized federal agencies, state and territory governments, cities, and national and local corporate sponsors to support a hackathon for citizen participants. In keeping with the large geographic distance between cities, regions host their own local contests, and the winners of each contest go on to compete at the national level. Solutions generated by the most recent GovHack include an energy calculator for appliances, an interactive water quality map, and an online interactive portal for children to interact with OD called "Mash Academy."

Participation in OD has been a challenge for Australia. There is a strong government mandate for OD with a creative commons license, yet participation has been disjointed, with territories and cities maintaining autonomy over the distribution of their data. One area of concern is citizen privacy with respect to the data released, and one Senator noted that "Having it fragmented is actually a defense against lack of security and privacy" (Timson 2014). Citizen participation in hackathons appears to be growing year on year, and the solutions generated are freely available for public use. Local and national industry partners are also increasing their support of annual hackathons, and the diversity of awards helps to ensure that industry partners see solutions relevant to their interests. Academic partnerships are surprisingly not mentioned on GovHack, however there is a strong Open Access movement in Australia, and future partnerships are planned (Picasso and Phelan 2014).

In keeping with the fragmented distribution of OD, we have found over a dozen smaller communities that concentrate on OD from their territory or city. GovHack maintains a hackerspace where participants can view projects and participate on a forum, but the focus is on the annual hackathon, rather than OD in general. Because of its dedication to GovHack, Australia stands out from many other OD initiatives by not offering a regular stream of highly financed contests throughout the year. Instead, they offer many small prizes in a variety of categories, attracting regular citizens to contribute their ideas and together to hack solutions from OD.

Asia

South Korea and Japan are two of the leading OD countries in the world, both in terms of maturity and capability, and their development has been well publicized in other outlets. Since the focus of this chapter is contrasting governmental OD developments, we have selected a different OD movement to reveal how OD can take root without overt government support. China offers an interesting insight into a technologically advanced country that has a long history of rigid data and communication standards, yet is slowly approving data for release to the public. Interestingly, the motivation to open this data comes as a response to the US and European OD movement, to capture China's portion of the innovation, competition, and productivity gains (Yang 2013), rather than a call for governmental transparency and accountability.

Starting in 2011, the Shanghai Internal Data Directory became the first OD repository in China (Chao 2014), which was quickly followed by Beijing in 2012, and the National Bureau of Statistics in 2013. Though OD portals are few in number, the data released has been substantial, with an initial 425 datasets from Shanghai and over 4000 from Beijing. In 2013, there was a "hackathon-type" event on climate change, where the government released climate data for analysis (Chao 2013), though there is no information on prizes. Some OD is even considered controversial, and the Chinese government has requested that foreign embassies in Beijing stop releasing air quality data (Bradsher 2012). Recently, with a change to the environmental law by the Chinese Parliament, the Beijing Institute of Public and Environmental Affairs created a public app that uses factory emission data to enable non-government activist groups to identify and penalize factories that are exceeding the pollution standards (Carannante 2014).

Without a stated mandate to governmental OD, China has managed to adopt several of the same principles that guide participation and innovation in other national contexts. One of the largest barriers to OD resides in a government that has traditionally held information very close to the chest, and censored Internet and public speech are a staple of Chinese policy (Qiang 2008). Citizen reaction to this has been mixed, with several groups calling for more OD, however these requests are tempered with caution to operate within the guidelines of the state to avoid arrest or persecution (Chao 2013). Not to be denied, civic hacking groups have scraped government websites and the social networking platform Weibo to generate their own datasets for analysis. Additionally, outside efforts to delve into the hidden wealth of governmental data have been engaged in by Western media, but these activities may be too risky for Chinese citizens (Barboza 2012). Industry in China may have the most to gain from this tightly controlled data release, because there is growing governmental support for entrepreneurship and innovation with data. Academia also plays a significant role in Chinese OD, with Fudan University holding an international meeting on e-governance in 2013, with a number of sessions dedicated to OD.

Thus, organization of governmental OD in China is largely a collaborative effort in the hands of a community of civic hackers who are willing to find or ask for the data that they need to create apps. Contests remain rare, but the initial success may inspire the Chinese government to release additional data in the future. China may be on the threshold of a governmental OD revolution, if their civic hacking community can demonstrate the value of OD to a political body that is historically skeptical of the free release and public analysis of information.

Concluding Remarks

In this chapter, we introduce the notion of Governmental Open Innovation (GOI), which is evolving from the global movement toward governmental OD. Since 2008, governments around the world have released machine-readable datasets that (potentially) are of general civic interest, which were previously inaccessible by the general public and tightly controlled by governmental rules and regulations. While the discussion on governmental OD is predominantly concerned with information transparency and the effect of OD on civic deliberation, inclusion, and democracy, we show that governmental OD sets the stage for a new model of how governmental actors "open" up for innovation. This makes use of three "revealing strategies" to shape how OD may turn into novel applications and solutions, namely OD release, problem revealing, and solution revealing. While OD release alone simply represents a collective-action effort, the other two strategies shift toward a model of GOI in which they shape the innovation activities in both a direct and indirect manner. Many GOI efforts make use of problem revealing to shape the agenda of innovators; however, solution revealing is a case-by-case decision. Some governments opt for a privatization of solutions that are developed by innovators – ranging from citizens, civic entrepreneurs, and start-ups, toward large integrated organizations. However, others support the idea of making solutions available to the public, for example, through creative commons licensing. The latter shift highlights the paradoxical nature of global policies for innovation in terms of control over innovation-related knowledge on the one hand, and free access to a public good and collective-action innovation (Dosi and Stiglitz 2014; Ostrom 1990; Hess and Ostrom 2005; Archibugi and Filippetti in this volume, Chapter 23). We are witnessing a rather contradictory logic of policies for innovation at the global scale. While governments strengthen IPR to foster innovation, they encourage innovation to governmental OD and "selective" revealing. Indeed, our discussions suggest that the governance responses to this paradoxical nature of innovation need to be made on a case-by-case basis, considering multiple local and global boundary conditions.

Today, GOI is in its infant stages, and our historical case analysis illustrates that all governments are experimenting with different revealing strategies, are starting to engage the civic crowd, and are making use of a variety of practices and governance forms to ensure that OD efforts are supporting governmental innovations. Across all GOI efforts, motivating participation is a major challenge for governmental agencies. Our historical analysis suggests that the evolution of governmental OD and the adoption of GOI are partly shaped by the "governmental" context, such as the IPR scheme of a particular country, the skill level, and the entrepreneurial capital in a particular region. Our chapter shows that GOI is a global phenomenon and introduces the key characteristics of GOI. However, it is too early to draw any conclusions about the economic and social welfare effects of GOI at a regional, national, or global scale, and the successful governance modes for supporting it. It sets the stage for future research at the intersection of innovation, public policy, and technology. We invite scholars to take up the challenge to further explore the particular characteristics of GOI and perform empirical research to advance our understanding on the role of "openness" in governmental innovation.

With this chapter, we also want to encourage policymakers to pursue OD efforts and shift toward GOI. Even though many questions and problems of designing and realizing GOI for greater economic and social welfare are not yet solved, we are convinced that transparency and openness may help us to solve complex societal problems which require the participation of the greater society including the citizen crowd.

References

Afuah, A., and C.L. Tucci. 2012. "Crowdsourcing as a Solution to Distant Search." *Academy of Management Review* 37(3): 355–375.

AGIMO. 2014. Australian Government Information Management Office (AGIMO). http://www.finance.gov.au/agimo (accessed January 8, 2015).

Alexy, Oliver, and Markus Reitzig. 2013. "Private–Collective Innovation, Competition, and Firms' Counterintuitive Appropriation Strategies." *Research Policy* 42(4): 895–913.

Alexy, Oliver, Gerhard Goerge, and Ammon Salter. 2013. "Cui Bono? The Selective Revealing of Knowledge and Its Implication for Innvation." *Academy of Management Review* 38(2): 270–291.

Almirall, Esteve, Melissa Lee, and Ann Majchrzak. 2014. "Open Innovation Requires Integrated Competition-Community Ecosystems: Lessons Learned from Civic Open Innovation." *Business Horizons* 57(3): 391–400.

Arora, A., and A. Gambardella. 2010. "Ideas for Rent: An Overview of Markets for Technology." *Industrial and Corporate Change* 19(3): 775–803.

Arora, A., A. Fosfuri, and A. Gambardella. 2001. "Markets for Technology and Their Implications for Corporate Strategy." *Industrial and Corporate Change* 10(2).

Barboza, David. 2012. "Obtaining Financial Records in China." *The New York Times*, October 26. http://www.nytimes.com/2012/10/27/business/global/obtaining-financial-records-in-china.html?ref=global (accessed January 8, 2015).

Bertot, John C., Paul T. Jaeger, and Justin M. Grimes. 2010. "Using ICTs to Create a Culture of Transparency: E-Government and Social Media as Openness and Anti-Corruption Tools for Societies." *Government Information Quarterly* 27(3): 264–271.

Boudreau, Kevin J., and K.R. Lakhani. 2013. "Using the Crowd as an Innovation Partner." *Harvard Business Review* 91(4): 61–69.

Bradsher, Keith. 2012. "China Asks Other Nations Not to Release Its Air Data." *The New York Times*, June 5. http://www.nytimes.com/2012/06/06/world/asia/china-asks-embassies-to-stop-measuring-air-pollution.html?_r=2& (accessed January 8, 2015).

Brunswicker, Sabine, and Vareska van de Vrande. 2014. "Exploring Open Innovation in Small and Medium-Sized Enterprises." In *New Frontiers in Open Innovation*, ed. Henry Chesbrough, Wim Vanhaverbeke, and Joel West. Oxford: Oxford University Press.

Brunswicker, Sabine, Esteve Almirall, and Melissa Lee. 2013. "*Open Public Policy Innovation (OPPI)*." ESADE Working Paper, Barcelona.

Burnett, Gary, Paul T. Jaeger, and Kim M. Thompson. 2008. "Normative Behavior and Information: The Social Aspects of Information Access." *Library & Information Science Research* 30(1): 56–66.

Byrne, Liam. 2009. *Putting the Frontline First: Smarter Government*. London: The Stationery Office, Treasury White Paper, Cm 7753.

Carannante, Thomas. 2014. "Chinese Environmental Group Develops Mobile App to Monitor Air Pollution." *Scienceworldreport.com*, June 9. http://www.scienceworldreport.com/articles/15282/20140609/chinese-environmental-group-develops-mobile-app-monitor-air-pollution.htm (accessed January 8, 2015).

Chan, Calvin M.L. 2013. "From Open Data to Open Innovation Strategies: Creating E-Services Using Open Government Data." Paper presented at HICSS 2013, 46th Hawaii International Conference on System Sciences.

Chan, Calvin M.L., and Shan L. Pan. 2008. "User Engagement in E-Government Systems Implementation: A Comparative Case Study of two Singaporean E-Government Initiatives." *Journal of Strategic Information Systems* 17(2): 124–139.

Chao, Rebecca. 2013. "The Hunt for Open Data in China." *Techpresident.com*, September 11. http://techpresident.com/news/wegov/24332/hunt-open-data-china (accessed January 8, 2015).

Chao, Rebecca. 2014. "In China, an Open Data Movement Is Starting to Take Off." *Techpresident.com*, April 24. http://techpresident.com/news/wegov/24940/China-Open-Data-Movement-Starting-Take-Off (accessed January 8, 2015).
Chesbrough, Henry W. 2003a. "A Better Way to Innovate." *Harvard Business Review* 81(7): 12–13.
Chesbrough, Henry W. 2003b. "The Era of Open Innovation." *MIT Sloan Management Review* 44(3): 35–41.
Chesbrough, Henry William. 2006. *Open Innovation: The New Imperative for Creating and Profiting from Technology*. Boston, MA: Harvard Business School Press.
Chesbrough, Henry. 2014. "Open Social Innovation." In *New Frontiers in Open Innovation*, ed. H. Chesbrough, W. Vanhaverbeke, and J. West. Oxford: Oxford University Press.
Chesbrough, H., and M. Bogers. 2014. "Explicating Open Innovation: Clarifying an Emerging Paradigm for Understanding Innovation." In *New Frontiers in Open Innovation*, ed. H. Chesbrough, W. Vanhaverbeke, and J. West. Oxford: Oxford University Press.
Chesbrough, Henry, and Sabine Brunswicker. 2014. "A Fad or a Phenomenon? The Adoption of Open Innovation Practices in Large Firms." *Research-Technology Management* 57(2): 16–25.
Chopra, Aneesh. 2011. "*Winning the Future through Open Innovation: A Progress Report on Our Open Government Initiative*." Washington, DC: Executive Office of the President.
Codeando México. 2013. "The 115 Million Peso App." http://codeandomexico.org/proyectos/21 (accessed January 8, 2015).
Danish Agency for Digitisation. 2013. "Open Data Innovation Strategy (ODIS)." http://www.digst.dk/Servicemenu/English/Policy-and-Strategy/Open-Data-Innovation-Strategy-ODIS (accessed January 8, 2015).
Dorgelo, Cristin. 2014. "Challenge.gov Wins Innovations in American Government Award." Executive Office of the President, Office of Science and Technology Policy, January 23. http://www.whitehouse.gov/blog/2014/01/23/challengegov-wins-innovations-american-government-award (accessed January 8, 2015).
Dosi, Giovanni, and Joseph Stiglitz. 2014. "The Role of Intellectual Property Rights in the Development Process, with Some Lessons from Developed Countries: An Introduction." *Intellectual Property Rights: Legal and Economic Challenges for Development* 1.
Federal Institute for Access to Information and Data Protection. 2014. "Transparency Obligations Portal." Government of Mexico. http://portaltransparencia.gob.mx/buscador/search/search.do?method=begin (accessed January 8, 2015).
Felin, Teppo, and Todd R. Zenger. 2014. "Closed or Open Innovation? Problem Solving and the Governance Choice." *Research Policy* 43(5): 914–925.
Fioretti, Marco. 2011. "Open Data: Emerging Trends, Issues and Best Practices." Laboratory of Economics and Management at the Sant'Anna School of Advanced Studies.
GovHack.org. 2014. "GovHack 2014 Report – A Nation of Civic Hackers." http://www.govhack.org/govhack-2014-report-a-nation-of-civic-hackers/ (accessed January 8, 2015).
Guijarro, Luis. 2007. "Interoperability Frameworks and Enterprise Architectures in E-Government Initiatives in Europe and the United States." *Government Information Quarterly* 24: 89–101.
Henkel, Joachim. 2006. "Selective Revealing in Open Innovation Processes: The Case of Embedded Linux." *Academy of Management Proceedings* 2006(1): J1–J6.
Henkel, Joachim, Simone Schöberl, and Oliver Alexy. 2014. "The Emergence of Openness: How and Why Firms Adopt Selective Revealing in Open Innovation." *Research Policy* 43(5): 879–890.
Hess, Charlotte, and Elinor Ostrom. 2005. "A Framework for Analyzing the Knowledge Commons: A Chapter from Understanding Knowledge as a Commons: From Theory to Practice." *Library and Librarians' Publication*, Paper 21.
Hilvert, John. 2013. "Slow Progress on Government's Open Data Effort." IT News for Australian Business, February 23. http://www.itnews.com.au/News/334212,slow-progress-on-governments-open-data-effort.aspx (accessed January 8, 2015).

Hippel, E. von. 1994. *The Sources of Innovation*. Oxford: Oxford University Press.

Hippel, Eric von, and Georg von Krogh. 2003. "Open Source Software and the 'Private-Collective' Innovation Model: Issues for Organization Science." *Organization Science* 14(2): 209–223.

Howells, Jeremy. 2006. "Intermediation and the Role of Intermediaries in Innovation." *Research Policy* 35(5): 715–728.

Huijboom, Noor, and Tijs van den Broek. 2011. "Open Data: An International Comparison of Strategies." *European Journal of ePractice* 12: 1–13.

Innovate UK. 2014a. "Innovate UK _connect." https://connect.innovateuk.org/ (accessed January 8, 2015).

Innovate UK. 2014b. "Innovate UK Technology Strategy Board." http://www.gov.uk/government/organisations/innovate-uk (accessed January 8, 2015).

Jaeger, Paul T., and John Carlo Bertot. 2010. "Transparency and Technological Change: Ensuring Equal and Sustained Public Access to Government Information." *Government Information Quarterly* 27(4): 371–376.

Janssen, Katleen. 2011. "The Influence of the PSI Directive on Open Governmental Data: An Overview of Recent Developments." *Government Information Quarterly* 28: 446–456.

Janssen, Marijn, Yannis Charalabidis, and Anneke Zuiderwijk. 2012. "Benefits, Adoption Barriers and Myths of Open Data and Open Government." *Information Systems Management* 29(4): 258–268.

Jeppesen, Lars Bo, and Karim R. Lakhani. 2010. "Marginality and Problem-Solving Effectiveness in Broadcast Search." *Organization Science* 21(5): 1016–1033.

Lathrop, Daniel, and Laurel Ruma. 2010. *Open Government: Collaboration, Transparency, And Participation in Practice*. Sebastopol, CA: O'Reilly Media, Inc.

Medaglia, Rony. 2012. "eParticipation Research: Moving Characterization Forward (2006–2011)." *Government Information Quarterly* 29(3): 346–360.

Meijer, A.J., D. Curtin, and M. Hillebrandt. 2012. "Open Government: Connecting Vision and Voice." *International Review of Administrative Sciences* 78(1): 10–29.

Mergel, Ines, and Kevin C. Desouza. 2013. "Implementing Open Innovation in the Public Sector: The Case of Challenge.gov." *Public Administration Review* 73(6): 882–890.

Mexico. 2014. "National Digital Strategy." Office of the Presidency, Mexico. http://www.presidencia.gob.mx/edn/en/ (accessed January 8, 2015).

Miragliotta, Narelle. 2012. "From Local to National: Explaining the Formation of the Australian Green Party." *Party Politics* 18(3): 409–425.

Nam, Taewoo. 2012. "Suggesting Frameworks of Citizen-Sourcing via Government 2.0." *Government Information Quarterly* 29(1): 12–20.

Nelson, Richard R., and Bhaven N. Sampat. 2001. "Making Sense of Institutions as a Factor Shaping Economic Performance." *Journal of Economic Behavior & Organization* 44(1): 31–54.

Nelson, Richard, and Sidney Winter. 1977. "In Search of the Theory of Innovation." *Research Policy* 6: 36–76.

Obama, B. 2009. "Open Government Directive." Executive Office of the President, 8 December. http://www.whitehouse.gov/open/documents/open-government-directive (accessed January 8, 2015).

Open Data 100 Mexico. 2014. "About the Open Data 100 Mexico." http://www.opendata500.com/mx/about/?lan=en (accessed January 8, 2015).

Open Government Partnership. 2013. "Open Government in Denmark." http://www.opengovpartnership.org/country/denmark/ (accessed January 8, 2015).

Opening Up Government. 2014. "Search Latest Apps." http://data.gov.uk/apps (accessed January 8, 2015).

OSTP. 2013. "Implementation of the Federal Prize Authority: Fiscal Year 2012 Progress Report." In *Report from the Office of Science and Technology Policy*. Washington, DC: Executive Office of the President.

Ostrom, Elinor. 1990. *Governing the Commons: The Evolution of Institutions for Collective Action*. Oxford: Oxford University Press.

Peled, Alon. 2011. "When Transparency and Collaboration Collide: The USA Open Data Program." *Journal of the American Society for Information Science and Technology* 62(11): 2085–2094.

Picasso, Vicki, and Liam Prior Phelan. 2014. "The Evolution of Open Access to Research and Data in Australian Higher Education." *RUSC: Universities and Knowledge Society Journal* 11(3): 122–133.

Qiang, Xiao. 2008. "How China's Internet Police Control Speech on the Internet." Radio Free Asia, November 24. http://www.rfa.org/english/commentaries/china_internet-11242008134108.html (accessed January 8, 2015).

Sheridan, John. 2013. Why so low data.gov.au? Australian Government Department of Finance, November 11. http://www.finance.gov.au/blog/2013/11/11/why-so-low-datagovau/ (accessed January 8, 2015).

Simcoe, Timothy S., Stuart J.H. Graham, and Maryann P. Feldman. 2009. "Competing on Standards? Entrepreneurship, Intellectual Property, and Platform Technologies." *Journal of Economics & Management Strategy* 18(3): 775–816.

Smart Aarhus. 2014. "Aarhus Challenges." http://www.smartaarhus.eu/projects/aarhus-challenges/ (accessed January 8, 2015).

Susha, Iryna, and Åke Grönlund. 2012. "eParticipation Research: Systematizing the Field." *Government Information Quarterly* 29(3): 373–382.

Tauberer, Joshua. 2014. *Open Government Data: The Book*.

Teece, D., G. Pisano, and A. Shuen. 1997. "Dynamic Capabilities And Strategic Management." *Strategic Management Journal* 18(7): 509–533.

Timson, Lia. 2014. "Privatise NBN, Cut Start-Up Funding and Appoint a Chief Digital Officer: Commission of Audit." *Sydney Morning Herald*, May 2. http://www.smh.com.au/it-pro/government-it/privatise-nbn-cut-startup-funding-and-appoint-a-chief-digital-officer-commission-of-audit-20140502-zr3h7.html (accessed January 8, 2015).

US Department of Commerce. 2014. "America COMPETES." http://www.commerce.gov/americacompetes (accessed January 8, 2015).

US Department of State. 2014. "The Open Government Partnership." http://www.state.gov/j/ogp/ (accessed January 8, 2015).

US General Services Administration. 2014. "Data.Gov Impact Apps." http://www.data.gov/impact/ (accessed January 8, 2015).

Vega, Ana. 2012. "Can a Former Authoritarian State Embrace Open Government?" *Slate.com*, September 5. http://www.slate.com/articles/technology/future_tense/2012/09/open_government_and_data_movement_is_making_progress_in_mexico_.html (accessed January 8, 2015).

Veit, Daniel, and Jan Huntgeburth. 2014. "Open Government." In *Foundations of Digital Government*, 87–100. New York: Springer.

West, Joel, and Marcel Bogers. 2014. "Leveraging External Sources of Innovation: A Review of Research on Open Innovation." *Journal of Product Innovation Management* 31(4): 814–831.

West, Joel, and Scott Gallagher. 2006. "Challenges of Open Innovation: The Paradox of Firm Investment in Open-Source Software." *R&D Management* 36(3): 319–331.

Yang, Wang. 2013. "Data is King: Financial Work is Inseparable from Large Data." Forbes China.

Chapter 25

Serendipity and Chance in Scientific Discovery: Policy Implications for Global Society

Donald Gillies

Introduction

Many researchers have found the concepts of serendipity and chance (or accident) useful for analyzing discovery in science and technology. Here are a few examples. Royston M. Roberts published in 1989 a book entitled *Serendipity: Accidental Discoveries in Science*. Kantorovich and Ne'eman in 1989, and van Andel in 1994 published papers on serendipity. In 2004, there was published, for the first time in English, a long book by Merton and Barber on serendipity. This book had originally been written in 1958, but did not appear in print until 2002, when Il Mulino published an Italian translation. In 2010, Norrby published *Nobel Prizes and Life Sciences*, and chapter 2 of this book is entitled "Serendipity and Nobel Prizes." Anguera de Sojo et al. (2014) discuss "Serendipity and the Discovery of DNA." What they deal with is not the discovery of the double helix structure of DNA by Watson, Crick, Franklin, and Wilkins in 1953, but the much earlier discovery of the molecule itself, originally called nuclein. This discovery was made by the Swiss chemist Miescher in 1869.

This paper fully endorses the importance which has been given to the role of serendipity and chance in scientific discovery. However, the concept of serendipity is not entirely clear, and indeed has been used in different senses by different authors. Moreover, the relationship between serendipity and chance (or accident) is also in need of clarification. I will therefore begin by discussing in the next section the concepts of serendipity and chance.

The Handbook of Global Science, Technology, and Innovation, First Edition.
Edited by Daniele Archibugi and Andrea Filippetti.

Serendipity and Chance

There is one definite fact about the term "serendipity" – it was coined by Horace Walpole in a letter to his friend Horace Mann, dated January 28, 1754. However, Merton and Barber (2004) show that the subsequent history of the term was tortuous and complicated. It has been, and still is, used in a variety of different senses. In this paper, I will use a definition of serendipity which, according to Merton and Barber (2004: 112), was first formulated by Edward Solly in 1880 and is cited in the Oxford English Dictionary. According to this definition, serendipity consists in "looking for one thing and finding another." This definition does have some basis in Horace Walpole's original letter, for he says there: "you must observe that *no* discovery of a thing you *are* looking for comes under this description" (Merton and Barber 2004: 2). However, as we shall see, serendipity is sometimes used in broader senses, which also have some basis in Horace Walpole's original letter.

The classic instance of serendipity, as it has just been defined, is Columbus's discovery of America. Columbus was looking for a sea route to the East Indies obtained by sailing west. However, what he actually found was a new continent, whose existence was unknown to European geographers.

The next step in clarifying serendipity in the sense to be used in this paper is to distinguish between serendipitous discoveries and accidental or chance discoveries. A discovery can be said to be accidental or to involve chance if some accidental or chance occurrence plays an important role in the discovery. The classic instance of an accidental or chance discovery is Fleming's discovery of penicillin. Fleming was carrying out a routine experiment when one of his Petri dishes became contaminated with a mold. Fleming noticed that the colonies of *staphylococcus aureus* (a pathogenic bacterium), which would normally have covered the Petri dish, appeared to have dissolved in the vicinity of the mold. He concluded that the mold must be producing an anti-bacterial agent, which he later named penicillin. There can be no doubt that chance (or accident) played a large part in this discovery, but was the discovery serendipitous?

In his 1944 account of the discovery of penicillin, Fleming begins by sketching the background of his researches, first into septic wounds and then into lysozyme. He then writes (4):

> It was however, fortunate that, with the background I have briefly sketched, *I was always on the lookout for new bacterial inhibitors*, and when I noticed on a culture plate that the staphylococcal colonies in the neighborhood of a mold had faded away I was sufficiently interested in the antibacterial substance produced by the mold to pursue the subject. (Fleming 1944: 4; my italics)

At the time Fleming was performing a routine experiment, but he says explicitly that he was always on the lookout for new bacterial inhibitors, and attributes his further investigation of the mold to this fact. Thus Fleming found something he was looking for, and so his discovery, according to the above definition, was not serendipitous. Despite this, Fleming's discovery of penicillin, and indeed accidental discoveries in general, are often described as serendipitous.

Van Andel also quotes from Fleming, but suggests (1994: 639) a compromise position according to which Fleming's discovery is described as an instance of *pseudoserendipity*. The concept of pseudoserendipity comes from Roberts, who defines it as follows:

> I have coined the term *pseudoserendipity* to describe accidental discoveries of ways to achieve an end sought for, in contrast to the meaning of (true) *serendipity*, which describes accidental discoveries of things not sought for. (1989: x)

This point of view is expressed also in the title of Roberts's book *Serendipity: Accidental Discoveries in Science*. My own view, however, is that it makes for greater clarity to distinguish the two concepts "serendipitous discovery" and "accidental discovery" rather than running them together in a portmanteau concept of "pseudoserendipity." One reason for this view is that many serendipitous discoveries are not in fact accidental discoveries. The standard example of Columbus discovering America is an illustration of this. Columbus's discovery did not involve any chance events or accidents like the contamination of a Petri dish. Anyone who followed Columbus's strategy of sailing west in order to discover a sea route to the East Indies would inevitably, if his ship hadn't accidentally sunk, have discovered America. America was simply there blocking the path, and any mariner following Columbus's search strategy would inevitably have run into it.

Let us further illustrate this idea by another discovery, which is acknowledged by all to have been serendipitous, namely Röntgen's discovery of X-rays. When he made the discovery, Röntgen was carrying out an investigation into whether cathode rays could penetrate the glass walls of the cathode ray tube in which they were produced. However, when he switched on his cathode ray tube, he noticed that a barium platinocyanide screen more than a meter from his apparatus started to glow with a shimmering light. He investigated the matter further and concluded that the glow was caused by some rays produced by the cathode ray tube which were different from cathode rays. These rays he named X-rays, because their nature was unknown.

Now Röntgen's discovery was clearly serendipitous. He was looking for one thing (cathode rays which penetrated the walls of a cathode ray tube), and found another (the new and mysterious X-rays produced by the cathode ray tube). It also seems that accident played a key role in the discovery. Röntgen happened to notice the glowing of the barium platinocyanide screen. But why was there a barium platinocyanide screen in the laboratory? The answer is that this was the fluorescent screen which Röntgen had planned to use as a detector of any cathode rays which penetrated the glass walls of the cathode ray tube. Now, suppose Röntgen had failed to notice the glowing of the screen. He would have continued his experimental protocol and used the screen to attempt to discover cathode rays penetrating the glass walls, and, in this investigation (if carried out systematically as it would have been), Röntgen would have undoubtedly discovered X-rays. So, although there was an accidental observation in the discovery of X-rays just as there was in the discovery of penicillin, the two cases are different. The accidental observation was essential to the discovery of penicillin which would not have occurred without it, but the accidental observation was only incidental to the discovery of X-rays which would almost certainly have occurred without it.

For these reasons, I favor distinguishing between serendipitous and accidental or chance discoveries. Of course it could be the case that a serendipitous discovery is also accidental, or, at least, involves chance elements. An example of this is Perkin's discovery of the first synthetic dye (Roberts 1989: 66–70). Perkin set out to try to synthesize quinine from coal tar products. He first used toluidine for his starting point, but did not meet with success. He then tried using aniline as his starting point, and was no more successful. He obtained, as a product of his chemical reactions, a black solid. Another investigator might have thrown this away, but Perkin noticed that the water or alcohol used to wash it out of the flask turned purple. This gave him the idea that the black solid might yield a purple dye: an idea which proved to be correct.

This is clearly an instance of serendipity. Perkin was looking for one thing (a method for synthesizing quinine), and found another (the first artificial dye). However, the case of Perkin differs from those of Columbus and Röntgen. As I have argued, it was almost inevitable that Columbus and Röntgen would have made their discoveries just by following the search strategy which they had laid down for themselves in advance. This was not so in the case of Perkin. He, like Fleming, needed to have the good judgment to change his search strategy in quite a radical way. Perkin when he obtained an unpromising solid black residue had to start investigating the new idea that it might yield a purple dye. He could just as easily have thrown it down the drain, and started another attempt to synthesize quinine. However, Perkin also benefited from a lucky chance. As Roberts says (1989: 67): "Actually, the aniline Perkin used contained small amounts of toluidine, which was essential to the formation of the purple dye." This all agrees surprisingly well with Walpole's original definition of serendipity, for Walpole says of the three princes of Serendip (quoted from Merton and Barber 2004: 2): "they were always making discoveries, by accidents and sagacity, of things which they were not in quest of." Perkin discovered something he was not in quest of partly by accident (the aniline he used was contaminated with toluidine), but also by sagacity. He showed considerable sagacity in reaching the conclusion that his black solid residue might yield a purple dye.

That concludes my main analysis of the concept of serendipity, and its relation to chance (or accident). However, it will be useful to make one further point about the concept. The word "serendipity" suggests something strange and mysterious. Yet there is really nothing mysterious about many instances of serendipity as it has been defined in this paper. Suppose I have mislaid my pen and search for it in a drawer filled with miscellaneous objects. The pen is not there, but, at the bottom of the drawer, I find an old notebook which had gone a-missing more than a year before. This is an example of serendipity since I was looking for one thing, and found another. However, there is nothing strange or mysterious about it. Some of the famous examples of serendipity such as Columbus's discovery of America, or Röntgen's discovery of X-rays, are scarcely more mysterious than my everyday example of the pen and the notebook.

Implications of Serendipity and Chance for the Individual Researcher

There are a number of clear implications of serendipity and chance for the individual researcher who wants to make a discovery in science or technology. First of all, such a researcher should always keep an eye open for something unexpected, and reflect as to whether this unexpected occurrence might have some significance. This was already stressed by the pre-Socratic philosopher Heraclitus, who wrote: "If one does not expect the unexpected one will not find it out" (Kirk and Raven 1957: 195). Second, a researcher should always have a flexible attitude toward his or her research plans, and be prepared to change them in the light of unexpected developments. This implication comes out very clearly from the cases of Fleming and Perkin. Fleming was conducting a fairly routine investigation of the pathogenic bacterium *staphylococcus aureus*, when he noticed the Petri dish contaminated by a mold. Now such contaminations were a frequent occurrence in the bacteriological laboratories of the time. Normally the plates so contaminated were of no further use, and were simply discarded. However, Fleming had the insight to realize that this contaminated plate had unexpected features which were worthy of further investigation, and he had the flexibility to abandon his planned investigation, and to start a new and different investigation of the contaminated plate. The case of Perkin is very similar. Perkin was trying to synthesize quinine, but instead obtained an unpromising black solid. However, noticing the unexpected fact that this black solid turned water and alcohol purple, Perkin had the flexibility to give up his attempt to synthesize quinine, and to start a new research program designed to synthesize an artificial purple dye. So, to sum up, the implication of serendipity and chance is that the individual researcher should have the flexibility to alter his or her plans for research in the light of unexpected developments.

At first sight this may seem just a matter of training researchers to acquire some psychological characteristics. However, this is to ignore the importance of the social and economic environment in which the researcher is working. This environment was very favorable for Fleming. He was the deputy of the head of the laboratory (under Sir Almroth Wright). Sir Almroth had a great admiration for Fleming's abilities as researcher, and did not interfere in any way with Fleming's research. Thus Fleming was free to carry out whatever research he liked. Other researchers are not in such a happy position. Perkin, for example, at the time of his discovery, was a student of the Royal College of Chemistry (now part of Imperial College London). His supervisor was a distinguished professor of chemistry (von Hofmann), who had come to London from Germany. He set Perkin the task of trying to synthesize quinine. Quinine was the only effective treatment for malaria, but it was an expensive natural product, and there could be considerable commercial advantages in finding how to make it in the chemical laboratory. When Perkin decided that it would be more interesting to try to create an artificial dye instead of synthesizing quinine, he was afraid that von Hofmann would not approve. Luckily von Hofmann was away on a visit to Germany at the time. So Perkin carried out his new research program secretly in a hut in his garden with the help of his brother Thomas and a friend, Arthur Church. However, if Perkin had been working in an official laboratory, under the watchful eye of von Hofmann, he might not have been able to pursue his research plans, and create the first synthetic dye. All this shows that a consideration of the psychology of the individual researcher is not enough, and we must also examine policy implications about the way in which research should be organized and funded. We will make a start with this investigation in the next section.

Policy Implications of Serendipity and Chance

The policy implications of serendipity and chance in scientific and technological discovery are clear. Research and its funding should be organized so that the researcher is encouraged to adopt a flexible attitude toward his or her research plans, and to be prepared to alter those plans in the light of unexpected developments. I will call this the principle of encouraging flexibility. Unfortunately, this principle is negated rather than adopted by most existing systems of research funding. In this section, I will give a particular striking example of a research funding system which negates the principle of encouraging flexibility,[1] and in the next section, I will discuss the problem in more general terms in the context of globalization.

The funding scheme, which I will now describe, was introduced in 2001 by the Arts and Humanities Research Board (AHRB), one of the leading UK funding bodies. The scheme was designed to provide some research leave for researchers to pursue their research free from teaching and administrative duties. In September 2001, the AHRB issued a Guide to the Research Leave Scheme (AHRB 2001). This guide explains the details of the new scheme in a document of 28 pages. It has to be said that such bureaucratic documents do not in general make very enjoyable reading. However, I will try to show that the judicious mind can find much of interest in the pages of AHRB (2001).

Let me begin by describing the application procedure, which is very complicated. At the end of AHRB (2001), there is a copy of the application form with a further three pages of guidance notes on how to fill it in. The heart of the application is of course the description of the research which the candidate wants to carry out. However, the candidate is certainly not allowed to use his or her initiative in describing the research. The description must be given in terms of a scheme devised by the AHRB. The relevant section begins: "You should read the guidance notes provided before completing this section," and the applicant is then told to use the following sub-headings:

- Research question(s)
- Aims and objectives
- Research context
- Research methods
- Timetable for completion within the period of leave (please refer to guidance notes for definition of completion)
- Plans for public dissemination (e.g., publication, exhibition, performance)

Of course the applicant might very well think that the scheme devised by the AHRB is not an appropriate one for describing his or her research, but then "rules are rules." Moreover the instructions conclude with the following instruction: "Please complete the word-count box provided: if you exceed the word limit, your application will be deemed ineligible for funding and will be returned to you."

Filling in an application form is, however, only the beginning. The candidate has to find two external assessors who are then sent a complicated form with guidelines about how to fill it in. They have to use this form to give an assessment of the candidate's application. After that "The external assessors' reports together with your application form will be sent for assessment to peer reviewers" (AHRB 2001: 13) because, as we would expect, "The Board is committed to the principle of peer review in its Advanced Research competitions" (AHRB 2001: 12).

So far the AHRB has introduced quite a lot of complexity into the application procedure. However, the most striking innovation of the AHRB is not here, but rather in what happens after the completion of the period of research leave. This is described in the sections of AHRB (2001) dealing with monitoring, and we will now give a brief account of the new procedures introduced.

Addressing the candidate, AHRB (2001: 19) says:

> You must submit an end-of-award report within three months of the end of the period covered by the award. ... In the report you will be asked to provide a self-assessment of the extent to which you have met the original aims and objectives of the research ... The report will normally be assessed through peer review.

The more detailed instructions require that the scheme of research from the original application be attached to the end-of-award report and referred to when completing the end-of-award report. The end-of-award report with its attachment is then sent for peer review so that the peer reviewers can judge to what extent the original plan has been carried out.

What is striking here is that the AHRB are imposing a particular strategy for carrying out research. This strategy has two parts. (1) A detailed plan for future research must be drawn up. Moreover, this plan must use categories imposed by the AHRB, and not categories selected by the researcher himself or herself. (2) This plan must be strictly adhered to. We can summarize this strategy as that of strict adherence to earlier detailed research plans. It is immediately obvious that this completely contradicts the principle of encouraging flexibility.

The AHRB makes clear that they fully endorse this strategy of strict adherence when they go on to describe how the peer reviewers will be required to assess the end-of-award report, and also the penalties which will fall on the heads of those researchers who have bad end-of-award reports. The peer reviewers are required to assess the end-of-award report using the following three categories: satisfactory, problematic, and unsatisfactory. The authors of AHRB (2001) go on to say:

> *unsatisfactory* – indicates a project that ... has failed to conduct the research as agreed at the time of the award (and any subsequent agreed changes to the plan of research), and which therefore does not meet the regulations and the aims and objectives of the particular scheme of awards. (20)

They then list the penalties to which the unfortunate researcher who has an unsatisfactory end-of-award report will be subjected. A selection from these penalties runs as follows:

> If your end-of-award report is assessed as unsatisfactory:
>
> - we will write to you, and to your host institution, to inform you of the status of the report.
> - you will be debarred from making applications to any of the AHRB's research schemes for two years ... We will write to you, and to the Head of your institution, confirming the penalty ...
> - we will keep a record of the unsatisfactory assessment on file, and this will be taken into consideration when you make further applications to the AHRB. (20)

These penalties are very harsh, particularly for a young researcher. Such a researcher, on receiving an unsatisfactory assessment, is debarred from applying for any research funding from the AHRB for two years, and, since the unsatisfactory report will be kept on file, it becomes very unlikely that he or she will get any further research funding from the AHRB even after two years. As the AHRB might well be one of the few sources for research funding, this is a heavy blow to the young researcher's hopes of a research career. In addition, there will be a letter to the Head of the researcher's institution confirming the penalty, and this makes it unlikely that the unfortunate researcher will get further research leave from that institution, as well as damaging promotion prospects. The curious thing about these heavy penalties is that they are quite compatible with the researcher having worked hard during the period of research leave, and even with the researcher having done what is generally agreed to be brilliant research. This is because one of the criteria for getting an unsatisfactory assessment is explicitly stated to be failure "to conduct the research as agreed at the time of the award."

It is now an easy matter to show that if the research of Fleming and Perkin had been funded by a scheme similar to AHRB (2001) (which was not of course the case!), then, if the two researchers had followed the requirements of the scheme, Fleming would have failed to discover penicillin, and Perkin would have failed to discover the first synthetic dye. Let us take the case of Fleming first.

Fleming made his discovery when engaged in a piece of fairly routine research into the staphylococcus bacterium. Staphylococci are responsible for a variety of infectious diseases – some quite serious. The most virulent form of the bacterium is the golden-colored *staphylococcus aureus*. There are also staphylococci with other colors, such as white, which are much less virulent. In 1927 or 1928, Fleming read a paper claiming that colonies of staphylococci changed color if they were kept at room temperature for several days. Fleming decided to check the claims of the paper he had read by conducting a program to investigate color changes in staphylococci. His procedure was very simple. He prepared colonies of staphylococci in Petri dishes, and left these dishes on his bench, examining them every few days to see if changes in the color of some of the staphylococci could be observed.

It should be pointed out that this research program was in line with a well-known principle introduced by Pasteur in the nineteenth century, that of *oxygen attenuation*. Pasteur discovered that many pathogenic bacteria lose some of their virulence if exposed to oxygen for lengthy periods. The bacteria might in some cases be treated with oxygen gas, but in others a simple exposure to air for a protracted period could suffice. Pasteur used this principle to create vaccines. So Fleming's research was a typical instance of what Kuhn would call normal science. If successful, the research might establish another instance of Pasteur's well-known principle of oxygen attenuation.

When Fleming noticed the contaminated plate, however, he immediately abandoned this rather routine bit of research, and devoted his energies to a quite new research program, namely that of investigating the mold which had contaminated his Petri dish, and which was apparently producing a bacterial inhibitor which acted on *staphylococcus aureus*. As we have already mentioned, Fleming's position in the laboratory meant that he was quite free to pursue whatever research he liked. Suppose, however, that he had been obliged to obtain funding for his research through a funding scheme similar to AHRB (2001). For his first, rather routine piece of research into color change in staphylococci, he would have had to fill in a complicated application form and find two assessors to support him. Both Fleming and his assessors would have had to spend some time on this form-filling, which would have been time deducted from their research. Fleming's application would then have been sent to a set of peer reviewers who would have had to take time away from their own research in order to review it. However, let us suppose that Fleming was finally awarded the research funds and began his research into color changes into staphylococci. Now, however, Fleming is financed by a research grant like AHRB (2001), which, as we have seen, imposes on researchers the strategy of strict adherence to earlier detailed research plans by threatening heavy penalties to anyone who fails to follow this strategy. If Fleming had been pressurized by such a grant-giving body to follow their directives, he would have continued his routine normal science investigation of color changes in staphylococci, and failed to discover penicillin.

The case of Perkin is of course just the same as that of Fleming. Perkin was given the task of trying to synthesize quinine by his professor (von Hofmann). However on the appearance of a black residue which stained water and alcohol purple, Perkin decided to change his research program to that of trying to synthesize a purple dye. Perkin was worried that von Hofmann would not approve of this plan, but luckily von Hofmann was away in Germany, and so Perkin went ahead with his plans, carrying them out secretly in a hut in his garden. But suppose Perkin had been financed by a research grant like AHRB (2001). His change of plan would then have risked the heavy penalties which, as we have seen, that scheme imposes on any researcher who does not carry out his original plan. As we remarked these penalties are particularly harsh for a young researcher whose whole research career might be ended by their imposition. Thus in an AHRB (2001) regime, Perkin might not have dared to change his research plans and so would have failed to discover the first artificial dye.

What is so remarkable here is that a leading UK research funding body should devise a research funding scheme based on principles which would probably have prevented two of Britain's most famous scientists making their most famous discoveries. This clearly shows that the role of serendipity and chance in scientific discovery has very important policy implications, and that at least some members of important funding bodies are quite unaware of these implications. Still, it could be that AHRB (2001) is just one isolated case. I will therefore, in the next section, discuss the problem in more general terms in the context of contemporary global science.

The Effects of Globalization on Research

Globalization has transformed the world economy, and it has similarly transformed the field of research. In the late 1960s and early 1970s, research groups were characteristically located in a single country, and often within a single university in that country. Nowadays, however, members of the same research group are typically scattered throughout the world. They can of course still communicate on a day-to-day basis by email, and meet regularly at international conferences. Virtual presence via Skype at conferences is now becoming more and more common. The communication possibilities of modern technology have been rendered effective by the adoption of a common language for research (English).

Another factor which favors the globalization of research is the increasingly easy availability of information via the Internet. In the late 1960s and early 1970s, it was very difficult for anyone to do research who did not live near a large research library, containing the necessary papers and books. This tended to concentrate research groups in a few major universities. Now, however, researchers with an Internet connection and an iPad can download nearly all the papers they need for their research, and even much of the data they require. There are still some blocks caused by attempts to preserve copyright legislation, and government secrecy, but it looks as if even these blocks will be removed in the coming years. So we already have a research community dispersed throughout the globe, but yet in constant communication with each other; and this community is likely to extend and become more and more the basis of research in the future.

Because it is based on the advances of contemporary technology, globalization is well grounded, and much more likely to increase than diminish. This does not mean, however, that we have to accept all the features of contemporary globalization as inevitable and fully determined by technology. There are both bad and good features of contemporary globalization, and, it may be perfectly possible to develop globalization in such a way that the good features are strengthened, while the bad features are eliminated or at least minimized.

In general we can say that, paradoxically, two exactly opposite tendencies can be detected in contemporary globalization. One is a good tendency and the other a bad tendency. The good tendency is toward a lively and beneficial variety. The bad tendency is toward a dull and harmful uniformity. These two tendencies can be illustrated by the example of food production and consumption under globalization. The tendency toward a lively and beneficial variety is clearly marked. In the past, the members of a particular nation would, in general, eat only dishes of the national cuisine, whatever that was, and would be completely unfamiliar with the cuisines of other countries. Nowadays, in the big cities at least, most people can enjoy a variety of different national cuisines in restaurants, and obtain the ingredients needed to make these dishes in their local supermarket. This is surely an example of a lively and beneficial variety.

However, there is also a tendency in the opposite direction through the increasing prevalence of fast food consisting of a limited number of dishes (the same throughout the world) delivered by a limited number of multinational chains. This is dull uniformity, and also harmful. Standard fast food is well known to be harmful to the health because of its high content of sugar, salt, and animal fats, and the large-scale production of fast food is also damaging to the environment.[2] The increasing international diffusion of standard fast food is a consequence of globalization, but not an inevitable one. It is a feature of contemporary globalization which could be eliminated.

Research production would seem to be very different from food production, and yet, strange to say, the two tendencies at work in the globalization of food also seem to be at work in the globalization of research. Let us start with the good tendency. The increasing number of researchers, and their wide diffusion throughout the globe, makes it possible for a considerable number of different research programs to be pursued in any field. Particularly in the light of serendipity and chance, the pursuit of a large number of different approaches to any given problem makes it more likely that significant scientific discoveries will be made. This was realized by Bacon as early as 1620, when he wrote: "For then only will men begin to know their strength when instead of great numbers doing all the same things, one shall take charge of one thing and another of another" (Bacon 1620: 293). Globalization of research could potentially encourage the pursuit of a considerable number of different research programs in any field, and this would certainly be a lively and beneficial variety.

Unfortunately the opposite tendency toward a dull and harmful uniformity also exists in globalized research. This occurs when a small group (the trendsetters) start a particular research program, and everyone else simply imitates them by working on the same research program. The situation becomes one of "great numbers doing all the same things." Possibly the research program which sets the trend does lead to some good results, but these would have been discovered anyway, even if pursuit of the program had been confined to a relatively small group of researchers. What is lost is the valuable results, arising directly or through serendipity and chance, which might have been produced by some groups pursuing a variety of alternative research programs.

But why does the tendency to a dull and harmful uniformity arise in globalized research? I will next argue that there are two factors principally responsible for this tendency, namely (1) the assignment of research funding by a competition between projects which is decided using anonymous peer review (I will refer to this, for brevity as "assignment by peer review"), and (2) the establishment of an alleged hierarchy of excellence among research groups, journals, and so on, and the use of this alleged hierarchy to evaluate research. I will now show that these two factors lead, in any specific area of research, to the predominance of one particular research program, and the elimination of alternatives, that is, to the exact opposite of what is desirable.

Let us start then by considering the assignment of funding by anonymous peer review. It will be useful to see how this would have affected the research in the four striking cases of serendipity and chance which we discussed earlier. Now Columbus had to raise funding for his expedition which resulted in the discovery of America. He applied first to the King of Portugal who turned him down. Davis (1973: 13) explains why as follows:

> Columbus failed to get Portuguese support because the king could muster advisers with a sound knowledge of world geography. The plans of Columbus were based on nonsensical geographical notions that grossly understated the distance westward from Europe to Japan, which was his intended destination, and the Portuguese experts exposed them.

It is perhaps unfair to describe Columbus's geographical notions as "nonsensical." Columbus based his estimate of the circumference of the Earth on the writings of Aristotle and Ptolemy which were generally accepted at the time. The Aristotelian-Ptolemaic paradigm in astronomy was only overthrown by the Copernican revolution which started in 1543, 51 years after Columbus's first voyage of 1492. However, Columbus modified Ptolemy's geography in the light of Marco Polo's travels. These had shown the extent of Cathay (China), and the existence of an island Zipangu (Japan) off the coast of China. Columbus calculated that Zipangu must be 4000 miles to the west of Portugal. This estimate agreed with a map produced by Toscanelli in 1474, and with Behaim's globe of 1492. Columbus reckoned that it would take him about 33 days to sail from the Azores to Zipangu, and, ironically, he sighted land after 35 days of sailing.

But although Columbus's geography was not nonsensical, the Portuguese experts knew better. This is a most instructive example because the experts who criticized Columbus were indeed correct. Yet Columbus's expedition was worthwhile all the same because of serendipity. This shows clearly that, once we take account of the important role of serendipity and chance in scientific discovery, then experts are no longer able to judge whether a proposed research project will prove fruitful or not.

The case of Perkin is also interesting from this point of view. The original project was to synthesize quinine. In fact we now know that this is a very difficult task, which could not have been carried out at the time (1856). Synthetic quinine has never been produced industrially as a substitute for naturally occurring quinine, and there has even been a debate about when the first laboratory synthesis of quinine occurred. One side argues for 1944, and the other for 2001. In the light of all this, experts in 1856 might well have argued correctly that the project for synthesizing quinine was not feasible with the methods available at the time, and so should not be funded. However, if the project had not been carried out, the first artificial dye would not have been discovered.

The other two cases are rather different. They were both rather routine pieces of research which might well have been refused funding on the grounds that they would not produce any very new or exciting results. Fleming's research was into a possible instance of the well-established principle of oxygen attenuation of pathogenic bacteria. Röntgen set out to discover whether cathode rays penetrated glass. A contemporary expert might very reasonably have argued that this was rather a boring project, and unlikely to produce any result of interest. After all, at that time (1895), cathode rays had been studied for over 20 years by means of cathode ray tubes. The expertise acquired made it unlikely that cathode rays would penetrate the glass walls of the tubes used, and, even if some did, this could almost certainly be corrected by making the glass wall a bit thicker. Yet this seemingly dull and routine research project yielded one of the most striking discoveries ever made in physics.

Let us now return to the question of whether the policy of assigning funds to research projects assessed by anonymous peer review is a good one. How likely is it that peer reviewers, even if they are experts in the field, will deal well with the very hard problem of assessing the potential of research projects? My own view is that they are not likely to make a very good job of it, because considerations of serendipity and chance show that no one really knows in advance what research projects are going to succeed. In particular, peer reviewers, even if experts in the field, do not really know which research projects are likely to succeed, but they do often have quite strong prejudices about the matter. These prejudices, taken in aggregate, lead, as I will now show, to sub-optimal decision-making.

The root of the problem is what I will call *researcher narcissism*. This is a condition which affects nearly all researchers (including the author of the present paper). It consists in an individual researcher believing quite strongly that his or her approach to research in the field is the best one, and the one most likely to produce good results; while the other approaches are less good and less likely to produce any good results. The existence of researcher narcissism is not surprising. Most researchers will spend some time thinking about which approach to adopt to their research, and, when they opt for a particular approach, there will be reasons for their decision. Moreover, once having made the decision, it tends to get reinforced by the fact that they mix a great deal with others working along similar lines, all of whom are convinced that what they are doing is right. In addition, any researcher has a strong interest in his or her approach proving to be the successful one. If it is successful, the researchers working on it will be more likely to get promotions and perhaps prizes. Their PhD students are more likely to get jobs, and so on. Conversely, if their approach to research proves unsuccessful, all these desirable consequences are much less likely to follow, and they are likely to see the rewards going to those working on a rival research program. Now it is a universal characteristic of humans to believe what is in their interest, and, in the present case, this amounts to the belief of researchers that they have adopted the best approach to research in the subject.

Let us assume therefore that we have community of peer reviewers who are all researchers who suffer from researcher narcissism. How will this affect the way in which research projects are selected for funding by peer review? Because of researcher narcissism, a peer reviewer is likely to take a favorable view of a research project which adopts the same general approach to research as that of the peer reviewer, and an unfavorable view of a research project which adopts a different approach. Now let us suppose that peer reviewers are chosen more or less at random. Very few of these peer reviewers will, by definition, be working on a minority research program. If the research project adopts the approach of this minority research program, therefore, it is likely to be viewed with disfavor by the majority of the peer reviewers and so won't be funded. Conversely a research project which adopts a majority or mainstream approach is much more likely to be funded. Hence, funding based on assessment of research projects by peer review will lead to most of the funding going to whatever is the majority research program, and minority research programs receiving little or no funding. Thus the system results in the dull and harmful uniformity, which was mentioned earlier.

The tendency is reinforced if attention is paid to the alleged hierarchies of excellence in research. The existence of such hierarchies is hardly in doubt since league tables for institutions, journals, and so on are published regularly. The problem, however, with laying too much weight on such tables is that, even if they are entirely correct at present, which is much to be doubted, it by no means follows that they will continue to be in the future. Often approaches to research are very fruitful for a while, but are then superseded by new approaches. Newton may have been the most brilliant mathematician in the world in the 1670s, but in the eighteenth century Newton's approach to the calculus was definitely superseded by the approach developed in continental Europe. Those who stuck to the Newtonian approach because of its prestige were onto a loser.

Let us consider a country (X say) whose research groups are relatively low down in the alleged hierarchies of research excellence. The government of X will now ask: "How can we get our researchers to improve?" The usual answer is the following. A group of American universities are at the top of the league at the moment. We must therefore get our researchers to imitate them. In particular we must rate very highly and reward those of our researchers who succeed in publishing in those journals controlled by these top universities. Now these journals will select papers by anonymous peer review, and so to get a publication in them, it will be necessary to participate in the research program supported by that journal, that is, the research program supported by the group at the top of the league. Hence, such research evaluations, which are very common, will result in the researchers of X imitating the research which is being done in the groups at the top of the league. This again reinforces the tendency to a dull and harmful uniformity. Of course, the considerations of serendipity and chance show that common government policies of the type we have just described are quite wrong, and the researchers of X would be better advised to pursue research programs which are different from those being followed by groups who are currently at the top of the league. There could indeed be intellectual and cultural traditions specific to X, which might suggest approaches different from those of researchers at the top of the league, and these different approaches could very well lead to interesting results. Yet instead of building on valuable local traditions, governments often just encourage imitation of whatever those currently at the top of the league are doing.

Having pointed out some features of present-day research organization which push in the undesirable direction of a dull and harmful uniformity, the question naturally arises of whether there are alternative policies which could be introduced and which would encourage lively and beneficial variety instead. Now this is obviously a complicated problem, which cannot be dealt with at an appropriate length in the present chapter. However, a couple of preliminary suggestions can be made. The first suggestion is that there should be a shift away from attempting to make any immediate evaluation of a researcher's work. The history of science shows over and over again that developments which in the long run are seen as major advances can be judged initially by contemporary researchers to be valueless. I give three examples of this typical phenomenon in Gillies (2008: 14–27). Very often 20 or 30 years need to elapse before a sound conclusion as to the value of a piece of research can be reached. Of course it is perfectly reasonable to check that researchers are working, but this can easily be done by counting their publications. The mistake is to evaluate a publication in a high-status journal as being of higher quality than one in a low-status journal. It could well be that in 20 or 30 years, the publication in the low-status journal will be considered as the one which really made the important breakthrough. This policy would allow local traditions with their journals to flourish, adding to the variety of the system.

The second suggestion is to set aside some funds to encourage the formation of international research teams. If we suppose that the members of such a team receive their main funding from their home university or research institute, the extra funding would be needed only to pay for communication, through travel and workshops, and for the publication of results. These items are not very costly. The formation of such international networks could be very helpful to researchers who are trying to develop minority approaches. In each individual country, a minority approach is liable to be crushed, but, if there are minorities who adopt the approach in many countries, together they can form an international group of researchers of a viable size. In this way a minority approach, which would be liable to disappear in every individual country, could flourish at the global level.

Conclusions

In this paper I have argued that the role of serendipity and chance in scientific and technological discovery does have important policy implications. It shows that researchers should be encouraged to have the flexibility to change their research plans in the light of unexpected developments, and that in every area of research a variety of different approaches and research programs should be encouraged. Existing systems of research organization, and, in particular, the method of assigning research funding on the basis of anonymous peer review, have exactly the opposite effect. These contemporary systems of research organization are therefore in need of considerable changes. A couple of suggestions have been made for changes, which could prove beneficial, but more study of this problem is clearly needed.

Notes

1 This example is taken from Gillies (2008: 101–114).

2 Two classic works about the harmful effects of standard fast food on health and the environment are Schlosser (2001) and Pollan (2006).

References

AHRB. 2001. "Guide to the Research Leave Scheme." Arts and Humanities Research Board.

Anguera de Sojo, A., J. Ares, M.A. Martínez, J. Pazos, S. Rodríguez, and J.G. Zato. 2014. "Serendipity and the Discovery of DNA." *Foundations of Science* 19: 387–401.

Bacon, F. 1620. *Novum Organum*. English translation in *The Philosophical Works of Francis Bacon*, ed. R.L. Ellis and J. Spedding, 212–387. London: Routledge, 1905.

Davis, R. 1973. *The Rise of the Atlantic Economies*. London: Weidenfeld and Nicolson.

Fleming, A. 1944. "The Discovery of Penicillin." *British Medical Bulletin* 2: 4–5.

Gillies, D. 2008. *How Should Research Be Organised?* London: College Publications.

Kantorovich, A., and Y. Ne'eman. 1989. "Serendipity as a Source of Evolutionary Progress in Science." *Studies in the History and Philosophy of Science* 20(4): 505–529.

Kirk, G.S., and J.E. Raven. 1957. *The Presocratic Philosophers*. Cambridge: Cambridge University Press.

Merton, R.K., and Barber, E. 2004. *The Travels and Adventures of Serendipity: A Study in Sociological Semantics and the Sociology of Science*. Princeton, NJ: Princeton University Press.

Norrby, E. 2010. *Nobel Prizes and Life Sciences*. Singapore: World Scientific.

Pollan, M. 2006. *The Omnivore's Dilemma: The Search for a Perfect Meal in a Fast-Food World*. London: Bloomsbury.

Roberts, R.M. 1989. *Serendipity: Accidental Discoveries in Science*. Hoboken, NJ: John Wiley & Sons, Inc.

Schlosser, E. 2001. *Fast Food Nation: What the All-American Meal Is Doing to the World*. London: Penguin.

van Andel, P. 1994. "Anatomy of the Unsought Finding: Serendipity: Origin, History, Domains, Traditions, Appearances, Patterns and Programmability." *British Journal for the Philosophy of Science* 45(2): 631–648.

Chapter 26

Global Climate Change and the Direction of Technological Change

Andrew Tylecote

Introduction

This chapter presents an unfamiliar approach to a familiar issue: man-made climate change. It is known that the emission of greenhouse gases tends to cause global warming, that the main greenhouse gas, carbon dioxide, persists in the atmosphere for centuries, and that emissions of this and other greenhouse gases are amply high enough to cause substantial and continuing warming. Since most of the world's population lives in areas in which excessive heat already presents problems, this must be of deep concern, except to those who disbelieve the science. It is also clear that rising temperatures will lead to shifting rainfall patterns and other meteorological miseries ("extreme weather events"). It follows that mankind needs to take action to reduce net greenhouse gas emissions (GGe) quickly and far – indeed to zero: the more challenging because of rising population and rising income per head, which must tend to push them up. Technological change across a wide front will play a crucial role. A wide range of economic (and other) incentives will be required both to drive such change and to get GGe-saving ("carbon-saving") technologies taken up as fast as possible.

This is all so familiar as to be almost banal – see Stern (2009) and next section. And yet GGe are not only continuing, but continuing to rise, because few countries are taking action which matches the magnitude of the challenge. The countries doing least are less developed countries, whose rising share of world population and income now accounts for a majority of GGe. The political inconveniences of action are clear, immediate, and specific to the decision-taker; the benefits from action will be spread across the world in general and centuries to come, and are most unlikely to be credited to any particular politician or businessman.

The Handbook of Global Science, Technology, and Innovation, First Edition.
Edited by Daniele Archibugi and Andrea Filippetti.

It follows that effective action against climate change will be taken much later than the good and wise have recommended. It is argued in this chapter that this has striking implications for the technologies which will need to be adopted and further developed once real action begins. It consequently has implications too for the technological development which should be done in the meantime.

I start by setting out the scientific consensus on greenhouse gases and climate change, and the policy position as of 2012/13. I then consider the process, and the technologies, by which greenhouse gas emissions might be reduced – to zero, net. It appears that the required speed of transition to a "carbon-neutral" world has a very important effect on the appropriate means of getting there. Where, as predicted, a very abrupt transition will be needed, *roundaboutness* in the expansion of a technology turns out to be a problem; and the best protection against this is the development of general and flexible technological competences. It is also necessary to take account of the likely positions of government. Even when the "real action" has begun, full commitment from most of the less developed countries' governments is unlikely, and I therefore set out means by which their economies might be brought toward carbon neutrality nevertheless.

The Oncoming Storm

In 2012 greenhouse gas concentrations in the atmosphere were ~476 ppm CO_2e (carbon dioxide equivalent, allowing for other GGs such as methane). They were rising at just over 1.4% per year, a rate which has been maintained for many years (it is the 1950–1990 average, and that for 1990–2012). Carbon dioxide emissions account for about two thirds of net GGe, and CO_2 concentrations were 393 ppm, rising at 1.7% per year, 1990–2012, and substantially faster than that since 2000 (NOAA 2013).

The Intergovernmental Panel on Climate Change suggests that a constant concentration of CO_2 alone at 550 ppm, 40% above 2012 levels (leaving the other GGs at 2012 levels) would lead in time to an average increase in Earth's temperature of ~3°C over pre-industrial levels (NOAA 2013). At recent rates of emissions we shall reach this level in the early 2030s. This prospect would be alarming enough, given its implications for climate change more broadly, and the likely increase in extreme weather events such as drought, floods, and typhoons. However the further GG concentrations are to rise, the greater the uncertainties about their effects. Moreover, "if this trend continues, it will put emissions on a trajectory corresponding to an average global temperature increase of around 6°C in the long term" (IEA 2013: 7). Consider, en route, the effects of a 4°C rise:

> At an average global temperature rise of 4°C, the hottest days experienced would be 6°C to 12°C hotter, ... yields for key crops such as maize, wheat and rice would drop by as much as 40%. Four degrees could be incompatible with organised global community and would inevitably lead to conflict and disruption and could potentially be beyond adaptation. (Dr Alice Bows, University of Manchester, speaking at the Warsaw meeting: Redfearn 2013)

The least that might be expected of governments, facing oncoming climate disaster, is to abstain from encouraging fossil fuel use through subsidy. In 2009, G20 leaders committed to "rationalize and phase-out over the medium term inefficient fossil fuel subsidies that encourage wasteful consumption" (Schwanitz et al. 2014: 882). Nonetheless, subsidies to fossil fuels continued to grow, to ~$523bn in 2011. Less developed countries, which are now responsible for most GG emissions, are also responsible for most fossil fuel subsidies. (As of 2005, consumer energy subsidies on fossil fuels in less developed economies amounted to ~$350bn per year (Schwanitz et al. 2014).) In energy-importing countries like India, and even in exporting ones like Nigeria, this is extremely expensive for the government; but it is done nonetheless (Siddig et al. 2014). Developing countries' economic growth is decidedly faster than advanced countries (higher per capita, and their populations are growing faster), and they are moving into the stage of growth which tends to be resource-intensive (expanding manufacturing; massive spread of consumer durables including cars). Less developed economies ("non-OECD") are currently expected to account for more than 75% of world primary energy demand by 2035 (Channell, Jansen, et al. 2013: Figure 5). They depend now overwhelmingly on fossil fuels to satisfy that demand, and between now and 2030 are expected to do so to satisfy the large bulk (over 75%) of the increment of demand (Channell, Jansen, et al. 2013: Figure 9). Meanwhile the OECD economies are expected to keep their fossil fuel consumption roughly static.

The estimates of Channell et al. are of particular interest, as the best recent guesses of experienced energy industry analysts, not as to what will happen on current policies, but as to what will happen, *given currently expected policy changes*. They chart the currently expected road to hell. (In principle any amount of fossil fuel consumption could be tolerated if it were neutralized by effective carbon capture and storage (CCS). As we see below, little CCS is currently in prospect.)

The need to address GGe in less developed economies was understood when the Kyoto Treaty of 1997 was negotiated, and a Clean Development Mechanism (CDM) was devised, under a UN umbrella, to come into force when the Treaty did, in 2005. The CDM is designed to provide funding for GGe-reducing projects in less developed economies, through the purchase of emissions permits in developed economies, by governments and businesses, as a partial alternative to cutting their own emissions. The United States, not having ratified Kyoto, has not participated; the European Union has, but made the cardinal error of giving out large numbers of permits to existing polluters free. Consequently the demand for permits is so low that the price has collapsed. The EU price of carbon, for those that must pay it through permits, dipped below €3 per tonne in early 2013, and as of early 2014 had been for years below $10 – insignificant (Hope 2014).

Even with brisk demand for permits the system would have been flawed. It gives incentives for particular groups of polluters in less developed economies to pollute less than they (allegedly) otherwise would have done. The incentive effect can easily be perverse – make sure you don't reduce your pollution until someone pays you to do so – and it does not apply to behavior in general. Its effects are therefore trivial compared to those of the subsidies for fossil fuel use already mentioned (Latin 2012).

The CDM is also feeble compared to the fossil fuel preference shown by those responsible for electricity generation in less developed economies. As we see below, the most attractive alternatives – wind and solar – are troublesome: because they are intermittent, they require back-up or storage capacity, and it is technically difficult, and expensive, to maintain grid stability; they are inevitably scattered across the country, involving expensive transmission lines and needing to be protected from theft and disorder (Channell, Jansen, et al. 2013). Easier to order standard coal-fired generating plant and put it next to a coalmine or port. A modest length of high-voltage transmission lines will then get the electricity everywhere it is really needed, politically: in the *least* developed countries, typically the main urban areas plus the president's home district. Even if all the pledges made by less developed countries for action to reduce emissions are acted on (together with those by developed countries), they will go nowhere near what is required to – probably – hold warming to 2°C (den Elzen, Hof, and Roelfsema 2013).

How will policy change commensurate with the need be brought about? Since the expectation of damaging climate change is not enough, it seems we must wait for the reality. Once we have, let us say, roughly as predicted, 2°C of warming above pre-industrial levels, and the extreme weather events expected to accompany that, public opinion at least in developed countries and China, may take the scientific consensus – that there is much worse to come – seriously. And these countries at least – countries with a coherent and effective state – will at last take action commensurate with the danger.

The nearer we are to the mouth of hell when we awake to our danger, and the faster we are moving, the more sharply we must brake. The implications of this are explored in the next section.

Getting Away from the Mouth of Hell

The Shift to loGGe and the Roundaboutness Problem

As each year passes without radical action, it creates another *fait accompli*: a further step toward intolerable climate change. First, the GG concentration rises further: and most of the greenhouse gases are very long-lived. This is a ratchet rather than a dial which can be turned back. Second, all kinds of long-lived equipment and infrastructure are created or maintained, which are designed for, or naturally suited to, high GG emissions (hiGGe). There are:

- hiGGe patterns of settlement – sprawling suburbs and exurbs: conventional European-style cities and towns are relatively loGGe, favoring walking, cycling, and public transport (Wegener 1996; Rickaby 1987);
- hiGGe buildings, poorly insulated, inefficiently heated and cooled;
- hiGGe machinery, vehicles, and electricity generation using fossil fuels, inefficiently;
- hiGGe agricultural equipment, using large quantities of nitrogenous fertilizers, pesticides and herbicides, made with fossil fuels.

And third, there is a limited range of known technologies to draw on when making new equipment and infrastructure. That at least will have moved in the right direction – but not as far as it might have done.

The obvious urgent need is to replace hiGGe equipment and infrastructure by loGGe alternatives. Even our current limited range of known loGGe technologies allows us to do that. Suppose it is made profitable or mandatory to do so, through carbon taxation, emissions trading, or regulation. The process by which hiGGe gives way to newly-made loGGe then has (in the last analysis) three elements: labor, time, and consumption of natural resources – this last, involving emissions. Labor is required to make the new equipment; this takes time, and inevitably the making involves extra emissions. (In fact of course physical capital – equipment and infrastructure – is also required in the manufacture of new equipment and infrastructure. But this is, or was, made by labor, over time, with emissions.)

The way the replacement process unfolds depends very much on the advantage that loGGe now has. If there is little urgency, so the advantage given is small, loGGe may simply be bought instead of new hiGGe, all existing hiGGe being left in place until no longer fit for use. We suppose that the loGGe type of capital is more expensive to produce than hiGGe. So our route to lower GGe requires extra labor, and *initially* extra emissions (implied in the greater capital expenditure). It also takes time, of course – which in the circumstances assumed doesn't seem to be a problem. But now let us assume that time is of the essence. Assume this urgency is transmitted to the decision-takers – through high carbon taxes, scarce emissions permits, or tight regulations. With a sudden much larger advantage of loGGe, even newly installed hiGGe may be scrapped, to make way for new loGGe. Then the net cost of loGGe, in labor and initial emissions, will be not much less than its gross costs – one cannot subtract the set-up costs saved by not installing hiGGe, because it has *just* been installed.

Then we shall hit the roundaboutness problem. Roundaboutness, or *roundabout methods of production*, is the process whereby goods are produced from or with the help of other goods (components and capital goods), which in their turn have been produced ditto ditto. A steel ladder, for example, will be produced only after the digging of iron ore, the smelting of steel, the making of machines that press steel into shape, the making of machines that make and help maintain those machines, and so on. As pointed out above, this requires labor, emissions, and time – the role of time being the key element of roundaboutness, for the economist who introduced the concept, Böhm-Bawerk (Buechner 1989). Roundaboutness is only problematic when we decide to make A (that ladder) and find that we have to make B first (the pressed steel), but can't until we have begun by making C (the pressing machines and the steel), for which we need some D (iron ore, perhaps), and find we have none to spare. In other words it is the *roundaboutness of newness or rapid change* that we have to worry about.

Let me make the point by contrast, with an example where roundaboutness would be minimal. Suppose a developing country decides to shift to renewables by building a lot of small earth dams to generate hydro-electricity. Suppose the dams are made by fairly low-skilled workers with general purpose tools – picks, shovels, barrows. Assume that almost all the rest of the investment is in the making of the turbines, which are an established type, for which there is spare capacity. So, as soon as the decision is made, they can proceed with designing and building dams and making turbines, neither of which will take very long: *they do not have to do anything else first.*

The example has three helpful ingredients:

1. No innovation is required: the technology is familiar.
2. Most of the work can be done with general purpose tools and indeed general purpose workers – with physical and human capital which already exists in abundance and can be switched from some other use, or idleness.

3. Where specific physical and human capital is required (probably in the manufacture of the turbines), there is spare capacity.

In consequence this route to renewable energy has very low roundaboutness.

Contrast (let us say) a decision to produce a very much larger amount of electricity, by windpower – if and when the windpower capacity of the world is fully stretched already, as it was in 2008. The 8000 distinct components required for a wind turbine include many that have to be manufactured to aerospace tolerances – but are much bigger – and must be both lightweight and robust (Ogando 2008). Such components need to be manufactured and put together with specialized equipment, used by skilled labor which needs to be trained accordingly. Some of them will require materials which are, or could quickly become, in short supply. Here we see a high degree of roundaboutness, implying extra emissions and long delays. Accordingly in 2008 some wind component manufacturers were quoting delivery dates of 2012 (Ogando 2008).

Note that roundaboutness is not simply a technical issue: it is also a function of quantities. Even those simple hydro-electric dams would be more roundabout if there were no spare turbine-making capacity, because then more capacity would have to be built: and there *would be* none if there were a lurch to hydro-electric power generation across the world.

The Route to Low-Carbon Electricity

There has been extensive research on the emissions costs of renewable energy, the most important means (potentially) of reducing emissions. In 2012

> For the first time ... the electricity generated by all the world's installed solar photovoltaic (PV) panels ... probably surpassed the amount of energy going into fabricating more modules ... With continued technological advances, the global PV industry is poised to pay off its debt of energy as early as 2015, and no later than 2020. (Golden 2013)

If the energy it used was from fossil fuels, thus extra GGe; and the energy it generated caused a corresponding reduction in GGe; then soon the global solar photovoltaic industry, cumulating its effects over its lifetime, will cease to be a net *contributor* to global warming! This relates to *energy cannibalism*: an effect where rapid growth of an entire energy-producing or energy efficiency industry creates a need for energy that uses ("cannibalizes") the energy of existing power plants. With some adjustment, such life cycle analysis is also valid for emissions. In fact we require *dynamic carbon life cycle analysis* (Kenny, Law, and Pearce 2010), which takes into account not simply the GGe caused, and saved, during the life of equipment of technology X, but their timing. With "renewable" energy-producing technologies the emissions caused are overwhelmingly "up-front," while all the savings are during the equipment's life. One can thus calculate "carbon-neutral" growth rates for a technology, above which it increases rather than reduces GGe (see Table 26.1).

The Table 26.1 figures assume the efficiencies of each technology to be fixed at the level known at the time of writing, ~2008. For wind and (above all) solar PV this is far from true. The rate of cost reduction of a technology depends on two factors: the rate of increase of installed capacity, worldwide, and the "learning rate" – the reduction in

Table 26.1 Rankings of current electrical energy technologies according to carbon neutral growth rates.

Electrical Energy Technology	*Carbon-Neutral Growth Rate (%)*
Geothermal	249
Wind	91
Biomass	50
Concentrating Solar Thermal (CST)	43
Small Hydro	43
Solar Photovoltaic	41
Nuclear	22
Hydro	5

Source: Kenny et al. (2010: Table 3).

cost for each doubling of installed capacity. As of 2012, wind capacity was doubling roughly every three years, and its learning rate was 7.4%. Solar PV's learning rate was 30% and its capacity was doubling every two years (Channell, Phuc, et al. 2013). (The average cost of a solar panel fell by 75% from 2009–2013: BP 2013.) Both would therefore have carbon-neutral growth rates over 100% – no serious constraint at all. (On the other hand, CST and geothermal are at an insignificant scale thus far. Nuclear and hydro have very slow capacity growth and low learning rates.)

Solar PV and wind have already reached "grid parity" with fossil fuel-generated electricity in a wide range of locations (Channell, Lam, and Pourreza 2012). (The main determining factors are, for PV, level of insolation (sunniness) and retail electricity price, including "feed-in tariff"; for wind, percentage of capacity utilization (roughly, windiness) and wholesale price of electricity.) As of 2012 wind power generated 521 TWh of electricity, 2.3% of total world electricity generation, which gave it 49% of renewable power generation, with solar at 8.9%, solar PV somewhat less (BP 2013); but as we saw above, solar PV was closing fast.

Wind and solar are however *intermittent* technologies. There is a rule of thumb that up to 20% of total electricity production (within a given grid area) may be supplied by intermittent technologies without requiring extra back-up production (or storage) capacity for windless/sunless periods (Kenny et al. 2010). Above that proportion, back-up or storage capacity must be costed in – and so the GGe cost, up-front, must rise sharply. Here is an important current constraint on the expansion of wind and solar. But it is not God-given. Storage technologies are diverse and much work is being done to develop and cheapen them (Channell, Phuc, et al. 2013; Economist 2014a; Butler 2014). There are also fast developing alternatives (and complements):

- Smart grids and smart appliances: means to reduce demand, for shorter or longer periods, in response to intermittency in supply (Kempener, Komor, and Hoke 2013).
- Long-distance high-voltage direct-current transmission – the cheapest means to move electricity (Kempener et al. 2013). The wider the connected area, the less the windless/sunless intermittency problem, and so less storage or back-up capacity is needed.

Moreover wind and solar are complementary more than competitive. Wind provides base-load (blowing around the clock); solar meets the daytime peak. There is more sun in summer, and (in Europe at least) more wind in winter.

We can thus expect solar and wind to be highly competitive on cost by the beginning of the assumed "climate catch-up period," so long as they are allowed to make the pace before it starts: that is, if the producers of solar PV electricity (mainly householders) and wind power (firms) are allowed to supply the grid at market rates, with no subsidy except that implied by the grid authority taking responsibility for back-up. The solar PV and wind industries will continue to expand rapidly, and so, pulled by them, will the back-up technologies, storage, smart grids/appliances, and long-distance transmission lines. (We return to the issue of renewable technologies in power generation in less developed economies below.)

But the challenge will be not simply to dominate the market for *new* power-generating equipment: it will be to replace the output of existing coal- and gas-fired power-stations. The total cost of the new renewable power generation – financially, and in emissions – will need to be less than the variable cost of the fossil fuel generation. This will become steadily harder, through the "low-hanging fruit" effect: the equipment that is replaced first, by expanding renewables production, is (presumably) the worst in GG emissions. The more is replaced, the lower the GGe of the capacity to be replaced next: thus conventional coal would go first, combined cycle gas with district heating last. And fossil fuel prices will fall, as the existing market shrinks, and it becomes clear that there will be little demand for them in future. At this stage the prior development of CCS will be of vital importance.

> If the world is to have a reasonable chance of limiting the global average temperature increase to 2°C … less than one-third of proven reserves of fossil fuels can be consumed prior to 2050, unless CCS technology is widely deployed. … If CCS is removed from the list of emissions reduction options in the electricity sector, the capital investment needed to meet the same emissions constraint is increased by 40% … CCS is currently the only large-scale mitigation option available to make deep reductions in the emissions from industrial sectors such as cement, iron and steel, chemicals and refining. Today, these emissions represent one-fifth of total global CO_2 emissions, and the amount of CO_2 they produce is likely to grow over the coming decades. … However, given today's level of fossil fuel utilization, and that a carbon price as a key driver for CCS remains missing, the deployment of CCS is running far below the trajectory required to limit long-term global average temperature increases to 2°C. (IEA 2013: 7–8)

The neglect of CCS may partly arise from its need for *extra* fossil fuel. Renewable energy sources reduce not only emissions, but also the need for energy imports. Fit CCS to a coal- or gas-fired power plant, and it needs *more* coal or gas for the same net electricity output.

The Route to Zero-Carbon Transport and Industry

By 2030, Channell, Jansen, et al. (2013) predict 7.7 btoe (billion tonnes of oil equivalent) for power generation, 4.7 btoe for industry, 2.8 btoe for transport, and 1.5 btoe for others. Unfortunately power generation is unusual in having renewable technologies with such immediate cost-competitiveness as solar and wind.

Transport exemplifies the roundaboutness problem: great economies in emissions could be made with railways and trams (using low-carbon electricity), cycling and walking. But that requires classic European patterns of settlement, with compact cities and towns – and rail infrastructure. For countries like the United States, to get from here to there would require much money, time, and emissions. Electric buses, trucks, and cars will be well suited to the existing infrastructure, plus charging facilities, but again will require much money, time and emissions, partly to improve the technologies. Biofuels can be used without delay for road vehicles, and indeed aircraft. However the "first generation" biofuels currently available require fertile land. The effective emissions cost of producing them must include the emissions needed to intensify production on the remaining agricultural land, to compensate for the land used for biofuels. Second generation biofuels will be produced on marginal land and from waste, and with higher levels of efficiency (see below) – but not yet.

The only transport mode which can produce electricity renewably, then use it, is the airship (pure or hybrid). Its large surface area is ideal for solar photovoltaic cells (Kalan 2013). Unlike all land transport, it requires no infrastructure en route; unlike conventional aircraft, it requires no large space and expensive ground infrastructure for take-off and landing. (Hybrid airships require a very short runway; pure airships none.) So now that its buoyancy management and control problems have essentially been solved (Stewart 2012), it could go point to point for most journeys, making it an attractive alternative to airfreight in the first instance (Jowit 2010). Unfortunately it is a striking example of path dependence: the airship *is* scarcely used, because it *has* scarcely been used since the 1930s. The faults which doomed the Hindenburg then – flammable lifting gas (hydrogen) and envelope materials – were dealt with long ago. But the competing modes have 80 years' lead on their learning curves.

As IEA (2013) points out, industrial sources of CO_2 provide excellent scope for the use of CCS; solar and wind can do little to help here. CO_2 emissions from the global *cement* industry alone account for ~5% of total anthropogenic emissions (Worrell et al. 2001). A long list of feasible measures for reducing emissions (Liska et al. 2012; Palmer 2012) is headed by a switch from calcium to magnesium minerals as base. Magnesium carbonate parts with its CO_2 at a much lower temperature than calcium carbonate does; magnesium-based concretes quickly reabsorb the CO_2 driven off in cement production; are stronger; and bind with, rather than repel, cellulose-based filler (e.g., wood chips): four big steps toward carbon neutrality. All that is lacking is the incentive to switch.

Most cement is of course used in construction, which is a major user also of *iron and steel* (similar to cement in its total CO_2 emissions: Gielen, Newman, and Patel 2008). The scope for saving emissions per ton of steel produced is, at the margin, far less than for concrete, but the scope for saving on total steel production is great: thus the construction industry, even for multistorey buildings, could switch to cross-laminated wood – potentially a carbon-negative material (Economist 2012b).

Chemicals and petrochemicals come third in emissions: 1.0 gigatonnes/year of CO_2 emissions in 2004 against 1.7 Gt/year for iron and steel. The rise of modern biotechnology has opened the way to huge reductions in carbon emissions here. "Technically speaking, the overwhelming share of the total demand for organic chemicals and polymers [accounting for the bulk of emissions from this industry] could be covered from bio-based feedstocks" (Gielen et al. 2008: 476). Even with current technology, tropical sugar cane can be made into ethanol and thus ethylene and its many derivatives, with carbon neutrality. Much work is being done to allow the same route to be followed, starting from bio-waste materials containing cellulose (e.g., BBSRC 2012).

Inorganic chemicals are less important, but nonetheless 20% of all energy use in chemicals and petrochemicals goes to produce ammonia, most of which is used to produce nitrogen fertilizer. Its use causes about 0.8 Gt/year of emissions of nitrous oxide (N_2O), also a greenhouse gas. Most of the biosphere's available nitrogen is already provided by biological fixation, mostly through leguminous plants (such as beans and other pulses) with symbiotic bacteria in root nodules (Gielen et al. 2008). We shall see below that this could be greatly extended.

Reflections on the Technologies Discussed

The Uses and Limitations of the Market Mechanism

A close examination of the economics of renewable energy shows the importance of using carbon taxes or similar uses of the price mechanism to drive the transition:

- Technologies with relatively heavy up-front emissions costs, such as nuclear, will incur relatively high carbon taxes, to the advantage of their rivals. So they should.
- Typically the sources cited in the last section say of disruptive carbon-reducing technologies, as concrete manufacturers do of magnesium concrete, "interesting, but without carbon taxes it is not worth the trouble." And CCS *cannot* be worth the trouble without carbon taxes, or similar incentives.
- Carbon-reducing technologies which are advancing rapidly would reduce carbon faster with carbon taxes. Thus the net gain from solar PV is significantly reduced by the electricity required to reduce silica to silicon (Golden 2013). Carbon taxation would steer this process to the lowest-emission sources.

There are however limits to the effectiveness of the "environmentally-adjusted" market mechanism. British experience shows how householders may spend far less on the thermal insulation of dwellings than is financially rational even without carbon taxes. If certain general practices are clearly carbon-reducing at modest cost, government should enforce them through regulation.

In the insulation case market pressures are baulked by householders' ignorance, inertia, and poor access to capital. But the market mechanism has more fundamental limitations. It works superbly in allocating resources, over space – far less well, over (long periods of) time (Richardson 1960). Thus according to recent projections, wind turbines and electrical vehicles will require an increase in supply of neodymium and dysprosium of over 700% and 2600% respectively, during the next 25 years. It is not increasing so fast; once a shortage appears, it will take a long time to put right (Economist 2012a): the roundaboutness of the production process will see to that.

The investment problem with rare metals is that the investors in mines and refining capacity lack information about demand, and competing supply, far enough ahead. At all events, the market will ensure there will be a substantial demand (barring the unexpected triumph of technologies not requiring these metals), and therefore that there will be a substantial – though perhaps inadequate – supply. Airships are an example of a carbon-reducing technology which poses even deeper problems for the market mechanism. As pointed out above, path dependence and learning curves have meant that this transport mode, outside a few niches, is highly uncompetitive. Only large-scale government intervention can get this transport mode far enough down its learning curve to, so to speak, take off.

The Key Role of General Purpose Technologies: ICT and Biotechnology

Specific products and technologies draw on progress made by, and for, other products and technologies. Solar PV luckily shares its dependence on silicon and similar materials with an older and far larger complex of industries: the information and communication technology (ICT) sectors. The learning by and for these sectors about these materials, their capabilities, and production processes, has been of great value to solar PV.

The moral: Develop technological competences which have a diversity of uses, and can be switched to emissions-saving innovation once this is made profitable. Do this particularly where those innovations will/can use general purpose equipment.

The ICT sectors produce, and use, a cluster of technologies which constitute what is now called a "general purpose technology" (Jovanovic and Rousseau 2005). This GPT has myriad uses for the purposes of control, which allows the more economical use of energy and materials. Thus integrated crop management (ICM) uses ICT to optimize the quantities and location of water, fertilizers, and pesticides, giving large reductions (Lancon et al. 2007). Intensive use of ICT will be required to develop smart grids (US Department of Energy 2013); also road pricing systems, vital to manage demand for road space when new roads and railways cannot be built (Glaister and Graham 2004).

Happily ICT has already thrown up two types of general purpose capital equipment, the first now well established: computer numerically-controlled (CNC) machine tools, now subsumed into CAD-CAM: computer-aided design linked to computer-aided manufacturing. Next is 3D printing: a process of creating three-dimensional objects from digital instructions using materials printers. The term is roughly synonymous with "additive manufacturing." The raw materials "added," in layers, were initially powders of metals such as titanium, or of thermopolymers (Excell and Nathan 2010) but their range is expanding (LaMonica 2013): indeed the basic technology is being extended to tissue engineering to build body parts, "bio-printing," by depositing layers of living cells onto a gel medium (Nathan 2010).

It is hard to envisage 3D printing being able to produce many established products as cheaply as they can be mass-produced with CAD-CAM. But it must help in unblocking capacity bottlenecks in the rapid ramp-up of production, thus in dealing with our roundaboutness problem.

Biotechnology is clearly a GPT of great potential in three key sectors: chemicals, agriculture, and medicine (Langeveld, Dixon, and Jaworski 2010).

As mentioned above, chemicals traditionally made by synthetic chemistry can be made, with far lower GGe, by micro-organisms (or plants), genetically modified to suit. Production is in fermentation tanks or in fields, where plants have been engineered to produce pharmaceuticals, for example: so, broadly speaking, in general purpose equipment. Develop the organisms first, learn how to use them, and production can be ramped up at high speed. Thus feces (farm and human) mixed with other organic waste can generate huge quantities of methane renewably, with high-grade organic fertilizer as by-product (Anaerobic Digestion 2014). Indigenous biomass resources could meet up to 44% of UK energy demand by 2050 without impacting food systems (Welfle, Gilbert, and Thornley 2014).

In agriculture, there is enormous potential for GGe savings by moving away from the established model, of annual crop plants grown with lavish watering, fertilizer, pesticides, and so on (Hazell 2009). The "bio-fixing" of nitrogen by leguminous plants was discussed above. The main crop plants (wheat, maize, rice, sugar cane, etc.) are grasses rather than leguminosae, but they too can benefit through interplanting with perennial leguminosae (Jordan 2013), a low-technology solution suitable for tropical small farmers. The grass family happily has a great deal of genetic variation. That within species can be exploited by conventional breeding. That among species makes them strong candidates for genetic modification. Thus some sugar cane varieties are capable of fixing nitrogen in association with the bacterium *Glucoacetobacter diazotrophicus* (Boddey et al. 1991). Maize (corn) and sugar cane use C4 photosynthesis, the others the less efficient C3 type. There are grasses which are tolerant of salty water (Munns, James, and Laeuchli 2006). No grasses appear to be insecticidal, but this trait has already been successfully transferred, from *Bacillus thuringiensis* (Bt) into cotton (Kingsbury 2009).

Many grasses, including sugar cane, are perennial. "Perennialization" of grain (and other) crops is potentially one of the largest contributors to GGe reductions. Since the root and part of the perennial plant's stem remains, year after year, the expenses of annual growth are reduced; and the plant is more drought-resistant, because it has larger reserves. It is also more flood-resistant, partly because its larger roots stabilize the ground, giving erosion-resistance (Bell, Wade, and Ewing 2010). The large perennial root gives free biological CCS (Paustian et al.1997).

Again, general-purpose equipment is available for this transformation – for new crops can thrive in old fields, with existing tractors, harvesters, barns. And once the genetic modification has been achieved, the competences required are general purpose too.

Yet Bell et al. present the perennialization of wheat in Australia, through genetic engineering, as likely to take around 18 years from project inception to the point of reaching 25% of the winter crop area – and with a 50% chance of failure. Compare that with the speed of progress in the digital realm! This slowness and uncertainty of pay-off reflects the immaturity of the biotechnology GPT; but even as Bell et al. (2010) wrote, it was maturing. In 2005 the *sequencing* of the first of the grain *genomes* was announced – rice (the easiest, being smallest; soya followed in 2008, maize in 2009). International Rice Research Institute researchers were then able to enhance their conventional breeding methods with "marker-assisted back crossing" which allowed much faster breeding-in of valuable traits found in unusual rice strains. Flood-tolerant rice, its first fruit, went from experimental use in 2008 to being planted by five million farmers across the world five years later. IRRI promises drought- and heat-tolerance next (Economist 2014b). Similar acceleration may be expected with genetic engineering of new traits. The expectations expressed for soya are typical:

> Right now it takes 15–20 years to get a new variety; we are hoping to make that 5–7 with the help of the genome. Markers identifying genes that breeders want can be used to find those plants that contain the genes without having to grow the plant to maturity. Or biotechnology can slot hot genes straight into plant cells. (Marris 2008)

Now *synthetic biology* aims to apply engineering principles to genetic modification. "Synthetic biology applies the knowledge and tools developed in analytical biology to synthesise biological entities ... drawing on expertise in molecular biology, computer science, chemistry, and engineering" (Nuffield Council on Bioethics 2012: 59.)

> Whereas traditional genetic engineering might involve transferring a gene from one organism to another and hoping it will do what it is supposed to, synthetic biology is about designing organisms to solve specific problems. It entails a more precise level of control: as well as inserting a gene you are also tinkering with cellular machinery that governs when and how much of that gene gets expressed. (Griggs 2013)

Since the building blocks of cells are fundamentally the same – DNA, RNA, amino-acids combined in proteins – the basic competences of synthetic biology have a very wide application. So here again is a (developing) body of general technological competences analogous to IT software, and indeed drawing heavily upon it. The money now is mostly on the pharma side, for example, engineering a phage (virus) to knock out drug-resistant staphyllococci (Nuffield Council on Bioethics 2012). But general competences built in one sector can be switched to another: to produce new species of algae, for example, which will (jointly) photosynthesize across the spectrum of visible light and thrive while doing so in the CO_2-rich exhaust gases from coal- and gas-fired power stations, yielding oils for biofuel, and much else (Global CCS Institute 2011).

To conclude: biotechnology "upstream," in agriculture very broadly defined, and "downstream," in chemicals, is becoming strongly "carbon-capable" by transforming agriculture and chemicals into net capturers of carbon acquired by photosynthesis – though in agriculture it will have work to do merely to adapt to, protect against, and compensate for climate change. Agriculture can be made vastly more productive, and thus in principle able to go beyond feeding the human and other animal population to produce a large "carbon surplus": enough to power its own requirements (e.g., biofuels for farm machinery), provide the feedstock for chemicals, and generate wood to be "locked up" in new construction, while increasing soil carbon (as in perennial roots) and above-soil carbon (new forest).

The Role of Less Developed Countries in Carbon Reduction

We saw above that less developed countries are increasingly the predominant source of GGe. So they must become the predominant location in which loGGe technologies are diffused, if not developed. On the other hand, we may expect most of their governments to remain laggards in the implementation of carbon taxes and similar emissions reduction measures. That is implied in the gloomy forecasts of Channell,

Jansen, et al. (2013). A nightmare scenario can be imagined, in which the governments of the "carbon-reducing North" try to impose their policies – such as carbon taxes – on recalcitrant governments in the "carbon-increasing South": "Do as we are now doing, not as we have done." "Our people are suffering from drought, floods, and above all from the heat – and you want us to tax the fuel that makes the electricity that powers their air-conditioners? How much will you pay us?"

However the developments envisaged above open up a very different possibility: a split within the South. Less developed countries are virtually all rich in one resource whose value can be expected to rise – sunshine; and the part of their population with privileged access to this resource is that which lives in the countryside: at least until recently, the majority of the population. In virtually all of them, country people have inferior access to every other kind of resource: transport, education, medical care, electric power, and political and economic power. But the mobile phone already shows what the right technologies can do to even the scales between "country" and "town." The biotechnological revolution can allow the "country" to generate an agricultural surplus without depending on the "town" to provide it with power, fertilizers, and pesticides; solar PV and other renewable energy technologies can yield a surplus of electric power available to develop rural industries, power country transport, and sell to the "town."

For a time, there will be two competing models of energy production and use in less developed countries. As we have seen, the most convenient model for a non-OECD government whose priority is to provide electricity cheaply and fairly reliably to the major urban areas, is to generate it in large conventional power stations, probably from coal. Similarly, the simplest means of making transport fuel available is from its own or imported petroleum. Renewable technologies are distributed, scattered: hard to organize, control, protect.

So cities in such countries might mostly be trapped in the hiGGe model. But loGGe would touch them nevertheless. Apartment blocks have roofs; not all city buildings are high-rise; shanty towns could generate solar PV; nearby villages might export it. Even without feed-in tariffs, it would be attractive to have one's own source of solar PV electricity, alongside the grid, if only for when the grid went down. But the more solar PV was used, the more strain there would be on the grid – just after sunset.

Picture a city in the (less developed) tropics, circa 2030. The summer heat has always been oppressive in that area; worse in the city; worst in apartment blocks. It is now a degree or two worse still, on average, due to climate change, and heat waves which push it up several degrees further have become more common. Air conditioning is no longer a luxury of the rich: it is necessary for survival. Grid electricity has become more expensive, if only because international pressure to remove subsidy has become irresistible. And *then* in a heatwave the grid goes down.

In the countryside a quite different model has become established. Each village is now at least self-sufficient in electricity and in biofuel for transport. Private or community entrepreneurs have provided the equipment for the solar PV and biofuel at market rates – as they had for 2000 Indian villages in one program by 2013 (Vidal 2013). So who needs a grid connection? Dirt roads are cheap, and can be built with village labor and initiative. And with 3D printing, and modern telecommunications, there is a greater degree of local self-sufficiency, whether for schooling, medicine, or repair of equipment.

The renewable village model triumphs, and the cities begin to empty.

Conclusion

This chapter started from a bleak assumption: that the world will, in a decade or two, be conscious of the danger of global environmental disaster due to climate change induced by GGe; and that policymakers at least in advanced countries will be, belatedly, prepared to take drastic action to avert – or rather mitigate – disaster. It showed that the *very rapid* expansion of a technology which would otherwise be an excellent means of reducing GGe, would introduce a number of problems which go under the heading of "roundaboutness." Roundaboutness causes delay, particularly where it is necessary to innovate before investment; and it causes much higher GGe during the period of heavy investment than during the "steady state" later.

We considered what might be the most effective technologies in this situation, for energy production and energy saving, and the conditions under which they could be "ramped up" most quickly. Two key requirements emerged:

- the development of general competences and flexible equipment which would cut down roundaboutness at source, so to speak.
- precautionary investment in specific competences and capital which might otherwise become bottlenecks.

Two "general purpose" technologies looked like playing a vital part: ICT and biotechnology. ICT has reached a very high level of development, though applied relatively little in the key sectors of energy production and agriculture. Biotechnology has delivered much less. It could give enormous savings in GGe in chemicals production, and in agriculture and its feeder industries. Indeed in agriculture it offers opportunities for biological carbon capture and storage on a massive scale through enhanced plant growth and a switch from annual to perennial crops. Such changes would seem to have little roundaboutness in emissions, but potentially much roundaboutness in the development process. Genome sequencing has already reduced this; synthetic biology offers the prospect of reducing it further.

Probably most less developed countries will be slow, at government level, to join in what must be a crash program of carbon reduction. One should not despair of negotiation at global level, with a combination of positive incentives and pressure, but it was argued in the last section that there is a model of carbon reduction which can work at village level, with very little help from national government. The "North" can help this process by developing the necessary technologies and assisting in their diffusion and adaptation in the "South."

References

Anaerobic Digestion. 2014. http://www.biogas-info.co.uk (accessed January 9, 2015).

BBSRC. 2012. "Big Score for British Biofuel Technology: Insight, Know-How and Collaboration Lead to Multi-Million Deal." BBSRC, January 27. http://www.bbsrc.ac.uk/news/industrial-biotechnology/2012/120127-f-british-biofuel-technology.aspx (accessed January 9, 2015).

Bell, L.W., L.J. Wade, and M.A. Ewing. 2010. "Perennial Wheat: Environmental and Agronomic Prospects for Its Development in Australia." *Crop and Pasture Science* 61: 679–690.

Boddey, R.M., S. Urquiaga, V. Reis, and J. Döbereiner. 1991. "Biological Nitrogen Fixation Associated with Sugar Cane." *Plant and Soil* 137: 111–117.

BP. 2013. *BP Statistical Review of World Energy June 2013*. London: BP.

Buechner, M. Northrup. 1989. "Roundaboutness and Productivity in Böhm-Bawerk." *Southern Economic Journal* 56: 499–510.

Butler, Nick. 2014. "A Ray of Sunshine – Breakthroughs on Storage Can Change the Game for Solar Power." Financial Times, May 4.

Channell, Jason, Timothy Lam, and Shahriar Pourreza. 2012. *Shale and Renewables: A Symbiotic Relationship*. New York: Citigroup Global Markets.

Channell, Jason, Nguyen Phuc, Shahriar Pourreza, and Timothy Lam. 2013. *Battery Storage: The Next Solar Boom?* New York: Citigroup Global Markets.

Channell, Jason, Heath Jansen, Alastair Syme, Sofia Savvantidou, Edward Morse, and Anthony Yuen. 2013. *Energy Darwinism: The Evolution of the Energy Industry*. New York: Citigroup Global Markets.

den Elzen, Michel, Andries Hof, and Mark Roelfsema. 2013. "Analysing the Greenhouse Gas Emission Reductions of Mitigation Action Plans by Non-Annex I Countries by 2020." *Energy Policy* 56: 633–643.

Economist. 2012a. "Rare Earths and Climate Change: In a Hole?" *The Economist,* March 17, 83.

Economist. 2012b. "Building Materials: Wooden Skyscrapers." *The Economist*, August 30.

Economist. 2014a. "Electricity Storage: Pumping Heat." *The Economist*, March 12.

Economist. 2014b. "The New Green Revolution: A Bigger Rice Bowl." *The Economist*, May 10.

Excell, Jon, and Stuart Nathan. 2010. "The Rise of Additive Manufacturing." The Engineer, May 24.

Gielen, Dolf, John Newman, and Martin Patel. 2008. "Reducing Industrial Energy Use and CO_2 Emissions: The Role of Materials Science." *MRS Bulletin* 33: 471–477.

Glaister, Stephen, and Dan Graham. 2004. *Pricing Our Roads: Vision and Reality*. London: Institute of Economic Affairs and Profile Books.

Global CCS Institute. 2011. *Accelerating the Uptake of CCS: Industrial Use of Captured Carbon Dioxide*. Docklands, Victoria, Australia: Global CCS Institute.

Golden, Mark. 2013. "Global Solar Photovoltaic Industry Is Likely Now a Net Energy Producer, Stanford Researchers Find." *Stanford Report*, April 2.

Griggs, Jessica. 2013. "The Odd Couple." *New Scientist December* 7, 46–49.

Hazell, Peter. 2009. "The Asian Green Revolution." Washington, DC: International Food Policy Research Institute Discussion Paper.

Hope, Mat. 2014. "Experts Unconvinced Latest Reforms Will Save the European Carbon Market." Carbon Brief, January 29.

IEA. 2013. *Technology Roadmap: Carbon Capture and Storage*. Prepared by Ellina Levina, Simon Bennett, and Sean McCoy. Paris: International Energy Agency.

Jordan, Carl. 2013. *Working with Nature: Resource Management for Sustainability*. New York: Routledge.

Jovanovic, Boyan, and Peter Rousseau. 2005. "General Purpose Technologies." In *Handbook of Economic Growth*, vol. 1B, ed. Philippe Aghion and Steven Durlauf, 1182–1224. Amsterdam: Elsevier.

Jowit, Juliette. 2010. "Blimps Could Replace Aircraft in Freight Transport, Say Scientists." *The Guardian*, June 30. http://www.guardian.co.uk/environment/2010/jun/30/blimps-aircraft-freight (accessed January 9, 2015).

Kalan, Jonathan. 2013. "Solar Ship Aims to Soar." *BBC.com*, January 13. http://www.bbc.com/future/story/20130111-solar-ship-set-to-soar (accessed January 9, 2015).

Kempener, Ruud, Paul Komor, and Anderson Hoke. 2013. *Smart Grids and Renewables: A Guide for Effective Deployment*. Abu Dhabi: International Renewable Energy Agency, Working Paper.

Kenny, R., C. Law, and J.M. Pearce. 2010. "Towards Real Energy Economics: Energy Policy Driven by Life-Cycle Carbon Emission." *Energy Policy* 38: 1969–1978.

Kingsbury, Noel. 2009. *Hybrid: The History and Science of Plant-Breeding*. Chicago, IL: University of Chicago Press.

LaMonica, Martin. 2013. "10 Breakthrough Technologies 2013: Additive Manufacturing." *MIT Technology Review* May–June.

Lancon, J., J. Wery, B. Rapidel, M. Angokaye, E. Gerardeaux, C. Gaborel, D. Ballo, and B. Fadegnon. 2007. "An Improved Methodology for Integrated Crop Management Systems." *Agronomy for Sustainable Development* 27: 101–110.

Langeveld, J.W.A., J. Dixon, and J.F. Jaworski. 2010. "Development Perspectives of the Biobased Economy: A Review." *Crop Science* 50: 142–151.

Latin, Howard. 2012. *Climate Change Policy Failures: Why Conventional Mitigation Approaches Cannot Succeed*. Singapore: World Scientific.

Liska, Martin, Abir Al-Tabbaa, Kneale Carter, and John Fifield. 2012. "Scaled-Up Commercial Production of Reactive Magnesium Cement Pressed Masonry Units. Parts I and II." *Proceedings of the ICE - Construction Materials* 165: 211–223 and 225–243.

Marris, Emma. 2008. "Soya Genome Sequenced." *Nature*, December 10.

Munns, Rana, Richard A. James, and André Laeuchli. 2006. "Approaches to Increasing the Salt Tolerance of Wheat and Other Cereals." *Journal of Experimental Botany* 57: 1025–1043.

Nathan, Stuart. 2010. "Building Body Parts with 3D Printing." *The Engineer*, May 24.

NOAA. 2013. "NOAA's Annual Greenhouse Gas Index (An Introduction)." National Oceanic and Atmospheric Administration. http://www.esrl.noaa.gov/gmd/aggi/ (accessed January 9, 2015).

Nuffield Council on Bioethics. 2012. *Emerging Biotechnologies: Technology, Choice and the Public Good*. London: Nuffield Council on Bioethics.

Ogando, Joseph. 2008. "Wind Energy's Manufacturing Crunch." *Design News*, September 17.

Palmer, Bill. 2012. "Future Cement Part II: Carbon-Negative Cement, Converting Carbon, and Recycled Materials." *The Concrete Producer*, October.

Paustian, K., O. Andrén, H.H. Janzen, R. Lal, P. Smith, G. Tian, H. Tiessen, M. Van Noordwijk, and P.L. Woomer. 1997. "Agricultural Soils as a Sink to Mitigate CO_2 Emissions." *Soil Use and Management* 13: 230–244.

Redfearn, Graham. 2013. "Warsaw's Widening Climate Chasm Could Lead to 4C Warming." *The Guardian*, November 21. http://www.theguardian.com/environment/planet-oz/2013/nov/21/warsaw-climate-change-conference-global-warming (accessed January 9, 2015).

Richardson, G.B. 1960. *Information and Investment*. Oxford: Oxford University Press.

Rickaby, P.A. 1987. "Six Settlement Patterns Compared." *Environment and Planning B: Planning and Design* 14: 193–223.

Schwanitz, Valeria, Franziska Piontek, Christoph Bertram, and Gunnar Luderer. 2014. "Long-Term Climate Policy Implications of Phasing Out Fossil-Fuel Subsidies." *Energy Policy* 67: 882–894.

Siddig, K., A. Aguiar, H. Grethe, P. Minor, and T. Walmsley. 2014. "Impacts of Removing Fuel Import Subsidies in Nigeria on Poverty." *Energy Policy* 69: 165–178.

Stern, Nicholas. 2009. *Blueprint for a Safer Planet*. Oxford: Oxford University Press.

Stewart, Jon. 2012. "Lighter-Than-Air Craft Rise Again." *BBC.com*, September 24. http://www.bbc.com/future/story/20120921-lighter-than-air-craft-rises (accessed January 9, 2015).

US Department of Energy. 2013. Renewable Energy and a Smart Grid. http://www.hostgeni.com/docviewer?filePath=http%3A%2F%2Fenergy.gov%2Fsites%2Fprod%2Ffiles%2FOE_Smart_Grid_Talking_Points.pdf (accessed January 9, 2015).

Vidal, John. 2013. "India's Villagers Reap Visible Benefits from Solar Electricity Scheme." *The Guardian*, March 6.

Wegener, Michael. 1996. "Reduction of CO_2 Emissions of Transport by Reorganisation of Urban Activities." In *Transport, Land-Use and the Environment*, ed. Y. Hayashi and J. Roy, 103–124. Dordrecht: Kluwer.

Welfle, Andrew, P. Gilbert, and P. Thornley. 2014. "Securing a Bioenergy Future Without Imports." *Energy Policy* 68: 1–14.

Worrell, Ernst, L. Price, C. Hendricks, and L. Ozawa Meida. 2001. "Carbon Dioxide Emissions from the Global Cement Industry." *Annual Review of Energy and Environment* 26: 303–329.

Chapter 27

Global Risks: Cause and Consequence of the New Interactions Between Science, Technology, and Society

Jean-Yves Heurtebise

Introduction: The "Anthropogenicity" of Global Social-Environmental Risks

This paper aims at understanding in what sense Global Risks can be understood as being the cause as well as the consequence of the new interactions between science, technology, and society that shaped contemporary societies.[1] Let us consider Climate Change.

Climate Change is widely recognized today as being caused by human activities: "Emissions from motor vehicles, power plants, deforestation, and other human sources are warming the Earth and damaging ecosystems and human well-being" (Gonzales 2010). Most scientists agree that Climate Change is among the negative ecological consequences of the diffusion to the whole planet of the patterns of development that emerged with the industrial revolution. The industrial revolution itself was rooted in the scientific revolution of the seventeenth century:[2] the application of the principles of mechanics to industrial processes in capitalistic economies lead to the invention of engines multiplying human workload. The engines producing this surplus of workload needed the fossil fuels (coal and oil) whose very consumption contributed to the emission of greenhouse gases from which Climate Change mostly derives.[3]

The industrialization process changed the whole social landscape. By attracting people from countryside to cities, where factories were located, it induced a new trend toward urbanization. Urbanization, by gathering a large number of people in the same place, creates new problems of hygiene and sanitation, whose resolution induced the medical revolution from which resulted an unprecedented demographic expansion. This demographic expansion in turn supported the global trend toward world industrialization by providing both the sufficient workforce for production and the necessary subjects for consumption.

The Handbook of Global Science, Technology, and Innovation, First Edition.
Edited by Daniele Archibugi and Andrea Filippetti.

Thus Climate Change illustrates that Global Risks come, in the first place, as a *consequence* of new interactions between science (experimental physics), technology (thermodynamic and cybernetic engines), and society (urbanization, medicine, and capitalism). However, Climate Change demonstrates today that the Global Risks it induced are also among the *cause* of the establishment of new interactions between science, technology, and society: in order to assess the gravity as well as the propensity of Climate Change, substantial progress will be needed both in ecological and meteorological sciences;[4,5] to mitigate Climate Change, innovation to improve the efficiency of renewable energies and the reliability of geo-engineering tools are required[6] (Robock 2008); finally, new economic taxes on carbon-emitting industries and new financial incentives for eco-friendly devices shall be implemented[7] and a new "behavioral awareness" should be developed for world citizens to adopt new ways of living that contribute to save energy and resources.[8,9]

Saying that Global Risks are at the same time cause and consequence of new interactions between science, technology, and society means that human beings are both their actual producers and their potential resolvers. Risk is indeed an anthropocentric concept. As Ortwin Renn said in the introduction to his co-edited book *Global Risk Governance*: "Risks are created and selected by human actors" (Renn 2008). However, stating that risks are anthropocentric in order to emphasize their "anthropogenicity" does not equate with saying that risks are subjective. Risks and the perception of risks are not the same. Indeed what we consider to be risky, for personal, social, cultural, economic and/or scientific reasons, does frame our assessment and management of risks;[10] but it is equally obvious that the absence of the perception of risks does not equate to the absence of risks. Thus if risks are created by human beings, this does not imply that risks are simply a social construct but it means that they are also and foremost a technological product.

This paper will start by discussing Ulrich Beck's and Mary Douglas's conception of risks and then will move on redefining risk from a naturalistic perspective (neither realist nor constructivist). On the basis of this redefinition, it will analyze the interactions between science, technology, and society and, finally, the political consequences for risk management.

Ulrich Beck's and Mary Douglas's Understanding of Risks: A Critical Review

One of the first academic accounts on the importance of risks in contemporary societies comes from the works of Ulrich Beck (influenced by the German school of sociology) alongside those of Mary Douglas (influenced by the American school of psychology).

Ulrich Beck's concept of Risk Society (*Risikogesellschaft*) stems from an historical reflection on modernity and "post-modernity" or "second modernity." Modernity means two things in Beck's interpretation: Enlightenment and Industrialization. Enlightenment implies a renovation of humanity's cultural agenda defined in two points: deconstruction of regional traditional beliefs and reconstruction of human hopes on the rational basis of techno-scientific progress. Industrialization applies Enlightenment theoretical agenda: peasants' agrarian life tuned by natural cyclical seasons is replaced by workers' existence shaped by the continuous rhythm of industrial production.

Second modernity emerged as an "unintended consequence" of modernity; it denotes "a social transformation within modernity, in the course of which people will be set free from the social forms of industrial society" (Beck 1992a: 87). The industrial economy ruled by patriarchal stratification is overturned by the service economy shaped by individual innovations. Politically speaking, the struggle for (majority) political rights is replaced by the fight for (minority) civil rights. But the most important change concerned risk insurance. During the period of the first modernity, industrial risks were covered by the state and private companies as a collective compensation for workers' labor. However, with scientific-technological progress giving to men a capacity of action exceeding their capacity of control and prediction, global risks emerged as uninsurable: "The residual risk society has become an uninsured society, with protection paradoxically diminishing as the danger grows" (Beck 1992b). Thus the struggle for sharing industrial goods and social benefits (positives) gives place, in the age of *second* modernity, to the debate about allocating ecological risks and financial "bads" (negatives).[11]

In Mary Douglas's interpretation, risks are definitively a social-cultural construction: "There could be no risks, illnesses, dangers, or any reality, knowledge of which is not construed" (Douglas 1997). To illustrate this point, in her book *Purity and Danger*, she draws a parallel between taboo and pollution: "pollution is a type of danger which is not likely to occur except where the lines of structure, cosmic or social, are clearly defined" (Douglas 1984: 117). Both taboo and pollution expressed a symbolic breaching in social-natural laws: they are contaminating powers, conveying ideas of fault and abnormality. The difference between taboo and pollution is that taboo is meant to protect the community while risks are meant to protect individuals (Douglas 1992: 28). But, precisely because they aim at protecting individuals, risk policies are paving the way of their own impossible resolutions: "Risk is … immeasurable and its unacceptability is unlimited … There can never be sufficient holiness and safety" (Douglas and Wildavsky 1983: 184). For Mary Douglas, the more society agrees to implement policies to deal with risks, the more the "sectarian groups" of political ecologists and environmental activists will raise new demands. Her view expresses a strong criticism of the ethical framework of risk analysis "using nature in the old primitive way: impurities in the physical world or chemical carcinogens in the body are directly traced to immoral forms of economic and political power" (Douglas and Wildavsky 1983: 47).

Douglas's analysis is based on a typology of social behaviors, named "grid-group typology" ("group" denoting the degree of social participation, and "grid" denoting the degree of social obedience), expressed by four profiles: the individualist (concerned with personal freedom), the isolated (who primarily cares for not being involved in anything whatsoever), the 'hierarchist' (believing that authority stands for truth), and the sectarian (thinking that truth alone holds real authority) (David 2005: 41). As for environmental policies, the individualist believes in regulations driven by market economy; the isolated thinks that things are too complicate to be involved with; the hierarchist affirms that administrative regulation supported by rational decisions will provide proper solutions; the sectarian assumes that changing the whole mindset of the dominant group is urgently needed.

It is common to oppose Mary Douglas's "subjective" conception of risks framed by psycho-social deviant behaviors to Ulrich Beck's "objective" interpretation of risk rooted in "techno-economic development."[12] However, this opposition is largely overstated. Beck frequently stresses that there is no increase of actual risks but rather an increase of "risk consciousness."[13] In "Risk Society Revisited" he clearly said: "*it is cultural perception and definition that constitutes risks.* 'Risks' and 'the (public) definition of risks' are one and the same" (Beck 2000).

Interestingly enough, critics of both Douglas and Beck all stress that their constructivist or culturalist understanding of risks neglects the sociological underpinnings of risks. Robbins, Hintz, and Moore (2010: 88) rightly pointed out the limits of Douglas's psycho-social typological definition of risks: "Confirming the actual real-world existence of these ideal types of groups has proven difficult." Similarly Engel and Strasser (1998) disagree with Beck's hypothesis that risk conflicts have replaced class conflicts and his idea that "risks display an equalizing effect within their scope and among those affected by them": risk conflicts are not replacing class conflicts but are a new form of their complex materialization.

On this account, Avner de-Shalit showed clearly that the distribution of bads (pollution and toxics) is as unequal as the distribution of goods (income and wealth) and that sociological and ecological unevenness mutually reinforce each other.[14] In "risk societies" social inequalities have become "embodied" in the food people from different ethnical groups and social classes eat, in the water to which they have access, in the air they breathe, and in the places where they work and dwell, so that poor and minorities are not only more exposed to risks but have also less means to avoid them (Athanasiou 1996). Defining risks as a social *construct* is correct so far as this "construction" is not simply a matter of mental perception but rather of material production: the point is not that different people *see* risks diversely but that different people are *affected* by them unevenly.

Redefining Risks from an Epistemological Point of View

It is necessary thus to propose a redefinition of the notion of risk. According to the editors of *Global Environmental Risk*, "Global environmental risk refers to threat (to human beings and what they value) resulting from human-induced environmental change, either systemic or cumulative, on the global scale" (Kasperson, Kasperson, and Dow 2001). This definition can be misleading because we need to distinguish between risk and hazard: hazard is "an object, condition, or process that threatens individuals and society in terms of production or reproduction" while risk is "the known (or estimated) probability that a hazard-related decision will have a negative consequence" (Robbins et al. 2010: 81) If this latter definition is correct in differentiating hazard from risk, however it still lacks an important element: a risk is the assessment about an event which is considered to be hazardous *so far as* it endangers human or social life.

Eugene A. Rosa rightly defined risk as "a situation or an event where something of human value (including humans themselves) is at stake and where the outcome is uncertain" (Rosa 2003). Rosa's definition stresses the fact that the notion of risk is anthropocentric: an earthquake occurring in the depths of the Pacific Ocean is a natural event that may be of interest for the geologist; it does not constitute a risk so far as it does not constitute a threat for human society. The definition proposed by Ortwin Renn emphasizes this anthropocentric nature of risk: "risk is an uncertain consequence of an event or an activity with respect to something that humans value" (Renn 2008). But the limits of these two definitions is that the notion of uncertainty and the notion of risk are not clearly distinguished. Actually, as noticed by Frank H. Knight, uncertainty and risk are different concepts that require different kinds of probabilistic tools.[15] Risk refers to the probability of the occurrence of the (hazardous) event (objective or frequency probability), while uncertainty refers to the quantity of knowledge we have access to for supporting our belief in its happening (subjective or Bayesian probability). As Terje Aven rightly noticed, it is important to not make a confusion between the notion of subjective uncertainty defined by probabilities and the constructivist interpretation of "perceived risk" illustrated by Ulrich Beck or by Paul Slovic:[16]

> subjective probabilities and related risk assignments are not the same as risk perception. The main difference is that risk perception is based on personal beliefs, affects and experiences irrespective of their validity. Subjective probabilities used in risk assessments are representations of individual and collective uncertainty assessments based on available statistical data, direct experience, models and theoretical approximations which all need justification that must be plausible to others. (Aven 2010: 87)

There is definitively in the construction of the risk assessment *also* a part which is coming from the cultural *and* sociological environment of the observer.[17] Risk is defined by our perception of it but it cannot be equated to it (unless we think, with Berkeley, that *esse est percipi*). The *dual* nature of "risks," material and socio-psychological, is something that should be acknowledged: "environmental risks are viewed as partly biophysical threats of harm to people and partly as socially constructed, a product of culture and social experience" (Kasperson et al. 2001). More precisely, risk is threefold: it refers to the *severity*, the *probability*, and the *uncertainty* of a hazardous event impacting human life.

The *severity* of a risk depends on the spatial scope (area of effect) and on the temporal impact (duration of effect) of the hazard. Defining risk as the same as the perception of the risk is true as far as one limits its definition to the individual or collective perception of the *severity* of a hazard. The *probability* of a risk refers to the likelihood of a hazardous event and, more precisely, to the frequency of its (more or less inevitable) occurrence: the realist conception of risk focuses on this aspect of risk assessment. The *uncertainty* of a risk refers usually to the amount of information a subject has about a potentially hazardous event. It is commonly assumed that "if a person's knowledge was complete, that person would have no uncertainty" (Windschitl and Wells 1996). However, as far as risk as uncertainty is concerned, one must take into account not only the subjective uncertainty that depends on the degree of our knowledge but also the objective uncertainty that is entailed by the inherent "chaotic" complexity of the world. Contemporary sciences have shown that non-linear deterministic causes can have unpredictable consequences because of interdependent processes of actualization.[18] Thus "complete" knowledge of the present does not imply total certainty about future outcomes.[19]

Such an intrinsic uncertainty is both a specific feature of risks and a factor qualifying its other components: risk-uncertainty means uncertainty regarding both the likelihood and the severity of risks. Subjective uncertainty affects risk-likelihood by implying that there exists an irreducible margin of errors regarding the probability of the occurrence of a risk; objective uncertainty affects risk-severity by implying that it is not possible to state a priori whether a risk will be global or local, temporary or durable. Because of non-linear chains of causality, a local event can have global consequences: a risk which is thought to be local can become global, or "glocal" (Beck 2000), that is, have different local consequences at diverse places all over the world. In this sense, it comes from the very definition of risk that any risk is virtually global. Thus Global Risks are not a specific kind of risk but the potential feature of risk itself.

Finally, we notice that common definitions of risk define it as something "bad" *happening to* human beings. However, risks cannot be equated to harming hazards and "bads" alone, so far as they also represent opportunities of change: in Mary Douglas's typology, seeing risk as opportunity is the psychological characteristic of "individualists" (Palmer 1996); in terms of management, it is frequently said that the public sector is risk-averse while the private sector sees risk as an opportunity for highest return in investments (Cowan 2005: 66). More importantly, defining global risks as both cause and consequence of science, technology, and society interactions implies that *human beings are not only the victims of hazards that they try to prevent and manage but also the very producers of these hazards.* As Nick Bostrom and Milan M. Ćirković said in the introduction to *Global Catastrophic Risks*: "The most likely global catastrophic risks all seem to arise from human activities, especially industrial civilization and advanced technologies" (Bostrom and Ćirković 2008).

Thus we will propose to define risk as: *the severity, probability, and uncertainty of a hazardous event whose management aims at preventing this hazard harming (unevenly) social collectivities and/or resulting (involuntarily) from human activities.*

Consequences of the Definition of Risk for Understanding the New Interactions Between Science, Technology, and Society

The first consequence of our definition of risk is that risk is structural. Indeed, different social, geographical, and cultural backgrounds induce in different groups or in different persons different perceptions of the severity of its outcome (Sivak et al. 1989). But whatever the probability of their occurrence, whatever the uncertainty of their nature, risks are unavoidable in so far as the planet in which we live is not a mechanistic engine but an evolving ecosystem.

This paradigmatic shift from a mechanistic vision of the world to a post-Newtonian, ecological one[20,21,22] changes not only our understanding of reality but also our perception of human relation to nature as well as our definition of science. As Philippe Descola (2006) and Bruno Latour demonstrated, the emergence of classical physics went with the dissociation of the domain of nature from the dominion of culture, and the separation of the scientific from the sociologic realm. Nature was equated to the universality of (single) objective physical laws while culture was equated to the relativity of (multiple) subjective human customs.[23] As a consequence, scientists should be isolated from social interferences to understand an objective nature separated from subjective cultures. This dual separation (of nature from culture and science from society) was obtained at the price of mechanizing the world. Conversely, understanding the world as an open and interconnected ecosystem leads us to realize that human cultures do have an impact on natural systems and that scientific experimentations and theories are embroiled in social experiences and practices.[24]

Man is indeed one essential part of our continuously evolving planetary ecosystem: "Even in pre-modern societies, human activities may have affected global environment and participated in the shaping of actual landscape and actual atmospheric composition" (Hibbard et al. 2007). Furthermore, with the development of science and technology and with their industrial application to the fabric of modern societies, human actions did have non-linear, irreversible impacts on environment.[25] This new state of things creates the condition for global risks to emerge: "the birth of risk analysis lay in the systematic appraisal of highly complex problems involving large uncertainties associated with human interactions with nature and technology" (Kasperson et al. 2001).

Our understanding of the function and position of science in society is thus transformed. The image of science expressed by Robert Merton's *normal science values* seems today largely mythical (Bell 2006: 25): science *universalism*, the first "Mertonian value," is largely criticized by feminists pointing out the patriarchal organization of the scientific community (Glover 2002); science *communalism* seems to be difficult to conciliate with the privatization of public research; science *disinterestedness* is undermined by conflicts of interest between laboratories and companies (Frankel 1996); and science *organized skepticism* is shadowed by the surge of fraudulent papers (Fanelli 2009).

Because of the limits of Merton's normal science, Funtowicz and Ravetz (1993) developed the concept of post-normal science. Post-normal science implies that since science is rooted on objective uncertainty, it cannot provide factual predictions but probabilistic forecasts. Since science is rooted on subjective probability, it needs to include social stakeholders in its deliberative process (extended peer-communities). Moreover, since science participates to shape society through technology, it can have both positive and negative impacts; it can not only *reduce* but also *produce* risks: "scientific and technological practices are among the main world uncertainty producers, introducing novel and emergent technologies, organisms and forms of life" (Funtowicz, and Strand 2007).

Science is not only a problem-solver but can also be a risk-producer for structural as well as conjectural reasons. The fact that science is a risk-producer is perhaps not something that can simply be avoided by "rationalizing" its practice since it is structurally linked to the fact that important science discoveries can also be made by chance; science as a risk-producer is the negative counterpart of science serendipity (see Gillies in this volume, Chapter 25). *Both serendipity and risks illustrates the fact that the parameters that define science mechanisms are not themselves mechanistic*. Science being a risk-producer is conjecturally a result of the reshaping of the relation between science, technology, and society in the contemporary world economy. Moreover, and as a direct consequence, when science is no more linked to close mechanistic systems but to open ecosystems, when science is no more separated from society but becomes a part of its edification and a condition of its development, experiments are no longer confined in closed laboratories submitted to controlled parameters but can take place in open fields, conducive to incontrollable effects:

> "One of the main feature of contemporary high-power technoscience is that research and implementation are undertaken outside of the secured, controlled and simplified setting of the laboratories, and directly experimented – or we should say *experienced* – on the socioenvironmental systems of the planet" (Benessia 2009).

This point is particularly obvious in the case of Genetically Modified Organisms (GMO) (Berlan et al. 2001: 74–75). According to the 2003 Cartagena Protocol on Biosafety, "an estimation of the overall risk posed by the living modified organism based on the evaluation of the likelihood and consequences of the identified adverse effects" should be done to approve its large-scale commercial use. However, on one hand scientific expertise regarding human and ecological direct or indirect adverse impacts of GMO is *still ongoing*; on the other hand, GMO are *already* massively produced, planted, and consumed in the United States, Argentina, Brazil, India, and elsewhere. As Angelica Hilbeck et al. (2011) said: "despite over 10 years of large scale commercial production of GM crops in at least five countries, no consensus on the applied ERA [Environmental Risk Assessment] methodologies, let alone agreed standardised testing procedures exist." Thus, whether GMO use proves to be innocuous or not for either human health or ecosystem equilibrium, in any case, the example of GMO reflects the actual fashion of science, technology, and society new interactions where *the use is contemporary to the test and consequences may come before evidence*.[26]

It is for this reason that, on the basis of the acknowledgement of the inherent complexity of social-ecological systems and of the unquantifiable degree of uncertainty attached to current scientific knowledge, the precautionary principle was promoted to *minimize risks* by implementing measures to *prevent* (human-generated) *hazards* from happening: "The emergence of the Precautionary Principle has marked a shift from *postdamage* control (civil liability as a curative tool) to the level of a *pre-damage* control (anticipatory measures) of risks" (COMEST 2005). However, the application of the precautionary principle has proved difficult for at least two reasons: first, different interpretations of its meaning can lead to opposite safety policies; second, the scientific expertise involved in a risk assessment may add to its irresoluteness (the precautionary principle acts upon the acknowledgement of scientific uncertainty but the degree to which science is uncertain depends itself on prior scientific experimental results).[27]

The dispute between the United States and the EU about the use of hormonal substances for the rearing of beef illustrates the difference of interpretations of the Precautionary Principle (PP). Preventing damage can have a strong or a weak interpretation. The PP can mean: (1) it is not possible to do *y* without the proof that no damage *will* emerge from doing *y* (strong interpretation); (2) it is possible to do *y* unless we have *already* a proof that damage will result from doing *y* (weak interpretation). Regarding the beef hormone case, the United States endorsed a weak interpretation of the PP while the EU endorsed a strong one (Rakel 2004). On the basis of the "belief" that hormone beef was safe since there was no proof of damage reported from its consumption, the United States attacked the EU import ban on hormone-raised beef at the WTO. Eventually, the WTO supported US claims on the basis that EU scientists failed to provide evidence "that an identifiable risk arises from the use of any of the hormones at issue for growth promotion purposes in accordance with good practice" (WTO 1997). It is interesting to note that the WTO's condemnation of the EU import ban clearly shows that the WTO opts for the US interpretation of the PP and thus adopts the procedural culture of its most powerful member. Besides, it is also likely that the EU ban on import of hormone-raised beef could have been motivated not only for the safety of its consumers but also for the protection of its producers. In any case, this example shows that risk assessments are not simply a matter of scientific knowledge influenced by "cultural worldviews" but also a matter of power relations influenced by political and economic factors.

Another difficulty with the precautionary principle comes from the nature of science involvement in risk assessments. Risk assessment is rooted on policymakers' belief that scientific expertise will reduce the uncertainties that plague the decision. However, as Dale Jamieson (1996) said, "rather than being a cause of controversy, scientific uncertainty is often a consequence of controversy," since "scientific uncertainty is constructed by both science and society in order to serve certain purposes." Science can reduce uncertainty in so far as there is no uncertainty about the social function of scientists and their interests in providing the risk assessment. *Science cannot help in reducing uncertainties that are not coming from objective causes but from the socio-economic factors shaping the uneven distribution of risks.*[28] Thus, finally, it is not so much *risks* but *uncertainties* that are the target of power relations among different social actors: either uncertainty is convoked by environmental activists to prohibit the use of a new technology in the name of partial knowledge; or it is used by industry to postpone the legal resolution of trials by asking the plaintiffs to provide always more information in the name of incomplete data (Jobin and Tseng 2014).

Redefining Risks and Risk Society: Political Challenges of Risk Management

The fact that Global Risks are at the same time cause and consequence of the new modes of interactions between science, technology, and society implies that progress is non-linear (see Figure 27.1). On one hand, the progress of science and technology increased the ability to predict and, to a certain extent, *to prevent global natural risks*; on the other, it increased the possibility *to generate*, either intentionally or unintentionally, *global anthropological risks* (Goussot 2009). The risk of a nuclear war or a terrorist use of nuclear waste comes from the invention of nuclear power and human needs of energy; the risk of global epidemic disease comes from the overuse of synthetic antibiotics and the potential use of sequenced viruses stored in open database (Kilbourne 2008). Risk as a concept refers to the analysis of the severity, probability, and uncertainty of hazards. Risk as an outcome proceeds from the irreducible indeterminacy between real and possible, true and false, and safe and unsafe.

Because events are unpredictable due to objective world complexity and data are uncertain due to subjective mind limitation, political decisions have to deal with ambivalent risks. Moreover, not only complexity and uncertainty put decision-making at stake, but socio-political disputes themselves reinforce uncertainty (by

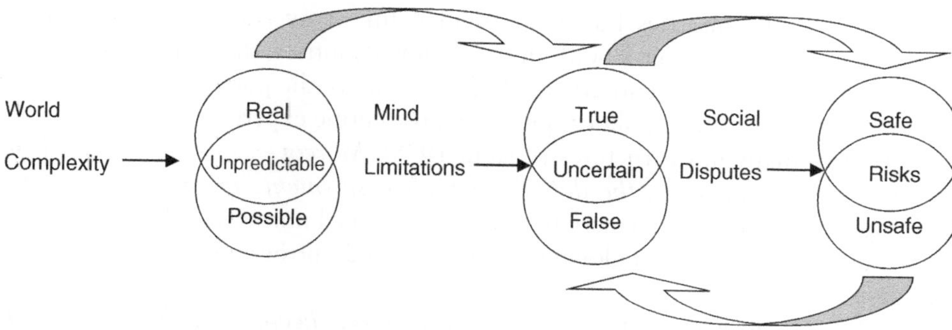

Figure 27.1 The ontology of social risks and its cognitive basis.

opening new debates about the value of scientific expertise) and complexity (by creating a new technology of management). Uncertainty is not only the reason for the need of science-based risk assessment but also the consequence of value-based risk management (Di Lucia, Ahlgren, and Ericsson 2012).

This dual nature of uncertainty leads to intractable difficulties for risk policies. Risk policies based on post-normal science extended peer communities (i.e., including stakeholders) can have contradictory results (Friedrichs 2011): on one hand, it may broaden the awareness of the general public (and potentially reduce the amount of involuntary risks); on the other, it can weaken people's belief in the objective nature of scientific expertise (and thus reinforce mistrust about recommendations to avoid risks). Similarly, risk policies based on the precautionary principle lead to political dilemma: either governments prohibit the use of potentially hazardous ("risky") new scientific-technological products or services, but at the expense of the research opportunities and economic benefices that can come from it; or they allow the use of potentially hazardous ("risky") new scientific-technological products or services, but at the expense of social-political stability and with the irreversible bio-ecological consequences that may result from them.

The problems of risk societies cannot be resolved only by participatory mechanisms and access to information. First, because these participatory debates are limited in both scope and impact (Pettsa 2004); second, because thinking that the more people will be informed by enlightened experts, the more they will accept governmental policies, supposes that public opposition comes from ignorance; however "conflicts and controversies surrounding risk management are not due to public ignorance or irrationality but, instead, are seen as a side effect of our remarkable form of participatory democracy" (Slovic 1993). The increase of protests against perceived risky facilities (from nuclear power plants to wind turbines to cell sites) are one of the consequences of "democracy" and access to information – and *not* of their (discursive) deficit.

But that does not mean that "the only solution is a sufficient measure of coercion" (Ophuls 1977: 150). On the contrary, good governance is the institutional condition of possibility of efficient risk management: openness, transparency, and accountability are necessary to identify risks, to integrate the information coming from both civil society and scientific experts and to react to them in a responsible manner (Heldeweg 2005). Ortwin Renn righty stressed that "these requirements are important for all countries but, in particular, for many transitional and most developing countries" (Renn 2008). In developing countries, the fact that the rules of good governance are not strictly applied can amplify the potential emergence of risks: in the case of China, political pressure on scientific expertise can bias the risk assessment of many infrastructures (Boland 1998). *Moreover, since local risks can have global consequences, the default of risks assessments in one country can affect the others.* Hence the emerging need for universal norms in risk assessments. But is a global governance of global risks possible and, furthermore, if possible, is it even desirable?

First, *the systemic characteristic of risks prevents us believing that global risks will be addressed by simply targeting individual risky behaviors.* As Elizabeth Shove said regarding policies of Climate Change, the problem is not to persuade individual actors to "behave responsibly" (either according to "universal" criteria of economic efficiency or to moral standards of generational responsibility) but to change the rule of the game in which they define their conducts: "one key condition is to shift the focus away from individual choice and to be explicit about the extent to which state and other actors configure the fabric and the texture of daily life" (Shove 2010). Risks are not subjective in the sense that risky individual behaviors are not the main factor of their emergence and transmission. Social collectives – either public (states and governments at the regional and national levels) or private (national and international profitable organizations) – are primary producers of risks.

Second, as shown by the cases of GMO and hormone beef, different decisions can come from a similar epistemological basis: it is ultimately for the political power to decide whether the risk should be "taken" (i.e., supported by society) or not. This is why global governance is not the ideal solution since it can become itself a vector of global risks. Global risk management can induce the establishment of new mechanisms of global control of society (*risk-panopticon*)[29] that could lead to global civil disobedience. Moreover, as demonstrated by the recent "panarchy theory" (theory of complexity), the more "organized" a system, the less *resilient* it is. A global governance of risks based on a continuous world survey of pre-identified loci of vulnerability not only may not be able to identify risks coming from unexpected places but will imply a bureaucratic, centralized structure of decision highly vulnerable itself to external disturbance: "part of the puzzle of adaptive management is how to build a non-bureaucratic bureaucracy" (Pritchard and Sanderson 2002).

Finally, since risks are a social and technological construct and product, they cannot be addressed without deeply amending what is dysfunctional in the social-economic structure in and from which they emerge: "Wealth production within a 'risk society' typically depends on production technologies that expose citizens to dangerous substances" (Cable, Shriver, and Mix 2008). Addressing the consequences without rectifying the causes will only raise the level of risks. Democratizing *discourses about risks* (rationally and openly discuss risk potential severity) is the necessary but not sufficient condition of global risk management; democratizing the *economy of risks* (fairly distribute hazards' actual impacts) will be more efficient in the long term (if everyone bears the same amount of risk, there will be no more incentive for their asymmetric accumulation in the social-ecological system).

Notes

1 This article is one of the results of the EU Commission Project IRSES "Evaluating Policies for Sustainable Energy Investments: Towards an Integrated Approach on National and International Stage (EPSEI)" coordinated by the University of Turin.

2 Mokyr (2005: 29): "the Industrial Revolution's timing was determined by intellectual developments, and the true key to the timing of the Industrial Revolution has to be sought in the scientific revolution of the seventeenth century and the enlightenment movement of the eighteenth century."

3 Dodson (2010: xix): "The science is clear that a significant part of recent global warming is driven by human activities. One of the main components of this is the accumulation of anthropogenically derived greenhouse active gases in the Earth's atmosphere. This gives rise to an example of a major feedback loop. In it, human economic activity (energy and resource use, agriculture, forestry and more) leads to changes in Earth system elements (climate, land cover, sea level change and more) which feedback into the distribution and kinds of human activities which are economic."

4 Carpenter et al. (2006): "The most catastrophic changes in ecosystem services identified in the MA (*Millennium Assessment*) involved nonlinear or abrupt shifts. We lack the ability to predict thresholds for such changes, whether or not such a change may be reversible, and how individuals and societies will respond."

5 Reid et al. (2010): "We need to develop what amounts to an enhanced Earth system simulator to improve our ability to anticipate impacts of a given set of human actions or conditions on global and regional climate and on biological, geochemical, and hydrological systems on seasonal to decadal time scales."

6 Wigley (2006): "Future climate change may be reduced through mitigation (reductions in greenhouse gas emissions) or through geoengineering."

7 Stern and Tubiana (2008: 75–107): "In the energy sector, energy efficiency is the first requirement of reducing emissions, since low historic prices have tended to make energy very wastefully used, and more efficient use can bring net positive economic benefit. However, as economies grow, there will be both thermodynamic and economic limits to ever increasing energy efficiency. So ultimately the control of energy emissions must come through the decarbonisation of energy sources – moving from unabated fossil fuels to renewables, nuclear and the capture and storage of fossil carbon. ... And here both energy efficiency and decarbonisation require the same kind of instrument: the pricing of carbon through taxes, cap and trade schemes (emissions trading), and regulations and standards."

8 Semenza et al. (2008): "Mitigating global climate change requires not only government action but also cooperation from consumers."

9 Shove (2010): "It is now widely agreed that the challenges of climate change are such that many familiar ways of life and many of the patterns of consumption associated with them are fundamentally unsustainable. If there is to be any effective response, new forms of living, working, and playing will have to take hold across all sectors of society."

10 Latour (2004: 124): "This is the case ... of the eight thousand people who die each year from automobile accidents in France: no way was found to keep them as full-fledged – and thus living! – members of the collective. In the hierarchy that was set up, the speed of automobiles and the flood of alcohol was preferred to highway deaths. ... for the time being, the rapid use of cars is 'worth' much more in France than eight thousand innocent lives per year."

11 Beck (1995: 3): "What was at stake in the older industrial conflict of labor against capital were positives: profits, prosperity, consumer goods. In the new ecological conflict, on the other hand, what is at stake are negatives: losses, devastation, threats."

12 Alexander and Smith (1996): "Beck wants to portray the risk society as an objective fact, both ontologically, in the sense that it exists as such, in a cold, hard and material way, and epistemologically, in the sense that these objective facts are perceived directly and accurately in the minds of citizens themselves."

13 Slater and Ritzer (2001): "The world has not grown unconditionally more risky. Rather it is the systemic lack of trust that makes consumers so quick to see risks everywhere. The less trust, the more risk."

14 de-Shalit (2000: 182): "Some gross inequalities related to ecological distribution are: dumping toxic and hazardous waste in locations inhabited by poorer or less advantaged ethnic groups or nations, often accompanied by a latent threat (if the poor oppose dumping they stand to lose their jobs, or, in the case of international relations, to lose military and economic aid); unequal vulnerability to heat waves; ecologically unequal exchange (importing products from poor countries or regions at prices that do not take into account the exhaustion or depletion of resources there); and exploitation of indigenous knowledge without adequate payment."

15 Knight (1921: 233): "The word 'risk' is ordinarily used in a loose way to refer to any sort of uncertainty viewed from the standpoint of the unfavorable contingency, and the term 'uncertainty' similarly with reference to the favorable outcome; we speak of the 'risk' of a loss, the 'uncertainty' of a gain. But if our reasoning so far is at all correct, there is a fatal ambiguity in these terms, which must be gotten rid of, and the use of the term 'risk' in connection with the measurable uncertainties or probabilities of insurance gives some justification for specializing the terms as just indicated. We can also employ the terms 'objective' and 'subjective' probability to designate the risk and uncertainty respectively."

16 Slovic (1992): "Risk does not exist 'out there', independent of our minds and cultures, waiting to be measured."

17 Jasanoff (1999): "the social sciences have deeply altered our understanding of what 'risk' means – from something real and physical if hard to measure, and accessible only to experts, to something constructed out of history and experience by experts and laypeople alike."

18 Capra (1996: 120): "Detailed models of self-organizing systems could be formulated only very recently when new mathematical tools became available that allowed scientists to model the non-linear interconnectedness characteristic of networks."

19 Emmeche, Køppe, and Stjernfelt (1997): "Something which is unpredictable does not have to be unexplainable. The phenomenon of deterministic chaos in complex dynamical systems has shown how futile it is to demand that the competence of prediction in physics should be complete, when the development of systems in phase space may be extremely sensitive on the initial conditions."

20 Capra (1996: 61): "Newtonian mechanics was a science of forces and trajectories, evolutionary thinking – thinking in terms of change, growth, and development – required a new science of complexity."

21 Smolin (2009): "We also no longer have any reason to suspect that time is an illusion because, as outlined above, the main arguments from physics for time being emergent and not fundamental come from the misapplication of the Newtonian schema to the universe as a whole."

22 Ulanowicz (2000): "one must abandon the [Newtonian] assumptions of closure, determinism, universality, reversibility, and atomism and replace them by the ideas of openness, contingency, granularity, historicity, and organicism, respectively."

23 Latour (1993: 31): "there shall exist a complete separation between the natural world (constructed, nevertheless, by man) and the social world (sustained, nevertheless, by things)."

24 Latour (2004: 25): "Political ecology does not shift attention from the human pole to the pole of nature; it shifts from certainty about the production of risk-free objects (with their clear separation between things and people) to uncertainty about the relations whose unintended consequences threaten to disrupt all orderings, all plans, all impacts."

25 Rockström et al. (2009): "Many subsystems of Earth react in a nonlinear, often abrupt, way, and are particularly sensitive around threshold levels of certain key variables. If these thresholds are crossed, then important subsystems, such as a monsoon system, could shift into a new state, often with deleterious or potentially even disastrous consequences for humans."

26 Jasanoff (2003): "Only in the wake of environmental disasters involving dioxin, methyl isocyanate, and PCBs, and only after the accidental exposure of populations and ecosystems, were gaps discovered in the information available about the long-term effects of many hazardous chemicals."

27 Harrison, McBride, and Bouldin (1999): "The uncertainties inherent in a risk assessment, originating from both missing data and a lack of understanding of how available data should be applied to complex systems with inherent biological variation, argue for the application of safety factors."

28 Robbins et al. (2010: 93): "The difference between state regulators/ power companies who may seek to develop nuclear waste and related waste storage, for example, and average citizens who may invoke precaution, may not be simply 'cultural' in the sense Douglas suggests, therefore. It may instead reflect fundamental differences in where each group sits in a complex political economy. Both can make rational risk decisions and still not agree. Risk assessment can no more resolve such differences than it can change the nature of bureaucratic government or the capitalist economy."

29 Luke (1995): "The individual human subject of today, and all of his or her unsustainable practices, would be reshaped through this environmentality, redirected by practices, discourses, and ensembles of administration that more efficiently synchronize the bio-powers of populations with the geo-powers of environments. Traditional codes defining human identity and difference would be reframed by systems of environmentality in new equations for making comprehensive global sustainability calculations as the bio-power of populations merges with the ecopower of environments."

References

Alexander, Jeffrey C., and Philip Smith. 1996. "Social Science and Salvation: Risk Society as Mythical Discourse." *Zeitschrift für Sociologie* 25(4): 251–262.

Athanasiou, Tom. 1996. *Divided Planet: The Ecology of Rich and Poor*. Boston, MA: Little, Brown and Company.

Aven, Terje. 2010. *Misconceptions of Risk*. Chichester: John Wiley & Sons Ltd.

Beck, Ulrich. 1992a. *Risk Society: Towards a New Modernity*. London: Sage.

Beck, Ulrich. 1992b. "From Industrial Society to the Risk Society: Questions of Survival, Social Structure and Ecological Enlightenment." *Theory, Culture and Society* 9(1): 97–123.

Beck, Ulrich. 1995. *Ecological Enlightenment: Essays on the Politics of the Risk Society*. Atlantic Highlands, NJ: Humanities Press International.

Beck, Ulrich. 2000. "Risk Society Revisited." In *The Risk Society and Beyond: Critical Issues for Social Theory*, ed. Barbara Adam, Ulrich Beck, and Joost Van Loon, 211–229. London: Sage.

Bell, David. 2006. *Science, Technology and Culture*. Maidenhead, UK: Open University Press.

Benessia, Alice. 2009. "From Certainty to Complexity." In *Science, Society and Sustainability*, ed. Donald Gray, Laura Colucci-Gray, and Elena Camino, 10–26. New York: Routledge.

Berlan, Jean-Pierre, et al. 2001. *La guerre au vivant*. Marseilles: Agone.

Boland, Alana. 1998. "The Three Gorges Debate and Scientific Decision-Making in China." *China Information* 13: 25–42.

Bostrom, Nick, and Milan M. Ćirković (eds.). 2008. *Global Catastrophic Risks*. Oxford: Oxford University Press.

Cable, Sherry, Thomas E. Shriver, and Tamara L. Mix. 2008. "Risk Society and Contested Illness: The Case of Nuclear Weapons Workers." *American Sociological Review* 73(3): 380–401.

Capra, Fritjof. 1996. *The Web of Life: A New Scientific Understanding of Living Systems*. New York: Anchor Books.

Carpenter, Stephen R., et al. 2006. "Millennium Ecosystem Assessment: Research Needs." *Science* 314(5797): 257–258.

COMEST. 2005. *The Precautionary Principle*. Paris: UNESCO, World Commission on the Ethics of Scientific Knowledge and Technology.

Cowan, Neil. 2005. *Risk Analysis and Evaluation*. Canterbury: Institute of Financial Services.

David, Matthew. 2005. *Science in Society*. London: Palgrave Macmillan.

Descola, Philippe. 2006. *Par-delà Nature et Culture*. Paris: Gallimard.

de-Shalit, Avner. 2000. *The Environment Between Theory and Practice*. Oxford: Oxford University Press.

Di Lucia, Lorenzo, Serina Ahlgren, and Karin Ericsson. 2012. "The Dilemma of Indirect Land-Use Changes in EU Biofuel Policy – An Empirical Study of Policy-Making in the Context of Scientific Uncertainty." *Environmental Science Policy* 16: 9–19.

Dodson, John. 2010. "Introduction." In *Changing Climates, Earth Systems and Society*, ed. John Dodson, xix–xx. New York: Springer.

Douglas, Mary. 1984. *Purity and Danger: An Analysis of the Concepts of Pollution and Taboo*. New York: Routledge.

Douglas, Mary. 1992. *Risk and Blame: Essays in Cultural Theory*. London: Routledge.

Douglas, Mary. 1997. "The Depoliticisation of Risk." In *Culture Matters: Essays in Honor of Aaron Wildavsky*, ed. R.J. Ellis and M. Thompson, 121–132. Boulder, CO: Westview Press.

Douglas, Mary, and Aaron Wildavsky. 1983. *Risk and Culture: An Essay on the Selection of Technological and Environmental Dangers*. Berkeley and Los Angeles: University of California Press.

Emmeche, Claus, Simo Køppe, and Frederik Stjernfelt. 1997. "Explaining Emergence: Towards an Ontology of Levels." *Journal for General Philosophy of Science* 28: 83–119.

Engel, Uwe, and Hermann Strasser. 1998. "Global Risks and Social Inequality: Critical Remarks on the Risk-Society Hypothesis." *The Canadian Journal of Sociology/Cahiers canadiens de sociologie* 23(1): 91–103.

Fanelli, Danielle. 2009. "How Many Scientists Fabricate and Falsify Research?" *PLoS One* 4(5): e5738.

Frankel, Mark S. 1996. "Perception, Reality, and the Political Context of Conflict of Interest in University-Industry Relationships." *Academic Medicine* 71(12): 1297–1304.

Friedrichs, Jörg. 2011. "Peak Energy and Climate Change: The Double Bind of Post-Normal Science." *Futures* 43(4): 469–477.

Funtowicz, Silvio, and Jerome R. Ravetz. 1993. "Science for the Post-Normal Age." *Futures* 25(7): 735–755.

Funtowicz, Silvio, and Roger Strand. 2007. "Models of Science and Policy." In *Biosafety First: Holistic Approaches to Risk and Uncertainty in Genetic Engineering and Genetically Modified Organisms*, ed. Terje Traavik and Lim Li Ching. Trondheim: Tapir Academic Press.

Glover, Judith. 2002. "Women and Scientific Employment: Current Perspectives from the UK." *Science Studies & Technology Studies* 15(1): 29–45.

Gonzalez, Patrick. 2010. "Impacts of Climate Change on Terrestrial Ecosystems and Adaptation Measures for Natural Resource Management." In *Changing Climates, Earth Systems and Society*, ed. John Dodson, 5–20. New York: Springer.

Goussot, Michel. 2009. "La représentation du changement." In *Le changement en environnement: Les faits, les représentations, les enjeux*, ed. Martine Tabeaud, 69–80. Paris : Publications de la Sorbonne.

Harrison, Ellen Z., Murray B. McBride, and David R. Bouldin. 1999. "Land Application of Sewage Sludges: An Appraisal of the US Regulations." *International Journal of Environment and Pollution* 11(1): 1–36.

Heldeweg, Michiel. 2005. "Towards Good Environmental Governance in Europe." *European Energy and Environmental Law Review* 14(1): 2–24.

Hibbard, Kathy A., et al. 2007. "Group Report: Decadal-Scale Interactions of Humans and the Environment." In *Sustainability or Collapse? An Integrated History and Future of People on Earth*, ed. Robert Constanza, Lisa J. Graumlich, and Will Steffen. Cambridge, MA: MIT Press.

Hilbeck, Angelica, et al. 2011. "Environmental Risk Assessment of Genetically Modified Plants – Concepts and Controversies." *Environmental Sciences Europe* 23: 13.

Jamieson, Dale. 1996. "Scientific Uncertainty and the Political Process." *Annals of the American Academy of Political and Social Science* 545(1): 35–43.

Jasanoff, Sheila. 1999. "The Songlines of Risk." *Environmental Values* 8(2): 135–152.
Jasanoff, Sheila. 2003. "Technologies of Humility: Citizen Participation in Governing Sciences." *Minerva* 41: 223–244.
Jobin, Paul, and Tseng Yu-Hwei. 2014. "Guinea Pigs Go to Court: Epidemiology and Class Actions in Taiwan." In *Powerless Science? The Making of the Toxic World in the Twentieth Century*, ed. Soraya Boudia and Nathalie Jas. Oxford: Berghahn Books.
Kasperson, Jeanne X., Roger Eugene Kasperson, and Kirstin Dow. 2001. "Global Environmental Risks and Society." In *Global Environmental Risk*, ed. Jeanne X. Kasperson and Roger Eugene Kasperson, 1–48. London: United Nations University Press.
Kilbourne, Edwin Dennis. 2008. "Plagues and Pandemics: Past, Present and Future." In *Global Catastrophic Risks*, ed. Nick Bostrom and Milan M. Ćirković, 218–231. Oxford: Oxford University Press.
Knight, Frank H. 1921. *Risk, Uncertainty, and Profit*. Boston, MA: Houghton Mifflin Company.
Latour, Bruno. 1993. *We Have Never Been Modern*. Cambridge, MA: Harvard University Press.
Latour, Bruno. 2004. *Politics of Nature*. Cambridge, MA: Harvard University Press.
Luke, Timothy W. 1995. "On Environmentality: Geopower and Ecoknowledge, in the Discourses of Contemporary Environmentality." *Cultural Critique* 31: 57–82.
Mokyr, Joel. 2005. *The Gifts of Athena: Historical Origins of the Knowledge Economy*. Princeton, NJ: Princeton University Press.
Ophuls, William. 1977. *Ecology and the Politics of Scarcity*. San Francisco, CA: W.H. Freeman and Company.
Palmer, Christina G.S. 1996. "Risk Perception: An Empirical Study of the Relationship Between Worldview and the Risk Construct." *Risk Analysis* 16(5): 717–723.
Pettsa, Judith. 2004. "Barriers to Participation and Deliberation in Risk Decisions: Evidence from Waste Management." *Journal of Risk Research* 7(2): 115–133.
Pritchard, Lowell Jr., and Steven E. Sanderson. 2002. "The Dynamics of Political Discourses in Seeking Sustainability." In *Panarchy: Understanding Transformations in Human and Natural Systems*, ed. Lance H. Gunderson and C.S. Holling, 147–172. Washington, DC: Island Press.
Rakel, Horst. 2004. "Scientists as Expert Advisors: Science Cultures Versus National Cultures?" In *Experts in Science and Society*, ed. Elke Kurz-Milcke and Gerd Gigerenzer, 3–26. New York: Springer.
Reid, W.V., et al. 2010. "Earth System Science for Global Sustainability: Grand Challenges." *Science* 330(6006): 916–917.
Renn, Ortwin. 2008. "White Paper on Risk Governance: Towards an Integrative Framework." In *Global Risk Governance: Concept and Practice Using IRGC Framework*, ed. Ortwin Renn and Katherine D. Walker, 3–73. Berlin and Heidelberg: Springer.
Robbins, Paul, John Hintz, and Sarah A. Moore. 2010. *Environment and Society: A Critical Introduction*. Oxford: Wiley-Blackwell.
Robock, Alan. 2008. "20 Reasons Why Geoengineering May Be a Bad Idea." *Bulletin of the Atomic Scientists* 64(2): 14–18.
Rockström, Johan, et al. 2009. "A safe Operating Space for Humanity." *Nature* 461(7263): 472–475.
Rosa, Eugene A. 2003. "The Logical Structure of the Social Amplification of Risk Framework (SARF): Metatheoretical Foundations and Policy Implications." In *The Social Amplification of Risk*, ed. Nick Pidgeon, Roger E. Kasperson, and Paul Slovic, 47–79. Cambridge: Cambridge University Press.
Semenza, Jan C., et al. 2008. "Public Perception of Climate Change: Voluntary Mitigation and Barriers to Behavior Change." *American Journal of Preventive Medicine* 35(5): 479–487.
Shove, Elizabeth. 2010. "Beyond the ABC: Climate Change Policy and Theories of Social Change." *Environment and Planning A* 42: 1273–1285.

Sivak, M., J. Soler, U. Trankle, and J.M. Spagnhol. 1989. "Cross-Cultural Differences in Driver Risk-Perception." *Accident Analysis and Prevention* 21(4): 355–362.
Slater, Don, and George Ritzer. 2001. "Interview with Ulrich Beck." *Journal of Consumer Culture* 1(2): 261–277.
Slovic, Paul. 1992. "Perception of Risk: Reflections on the Psychometric Paradigm." In *Social Theories of Risk*, ed. S. Krimsky and D. Golding, 117–152. Westport, CT: Praeger.
Slovic, Paul. 1993. "Perceived Risk, Trust, and Democracy." *Risk Analysis* 13(6): 675–682.
Smolin, Lee. 2009. "The Unique Universe: Against the Timeless Multiverse." *Physics World June:* 21–26.
Stern, Nicholas, and Laurence Tubiana. 2008. "A Progressive Global Deal on Climate Change." In *A Progressive Agenda for Global Action*, ed. Policy Network. London: Policy Network.
Ulanowicz, Robert E. 2000. "Life after Newton: An Ecological Metaphysic." In *The Philosophy of Ecology: From Science to Synthesis*, ed. David R. Keller and Frank B. Golley, 81–100. Athens and London: University of Georgia Press.
Wigley, T.M.L. 2006. "A Combined Mitigation/Geoengineering Approach to Climate Stabilization." *Science* 314(5798): 452–445.
Windschitl, P.D., and G.L. Wells. 1996. "Measuring Psychological Uncertainty: Verbal Versus Numeric Methods." *Journal of Experimental Psychology: Applied* 2(4): 343–364.
WTO. 1997. *EC Measures Concerning Meat and Meat Products (Hormones): Complaint by the United States: Report of the Panel*, Geneva:.World Trade Organization (WT/DS26/R/USA).

Chapter 28

Globalization, Regionalization, and Technological Change

Frederick Guy

Introduction

National economies have become, over the past half century, radically more integrated. The process of international economic integration is viewed by many as being *global* (hence the term "globalization"). In this chapter I will argue that regional economic integration – regions here in the sense of sub-continental mega-regions, such as Europe and China – is likely to eclipse global integration in coming decades.

Before going further, it is necessary first to unpack the term "international economic integration" which, if it can be considered a single phenomenon at all, is certainly one of many distinct dimensions. A simple categorization of dimensions includes (1) trade in goods and services; (2) the organization of production processes across borders – something that requires trade, but also requires far more by way of cross-border corporate activity than simple trade does; (3) production in multiple countries by multinational corporations (MNCs), which requires international capital flows in the forms both of foreign direct investment (FDI), and eventual repatriation of profit or disinvested capital; (4) the international mobility of capital for other purposes; (5) the transmission of economically useful knowledge – scientific discoveries, new technologies, designs, business know-how – through the activities of MNCs, through migration, licensing, or simple copying, and through various forms of public information dissemination (publishing); (6) migration in various forms – for work, study, or tourism, legal or illegal, permanent or temporary. Each of these dimensions may be understood both in terms of activity (e.g., trade), and of institutions (e.g., rules governing trade and trade barriers). Finally, and central to the argument of this chapter, these dimensions can manifest on different geographical

The Handbook of Global Science, Technology, and Innovation, First Edition.
Edited by Daniele Archibugi and Andrea Filippetti.

scales: "international" can range from integration with a single neighboring country, to throwing one's doors open to the world.

Seen thus, it becomes difficult to say simply "international economic integration has increased (or decreased)," because it may increase in one dimension while decreasing in another; or it may increase globally in trade but regionally in production or migration. It is therefore necessary to refine the thesis stated above: global integration will be increasingly concentrated in the transmission of knowledge, while the integration of production, and direct investment, will be largely regional phenomena.

To establish this, we need to consider first how technology provides means and motive for international economic integration. Here we need to distinguish between the effects of particular technologies, and the effects of technological change itself.

Particular technologies – containerization, jumbo jets, the Internet – are *distance-shrinking*, and because of this they facilitate international integration. Once such a technology has been invented and adopted, it can help sustain a certain level of integration even if the technology changes no further.

The ongoing process of technological change creates new differences in what is known in one place and another. Most international economic integration can be understood as a response to these differences in knowledge. Comparative advantage, which motivates trade, is largely rooted in differences in knowledge. Modern trade theory (the "new" trade theory of the 1980s and after) explains trade in produced goods and services between countries at similar levels of development in terms of localized increasing returns, which are in turn rooted in locally sticky knowledge – learning-by-doing, local knowledge spillovers, organizational capabilities. The older Heckscher–Ohlin theory located national comparative advantage in different levels of development; in simple formal terms, this meant different capital to labor ratios (high in rich countries, low in poor ones), but modern understandings of that theory would locate much of the difference in the quality, experience, and accumulated knowledge of the labor force – human capital. Only when comparative advantage is entirely a product of natural endowments – Ecuador has a comparative advantage in bananas, Kuwait in oil – do differences in knowledge cease to be an important motivation for trade.

Cross-border production networks allow comparative advantage to be realized on a more fine-grained basis – it is now necessary to have a comparative advantage only in one stage of production, rather than the entire product. Such networks require management, which is provided by MNCs. Most of what the MNCs bring to this process is, again, knowledge: knowledge of markets, production processes, or management methods (see this volume, Ietto-Gillies, Chapter 6 and Iammarino and McCann, Chapter 14).

MNCs, however, have a much bigger role in the generation and transfer of knowledge than what is needed to coordinate cross-border production. They develop new products and processes, and they carry that knowledge across borders – often, to set up production in foreign countries. Such ventures then require considerable international capital flows, and in particular FDI. MNCs invest in foreign operations – in short, they exist as an MNC – largely to profit from knowledge they have, or to profit from the acquisition of knowledge in the foreign country. Knowledge the MNC has may take the form of organizational competencies or of intellectual property. The former is the firm's collective know-how, the visible hand capabilities of its organizational hierarchy (Chandler 1977, 1992). These competencies may enable the MNC to produce, even in a foreign environment, at a lower cost than its competitors. Alternatively, the MNC may enter a market to preserve or extend market power, by preventing the emergence of competitors – that is, firms with comparable competencies – and by protecting its intellectual property by keeping functions internal. And, finally, MNCs locate their investments – in particular, R&D facilities and joint ventures – with the strategic aim of acquiring knowledge from other firms (Cantwell and Iammarino 2001, 2003).

Commercially useful knowledge, of course, gets transferred in other ways as well: designs and methods are reverse engineered; academic papers are published; patents are filed, and then read by others; knowledgeable workers migrate; programmers from around the world collaborate on open-source software. Just as we understand large elements of international economic integration as responses to differences between countries in what is known, we regard the use of these varied mechanisms to spread or share knowledge across borders, as forms of international economic integration.

At what scale, geographically, does integration take place, and at what level is it governed? Global? National? Something in between? The answers lie in countless choices of policy, and of institutional design, which affect the scale of integration: trade barriers, rules governing international investment, technical standards, intellectual property, and so on. The central arguments here are two, one concerning the properties of global and sub-global economic governance, and the other the implication that those properties have for the timing of their adoption. First: global integration tends to be institutionally minimalist, such as we see reflected in simple World Trade Organization (WTO) market liberalization. Economic processes integrated at a sub-global level can be associated with deeper, more robust, and (potentially) more accountable regulation. Second: following a shock which leads to a restructuring of the institutions governing economic integration, the institutional simplicity of global integration allows for relatively rapid adoption, but global integration will be rolled back as sub-global integration progresses. Global integration is the hare, regional integration the tortoise.

Large markets offer economic advantages – notably, an elaborate division of labor, and competition even in the face of considerable returns to scale. Since the division of labor becomes more elaborate as economic development progresses (Jonscher 1994; Smith 1991), and since distance-shrinking technologies reduce the costs of operating in large markets, it is often argued that the appropriate scale for most forms of economic integration has become global, and that economic governance should be global as well. Yet others see advantages to regulating markets at sub-global levels, and sub-global regulation generally limits global integration.

One way to approach the question of what level of governance is Rodrik's formulation of "the inescapable trilemma of the world economy" (Rodrik 2007). He argues that the nation-state, deep global economic integration, and democracy are not simultaneously possible. For nation-state, we can read *any* sub-global sovereign entity. Deep international economic integration puts the nation-state in what Rodrik calls a golden straitjacket, which is inconsistent with democratic control. Rodrik makes this point out of concern for democracy, but here it is useful to frame it in a more general way: deep international economic integration is inconsistent with *sub-global* control, whether democratic or not: a fascist dictator or a traditional oligarchy could find itself uncomfortably constrained by the rules of the WTO, just as easily as a democratic polity. The alternatives, Rodrik points out, are deeper global governance (democratic and federal, in his account), or shallower international economic integration. Deeper global governance, however, is unlikely, for two reasons. Frst, there is the long and problematic process required to develop the legitimacy and institutions required for governance (especially democratic governance) at any level; seeking these on a global scale is a worthy project, but there is little reason to believe that anybody alive today will see it. Second, deeper institutions of global governance, especially if democratic, would vitiate one of the virtues global economic integration holds for its most fervent and powerful advocates, which is precisely that it does put a set of golden handcuffs on states. The minimalist institutions of the WTO, run from a small secretariat in Geneva, can be seen as aspiring to impose an economically liberal order of property and free contracts, à la Buchanan and Tullock (1962). For this constituency, deep and accountable institutions of global governance would offer no real advantages over national or regional governance.

Because deeper global governance is unlikely, those resisting golden handcuffs will find it necessary to focus on renewing institutions of economic regulation at a sub-global level. Given the scale properties of today's production systems and the distance-shrinking technologies in use, it is likely that that level will be regional.

It is not necessary here to judge whether these national and regional political actors can produce any genuine public advantages, or will simply use sub-global regulation to protect rents for some small number of the powerful. Nor is it necessary to go into the long menu of regulation – from macroeconomic stabilization to various forms of social policy – that deeper institutional integration can offer; nor to consider the decision-making frameworks, democratic or otherwise, within which sub-global economic governance can be held accountable; nor the bases of identity and legitimacy attached to all of these institutions. What is important here is only that there are many who do value economic governance at some sub-global level, and so weigh the benefits of the global and the sub-global in some sort of balance. This chapter is about how that balance can shift.

The "First Globalization" of the Nineteenth Century, and the Emergence of National Economies

This is not a new story: the parallel processes of global and sub-global (national or regional) integration, with global integration racing ahead only to be contained by the slower but stronger force from below, is not a new one. It recapitulates a process that took place between the mid-nineteenth and early twentieth centuries. That earlier case also saw the rapid liberalization of international trade in Europe, in parallel with the growth and integration of national markets. In a few years following 1850, trade barriers between European countries were abruptly removed in a great wave of international trade liberalization. Then, from about 1879, the trade liberalization process reversed; tariff barriers were restored; following 1914, international capital flows were curtailed as well, by many measures, and it was not until 1980 that the world economy achieved its pre-1914 measures of integration.

The countries that closed their borders in the late nineteenth and early twentieth centuries were not at all the same as the countries that opened them in the 1850s and 1860s. The states now controlled larger territories, and consequently were fewer in number (empires grew; Germany and Italy were created out of numerous small states). The domestic markets controlled by the states had each become more closely integrated, both because of investment in infrastructure based on new technologies (railways, telegraph), and because of institutional change (the elimination of local monopoly privileges and internal trade barriers). The newly industrialized countries (NICs) – Germany, the United States, France, Japan, and so on – had greatly narrowed the gaps in technological capability and industrial output between themselves and Britain.

All of these factors – the growth of states, the improved integration of domestic markets, and industrial and technological catch-up – changed the relative pay-offs from international liberalization and national economic management.

Trade Liberalization Comes and Goes

The label "first globalization" has been applied, tellingly, to two distinct periods: the flowering of international free trade in goods within Europe between 1850 and 1879; and the flowering of international investment, trade in intellectual property, and high-speed communications (telegraph) and transport (steam ships) between 1870 and 1914. The latter period was also a period of historically high levels of international trade, but in contrast to the 1850–1870 period it was marked by the systematic use of tariffs to protect heavy industry and agriculture in most industrial countries, and by the extension of systems of imperial preference within the newly extended overseas empires. In that sense, the two periods illustrate the importance of viewing international economic integration as multidimensional.

Either period, or the two together, can be seen as a precursor to today's "globalization," and also as a cautionary tale of the ways in which global liberalization can retreat (Findlay and O'Rourke 2007; Frieden 2006). The beginning and the end of the first phase – the dramatic liberalization of trade in Europe between 1850 and 1879 – has also been a puzzle: why did so many countries suddenly eliminate trade barriers? Why, given the mutual benefits of trade, did the barriers so suddenly return?

An influential theory links the coming and going of free trade to the coming and going of British hegemony. Krasner (1976) provides the classic account of this. The idea here is that when a single state exercises unchallenged power within the international system, it is able to force open trade, establishing and enforcing a common set of rules which in some respects solve a collective action problem among states, and in others acts to the benefit of the hegemon. In the absence of a hegemon, trade barriers rise. Britain, by this account, was the hegemon during the nineteenth century, while the United States played that role with the establishment of the Bretton Woods system after the Second World War.

Krasner's theory faces two empirical objections. One, which Krasner discussed, concerns timing: Britain's hegemony in Europe surely dated from the defeat of France at Waterloo in 1815, yet Europe's avalanche of bilateral free-trade deals did not begin until after Britain repealed the Corn Laws (barriers to grain imports) in 1846. The second objection is that it is not clear how Britain's supposed hegemonic power actually figured in the establishment and enforcement of trade liberalization. It is clear that, following 1846, Britain's longstanding intellectual enthusiasm for free trade (free, that is, with other sovereign states – colonies would continue to play by asymmetric rules internal to the empire) was finally backed up by action, with the repeal of the Corn Laws and the removal of restrictions on machinery exports and on the migration of people with technical skills. Yet there is little evidence of the policy's imposition by threat or force within Europe (McKeown 1991). Free trade in Europe was realized not through a grand brokered deal, like the later carve-up of Africa or the intellectual property conventions of the 1880s; rather, it came about through a series of bilateral deals between numerous states. This suggests that, during the free-trade episode, the various national governments found free trade to be in their mutual interest, and were able to overcome collective action problems without a hegemonic enforcer.

Kindleberger (1975), attributes the prevalent mid-nineteenth-century free-trade practice to a contagion of free-trade ideology.[1] He offers ample evidence that free-trade beliefs were widely held among elites at the time, and that they figured prominently in the rationale for reducing trade barriers. His theory is silent on why, after being so powerful from the 1850s and through the mid-1870s, this ideology went into headlong retreat. It is an interesting argument, but offers nothing to counter the objection that free-trade doctrine may have become intellectually hegemonic simply because it reflected the preferences of numerous influential actors. It is more plausible, perhaps, to think of the free-trade ideology as a coordinating mechanism, providing a focal point for policies and thus helping those actors overcome the considerable collective action problems between states. Certainly, it would be hard to understand the sudden retreat of free trade in the 1870s as the result simply of an abrupt change in intellectual fashion, while there are clearly ways in which the interests of economic and political actors had changed.

What did change to end free trade? There was, of course, the end of British hegemony, with the rise of Germany and the United States, and several other powers close behind. We also have changes in domestic political alignments brought on by the impact of cheap New World grain on unprotected markets – a classic example of how different dimensions of integration need not move in the same direction: here, global grain markets, the product of settlement by European emigrants together with steamships, railways, and canals, brought price changes that contributed to a protectionist political response. Yet, as Gourevitch (1977) shows, the protection was never for agricultural interest alone; indeed, farmers got protection only if they managed to strike a deal with heavy industry, as in Bismarck's "marriage of iron and rye." This raises the question of why industry, seemingly a beneficiary of the free-trade system, was now behind national policies of protection. One explanation is that the character of industry had changed; the Bessemer process and a cluster of other steel manufacturing and chemical inventions had come about in the 1860s; this led, from the 1870s onward, to the machine age, an outpouring of new steel- and chemical-based industrial processes with high output capacities and high fixed costs. High fixed costs and low marginal costs brought an unstable industrial structure, which was answered within states by the creation of giant firms or cartels and between states by protection against imports (Lamoreaux 1985; Piore and Sabel 1984; Guy 2009: chs. 6 and 8).

Technological Catch-Up

Another critical factor in the renewal of trade barriers was a reduction in the gaps between Britain and her competitors, in technological capability and industrial capacity. Thompson (1990) and Thompson and Vescera (1992) use this fact to reconcile the anomalies in Krasner's theory. They argue that openness to trade comes about while rivals are beginning to close the technological gap on the hegemon. There is a period during such a "catch-up" episode in which international liberalization is beneficial both to the technological leaders and to those catching up. Between the leader and all the followers, the differences in technological capabilities are such that comparative advantage is unambiguous – there is little immediate rivalry. One might think that the same did not apply between the NICs themselves, but countries at lower levels of development are typically more specialized. In today's terms, development continues to lead to reduced specialization (i.e., increased diversification) at the national level, up to about the per capita GDP of Ireland (see Rodrik 2004), so NICs can gain a great deal by trading with each other.

Catch-up, by bringing parity of power, ended British hegemony. But how could it have ended free trade? Thompson and Vescera say catch-up brought excess capacity, leading to political pressure for protection from foreign competition. Excess capacity could be seen as a consequence of slower growth, which was inevitable in the long-wave framework (Freeman and Perez 1988; Freeman and Soete 1997) that Thompson and Vescera adopt. Yet is does not seem plausible to attribute the comings and goings of protected markets entirely – or even largely – to shifts between constrained and excess capacity: was production capacity really constrained from 1850–1879, then in surplus for a century after that?

There were other ways, however, in which the technological and industrial convergence documented by Thompson and Vescera changed the tradeoff between openness to trade and governance. Having come closer to the technological frontier, Germany, the United States, and the other NICs of the day had greatly improved what Cohen and Levinthal (1990) call "absorptive capacity": the ability to absorb, understand, and make use of new knowledge. In, say, 1850, British manufacturers often could put cutting-edge knowledge to use far more quickly and more effectively than anyone else, and in such cases Britain's comparative advantage was clear. By 1880, manufacturers in several other countries were sufficiently sophisticated that they could just as well learn (or improve) the new process themselves. 1850 was a Heckscher–Ohlin world, where comparative advantage was rooted in different levels of economic development, and the mutual advantages to trade were clear. 1880 was, for Britain and its new competitors, a New Trade Theory world, where comparative advantage grows out of increasing returns to knowledge. Gomory and Baumol (2000) show how in such a world free trade may not be a win-win proposition, but a theatre of battle between contending national industrial policies.

The renewal of trade barriers around 1880 did not end international economic integration; what it did is to change the form of integration, and did so in a way that facilitated national economic management. Many national industrial and agricultural markets enjoyed substantial protection from import competition, but on the whole trade thrived, and international capital flows increased. Knowledge and technology, in particular, flowed more freely than ever. Historically, technical knowledge had traveled mostly through reverse engineering, or through the migration of skilled workers. Now those sources were supplemented by more organized international markets in intellectual property, following the Paris Convention on Patents, 1883, and the Berne Convention on Copyright, 1886 (Sell and May 2001); by a rise in international capital flows; and by the tentative steps of early MNCs.

Larger Domestic Markets: Bigger States, Bigger Empires

While Britain's new competitors were catching up technologically, their domestic markets were growing as well. This growth was not an accident, but the result of policies of nation (and empire) building carried out systematically over many decades, both during the free-trade period and beyond.

Nation-states took various paths to expansion. In Germany and Italy, new and much larger states were forged out of numerous smaller ones. It is important to note that neither of these unifications was a simple or abrupt event. The modern nation-state of Germany grew from the *Zollverein*, a customs union organized under Prussian leadership in 1829. The *Zollverein* grew in membership, but took a long time to include all of what later became Germany – Bremen and Hamburg were notable holdouts – and its members were not above fighting wars with each other. The *Zollverein* was succeeded by the modern German nation-state only in 1871. Italy's *Risorgimento* – its transition from a maze of principalities and foreign dominions into a single state – occupied the time from the end of the Napoleonic wars to the unification of the peninsula in 1870. The creation of these states, and the integration of the markets they governed, were, then, long-drawn and hard-fought political processes.

In the same period, the United States and Russia expanded across their respective continents; a few years later, Japan acquired colonies that were across seas, but close to home (Taiwan, Korea, Manchuria). Most of the industrializing countries of the day also created, enlarged, or consolidated, overseas empires: before the rise in tariffs, this list would include Britain, France, and the Netherlands; later in the nineteenth century, the United States, Belgium, Japan, Germany, and Italy joined the list of industrial countries with substantial overseas possessions. The population of the United States – and that of all European settler colonies, several of which went on to become industrial countries – grew rapidly due to migration.

European overseas empires had been important since the 1500s. These empires had retrenched considerably in the late eighteenth and early nineteenth centuries – though over the first half of the nineteenth century European powers did consolidate control of southern Asia. A great renewal of imperial expansion then took place late in the later nineteenth century, after the free-trade era: various European powers and the United States carved out concessions in China; at the Berlin Conference of 1884–1885 they carved up the continent of Africa; the United States conquered Hawaii, Cuba, Puerto Rico, and the Philippines; Japan conquered Taiwan, Korea, and Manchuria; the United States asserted its hegemony in Latin America. The new imperial ventures served to expand the coverage of protected markets for national manufactures, in addition to their primary function of securing supplies of raw materials for those same manufactures (Frieden 1994).

Breaking Down Internal Trade Barriers

The enlargement of markets controlled by individual nation-states would be of little relevance here had it not been accompanied by a substantial improvement in the internal integration of national markets, both because of institutional changes and because of improved transport and communications infrastructure.

Large countries are not necessarily large markets. In 1750 the population of the British Isles was only about 11 million – less than half that of France. Britain, however, had a much better integrated domestic market, and this was an important factor in the progress of the first industrial revolution (Landes 1969). Trade within France was hindered by internal tariffs and tolls until after the revolution. In addition, in Britain after 1624 the Statute of Monopolies had curtailed (at the domestic level, if not in the empire) the common European practice of raising state revenues through the sale of monopoly rights (Sell and May 2001); since these monopolies were not always national in extent, they could have the effect of restricting market size as well as imposing the usual costs of monopoly. The Netherlands likewise had an integrated domestic market, and the French Revolution together with Napoleon had largely succeeded in creating a single market within France by the early nineteenth century. Napoleonic reforms were also influential in much of the rest of Europe. Still, the integration and liberalization of domestic markets remained contested. The process of domestic liberalization picked up speed in 1848, just before free international trade gained traction. Switzerland, for instance, began dismantling its internal trade barriers in 1848, completing the process in 1874. Austria dismantled internal trade barriers in 1850 (Pollard 1974: 109–115). A few years later, in Japan, feudal relationships that constrained the growth of domestic markets were uprooted following the collapse of Tokugawa rule, and the Meiji "restoration" of the 1860s (Gordon 2003: 46–76).

Transport and Communications Infrastructure

Institutional integration of a market may have limited effect if transport and communication are inadequate. The period of catch-up in the middle and late nineteenth century was a period of furious railway building. Prior to railways, most long-distance and bulk transport was water-borne. While canals had done much to integrate some domestic markets, they were slower and of more limited coverage. On the whole water transport exhibited a bias toward foreign trade rather than national integration: it connected port to port, and many countries enjoyed rich trade overseas but only limited links with their own interiors. This outward bias held not only for mountainous countries with long seacoasts like Italy, Spain, Greece, and Norway (Japan would be included here, but for the Tokugawa Shogunate's uncommonly severe restrictions on foreign trade): before the railway, US population and industry were concentrated around seaports on the Eastern seaboard, with long travel times between its isolated inland communities (Chandler 1977; Gordon 1996).

In 1840, almost 70% of the railway mileage in Europe was in the United Kingdom; at that time Germany, Italy, Austria, and Russia had essentially no railways. Over the next 30 years, Britain's railway mileage increased tenfold, but even so fell to about 24% of Europe's total. As with the enlargement of states and the removal of internal trade barriers, the development of railway networks was a fraught process; in industrial countries, it was also one directed at the creation of national markets (Pollard 1974: 42–46); this is in marked contrast with the construction of railways in colonies and other countries specializing in natural resource exports, across Asia, Africa, and Latin America, where railways were designed to get raw materials from the interior to the sea. In 1846, 24 years before leading Italy's first national government, Cavour drew a railway map for his imagined country which ignored existing political boundaries, existing commercial relationships, and some ongoing wars: that fantasy map turns out to be quite close to the actual network built after the unification of the country (Hooper 2012). In the United States, the construction of railroads and other "internal improvements" (which meant mainly transportation infrastructure – roads, canals, bridges, railways) was a major bone of contention between slave states and free states prior to the Civil War: the free states, which were industrializing, wanted the improvements, and the market integration that went with them; the slave states wanted only the minimal infrastructure required to get cotton to coastal ports. The industrial side won the war in 1865, and the railroad boom redoubled. Japan's railway expansion followed close on the fall of the Tokugawa; its first railway, connecting Tokyo with the new port city of Yokohama, opened in 1872. From that starting point, the Japanese government promoted the rapid development of a national railway network, converting many local markets to a national one (Gordon 2003: 71).

Lessons from the First Globalization

What do we learn from this? Two things. First, that it would be a serious mistake to think in terms of a fixed thing called a "nation-state" or a "national government" switching between protectionist and free-trade policies. The nation-states that raised tariff barriers in 1879 and later were entirely different creatures from those that lowered them between 1846 and 1865. They had much larger and better integrated internal markets, much more capable of exploiting economies of scale; and they were closer to the technological leaders, and so were better able to absorb new technologies rather than importing products.

Second, the flowering of free trade began, and also ended, *during* this period of nation-state growth and national catch-up. During the free-trade period, the two processes occurred in parallel, but national market expansion and integration continued after trade liberalization went into reverse. This should not be surprising. Compared with national integration, international liberalization was institutionally minimal: certain import tariffs and export taxes were no longer collected at ports and border crossings. Because international commerce was often seaborne traffic between port cities, the infrastructure required for international integration was similarly minimal. Expanded national markets, on the other hand, required not only large investments in railways, canals, roads, and telegraphs, but decades of political, military, and bureaucratic work and strife. Indeed, the construction of national markets required the construction of national identity itself – something in which the school system and the organs of public administration played large parts (Gellner 1983). By the 1870s, these efforts had defined the national markets that were to serve largely unchanged for over 100 years. Soon after that, tariff barriers between those national markets were raised. A century of national economies had begun.

Global Liberalization and Regional Integration Today

After Britain, of course, came America. From the late nineteenth century onward, the United States claimed ascendancy in new production methods, in science and technology, in military might, and in finance. Its dominance was settled by the defeat of Germany and Japan in the Second World War, and by the inability of the USSR to match the United States in innovation and growth. Under American dominance, international markets were gradually opened – first in trade in goods and in FDI.

US ascendancy seemed to falter in the 1970s, with a combination of military defeat in Vietnam, persistent Soviet influence, and the renewed commercial parity of Western Europe and Japan. But then, the USSR collapsed. Moreover, the USSR's collapse was only the most extreme manifestation of a more general crisis in the mass production approach to industry, and the systems of regulation and governance that went with it (Guy 2009: chs. 9–10). Flexible production methods, network products, and information products between them produced market structures that were incompatible with the systems of post-war regulation in much of the capitalist world (Aglietta 2001). The principal political response to this situation was neo-liberalism. In this setting, the United States was more dominant than ever, and with that dominance came the more rapid liberalization of international markets – broad international capital flows, trade in services, and more standardized and extensive intellectual property markets, all under the WTO framework in the 1990s. These changes in the institutional framework were accompanied by changes in the geography of economic activity: the new international division of labor (NIDL) in manufacturing, the burgeoning of cross-border production networks, and corporations claiming to be "global."

The late twentieth-century liberalization was accompanied by the arrival of the first major new aspirants to the club of developed industrial nations since the modern system of national economies had consolidated in the late nineteenth century. Before that, in the first industrial revolution, Britain's initial industrial growth had been so exceptional that it brought a "great divergence" in wealth, technology, and power between Britain and most of the world, including, most shockingly, China (Pomeranz 2000). The countries catching up with Britain in the nineteenth century were (with the exception of Japan) blessed to be either Britain's neighbors in Europe, or its settler colonies. From the development of the system of national economies in the late nineteenth century until the 1990s, the world saw what Pritchett (1997) called "divergence, big time": between those developed countries and the "developing" ones, the gaps in wealth and in technological capability grew and grew. The few late exceptions – the Tigers of East Asia who joined the ranks of the global middle class so rapidly between 1970 and 1990 – were anomalies that stood out so starkly in that divergent world that they became the subjects of endless study, and of largely fruitless emulation (Amsden, Kochanowicz, and Taylor 1994; Gereffi and Wyman 1990; Rodrik 2004; Wade 1990).

In the 1990s, suddenly, vast China was joining the ranks of the industrial and technologically sophisticated; most of Southeast Asia seemed to be following in its wake; across Latin America, South Asia, and, in the 2000s, Africa, numerous other countries followed. The relationship of these NICs to the old OECD countries is the same as that of Germany, the United States, France, and Japan to Britain in the 1870s. And, while the new wave of industrialization and development owes much to international liberalization (Williamson 1996), we should not let that debt blind us to two facts. One is that development – the reduction in the huge differentials of wealth, power, and in particular scientific and technological capability between the leading economies and the more successful "emerging" markets – will change the payoffs that different forms of international economic integration provide to different actors in both sets of countries; players who now find global liberalization beneficial may not continue to do so. The other is that at the same time global institutions of liberalization and apparently global production systems have been put in place, regional systems have developed: regional systems of production, markets for goods and services, transport networks, and institutions of governance. Such regional integration has not occurred to the same extent or in the same way everywhere, but it has been substantial. Two regions in particular, Europe and China (the latter being a nation-state which, for reasons that will become clear, I treat as a region), are well along in the process of recapitulating, on a larger scale, the creation of national economies in the late nineteenth century.

Regional Production

However much we talk of a globalized economy, the fact is that a disproportionate share of international economic integration has been regional. In the midst of the globalist euphoria of the 1990s, Mansfield and Milner (1999) noted that the growth of international trade between 1948 and 1990 had actually been higher at the intra-regional level than at the inter-regional. In aggregate, this was so whether the regions were defined politically (as regional trade blocs) or by the more basic geography of continents and sub-continents. In many regions, however, Mansfield and Milner found a decline in the relative share of regional trade in the 1980s. That could be interpreted in the conventional globalization lens: after 1980, real international production kicks into gear. Yet, if we update the numbers we see that intra-regional trade has, on the whole, continued to grow more quickly than inter-regional trade. Figure 28.1 takes as examples a number of trade blocs, large and small, on different continents. The data is the ratio between foreign trade within the bloc, and foreign trade between members of the bloc and the rest of the world, from 1954 to 2012. To avoid bias from the changing (usually, growing) membership of the blocs, countries are included if and only if they were members at the end of the data series – indeed, in the early years of the series, many of these blocs did not formally exist, in which case the bloc is just a set of independent countries or colonial territories from that time. The results vary considerably from year to year; events leave their mark: the debt crisis in Argentina in the late 1990s was a big setback for the Mercado Común del Sur (Mercosur), while the 15 southern and eastern African countries which are now the Southern African Development Community (SADC) did little trade with each other during the later years of apartheid and Portugal's colonial wars; the overall trend, however, is upward, not only in huge blocs such at the EU and the Association of Southeast Asian Nations (ASEAN) – 10 countries with a combined population of over 600 million – but in small ones such as the Caribbean

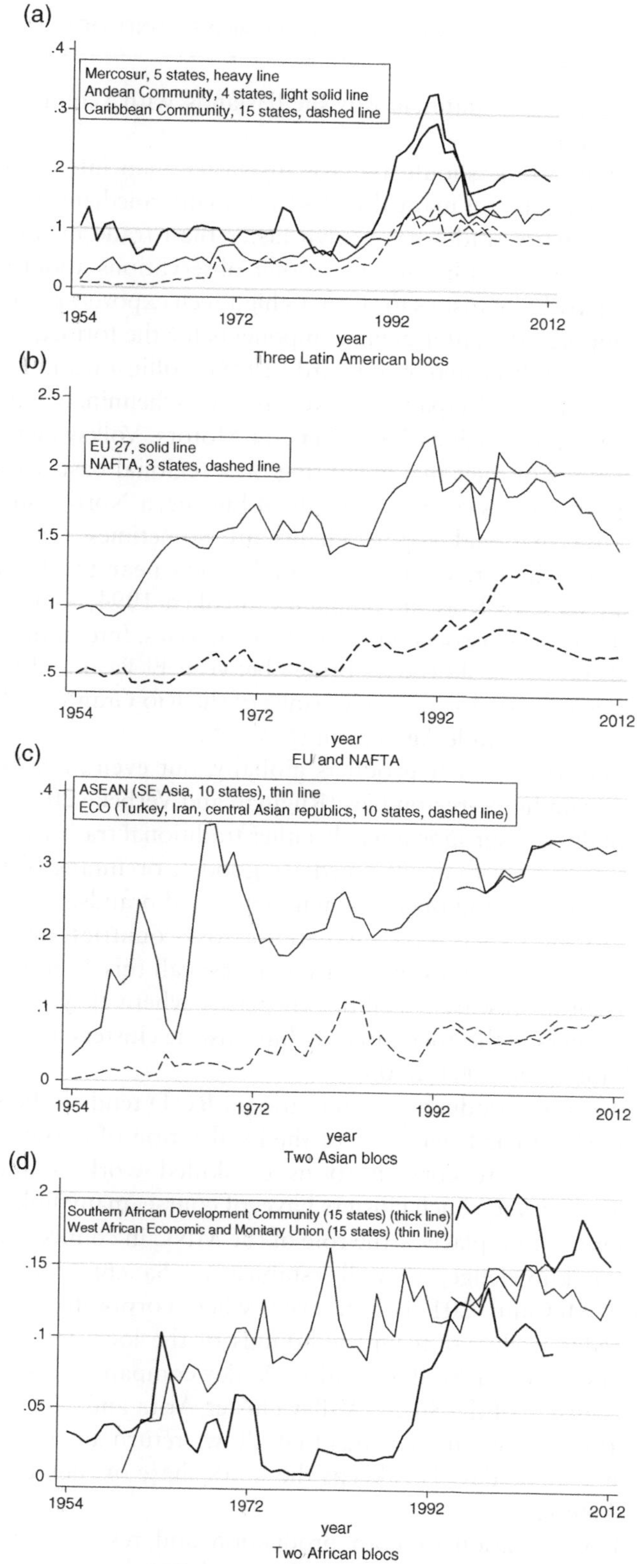

Figure 28.1 Selected regional trade agreements. Discontinuity and overlapping lines are due to differences between the two UNCTAD data sets used for the 1954–2006 and 1995–2012 series. Source: based on data from UNCTAD.

Community (Caricom) of 15 states, mostly small islands, with a combined population of less than 17 million.

That intra-regional trade should be growing faster than inter-regional trade is consistent with the fact that for many decades trade in intermediate goods (sometimes proxied by intra-industry trade) has grown faster than trade in raw materials or finished goods. It also fits with numerous other observations about supply chains and production networks. Statistics for China's high-tech exports are closely mirrored by its high-tech imports, the latter being components for the former, and originating mostly within East Asia (Branstetter and Lardy 2008). Collinson and Rugman (2008) find that the foreign assets of Japanese MNCs are overwhelmingly within East Asia. "Global" auto companies such as Ford, General Motors, Volkswagen, and Nissan/Renault organize most of their manufacturing on a regional basis to serve regional markets: a European Ford is essentially made in Europe, a North American Ford in North America. International supply chains are sometimes very short indeed. Although some maquiladoras – factories in Mexico near the US border – are affiliated with Japanese or Korean companies (Shaiken 1994; Kenney and Florida 1994), they are more often associated with American ones. International integration can occur over distances as short as a bridge between El Paso and Ciudad Juarez, with different wage rates on the opposite banks of the Rio Grande – all governed by the North American Free Trade Agreement (NAFTA).

MNCs do aspire to sell their products globally, but even in sales they seldom overcome a significant home-region bias (Rugman and Verbeke 2004). What MNCs move inter-regionally is disproportionately either traditional trade (finished products and raw materials, rather than intermediate goods), or intangibles: technology, organizational know-how and operating templates, and brands.

MNCs also distribute their R&D activities across countries, or engage in joint R&D projects with corporations in other countries: call this "international R&D." MNCs typically locate research facilities in places where they can pick up new knowledge from others in the area, often within existing clusters in foreign countries (Cantwell and Iammarino 2001, 2003).

Yet like international production, international R&D tends to be more regional than it is global. Much has been made of the localization of knowledge exchange, in localized clusters where concentrations of skilled workers and face-to-face contact can foster discovery. Such clusters have, of course, many links to people and companies in more distant places – the role of the MNC in connecting clusters, and mining them for knowledge, is well established (Saxenian and Sabel 2008; Iammarino and McCann 2013). Yet, even when corporate organization has liberated knowledge exchange from the bounds of the local travel-to-work area, distance continues to matter. Studying Silicon Valley companies' research collaborations with firms *outside* of the Silicon Valley cluster, Arita and McCann (2000) find them disproportionately within a radius that allows return air travel, and time for meetings, within a single day. This keeps the lion's share of such collaborations in western North America.

The regional organization of both production and research within MNCs is reflected in the growing decentralization of control (Iammarino and McCann, this volume, Chapter 14). As modern corporations are essentially portfolios of operational and intellectual assets, decentralization of control presents a situation in which a global corporation could be easily disassembled into a number of regional ones, should that prove the more profitable course.

Regional Economic Integration: Europe, China, and Others

The European Project

If we essentialize the nation-state, we could mistake the European project for a miniature version of the global liberalization project. The EU is, after all, characterized by the elimination of trade barriers and the liberalization of markets. Perhaps we should see the EU just as 28 nation-states with a combined population of over 500 million, enjoying not only the free flow of goods, services, and capital, but also freedom of movement, to live and to work. Is that, perhaps, simply a transitional step toward the *global* erasure of economic borders?

What most clearly distinguishes the European project from the global one is that it has involved the incremental creation of an elaborate structure of economic governance at the European level. Over the decades, aspects of consumer protection, employment relations, competition (anti-trust) policy, environmental protection, regional (i.e., micro-regional) economic development, community-wide collaboration in research and education, agricultural policy, human rights law, and (within the eurozone) monetary policy have come under the European umbrella. This is not to say that anything like a comprehensive central government has developed; even the term "federal" remains, depending on one's viewpoint, an aspiration or an accusation rather than a description of the status quo, and most functions of both policy and administration remain primarily in the hands of nation-states. The financial crisis of 2008 and its aftermath have pushed the eurozone toward tighter integration, although the political will has often been lacking to deal effectively with community deficits in fiscal policy, banking supervision, and the paradox of maintaining community-wide price stability when productivity growth varies greatly from one part of the union to another. Still, together with its functions of economic regulation, the EU has acquired many conventional institutions of government: a growing bureaucracy, an executive, and an elected parliament. Finally, among the many infrastructure projects sponsored or promoted by the EU has been an ongoing effort to integrate rail transport systems across the continent, harmonizing systems which had been developed primarily to serve national markets, and linking them with tunnels though mountain ranges, and tunnels and bridges across straights.

We cannot know the future of this European state, but its development to date surely recapitulates the emergence of Germany and Italy in the nineteenth century. It has been a long, slow process precisely because it goes deep.

The Single Chinese Market

China is treated here as a case of regional integration for two reasons. First, in terms of population it far outstrips any international regional grouping. The other is that China's internal market has been undergoing a process of integration, in parallel to its international opening, much as the European powers did in the nineteenth century.

China's transition from a centrally planned economy to a market economy began in the early 1980s. Many businesses in China had historically been controlled by local or provincial governments; the more so, following the early years of economic reforms in which the central planners loosened their grip, giving local and provincial governments more autonomy. In any case, the nation's transportation infrastructure was rudimentary. Railway construction in China had been negligible before the 1912 revolution, and made slight progress during the ensuing decades of political instability and war; after 1948 the Communist government did invest in railways, but this investment suffered severe reversals during the Cultural Revolution (1966–1976). Highway capacity was even more constrained. Early experiments in economic liberalization were in selected coastal zones, which had ready access to ports and thus international markets; development of the interior, and improvement of links between different parts of China, came later.

Thus, in the early years of its economic reforms, China was not only a poor country but one with a seriously balkanized internal market: local and provincial governments controlled monopolies in many lines of business, and did not welcome competition; movement of goods and people from one part of China to another was slow.

A central task of the reform process has therefore been to create, within this vast nation-state, a single market. Montinola, Qian and Weingast (1995) argue that China's growth is underpinned by the development of "market-sustaining federalism" in which the national government takes responsibility for establishing competition among sub-national units. Branstetter and Lardy (2008) show how China used its negotiations for entry into the WTO to impose internal market integration. Hsueh (2011, 2012) describes the "de-regulation re-regulation two-step" whereby the old rules are swept aside, then replaced by new ones. Notably, in Hsueh's account, the state keeps control of industries whose technologies it regards as strategic. These institutional developments have been accompanied by the frenzied expansion of the internal transportation infrastructure.

Other Regions

The European and Chinese cases are the furthest advanced and largest examples of regional integration from, respectively, the developed and developing worlds. On their own, the integration and consolidation of these two huge regional economies would represent a substantial restructuring of the world economy. There is good reason, however, to believe that they reflect a broader trend. Even within regional trade blocs that appear institutionally weak – ASEAN, Mercosur, NAFTA, Caricom, and others – the growth of intra-region trade has far outstripped that of international trade generally. In many regions of the developing world, it is apparent that global liberalization has spurred the development of regional institutions. This appears to be both because it has encouraged economic modernization and development, thus enhancing the gains to be had from intra-regional trade; and because liberalization has exposed national economies to the vagaries of the global market, and regional integration is seen as a way to restore both stability and accountability to actors within the region. Moreover, other large nation-states in developing countries are seeing market integration analogous to China's. India, in particular, had been characterized by localized monopolies in many industries; internal liberalization and infrastructure improvements have, again, produced more integrated markets.

Will It Be Different This Time?

Are we, then, on the way to seeing regional blocs take the place of nation-states as the basic sub-global economic units? Or are there features of today's world that make this outcome unlikely? Let us consider three factors: differences between our "knowledge economy" and the manufacturing economy of the nineteenth century; the political power of MNCs; and the global environmental crisis.

It may be that the knowledge economy of today is less conducive to the creation of closed national or regional units than was the manufacturing economy of the late nineteenth and early twentieth centuries. There are two reasons this might be so. One is that information- based products and processes often exhibit unlimited increasing returns. To take an extreme case: all of the costs of making Microsoft Windows are sunk, marginal costs of production are zero, so the natural market would seem to be global; the same could be said of genetically modified seeds and any digitized information. These are extreme, if important cases, but their greater importance lies in illustrating a more general fact: most of the goods and services we use have a much greater share of code – sets of instructions that can be replicated at zero marginal cost – in their makeup than they used to.

The second reason is that the complexity of knowledge used in production, and the speed with which that knowledge changes, may mean that the global circulation of that knowledge requires the visible hand capabilities, the dynamic competencies, of big corporations: that is, the argument goes, the problems of knowledge management and technology integration have become too complex to be left to the unregulated market or the free exchange of knowledge; in the international sphere that puts the MNC, with its internalization advantages, front and centre. It may be ironic that one of the key arguments for a regime of global liberalism is rooted in the inability of markets to match the performance of corporate bureaucracies, but irony alone does not invalidate an argument.

The apparent scale economies associated with information products, however, are legal constructs, not physical properties of the production process. Certain countries – notably the United States, United Kingdom, France, and Sweden – have become major net exporters of information products and other intellectual property, and have come to rely on them for foreign exchange; moreover, many of the world's great fortunes are built on intellectual property monopolies, and those holding these fortunes fight to keep them. Consequently, the enforcement and extension of these rights has become a priority in the trade policies of information product exporters. The volumes of trade involved are large, and if we take enforcement of the rights as given the natural markets are indeed global. Yet this is a precarious foundation for any form of global integration. The legal rights which underlie the appropriation of rents from information products are contested, through legal and political processes, through the development of open source alternatives, through simple violation of the rights, and through engineering around the protected properties. It is not to be taken for granted that the actors who are on the deficit end of transactions in intellectual property – actors who constitute most of the world's countries, most of the world's population, and most of the world's industries – will tolerate the current intellectual property regime indefinitely (Guy 2007).

The apparent superiority of MNCs in producing, transferring, and applying new technical knowledge is, like infinite economies of scale in software, an illusion borne of naïve extrapolation of recent trends. The role MNCs play in transferring knowledge today is analogous to the role played by machinery exports from Britain in the mid-nineteenth century: alternatives are quickly catching up. As the new industrial countries improve their absorptive capacity, they will be able to rely more on copying (even now, not a rarity) and on licensing (Athreye and Kapur, this volume, Chapter 9), rather than inviting global MNCs in. As a result, the overseas presence of corporations will shrink, along with FDI.

MNCs now wield considerable political power – again, a difference from the situation in 1880, when MNCs scarcely existed. Where does the power come from? Like the Pope in Stalin's famous question, MNCs have no armored divisions. Strange (1992) explained their power as a consequence of their technological capabilities together with their ability to provide access to consumer and capital markets in rich countries: with these levers, they could demand from poor countries the liberalization of international trade and investment. These levers will lose their force as regional markets are consolidated and the absorptive capacity of the new industrial countries improves.

The era of national economic management ran for 100 years, from 1880 to 1980. It brought with it some exceedingly ugly and tragic times, but its last three decades were ones of unprecedented peace, prosperity, and equality within the developed world – what has been called the Golden Age of Capitalism (Marglin and Schor 1990). The neoliberal program, of which global economic integration must be reckoned a part, has served to prevent any return to the relatively egalitarian policies of the Golden Age. While regional integration has often been part of this neo-liberal agenda (NAFTA being a case in point), robust and democratically accountable regional institutions would be anathema to it. The development of such regional institutions could form part of a society's latest defense against the destabilizing depredations of the market (Polanyi 1957). Or not – perhaps we're simply at the end of democracy and dreams of a fair society (Crouch 2004).

Finally, decisions about international economic integration are taking place in a world threatened by climate change – an irreducibly global problem. They involve extremely difficult global collective action problems, and it seems unlikely these can be solved without substantial strengthening of global governance (Gardiner 2011). Within such a framework, would more extensive global economic integration not be likely?

To answer this we must return to the taxonomy of types of integration with which this chapter began, and ask how an effective system of global environmental governance would affect them. Any such system would bring with it some comprehensive scheme for reducing release of carbon into the atmosphere, probably through pricing. This would have a pronounced effect on the cost of high-speed air and sea transport – the two major carbon sources for which feasible low-carbon close substitutes are not available. It is increases in speed, rather than reductions in transportation cost, which have facilitated far-flung production networks (Hummels 2006). Sea and air transport can be decarbonized using existing technologies, but only at a great sacrifice of speed – sail-assisted freighters, and dirigibles. The weight of increased air and sea transport costs (or reduced speed) would fall on long-distance supply chains; and on various forms of short-term migration, including those which hold together far-flung corporate operations, tourism, and, it must be said, academic conferences.

Moreover, at present the new international division of labor receives a substantial subsidy from the fact that environmental regulations, including the carbon credit market under the Kyoto protocol, are not global. Factories have shifted to countries in which carbon pollution and other forms of air pollution are lightly regulated; the goods are then exported to countries with tighter controls on emissions (Davis and Caldeira 2010; Caldeira and Davis 2011). Global internalization of carbon costs would end this subsidy to offshoring. Without effective global environmental regulation, of course, global supply chains and profligate use of air travel might continue unabated. Such a world is probably not one in which we want our grandchildren to live.

If we combine technological catch-up by new industrial countries, the gradual rescaling of deep economic governance from the national to the regional level, and the civilizational imperative to restrict carbon emissions, we are likely to see global economic integration becoming shallower. We will see thriving global markets in intellectual property (licensing, in place of the direct operation of MNCs); global technical standards (which, again, substitute for the operation of MNCs by allowing production of interchangeable components by various producers); continued global use and development of open source software, with local customization; global environmental standards and carbon taxation guidelines; global trade deals and efforts to maintain a stable system of currencies and credit. Sub-global governments – nation-states, regional blocs – will likely reassert control over capital flows, and end the rush to protect foreign investment through an extra-territorial legal system.

Note

1 In other writings Kindleberger offered his own theory of hegemonic stability (Kindleberger 1981, 1986), somewhat different from Krasner's. That is beyond the scope of this chapter, but is noted here because some readers will associate Kindleberger with that and not with his argument about the role of ideology.

References

Aglietta, Michel. 2001. *A Theory of Capitalist Regulation: The US Experience*. Translated by D. Fernbach. London: Verso. Original edition, 1979.

Amsden, Alice H., Jacek Kochanowicz, and Lance Taylor. 1994. *The Market Meets Its Match: Restructuring the Economies of Eastern Europe*. Cambridge, MA: Harvard University Press.

Arita, Tomokazu, and Philip McCann. 2000. "Industrial Alliances and Firm Location Behaviour: Some Evidence from the US Semiconductor Industry." *Applied Economics* 32: 1391–1403.

Branstetter, Lee, and Nicholas Lardy. 2008. "China's embrace of globalization." In *China's Great Economic Transformation*, ed. L. Brandt and T.G. Rawski. Cambridge: Cambridge University Press.

Buchanan, James M., and Gordon Tullock. 1962. *The Calculus of Consent: Logical Foundations of Constitutional Democracy*. Indianapolis, IN: Liberty Fund.
Caldeira, Ken, and Steven J. Davis. 2011. "Accounting for Carbon Dioxide Emissions: A Matter of Time." *Proceedings of the National Academy of Sciences* 108(21): 8533–8534.
Cantwell, John, and Simona Iammarino. 2001. "EU Regions and Multinational Corporations: Change, Stability and Strengthening of Technological Comparative Advantages." *Industrial and Corporate Change* 10(4): 1007–1037.
Cantwell, John, and Simona Iammarino. 2003. *Multinational Corporations and European Regional Systems of Innovation*. London: Routledge.
Chandler, Alfred D. Jr. 1977. *The Visible Hand: The Managerial Revolution in American Business*. Cambridge, MA: Harvard University Press.
Chandler, Alfred D. Jr. 1992. "Organizational Capabilities and the Economic History of the Industrial Enterprise." *Journal of Economic Perspectives* 6(3): 79–100.
Cohen, Wesley M., and Daniel A. Levinthal. 1990. "Absorptive Capacity: A New Perspective on Learning and Innovation." *Administrative Science Quarterly* 35: 128–152.
Collinson, Simon, and Alan M. Rugman. 2008. "The Regional Nature of Japanese Multinational Business." *Journal of International Business Studies* 39: 215–230.
Crouch, Colin. 2004. *Post-Democracy*. Cambridge: Polity.
Davis, Steven J., and Ken Caldeira. 2010. "Consumption-Based Accounting of CO2 Emissions." *Proceedings of the National Academy of Sciences* 107(12): 5687–5692.
Findlay, Ronald, and Kevin H. O'Rourke. 2007. *Power and Plenty: Trade, War, and the World Economy in the Second Millennium*. Princeton, NJ: Princeton University Press.
Freeman, Christopher, and Carlota Perez. 1988. "Structural Crises of Adjustment: Business Cycles and Investment Behaviour." In *Technical Change and Economic Theory*, ed. G. Dosi, C. Freeman, R. Nelson, G. Silverberg, and L. Soete. London: Pinter.
Freeman, Christopher, and Luc Soete. 1997. *The Economics of Industrial Innovation*, 3rd edn. London: Pinter.
Frieden, Jeffry A. 1994. "International Investment and Colonial Control: A New Interpretation." *International Organization* 48(4): 559–593.
Frieden, Jeffry A. 2006. *Global Capitalism: Its Fall and Rise in the Twentieth Century*. New York: Norton.
Gardiner, Stephen M. 2011. *A Perfect Moral Storm: The Ethical Tragedy of Climate Change* Oxford: Oxford University Press.
Gellner, Ernest. 1983. *Nations and Nationalism*. Oxford: Blackwell.
Gereffi, Gary, and Donald L. Wyman (eds.). 1990. *Manufacturing Miracles: Paths of Industrialization in Latin America and East Asia*. Princeton, NJ: Princeton University Press.
Gomory, Ralph E., and William J. Baumol. 2000. *Global Trade and Conflicting National Interest*. Cambridge, MA: MIT Press.
Gordon, Andrew. 2003. *A Modern History of Japan: From Tokugawa Times to the Present*. Oxford: Oxford University Press.
Gordon, Sarah H. 1996. *Passage to Union: How the Railroads Transformed American Life, 1829–1929*. Chicago, IL: Ivan R. Dee.
Gourevitch, Peter A. 1977. "International Trade, Domestic Coalitions, and Liberty: Comparative Responses to the Crisis of 1873–1896." *Journal of Interdisciplinary History* 8(2): 281–313.
Guy, Frederick. 2007. "Strategic Bundling: Information Products, Market Power, and the Future of Globalization." *Review of International Political Economy* 14(1): 26–48.
Guy, Frederick. 2009. *The Global Environment of Business*. Oxford: Oxford University Press.
Hooper, John. 2012. "Italy's High-Speed Train Line Under the Alps Gathers Pace." *The Guadian*, April 9.
Hsueh, Roselyn. 2011. *China's Regulatory State: A New Strategy for Globalization*. Ithaca, NY: Cornell University Press.
Hsueh, Roselyn. 2012. "China and India in the Age of Globalization: Sectoral Variation in Post-Liberalization Reregulation." *Comparative Political Studies* 45(1): 32–62.

Hummels, David. 2006. "Transportation Costs and International Trade in the Second Era of Globalization." *Journal of Economic Perspectives* 21(3): 131–154.

Iammarino, Simona, and Philip McCann. 2013. *Multinationals and Economic Geography: Location, Technology and Innovation*. Cheltenham: Edward Elgar.

Jonscher, Charles. 1994. "An Economic Study of the Information Technology Revolution." In *Information Technology and the Corporation of the 1990s: Research Studies*, ed. T.J. Allen and M.S. Scott Morton. Oxford: Oxford University Press.

Kenney, Martin, and Richard Florida. 1994. "Japanese Maquilddoras – Production Organization and Global Commodity Chains." *World Development* 22(1): 27–44.

Kindleberger, Charles P. 1975. "The Rise of Free Trade in Western Europe." *Journal of Economic History* 35(1): 20–55.

Kindleberger, Charles P. 1981. "Dominance and Leadership in the International Economy: Exploitation, Public Goods, and Free Rides." *International Studies Quarterly* 25: 242–254.

Kindleberger, Charles P. 1986. "Hierarchy vs. Inertial Cooperation." *International Organization* 40: 841–847.

Krasner, Stephen D. 1976. "State Power and the Structure of International Trade." *World Politics* 28(3): 317–347.

Lamoreaux, Naomi R. 1985. *The Great Merger Movement in American Business, 1895–1904*. Cambridge: Cambridge University Press.

Landes, David S. 1969. *The Unbound Prometheus: Technological Change and Industrial Development in Western Europe from 1750 to Present*. Cambridge: Cambridge University Press.

Mansfield, Edward E., and Helen V. Milner. 1999. "The New Wave of Regionalism." *International Organization* 53(3): 589–627.

Marglin, Stephen A., and Juliet B. Schor (eds.). 1990. *The Golden Age of Capitalism*. Oxford: Oxford University Press.

McKeown, Timothy J. 1991. "A Liberal Trade Order? The Long-Run Pattern of Imports to the Advanced Capitalist States." *International Studies Quarterly* 35: 151–172.

Montinola, Gabriella, Yingyi Qian, and Barry R. Weingast. 1995. "Federalism, Chinese Style – The Political Basis for Economic Success in China." *World Politics* 48(1): 50–81.

Piore, Michael J., and Charles Sabel. 1984. *The Second Industrial Divide: Possibilities for Prosperity*. New York: Basic Books.

Polanyi, Karl. 1957. *The Great Transformation: The Political and Economic Origins of Our Time*. Boston, MA: Beacon Press. Original edition, 1944.

Pollard, Sidney. 1974. *European Economic Integration, 1815–1970*. London: Thames and Hudson.

Pomeranz, Kenneth. 2000. *The Great Divergence: China, Europe, and the Making of the Modern World Economy*. Princeton, NJ: Princeton University Press.

Prichett, Lant. 1997. "Divergence, Big Time." *Journal of Economic Perspectives* 11(3): 3–18.

Rodrik, Dani. 2004. "Industrial Policy for the Twenty-First Century." Paper prepared for UNIDO.

Rodrik, Dani. 2007. "The Inescapable Trilemma of the World Economy." Dani Rodrik's weblog: Unconventional thoughts on economic development and globalization.

Rugman, Alan M., and Alain Verbeke. 2004. "A Perspective on Regional and Global Strategies of Multinational Enterprises." *Journal of International Business Studies* 35: 3–18.

Saxenian, AnnaLee, and Charles Sabel. 2008. "Venture Capital in the 'Periphery': The New Argonauts, Global Search, and Local Institution Building." *Economic Geography* 84(4): 379–394.

Sell, Susan, and Christopher May. 2001. "Moments in Law: Contestation and Settlement in the History of Intellectual Property." *Review of International Political Economy* 8(3): 467–500.

Shaiken, Harley. 1994. "Advanced Manufacturing and Mexico: A New International Division of Labor?" *Latin American Research Review* 29(2): 39–71.

Smith, Adam. 1991. *The Wealth of Nations*. New York: Everyman's Library. Original edition, 1776.

Strange, Susan. 1992. "States, Firms and Diplomacy." *International Affairs* 68(1): 1–15.

Thompson, William R. 1990. "Long Waves, Technological Innovation, and Relative Decline." *International Organization* 44(2): 201–233.

Thompson, William R., and Lawrence Vescera. 1992. "Growth Waves, Systemic Openness, and Protectionism." *International Organization* 46(2): 493–532.

Wade, Robert. 1990. *Governing the Market: Economic Theory and the Role of Government in East Asian Industrialization*. Princeton, NJ: Princeton University Press.

Williamson, Jeffrey G. 1996. "Globalization, Convergence and History." *Journal of Economic History* 56(2): 277–306.

Index

The Handbook of Global Science, Technology, and Innovation, First Edition.
Edited by Daniele Archibugi and Andrea Filippetti.